PLANETARY TECTONICS

This book describes the tectonic landforms resulting from major internal and external forces acting on the outer layers of solid bodies throughout the solar system. It presents a detailed survey of tectonic structures at a range of length scales found on Mercury, Venus, the Moon, Mars, the outer planet satellites, and asteroids.

A diverse range of models for the sources of tectonic stresses acting on silicate and icy crusts is outlined, comparing processes acting throughout the solar system. Rheological and mechanical properties of planetary crusts and lithospheres are discussed to understand how and why tectonic stresses manifest themselves differently on various bodies. Results from fault population data are assessed in detail.

The book provides methods for mapping and analysing planetary tectonic features and is illustrated with diagrams and spectacular images returned by manned and robotic spacecraft. It forms an essential reference for researchers and students in planetary geology and tectonics.

THOMAS R. WATTERS is a Senior Scientist in the Smithsonian Institution's Center for Earth and Planetary Studies at the National Air and Space Museum and Director of the Smithsonian's Regional Planetary Image Facility. He is involved in three planetary missions: Mars Express (MARSIS Radar Sounder team), MESSENGER (MDIS imaging team), and the Lunar Reconnaissance Orbiter (LROC imaging team). He has served as Chair of Planetary Geology Division of the Geological Society of America, guest editor for *Geophysical Research Letters*, and on the editorial board of the journal *Geology*. His research interests are planetary tectonics, and planetary geology, geophysics and remote sensing.

RICHARD A. SCHULTZ is Professor of Geological Engineering and Geomechanics at the University of Nevada, Reno, where he teaches courses in geology, rock mechanics, and planetary science. He has served on the IUGS Commission on Comparative Planetology, as associate editor for the *Journal of Geophysical Research*, guest editor for the *Journal of Structural Geology*, and co-editor of the 35th U.S. Symposium on Rock Mechanics and the 33rd Symposium on Engineering Geology and Geotechnical Engineering, with visiting professorships at Woods Hole Oceanographic Institution and the University of Paris. His research interests include the evolution of fault systems on various planetary bodies.

Cambridge Planetary Science

Series Editors Fran Bagenal, David Jewitt, Carl Murray, Jim Bell, Ralph Lorenz, Francis Nimmo, Sara Russell

Books in the series

1. *Jupiter: The Planet, Satellites and Magnetosphere*[†]
   Edited by Bagenal, Dowling and McKinnon
   978-0-521-03545-3

2. *Meteorites: A Petrologic, Chemical and Isotopic Synthesis*[†]
   Hutchison
   978-0-521-03539-2

3. *The Origin of Chondrules and Chondrites*
   Sears
   978-0-521-83603-6

4. *Planetary Rings*
   Esposito
   978-0-521-36222-1

5. *The Geology of Mars*
   Edited by Chapman
   978-0-521-83292-2

6. *The Surface of Mars*
   Carr
   978-0-521-87201-0

7. *Volcanism on Io: A Comparison with Earth*
   Davies
   978-0-521-85003-2

8. *Mars: An Introduction to its Interior, Surface and Atmosphere*
   Barlow
   978-0-521-85226-5

9. *The Martian Surface: Composition, Mineralogy and Physical Properties*
   Edited by Bell
   978-0-521-86698-9

10. *Planetary Crusts: Their Composition, Origin and Evolution*
    Taylor and McLennan
    978-0-521-84186-3

11. *Planetary Tectonics*
    Edited by Watters and Schultz
    978-0-521-76573-2

[†] Issued as a paperback

PLANETARY TECTONICS

Edited by

THOMAS R. WATTERS
Smithsonian Institution, Washington, DC

RICHARD A. SCHULTZ
University of Nevada, Reno

CAMBRIDGE UNIVERSITY PRESS
Cambridge, New York, Melbourne, Madrid, Cape Town, Singapore,
São Paulo, Delhi, Dubai, Tokyo

Cambridge University Press
The Edinburgh Building, Cambridge CB2 8RU, UK

Published in the United States of America by Cambridge University Press, New York

www.cambridge.org
Information on this title: www.cambridge.org/9780521765732

© Cambridge University Press 2010

This publication is in copyright. Subject to statutory exception
and to the provisions of relevant collective licensing agreements,
no reproduction of any part may take place without the written
permission of Cambridge University Press.

First published 2010

Printed in the United Kingdom at the University Press, Cambridge

A catalogue record for this publication is available from the British Library

ISBN 978-0-521-76573-2 Hardback

Cambridge University Press has no responsibility for the persistence or
accuracy of URLs for external or third-party internet websites referred to
in this publication, and does not guarantee that any content on such
websites is, or will remain, accurate or appropriate.

Contents

List of contributors		*page* vii
Preface		xi
1	Planetary tectonics: introduction	1
	Thomas R. Watters and Richard A. Schultz	
2	The tectonics of Mercury	15
	Thomas R. Watters and Francis Nimmo	
3	Venus tectonics	81
	George E. McGill, Ellen R. Stofan and Suzanne E. Smrekar	
4	Lunar tectonics	121
	Thomas R. Watters and Catherine L. Johnson	
5	Mars tectonics	183
	Matthew P. Golombek and Roger J. Phillips	
6	Tectonics of small bodies	233
	Peter C. Thomas and Louise M. Prockter	
7	Tectonics of the outer planet satellites	264
	Geoffrey C. Collins, William B. McKinnon, Jeffrey M. Moore, Francis Nimmo, Robert T. Pappalardo, Louise M. Prockter and Paul M. Schenk	
8	Planetary structural mapping	351
	Kenneth L. Tanaka, Robert Anderson, James M. Dohm, Vicki L. Hansen, George E. McGill, Robert T. Pappalardo, Richard A. Schultz and Thomas R. Watters	
9	Strength and deformation of planetary lithospheres	397
	David L. Kohlstedt and Stephen J. Mackwell	

10	Fault populations *Richard A. Schultz, Roger Soliva, Chris H. Okubo and Daniel Mège*	457

Index 511

Color plates are located between pages 212 and 213.

Contributors

Robert Anderson
Jet Propulsion Laboratory
California Institute of Technology, Pasadena, California

Geoffrey C. Collins
Wheaton College, Norton, Massachusetts

James M. Dohm
Department of Hydrology and Water Resources
University of Arizona, Tucson, Arizona

Matthew P. Golombek
Jet Propulsion Laboratory
California Institute of Technology, Pasadena, California

Vicki L. Hansen
Department of Geological Sciences
University of Minnesota Duluth, Minnesota

Catherine L. Johnson
Earth and Ocean Sciences
University of British Columbia, Vancouver, Canada

David L. Kohlstedt
Department of Geology and Geophysics
University of Minnesota, Minneapolis, Minnesota

Stephen J. Mackwell
Lunar and Planetary Institute, Houston, Texas

George E. McGill
University of Massachusetts, Amherst, Massachusetts

William B. McKinnon
Washington University, Saint Louis, Missouri

Daniel Mège
Laboratoire de Planetologie et Geodynamique
UFR des Sciences et Techniques Université de Nantes, France

Jeffrey M. Moore
NASA Ames Research Center, Moffett Field, California

Francis Nimmo
University of California, Santa Cruz, California

Chris H. Okubo
Lunar and Planetary Laboratory
The University of Arizona, Tucson, Arizona

Robert T. Pappalardo
Jet Propulsion Laboratory
California Institute of Technology, Pasadena, California

Roger J. Phillips
Planetary Science Directorate,
Southwest Research Institute, Boulder, Colorado

Louise M. Prockter
Applied Physics Laboratory, Laurel, Maryland

Paul M. Schenk
Lunar and Planetary Institute, Houston, Texas

Richard A. Schultz
Geomechanics – Rock Fracture Group
Department of Geological Sciences and Engineering
University of Nevada, Reno, Nevada

Suzanne E. Smrekar
Jet Propulsion Laboratory
California Institute of Technology, Pasadena, California

Roger Soliva
Université Montpellier II
Département des Sciences de la Terre et de l'Environnement, France

Ellen R. Stofan
Proxemy Research, Laytonsville, Maryland

Kenneth L. Tanaka
U.S. Geological Survey, Flagstaff, Arizona

Peter C. Thomas
Center for Radiophysics and Space Research
Cornell University, Ithaca, New York

Thomas R. Watters
Center for Earth and Planetary Studies, National Air and Space Museum
Smithsonian Institution, Washington DC

Preface

There are few periods in the history of science that compare to the explosion of knowledge from robotic and manned exploration of the bodies of our solar system over the last 50 years. In this golden age of planetary exploration, hundreds of thousands of detailed images of the terrestrial planets, the outer planets and their icy satellites, and many asteroids and comets have been obtained by manned and unmanned spacecraft. In the near future, spacecraft already in flight will complete surveys of our innermost planet, Mercury, and provide the first high-resolution images of outermost Pluto.

In the pursuit of understanding the origins and geologic evolution of the solid bodies in the solar system, many similarities and differences have emerged in the processes that shaped their landscapes. One of the most fundamental of these processes is tectonics. The number and diversity of tectonic landforms is truly remarkable. The investigation of these tectonic landforms has stimulated an equally diverse range of studies, from the characterization and modeling of individual classes of tectonic landforms to the assessment of regional and global tectonic systems. These investigations expose the complex interplay between the forces that act on planetary crusts, both internal and external, and the mechanical properties of crustal material.

Over the past several decades, planetary tectonics has become an important component at geoscience and planetary science meetings, conferences, and workshops worldwide. In the year of the new millennium, a topical session entitled "Structure and Tectonics of Planets and Satellites" was held at the Geological Society of America Annual Meeting in Reno, Nevada (November 9–18, 2000). The overview talks presented in that session were the impetus for this book, the first comprehensive treatment of planetary tectonics.

We wish to thank the authors for their considerable work in collecting, synthesizing, and presenting the vast amount of material on planetary tectonics in their respective chapters, and the many reviewers for their detailed, thoughtful, and insightful evaluations of the chapters. As a result of this process, many new innovations and creative insights appear in these pages that have not been presented or published elsewhere. Finally, we gratefully acknowledge and thank the staff at Cambridge University Press for their expert assistance from inception through completion in the production of this book.

Thomas R. Watters
Richard A. Schultz

1
Planetary tectonics: introduction

Thomas R. Watters
*Center for Earth and Planetary Studies, National Air and Space Museum,
Smithsonian Institution, Washington, DC*

and

Richard A. Schultz
*Geomechanics – Rock Fracture Group, Department of Geological Sciences
and Engineering, University of Nevada, Reno*

Summary

The geocentric realm of tectonics changed with the dawn of robotic exploration of the other bodies of the solar system. A diverse assortment of tectonic landforms has been revealed, some familiar and some with no analogues to terrestrial structural features. In this chapter, we briefly introduce some of the major topics in the book. The chapters review what is known about the tectonics on Mercury, Venus, the Moon, Mars, the outer planet satellites, and asteroids. There are also chapters that describe the mapping and analysis of tectonic features and review our understanding of the strength of planetary lithospheres and fault populations.

1 Introduction

At the most basic level, tectonics concerns how landforms develop from the deformation of crustal materials. The root of the word "tectonics" is the Greek word "tektos," meaning builder. The building of tectonic landforms is in response to forces that act on solid planetary crusts and lithospheres. Tectonic landforms in turn provide a wealth of information on the physical processes that have acted on the solid-surface planets and satellites.

Until little more than a century ago, the study of tectonics and tectonic landforms was limited to those on the Earth. This changed in the early 1890s when G. K. Gilbert began to study the lunar surface with a telescope. He described sinuous ridges in the lunar maria and interpreted them to be anticlinal folds (see Watters and Johnson, Chapter 4). This marked the beginning of the scientific investigation of

Planetary Tectonics, edited by Thomas R. Watters and Richard A. Schultz. Published by Cambridge University Press. Copyright © Cambridge University Press 2010.

tectonic landforms on planetary surfaces. In the decades that followed, during the era when planetary exploration was limited to telescopic observations, the Moon was the only other object in the solar system that was known to have tectonic landforms.

In the 1960s, with the first successful launches and operations of lunar and interplanetary spacecraft, a new era of planetary exploration began. The Mariner 4 spacecraft returned the first high-resolution ($\sim$3 km/pixel) images of the surface of Mars in 1965. In one of the 22 images, a poorly defined linear feature that occurs in the area of a mapped "canal" was identified and interpreted to be evidence of escarpments associated with an eroded rift valley. Higher resolution images from subsequent missions (Mariner 9 and the Viking Orbiters) revealed the areas imaged by Mariner 4 were on the western and southern flanks of the Tharsis volcanic and tectonic province. The linear feature was actually a segment of Sirenum Fossae, a prominent extensional trough that is part of the Tharsis radial graben system (see Golombek and Phillips, Chapter 5; and Tanaka *et al.*, Chapter 8).

Since the 1960s, an armada of exploratory spacecraft have identified widespread evidence of tectonism on all the terrestrial planets, most of the satellites of the outer planets, and on a number of asteroids. Tectonic landforms on large and small solid bodies in the solar system are as ubiquitous as impact craters. They express crustal deformation across a very large range in length-scale, from a few tens of kilometers up to several thousand kilometers (see Schultz *et al.*, Chapter 10; Watters and Nimmo, Chapter 2; and Watters and Johnson, Chapter 4). Deformation occurs in a wide range of crustal materials, from silicates to water ice to exotic ices, and exhibits local, regional, and global distributions.

The chapters in this book explore the diversity and similarity of tectonic landforms known to exist in the solar system. The characterization of tectonic landforms, methods of mapping the spatial distribution and bounding the age of structures and deformed terrains, the relationship between individual faults and fault populations, the kinematics and mechanics of crustal deformation, the thermal and mechanical properties of deformed lithospheres, and models for the origin and evolution of tectonic stresses are all described.

2 Terrestrial planets

The foundation for the interpretation of tectonic landforms on the terrestrial planets, and also on icy satellites in the outer solar system, is inexorably rooted in our understanding of tectonic landforms and processes on the Earth. On our planet, the vast majority of tectonic landforms occur within the context of plate tectonics. The largest scale tectonic landforms on Earth are the direct result of the relative motion of the lithospheric plates. Thus, most of Earth's tectonic landforms can be described as localized, occurring in or near current or past plate margins. Ironically, the plate tectonics paradigm was gaining acceptance during the same period, the 1960s to 1970s, when the exploration of the Moon and other terrestrial planets was gaining momentum. Yet, the large-scale manifestations of plate tectonics – mountain belts, spreading centers, and ocean trenches – now appear to be unique to the Earth.

In contrast to Earth, tectonism on Mercury, Venus, the Moon, and Mars is generally more distributed in character. Rather than a lithosphere comprised of a mosaic of individual plates that can shift and move relative to one another, the lithospheres of the other terrestrial planets and Earth's Moon behave largely as a single continuous shell. This is particularly puzzling in the case of Venus, a tectonically dynamic planet almost identical to the Earth in size and bulk density (see McGill *et al.*, Chapter 3). The relatively small sizes of the Moon and Mercury might be expected to work against the development of plate tectonics as a dominant interior cooling mechanism given the tenet that a planet's tectonic vigor and longevity scale approximately with its volume. On Mercury, contractional deformation (i.e., surface-breaking thrust faults) is broadly distributed in the regions of the planet imaged by spacecraft, possibly caused by contraction of the lithosphere from interior cooling (see Watters and Nimmo, Chapter 2). Tectonism on the Moon, on the other hand, is generally localized in the mare impact basins, attributed to subsidence and flexure-related deformation (see Watters and Johnson, Chapter 4). The bimodal distribution of topography and crustal thickness on Mars is strikingly similar to that on the Earth. However, tectonics on Mars, a planet intermediate in size between Mercury and Earth, occurred predominantly in the Tharsis volcanotectonic province and in only a few other major areas (see Golombek and Phillips, Chapter 5). Thus, each of the terrestrial planets and Earth's Moon exhibit a distinctive style of tectonics, expressing different interior structures, different mechanisms of heat loss, and different paths of crustal evolution.

2.1 Mercury

Mercury has many unusual characteristics that make it stand out in the spectrum of terrestrial planets in the solar system. As first in order from the Sun, Mercury is located in a harsh environment, exposed to intense solar radiation, flares, and tidal forces. It is the smallest terrestrial planet, about one-third the size of the Earth and only slightly larger than Earth's Moon. Yet, Mercury's mean density is comparable to that of the Earth, indicating that the proportion of its core volume to that of its mantle and crust is much larger than that of the other terrestrial planets.

Chapter 2 by Watters and Nimmo describes tectonic landforms imaged by Mariner 10 in three flybys of Mercury during the mid 1970s, along with some newly discovered tectonic landforms imaged by the MESSENGER spacecraft in its first flyby of Mercury. New images and data obtained by MESSENGER in two future flybys of Mercury and during the orbital phase of the mission will revolutionize our understanding of this intriguing planet.

Landforms indicative of both crustal shortening and extension are evident on Mercury. The number and scale of tectonic landforms was one of the surprising aspects of Mercury first revealed by Mariner 10. Lobate scarps are the most widely distributed tectonic landforms, interpreted as thrust faults that cut the oldest intercrater plains and the youngest smooth plains (see Watters and Nimmo, Chapter 2). Associated with lobate scarps are tectonic landforms known as high-relief ridges. Their maximum relief can exceed 1 km. In some cases, high-relief ridges transition along-strike into lobate scarps, suggesting they too are an expression of crustal shortening. Wrinkle ridges are another common tectonic landform on Mercury. These complex, positive relief structures occur predominantly in smooth plains in the interior of the Caloris impact basin and in the exterior smooth plains to the east. Wrinkle ridges, another result of crustal shortening, are structural anticlines formed by thrust faulting and folding.

Also surprising is the near absence of extensional landforms outside the interior smooth plains materials of the Caloris basin. Extensional deformation of the interior smooth plains is expressed by a complex and widespread network of basin radial and basin concentric troughs formed by graben that crosscut the wrinkle ridges within the Caloris basin. These extensional troughs appear to be among the youngest endogenic features known on Mercury. The origin of the stresses that form the graben are likely due to exterior loading of the Caloris basin or lateral crustal flow causing uplift of the basin floor (see Watters and Nimmo, Chapter 2).

Global tectonic stresses on Mercury may have arisen from a number of sources. Cooling of Mercury's interior is thought to have resulted in global contraction and widespread thrust faulting. Tectonic stresses may also have occurred from slowing of Mercury's rotation by despinning due to solar tides, and relaxation of an early equatorial bulge. A combination of global contraction and tidal despinning or a combination of global contraction and the formation of the Caloris basin have also been suggested. Mantle convection may have also been a source of stress. Each model predicts distinctive spatial and temporal distributions of lobate scarps, and therefore a global investigation of Mercury's tectonic landforms is needed for these models to be fully evaluated.

2.2 Venus

Venus also occupies a distinctive place in the spectrum of terrestrial planets in the solar system. In some ways it is strikingly similar to the Earth, but the difference between Venus and Earth are equally perplexing. The two planets could pass as twins with respect to diameter and bulk density. However, Venus and Earth have little in common when it comes to the composition and pressure of their atmospheres, their surface temperatures, and their tectonics. Chapter 3 by McGill *et al.* describes the tectonic landforms and terrains on Venus.

The exposed surface of Venus is relatively young, with the total population of impact craters of less than 1000 (see McGill *et al.*, Chapter 3). Although volcanic plains are the dominant terrain on Venus, features analogous to spreading centers, subduction zones, oceanic transform faults, and other diagnostic plate-boundary signatures found on Earth are absent. This suggests that Earth-like plate tectonics has not developed on Venus, at least during the recent period of time preserved in the crust (less than ~700 Ma; McGill *et al.*, Chapter 3; see Figure 8.1 in Tanaka *et al.*, Chapter 8).

Even in the absence of plate tectonics, Venus displays a surprising number of tectonic landforms and terrains. McGill *et al.* (Chapter 3) describe the several classes of tectonic terrains on Venus that include plains, tesserae, coronae, and chasmata. The expansive plains of Venus display a diverse suite of tectonic landforms, including wrinkle ridges, fractures and graben, polygonal terrains, ridge belts, fracture belts, and broad-scale uplifts and ridges. Some of these landforms are also associated with either coronae (large subcircular volcanotectonic complexes) or chasmata (lithospheric rift valleys). Coronae vary greatly in size, topographic relief, and in the extent of associated volcanism, and are generally associated with distinctive curvilinear fracture and graben systems. They are thought to have formed over thermal mantle plumes and may account for a significant fraction of the total planetary heat loss on Venus, thus replacing plate tectonics as the dominant interior heat loss mechanism. The large-scale graben that make up chasmata can have kilometer-scale relief and may be thousands of kilometers long. Tesserae are complex terrains consisting of multiple sets of ridges or graben that intersect at high angles, indicating a sequence of polyphase folding and extension over broad regions of Venus.

McGill *et al.* (Chapter 3) discuss the two principal hypotheses for the timing of the formation of the tectonic terrains. One hypothesis proposes that similar tectonic landforms formed at the same time globally, while the competing hypothesis postulates that tectonic landforms formed at different times in different places. In the absence of plate tectonics and liquid water, the tectonic history of Venus could conceivably involve a combination of both scenarios. An important question is how Venus and Earth, with nearly the same bulk properties, evolved into two distinctly different planets. Significant differences in tectonic evolution are predicted by geophysical models that involve a high-viscosity crust, asthenosphere, and mantle due to a lack of liquid water, motivating tests of these models constrained by the tectonic landforms.

2.3 The Moon

Among the diverse bodies in the solar system that display evidence of tectonism, Earth's Moon stands out because most of the tectonic landforms are found in and around the nearside lunar maria, impact basins subsequently flooded by basalt. Chapter 4 by Watters and Johnson describes the tectonic landforms that have been recognized on the Moon. The major tectonic landforms are wrinkle ridges, that occur exclusively in mare basalts, and rilles, narrow troughs that occur along basin margins and the adjacent highlands. Lunar rilles are the result of extension, formed by graben. The lunar farside, however, is not without tectonics. Lobate scarps are tectonic landforms that occur primarily in the farside highlands (Watters and Johnson, Chapter 4). Lunar lobate scarps are small-scale landforms relative to the nearside wrinkle ridges and rilles. They, like wrinkle ridges, are the surface expression of thrust faults.

The timing of tectonic events on the Moon suggests an interplay between periods of extension and compression. The crustal extension that formed the basin-localized graben ceased at $\sim$3.6 Ga (see Figure 8.1 in Tanaka *et al.*, Chapter 8), while crustal shortening in the maria continued to $\sim$1.2 Ga ago (Watters and Johnson, Chapter 4). The much earlier termination of extension on the Moon may have occurred when compressional stresses from global contraction were superposed on the basin-localized extensional stress, marking a shift from net expansion of the lunar interior to net contraction. Lobate scarps may be the youngest tectonic landforms on the Moon, forming less than 1 Ga ago. Young lobate scarp thrust faults indicate late-stage compression of the lunar crust.

One of the important accomplishments of the Apollo missions was deployment of a seismic network on the Moon. The deepest moonquakes ($\sim$800–1000 km) are spatially associated with nearside maria (Watters and Johnson, Chapter 4). These deep moonquakes may be related to the effects of impact basin formation and the production of mare-filling basalts. Shallow-depth moonquakes, some of which may occur within the lunar crust, are also associated with the nearside maria. This suggests that some moonquakes may be associated with nearside lunar faults and that the Moon may still be tectonically active.

Lunar gravity data show that many of the maria have large positive anomalies indicative of mass concentrations ("mascons") due to the mare basalts on an impact-induced thinned lithosphere. Watters and Johnson (Chapter 4) note that the shoulders of the positive anomalies, that are spatially correlated with basin-interior wrinkle-ridge rings in some maria, suggest a pan- rather than a bowl-shaped mare-fill geometry. Models for the origin of stresses in mascons suggest that the spatial distribution of the wrinkle ridges and rilles is best fit by a relatively uniform thickness of mare basalts. Watters and Johnson (Chapter 4) conclude that the stresses and the estimated contractional strain expressed by the lobate scarps are consistent with thermal history models that predict a small change in lunar radius over the last 3.8 Ga.

2.4 Mars

Mars is intermediate in size in the spectrum of the terrestrial planets. By comparison with smaller bodies, the Moon and Mercury, the size of Mars appears to have been favorable to sustain tectonic activity throughout much of its geologic history. Although smaller in size than Venus and the Earth, Mars has some of the largest tectonic landforms and the most intensely deformed terrains in the solar system. Golombek and Phillips (Chapter 5) describe the principal tectonic landforms and provinces on Mars.

Two large-scale physiographic provinces dominate the planetary landscape, and their development and evolution are directly responsible for, or have influenced the formation of, most of the tectonic landforms on Mars (Golombek and Phillips, Chapter 5). These are the Tharsis province and the highland–lowland dichotomy. The highland–lowland dichotomy boundary separates the sediment-covered northern lowland plains from the more rugged and heavily cratered southern highlands. The Tharsis province is a broad, topographic rise superimposed on the dichotomy boundary in the western hemisphere. Its massive volcanoes and large expanses of volcanic plains are punctuated by a vast array of extensional and compressional fault populations that include the striking Valles Marineris rift system.

Extensional tectonic landforms on Mars vary greatly in scale, from narrow fractures and graben to rift valleys up to 100 km wide and kilometers deep. The most common compressional tectonic landform on Mars is wrinkle ridges. These anticlines occur in broadly distributed plains units in the Tharsis province, and elsewhere in the highlands and the northern lowlands. Another common compressional tectonic landform is lobate scarps. Many of the lobate scarps on Mars occur in the highlands along the dichotomy boundary in the eastern hemisphere. Large-scale ridges, landforms that can have up to several kilometers in relief, have also been identified on Mars.

Golombek and Phillips (Chapter 5) review two predictive models for the gravity field and topography associated with Tharsis: elastic spherical-shell loading and plume-related uplift of a lithospheric shell. In elastic spherical-shell loading models, Tharsis constitutes a massive lithospheric load supported by membrane (bending) stresses in a rigid lithospheric shell. In plume-related models, the long-wavelength topography of Tharsis supported by a mantle plume and associated radial dikes are responsible for forming the graben system. Outside of the Tharsis province, the broad distribution of wrinkle ridges on Mars indicates widespread contractional deformation across the planet. While stresses from the Tharsis load appear to have had a global influence, many wrinkle ridges in the eastern hemisphere highlands and in the northern lowlands are not obviously related to Tharsis, implying additional controls on the planet's tectonic history. Golombek and Phillips (Chapter 5) suggest that a globally isotropic contractional strain event, modulated by Tharsis stresses in the western hemisphere, can account for the distribution and orientation of wrinkle ridges.

3 Small bodies of the solar system

It might be expected that the large number of smaller bodies of the solar system would not display evidence of tectonics, simply because their size would limit the number of mechanisms that generate significant stress in larger bodies. However, a surprising number of small bodies have tectonic landforms. Small bodies, defined as objects with radii less than about 200 km, include the satellites of some planets, along with many of the known asteroids, cometary nuclei, and Kuiper-belt objects. In the spectrum of objects in the solar system, they are the remnants of the building blocks of the planets and larger satellites, and perhaps fragments of large bodies that were disrupted. A variety of spacecraft have now imaged small satellites, asteroids, and cometary nuclei, providing insight into tectonic processes on these bodies. Perhaps the most dramatic to date was the Near Earth Asteroid Rendezvous (NEAR) mission. After orbiting 433 Eros, the NEAR Shoemaker spacecraft touched down on the asteroid. Thomas and Prockter (Chapter 6) describe tectonic landforms that have been discovered on the solar system's small bodies.

The most common tectonic landforms on small bodies are grooves, troughs, and ridges. Although these landforms deform the regolith of the bodies, the underlying fractures and faults are likely rooted in more competent substrate materials (Thomas and Prockter, Chapter 6). The scale of tectonic landforms on small bodies varies as greatly as their size. Grooves are generally small-scale features that vary morphologically from linear, straight-walled depressions to rows of coalesced pits, and are common tectonic landforms on Gaspra, Ida, Eros and Phobos. Grooves likely result from extension, expressed by collapse of the regolith into near-vertical subsurface fractures or cracks. Troughs are larger-scale landforms than grooves, typically exhibiting greater widths and depths. An example is Calisto Fossae, which is made up of some of the largest tectonic landforms on Eros. Calisto Fossae and troughs on other small bodies are extensional landforms formed by graben. Ridges are positive-relief landforms found on Eros and Gaspra. One of the most striking examples of these is Rahe Dorsum on Eros, an asymmetric ridge that extends over a significant portion of the length of the asteroid. Rahe Dorsum appears to be the surface expression of thrust faulting.

Thomas and Prockter (Chapter 6) review models for the origin of structure-forming stresses on small bodies. Internal thermal activity on small bodies, although limited, may have been significant enough to induce tectonic stresses. External thermal stresses are created by temperature variations due to short- and long-period changes in rotation or distance from the Sun. Significant stress can also be induced externally from collisions with other small bodies. Sufficiently high-velocity impacts will result in stresses that far exceed the yield strengths of the small body, leading to the growth of tectonic landforms. Another external source of stress is tidal forces, which are induced by close encounters of small bodies with larger bodies and planets.

4 Outer planet satellites

The satellites of the outer planets (Jupiter, Saturn, Uranus, and Neptune) are diverse bodies that rival the inner planets and Earth's Moon in size and the range of tectonic complexity. Jupiter's moon Ganymede, and Saturn's moon Titan, are both larger than the planet Mercury. Water ice, rather than silicates, is the dominant constituent of the near-surface crusts of most of the larger outer planet satellites. Jupiter's Io is a notable exception: its widespread volcanism and isolated giant mountains present a sharp contrast to the icy satellites. Most of the tectonic landforms on the outer planet satellites are indicative of crustal extension. Landforms that express crustal shortening are rare. Collins *et al.* (Chapter 7) describe the vast array of familiar to enigmatic tectonic landforms on Io and the icy satellites. Just a few examples of these are summarized below.

The crust of Europa records a long-lived and complex history of tectonic activity. Common tectonic landforms there include troughs, ridges, bands, and undulations. Troughs on Europa are linear to curvilinear in plan form and often V-shaped in cross section, and are interpreted to be tension fractures analogous to terrestrial crevasses. Ridges are the most common landform on Europa. They often occur in pairs forming a double ridge with a medial trough. Numerous hypotheses have been proposed for Europa's ridges, many involving extensional stresses generated by mechanisms such as tidal squeezing, volcanism, dike intrusion, or diapirism. The diversity of hypotheses attests to the continuing difficulty in interpreting this exotic structural landform.

Another unusual tectonic landform on Europa is bands, which are polygonally shaped areas with sharp boundaries formed by crustal extension that appear to be offset laterally and displaced from adjacent bands. Bands can be reconstructed geometrically by using rigid-block rotations to restore originally contiguous landforms, suggesting significant strike-slip motion on the body. Subtle shading variations in high-resolution Galileo images revealed undulations comprised of alternating zones of small-scale ridges and troughs. These undulations are interpreted to be folds with bending-related fractures along the inferred anticlinal rises and compressional ridges in the synclinal valleys. Such folds may accommodate some part of Europa's widespread crustal extension.

A principal source of tectonic stress on Europa arises from nonsynchronous rotation which causes a shift in the moon's surface relative to its fixed tidal axes. Diurnal tidal variations also induce a daily rotating stress field in the icy crust that changes the orientation of cracks developing in response to stresses from nonsynchronous rotation.

One of Saturn's smaller satellites, Enceladus, has some of the most remarkable tectonic landforms found to date on icy bodies. The south polar region of Enceladus has a series of subparallel troughs flanked by ridges. Informally described as "tiger stripes," these features are likely associated with thermal anomalies and active plumes of water vapor (Collins *et al.*, Chapter 7). Arcuate scarps in the south polar region are interpreted to be evidence of thrust faults. In contrast to the south polar region, the more ancient north polar cratered terrain appears undeformed. The deformation on Enceladus is likely to have arisen from a number of sources, including tidally induced stress, diapirs in a subsurface ocean, crustal subsidence, poleward reorientation, and ice shell thickening above a subsurface ocean.

The largest of Neptune's moons is Triton. Although its crust is thought to be composed mainly of water ice or ammonia hydrate ice, evidence of other exotic ices has been found. The dominant tectonic landform on Triton is double ridges similar to those found on Europa. These sinuous features are characterized by continuous, along-strike depressions that are flanked by ridges. The stresses that formed these ridges may reflect tidally induced shear heating generated during the highly eccentric phase of Triton's orbital evolution. Triton has one of the most puzzling surface features known on icy bodies, the "cantaloupe terrain." Quasi-circular shallow depressions with raised rims give the appearance of a cantaloupe rind. This terrain may result from diapirism and gravity-driven overturn due to an instability caused by more dense ices emplaced on less dense ice layers.

5 Structural mapping on planetary bodies

Tectonic landforms on planetary bodies beyond the Earth have been identified and analyzed using remote sensing data. By and large, the tectonic landforms discussed in this book have been characterized using images and topographic data obtained by spacecraft observations. The attributes of a dataset determine what scales and aspects of tectonic landforms can be identified and mapped. Tanaka *et al.* (Chapter 8) describe the principal datasets and methods utilized in the recognition and geologic mapping of tectonic landforms. Chapter 8 and the two that follow contain material that connects the previous chapters, providing a synthesis of concepts and processes common to planetary tectonics on any object in the solar system.

Planetary geologic mapping relies on spacecraft data and generally lacks ground truth. One of the cornerstones of planetary structural mapping is imaging. Spacecraft-borne imaging systems are usually designed to obtain images in the visible and infrared part of the electromagnetic spectrum. The spatial resolution and image quality are governed by the characteristics of the camera(s), the mission parameters, and the properties of the planetary target. Images recorded on film were replaced by digital images obtained by vidicon cameras. Charge-coupled device (CCD) cameras are the current standard for planetary imaging. Multiple images are combined into mosaics that can provide regional and even global-scale image maps. Essential to the identification and morphologic analysis of tectonic landforms is slope information, where brightness variations are related to the slope orientation and incident angle. Albedo variations of surface materials may also highlight tectonic landforms and aid in their geologic interpretation.

An important alternative to passive visible and infrared imaging is radar. Radar uses the microwave part of the spectrum and is capable of obtaining image and altimetry data independent of solar illumination or atmospheric density. Factors that influence radar return are surface slope, surface roughness, and the electrical properties of the surface materials. For objects in the solar system with dense, opaque (at visible wavelengths) atmospheres such as Venus and Titan, radar has been utilized to obtain images of their surfaces. The synthetic aperture radar (SAR) instrument on the Magellan spacecraft returned a global set of high-resolution images of the surface of Venus, revealing the planet's complexly deformed terrains.

Another of the cornerstones of planetary mapping is topography. Accurate characterization of the morphology and the structural relief of tectonic landforms are often impossible without topographic data. Topography is also critical to estimates of strain and in quantitatively evaluating kinematic and mechanical models for tectonic features (see Tanaka *et al.*, Chapter 8; and Schultz *et al.*, Chapter 10). As well as an important adjunct to imaging, topography has been used to identify and define the extents of tectonic landforms not clearly resolved in images. Topography of planetary bodies has been obtained from a variety of sources. Methods to derive topographic data from images includes shadow measurements, photoclinometry (shape from shading), and stereo imaging. Earth-based radar altimetry and radar interferometry are also an important independent source of topographic data. Finally, radar and laser altimeters on orbiting spacecraft have been used to obtain global topographic datasets.

Structural and tectonic maps are invaluable in deducing the tectonic evolution of a body. Combined with appropriate analysis techniques, such maps can provide insight into the timing of deformation, sources of stress, the mechanical properties of deformed materials, and the thermal and interior evolution of the body. Tanaka *et al.* (Chapter 8) review the major tectonic landforms identified and mapped on the terrestrial planets, the Moon, and the satellites of the outer planets.

6 Planetary lithospheres

The mechanical properties of a lithosphere have a major influence on the tectonic processes that shape it. As in the analysis of most planetary tectonic landforms, the understanding of planetary lithospheres is strongly dependent on the understanding of mechanical properties of Earth's lithosphere. This is informed primarily by experimental investigations of the properties of minerals and rocks at elevated temperatures and pressures. Recent experimental studies have focused on the question of the rock strength of other terrestrial planets and outer planet satellites. Kohlstedt and Mackwell (Chapter 9) review work on lithospheric mechanical properties of the planets Venus and Mars, along with Jupiter's moons Europa, Ganymede, and Io.

The exotic tectonic landforms on Europa and Ganymede are intimately related to the rheological properties of their water-ice crustal shells that likely overlie mantles of liquid water. These properties also determine the ability of an icy body to retain tectonic landforms. Deformation of ice, like other crystalline materials, occurs by diffusion of ions or diffusion creep and propagation along grain boundaries or dislocation creep. Modeling that incorporates dislocation creep, which is sensitive to grain size, predicts topography at wavelengths consistent with Ganymede's grooved terrain. Relaxation models for Ganymede suggest that crater topography can survive for billions of years, as observations suggest (Kohlstedt and Mackwell, Chapter 9).

Water, even in very small amounts, also has a major influence on the rheological properties of the silicate lithosphere of Earth. The influence of water or "water weakening" is likely all but absent, however, on some of the other terrestrial planets. The lithosphere of Venus, for example, is expected to be completely dehydrated because of the high surface temperatures and widespread volcanic activity (Kohlstedt and Mackwell, Chapter 9). The high strength of basaltic rocks under dry conditions means that most of the strength of Venus's lithosphere is in the crust. Such a strong crust is capable of supporting topography for periods of tens of millions of years. A low-strength contrast between the Venusian crust and mantle predicted by experimental work suggests that lithospheric deformation there is likely strongly coupled to mantle convection, unlike the Earth that has a wet, weak asthenosphere. Thus, regional- and global-scale tectonics on Venus may directly reflect mantle processes.

In contrast to Venus, an early wet Mars may have contributed to a lithosphere that was relatively weak. In addition, however, the Fe content of Martian mantle rocks is thought to be about twice that of Earth's mantle. Experimental data indicate that mantle strength decreases with increasing Fe content. Water-related weakening is also greater in Fe-rich rocks than in Fe-poor rocks. A weak early Martian mantle may have contributed to mantle dynamics controlled by stagnant-lid convection rather than plate tectonics (Kohlsted and Mackwell, Chapter 9). This would have resulted in interior heat loss through widespread volcanic activity and possibly through crustal delamination. The Tharsis volcanotectonic province and the crustal dichotomy may then be expressions of this stagnant-lid mode of interior cooling.

7 Planetary fault populations

The primary window into the state of stress in crustal materials of a planetary body, past and present, is its fault populations. Fault populations also express the magnitude of strain and reflect the mechanical properties and strength of the deformed rocks. The characteristics of faults, such as their length, height, displacement, and spacing are dependent on each other. Thus, knowledge of some of these characteristics can provide insight into the relationships of other characteristics and permit quantification of the fault-related deformation. Schultz *et al.* (Chapter 10) review the stress states in a planetary lithosphere, the characteristics of fault populations, the structural topography generated by faulting, and methods of determining strains from planetary fault populations.

The state of stress in the Earth's crust, as determined from *in situ* stress measurements, is generally compressive and controlled in magnitude by the frictional resistance of the fractured crust. Stress differences that are sufficiently greater than the lithostatic stress result in faulting. These values provide a basis for defining the stress states and failure criteria for the crusts and lithospheres of other planetary bodies.

Individual faults in a population can occur as isolated or linked structures in a wide range of sizes and geometrical configurations. Statistical analysis reveals a consistent scaling relationship between the maximum displacement and fault length. This relation describes, in part, how faults grow because displacement accumulates with lateral and/or down-dip growth of the fault (Schultz *et al.*, Chapter 10). The displacement–length (D/L) ratio describes the rate of displacement accumulation relative to the fault length. Analysis of multiple fault populations on the Earth and other terrestrial planets indicates that each population has a particular D/L ratio that is controlled by factors such as planetary gravity, lithology, and three-dimensional fault geometry.

Linkage of initially isolated faults is an important process in fault growth. As the faults grow and interact, shear stress around the relay zone increases, and transfer of displacement on the segments results in an increase in the displacement gradient at the ends of the faults. The interaction of fault segments can result in an increase of the D/L ratio by linkage of the temporarily over-displaced fault segments. After a period of fault displacement recovery, the resulting linked segmented fault can have the D/L ratio of single isolated faults. One of the exciting frontiers in the study of planetary fault populations is the recent identification of vertical restriction of faults on Mars by mechanical stratigraphy in the crust or lithosphere, which influences fault lengths, spacing, topography, and strain magnitudes in both space and time.

8 Conclusions

In this introduction to the chapters that follow, only a small number of the many and diverse aspects of planetary tectonics are highlighted. The planets, satellites, and small bodies of our solar system have a remarkable collection of tectonic landforms. Just as remarkable are the similarities and differences in the tectonics of bodies with silicate and non-silicate crusts. Water, whether enriched or absent in the crustal material of a terrestrial planet, or as the major constituent in the icy crust of an outer planet satellite, has an extraordinary influence on the style and magnitude of deformation. The sources of tectonic stresses are also remarkably diverse, arising from everything from interior heat loss, external tidal influences, or collisional interactions with other objects. We hope the following chapters will both inform and stimulate interest in planetary tectonics. Each chapter has been peer-reviewed and revised accordingly.

2
The tectonics of Mercury

Thomas R. Watters
*Center for Earth and Planetary Studies, National Air and Space Museum,
Smithsonian Institution, Washington, DC*

and

Francis Nimmo
Department of Earth and Planetary Sciences, University of California, Santa Cruz

Summary

Mercury has a remarkable number of landforms that express widespread deformation of the planet's crustal materials. Deformation on Mercury can be broadly described as either distributed or basin-localized. The distributed deformation on Mercury is dominantly compressional. Crustal shortening is reflected by three landforms, lobate scarps, high-relief ridges, and wrinkle ridges. Lobate scarps are the expression of surface-breaking thrust faults and are widely distributed on Mercury. High-relief ridges are closely related to lobate scarps and appear to be formed by high-angle reverse faults. Wrinkle ridges are landforms that reflect folding and thrust faulting and are found largely in smooth plains material within and exterior to the Caloris basin. The Caloris basin has an array of basin-localized tectonic features. Basin-concentric wrinkle ridges in the interior smooth plains material are very similar to those found in lunar mascon basins. The Caloris basin also has the only clear evidence of broad-scale, extensional deformation. Extension of the interior plains materials is expressed as a complex pattern of basin-radial and basin-concentric graben. The graben crosscut the wrinkle ridges in Caloris, suggesting that they are among the youngest tectonic features on Mercury. The tectonic features have been used to constrain the mechanical and thermal structure of Mercury's crust and lithosphere and to test models for the origin of tectonic stresses. Modeling of lobate scarp thrust faults suggests that the likely depth to the brittle–ductile transition (BDT) is 30 to 40 km. Plausible thermal and mechanical structures for Mercury at the time of faulting suggest an elastic thickness T_e of 25 to 30 km and a heat flux of roughly 30 mWm^{-2}. The thickness of the crust is also constrained to be <140 km. A combination of despinning and thermal contraction may account for equatorial N–S and polar E–W trending lobate scarp thrust faults,

Planetary Tectonics, edited by Thomas R. Watters and Richard A. Schultz. Published by Cambridge University Press. Copyright © Cambridge University Press 2010.

the latter by reactivation of normal faults. However, the observed spatial and temporal distribution of lobate scarps suggests other sources of stress may be important. Stresses resulting from mantle convection may have been significant, while those due to polar wander and buoyancy forces are currently uncertain. Loading in the Caloris basin from infilling of volcanic material likely produced significant compressional stresses that formed the wrinkle ridges. The Caloris graben are probably due to basin exterior loading or lateral flow of the lower crust causing uplift of the basin floor. The absence of widespread extensional faulting outside the Caloris basin is probably due to a compressive stress bias resulting from global thermal contraction. Modeling suggests that T_e during Caloris loading was $\sim$100 km, up to a factor of four larger than estimates of T_e based on lobate scarp thrust faults.

1 Introduction

Mercury is a planet with a number of important characteristics. It is the smallest of the terrestrial planets, not much larger than Earth's moon, and it is the closest planet to the Sun. In spite of its small size, Mercury's mean density is comparable to that of the Earth. This indicates that the proportion of Mercury's core volume to that of its mantle and crust is much larger than the other terrestrial planets. In three flybys during 1974 and 1975, Mariner 10 imaged about 45% of Mercury's surface ($\sim$90% of the eastern hemisphere). The Mercury Surface, Environment, Geochemistry, and Ranging (MESSENGER) mission successfully completed the first of three flybys in January 2008 and will become the first spacecraft to orbit Mercury in 2011 (Solomon et al., 2001, 2007, 2008). During the first flyby, images of 21% of the hemisphere unseen by Mariner 10 were obtained by MESSENGER's Mercury Dual Imaging System (MDIS) cameras (Hawkins et al., 2007). Although the summary presented in this chapter is based largely on the Mariner 10 view of Mercury, some of the early new findings from MESSENGER are described.

At first glance, the surface of Mercury looks much like Earth's moon. However, images revealed evidence of tectonism on a much larger scale than exists on the Moon. In fact, the scale of tectonic activity is comparable to that of the larger terrestrial planets. Preserved in Mercury's ancient crust is evidence of post late heavy bombardment crustal deformation. The tectonics of Mercury can be broadly described as either distributed or basin-localized. Distributed deformation on Mercury is dominantly contractional. Three landforms have been found on Mercury that reflect crustal shortening: lobate scarps, wrinkle ridges, and high-relief ridges. Lobate scarps are linear or arcuate features that are asymmetric in cross section, generally consisting of a steeply sloping scarp face and a gently sloping back scarp (Strom et al., 1975; Cordell and Strom, 1977; Dzurisin, 1978; Melosh and McKinnon, 1988; Watters et al., 1998). Where lobate scarps transect impact craters, there is clear evidence of offsets in wall and floor materials, suggesting they are the expression of surface-breaking thrust faults (Strom *et al.*, 1975; Cordell and Strom, 1977; Melosh and McKinnon, 1988; Watters *et al.*, 1998, 2001).

Wrinkle ridges, in contrast to lobate scarps, are generally more complex morphologic features. They typically consist of a broad, low-relief arch with a narrow superimposed ridge (Strom, 1972; Bryan, 1973; Maxwell *et al.*, 1975). A compressional origin involving a combination of folding and thrust faulting is supported by studies of terrestrial analogues (Plescia and Golombek, 1986; Watters, 1988; Schultz, 2000); however, there is no consensus on the geometry and number of faults or if the faults are surface breaking or non-surface breaking (blind) (see Schultz, 2000; Golombek *et al.*, 2001).

High-relief ridges are a much less common tectonic landform than lobate scarps and wrinkle ridges (Watters *et al.*, 2001). They are generally symmetric in cross section with greater relief than wrinkle ridges. Like lobate scarps, when high-relief ridges transect impact craters they deform the walls and floors. Some high-relief ridges transition into lobate scarps, suggesting that the origins of the two structures are related.

Unlike the other terrestrial planets, evidence of crustal extension on Mercury is nearly absent. The only extensional deformation on Mercury found thus far occurs in the interior plains material of three impact basins (Solomon *et al.*, 2008; Watters *et al.*, 2009a, b). Extension in the Caloris basin is manifested by a complex network of linear and sinuous narrow troughs that are interpreted to be graben (Strom *et al.*, 1975; Melosh and McKinnon, 1988; Watters *et al.*, 2005). No graben are found in the annulus of materials exterior to the Caloris basin.

The array of tectonic features on Mercury provides important clues to the properties and evolution of the planet's crust and lithosphere and its thermal history. The spatial distribution and contractional strain expressed by Mercury's lobate scarps are important to constrain thermal history models and models for the origin of tectonic stresses (see Melosh and McKinnon, 1988). Models for the origin of lobate scarps that emerged from the post Mariner 10 era studies of Mercury are global contraction due to secular cooling of the interior, tidal despinning, a combination of thermal contraction and tidal despinning, and a combination of thermal contraction and stresses related to the formation of the Caloris basin (Strom *et al.*, 1975; Cordell and Strom, 1977; Melosh and Dzurisin, 1978a; Pechmann and Melosh, 1979; Melosh and McKinnon, 1988; Thomas *et al.*, 1988; Dombard and Hauck, 2008).

Interest in Mercury's geologic and tectonic evolution has been rekindled by the new planetary missions to the iron planet. Meantime, new insights into the nature of the tectonic features, their spatial and temporal distribution, the origin of the tectonic stresses, and the mechanical and thermal structure of Mercury's crust and lithosphere have been gained through the continued analysis of Mariner 10 images. In the following sections, we will describe the current understanding of Mercury's tectonic features and the constraints they provide on existing geophysical and thermal history models.

2 Tectonic features of Mercury

2.1 Topographic data

A major challenge to the identification and analysis of tectonic features on Mercury has been the paucity of topographic data. Topographic data for Mercury have been derived from shadow measurements (Strom *et al.*, 1975; Malin and Dzurisin, 1977; Pike, 1988; Watters, 1988), photoclinometry (Hapke *et al.*, 1975; Mouginis-Mark and Wilson, 1981), point stereoanalysis (Dzurisin, 1978), and Earth-based radar altimetry (Harmon *et al.*, 1986; Harmon and Campbell, 1988; Clark *et al.*, 1988).

Earth-based radar altimetry is an important source of reliable topographic data for Mercury. Altimetry data is determined by transmitting short pulses (12.5 cm or S-band wavelength) and analyzing the radar echoes using the delay-Doppler method (see Harmon *et al.*, 1986). The limitation of Earth-based radar altimetry is that, because of Mercury's 7° orbital inclination, only a band ±12° of the equator can be covered (Harmon *et al.*, 1986; Harmon and Campbell, 1988). Radar altimetry data obtained by the Arecibo antenna between 1978 and 1984 have a resolution cell of 0.15° longitude by 2.5° latitude or 6 km by 100 km and an altitude resolution of ∼100 m (Harmon *et al.*, 1986). Data obtained by the Goldstone antenna between 1972 and 1974 have a latitude resolution of ∼65 km, a longitude resolution that varies from 10 to 20 km, and an altitude resolution of ∼150 m (Clark *et al.*, 1988).

Topographic data with high spatial resolution can be obtained for relatively small landforms with gentle slopes using monoscopic (one dimensional) photoclinometry. The spatial resolution of the topography is only limited by the spatial resolution of the image, and the method is sensitive to small changes in slope (Davis and Soderblom, 1984). Using a photometric function appropriate for Mercury, such as the Lommel-Seeliger/Lambert function (see McEwen, 1991) that describes the photometric properties of the surface, the slope along a defined line is recovered and the relative elevation calculated between adjacent pixels (Davis and Soderblom, 1984; Tanaka and Davis, 1988; Watters and Robinson, 1997). Two-dimensional photoclinometry can be used to generate digital elevation models from an image (Kirk *et al.*, 2003). The method involves generating a topographic surface that approximates the image based on its viewing geometry (Kirk *et al.*, 2003).

Recent efforts to update Mariner 10 camera orientations (Robinson *et al.*, 1999) and improve radiometry (Robinson and Lucey, 1997) have greatly enhanced the accuracy of topographic data derived from the images. Topographic data for Mercury has been obtained using Mariner 10 stereo images and modern digital stereo methods (Watters *et al.*, 1998, 2001, 2002; Cook and Robinson, 2000; André *et al.*, 2005). The spatial resolution and height accuracy of stereo derived topography is limited by the resolution of the images. Stereo derived topography will be important in obtaining a global topographic map of Mercury. Because MESSENGER will be in an elliptical orbit, the Mercury Laser Altimeter (MLA) will not return ranging data for much of the southern hemisphere (Solomon *et al.*, 2007). The MDIS cameras, however, will obtain global stereo image coverage of Mercury (Hawkins *et al.*, 2007).

2.2 Mapping tectonic features

Tectonic features (i.e., lobate scarps, high-relief ridges, wrinkle ridges, graben) imaged by Mariner 10 have been identified and their locations digitized (Plate 1). The tectonic features were digitized directly from Mariner 10 image mosaics. Where available, topographic data from Mariner 10 stereo coverage and radar altimetry was used to aid in the identification of tectonic landforms (Watters *et al.*, 2004). The stereo topography is particularly useful in identifying tectonic features in areas where the lighting geometry is not optimal for morphologic analysis (incidence angles <45°).

2.3 Lobate scarps

Among the most important discoveries of the Mariner 10 mission was the number and scale of features described by Strom *et al.* (1975) as lobate scarps. The most convincing evidence for a tectonic origin is the expression of a surface-breaking thrust fault in the walls and floors of large impact craters cut by these structures (Figure 2.1). Discovery Rupes (54°S, 37°W), one of the most prominent lobate scarps imaged by Mariner 10, is over 450 km long (Figure 2.1B). New images obtained by MESSENGER confirm that there are lobate scarps in the hemisphere unseen by Mariner 10 (Solomon *et al.*, 2008). One of these scarps, Beagle Rupes (2°S, 259°W), is over 600 km long, greater than the length of Discovery Rupes (Figure 2.2). Sometimes characterized as features unique in the solar system (Thomas *et al.*, 1988), landforms virtually indistinguishable from Mercurian lobate scarps are found in highland intercrater plains on Mars (Watters, 1993, 2003; Watters and Robinson, 1999). Small-scale lobate scarps have also been found in the highlands of the Moon (Howard and Muehlberger, 1973; Lucchitta, 1976; Binder, 1982; Binder and Gunga, 1985; Watters and Johnson, Chapter 4). Reasonable terrestrial analogues to planetary lobate scarps may be Rocky Mountain foreland thrust faults

20   *Planetary Tectonics*

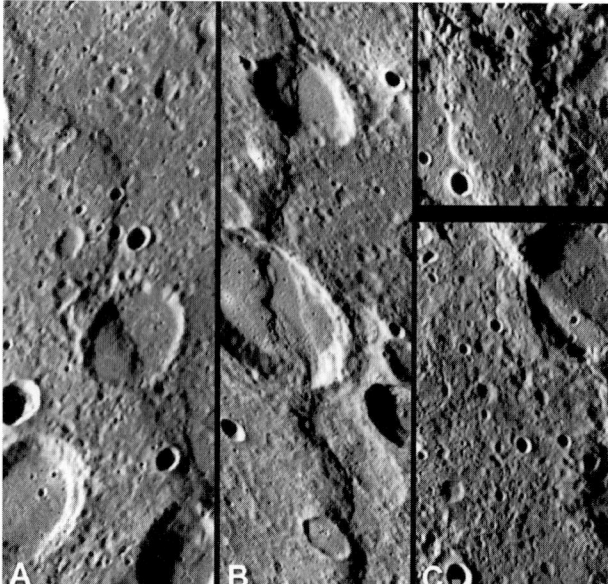

Figure 2.1. Prominent lobate scarps on Mercury. Santa Maria Rupes (A), Discovery Rupes (B), and Vostok Rupes (C) are landforms interpreted to be the surface expressions of thrust faults. The images are approximately 65 km across.

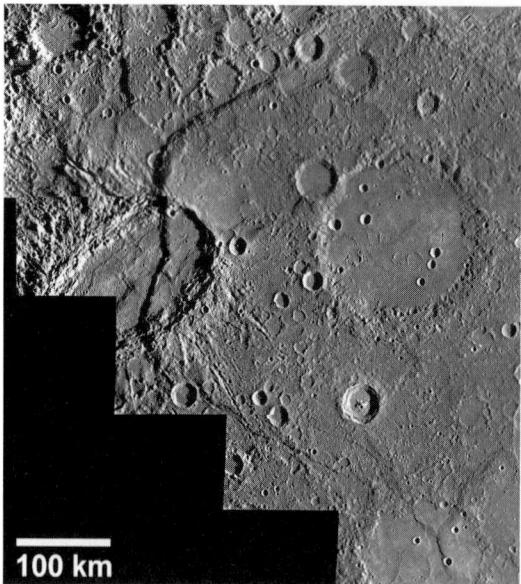

Figure 2.2. Beagle Rupes, a large-scale lobate scarp in the hemisphere unseen by Mariner 10. Beagle is an arcuate lobate scarp over 600 km long, making it one of the longest lobate scarps yet found on Mercury. The walls and floor of the ~220-km diameter, elliptically shaped impact basin named Sveinsdóttir are offsets. The image mosaic was obtained by the MDIS Narrow Angle Camera on the MESSENGER spacecraft.

in Wyoming that cut Precambrian basement and Paleozoic sedimentary sequences (Watters and Robinson, 1999; Watters, 2003).

2.3.1 Topography of lobate scarps

Topographic data for lobate scarps has been obtained from a variety of sources (see Section 2.1). Mariner 10 stereo derived topographic data for Discovery Rupes, Resolution Rupes (63°S, 50°W), and Adventure Rupes (65°S, 63°W) clearly show

The tectonics of Mercury 21

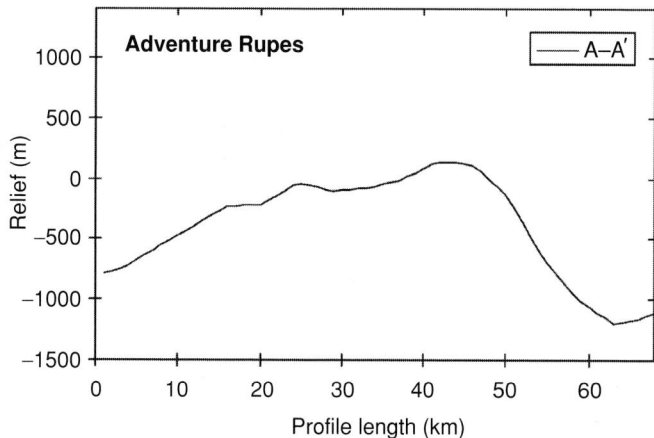

Figure 2.3. Topographic profile across Adventure Rupes obtained from a digital stereo derived DEM (Watters *et al.*, 2001). Profile location is shown in Plate 2. Elevations are relative to the 2439.0 km Mercury radius reference sphere (vertical exaggeration is ∼15:1).

the morphology of lobate scarps consists of a relatively steep scarp face and rise and a gently sloping back scarp (Plate 2, Figure 2.3) (Watters *et al.*, 1998, 2001). Discovery Rupes has the greatest relief (∼1.5 km) of the lobate scarps measured (Watters *et al.*, 1998, 2001). Adventure Rupes and Resolution Rupes have a maximum relief of ∼1.3 km and ∼0.9 km respectively.

Photoclinometric data provide an independent check of the accuracy of the stereo derived digital elevation model. Monoscopic photoclinometric profiles across Discovery Rupes compared to stereo derived profiles across the same area of scarp indicate that the difference between the two methods is <10% (<100 m) (Watters *et al.*, 1998). Topographic data for three other prominent lobate scarps, Santa Maria Rupes (3.5°N, 19°W), Endeavour Rupes (38°N, 31°W), and a lobate scarp designated S_V1 (30°N, 28°W) were also obtained using photoclinometry. Of the three, Endeavour Rupes has the greatest relief at ∼830 m, followed by Santa Maria Rupes at ∼710 m and S_V1 at ∼260 m.

Earth-based radar altimetry profiles cross several lobate scarps near the equator. The most prominent is Santa Maria Rupes (Figures 2.1, 2.4). A profile obtained by Arecibo shows that the scarp has a relief of about 700 m (Harmon *et al.*, 1986) (Figure 2.5), in good agreement with the relief estimated from co-located photoclinometric profiles (Watters *et al.*, 1998). The same Arecibo profile indicates that another lobate scarp (4°N, 15°W) to the east of Santa Maria Rupes (designated S_K4) has a relief of ∼700 m (Harmon *et al.*, 1986). An Arecibo profile also crosses a lobate scarp (8°N, 13°W) to the northeast of Santa Maria Rupes. This lobate scarp (designated S_K3) has ∼470 m of relief.

22   *Planetary Tectonics*

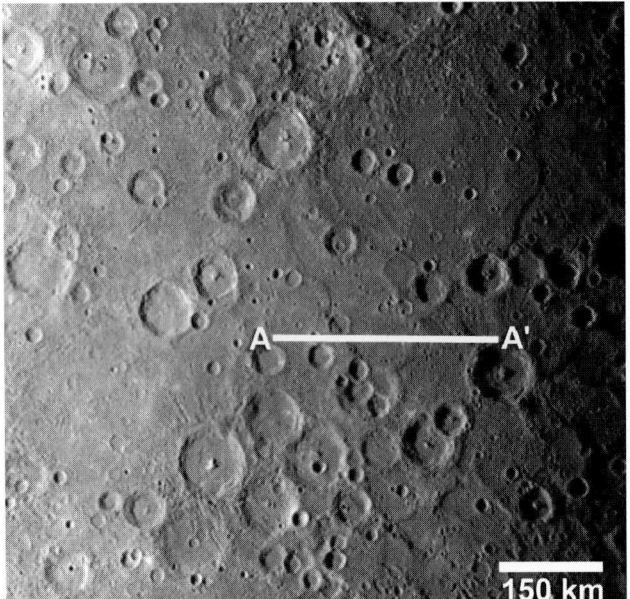

Figure 2.4. Mariner 10 mosaic of part of the Kuiper quadrangle area in the equatorial zone of Mercury. Santa Maria Rupes and other lobate scarps occur in smooth and intercrater plains (see Plate 1). Mosaic covers the region from about 5°S to 15°N and 10°W to 30°W.

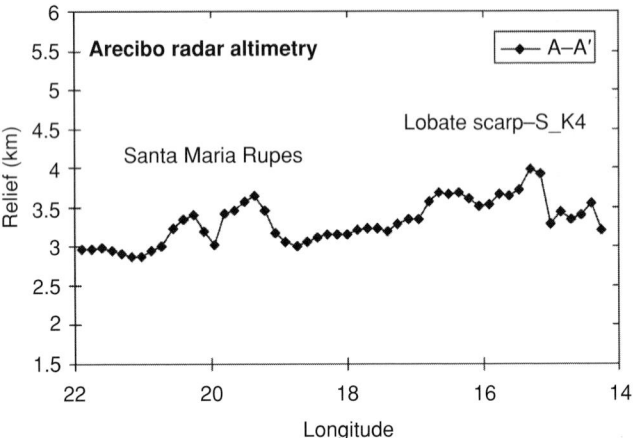

Figure 2.5. Arecibo radar altimetry profile crossing Santa Maria Rupes and another lobate scarp to the east (S_K4). These data are at latitudes of 3.9° to 4.0°N and elevations are relative to the Mercury 2439.0 km radius reference sphere (Harmon et al., 1986). Profile location is shown in Figure 2.4.

The maximum relief of the lobate scarps in the Mariner 10 hemisphere measured with topographic data obtained from digital stereo, monoscopic photoclinometry, and radar altimetry ranges from ∼0.26 to 1.5 km ($n = 8$) (Table 2.1). Although based on a small number of measurements, this range in maximum elevation is remarkably consistent with that of lobate scarps on Mars (∼0.13 to 1.2 km) (Watters, 2003).

Table 2.1. *Dimensions of lobate scarps imaged by Mariner 10*

Index	Latitude	Longitude	Maximum Relief (m)	Length (km)	D $\theta = 30°$ (m)
DR*	54°S	37°W	1500	457	3000
AR*	65°S	63°W	1300	261	2600
RR*	63°S	50°W	900	127	1800
ER[†]	38°N	31°W	830	221	1660
SMR[‡]	5°N	20°W	700	227	1400
S_K4[‡]	4°N	15°W	700	239	1400
S_K3[‡]	8°N	13°W	470	163	940
S_V1[†]	30°N	28°W	260	203	520

The lobate scarps Discovery Rupes (DR), Adventure Rupes (AR), Resolution Rupes (RR), Endeavour Rupes (ER), Santa Maria Rupes (SMR), S_V1, S_K3, and S_K4 are in the digitized database shown in Plate 1. Topographic data was obtained from digital stereo (*), monoscopic photoclinometry (†), and radar altimetry (‡).

2.3.2 Spatial and temporal distribution of lobate scarps

In an analysis of the spatial distribution of lobate scarps imaged by Mariner 10 (Plate 1) the cumulative lengths were plotted in 20° latitude bins, normalized by area (less the portion of the hemisphere not imaged by Mariner 10) (Watters *et al.*, 2004, Fig. 3). The distribution of the mapped lobate scarps does not appear to be uniform in the Mariner 10 hemisphere. It is clear that a relatively large number of lobate scarps occur in the southern hemisphere. In fact, over 50% of the area-normalized cumulative length of the lobate scarps mapped by Watters *et al.* (2004) occurs below 30°S, with the greatest cumulative length between 50°S to 90°S. In the northern hemisphere, there is a gradual decrease in the area-normalized cumulative length from the equator toward higher latitudes, with the exception of the north polar region (70°N to 90°N)(see Watters *et al.*, 2004, Fig. 3).

An inherent weakness of any analysis of the distribution of lobate scarps or other tectonic features on Mercury is the effect of observational bias. Significant observational bias may be caused by variations in lighting geometry (see Cordell and Strom, 1977; Melosh and McKinnon, 1988; Thomas *et al.*, 1988). The incidence angle of the images acquired by Mariner 10 changes from 90° at the terminator to 0° at the subsolar point (0°N, 100°W). This means that only a small percentage of the hemisphere imaged by Mariner 10 has an optimum lighting geometry for the identification of tectonic and other morphologic features. Many of the lobate scarps mapped using Mariner 10 images occur where the incidence angle is >50° (~40° from the terminator), suggesting that the apparent distribution of tectonic features may be significantly biased by lighting geometry. However, the distribution of the lobate scarps imaged by Mariner 10 is not uniform, even in areas where the incidence angle is >50°. In the intercrater plains between 10°W to 40°W and about 5°S to 30°S (~60° to 90° incidence angle), for example, no lobate scarps are found (Plate 1). It should be noted that lobate scarps are not absent where the incidence angle is <50° and some are found even near the subsolar point (Watters et al., 2004).

Nevertheless, it is clear from an examination of Plate 1 that the number of lobate scarps is much less near the Mariner 10 subsolar point. On approach to Mercury during its first flyby, MESSENGER imaged a portion of the Mariner 10 hemisphere. The location of the terminator was close to the Mariner 10 subsolar longitude. MESSENGER has revealed lobate scarps that were previously undetected in Mariner 10 images of the region (Solomon et al., 2008; Watters et al., 2009a). This clearly demonstrates that Mariner 10-based maps of the lobate scarps are not complete and are strongly influenced by lighting geometry.

To determine the possible effects of lighting geometry on the spatial distribution of the lobate scarps mapped using Mariner 10 images, Watters et al. (2004) removed areas of low-incidence angle from the analysis. Lobate scarps within 50° of the subsolar point were eliminated and the corresponding area subtracted from the area-normalized cumulative length plots. Eliminating the area within 50° of the subsolar point (a survey area of ~23% of the surface) does increase the cumulative length of lobate scarps at the equator and in the northern mid-latitudes, but the distribution is far from uniform (Watters et al., 2004, Fig. 3). Thus, the structures identified in Mariner 10 images are not uniformly distributed even in the areas where the illumination geometry is the most favorable (incidence angle >50°) (Watters et al., 2004).

The true distribution of orientations of the lobate scarps imaged by Mariner 10 has been the subject of some debate (Melosh and McKinnon, 1988). Cordell and Strom (1977) first noted a concentration of lobate scarp orientations between ±45° of north. However, they argued that any apparent preferred orientation of the lobate scarps was an artifact of observational bias due to the lighting geometry of the Mariner 10 images. Watters et al. (2004) analyzed the distribution of orientations of the mapped lobate scarps by plotting the length-weighted azimuths of the digitized segments and concluded that there is a strong preferred orientation (Watters et al., 2004, Fig. 4). The majority of the segments have orientations that fall between ±50° of north, with the most frequently occurring orientations between N30°W and N40°W and the mean vector at ~N12°E with a circular variance of 0.72 (a uniform distribution has a circular variance of 1). Eliminating the lobate scarps in areas within 50° of the subsolar point does not significantly change the distribution of orientations or the mean vector (Watters et al., 2004, Fig. 4). North of 50°S, the majority of the lobate scarps have NNE, NNW, or N–S orientations. Where the largest area-normalized cumulative length of lobate scarp segments occur, south of 50°S, the preferred orientations are WNW, ESE, or E–W.

Watters et al. (2004) examined the possibility that the orientations of lobate scarps imaged by Mariner 10 are a sample of a larger uniformly distributed population using the Kuiper's Test (Fisher, 1993) for statistical randomness. The Kuiper's V statistic, a measure of the largest vertical deviations above and below the diagonal line representing a uniform distribution, for the lobate scarp orientations is large. Based on the deviation of the mapped orientations from a uniform distribution, the hypothesis that the sample orientations are drawn from a uniform distribution was rejected (Watters et al., 2004).

The timing of formation of the lobate scarp thrust faults can be constrained by the age of the materials they deform. The oldest crustal material deformed by lobate scarp thrust faults imaged by Mariner 10 is pre-Tolstojan intercrater plains (see Tanaka et al., Figure 8.1, Chapter 8) emplaced near the end of the period of heavy bombardment (Strom et al., 1975). Younger Tolstojan and Calorian aged smooth plains units (see Tanaka et al., Figure 8.1, Chapter 8) are also deformed by lobate scarps (Plate 1) suggesting that thrust faulting continued after the formation of the Caloris basin and the emplacement of the youngest smooth plains (Strom et al., 1975; Melosh and McKinnon, 1988; Watters et al., 2001, 2004). The absence of lobate scarps in hilly and lineated terrain antipodal to the Caloris basin, where landforms may have been disrupted by basin formation-induced seismic shaking, would suggest that most of the scarps are pre-Caloris in age (Cordell and Strom, 1977), assuming scarps were uniformly distributed. However, a post-Calorian age of formation for many of the lobate scarps imaged by Mariner 10 is suggested by several lines of evidence. One is the crosscutting relations between lobate scarps and impact craters. Lobate scarps often crosscut impact craters (Figure 2.1), some with diameters >60 km. The largest impact feature crosscut by a lobate scarp observed thus far is the ~220-km diameter Sveinsdóttir basin cut by Beagle Rupes (Figure 2.2). Conversely, there are no clear examples of large-diameter impact craters (>40 km in diameter) superimposed on lobate scarps. Further, there is no apparent degradation or partial burial of lobate scarps by Caloris ejecta in the northern hemisphere (Watters et al., 2004). MESSENGER images, however, do show evidence of smaller-diameter impact craters superimposed on lobate scarps (Solomon et al., 2008; Watters et al., 2009a). An example is where a ~30-km diameter impact crater crosscuts the southern segment of Beagle Rupes (Figure 2.2). Another line of evidence is embayment relations between lobate scarps and plains material. Younger Tolstojan and Calorian aged smooth plains overlay intercrater plains (Plate 1), covering ~40% of the hemisphere imaged by Mariner 10 (Spudis and Guest, 1988). In a survey of tectonic features imaged by Mariner 10, no evidence of embayment of lobate scarps by Tolstojan and Calorian smooth plains materials was found (Watters *et al.*, 2004). The southern segments of Discovery Rupes, for example, deform intercrater plains while the northern segments cut Calorian smooth plains (Plate 1). However, in the hemisphere unseen by Mariner 10, MESSENGER images show the evidence of a lobate scarp embayed by smooth plains material (Solomon *et al.*, 2008; Watters, *et al.*, 2009a). Low-relief ridges in the smooth plains flanking the scarp suggest movement on the thrust fault scarp may have continued after the emplacement of the smooth plains. Thus, these observations suggest that lobate scarp formation developed after the emplacement of the intercrater plains (∼4 Ga) and before the end of smooth plains emplacement, continuing after the emplacement of the youngest smooth plains material (Watters *et al.*, 2004; Solomon *et al.*, 2008; Watters *et al.*, 2009a). Some lobate scarps appear to be relatively young and are perhaps the youngest endogenic landforms on Mercury.

2.3.3 Geometry of lobate scarp thrust faults

There is general agreement that Mercury's lobate scarps are the surface expression of thrust faults. However, the fault geometry, fault plane dip, and depth of faulting are unknown. Forward mechanical modeling of lobate scarp thrust faults has helped to constrain the characteristics of these faults on Mercury and Mars (Schultz and Watters, 2001; Watters *et al.*, 2002). Using elastic dislocation modeling, Watters *et al.* (2002) examined the fault geometry, sense of slip, and the amount of displacement D on the thrust fault underlying Discovery Rupes, where the relief on the scarp is the greatest. The greatest relief occurs just south of the Rameau crater (60 km in diameter) where the average relief is ∼1.3 km (Watters *et al.*, 2002, Fig. 1). A broad, shallow trough about 40 km wide and several hundred meters deep occurs roughly 90 km west of the base of the scarp. The axis of the trough roughly parallels the strike of the scarp face of Discovery Rupes. The trough is interpreted to be evidence of a trailing syncline, and the distance between it and the surface break defines the cross-strike dimension of the upper plate of the thrust fault (Watters *et al.*, 2002). Other back scarp topographic troughs are associated with Adventure Rupes and an analogous lobate scarp on Mars called Amenthes Rupes (Schultz and Watters, 2001; Watters *et al.*, 2001). In the model, the lower tip of the thrust fault is placed below the trailing syncline and its upper tip at the surface break. The initial amount of slip is estimated from the kinematic offset calculated from the height of the scarp adjacent to the surface break. Mercury's seismogenic lithosphere is approximated by an elastic halfspace with a Young's modulus E of 80 GPa and Poisson's ratio v of 0.25. The best fits to the topography across Discovery Rupes are for a depth of faulting $T = 35$ to 40 km, a fault dip $\theta = 30°$ to $35°$, and $D = 2.2$ km (Watters *et al.*, 2002, Fig. 3). It is possible that other non-planar fault geometries, particularly a listric geometry, could be involved in the formation of lobate scarps. Elastic dislocation modeling of listric thrust faults, approximated by linear segments with varying dips, has been modeled (Watters *et al.*, 2002; Watters, 2004). None of the listric geometries modeled produce fits to the topographic data as good as the planar geometry.

2.3.4 Displacement–length relationship of lobate scarp thrust faults

Studies of terrestrial faults show that the maximum displacement on a fault D scales with the planimetric length of the fault L (e.g., Walsh and Watterson, 1988; Cowie and Scholz, 1992b; Gillespie *et al.*, 1992; Cartwright *et al.*, 1995) and the relationship holds for planetary faults (Schultz and Fori, 1996; Schultz, 1997, 1999; Schultz *et al.*, Chapter 10; Watters *et al.*, 1998, 2000; Watters and Robinson, 1999; Watters, 2003). It has been proposed that D is related to L by $D = cL^n$, where c is a constant related to material properties and n is estimated to be >1 (Walsh and Watterson, 1988; Marrett and Allmendinger, 1991; Gillespie *et al.*, 1992). A linear relationship $D = \gamma L$, where γ is a constant determined by rock type and tectonic setting (where $n = 1$, $\gamma = c$) (Cowie and Scholz, 1992b), is supported by studies of faults in populations formed in uniform rock types (Dawers *et al.*, 1993; Clark and Cox, 1996). Evidence also indicates that the scaling relationship between D and L generally holds for all the fault types (i.e., normal, strike-slip, and thrust) in a wide variety of tectonic settings and eight orders of magnitude in length scale (Cowie and Scholz, 1992b). The ratio of displacement to fault length γ for the fault populations analyzed by Cowie and Scholz (1992b) ranges between 10^0 and 10^{-3}. Scatter in the D–L data can arise from several sources, including fault segmentation, uncertainties in fault dip and depth or aspect ratio, interaction with other faults, and ambiguities in determining the maximum value of D along the scarp trace (Cartwright *et al.*, 1995, 1996; Dawers and Anders, 1995; Wojtal, 1996; Schultz, 1999; Schultz *et al.*, Chapter 10).

The amount of displacement on some lobate scarp thrust faults imaged by Mariner 10 has been estimated using a kinematic model that assumes the total slip is a function of the relief of the lobate scarp and the dip of the surface-breaking fault plane. Given the measured relief of the scarp h and the fault plane dip θ, the displacement necessary to restore the topography to a planar surface is given by $D = h/\sin\theta$ (see Wojtal, 1996; Watters *et al.*, 2000). The optimum angle θ at which faulting will occur, the angle for which the differential horizontal stress necessary to initiate faulting is a minimum, is given by $\tan 2\theta = 1/\mu_s$ where μ_s is the coefficient of static friction (see Jaeger and Cook, 1979; Turcotte and Schubert, 2002). Laboratory data on the maximum shear stress to initiate sliding for a given normal stress indicates a range in maximum coefficient of static friction of 0.6 to 0.9 (Byerlee, 1978) corresponding to thrust faults with dips from about 24° to 30° with an optimum $\mu_s = 0.65$ resulting in faults with dips of 28.5° (Scholz, 2002).

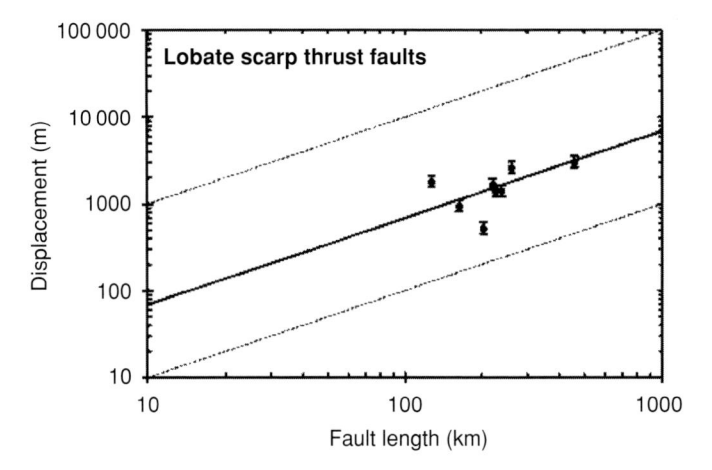

Figure 2.6. Log–log plot of maximum displacement as a function of fault length for eight lobate scarps on Mercury. The ratio γ ($\sim 6.9 \times 10^{-3}$ using estimates of D based on fault plane dips $\theta = 30°$) for the thrust faults (Table 2.1) were obtained by a linear fit to the D–L data with the intercept set to the origin (Cowie and Scholz, 1992b).

The range of expected dips is consistent with field observations of θ for thrust faults that typically range between 20° to 35° (e.g., Jaeger and Cook, 1979; Brewer et al., 1980; Gries, 1983; Stone, 1985). As discussed in the previous section, elastic dislocation modeling of the Discovery Rupes thrust fault and the Amenthes Rupes thrust fault on Mars suggests that the faults dip from 25° to 35° (Schultz and Watters, 2001; Watters et al., 2002).

The estimated displacement on some large-scale lobate scarp thrust faults imaged by Mariner 10 range from 0.52 to 3.0 km, assuming fault plane dips of 30° (Table 2.1), with an average of $\sim$1.7 km ($n = 8$). The amount of displacement on the thrust fault associated with Discovery Rupes is on the order of 2.6 to 3.5 km for θ of 35° and 25° respectively. The value of γ is determined from a linear fit to D–L data for the eight large-scale lobate scarps (Figure 2.6). For a range in θ of 25° to 35°, the corresponding γ is $\sim 6.0 \times 10^{-3}$ to $\sim 8.1 \times 10^{-3}$ and for $\theta = 30°$, $\gamma \cong 6.9 \times 10^{-3}$. These values of γ are consistent with previous estimates for lobate scarps imaged by Mariner 10 (Watters et al., 1998, 2000) and with estimates of γ for lobate scarp thrust faults on Mars (Watters and Robinson, 1997; Watters, 2003).

2.3.5 Influence of buried basins on lobate scarps

Many of the known lobate scarps on Mercury are arcuate in plan view (Dzurisin, 1978). Prominent examples are Discovery Rupes, Hero Rupes ($\sim$59°S, 172°W), Adventure Rupes, Resolution Rupes (Plate 1) and Beagle Rupes (Figure 2.2). Spudis and Guest (1988) suggest that the population of pre-Tolstojan basins forms a structural framework that influenced the subsequent geologic evolution of the surface. They mapped 20 pre-Tolstojan multiring basins randomly distributed over the hemisphere imaged by Mariner 10. A number of the proposed remnant rings of the multiring basins have been identified on the basis of circular or arcuate patterns in tectonic features (Pike and Spudis, 1987; Spudis and Guest, 1988). The formation of many of the arcuate lobate scarps may have been influenced by preexisting mechanical discontinuities in the Mercurian crust (Watters et al., 2001, 2004). In an analysis of inferred stresses, where great circles are fit to the perpendiculars of digitized segments of Adventure Rupes, Resolution Rupes, and Discovery Rupes, the maximum concentration of intersection (27% per 1% area) was found to be located at ~48°S, 58°W (Watters et al., 2001, Fig. 7). A possible basin identified by Spudis and Guest (1988) is centered at 43°S, 49°W (Andal-Coleridge basin). The rings of the Andal-Coleridge basin (Spudis and Guest, 1988) are not parallel to the arcuate trend of Adventure and Resolution Rupes and are only subparallel to some northern segments of Discovery Rupes. However, a broad, shallow, roughly circular depression centered at approximately 43°S, 54°W was found in a digital stereo derived regional DEM (Figure 2.7) (Watters et al., 2001). This topographic feature is interpreted to be evidence of another ancient, pre-Tolstojan impact basin, informally named the Bramante-Schubert basin (~550 km in diameter). Segments of Discovery Rupes are roughly concentric to the Bramante-Schubert basin. This may indicate the presence of a ring of the basin that coincides with Discovery Rupes. Another ring of the Bramante-Schubert basin may coincide with two other lobate scarps in the region, Astrolabe Rupes and Mirni Rupes (Watters et al., 2001, Fig. 1). Segments of Adventure Rupes and Resolution Rupes, however, are not strongly concentric to this proposed basin. A topographic depression to the north of Adventure Rupes and Resolution Rupes, roughly centered on Rabelais crater, may indicate the presence of an ancient buried basin (Plate 2, Figure 2.7) that influenced the localization of these structures. It is also possible that the arcuate trend of Adventure and Resolution reflects the influence of an unidentified basin or basin ring not visible in the topography (Watters et al., 2001).

Topographic data suggest that other prominent lobate scarps in the Mariner 10 hemisphere, such as Hero Rupes (~59°S, 172°W) and Fram Rupes (~58°S, 94°W), may be associated with ancient impact basins (Watters et al., 2004). Many of the basins suggested by the topographic data correspond to previously identified pre-Tolstojan multiring basins (Spudis and Guest, 1988). Two other previously unrecognized buried impact basins in the Mariner 10 hemisphere are suggested by opposite-facing arcuate lobate scarps (Plate 1) that ring topographic depressions. They are centered at ~45°S, 142°W and ~10°N, 17°W and are ~300 km in diameter. Thus, mechanical discontinuities introduced by ancient buried impact basins may have influenced the localization of many prominent lobate scarps

30   *Planetary Tectonics*

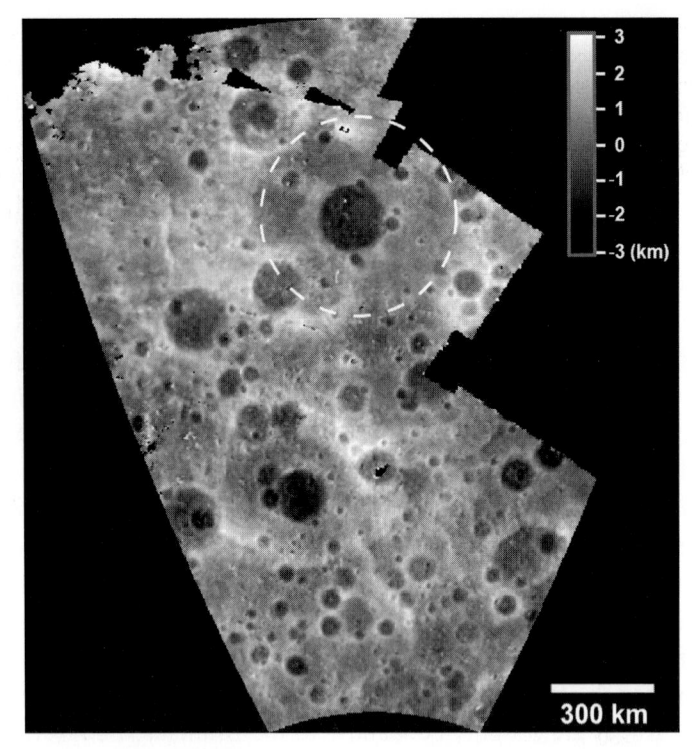

Figure 2.7. Regional-scale DEM of part of the Discovery quadrangle. The dashed line shows the proposed location of an ancient buried impact basin. The DEM covers an area from 25°W to 80°W, 30°S to 75°S and was generated using over 350 individual stereo pairs. Elevations are relative to the 2439.0 km Mercury radius reference sphere.

(Watters *et al.*, 2001, 2004). Many other lobate scarps, however, do not appear to have been influenced by buried impact basins.

2.4 High-relief ridges

High-relief ridges were first described by Dzurisin (1978) as ridges with significant relief found in intercrater plains (also see Melosh and McKinnon, 1988). The morphology and dimension of the high-relief ridges on Mercury is similar to high-relief ridges observed in highland material on Mars (Watters, 1993, Fig. 4a). Although rare in comparison to lobate scarps and wrinkle ridges, eight high-relief ridges imaged by Mariner 10 (Watters *et al.*, 2004) are prominent landforms (Plate 1). Antoniadi Dorsa (30°N, 31°W) is one of the most distinct examples of a high-relief ridge on Mercury (Figure 2.8). It is roughly symmetric in cross section, occurs in intercrater plains, and is ∼340 km in length. Antoniadi Dorsa crosscuts an ∼85-km diameter impact crater, clearly deforming the crater wall and floor materials.

Figure 2.8. Mariner 10 mosaic of the high-relief ridge Antoniadi Dorsa. Antoniadi Dorsa and other high-relief ridges on Mercury are interpreted to be the surface expression of high-angle reverse faulting.

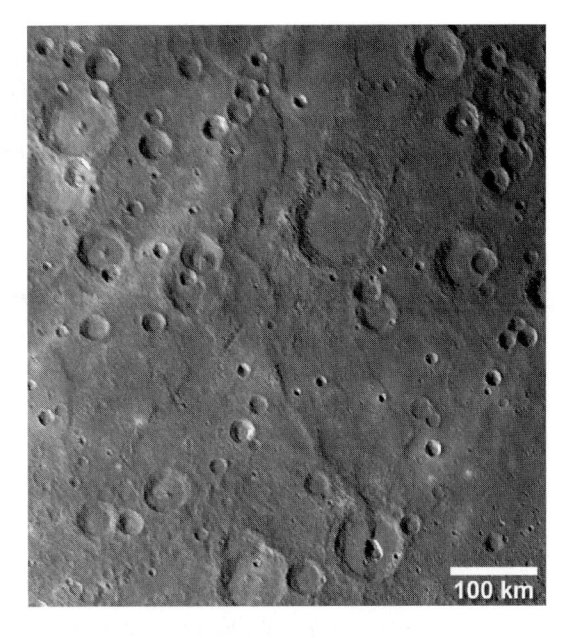

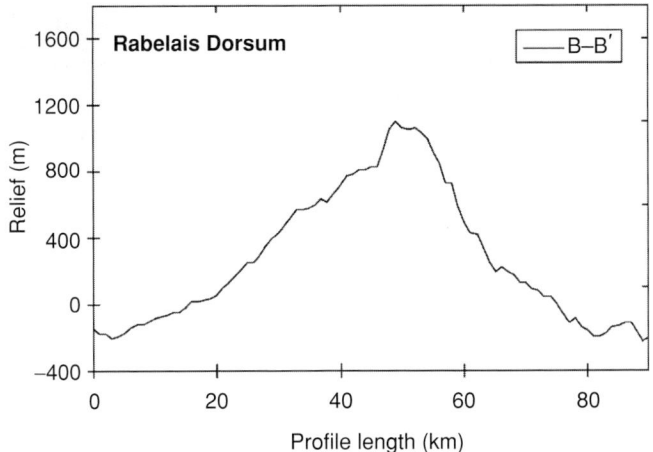

Figure 2.9. Topographic profile across the Rabelais Dorsum high-relief ridge obtained from a digital stereo derived DEM. Rabelais Dorsum is roughly symmetric in cross section with a maximum relief comparable to that of Discovery Rupes. Profile location is shown in Plate 2. Elevations are relative to the 2439.0 km Mercury radius reference sphere (vertical exaggeration is ~40:1).

A high-relief ridge, first described by Dzurisin (1978), is not readily apparent in monoscopic Mariner 10 images. The ridge, informally named Rabelais Dorsum after a nearby crater (Watters *et al.*, 2001), appears as a prominent landform in a Mariner 10 stereo derived DEM of the region (Plate 2). To the north, Rabelais Dorsum crosscuts the Adventure–Resolution Rupes trend (Plate 2). The topographic data show that the northern segment of Rabelais Dorsum is roughly symmetric with a maximum relief of ~1.3 km (Figure 2.9), comparable in relief to that of

Discovery Rupes. Rabelais Dorsum is ~370 km in length, comparable to Antoniadi Dorsa. Landforms described by Malin (1976) as lobate fronts are found on the eastern side of Rabelais Dorsum, superimposed on intercrater plains and the walls and floors of a number of craters with no evidence of significant offset. Malin (1976) suggests that these features are not fault scarps but rather formed through mass movement associated with seismic activity or tectonically controlled volcanic extrusion. To the south, the morphology of the southwest trending segment of the structure transitions from a high-relief ridge to a lobate scarp (Plate 2) (Watters *et al.*, 2001). The lobate scarp segment is ~140 km in length and offsets the walls and floor material of a crosscut impact crater (72°S, 47°W). Another prominent lobate scarp – high-relief ridge transition in the Mariner 10 hemisphere occurs in the intercrater plains south of Tir Planitia (22°S, 178°W) (Plate 1).

The transformation of some high-relief ridges into lobate scarps suggests that the origin of the two structures is related. One possible explanation for the contrast in morphology between high-relief ridges and lobate scarps, if both structures involve reverse faulting, is the dip of the fault plane. High-relief ridges may reflect deformation over high-angle reverse faults ($\theta > 45°$) rather than thrust faults ($\theta < 45°$). If this is the case, the change in morphology may reflect the transition from a high-angle reverse fault underlying the ridge to a low-angle thrust fault underlying the lobate scarp.

High-relief ridges appear to be comparable in age to lobate scarps (Melosh and McKinnon, 1988; Watters *et al.*, 2004). This interpretation is supported by lobate scarp – high-relief ridge transitions and a lack of superimposed large-diameter impact craters. The crosscutting relationship between Rabelais Dorsum and Adventure Rupes and Resolution Rupes, however, suggests that the ridge postdates the formation of these thrust fault scarps. The superposition of Rabelais Dorsum on the trend of these lobate scarps may indicate a local change in the orientation of the stress field over time (Watters *et al.*, 2001).

2.5 Wrinkle ridges

Wrinkle ridges are one of the most ubiquitous tectonic features on the terrestrial planets. Planetary wrinkle ridges are commonly found on topographically smooth material in two physiographic settings, the interior of large impact basins, or broad expansive plains (Watters, 1988). On the Moon, wrinkle ridges are confined to mare material known to be composed of basalt. Wrinkle ridges on Mars and Venus also appear to occur in basalt-like units (Watters, 1988, 1992, 1993). Terrestrial analogues to wrinkle ridges are also found in the continental flood basalts of the Columbia Plateau in the northwestern United States (Plescia and Golombek, 1986; Watters, 1988, 1992). The common association between wrinkle ridges and volcanic plains has led to the widely held conclusion that there is a genetic

Figure 2.10. Mariner 10 mosaic of the prominent wrinkle ridge, Schiaparelli Dorsum. Schiaparelli Dorsum exhibits the broad arch and superimposed ridge that is typical of planetary wrinkle ridges.

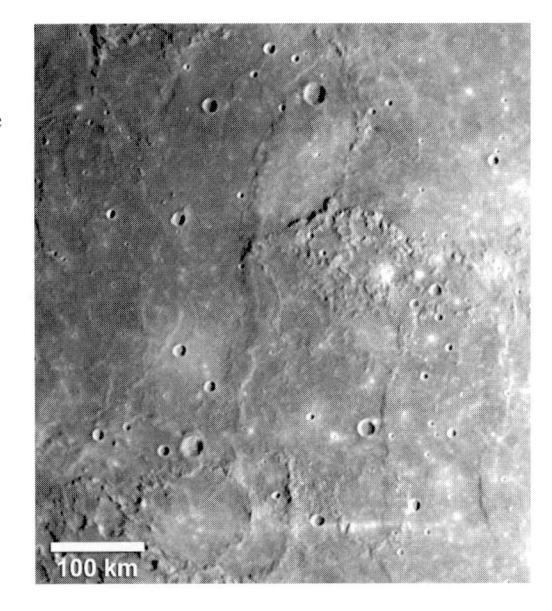

relationship between the two, although this interpretation has been challenged (see Plescia and Golombek, 1986; Schultz, 2000; Watters, 2004).

On Mercury, wrinkle ridges are found in both physiographic settings, basin interior plains and expansive plains outside of basins. By far the largest number of wrinkle ridges in the hemisphere imaged by Mariner 10 occur in the interior smooth plains material of Caloris and the exterior annulus of smooth plains (Plate 1). Some of Mercury's wrinkle ridges would be difficult to distinguish from those in lunar mare or those in ridged plains on Mars. Schiaparelli Dorsum (22°N, 164°W) on the eastern edge of Odin Planitia is a wrinkle ridge that exhibits the typical morphologic elements – a broad, low-relief arch and a superimposed ridge (Figure 2.10). One aspect of planetary wrinkle ridges, however, is that the two morphologic elements that comprise them can occur independently of one another. That is, the relatively narrow ridge and the broad arch may not be superimposed (Watters, 1988). This is the case for many of the wrinkle ridges imaged by Mariner 10 (Strom et al., 1975; Maxwell and Gifford, 1980).

2.5.1 Topography of wrinkle ridges

Little topographic data is available for Mercury's known wrinkle ridges. Estimates of the maximum relief of some wrinkle ridges imaged by Mariner 10 was obtained using poorly constrained shadow measurements (Watters, 1988). Although no Mariner 10 stereo image pairs cover wrinkle ridges, Earth-based radar altimetry profiles do cross the ridged plains of Tir Planitia (Figure 2.11) (Harmon et al., 1986; Harmon and Campbell, 1988; Clark et al., 1988). The relief of seven wrinkle ridges measured in Tir Planitia ranges from 200 to 730 m (Table 2.2). One of these wrinkle ridges (WR_T14), found near the northeastern edge of Tir Planitia (10°N,

Table 2.2. *Dimensions of wrinkle ridges in Tir Planitia*

Index	Latitude	Longitude	Maximum Relief (m)	Length (km)	D $\theta = 30°$ (m)
Wr_T14	10°N	176°W	400	24	800
Wr_T18	9°N	183°W	730	35	1460
Wr_T7	3°N	174°W	440	22	880
Wr_T9	2°S	172°W	200	15	400
Wr_T3	1°S	182°W	370	17	740
Wr_T8	6°N	175°W	250	22	500
Wr_T6	3°N	178°W	340	16	680

The wrinkle ridges are in the digitized database shown in Plate 1. Topographic data was obtained from radar altimetry. Widths were measured from Mariner 10 image mosaics.

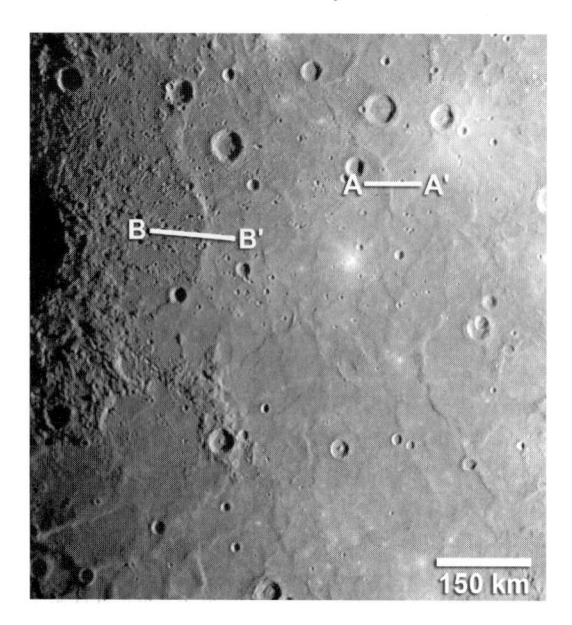

Figure 2.11. Mariner 10 mosaic of part of the Tir quadrangle in the equatorial zone. Wrinkle ridges occur in the smooth plains of Tir Planitia (see Plate 1). Mosaic covers the region from about 5°S to 15°N and 170°W to 190°W.

176°W), has a maximum relief of ~400 m and is ~130 km long (Figure 2.12) (Table 2.2). The altimetry data shows that the ridge is broad and asymmetric. The large width of the ridge, over 30 km, indicates that it is dimensionally more akin to an arch (Watters, 1988). This is the case for all the measured wrinkle ridges in Tir Planitia (Table 2.2). The arch-like structure (WR_T18) near the western edge of Tir Planitia (9°N, 183°W) has the greatest relief of the measured structures at ~730 m. The maximum width of the structure, as measured from Mariner 10 images, is also the largest at ~35 km (Table 2.2). However, the radar altimetry data indicates that this arch is much wider than it appears in images (Figure 2.13). The visible arch is superimposed on a broader landform that is over 50 km wide where the radar altimetry profile crosses it. A Goldstone radar altimetry profile across the

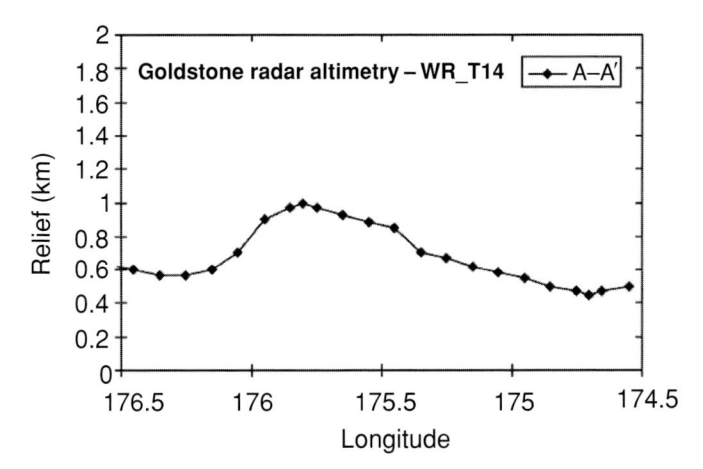

Figure 2.12. Goldstone radar altimetry profile across a wrinkle ridge in Tir Planitia (WR_T14). The width and morphology of the structure indicates that it is an arch. These data are at a latitude of ~9.8°N. Elevations are relative to the Mercury 2439.0 km radius reference sphere (Clark et al., 1988). Profile location is shown in Figure 2.11.

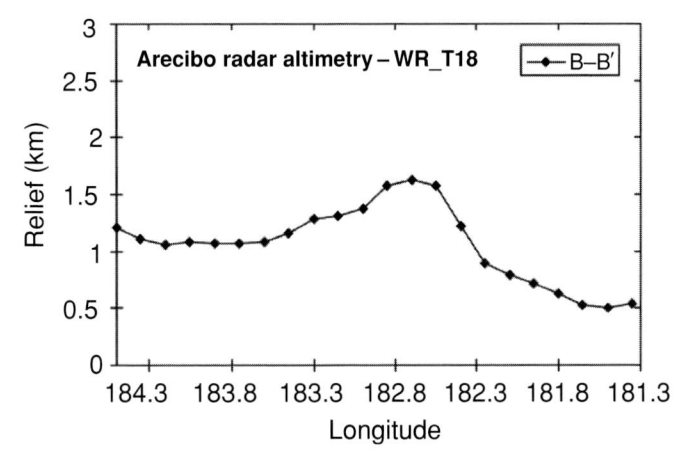

Figure 2.13. Arecibo radar altimetry profile across a high-relief arch-like structure in Tir Planitia (WR_T18). The morphology and high relief of the feature suggests it may be similar to high-relief ridges found in intercrater plains. These data are at a latitude of 8°N and elevations are relative to the Mercury 2439.0-km radius reference sphere (Harmon et al., 1986). Profile location is shown in Figure 2.11.

northern segment of WR_T18 indicates that it may be up to 80 km wide. Although the structure occurs in smooth plains (Plate 1), its large width and great relief make it more akin to a high-relief ridge than to a wrinkle-ridge arch. The high relief of the arches measured using the radar altimetry data, however, is in sharp contrast with the low-relief arches found on the Moon and Mars (see Watters, 1988, 2004).

2.5.2 Spatial distribution of wrinkle ridges on Caloris exterior plains

The proximity of the exterior smooth plains to Caloris, and a quick examination of the orientations of the wrinkle ridges (Plate 1), suggests a relationship between the basin and the stresses that formed the wrinkle ridges (Strom *et al.*, 1975; Melosh and McKinnon, 1988; Spudis and Guest, 1988). Many wrinkle ridges on the exterior plains east of Caloris have orientations that appear to be either roughly concentric or radial to the basin (Plate 1). However, many other wrinkle ridges on these smooth plains are not strongly concentric or radial to the Caloris basin. Examining the distribution of orientations of wrinkle ridges in Tir Planitia (~0°N, 180°W), Odin Planitia (~25°N, 170°W), and Suisei Planitia and adjacent plains (~60°N, 160°W) imaged by Mariner 10, suggests there are major trends that could be related to the Caloris basin (Plate 3). In Tir Planitia, the length-weighted azimuths of digitized wrinkle-ridge segments have a vector mean of ~N4°E with a circular variance of 0.62. Although the vector mean is not radial or concentric to Caloris, the trends between N10°W and N40°W are roughly radial, and the trends between N50°E and N60°E are roughly concentric (Plate 3). However, many of the wrinkle ridges in Tir Planitia appear to have been localized by shallow buried impact craters. The vector mean of wrinkle-ridge segments in Odin Planitia is also nearly N–S (~N7°E, circular variance = 0.45). Although this trend and a major trend between N20°E and N30°E are roughly concentric to Caloris, wrinkle ridges form a local concentric pattern in Odin Planitia, possibly reflecting the influence of a buried impact basin. Suisei Planitia has the most consistent trends of the three ridged plains provinces (Plate 3), with a vector mean of ~N38°E (circular variance = 0.33). This trend is perhaps the most clearly radial to the Caloris basin. However, as can be seen in the geologic map of the Mariner 10 hemisphere (Plate 1), this trend also parallels the orientation of the contact between the smooth plains in the Suisei Planitia region and the surrounding intercrater plains. The orientations of many of the wrinkle ridges in Tir, Odin, and Suisei Planitia can be correlated to the gross geometry of smooth plains units that presumably occupy regional-scale lowlands. Earth-based radar altimetry data show that the smooth plains of northern Tir Planitia occur in a broad, roughly N–S trending trough that at its lowest is more than 1 km below the adjacent intercrater plains to the west (Figure 2.14) (Harmon *et al.*, 1986; Harmon and Campbell, 1988). The occurrence of wrinkle ridges in plains units that occupy topographic lows is a characteristic shared by many ridged plains provinces on Mars (e.g., Hesperia Planum) (see Watters, 1993). The presence of wrinkle-ridge rings, wrinkle ridges with semicircular orientations that may have formed over buried impact craters, suggests the ridges are largely the result of subsidence of the plains material (Watters, 1988, 1993). Wrinkle-ridge rings have also been found in MESSENGER images of smooth plains in the hemisphere unseen by Mariner 10 (Head *et al.*, 2008; Watters *et al.*, 2009a). Their presence in volcanic plains on

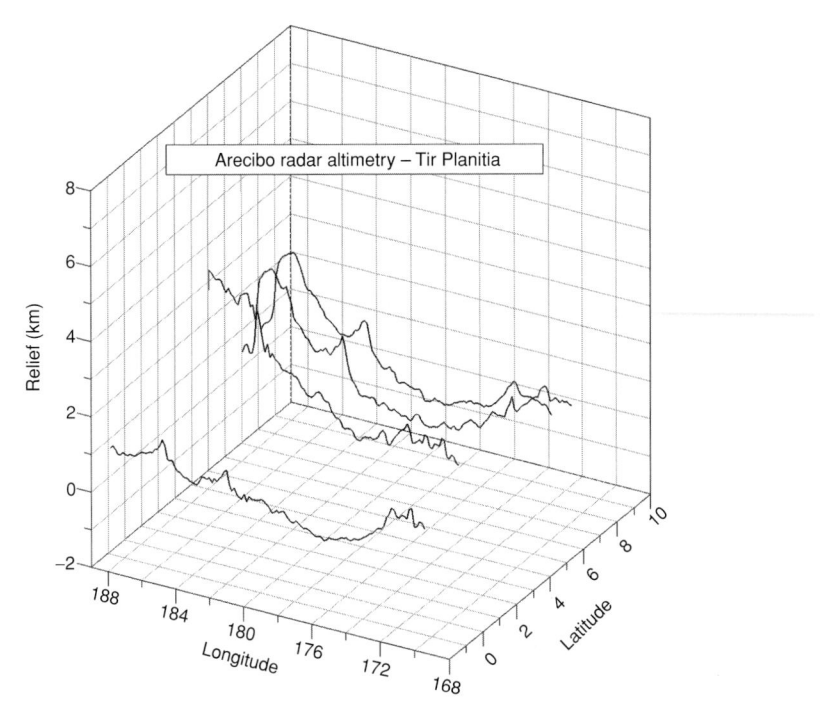

Figure 2.14. Arecibo radar altimetry profiles across Tir Planitia. The topographic data shows that the ridged plains of Tir Planitia occur in a broad trough more than 1 km below the adjacent intercrater plains. Elevations are relative to the Mercury 2439.0 km radius reference sphere (Harmon et al., 1986).

the Moon and Mars along with observed embayment relations supports a volcanic origin for the smooth plains material (Head et al., 2008).

Stresses resulting from subsidence are likely strongly influenced by the geometry of the subsiding plains. Thus, a correlation between the orientation of wrinkle ridges and the geometry of the lowland plains is expected, and stresses resulting from subsidence and the influence of local, shallow, mechanical discontinuities are expected to dominate over influences from the Caloris basin. It is possible, however, that the mechanical discontinuities in the crust introduced by the formation of the Caloris basin are reflected in the trends of wrinkle ridges in the surrounding smooth plains. Wrinkle ridges in Budh Planitia (~20°N, 150°W), to the southeast of Odin Planitia, trend roughly N–S (Plates 1, 3). Budh Planitia has been identified as the location of a possible ancient buried impact basin (Spudis and Guest, 1988). The ridges here, however, are not concentric to Caloris and do not appear to have been influenced by the geometry of the presumed lowland smooth plains. This is also the case for the only other major ridged plains province in the hemisphere imaged by Mariner 10, Borealis Planitia (~75°N, 80°W). Here the wrinkle ridges have predominantly E–W orientations (Plate 1).

Figure 2.15. Mariner 10 mosaic of the Caloris Basin. The main rim diameter of Caloris is estimated to be ~1300 km. The smooth plains floor material of the imaged side of the Caloris basin has been deformed by both compression and extension.

2.6 Wrinkle ridges in the Caloris basin

The Caloris basin is one of the largest basins in the solar system (Figure 2.15). Based on the portion of the main basin rim imaged by Mariner 10, the diameter of Caloris was estimated to be from ~1300 km (Strom *et al.*, 1975) up to ~1420 km (Maxwell and Gifford, 1980). Less than half of the basin was imaged by Mariner 10 (Figure 2.15). Images obtained by MESSENGER in its first flyby of Mercury cover the entire basin and the diameter of Caloris is now estimated to be about 1550 km (Plate 4) (Murchie *et al.*, 2008). Mariner 10 and MESSENGER images show that wrinkle ridges are found on the interior smooth plains throughout the basin (Plate 4). MESSENGER color data, observed embayment relations, and the discovery of rimless depressions interpreted to be volcanic vents around Caloris support a volcanic origin for the interior smooth plains material (Head *et al.*, 2008; Murchie *et al.*, 2008; Robinson *et al.*, 2008; Watters *et al.*, 2009c).

Wrinkle ridges in Caloris basin are found throughout the interior plains material (Murchie *et al.*, 2008; Watters *et al.*, 2009c). As can be seen in Plate 4, many more wrinkle ridges are found on the eastern interior plains than on the western plains. This is due, in part, to the much more optimal lighting geometry of the Mariner 10 images of the eastern portion of the basin. The Caloris ridges appear to exhibit only the narrow ridge morphologic element of wrinkle ridges; arches have not been observed (Strom *et al.*, 1975; Maxwell and Gifford, 1980). The widths of Caloris wrinkle ridges imaged by Mariner 10 range from 1 to 12 km, and their maximum relief is estimated at ~500 m (Dzurisin, 1978). This is consistent with the range of widths and maximum relief of wrinkle ridges measured from Mariner 10 images in smooth plains outside of the Caloris basin (Watters, 1988). The location and orientation of wrinkle ridges on the interior plains of Caloris is similar in many ways to lunar maria (Maxwell and Gifford, 1980). These ridges have two predominant orientations, basin concentric and basin radial (Figures 2.15, Plate 4). This pattern is common in lunar mare (Strom, 1972; Bryan, 1973; Maxwell *et al.*, 1975). What is in contrast to lunar mare is the complexity of the pattern of wrinkle ridges of Caloris (Figure 2.15, Plate 4). Crosscutting wrinkle ridges form what has been described as a crude polygonal network (Melosh and McKinnon, 1988). The spacing of parallel to subparallel trending ridges measured in the Mariner 10 portion of the basin ranges from under 10 km to up to 60 km. In some cases, segments of subparallel trending ridges converge and either become subparallel again at a reduced spacing or continue to converge until they merge into a single ridge. A number of wrinkle-ridge rings are also present in the portion of Caloris imaged by Mariner 10, suggesting the influence of shallow buried impact craters.

It has been noted that the basin-concentric wrinkle ridges in the portion of Caloris imaged by Mariner 10 fall along rough arcs in the interior smooth plains (Maxwell and Gifford, 1980; Melosh and McKinnon, 1988). These arcs have been associated with the location of possible interior rings in Caloris. Maxwell and Gifford (1980) found two rings, an innermost ring delineated by a gentle topographic rise at 800 km and a second ring corresponding to ridges at 1060 km diameter. Melosh and McKinnon (1988) determined the diameter of two ridge arcs at 700 km and 1000 km. As in the case of wrinkle ridges in mare basins (see Maxwell *et al.*, 1975), buried basin rings within Caloris may have strongly influenced the localization of the ridges.

2.7 Caloris basin graben

One of the more significant aspects of the tectonics of Mercury is the absence of clear evidence of extension. A network of lineaments may be the subtle expression of extension in the form of a fabric of fractures that make up what has been described as a tectonic grid (Dzurisin, 1978; Melosh and McKinnon, 1988; Thomas, 1997). This grid may reflect ancient lines of weakness in the lithosphere (Melosh and McKinnon, 1988). Clear evidence of extension on Mercury is found in the interior

Figure 2.16. Third-encounter Mariner 10 image mosaic of a portion of the interior plains of Caloris showing extensional troughs that form giant polygons (Mariner 10 images 0529055 and 0529056). One of the longest, continuous troughs in the mosaic (center) forms the boundary of a giant polygon, and cuts the flanks and along the crest of a wrinkle ridge. The maximum width of the mosaic is ~115 km.

Figure 2.17. Radial graben of Pantheon Fossae in the Caloris basin. The radial graben converge on an area near the center of the basin. The ~40-km diameter impact crater Apollodorus is located in the convergence area. The image mosaic was obtained by the MDIS Narrow Angle Camera on the MESSENGER spacecraft.

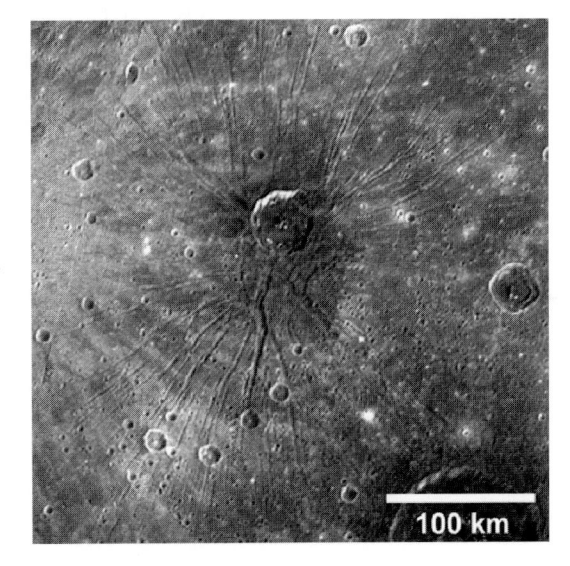

plains of the Caloris basin. In addition to the numerous wrinkle ridges, a population of narrow graben near the basin margin was imaged by Mariner 10 (Strom *et al.*, 1975; Dzurisin, 1978). In plan view these graben are highly variable; some segments are linear while others are very sinuous (Figures 2.15, 2.16). The basin-concentric graben account for the dominant trends, however, like the wrinkle ridges there are also basin-radial graben. The basin-concentric and basin-radial graben form polygonal patterns (Plate 4, Figure 2.16) (Dzurisin, 1978; Melosh and McKinnon, 1988; Watters *et al.*, 2005).

MESSENGER has revealed a remarkable new facet of the tectonics of the Caloris basin. A population of radial graben named Pantheon Fossae converge on an area near the center of the basin, close to a ~40-km diameter impact crater (Plate 4, Figure 2.17) (Murchie *et al.*, 2008; Watters *et al.*, 2009c). Other graben near the center of the complex form a polygonal pattern like that seen near the interior margin. MESSENGER images also confirm that basin-concentric graben near the margin extend beyond the Mariner 10 hemisphere, into the western half of the interior plains (Murchie *et al.*, 2008; Watters *et al.*, 2009c). Evidence of extensional faulting outside of Caloris has been found by MESSENGER in the interior plains of the ~250-km diameter Raditladi basin (~27°N, 241°W) (Solomon *et al.*, 2008; Watters *et al.*, 2009a) and in the ~715-km diameter Rembrandt basin (~33°S, 272°W) (Watters *et al.*, 2009b).

The polygonal pattern of the graben in Caloris strongly resembles giant polygons found in polygonal terrains on Mars (Pechmann, 1980; McGill, 1986; Hiesinger and Head, 2000) and Venus (Johnson and Sandwell, 1992; Smrekar *et al.*, 2002). The trough spacing of the largest polygons in the Mariner 10 portion of Caloris is up to 50 km (Figure 2.16), larger than the maximum size of giant polygons

Table 2.3. *Dimensions of troughs/graben in the Caloris basin*

Index	Latitude	Longitude	Maximum Relief (m)	Length (km)	D $\theta = 60°$ (m)
G_S40	27.7°N	184.7°W	220	6.9	254
G_S41	28.8°N	184.9°W	220	9.3	254
G_S42	29.4°N	184.4°W	180	6.8	208
G_S43	27.7°N	185.5°W	180	9.7	208
G_S44	27.5°N	185.1°W	100	6.8	116
G_S45	25.0°N	186.5°W	130	5.6	150
G_S48	28.3°N	185.9°W	160	5.7	185

The troughs/graben are in the digitized database shown in Plate 1. Topographic data were obtained from 2D photoclinometry on a third-encounter Mariner 10 image (0529055). Widths were measured from topographic profiles. The relief of a trough wall measured from the photoclinometrically derived DEM was compared to the relief determined using shadow measurements. They were found to agree by better than ±10 m.

observed on Venus (30 km) (Smrekar *et al.*, 2002), but comparable in scale to some of the largest found in Utopia Planitia on Mars. The mixture of highly sinuous and linear troughs, the wide variation in trough widths and depths, and the wide range in organization of the polygonal pattern are all characteristics strikingly similar to polygon terrain on Mars and Venus.

The widths of graben imaged by Mariner 10 are also highly variable, ranging from hundreds of meters up to ~10 km (Dzurisin, 1978) (Table 2.3). In the wider graben where the interiors are not obscured by shadows, the floors appear to be flat (Strom *et al.*, 1975; Dzurisin, 1978). Topographic data for some of the graben imaged by Mariner 10 have been obtained using 2D photoclinometry (Kirk *et al.*, 2003). Photoclinometry does not produce reliable topographic data in areas obscured by true shadows. Thus, measurements were made only where the majority of the trough floor and walls are without shadows. The topographic data confirm that many of the troughs have relatively flat floors (Watters *et al.*, 2005) (Figure 2.18). The depth of the measured graben in the Mariner 10 portion of the basin varies from ~100 m up to a maximum of ~220 m (Table 2.3) and the walls of the troughs have gentle slopes that are generally between roughly 4° to 6° and do not exceed 10° (Watters *et al.*, 2005). The topographic data show that many graben are flanked on both sides by gently sloping rises (Figure 2.18). These rises or rounded rims vary in relief and are often tens of meters above the surrounding plains.

Basin-radial and basin-concentric graben and wrinkle ridges is a tectonic pattern found in the Caloris basin (Plate 4) and in the newly discovered Rembrandt basin (Watters *et al.*, 2009b, Fig. 4). Similar patterns of tectonic features have not yet been observed in the interior fill of other large impact basins on Mercury, such

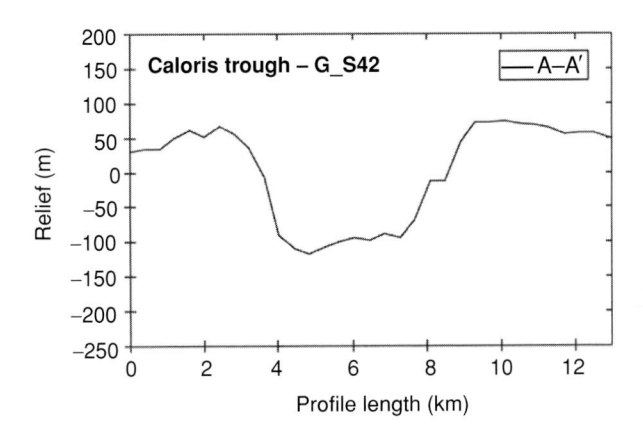

Figure 2.18. Topographic profile across a trough in the Caloris basin. The topographic data show that many of the troughs have relatively flat floors, gently sloping walls, and raised or rounded rims. The profile is from a DEM generated using 2D photoclinometry (see Kirk et al., 2003) on Mariner 10 frame 529055. Profile location is shown in Figure 2.16. Elevations are relative to an arbitrary datum.

as Beethoven and Tolstoj (André et al., 2005). The tectonic patterns in Caloris and Rembrandt are in sharp contrast to lunar mare, where radial and concentric wrinkle ridges occur in mare basalts, and circumferential trough-forming graben occur only near the basin margins (Strom, 1972; Bryan, 1973; Maxwell et al., 1975; Watters and Johnson, Chapter 4). Lunar graben form linear to arcuate troughs with flat floors and steep walls (Baldwin, 1963; McGill, 1971; Golombek, 1979; Watters and Johnson, Chapter 4). The graben are located along basin rims and can cut both mare and basin material (Wilhelms, 1987). Lunar graben often occur in parallel or echelon sets (Golombek, 1979; Wilhelms, 1987). The pattern of wrinkle ridges and graben in some maria are a result of mascon tectonics, where loading from uncompensated basalts causes downward flexure of the lithosphere, resulting in compressional stresses in the interior of the basin that form the wrinkle ridges and extensional stresses on the margins that form arcuate graben (Phillips et al., 1972; Melosh, 1978; Solomon and Head, 1979, 1980; Freed et al., 2001). It is clear that the pattern of compressional and extensional deformation observed in the Caloris and Rembrandt basins is not consistent with lunar-style mascon tectonics (Melosh and Dzurisin, 1978b; Melosh and McKinnon, 1988; Watters et al., 2005; Watters et al., 2009b,c).

An important constraint on the mode and timing of formation of the Caloris graben is the crosscutting relationship between them and the wrinkle ridges. Many of the wrinkle ridges are crosscut by the graben (Figure 2.19). The angle of intersection between the ridges and the crosscutting graben is highly variable. In the Mariner 10 portion of the basin, some graben crosscut wrinkle ridges at nearly orthogonal angles. In other cases, graben roughly parallel wrinkle ridges, cutting

44   *Planetary Tectonics*

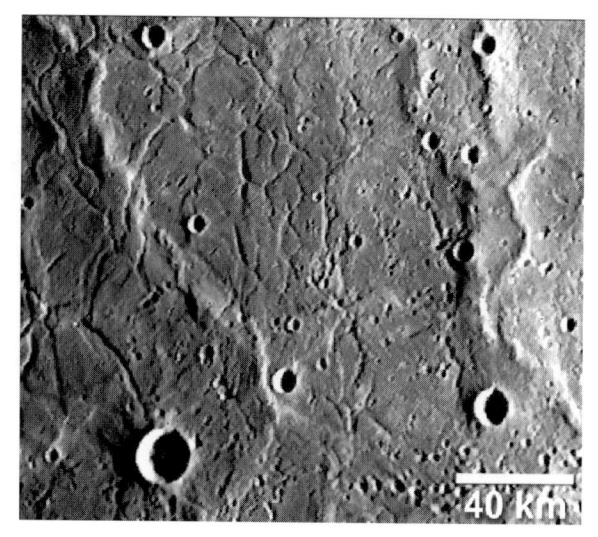

Figure 2.19. Mariner 10 image of troughs and wrinkle ridges on the floor of Caloris. The troughs that form the polygonal terrain crosscut wrinkle ridges suggesting that the troughs postdate the ridges. The areal extent of the troughs is not as great as that of the wrinkle ridges. Mariner 10 frame 106.

their flanks. The crosscutting relations indicate that the graben postdate the wrinkle ridges (Strom *et al.*, 1975; Dzurisin, 1978; Melosh and McKinnon, 1988; Watters *et al.*, 2005; Murchie *et al.*, 2008; Watters *et al.*, 2009c). Thus, compressional deformation of the interior plains materials predates the extension that resulted in the basin-radial and basin-concentric graben. This suggests that initial cooling of the Caloris basin floor material after its emplacement was probably not the cause of the extension, leaving either subsurface heating or uplift of the floor material after the formation of the wrinkle ridges as the explanation (see Section 4.3).

2.8 Summary

The tectonic features on Mercury are dominantly contractional. Lobate scarps, high-relief ridges, and wrinkle ridges are landforms that exhibit crustal shortening. Lobate scarps are the most widely distributed tectonic landform. They are the expression of surface-breaking thrust faults that deform old pre-Tolstojan intercrater plains and relatively young Calorian smooth plains material. Lobate scarps imaged by Mariner 10 are not uniformly distributed spatially and do not exhibit random orientations. They have regional preferred orientations and fault plane dip directions that indicate the influence of regional stresses. Embayment and superposition relations suggest that lobate scarp formation developed after the emplacement of the oldest intercrater plains and continued after the emplacement of the youngest smooth plains material. High-relief ridges, like lobate scarps, are found in intercrater plains and younger smooth plains and appear to be formed by high-angle reverse faults. Wrinkle ridges are complex landforms that reflect a combination of folding and thrust faulting. They occur exclusively in Calorian smooth plains material. The largest contiguous ridged plains units in the hemisphere imaged by Mariner 10 are in the Caloris exterior annulus. Topographic data indicate that some ridged plains of the eastern annulus occupy lowland areas and suggest that subsidence may have been an important source of compressional stress. Wrinkle ridges are also found in the Caloris interior plains material. The patterns of wrinkle ridges in Caloris are very similar to those found in lunar mascon basins where wrinkle ridges form in response to a super-isostatic load from infilling of mare basalts. Evidence of extension on Mercury is only found in the interior plains of Caloris and two other impact basins. Graben in the interior plains material of Caloris form a complex basin-radial and basin-concentric pattern. Some graben form giant polygons that occur along the interior margin and near the basin center. The complex pattern of graben in the Caloris interior plains material has no analogue in lunar mascons. The graben crosscut the wrinkle ridges and are thus among the youngest tectonic features on Mercury.

3. Mechanical and thermal structure of Mercury's crust and lithosphere

3.1 Introduction

Having discussed the characteristics of tectonic features on Mercury, it is now appropriate to discuss the geophysical implications of these structures. In general, the following sections will concentrate on issues which have changed since the comprehensive reviews of Melosh and McKinnon (1988) and Thomas *et al.* (1988). Few constraints presently exist on the mechanical and thermal structure of Mercury. However, based on the limited data and our experience with other terrestrial planets, Mercury is thought to have an iron core and a silicate mantle (Schubert *et al.*, 1988). Typically, the mantle is overlain by a crust which is compositionally different to the mantle, and is usually more silica-rich and of a lower density. Those parts of the crust and mantle which are relatively cold do not participate in mantle convection over geological timescales and together will be termed the lithosphere. The rigidity of this layer is often expressed as an effective elastic thickness, T_e, which is usually several times smaller than the lithospheric thickness. It is important to recognize that T_e does not give the depth to a particular isotherm or geological feature, though lower elastic thicknesses are usually associated with higher heat fluxes, and that the value of T_e recorded is the *lowest* since deformation occurred. Because the lithosphere usually becomes more rigid over time (due to cooling), the values of T_e estimated are generally relevant to the period of deformation and not the present day.

The heat flux or temperature gradient also determines the depth to the brittle–ductile transition (BDT). Above this level, deformation is typically accommodated by faulting (earthquakes), while below this level, it is accommodated in a ductile (aseismic) fashion. Because both the BDT depth or seismogenic thickness and elastic thickness are controlled by heat flux, on Earth there is generally a good correlation between these two quantities (e.g., Maggi *et al.*, 2000).

Mariner 10 flyby data give a density structure for Mercury consistent with a mantle approximately 600 km thick (Schubert *et al.*, 1988). Surface reflectance spectra (Vilas, 1988; Sprague *et al.*, 1997) and Mariner 10 color observations (Robinson and Lucey, 1997) suggested a surface composition deficient in opaque mineral and possibly similar to that of the lunar highlands crust. MESSENGER color observations, however, suggest Mercury's crust is largely volcanic in origin, with significant compositional heterogeneities, and that a low-reflectance component is enriched in iron- and titanium-bearing opaque materials (Robinson *et al.*, 2008; Denevi *et al.*, 2009). Thus, the existence of a crust on Mercury is very likely. However, the thickness of this crust is poorly constrained.

Mercury has a significant topographic equatorial ellipticity, based on radar ranging data (Anderson *et al.*, 1996). However, the equivalent gravity coefficient is much smaller than the value which would be inferred from this ellipticity, suggesting that the long-wavelength topography is compensated. The apparent compensation depth is 200 ± 100 km (Anderson *et al.*, 1996), which is interpreted to be the crustal thickness. On other planets, similar analyses at shorter wavelengths sometimes yield comparable apparent depths of compensation, which are usually interpreted as reflecting lithospheric thicknesses (e.g., Smrekar and Phillips, 1991). It will be argued below, on the basis of geological observations, that crustal thicknesses on Mercury are likely to be at or less than the bottom of the range proposed by Anderson *et al.* (1996).

On Earth, the base of the lithosphere is often defined as a particular isotherm, usually around 1100 to 1300 K (e.g., Parsons and McKenzie, 1978; Breuer *et al.*, 1993). The justification for this approach is that geological materials have strongly temperature-dependent viscosity, and thus below a particular temperature they will not participate in convective motions (e.g., Solomatov, 1995). Here, we will generally assume a base lithosphere temperature of 1400 K, both to be conservative and to take into account the likely dryness of the lithosphere on Mercury.

3.2 Thermal structure

The lobate scarps are perhaps the most distinctive class of tectonic feature on Mercury. Earlier in this chapter (subsection 2.3.4) it was shown that stereo topography across lobate scarps can be used to infer the approximate depth of faulting (30 to 40 km). On Earth, large earthquakes rupture all the way to the base of the seismogenic zone. Hence, the inferred vertical extent of the faults on Mercury places a crude constraint on the BDT depth and thus the thermal structure of

Mercury at the time of faulting (Watters *et al.*, 2002). This simple observation has been used to make several inferences about the thermal and mechanical properties of the lithosphere at the time that faulting occurred.

Watters *et al.* (2002) assumed that the BDT depth was in the range 300 to 600 °C and inferred a heat flux at the time of faulting of 10 to 43 mWm^{-2}. They assumed a thermal gradient in which little heat was produced within the crust. Nimmo and Watters (2004) obtained a subcrustal heat flux of 30 to 50 mWm^{-2} if little heat production took place within the crust, but obtained a range of 0 to 45 mWm^{-2} if crustal heat production rates were similar to lunar highland values. In either case, the total heat flux required to produce the correct BDT depth exceeded the likely total radiogenic heat flux by about 10 mWm^{-2}. A heat flux in excess of the heat production is characteristic of planets during their early thermal evolution (see Sections 5.2 and 5.5).

Given these approximate bounds on heat flux at the time of deformation, it is then possible to place constraints on the thickness of the crust. Topography which is supported by variations in crustal thickness decays with time as the lower part of the crust flows laterally; the rate of decay increases with crustal thickness and temperature (e.g., Nimmo and Stevenson, 2001). Based on the survival of Mercurian topography over 4 Gyr, and using a rheology appropriate to dry plagioclase and heat fluxes similar to those derived above, Nimmo (2002) concluded that the crust was <200 km thick. A different approach was presented by Nimmo and Watters (2004), who argued that the joint requirements of a 30- to 40-km BDT depth and an absence of melting at the base of the crust constrained the crustal thickness to be <140 km. Thicker crusts either produced BDT thicknesses which were too shallow or resulted in widespread crustal melting. These constraints, although still crude, are consistent with the geodetic estimate of Anderson *et al.* (1996) of 100 to 300 km.

Using the approach of Nimmo and Watters (2004), two hypothetical temperature profiles for Mercury which produce the correct BDT depth are shown in Figure 2.20A. Both cases have a relatively thin crust (60 km). The lithospheric thickness varies depending on the balance between heat production within the crust and heat flow into the base of the crust, but in either case is relatively thin (81 to 109 km). Crusts which were much thicker would result in the crustal thickness exceeding the lithospheric thickness. Figure 2.20B shows the resulting stress–depth distribution and demonstrates that the bulk of the strength of the lithosphere arises from the brittle portion.

3.3 Elastic thickness

The bounds on heat flux also allow constraints to be placed on the rigidity or elastic thickness T_e of the lithosphere. Nimmo and Watters (2004) used the

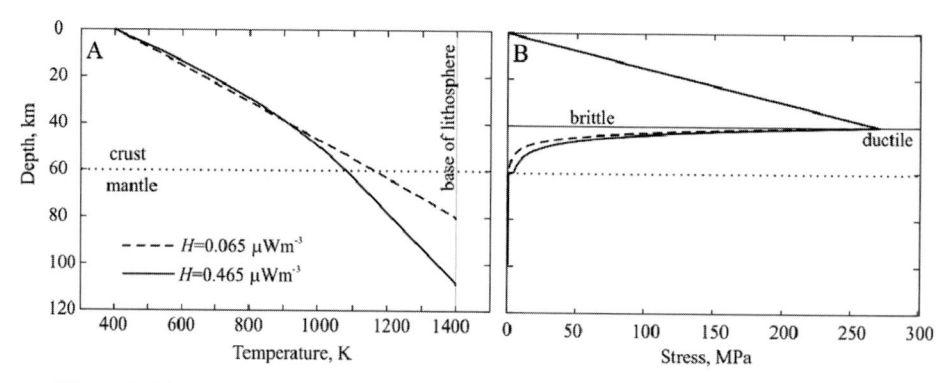

Figure 2.20. Hypothetical Mercury temperature profiles resulting in a BDT depth of 40 km, both with crustal thickness 60 km. Solid line has crustal heat generation rate $H = 0.465$ µWm^{-3} and subcrustal heat flux $F_b = 20$ mWm^{-2}; dashed line has $H = 0.065$ µWm^{-3} and $F_b = 36$ mWm^{-2}. Lithospheric thicknesses (defined as depth to 1400 K isotherm) are 109 km and 81 km, respectively. (B) Stress–depth profiles for temperature profiles shown in (A). Crustal rheology is dry plagioclase in the dislocation creep regime, mantle rheology is dry olivine. Strain rate is 10^{-17} s^{-1}, grain size 1 mm. For a curvature of 5×10^{-7} m^{-1}, the effective elastic thicknesses are 33 km and 25 km, respectively. Method and other parameters are given in Nimmo and Watters (2004).

yield-strength envelope approach of McNutt (1984) and showed that for the kind of stress–depth profiles shown in Figure 2.20B and for typical curvatures associated with the large-scale lobate scarps, T_e was typically about 75% of the BDT depth. Thus, the effective elastic thickness at the time of faulting is predicted to have been 25 to 30 km. The value of T_e at the present day on Mercury is undoubtedly larger than this value; however, the value recorded is the lowest since the deformation occurred. Thus, only the youngest deformation features on Mercury are likely to provide estimates of T_e relevant to the planet's present-day state. The issue of spatial and temporal variations in T_e is discussed further below (Section 5.2).

3.4 Fault properties

Both the depth to the BDT and the thickness of the crust are likely a factor >2 larger than typical terrestrial continental values. However, the gravitational acceleration on Mercury is only about 38% of the Earth's. Thus, the overburden pressure at the base of the BDT on Mercury is comparable to the terrestrial value, about 300 MPa (Figure 2.20B). Despite the large overburden pressures acting to prevent fault motion, terrestrial faults slip under low differential stresses (of order 10 MPa), probably because the presence of pore fluids reduces the overburden pressure and can lubricate fault planes (e.g., Rice, 1992). No such lubrication is likely on Mercury, which may help to explain why large-scale tectonic deformation is relatively rare. However, the fact that faults formed and accumulated significant displacements suggests that stresses could be quite large; potential sources of stress are discussed in Section 5.4 below.

Observations of lobate scarps may provide insight into fault properties on Mercury. A limited dataset ($n=8$) of lobate scarps suggests that the maximum *D–L* ratio γ is $6.9 \pm 1 \times 10^{-3}$ (subsection 2.3.5). It has been argued that γ contains information about the mechanical properties of the rock and the regional stresses (Cowie and Scholz, 1992a). Specifically, by adopting a model which takes into account inelastic deformation near the fault tip, it can be shown that

$$\gamma = \frac{(1-\nu)(\sigma_o - \sigma_f)}{\mu} f(\sigma_a, \sigma_o, \sigma_f), \qquad (2.1)$$

where ν is Poisson's ratio, σ_o is the shear strength of the rock, σ_f is the residual frictional stress on the fault, μ is the shear modulus, and $f(\sigma_a, \sigma_o, \sigma_f)$ is a factor of order 1 which increases as the remote stress σ_a increases.

Terrestrial faults show a wide variety of scatter in γ (e.g., Cartwright *et al.*, 1995; Dawers and Anders, 1995). Thus, equation (2.1) is undoubtedly a simplification of a more complex mechanical problem (see, for example, Cowie and Shipton, 1998; Schultz and Fossen, 2002; Schultz *et al.*, Chapter 10). Nonetheless, thrust faults in the Canadian Rocky Mountains ($n=29$), a likely terrestrial analogue to lobate scarps, have a γ of 8×10^{-2} (Watters *et al.*, 2000). This value of γ is significantly higher than the Mercury lobate scarps. Similarly, the value of γ obtained for Mercury is towards the low end of the range of values compiled by Cowie and Scholz (1992b). Equation (2.1) shows that a likely explanation for the low value of γ is that either μ or σ_f are larger on Mercury than on Earth, or that σ_o is smaller.

It is unlikely that σ_o is smaller than the terrestrial value, since dry rocks are typically stronger than wet rocks. Although μ might be higher on Mercury than on Earth, it is perhaps more likely that σ_f, the residual frictional stress on the fault, is larger. As argued above, terrestrial faults on Earth are weak because of the presence of water; on Mercury, there is no water available to lubricate the fault and reduce σ_f. Thus, a natural explanation for the low value of γ found for the Mercury lobate scarps is the high friction on faults on Mercury, due to the absence of water. Faults on Venus are also thought to be strong for similar reasons (Foster and Nimmo, 1996).

3.5 Summary

The observed and modeled characteristics of lobate scarp thrust faults have allowed the likely depth to the BDT transition (30 to 40 km) to be inferred. This BDT depth places constraints on the thermal structure at the time of faulting and also implies an elastic thickness of 25 to 30 km. The thickness of the crust at the time of faulting was probably <140 km, and the inferred heat fluxes exceed the likely amount from radiogenic sources 4 Ga. Plausible thermal and mechanical structures for Mercury 4 Ga are shown in Figure 2.20. Faults on Mercury are probably significantly stronger than their terrestrial counterparts, owing to the absence of water.

4. Mechanical and thermal structure of the Caloris basin

The tectonic features in the Caloris basin are among the most complex and interesting on Mercury, consisting of structures that express at least two distinct episodes of deformation (see Sections 2.6, 2.7). The interior of Caloris consists of early compressional wrinkle ridges, cut by later extensional troughs. Both tectonic landforms have implications for the local crustal structure of the basin.

4.1 Caloris wrinkle ridges and the brittle–ductile transition depth

Wrinkle ridges have been recognized on all the terrestrial planets (see Sections 2.5, 2.6). It has long been thought that the characteristic spacing of these features provides information about the thickness of the layer in which deformation occurs. However, wrinkle ridge spacing in ridged plains on Mars of ~30 km have been interpreted as indicating both thin (few km) and thick (few tens of kilometers) layers (Zuber and Aist, 1990; Watters, 1991). This is because ridge spacing can be accounted for with layer instability models where the lithosphere is free to deform (Zuber and Aist, 1990) or where the lithosphere is assumed to be rigid (Watters, 1991).

In a recent series of papers, Montesi and Zuber (2003a,b) have developed a model in which strain localization occurs in response to compression of a material which is strongly strain-weakening. The characteristic spacing of the zones of localization depends on the depth to which the localization extends; that is, the BDT depth. The model parameters are not easy to relate to measurable properties of rock specimens. Nonetheless, for likely parameters Montesi and Zuber (2003a,b) concluded that a typical ratio of ridge spacing to BDT depth was in the range of 0.25 to 1.15. Note that in this formulation, the BDT depth can exceed the ridge spacing, while in conventional buckling theory the buckling wavelength generally significantly exceeds the layer depth.

Wrinkle ridge spacing in the interior of the Caloris basin varies greatly from under 10 km to over 60 km with a median of 33 km ($n = 32$). The results of Montesi and Zuber (2003a,b) thus imply a minimum BDT thickness of 25 km and an upper bound of 120 km. Note that this estimate assumes that any rheological layering at the base of the basin infill does not have a significant effect. As noted above, the importance of the BDT depth is that it places constraints on both the thermal structure and effective elastic thickness of the lithosphere. Although the uncertainties are so large that further investigation is premature, it is nonetheless striking that the BDT depth range thus obtained is consistent with that inferred for the lobate scarps (subsection 2.3.4). It is important to note that ridge spacing in the Caloris basin may have been influenced by shallow buried impact craters and possibly by buried interior basin rings (see Section 2.6).

As Watters *et al.* (2000) point out, generating sufficient horizontal stresses to achieve motion on low-angle thrust faults is non-trivial. For a stress of 100 MPa, thrust faulting is restricted to depths shallower than about 10 km. Since stresses of this order are likely to have been present on Mercury (Section 5.4), invoking thrust faults extending to a few tens of kilometers is probably not unreasonable. Nonetheless, the possibility remains that wrinkle ridges are relatively shallow-rooted structures.

Wrinkle ridges are interpreted to be the surface expressions of folding and thrust faulting (e.g., Watters, 1988, 2004; Golombek *et al.*, 1991; Schultz, 2000). Assuming that the bulk of the strain is due to displacement on blind thrust faults, it is possible to estimate the regional strain if the D–L ratio γ for wrinkle ridges is known. The estimated displacement on Mercury's wrinkle ridge thrust faults range from 0.4 to 1.5 km with an average of ~0.8 km ($n = 7$), assuming fault plane dips of 30° (Table 2.2). A linear fit to the D–L data for the wrinkle ridges yields a value of $\gamma \cong 3.5 \times 10^{-3}$ for $\theta = 30°$. The Yakima folds in the Columbia Plateau basalts, which are a likely terrestrial analogue to wrinkle ridges (Watters, 1988, 1992), show a somewhat larger γ of $2.9 \pm 1.0 \times 10^{-2}$ (Mege and Reidel, 2001). This difference likely reflects the much larger accumulated shortening on the anticlinal ridges of the Columbia Plateau (Reidel, 1984) compared to planetary wrinkle ridges (Watters, 1988). The strain can be calculated by summing the lengths of the digitized wrinkle ridges within the Caloris basin (Plate 4) if the D–L relationship of the fault population is known (Scholz and Cowie, 1990; Cowie *et al.*, 1993). The strain for large faults ($L \geq$ the maximum depth of faulting) is given by

$$\varepsilon = \frac{\cos(\theta)}{A} \sum_{k=1}^{n} D_k L_k, \qquad (2.2)$$

where θ is the fault plane dip, A is the survey area, n is the total number of faults, and $D = \gamma L$ (Cowie *et al.*, 1993). The compressional strain over the entire area of the basin ($\sim 1.89 \times 10^6$ km^2) is roughly 0.07% (Watters *et al.*, 2009c). More sophisticated models which take into account inelastic effects will reduce the calculated strain somewhat (Mege and Riedel, 2001).

On Earth, fault length distributions can be either power-law or exponential (e.g., Hardacre and Cowie, 2003). For wrinkle ridges <150 km long, the distribution within Caloris is clearly exponential, with a slope of 0.03 km^{-1}. Such exponential fault distributions are also found at terrestrial mid-ocean ridges (Cowie *et al.*, 1993). Unfortunately, even on Earth, the significance of these distributions is unclear. For instance, although in some cases it appears that the thickness of the brittle layer can have an effect on fault population statistics (e.g., Ackermann *et al.*, 2001), numerical models suggest that such statistics depend in a complex fashion on the amount of strain, the degree of fault linkage and the amount of fault nucleation (Hardacre and Cowie, 2003). The significance of the Caloris wrinkle-ridge length distribution is thus not clear at the present time.

4.2 Strain and depth of faulting of extensional troughs

The extensional troughs in the Caloris basin, postdate the wrinkle ridges, are more closely spaced than the wrinkle ridges, and form polygonal patterns. As with the lobate scarp and wrinkle-ridge thrust faults, the regional strain can be estimated if the *D–L* ratio of the trough forming graben is known. The slope of the measured trough walls does not exceed 10°, which is much lower than likely fault plane dips on the bounding normal faults (>45°). The low trough wall slopes are presumably due to mass wasting, probably related to impact processes (Watters *et al.*, 2005). Assuming initial fault plane dips of ~60°, the estimated displacement on the normal faults ranges from 0.11 to 0.25 km with an average of ~0.2 km ($n = 7$) (Table 2.3). A linear fit to the *D–L* data for the graben yields a value of $\gamma \cong 2.2 \times 10^{-3}$ ($\theta = 60°$). Determination of γ is complicated by the variability of the trough width and the combination of highly sinuous and linear trough segments. The linkage relationship between fault segments forming the polygonal pattern is also not obvious. It is assumed that each trough segment is independent. Using Equation (2.2), the extensional strain over the total area of the Caloris basin is roughly 0.08% (Watters *et al.*, 2009c), comparable to the estimated contractional strain due to the wrinkle ridges.

The depth of trough-forming faults is not well constrained, but the relatively close spacing of the troughs and their narrow widths suggest relatively shallow-rooted faulting. If the fault planes actually intersect, they will do so at a depth of about 2 to 5 km. Although the faults may penetrate below the 2 to 5 km intersection depth, this depth of faulting is either comparable to or smaller than that estimated from the wrinkle ridges (see Section 4.1). The closer average spacing of the troughs compared to the wrinkle ridges suggests that the trough-forming faults are confined to the basin floor material.

It may also be possible to constrain the depth of faulting using the relationship between the spacing of the polygonal graben and the thickness of the deformed floor material. Based on analysis of the variation in the maximum depth of failure with temperature gradient, Johnson and Sandwell (1992) estimated that the range in ratio of polygon width to thickness on Venus varies from 0.5 to 16. They also noted that the ratio of the dominant wavelength of deformation to layer thickness in thin layers under uniaxial compression or tension is about 4, roughly the same order of magnitude. The range in widths of the most pronounced Caloris polygons varies greatly from ~7 to 50 km, with a median of 14 km ($n = 15$). A ratio of 4 corresponds to a range in depth of 1.8 to 12.5 km, in good agreement with the above estimate based on fault geometry.

4.3 Caloris loading history

The fact that Caloris experienced compression followed by extension places constraints on its geological history (e.g., McKinnon, 1980; Melosh and McKinnon, 1988; Thomas et al., 1988). Some insight may be gained by considering the likely events following a large impact. The initial impact creates a transient cavity which is modified over the course of a few minutes by collapse (Melosh, 1989). The vaporization and excavation of crustal material results in the crater being initially out of isostatic equilibrium; vertical mantle uplift results, with a timescale probably comparable to the ~10^3 year timescale characteristic of glacial rebound on Earth. At the end of this process, the basin will overlie thinned crust and be in isostatic equilibrium. The energy delivered by the impactor will result in raised temperatures surrounding the basin, which will decay over a timescale set by diffusion (at least a few Myr for Caloris).

4.3.1 Wrinkle ridges

On the Moon, mare volcanism flooded the nearside basins (Solomon and Head, 1979). Geological mapping and crater counting of the deposits in the Caloris basin suggest that they are probably volcanic in origin and postdate the formation of the basin (Spudis and Guest, 1988). Unfortunately, the thickness of the deposits is not well constrained (see Section 4.2) and the topography of the basin is currently unknown. If later volcanism does occur, the original floor of the basin will subside, but depending on the rigidity of the basin and the density of the fill, its new surface may be either higher or lower than the original level. Subsidence in response to loading can lead to compression of the basin-interior floor material. Melosh and McKinnon (1988) concluded that subsidence was the most likely cause of the Caloris wrinkle ridges.

If the basin maintains elastic strength, then the maximum strain ε caused by sagging or uplifting of the basin center by an amount w_0 is given by

$$\varepsilon \approx w_0 \left(\frac{2\pi}{\lambda}\right)^2 \frac{T_e}{2}, \tag{2.3}$$

where λ is the basin width (~1550 km for Caloris). An inferred compressive strain of 0.07% implied by the wrinkle ridges imaged by Mariner 10 and MESSENGER requires $w_0 \sim 3$ km for $T_e = 30$ km, $w_0 \sim 1$ km for $T_e = 100$ km, and $w_0 < 1$ km for $T_e = 150$ km. Alternatively, subsidence of the basin floor due to surface loading may lead to compressional thermal stresses at depth. A compressional strain of ~0.07% requires a temperature increase of 25 K. For the likely heat fluxes of 30 to 50 mWm^{-2}, such a temperature increase requires a depth increase of 1.5 to 2.5 km.

Cooling of an initially hot layer can also lead to compressional stresses. Parmentier and Haxby (1986) showed that a cooling layer which is free to contract horizontally will develop compressional stresses near the surface and tensional stresses at depth. We think it unlikely that this process is important for Caloris because it is not obvious that the basin fill is free to contract horizontally and because it is unlikely that the compressional features are as shallow as the theory requires.

For a basin initially in isostatic equilibrium which then has a thickness of material b added on top, it can be shown that the original floor subsides by an amount $b\rho_b/\rho_m$, where ρ_b and ρ_m are the densities of the added material and the mantle respectively. This subsidence is reduced if the elastic strength of the lithosphere is significant at the basin length-scale. Thus, to cause the inferred strains would probably require >5 km of fill material to the basin interior. This is a large amount of volcanism; for instance, estimated thicknesses of mare fill on the Moon range from 0.5 to 4.7 km (Williams and Zuber, 1998). It is, therefore, not yet clear whether subsidence alone can explain the observed wrinkle ridges. The apparent age of many of the known compressional tectonic features on Mercury suggests the Caloris wrinkle ridges formed at about the same time as wrinkle ridges and lobate scarp thrust faults in the smooth plains and intercrater plains (Watters *et al.*, 2004, 2005). A combination of subsidence and compressional stresses due to global contraction could account for the Caloris wrinkle ridges.

Loads which are partially rigidly supported produce positive gravity anomalies which can lead to reorientation of the planet; Murray *et al.* (1974) suggested that the present-day location of Caloris might be due to such a gravity anomaly. If the wrinkle ridges are evidence for subsidence, as seems likely, then it is important to examine whether the inferred subsidence and fill thickness are consistent with the predicted gravity anomaly. We discuss this issue below (Section 4.4).

4.3.2 Extensional troughs, exterior loading, and lateral crustal flow

The existence of radial and concentric graben that postdate the wrinkle ridges is puzzling, especially in view of the fact that similar features are not observed in mascons on the Moon. It is clear that a mascon tectonic model cannot account for the spatial and temporal distribution of tectonic features in Caloris. Thomas *et al.* (1988) suggested that inward motion towards Caloris caused updoming and extension. Melosh and McKinnon (1988) advocated exterior, annular loading of the Caloris basin causing basin interior uplift. This model invokes an annular load extending 650 to 1800 km radial distance from the basin center (Melosh, 1978) with a T_e from 75 km to 125 km (McKinnon, 1986). The Thomas *et al.* (1988) model does not really explain why similar features are not seen in similar basins on the Moon or Mars; while the Melosh and McKinnon (1988) basin exterior loading model depends on the poorly known thickness of the circum-Caloris deposits. A third possibility is that the Caloris graben are due to thermal contraction or subsurface heating of near-surface volcanic plains, as has been proposed for Venus (Johnson and Sandwell, 1992); however, both these explanations are problematic. Surface cooling is unlikely to produce features of the scale observed, while subsurface heating is not only ad hoc but also disagrees with the observations.

The Caloris graben may also be explained by lateral crustal flow (Watters *et al.*, 2005). After the formation of an impact basin pressure gradients exist that can drive flow in the lower crust, even if the crust is isostatically compensated. This lateral flow of the lower crust will result in uplift of the existing basin materials and occurs at a rate determined by the crustal thickness, rheology, and thermal gradient. The rate of flow is always much slower than the vertical mantle motion discussed above (Zhong, 1997; McKenzie *et al.*, 2000) but can still be significant. If such lateral flow does occur, the result is uplift by an amount depending on the initial thinning of the crust and the lithospheric rigidity. Although the topography of Caloris is currently not well determined, the basin is clearly relatively shallow (e.g., Schaber *et al.*, 1977), and thus uplift due to lateral crustal flow is very likely to have occurred. Melosh and McKinnon (1988) note that domical uplift would likely lead to radial graben, while a mix of radial and concentric graben is actually observed. An extensional strain of roughly 0.08% requires an uplift of about 3 km for $T_e = 30$ km, $w_0 \sim 1$ km for $T_e = 100$ km, and $w_0 < 1$ km for $T_e = 150$ km (Equation 2.3).

Lateral crustal flow causes thickening of the crust beneath the impact basin, which in turn generates uplift and extension (Watters *et al.*, 2005, Fig. 4). As lower crustal flow proceeds, uplift leads to extensional stresses. The main controls on the amount of lower crustal flow are heat flux, crustal thickness, and rheology. The degree of lower crustal flow can be parameterized by the basin relaxation, defined as the ratio of basin uplift:initial depth. If the heat flux is set by the likely radiogenic values, then for the assumed rheology, a crustal thickness of 90 to 140 km is required (Watters *et al.*, 2005). This derived range in crustal thickness is consistent with other estimates (see Section 3.2).

The maximum extensional stress occurs away from the basin center because Watters *et al.* (2005) assumed a flat-floored initial basin geometry. This geometry is in agreement with recent studies of gravity anomalies associated with lunar mascons (Watters and Konopliv, 2001). An initially Gaussian-shaped basin results in stresses of the same magnitude but maximized at the basin center. The existence of a radial graben complex that converges near the center of Caloris (see Section 2.7), suggests that a combination of the two basin geometries, or preexisting fractures, may be needed to account for the basin-radial and basin-concentric graben.

Kennedy *et al.* (2008) evaluated both the external loading and the lateral flow models for a range of lithospheric structures using finite element analysis. In agreement with Melosh and McKinnon (1988), they found that the distribution of basin-interior graben can be accounted for by an annular load extending 750 to 1800 km radial distance from the basin center. To induce the necessary load, the basin exterior smooth plains must be thick, 5 km or more, so that extensional stresses from flexural uplift exceed compressional stresses due to subsidence of interior plains and global contraction. Kennedy *et al.* (2008) also conclude that infilling of the basin interior must have occurred in stages spaced over a period of time that exceeds the timescale for basin subsidence from a single interior load, to reduce compressional stresses in the uppermost layer of the plains material. In their analysis of the lateral crustal flow model, Kennedy *et al.* (2008) found that although basin-interior uplift will result from late-stage flow, the extensional stress patterns from lateral flow-induced uplift are not consistent with the observed distribution of graben imaged by Mariner 10.

MESSENGER's discovery of Pantheon Fossae is a critical new constraint on models for the origin of stresses in the Caloris basin. Both the exterior loading model and the lateral crustal flow model must be reevaluated to determine if they can account for the evolution of stresses necessary to explain the spatial and temporal distribution of wrinkle ridges and basin-radial and basin-concentric graben in Caloris. It is also unclear whether the external loading model is consistent with existing constraints from gravity on the mass of loading material present (see below).

4.4 Elastic thickness

Since Caloris is located on the approximate longitude (although not the latitude) of one of Mercury's principal axes of inertia, it has been suggested (Murray *et al.*, 1974) that it may possess a positive gravity anomaly – a mascon. Mascons are formed on the Moon by dense mare basalts flooding the basin after sufficient time has elapsed that the lithosphere is highly rigid (Solomon and Head, 1979; see Watters and Johnson, Chapter 4). Melosh and Dzurisin (1978b) used the location of Caloris to place a lower bound on the gravity anomaly and hence the local thickness of uncompensated lavas. Willemann (1984) pointed out that the fact that Caloris is not on the equator also places an upper bound on the thickness of the interior plains material, though more recent work suggests that Willemann's (1984) theoretical approach is incomplete (Matsuyama *et al.*, 2006).

The models of Melosh and Dzurisin (1978b) and Willemann (1984) both used moment of inertia (MoI) values which were subsequently updated by Anderson *et al.* (1987). Furthermore, they assumed that the volcanic fill existed in an exterior annulus at a radial distance of 650 km to 1950 km from the center of the basin. Here we use the updated MoI and basin radius values, and also consider the effect of uncompensated volcanic fill forming a disk inside the basin (outer radius 775 km).

The revised upper bound (Anderson *et al.*, 1987) on J_2 of 8×10^{-5} gives an upper bound on the uncompensated exterior annular fill thickness of 1.8 km, slightly lower than the original Willemann (1984) estimate. If instead the volcanic fill forms a disk within the basin, the upper bound is 5.6 km. The lower bound on $(B - A)/C$ of 5×10^{-4} results in a lower bound on the annular fill thickness of 0.35 km, similar to the original Melosh and Dzurisin (1978b) estimate. The lower bound for the interior uncompensated fill thickness is 1.2 km. It should be noted that an upper limit on the uncompensated annular fill thickness of 1.8 km is significantly lower than the minimum thickness invoked by Kennedy *et al.* (2008) to account for the Caloris graben by exterior loading.

These results therefore suggest that roughly 1 km of annular uncompensated volcanic fill, or a few kilometers of uncompensated basin-filling volcanic material, can explain the position of Caloris. The actual thickness of the volcanic sequence could be much greater since wide loads will tend to be at least partially compensated. Basin fill thicknesses of >5 km were inferred above (Section 4.2) to explain trough and polygon widths and wrinkle-ridge strains. Thus, it appears that the basin fill is mostly, but not entirely compensated, which has implications for the elastic thickness of Caloris.

For a Cartesian situation, the degree to which a load of a particular wavelength is compensated, C, is given by (Turcotte and Schubert, 2002)

$$C = \frac{1}{1 + \frac{Dk^4}{\Delta \rho g}}, \quad D = \frac{ET_e^3}{12(1 - \nu^2)}, \quad (2.4)$$

where $0 \leq C \leq 1$, k is the wavenumber of the load, $\Delta \rho$ is the density contrast between mantle and crust, g is the acceleration due to gravity, D is the rigidity, E is the Young's modulus and ν is the Poisson's ratio. More complicated results exist for situations in which the curvature of the planet is important (e.g., Turcotte *et al.*, 1981; Willemann and Turcotte, 1982), but the basic dependence of the compensation factor on T_e remains the same. Assuming an effective load wavelength of 3100 km, values of C of 93%, 79% and 61% for T_e values of 100 km, 150 km and 200 km are obtained respectively. Since a mostly compensated situation is required

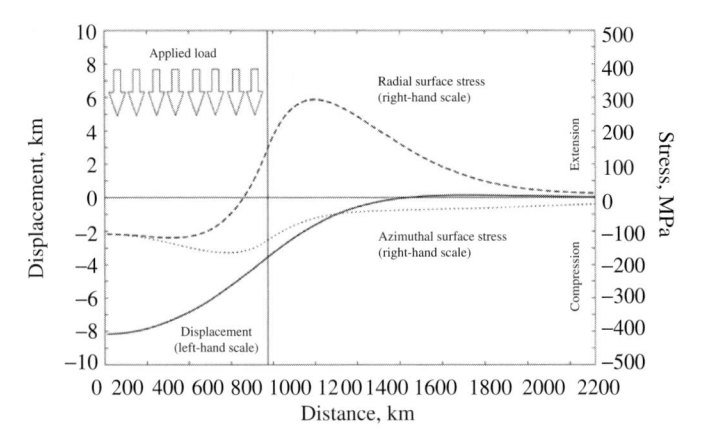

Figure 2.21. Displacement and stresses due to a radial load of 126 MPa (equivalent to a fill thickness of 10 km) and radius 775 km operating on an elastic shell with $T_e = 200$ km. Solid line is displacement, dotted line is surface azimuthal stress, dashed line is surface radial stress, arrows denote load, vertical line denotes edge of load. Values calculated using method of Solomon and Head (1979) using a planetary radius of 2440 km, Young's modulus of 100 GPa, Poisson's ratio 0.25, density 3400 kg m^{-3}, and gravity 3.7 m s^{-2}.

(C approaching 1) these results suggest that an elastic thickness on the order 100 km is appropriate.

A more precise estimate of T_e can be made by considering the case of a circular load imposed on a thin spherical shell. Figure 2.21 shows the deformation and stresses due to a load of radius 775 km on a shell with $T_e = 200$ km, calculated using the method of Solomon and Head (1979). The thickness of the load is 10 km if the load density is the same as that of the mantle (3400 kg m^{-3}); a higher density would result in a thinner load. The maximum subsidence is 8.2 km, indicating that the load is 82% compensated and the maximum compressional strain is of order 0.1%. The resulting gravity anomaly would be around 260 mGal. This model thus crudely agrees with both the geodetic constraints on the basin fill thickness and the requirements from the wrinkle-ridge strain estimates (Section 4.3). The same model with $T_e = 100$ km results in a load that is 99% compensated, inconsistent with the geodetic requirements. Thus, it appears that an elastic thickness of ∼200 km is required to explain the Caloris observations. While this T_e estimate may appear large, similar values have been obtained for the lithosphere supporting the large Martian shield volcanoes (Zuber et al., 2000).

4.5 Stresses

Figure 2.21 shows that substantial extensional stresses (∼200 MPa) develop exterior to the basin. Circumferential grabens are observed on the margins of lunar mascon basins, presumably in response to these extensional stresses. Solomon and Head (1979) showed that stresses of order 50 MPa were apparently enough to initiate lunar graben formation. Similar features have not been identified around the Caloris basin, as noted by Thomas *et al.* (1988) (see Section 2.7). The above calculations do not take into account sphericity, which may alter the resulting stresses somewhat (Janes and Melosh, 1990; Freed *et al.*, 2001). However, the relative magnitudes of the stresses for the Mercurian and lunar cases are unlikely to change, making the lack of observed grabens around Caloris puzzling. Possible explanations for their absence are: (1) the rock fracture strength is higher on Mercury; (2) the grabens do exist but have been buried by later resurfacing events; (3) the elastic thickness of Mercury's lithosphere prevented stresses from exceeding the rock strength; (4) the loading is sufficiently small that the fracture strength is not exceeded; (5) global compressional stresses reduced the local extensional stresses.

Possibility (1) seems unlikely; given that lunar rocks are likely at least as dry as rocks on Mercury, there seems little justification for assuming that their strengths are very different. Possibility (2) depends on whether volcanism exterior to Caloris postdated that of the interior. Crater density data indicates that the Caloris interior plains material may be older than the exterior annulus smooth plains material (Spudis and Guest, 1988, Table III; Strom *et al.*, 2008; Fassett *et al.*, 2009). Thus, it is unlikely that the exterior smooth plains material buried circumferential graben.

The stresses external to Caloris depend on both the elastic thickness and the amount of basin loading. Using the method of Solomon and Head (1979), the maximum extensional stress produced outside the Caloris basin as a function of basin fill thickness and lithospheric elastic thickness is shown in Figure 2.22. The location of this maximum stress point increases from $1.07 R_0$ to $1.30 R_0$ as T_e increases from 20 km to 200 km, where R_0 is the radius of the basin.

For a nominal rock strength of 50 MPa, Figure 2.22 shows that the basin fill thickness cannot exceed 1.7 km unless T_e exceeds 200 km. This fill estimate is substantially smaller than that derived from the wrinkle ridge observations. The basic problem is that the maximum compressional and extensional strains and stresses are always about the same (see Figure 2.21), and thus, if basin-interior compressional features are created, then it is likely that basin-exterior extensional features will also arise. The problem is exacerbated by the fact that crustal materials are generally weaker in tension than in compression, so lower stresses are required to produce high-angle normal faults rather than low-angle thrust faults (e.g. Turcotte and Schubert, 2002).

The most likely resolution of this problem is the possibility that global compressional stresses occurred simultaneously with the local loading stresses. A similar mechanism was advocated by Solomon and Head (1979) for the Moon. These global compressional stresses would have the effect of suppressing extensional

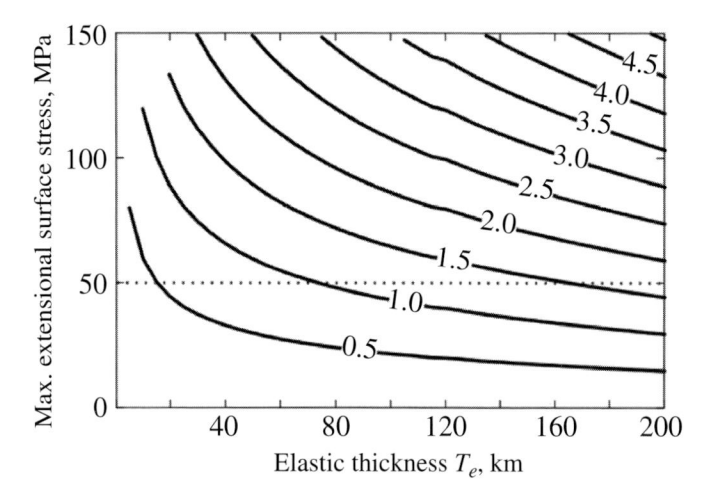

Figure 2.22. Maximum surface radial extensional stress as a function of fill thickness and elastic thickness. Load radius is 775 km, contours denote equivalent load thickness in kilometers, other parameters and method as in Figure 2.21. Dashed line denotes stress required to initiate extensional faulting on the Moon.

features and would thus permit thicker basin fills. As discussed below (Section 5.4), such stresses are likely to arise and can be comparable in magnitude to the flexural stresses discussed here.

4.6 Summary

The wrinkle ridges in Caloris are probably a result of subsidence due to infilling of the basin. This subsidence implies loading by a few to several tens of kilometers of basin fill material. Because Caloris is located at the longitude but not the latitude of one of Mercury's principal axes, both upper and lower limits can be placed on the thickness of uncompensated basin fill. Partially compensated loads can explain these constraints if the elastic thickness at the time of loading was ~150 to 200 km. The absence of circumferential graben about Caloris is probably because global contraction was occurring during basin loading and suppressed the formation of exterior extensional features. The interior extensional troughs, which overprint the wrinkle ridges, are probably due to subsequent uplift of the basin floor due to lateral crustal flow or basin-exterior loading.

5 Global implications

In the preceding sections, we have argued that the existence of lobate scarps and the characteristics of Caloris may be used to infer the properties of Mercury's crust and lithosphere. Here we investigate the pattern of deformation in the hemisphere

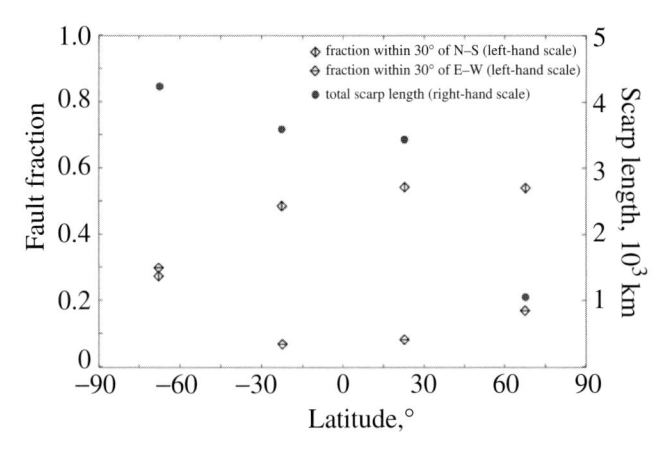

Figure 2.23. Lobate scarp orientations and distribution as a function of latitude. Diamonds indicate fraction of scarps within 30° of N–S or E–W orientations; contribution of each scarp segment is scaled by its length compared to the mean length. Circles indicate total length of scarps within each latitude bin. Mean orientations (standard deviations) of scarps within each latitude bin are (south to north): 45° (24°), 59° (19°), 60° (20°), 57° (22°).

imaged by Mariner 10 and the implications for the sources of stress and thermal evolution of the planet. As will become clear, a major difficulty in interpreting the different estimates of lithospheric properties is the poor understanding of both the relative and absolute timing of events.

5.1 Fault distribution

The distribution of orientations of lobate scarps mapped using Mariner 10 images is not uniform with latitude (see subsection 2.3.2). The fraction of lobate scarps which are within 30° of a N–S orientation and the fraction within 30° of an E–W direction as a function of latitude is shown in Figure 2.23. The plot was produced by combining mapped lobate scarp segments into longer straight line segments 20 to 30 km in length and obtaining the azimuth of each segment (see Section 2.2). The main trend is that equatorial lobate scarps are more likely to be N–S on average, while scarps at south latitudes have a stronger E–W component. The larger total length of lobate scarps towards the south pole is also evident (see Plate 1).

The predominance of N–S trending faults near the equator (Plate 1, Figure 2.23) is crudely consistent with the predictions of stresses caused by despinning (Melosh, 1977; Melosh and McKinnon, 1988, Fig. 3; Dombard and Hauck, 2008). However, the distribution of N–S trending lobate scarps imaged by Mariner 10 is not uniform across the equatorial zone (Plate 1), and the despinning model predicts no thrust faults in the mid-latitudes where lobate scarps are common (Watters *et al.*, 2004, Fig. 3). As explained below, the equatorial N–S thrust faults are diagnostic of either a relatively thick lithosphere or an additional isotropic compressive stress. The main disagreement is that the despinning model predicts high-latitude E–W normal faults, while E–W thrust faults are actually observed. This could be the result of reactivation of preexisting normal faults, a well-known phenomenon on Earth. For instance, thrust faulting beneath the Zagros Mountains, Iran, occurs on relatively steeply dipping planes (30 to 60°) which are likely to be reactivated normal faults (Jackson, 1980). It is, therefore, possible that polar normal faults were generated by (extensional) despinning stresses and with continued contraction eventually led to these features being reactivated as thrust faults.

The fact that the greatest area-normalized cumulative length of lobate scarps occurs below 50°S in the Mariner 10 hemisphere (Watters *et al.*, 2004, Fig. 3) and that all the thrust faults in this region have the same fault-plane dip direction (Plate 1) is not easily explained by either tidal despinning or global thermal contraction. One possibility is that it may reflect some kind of preexisting structure, mechanical inhomogeneities, or variations in crustal thickness; another is some process similar to convective downwelling (Watters *et al.*, 2004) or mantle convection (King, 2008). Many lobate scarps in the hemisphere image by Mariner 10 have preferred orientations and thrust slip directions. New MESSENGER images of Mercury suggest that lobate scarps have a similar distribution with latitude as those imaged by Mariner 10 (see Solomon *et al.*, 2008; Watters *et al.*, 2009a,b). The global view to be provided by MESSENGER will reveal if there are other patterns in the spatial distribution, preferred orientations, and preferred thrust slip directions of the lobate scarps.

5.2 Rigidity

Evidence derived from the lobate scarps (Section 3.3) has been used to argue that T_e values on Mercury were quite modest (25 to 30 km). Conversely, the data for Caloris (Section 4.4) imply T_e values around 200 km and, as will be shown below, other authors have obtained comparable numbers. Since T_e is closely linked to heat flux, an understanding of these T_e variations is necessary when considering a planet's thermal evolution.

For planets such as Mars or Mercury which likely started hot, initial surface heat fluxes tend to exceed the radiogenic heat production (e.g., Nimmo and Stevenson, 2000). Although Mercury's complement of radiogenic elements is unknown, for bulk silicate Earth or bulk Moon abundances, the radiogenic heat flux at 4 Ga was about 30 mWm^{-2}. As long as radiogenic elements are not greatly concentrated in the near-surface (see Figure 2.20), the thermal gradient implied is about 10 K/km. Since this gradient neglects secular cooling, it is likely to be an underestimate. A likely upper bound on the base of the lithosphere is 1400 K, so the maximum possible lithospheric thickness at 4 Ga is 100 km. As noted above (Section 3.2), much of the lithosphere retains little strength, so that the effective elastic thickness T_e is typically a factor of 3 to 4 lower than the lithospheric thickness. Thus, despite the uncertainties, our expectations of elastic thicknesses on Mercury at about 4 Ga are that they should be in the range 20 to 40 km. These expectations are consistent with the inferred BDT transition depth for the lobate scarps (see Section 3.2). However, they are inconsistent with other estimates, which we discuss below.

Melosh and McKinnon (1988) make three arguments that the elastic thickness of Mercury was ∼100 km during the period of deformation. Firstly, the stresses which occur due to a despinning planet only lead to equatorial, N–S trending thrust faults if the elastic thickness exceeds about 0.05 of the planet's radius (Melosh, 1977). Since such faults are indeed observed, it was inferred that $T_e > 100$ km at the time of deformation. Secondly, the observed lack of multiring basins on Mercury was used to infer a T_e roughly greater than 100 km at the time of heavy bombardment. Finally, Melosh and McKinnon (1988) argued that the extensional faulting within Caloris could be explained by the emplacement of annular circum-Caloris plains materials. For such materials to cause the observed pattern of faulting requires a $T_e < 125$ km, while it was argued that faults would not form at all for a $T_e < 75$ km (the stresses would be too low).

Each of these constraints has its problems. For instance, even in a thin lithosphere, N–S trending faults can form at the equator if both contraction and despinning occur simultaneously (Pechmann and Melosh, 1979; Dombard and Hauck, 2008). Theoretical arguments and observations both suggest compressional stresses of comparable magnitudes to the despinning stresses (Sections 4.5, 5.4). Neither process has a well-constrained timescale, but both are likely to have been important in the first Gyr of Mercury's history. Dombard and Hauck (2008) examined despinning in combination with 3 to 5.5 km of global contraction before the end of the period of late heavy bombardment. The record of the resulting N–S oriented thrust faults is assumed to have been lost and the present population of lobate scarps is the expression of reactivated faults by later global contraction. Thus, the observed equatorial thrust faults are not necessarily evidence for a thick lithosphere. Similarly, simulations of the effects of large impacts on planetary lithospheres have only become available recently. Although such simulations can now be performed (e.g., Turtle and Pierazzo, 1998), the quantitative effect of lithospheric thickness on multiring basin formation is still unknown. Furthermore, strain rates during impacts are so different to strain rates of geological processes that the effective elastic thickness during an impact may be very different to that measured over geological timescales. Finally, as was argued above, an annular load may not be required to explain the extensional features within Caloris and it is not clear that the exterior plains materials are thick enough to generate the stresses required.

Nonetheless, despite the uncertainties, it seems likely that T_e during the formation of Caloris was ~100 km. Even higher values of T_e (~200 km) were obtained in Section 4.4 by considering the amount of subsidence required and the thickness of uncompensated basin fill. The disagreement between these two values is probably not significant, particularly in view of the fact that T_e may have varied over the course of the history of Caloris. For example, Solomon and Head (1979) found that T_e of the Serenitatis basin increased from 25 to 50 km at 3.8 to 3.6 Ga to ~100 km at 3.4 to 3.0 Ga. It is clearly important to understand the loading history of Caloris in more detail, both the order in which events occurred and the absolute durations of the events. Although as mentioned previously, crater density data suggest that there was a significant interval between the formation of the Caloris basin and the emplacement of the interior plains material (Spudis and Guest, 1988; Strom et al., 2008; Fassett et al., 2009), the absolute age of the two events is not well constrained.

There is a significant difference between the 100 to 200 km values of T_e obtained for Caloris and the 25 to 30 km values obtained from the lobate scarps (Section 3.3) and simple heat flux predictions. Since the lobate scarps generally appear to postdate the emplacement of the Calorian smooth plains (Spudis and Guest, 1988; Watters et al., 2004), it is unlikely that Caloris loading occurred after the lobate scarps formed. Thus, the discrepancy cannot simply be due to an increase in T_e with time as expected for a cooling planet. One possibility is that Caloris is an area of thinned crust, resulting in a lithosphere containing more mantle material than usual. Since mantle material is intrinsically stronger than most crustal materials and contains fewer radiogenic elements, a thinned crust would likely lead to higher T_e values. Another possible explanation is that the lithospheric curvature associated with Caloris is much smaller than that of the short-wavelength scarps, resulting in a higher effective rigidity for the same thermal structure (cf. Watts, 2001). Regional variations in heat flux, either due to differing crustal concentrations (the lunar KREEP terrain; Wieczorek and Phillips, 2000) or mantle contributions, would need to have exceeded a factor of three to generate the T_e differences observed. None of these explanations is particularly satisfactory; the acquisition of gravity and topography data from the MESSENGER mission (Solomon et al., 2001, 2007, 2008) will allow direct measurements of T_e and help to resolve some of these issues.

5.3 Amount of strain and contraction

While it is clear that Mercury has undergone global contraction, the amount of contraction is unclear. The amount of shortening and strain recorded by Mercury's lobate scarps has been used to estimate the global crustal shortening and radius change due to global contraction (Strom et al., 1975; Watters et al., 1998). Strom et al. (1975) estimated the horizontal shortening due to the lobate scarps by assuming an average throw of 1 km and fault plane dips θ of 25° and 45°. The decrease in surface area expressed by the thrust faults was estimated by multiplying the average shortening by the total length of the lobate scarps mapped over an area covering ~24% of the surface. Extrapolating to the entire surface of Mercury, Strom *et al.* (1975) estimated a net decrease in area of up to 1.3×10^5 km². This corresponds to up to a 2 km decrease in the planet's radius (Strom *et al.*, 1975), consistent with thermal history models (Solomon, 1976; Schubert *et al.*, 1988).

The strain expressed by the lobate scarps can be determined using the D–L relationship of the faults (Equation 2.2) (Watters *et al.*, 1998). The lengths of the 82 mapped lobate scarps (Watters *et al.*, 2004) range from 36 km to 457 km, with a mean of 143 km and a standard deviation of 69 km. The longest lobate scarp in the hemisphere imaged by Mariner 10 is Discovery Rupes. Using the range of γ given in subsection 2.3.4 and the total length of the mapped lobate scarp fault segments (12 324 km), the areal strain is estimated to range from 0.035% to 0.053% (0.043% for $\theta = 30°$) for the 45% of the surface imaged by Mariner 10.

The radius change due to global contraction can be expressed by $\Delta R = R_d - R_u$ where R_u is the pre-deformation planetary radius and R_d is the post-deformation (current) planetary radius. R_u is related to the contractional strain by

$$R_u = \left[\frac{R_d^2}{\varepsilon + 1}\right]^{0.5}. \tag{2.5}$$

Assuming the hemisphere imaged by Mariner 10 is representative of the entire surface, this range in ε corresponds to a range in ΔR of approximately -0.43 km to -0.64 km (-0.52 km for $\theta = 30°$). This amount of radius change is consistent with estimates by Watters *et al.* (1998) for a survey of lobate scarps over about 19% of the surface. Images obtained by MESSENGER showing lobate scarps in the hemisphere unseen by Mariner 10 and previously undetected lobate scarps in the Mariner 10 hemisphere, however, indicate that the above estimates of strain are too low. New estimates of the average contractional strain based on a survey of lobate scarps imaged by both MESSENGER and Mariner 10 are about one third greater (Solomon *et al.*, 2008) corresponding to a decrease in radius of up to 0.8 km (Watters *et al.*, 2009a).

Estimates of strain and radius change are critical to constraining thermal history models for Mercury (Solomon, 1976, 1977, 1978, 1979; Schubert *et al.*, 1988; Phillips and Solomon, 1997; Hauck *et al.*, 2004). Existing thermal history models predict much more radial contraction than can be accounted for with the observed tectonic features (on the order of 5 to 6 km) (Dombard *et al.*, 2001). Recent modeling suggests 1 to 2 km of radial contraction since 4 Ga can only be achieved with a bulk core sulfur content >6.5 wt% and a dry-olivine mantle rheology (Hauck *et al.*, 2004). Complete solidification of Mercury's core would result in a reduction in planetary radius of about 17 km (Solomon 1976). However, core solidification may have begun before the crust was capable of recording any contraction. Thus, the observed lobate scarps only record the amount of planetary contraction that occurred after the period of late heavy bombardment and emplacement of the intercrater plains and thus likely only a fraction of the total contraction. The discrepancy between the observed and predicted strain has led to the suggestion that long-wavelength (>100 km), low-amplitude lithospheric folds, not visible in Mariner 10 images, may be present (Dombard *et al.*, 2001). Alternatively, since many of the known thrust faults do not appear to have formed prior to the emplacement of the youngest Calorian smooth plains, the strain recorded by the lobate scarps may represent only the last phase of planetary contraction.

On the Moon, global contraction is predicted on the basis of thermal evolution models, but lobate scarp thrust faults like those on Mercury are not observed (see Watters and Johnson, Chapter 4). Pritchard and Stevenson (2000) carried out a thorough investigation into this mismatch and concluded that it could be explained by: near-surface regolith being able to accommodate more contraction strain than intact rock before failing (e.g., Binder and Gunga, 1985); non-monotonic thermal evolution histories; or accommodation by numerous small faults (Weisberg and Hager, 1998). It is also likely that the crust can accommodate a small amount of elastic strain without undergoing observable brittle deformation. Any of these effects, including low-amplitude folds not detectable with the available data (Dombard *et al.*, 2001), could be important on Mercury. About all that can be said with certainty is that complete core solidification would produce more contraction than is observed and can be ruled out. This conclusion agrees with the observations of Peale (2003) and Margot *et al.* (2007) and new observations by MESSENGER (Solomon *et al.*, 2008) which suggest Mercury's core is at least partly liquid.

5.4 Sources of stress

5.4.1 Contraction

Both theoretical considerations and observations suggest that Mercury has undergone global contraction. The timing, duration, and magnitude of this contraction are unclear, but it seems likely that significant thermal contraction-related compressional stresses were present during the loading of Caloris (Section 4.5). The isotropic stresses σ resulting from global contraction (Melosh and McKinnon, 1988) are given by

$$\sigma = 2\mu \frac{(1+\nu)}{(1-\nu)} \frac{\Delta R}{R}, \qquad (2.6)$$

where μ is the shear modulus, ν is the Poisson's ratio, R is the planetary radius, and ΔR is the radial contraction. A change in radius of 2 km gives rise to stresses on the order of 100 MPa. If the upper limit of the radius change is on the order of 1 km (see Section 5.3), then the corresponding stresses are ∼55 MPa (Watters et al., 2000). This is comparable to the stresses derived from the inferred loading of the Caloris basin (Section 4.5).

5.4.2 Thermal stresses

Global thermal stresses arising from planetary cooling are closely related to planetary contraction. Models calculating vertically varying thermal stresses as the temperatures of the lithosphere and interior evolve have been constructed (e.g., Solomon and Head, 1979; Turcotte, 1983; Hauck et al., 2004). The importance of these models is that in some circumstances they predict extension as well as compression. The magnitude of the stresses is given by equations of the form

$$\sigma = \alpha E \frac{\Delta T}{(1-\nu)}, \qquad (2.7)$$

where σ is the thermal expansivity and ΔT is the temperature change. A temperature change of 100 K gives rise to stresses of order 300 MPa for a thermal expansivity of 3×10^{-5} K^{-1}.

Local thermal stresses may also arise due to cooling of volcanic plains, lithospheric reheating, or subsidence. While less important for global tectonic patterns, such stresses may be important in generating local features like polygonal troughs (Johnson and Sandwell, 1992).

5.4.3 Despinning

Observational arguments that despinning may have occurred are presented in Section 5.1. The maximum differential stress caused by despinning (Melosh and McKinnon, 1988; Matsuyama and Nimmo, 2009) is given by

$$\sigma = 5\left(\omega_1^2 - \omega_2^2\right) \frac{R^3}{GM} \mu \left(\frac{1+\nu}{5+\nu}\right), \qquad (2.8)$$

where ω_1 and ω_2 are the initial and final angular rotation velocities, R is the planetary radius, M is the planetary mass, G is the gravitational constant, and μ is the shear modulus. If the initial spin period of Mercury was 20 hr, the stresses developed during despinning are of order 200 MPa, comparable to stresses due to contraction.

5.4.4 True polar wander

The discussion of Caloris suggests that the orientation of the planet's surface may have shifted with respect to its pole of rotation (Section 4.4). This reorientation causes the rotational bulge to change its position in relation to the surface and leads to lithospheric stresses (e.g., Melosh, 1980; Leith and McKinnon, 1996; Matsuyama and Nimmo, 2009). The maximum differential stress σ which develops is given by (Melosh, 1980)

$$\sigma = \frac{4(1+\nu)}{(5+\nu)} \mu f \sin\theta, \qquad (2.9)$$

where f is the planetary flattening at the time of reorientation, and θ is the angle of rotation.

Mercury's rotation is slow, and so the flattening would be expected to be small. Furthermore, radar altimetry data are restricted to an equatorial band, so the planetary flattening is poorly determined. The best estimate of f is 0.0029 ± 0.0036 (Anderson et al., 1996) which is consistent with a value of zero. There is also a small equatorial ellipticity of 0.00054 ± 0.00005.

Taking $\mu = 40$ GPa, $\nu = 0.25$, and using $f = 0.003$ results in a differential stress of 20 MPa for a rotation of $10°$. These stresses are not negligible compared with the other likely contributions if the flattening really is this large. However, detecting planetary reorientation through the spatial distribution of fractures is difficult (see Leith and McKinnon, 1996; Sarid et al., 2002). Whether reorientation turns out to be an important effect will depend on better constraints on f and mapping of the global distribution of faults on Mercury.

5.4.5 Convection

Subsolidus mantle convection can potentially produce large stresses in the overlying lithosphere. These stresses depend on the temperature difference available to drive convection and for a strongly temperature-dependent viscosity (Solomatov and Moresi, 2000) are given by

$$\sigma \approx 0.03 \alpha \rho g \Delta T_{rh} \delta_{rh}, \qquad (2.10)$$

where σ is the thermal expansivity, ρ is the density, and ΔT_{rh} and δ_{rh} are the rheologically controlled temperature drop and boundary layer thickness. The value of δ_{rh} depends on the vigor of convection and is typically of order 100 km (see Solomatov and Moresi, 2000). The temperature contrast ΔT_{rh} for Newtonian rheologies is given by

$$\Delta T_{rh} \approx \frac{2.4 R T_i^2}{Q}, \qquad (2.11)$$

where R is the gas constant, T_i is the interior temperature, and Q is the activation energy. Assuming $Q = 250$ kJ/mol and $T_i = 1500$ K, we have $\Delta T_{rh} = 180$ K. The resulting stresses, from equation (2.10), are 6 MPa for a boundary layer thickness of 100 km. These stresses are small compared to the stresses arising due to other mechanisms.

Recent modeling, however, suggests that mantle convection may be an important source of tectonic stresses on Mercury. In three-dimensional convection simulations of Mercury's mantle, King (2008) found that a regularly spaced pattern developed. At low latitudes, mantle upwelling has a linear, sheet-like pattern creating dominantly E–W compression, while at the poles the pattern is almost hexagonal, creating generally N–S compression. These predicted stress patterns are roughly consistent with the orientations of the lobate scarps (see Plate 1). The modeled stresses are of the order of 500 MPa, which is much higher than the estimates derived from scaling arguments above, and in the absence of an elastic lithosphere would imply topographic variations with an amplitude of roughly 50 km. Further modeling is clearly required to resolve these issues.

5.4.6 Buoyancy forces

If isostatically compensated lateral crustal thickness contrasts exist, then buoyancy forces occur. The force (per unit length) (e.g., Buck, 1991) is given by

$$F \approx g \Delta \rho t_c \Delta t_c, \qquad (2.12)$$

where $\Delta \rho$ is the density contrast between crust and mantle, and t_c and Δt_c are the mean crustal thickness and the thickness variation, respectively. If these forces are distributed uniformly across the entire crust, the stress is simply F/t_c. The stresses will be larger if they are confined to the elastic layer.

To produce a 1 km variation in elevation requires a Δt_c of about 6 km if $\Delta \rho = 500$ kg m^{-3}. The resulting buoyancy force is 10^{12} Nm^{-1} for a crustal thickness of 100 km. The mean stress across the entire crust is 10 MPa but may be higher if the stresses only occur within the elastic portion of the lithosphere. In any event, these stresses are probably small compared to other likely sources of stress.

5.5 Thermal evolution

Various authors have modeled the thermal evolution of Mercury with time (e.g., Schubert *et al.*, 1988; Spohn, 1991; Solomatov and Reese, 2001; Hauck *et al.*, 2004; Williams *et al.*, 2007). Unfortunately, there are essentially only two constraints at present. Firstly, the existence of a present-day magnetic field suggests an active dynamo, which in turn places constraints on the present-day core heat flux (Schubert *et al.*, 1988; Hauck *et al.*, 2004; Williams *et al.*, 2007). Secondly, the amount of global contraction places constraints on the total cooling. Understanding how T_e has varied with time would add an important constraint on thermal evolution, as it has for Mars (e.g., Zuber *et al.*, 2000).

Existing thermal evolution models leave many questions unanswered. For instance, it is not actually clear that mantle convection is required, especially if the material deforms in a non-Newtonian fashion (Solomatov and Reese, 2001). Similarly, the apparently plagioclase-rich crust is not likely to be generated by simple melting of mantle material, and perhaps implies an early magma ocean (cf. Wetherill, 1988) with little melt generation subsequently (Jeanloz et al., 1995). This early event may have had important consequences for the transport of heat and radiogenic elements within Mercury. Finally, the recent discovery that Mercury's core is at least partially liquid (Peale, 2003; Margot et al., 2007) has implications for the composition of the core and the cooling of the mantle.

5.6 Summary

The distribution of lobate scarps in the hemisphere imaged by Mariner 10 is not consistent with thrust faults formed during a despinning or thermal contraction episode alone. A combination of tidal despinning and thermal contraction may account for equatorial N–S and polar E–W trending lobate scarp thrust faults in the regions by reactivation of normal faults. These two mechanisms generate comparable stresses, probably occurred over similar timescales, and are likely to be important stress-generating mechanisms on Mercury. Nonetheless, they do not by themselves account for the spatial and temporal distribution of all lobate scarps. Another important source of stress that may account for the distribution and orientation of the lobate scarps is mantle convection. Local loading in the Caloris basin produced significant compressional stresses, similar in magnitude to those produced by despinning and thermal contraction (Section 4.5). The loading may also have led to planetary reorientation. Several lines of evidence suggest that T_e during Caloris loading was ~100 km. The role of polar wander-induced stresses is uncertain, while stresses due to crustal thickness variations are likely to be small. The amount of global contraction is at present poorly constrained, but estimates will be greatly improved by global high-resolution imaging obtained by MESSENGER.

6 Conclusions

Tectonic features on Mercury are dominantly compressional in origin. The most prominent tectonic landforms are the lobate scarp thrust faults found in the oldest (pre-Tolstojan) and youngest (Calorian) plains on Mercury. The lobate scarps are not uniformly distributed and have preferred orientations and fault-plane dip directions. Although the lobate scarps deform the oldest plains units, the scarcity of large, superimposed impact craters and their relatively undegraded appearance suggest many formed after the emplacement of the youngest Calorian plains units. Wrinkle ridges are found exclusively in Calorian smooth plains units that generally occupy topographic lows, including the floor material of the Caloris basin. The wrinkle ridges in the Caloris basin form a circumferential pattern that is very similar to the pattern of wrinkle ridges found in lunar mascon basins. The Caloris basin is one of only three basins on Mercury with evidence of extensional deformation. Extension of the interior plains is expressed by a complex pattern of basin-radial and basin-concentric graben. The Caloris graben crosscut the wrinkle ridges and thus may be among the youngest tectonic features on Mercury.

Observations of tectonic features place constraints on geophysically important variables such as elastic thickness and heat flux. The lobate scarps provide constraints on T_e (~25 to 30 km), heat flux, and the amount of global contraction. Tidal despinning and planetary cooling and contraction are important sources of stress on Mercury, but by themselves or in combination do not account for all of the temporal and spatial distribution observed in the lobate scarp thrust faults. Local preferred orientations of the lobate scarps and uniform thrust slip dip directions suggest regional-scale stresses influenced the formation of the thrust faults. The wrinkle ridges in smooth plains outside of the Caloris basin are likely due to loading and subsidence of volcanic material that flooded lowland areas.

The wrinkle ridges in Caloris are probably due to subsidence following the infilling of the basin by volcanic material. The extensional polygonal troughs most likely result from either subsequent lateral crustal flow and uplift or basin-exterior loading. The absence of extensional features surrounding Caloris may be due to a background compressive stress resulting from global thermal contraction. Elastic thickness estimates based on Caloris structures imply $T_e \sim 100$ km, which is not consistent with estimates based on lobate scarp fault modeling. This mismatch cannot be explained by the lithosphere thickening with time, as Caloris formation appears to predate many of the lobate scarps. The explanation for this discrepancy is not currently obvious and will have to await the return of additional data from MESSENGER (Solomon et al., 2001, 2007, 2008).

Acknowledgments

We wish to thank Jay Melosh and John Guest for their reviews and very helpful comments and suggestions that greatly improved this chapter. We also wish to thank Mark Robinson for providing Mariner 10 images and mosaics and for valuable insights into their processing and interpretation. The original version of this chapter was completed in April, 2006. This work was supported by the National Aeronautics and Space Administration under Grants issued through the Office of the Planetary Geology and Geophysics Program.

References

Ackermann, R. V., Schlische, R. W., and Withjack, M. O. (2001). The geometric and statistical evolution of normal fault systems: An experimental study of the effects of mechanical layer thickness on scaling laws. *J. Struct. Geol.*, **23**, 1803–1819.

Anderson, J. D., Colombo, G., Esposito, P. B., Lau, E. L., and Tracer, G. B. (1987). The mass, gravity field and ephemeris of Mercury. *Icarus*, **71**, 337–349.

Anderson, J. D., Jurgens, R. F., Lau, E. L., and Slade, M. A. (1996). Shape and orientation of Mercury from radar ranging data. *Icarus*, **124**, 690–697.

André, S. L., Watters, T. R., and Robinson, M. S. (2005). The long wavelength topography of Beethoven and Tolstoj basins, Mercury. *Geophys. Res. Lett.*, **32**, L21202, doi:10.1029/2005GL023627.

Baldwin, R. B. (1963). *The Measure of the Moon.* Chicago, IL: University of Chicago Press.

Binder, A. B. (1982). Post-Imbrian global lunar tectonism: Evidence for an initially totally molten Moon. *Moon and the Planets*, **26**, 117–133.

Binder, A. B. and Gunga, H. C. (1985). Young thrust-fault scarps in the highlands: Evidence for an initially totally molten Moon. *Icarus*, **63**, 421–441.

Breuer, D., Spohn, T., and Wullner, U. (1993). Mantle differentiation and the crustal dichotomy of Mars. *Planet. Space Sci.*, **41**, 269–283.

Brewer, J. A., Smithson, S. B., Oliver, J. E., Kaufman, S., and Brown, L. D. (1980). The Laramide orogeny: Evidence from COCORP deep crustal seismic profiles in the Wind River mountains, Wyoming. *Tectonophysics*, **62**, 165–189.

Bryan, W. B. (1973). Wrinkle-ridges as deformed surface crust on ponded mare lava. *Geochim. Cosmochim. Acta*, **1**, (Suppl.), 93–106.

Buck, W. R. (1991). Modes of continental lithospheric extension. *J. Geophys. Res.*, **96**, 20 161–20 178.

Byerlee, J. (1978). Friction of rocks. *Pure Appl. Geophys.*, **116**, 615–626.

Cartwright, J. A., Trudgill, B. D., and Mansfield, C. S. (1995). Fault growth by segment linkage: An explanation for scatter in maximum displacement and trace length data from the Canyonlands grabens of SE Utah. *J. Struct. Geol.*, **17**, 1319–1326.

Cartwright, J. A., Mansfield, C. S., and Trudgill, B. D. (1996). The growth of normal faults by segment linkage. *Geol. Soc. Am. Spec. Publ.*, **99**, 163–177.

Clark, P. E., Leake, M. A., and Jurgens, R. F. (1988). Goldstone radar observations of Mercury. In *Mercury*, ed. F. Vilas, C. R. Chapman and M. S. Matthews. Tucson, AZ: University of Arizona Press.

Clark, R. and Cox, S. (1996). A modern regression approach to determining fault displacement-length scaling relationships. *J. Struct. Geol.*, **18**, 147–154.

Cook, A. C. and Robinson, M. S. (2000). Mariner 10 stereo image coverage of Mercury. *J. Geophys. Res.*, **105**, 9429–9443.

Cordell, B. M. and Strom, R. G. (1977). Global tectonics of Mercury and the Moon. *Phys. Earth Planet. Inter.*, **15**, 146–155.

Cowie, P. A. and Scholz, C. H. (1992a). Physical explanation for the displacement–length relationship of faults using a post-yield fracture-mechanics model. *J. Struct. Geol.*, **14**, 1133–1148.

Cowie, P. A. and Scholz, C. H. (1992b). Displacement-length scaling relationship for faults data synthesis and discussion. *J. Struct. Geol.*, **14**, 1149–1156.

Cowie, P. A. and Shipton, Z. K. (1998). Fault tip displacement gradients and process zone dimensions. *J. Struct. Geol.*, **20**, 983–997.

Cowie, P. A., Scholz, C. H., Edwards, M., and Malinverno, A. (1993). Fault strain and seismic coupling on mid-ocean ridges. *J. Geophys. Res.*, **98**, 17 911–17 920.

Davis, P. A. and Soderblom, L. A. (1984). Modeling crater topography and albedo from monoscopic Viking Orbiter images. *Icarus*, **89**, 9449–9457.

Dawers, N. H. and Anders, M. H. (1995). Displacement-length scaling and fault linkage. *J. Struct. Geol.*, **17**, 607–614.

Dawers, N. H., Anders, M. H., and Scholz, C. H. (1993). Growth of normal faults: displacement length scaling. *Geology*, **21**, 1107–1110.

Denevi, B. W., Robinson, M. S., Solomon, S. C., Murchie, S. L., Blewett, D. T., Domingue, D. L, McCoy, T. J., Ernst, E. M., Head, J. W., Watters, T. R., and Chabot, N. L. (2009). The evolution of Mercury's crust: A global perspective from MESSENGER. *Science*, **324**, 613–618.

Dombard, A. J. and Hauck, S. A. (2008). Despinning plus global contraction and the orientation of lobate scarps on Mercury. *Icarus*, **198**, 274–276.

Dombard, A. J., Hauck, S. A., Solomon, S. C., and Phillips, R. J. (2001). Potential for long-wavelength folding on Mercury (abs.). *Lunar Planet. Sci. Conf. XXXII*, 2035.

Dzurisin, D. (1978). The tectonic and volcanic history of Mercury as inferred from studies of scarps, ridges, troughs, and other lineaments. *J. Geophys. Res.*, **83**, 4883–4906.

Fisher, N. I. (1993). *Statistical Analysis of Circular Data*. Cambridge: Cambridge University Press, p. 277.

Foster, A. and Nimmo, F. (1996). Comparisons between the rift systems of East Africa, Earth, and Beta Regio, Venus. *Earth Planet. Sci. Lett.*, **143**, 183–195.

Freed, A. M., Melosh, H. J., and Solomon, S. C. (2001). Tectonics of mascon loading: Resolution of the strike-slip faulting paradox. *J. Geophys. Res.*, **106**, 20 603–20 620.

Gillespie, P. A., Walsh, J. J., and Watterson, J. (1992). Limitations of dimension and displacement data from single faults and the consequences for data analysis and interpretation. *J. Struct. Geol.*, **14**, 1157–1172.

Golombek, M. P. (1979). Structural analysis of lunar grabens and the shallow crustal structure of the Moon. *J. Geophys. Res.*, **84**, 4657–4666.

Golombek, M. P., Plescia, J. B., and Franklin, B. J. (1991). Faulting and folding in the formation of planetary wrinkle ridges. *Proc. Lunar Planet. Sci. Conf. 21*, 679–693.

Golombek, M. P., Anderson, F. S., and Zuber, M. T. (2001). Martian wrinkle ridge topography: Evidence for subsurface faults from MOLA. *J. Geophys. Res.*, **106**, 23 811–23 821.

Gries, R. (1983). Oil and gas prospecting beneath Precambrian of foreland thrust plates in Rocky Mountains. *Am. Assoc. Petrol. Geol. Bull.*, **67**, 1–28.

Hapke, B., Danielson, E., Klaasen, K., and Wilson, L. (1975). Photometric observations of Mercury from Mariner 10. *J. Geophys. Res.*, **80**, 2431–2443.

Hardacre, K. M. and Cowie, P. A. (2003). Controls on strain localization in a two-dimensional elastoplastic layer: Insights into size-frequency scaling of extensional fault populations. *J. Geophys. Res.*, **108**, 2529.

Harmon, J. K. and Campbell, D. B. (1988). Radar observations of Mercury. In *Mercury*, ed. F. Vilas, C. R. Chapman and M. S. Matthews. Tucson, AZ: University of Arizona Press, pp. 101–117.

Harmon, J. K., Campbell, D. B., Bindschadler, K. L., Head, J. W., and Shapiro, I. I. (1986). Radar altimetry of Mercury: A preliminary analysis. *J. Geophys. Res.*, **91**, 385–401.

Hauck, S. A., Dombard, A. J., Phillips, R. J., and Solomon, S. C. (2004). Internal and tectonic evolution of Mercury. *Earth Planet. Sci. Lett.*, **222**, 713–728.

Hawkins, S. E., Boldt, J., Darlington, E. H., Espiritu, R., Gold, R., Gotwols, B., Grey, M., Hash, C., Hayes, J., Jaskulek, S., Kardian, C., Keller, M., Malaret, E., Murchie, S. L., Murphy, P., Peacock, K., Prockter, L., Reiter, A., Robinson, M. S., Schaefer, E., Shelton, R., Sterner, R., Taylor, H., Watters, T., and Williams, B. (2007). The Mercury Dual Imaging System (MDIS) on the MESSENGER spacecraft. *Space Sci. Rev.*, **131**, 247–338.

Head, J. W., Murchie, S. L., Prockter, L. M., Robinson, M. S., Solomon, S. C., Strom, R. G., Chapman, C. R., Watters, T. R., McClintock, W. E., Blewett, D. T., and Gillis-Davis, J. J. (2008). Volcanism on Mercury: Evidence from the first MESSENGER flyby. *Science*, **321**, 69–72.

Hiesinger, H. and Head, J. W. (2000). Characteristics and origin of polygonal terrain in southern Utopia Planitia, Mars: Results from Mars Orbiter Laser Altimeter and Mars Orbiter Camera Data. *J. Geophys. Res.*, **105**, 11 999–12 022.

Howard, K. A. and Muehlberger, W. R. (1973). Lunar thrust faults in the Taurus-Littrow region. *Apollo 17 Prel. Sci. Rep., NASA Spec. Publ.*, **SP-330**, 31–22 – 31–25.

Jackson, J. A. (1980). Reactivation of basement faults and crustal shortening in orogenic belts. *Nature*, **283**, 343–346.

Jaeger, J. C. and Cook, N. G. W. (1979). *Fundamentals of Rock Mechanics*, 3rd edn. London: Chapman and Hall, p. 593.

Janes, D. M. and Melosh, H. J. (1990). Tectonics of planetary loading: A general model and results. *J. Geophys. Res.*, **95**, 21 345–21 355.

Jeanloz, R., Mitchell, D. L., Sprague, A. L., and DePater, I. (1995). Evidence for a basalt-free surface on Mercury and implications for internal heat. *Science*, **268**, 1455–1457.

Johnson, C. L. and Sandwell, D. T. (1992). Joints in Venusian lava flows. *J. Geophys. Res.*, **97**, 13 601–13 610.

Kennedy, P. J., Freed, A. M., and Solomon, S. C. (2008). Mechanisms of faulting in and around Caloris basin, Mercury. *J. Geophys. Res.*, **113**, E08004, doi:10.1029/2007JE002992.

King, S. D. (2008). Pattern of lobate scarps on Mercury's surface reproduced by a model of mantle convection. *Nature Geoscience*, **1**, 229–232.

Kirk, R. L., Barrett, J. M., and Soderblom, L. A. (2003). Photoclinometry made simple...? In *Advances in Planetary Mapping 2003*. Houston, TX: Lunar and Planetary Institute.

Leith, A. C. and McKinnon, W. B. (1996). Is there evidence for polar wander on Europa? *Icarus*, **120**, 387–398.

Lucchitta, B. K. (1976). Mare ridges and related highland scarps: Results of vertical tectonism. *Geochim. Cosmochim. Acta*, **3**, (Suppl.), 2761–2782.

Maggi, A., Jackson, J. A., McKenzie D., and Priestley K. (2000). Earthquake focal depths, effective elastic thickness, and the strength of the continental lithosphere. *Geology*, **28**, 495–498.

Malin, M. C. (1976). Observations of intercrater plains on Mercury. *Geophys. Res. Lett.*, **3**, 581–584.

Malin, M. C. and Dzurisin, D. (1977). Landform degradation on Mercury, the Moon, and Mars: Evidence from crater depth/diameter relationships. *J. Geophys. Res.*, **82**, 376–388.

Margot, J. L., Peale, S. J., Jurgens, R. F., Slade, M. A., and Holin, I. V. (2007). Large longitude libration of Mercury reveals a molten core. *Science*, **316**, 710–714.

Marrett, R. and Allmendinger, R. W. (1991). Estimates of strain due to brittle faulting: Sampling of fault populations. *J. Struct. Geol.*, **13**, 735–738.

Matsuyama, I. and Nimmo, F. (2009). Gravity and tectonic patterns of Mercury: The effect of tidal deformation, spin-orbit resonance, non-zero eccentricity, despinning and reorientation. *J. Geophys. Res.*, **114**, E01010.

Matsuyama, I., Mitrovica, J. X., Manga, M., Perron, J. T., and Richards, M. A. (2006). Rotational stability of dynamic planets with elastic lithospheres. *J. Geophys. Res.*, **111**, E2, E02003.

Maxwell, T. A. and Gifford, A. W. (1980). Ridge systems of Caloris: Comparison with lunar basins. *Proc. Lunar Planet. Sci. Conf. 11*, 2447–2462.

Maxwell, T. A., El-Baz, F., and Ward, S. W. (1975). Distribution, morphology, and origin of ridges and arches in Mare Serenitatis. *Geol. Soc. Am. Bull.*, **86**, 1273–1278.

McEwen, A. S. (1991). Photometric functions for photoclinometry and other applications. *Icarus*, **92**, 298–311.

McGill, G. E. (1971). Attitude of fractures bounding straight and arcuate lunar rilles. *Icarus*, **14**, 53–58.

McGill, G. E. (1986). The giant polygons of Utopia, northern Martian plains. *Geophys. Res. Lett.*, **13**, 705–708.

McKenzie, D., Nimmo, F., Jackson, J. A., Gans, P. B., and Miller, E. L. (2000). Characteristics and consequences of flow in the lower crust. *J. Geophys. Res.*, **105**, 11 029–11 046.

McKinnon, W. B. (1980). Large impact craters and basins: Mechanics of syngenetic and postgenetic modification. Ph.D. thesis, California Institute of Technology.

McKinnon, W. B. (1986). Tectonics of the Caloris basin, Mercury (abs.). In *Mercury: Program and Abstracts*. Tucson, AZ: Div. Planet. Sci., Amer. Astron. Soc.

McNutt, M. K. (1984). Lithospheric flexure and thermal anomalies. *J. Geophys. Res.*, **89**, 1180–1194.

Mege, D. and Reidel, S. P. (2001). A method for estimating 2D wrinkle ridge strain from application of fault displacement scaling to the Yakima folds, Washington. *Geophys. Res. Lett.*, **28**, 3545–3548.

Melosh, H. J. (1977). Global tectonics of a despun planet. *Icarus*, **31**, 221–243.

Melosh, H. J. (1978). The tectonics of mascon loading. *Proc. Lunar Planet. Sci. Conf. 9*, 3513–3525.

Melosh, H. J. (1980). Tectonic patterns on a tidally distorted planet. *Icarus*, **43**, 454–471.

Melosh, H. J. (1989). *Impact Cratering: A Geologic Process*. Oxford Monographs on Geology and Geophysics, No. 11. New York, NY: Oxford University Press, p. 245.

Melosh, H. J. and Dzurisin, D. (1978a). Mercurian global tectonics: A consequence of tidal despinning? *Icarus*, **35**, 227–236.

Melosh, H. J. and Dzurisin, D. (1978b). Tectonic implications for gravity structure of Caloris basin, Mercury. *Icarus*, **33**, 141–144.

Melosh, H. J. and McKinnon, W. B. (1988). The tectonics of Mercury. In *Mercury*, ed. F. Vilas, C. R. Chapman and M. S. Matthews. Tucson, AZ: University of Arizona Press.

Montesi, L. G. J. and Zuber, M. T. (2003a). Spacing of faults at the scale of the lithosphere and localization instability: 1. Theory. *J. Geophys. Res.*, **108**, 2110.

Montesi, L. G. J. and Zuber, M. T. (2003b). Spacing of faults at the scale of the lithosphere and localization instability: 2. Application to the Central Indian Basin. *J. Geophys. Res.*, **108**, 2111.

Mouginis-Mark, P. J. and Wilson, L. (1981). MERC: A FORTRAN IV program for the production of topographic data for the planet Mercury. *Comput. Geosci.*, **7**, 35–45.

Murchie, S. L., Watters, T. R., Robinson, M. S., Head, J. W., Strom, R. G., Chapman, C. R., Solomon, S. C., McClintock, W. E., Prockter, L. M., Domingue, D. L., and Blewett, D. T. (2008). Geology of the Caloris Basin, Mercury: A new view from MESSENGER. *Science*, **321**, 73–76.

Murray, B. C., Belton, M. J. S., Danielson, G. E., Davies, M. E., Gault, D. E., Hapke, B., O'Leary, B., Strom, R. G., Suomi, V., and Trask, N. (1974). Mercury's surface: Preliminary description and interpretation from Mariner 10 pictures. *Science*, **185**, 169–179.

Nimmo, F. (2002). Constraining the crustal thickness of Mercury from viscous topographic relaxation. *Geophys. Res. Lett.*, **29**, 1063.

Nimmo, F. and Stevenson, D. J. (2000). Influence of early plate tectonics on the thermal evolution and magnetic field of Mars. *J. Geophys. Res.*, **105**, 11 969–11 979.

Nimmo, F. and Stevenson, D. J. (2001). Estimates of Martian crustal thickness from viscous relaxation of topography. *J. Geophys. Res.*, **106**, 5085–5098.

Nimmo, F. and Watters, T. R. (2004). Depth of faulting on Mercury: Implications for heat flux and crustal and effective elastic thickness. *Geophys. Res. Lett.*, **31**, L02701.

Parmentier, E. M. and Haxby, W. F. (1986). Thermal stresses in the oceanic lithosphere. Evidence from geoid anomalies at fracture zones. *J. Geophys. Res.*, **91**, 7193–7204.

Parsons, B. and McKenzie, D. (1978). Mantle convection and thermal structure of plates. *J. Geophys. Res.*, **93**, 4485–4496.

Peale, S. J. (2003). Mercury's interior from geodesy of librations (abs.). *Eos Trans. Am. Geophys. Union*, **84**(46), (Fall Meet. Suppl.), G42C-03.

Pechmann, J. B. (1980). The origin of polygonal troughs on the northern plains of Mars. *Icarus*, **42**, 185–210, doi:10.1016/0019-1035(80)90071-8.

Pechmann, J. B. and Melosh, H. J. (1979). Global fracture patterns of a despun planet application to Mercury. *Icarus*, **38**, 243–250.

Pike, R. J. (1988). Geomorphology of impact craters on Mercury. In *Mercury*, ed. F. Vilas, C. R. Chapman and M. S. Matthews. Tucson, AZ: University of Arizona Press, pp. 165–273.

Pike, R. J. and Spudis, P. D. (1987). Basin-ring spacing on the Moon, Mercury, and Mars. *Earth, Moon, and Planets*, **39**, 129–194.

Phillips, R. J. and Solomon, S. C. (1997). Compressional strain history of Mercury (abs.). *Lunar Planet. Sci. Conf. XXVIII*, 1107–1108.

Phillips, R. J., Conel, J. E., Abbott, E. A., Sjogren, W. L., and Morton, J. B. (1972). Mascons: Progress toward a unique solution for mass distribution. *J. Geophys. Res.*, **77**, 7106–7114.

Plescia, J. B. and Golombek, M. P. (1986). Origin of planetary wrinkle ridges based on the study of terrestrial analogs. *Geol. Soc. Am. Bull.*, **97**, 1289–1299.

Pritchard, M. E. and Stevenson, D. J. (2000). Thermal aspects of a lunar origin by giant impact. In *Origin of the Earth and Moon*, ed. R. Canup and K. Righter. Tucson, AZ: University of Arizona Press, pp. 179–196.

Reidel, S. P. (1984). The Saddle Mountains: The evolution of an anticline in the Yakima fold belt. *Am. Jour. Sci.*, **284**, 942–978.

Rice, J. R. (1992). Fault stress states, pore pressure distributions and the weakness of the San Andreas fault. In *Fault Mechanics and Transport Properties of Rocks*, ed. B. Evans and T.-F. Wong. San Diego: Academic Press, pp. 475–503.

Robinson, M. S. and Lucey, P. G. (1997). Recalibrated Mariner 10 color mosaics: Implications for Mercurian volcanism. *Science*, **275**, 197–200.

Robinson, M. S., Davies, M. E., Colvin, T. R., and Edwards, K. E. (1999). A revised control network for Mercury. *J. Geophys. Res.*, **104**, 30 847–30 852.

Robinson, M. A., Murchie, S. L., Blewett, D. T., Domingue, D. L., Hawkins, S. E., Head, J. W., Holsclaw, G. M., McClintock, W. E., McCoy, T. J., McNutt, R. L., Prockter, L. M., Solomon, S. C., and Watters, T. R. (2008). Reflectance and color variations on Mercury: Indicators of regolith processes and compositional heterogeneity. *Science*, **321**, 66–69.

Sarid, A. R., Greenberg, R., Hoppa, G. V., Hurford, T. A., Tufts, B. R., and Geissler, P. (2002). Polar wander and surface convergence of Europa's ice shell: Evidence from a survey of strike-slip displacement. *Icarus*, **158**, 24–41.

Schaber, G. G., Boyce, J. M., and Trask, N. J. (1977). Moon-Mercury large impact structures, isostasy and average crustal viscosity. *Phys. Earth Planet. Inter.*, **15**, 189–201.

Scholz, C. H. (2002). *The Mechanics of Earthquakes and Faulting*. Cambridge: Cambridge University Press, p. 471.

Scholz, C. H. and Cowie, P. A. (1990). Determination of total strain from faulting using slip measurements. *Nature*, **346**, 837–839.

Schubert, G., Ross, M. N., Stevenson, D. J., and Spohn, T. (1988). Mercury's thermal history and the generation of its magnetic field. In *Mercury*, ed. F. Vilas, C. R. Chapman and M. S. Matthews. Tucson, AZ: University of Arizona Press.

Schultz, R. A. (1997). Displacement-length scaling for terrestrial and Martian faults: Implications for Valles Marineris and shallow planetary grabens. *J. Geophys. Res.*, **102**, 12 009–12 015.

Schultz, R. A. (1999). Understanding the process of faulting: Selected challenges and opportunities at the edge of the 21st century. *J. Struct. Geol.*, **21**, 985–993.

Schultz, R. A. (2000). Localizaton of bedding-plane slip and backthrust faults above blind thrust faults: Keys to wrinkle ridge structure. *J. Geophys. Res.*, **105**, 12 035–12 052.

Schultz, R. A. and Fori, A. N. (1996). Fault-length statistics and implications of graben sets at Candor Mensa, Mars. *J. Struct. Geol.*, **18**, 272–383.

Schultz, R. A. and Fossen, H. (2002). Displacement-length scaling in three dimensions: The importance of aspect ratio and application to deformation bands. *J. Struct. Geol.*, **24**, 1389–1411.

Schultz, R. A. and Watters, T. R. (2001). Forward mechanical modeling of the Amenthes Rupes thrust fault on Mars. *Geophys. Res. Lett.*, **28**, 4659–4662.

Smrekar, S. E. and Phillips, R. J. (1991). Venusian highlands geoid to topography ratios and their implications. *Earth Planet. Sci. Lett.*, **107**, 582–597.

Smrekar, S. E., Moreels, P., and Franklin, B. J. (2002). Characterization and formation of polygonal fractures on Venus. *J. Geophys. Res.*, **107**, 8–1 – 8–18.

Solomatov, V. S. (1995). Scaling of temperature-dependent and stress-dependent viscosity convection. *Phys. Fluids*, **7**, 266–274.

Solomatov, V. S. and Moresi, L. N. (2000). Scaling of time-dependent stagnant lid convection: Application to small-scale convection on Earth and other terrestrial planets. *J. Geophys. Res.*, **105**, 21 795–21 817.

Solomatov, V. S. and Reese, C. C. (2001). Mantle convection and thermal evolution of Mercury revisited (abs.). In *Workshop on Mercury: Space Environment, Surface and Interior*. Houston, TX: Lunar and Planetary Institute, 1097, 92–95.

Solomon, S. C. (1976). Some aspects of core formation in Mercury. *Icarus*, **28**, 509–521.

Solomon, S. C. (1977). The relationship between crustal tectonics and internal evolution in the Moon and Mercury. *Phys. Earth Planet. Inter.*, **15**, 135–145.

Solomon, S. C. (1978). On volcanism and thermal tectonics on one-plate planets. *Geophys. Res. Lett.*, **5**, 461–464.

Solomon, S. C. (1979). Formation, history and energetics of cores in the terrestrial planets. *Phys. Earth Planet. Inter.*, **19**, 168–182.

Solomon, S. C. and Head, J. W. (1979). Vertical movement in mare basins: Relation to mare emplacement, basin tectonics and lunar thermal history. *J. Geophys. Res.*, **84**, 1667–1682.

Solomon, S. C. and Head, J. W. (1980). Lunar mascon basins: Lava filling, tectonics, and evolution of the lithosphere. *Rev. Geophys. Space Phys.*, **18**, 107–141.

Solomon, S. C., McNutt, R. L., Gold, R. E., Acuña, M. H., Baker, D. N., Boynton, W. V., Chapman, C. R., Cheng, A. F., Gloeckler, G., Head, J. W., Krimigis, S. M., McClintock, W. E., Murchie, S. L., Peale, S. J., Philips, R. J., Robinson, M. S., Slavin, J. A., Smith, D. E., Strom, R. G., Trombka, J. I., and Zuber, M. T. (2001). The MESSENGER mission to Mercury: Scientific objectives and implementation. *Planet. Space Sci.*, **49**, 1445–1465.

Solomon, S. C., McNutt, R. L., Gold, R. E., and Domingue, D. L. (2007). MESSENGER mission overview. *Space Sci. Rev.*, **131**, 3–39.

Solomon, S. C., McNutt, R. L., Watters, T. R., Lawrence, D. J., Feldman, W. C., Head, J. W., Krimigis, S. M., Murchie, S. L., Phillips, R. J., Slavin, J. A., and Zuber, M. T. (2008). Return to Mercury: A global perspective on MESSENGER's first Mercury flyby. *Science*, **321**, 59–62.

Spohn, T. (1991). Mantle differentiation and thermal evolution of Mars, Mercury and Venus. *Icarus*, **90**, 222–236.

Sprague, A. L., Nash, D. B., Witteborn, F. C., and Cruikshank, D. P. (1997). Mercury's feldspar connection: Mid-IR measurements suggest plagioclase. *Adv. Space Res.*, **19**, 1507–1510.

Spudis, P. D. and Guest, J. E. (1988). Stratigraphy and geologic history of Mercury. In *Mercury*, ed. F. Vilas, C. R. Chapman and M. S. Matthews. Tucson, AZ: University of Arizona Press.

Stone, D. S. (1985). Geologic interpretation of seismic profiles, Big Horn Basin, Wyoming: Part I. East Flank. In *Seismic Exploration of the Rocky Mountain Region*, ed. R. R. Gries and R. C. Dyer. Denver, CO: Rocky Mountain Association of Geologists, pp. 165–174.

Strom, R. G. (1972). Lunar mare ridges, rings and volcanic ring complexes. *Mod. Geol.*, **2**, 133–157.

Strom, R. G., Trask, N. J., and Guest, J. E. (1975). Tectonism and volcanism on Mercury. *J. Geophys. Res.*, **80**, 2478–2507.

Strom, R. G., Chapman, C. R., Merline, W. J., Solomon, S. C., and Head, J. W. (2008). Mercury cratering record viewed from MESSENGER's first flyby. *Science*, **321**, 79–81.

Tanaka, K. L. and Davis, P. A. (1988). Tectonic history of the Syria Planum province of Mars. *J. Geophys. Res.*, **93**, 14 893–14 907.

Thomas, P. G. (1997). Are there other tectonics than tidal despinning, global contraction and Caloris-related events on Mercury? A review of questions and problems. *Planet. Space Sci.*, **45**, 3–13.

Thomas, P. G., Masson, P., and Fleitout, L. (1988). Tectonic history of Mercury. In *Mercury*, ed. F. Vilas, C. R. Chapman and M. S. Matthews. Tucson, AZ: University of Arizona Press.

Turcotte, D. L. (1983). Thermal stresses in planetary elastic lithospheres (Proc. Lunar Planet. Sci. Conf. 13). *J. Geophys. Res.* (Suppl. 88), A585–A587.

Turcotte, D. L. and Schubert, G. (2002). *Geodynamics: Application of Continuum Physics to Geological Problems*. Cambridge: Cambridge University Press, p. 450.

Turcotte, D. L., Willemann, R. J., Haxby, W. F., and Norberry, J. (1981). Role of membrane stresses in the support of planetary topography. *J. Geophys. Res.*, **86**, 3951–3959.

Turtle, E. P. and Pierazzo, E. (1998). Constraints on the size of the Vredefort impact crater from numerical modeling. *Meteorit. Planet. Sci.*, **33**, 483–490.

Vilas, F. (1988). Surface composition of Mercury from reflectance spectrophotometry. In *Mercury*, ed. F. Vilas, C. R. Chapman and M. S. Matthews. Tucson, AZ: University of Arizona Press.

Walsh, J. and Watterson, J. (1988). Analysis of the relationship between displacements and dimensions of faults. *J. Struct. Geol.*, **10**, 239–247.

Watters, T. R. (1988). Wrinkle ridge assemblages on the terrestrial planets. *J. Geophys. Res.*, **93**, 10 236–10 254.

Watters, T. R. (1991). Origin of periodically spaced wrinkle ridges on the Tharsis plateau of Mars. *J. Geophys. Res.*, **96**, 15 599–15 616.

Watters, T. R. (1992). A system of tectonic features common to Earth, Mars and Venus. *Geology*, **20**, 609–612.

Watters, T. R. (1993). Compressional tectonism on Mars. *J. Geophys. Res.*, **98**, 17 049–17 060.

Watters, T. R. (2003). Thrust faulting along the dichotomy boundary in the eastern hemisphere of Mars. *J. Geophys. Res.*, **108** (E6), 5054, doi:10.1029/2002JE001934.

Watters, T. R. (2004). Elastic dislocation modeling of wrinkle ridges on Mars. *Icarus*, **171**, 284–294.

Watters, T. R. and Konopliv, A. S. (2001). The topography and gravity of Mare Serenitatis: Implications for subsidence of the mare surface. *Planet. Space Sci.*, **49**, 743–748.

Watters, T. R. and Robinson, M. S. (1997). Radar and photoclinometric studies of wrinkle ridges on Mars. *J. Geophys. Res.*, **102**, 10 889–10 903.

Watters, T. R. and Robinson, M. S. (1999). Lobate scarps and the origin of the Martian crustal dichotomy. *J. Geophys. Res.*, **104**, 18 981–18 990.

Watters, T. R., Robinson, M. S., and Cook, A. C. (1998). Topography of lobate scarps on Mercury: New constraints on the planet's contraction. *Geology*, **26**, 991–994.

Watters, T. R., Schultz, R. A., and Robinson, M. S. (2000). Displacement–length relations of thrust faults associated with lobate scarps on Mercury and Mars: Comparison with terrestrial faults. *Geophys. Res. Lett.*, **27**, 3659–3662.

Watters, T. R., Robinson, M. S., and Cook, A. C. (2001). Large-scale lobate scarps in the southern hemisphere of Mercury. *Planet. Space Sci.*, **49**, 1523–1530.

Watters, T. R., Schultz, R. A., Robinson, M. S., and Cook, A. C. (2002). The mechanical and thermal structure of Mercury's early lithosphere. *Geophys. Res. Lett.*, **29**, 1542.

Watters, T. R., Robinson, M. S., Bina, C. R., and Spudis, P. D. (2004). Thrust faults and the global contraction of Mercury. *Geophys. Res. Lett.*, **31**, L04071 doi:10.1029/2003GL019171.

Watters, T. R., Nimmo, F., and Robinson, M. S. (2005). Extensional troughs in the Caloris basin of Mercury: Evidence of lateral crustal flow. *Geology*, **33**, 669–672.

Watters, T. R., Solomon, S. C., Robinson, M. S., Head, J. W., André, S. L., Hauck, S. A., and Murchie, S. L. (2009a). The tectonics of Mercury: The view after MESSENGER'S first flyby. *Earth Planet. Sci. Lett.*, doi:10.1016/j.epsl.2009.01.025.

Watters, T. R., Head, J. W., Solomon, S. C., Robinson, M. S., Chapman, C. R., Denevi, B. W., Fassett, C. I., Murchie, S. L., and Strom, R. G. (2009b). Evolution of the Rembrandt impact basin on Mercury. *Science*, **324**, 618–621.

Watters, T. R., Murchie, S. L., Robinson, M. S., Solomon, S. C., Denevi, B. W., André, S. L., and Head, J. W. (2009c). Emplacement and tectonic deformation of smooth plains in the Caloris basin, Mercury. *Earth. Planet. Sci. Lett.*, doi:10.1016/j.epsl.2009.03.040.

Watts, A. B. (2001). *Isostasy and Flexure of the Lithosphere*. Cambridge: Cambridge University Press.

Weisberg, O. and Hager, B. H. (1998). Global lunar contraction with subdued surface topography. In *Origin of the Earth and Moon*, Proceedings of the Conference Held 1–3 December, 1998 in Monterey, California. Houston, TX: Lunar and Planetary Institute, p. 54.

Wetherill, G. W. (1988). Accumulation of Mercury from planetesimals. In *Mercury*, ed. F. Vilas, C. R. Chapman and M. S. Matthews. Tucson, AZ: University Arizona Press.

Wieczorek, M. A. and Phillips, R. J. (2000). The Procellarum KREEP Terrane: Implications for mare volcanism and lunar evolution. *J. Geophys. Res.*, **105**, 20 417–20 430.

Wilhelms, D. E. (1987). *The Geologic History of the Moon*. Washington, DC: U.S. Government Printing Office.

Willemann, R. J. (1984). Reorientation of planets with elastic lithospheres. *Icarus*, **60**, 701–709.

Willemann, R. J. and Turcotte, D. L. (1982). The role of lithospheric stress in the support of the Tharsis Rise. *J. Geophys. Res.*, **87**, 9793–9801.

Williams, J.-P., Aharonson, O., and Nimmo, F. (2007). Powering Mercury's dynamo. *Geophys. Res. Lett.*, **34**, L21201.

Williams, K. K. and Zuber, M. T. (1998). Measurement and analysis of lunar basin depths from Clementine altimetry. *Icarus*, **131**, 107–122.

Wojtal, S. F. (1996). Changes in fault displacement populations correlated to linkage between faults. *J. Struct. Geol.*, **18**, 265–279.

Zhong, S. (1997). Dynamics of crustal compensation and its influences on crustal isostasy. *J. Geophys. Res.*, **102**, 15 287–15 299.

Zuber, M. T. and Aist, L. L. (1990). The shallow structure of the Martian lithosphere in the vicinity of the ridged plains. *J. Geophys. Res.*, **95**, 14 215–14 230.

Zuber, M. T., Solomon, S. C., Philips, R. J., Smith, D. E., Tyler, G. L., Aharonson, O., Balmino, G., Banerdt, B. W., Head, J. W., Johnson, C. L., Lemoine, F. G., McGovern, P. J., Neumann, G. A., Rowlands, D. D., and Zhong, S. (2000). Internal structure and early thermal evolution of Mars from Mars Global Surveyor topography and gravity. *Science*, **287**, 1788–1793.

3

Venus tectonics

George E. McGill
University of Massachusetts, Amherst

Ellen R. Stofan
Proxemy Research, Laytonsville

and

Suzanne E. Smrekar
Jet Propulsion Laboratory, Pasadena

Summary

Venus has a pressure-corrected bulk density that is only 3% less than that of the Earth. In contrast, the surface environments of these two planets are very different. At the mean planetary radius the atmospheric pressure and temperature on Venus are 95 bars and 737 K, respectively. Liquid water cannot exist on the surface, which implies the absence of the processes most effective for erosion and sediment transport on Earth. The planet is completely shrouded in clouds, and temperatures of the lower atmosphere do not vary much from equator to poles, resulting in winds not capable of significant erosion. Most of the materials exposed on the surface of Venus apparently formed during approximately the last 20% of solar system history, with no significant clues to conditions on the planet during prior eons. Because the dense atmosphere has destroyed small bolides, the smallest surviving impact craters have diameters of ∼2 km, and the total population of impact craters is less than 1000. The dominant terrain on Venus is plains, which constitute ∼80% of the planet's surface. Impact craters are randomly distributed on these plains, and thus differences in the relative age of surface materials based on differences in crater frequency are not statistically robust.

The global topography of Venus does not include the diagnostic plate-boundary signatures that are present on Earth, and thus plate tectonics has not been active on Venus during the time represented by the current surface materials and features. Plate-boundary processes are responsible for most heat loss on Earth, but there

Planetary Tectonics, edited by Thomas R. Watters and Richard A. Schultz. Published by Cambridge University Press. Copyright © Cambridge University Press 2010.

is significant uncertainty concerning how heat is lost on Venus in the absence of plate tectonics. Slow erosion and no plate tectonics result in a planet exposing a bewildering array of structural and volcanic features that very likely were formed over a significant time span.

The tectonic landforms and terrains on Venus are divided into five classes: plains, volcanic rises, crustal plateaus and tesserae, coronae, and chasmata. The plains host an interesting and important array of structures and landforms, including wrinkle ridges, fractures and graben, polygonal terrains, ridge belts, fracture belts, and very broad-scale uplifts and ridges. As the name implies, volcanic rises are associated with significant volcanism and large central volcanoes. Some also are associated with either coronae or chasmata. Tessera terrain occurs both on large crustal plateaus and as inliers in plains. Tessera materials are very bright in SAR images, and are generally deformed by at least two sets of ridges or fractures at high angles to each other. These structures are abruptly truncated at contacts with plains or other materials, implying that tessera terrain is locally the oldest material. There is significant uncertainty concerning both the origin of tessera material and the processes responsible for crustal plateaus. Coronae are circular to elliptical features ranging in diameter from 50 to 2600 km. They vary significantly in internal topography and in the amount of associated volcanism. Most likely, these features form over thermal plumes in the mantle, and may account for a significant fraction of the total planetary heat loss. Chasmata are large graben structures that form major rift systems up to thousands of km long and with up to 7 km of relief. Some chasmata are associated with volcanic rises, others cut across the Venusian plains.

A major issue is whether similar features formed at the same time globally, or if these features formed at different times in different places, which is a more Earth-like model. A planet with no liquid water and no plate tectonics could have a tectonic history involving some combination of these models. Global geophysical models primarily address how Venus can be so similar to Earth in bulk properties yet have a tectonic style that is so different. Most models predict significant changes in the style of tectonics due to the low water content and associated increase in viscosity of the crust, asthenosphere, and mantle. Solving these global-scale issues is not only important for understanding Venus, but also for understanding Earth.

1 Introduction

Venus is commonly referred to as Earth's twin planet. This stems in part from its proximity to Earth in the solar system, but mostly because the bulk properties of Venus are not much different from the bulk properties of Earth (Tables 3.1, 3.2). Of particular importance is the small difference in mean density, which implies that the bulk compositions of Venus and Earth must not differ much. The difference in mass

Table 3.1. *Venus vital statistics*

Radius	6051.9 Km
Mass	4.87×10^{24} kg
Sidereal day	243.0 Earth days
Solar day	116.75 Earth days
Year	224.7 Earth days
Obliquity	$-2.6°$
Mean radius of orbit	0.723 32 AU
Orbital eccentricity	0.006 787

Table 3.2. *Venus vs. Earth*

	Earth	Venus
Diameter	1	0.95
Mass	1	0.82
Mean density	1	0.95
Gravity	1	0.91

is largely due to Venus' slightly smaller radius, but the pressure-corrected density of Venus is about 3% less than the pressure-corrected density of Earth (Phillips and Malin, 1983). This may imply that the Mg/(Mg+Fe) ratio is somewhat larger than for Earth's mantle (Wood *et al.*, 1981). One bulk property of Venus that is very different from Earth's is the apparent absence of a magnetic field. Venus also lacks a satellite.

In contrast, the surface environment of Venus is not at all Earth-like. The atmospheric pressure and temperature are 95.0 bars and 737 K at the mean planetary radius (Donahue and Russell, 1997). These values vary significantly with elevation, and are ∼45 bars and ∼650 K at the top of Maxwell Montes (Seif *et al.*, 1980), the highest place on the planet. Clearly, water cannot exist on the surface of Venus and thus the most important erosional processes on Earth are not possible on Venus. Even aeolian erosion is much less effective than on Earth, in part because the hot, dry atmosphere is not conducive to the production of fine particles by weathering, and in part because the very sluggish winds near the surface, although capable of moving sand-sized particles, are not capable of extensive abrasion (Greeley *et al.*, 1997). Probably the most effective processes for degradation are slow weathering coupled with mass wasting.

In the absence of an aggressive regime of weathering and erosion, most landforms are primary features due directly to tectonic and volcanic processes. These landforms, and structures as well, are thus able to survive for geologically long

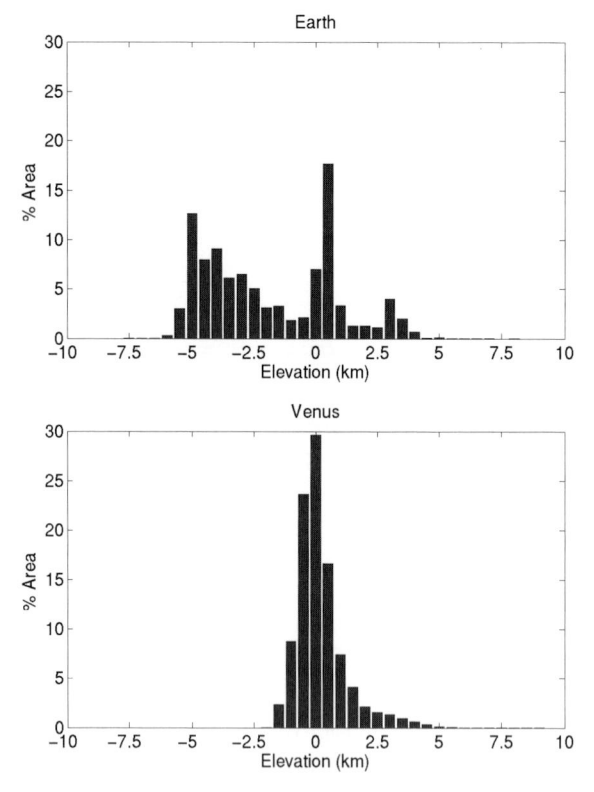

Figure 3.1. Comparison of hypsometric histograms for Venus and Earth. The pronounced bimodality of the Earth elevation distribution is a function of plate tectonics and continental freeboard. The Venus histogram is strongly unimodal, presumably due to no plate tectonics and no oceans.

periods of time. The main processes modifying tectonic landforms and structures are burial by lava flows, disruption by intense fracturing, and destruction and burial by impact craters and their ejecta.

A striking difference between Earth and Venus may be seen in their global topographies (Plate 5). The topography of an Earth with transparent oceans is dominated by very long, narrow, elevated regions, both within the ocean basins and along the margins of continental blocks (Plate 5a). These elevated regions are, of course, the divergent and convergent boundaries of our terrestrial plate-tectonic system. The global topography of Venus (Plate 5b) exhibits no comparable pattern, leading to the generally accepted view that the tectonic style of the preserved geological record on Venus does not involve plate tectonics (e.g., Kaula and Phillips, 1981; Solomon *et al.*, 1992). This contrast also is reflected in the hypsometric curves for both planets (Figure 3.1). The distribution of elevations on Earth is markedly bimodal due to the preservation of high-standing continental crust and the continuous formation of new low-standing oceanic crust. On Venus, the distribution of elevations is strongly unimodal, supporting the inference that Venus currently is not in a plate-tectonic regime. Also in contrast to Earth, large-scale strike-slip faults are rare or absent on Venus; the few inferred shear zones appear to be local features only.

Although difficult to date accurately, the preserved crustal record of Venus was most likely formed in the last quarter of the total life span of the solar system (e.g., McKinnon *et al.*, 1997). The number of impact craters is greatly reduced because the dense atmosphere screens out bolides capable of creating craters smaller than about 2 km in diameter, depriving us of the large numbers of craters needed to separate material units by age. The spatial distribution of the surviving craters on the plains of Venus cannot be statistically distinguished from a random distribution (Schaber *et al.*, 1992; Campbell, 1999). The small number of modified craters suggests that the rate of geologic activity declined following global resurfacing (Schaber *et al.*, 1992). The nature of Venus tectonics during earlier eons is entirely speculative.

To first order, the processes that together constitute a planet's tectonic style relate directly to how heat is lost. For Earth, it is estimated that ~70% of the planet's heat is lost at divergent and convergent plate boundaries (Sclater *et al.*, 1980). Because Venus has no plate tectonics, most of its heat must escape by means of other processes. These processes presumably create structural and topographic signatures on the crust of the planet.

One alternative to plate tectonics is what has been called "vertical tectonics," in which mantle upwelling is the dominant heat loss mechanism (Solomon and Head, 1982; Phillips and Malin, 1983; Morgan and Phillips, 1983). In addition to features that strongly resemble terrestrial hotspots (volcanic rises) (e.g., Smrekar and Phillips, 1991; McGill, 1994; Phillips and Hansen, 1994), Venus has plateaus several thousand km in diameter and quasicircular volcanotectonic features (coronae) typically several hundred km in diameter, both of which have been attributed to upwelling. Although all of these features may contribute to heat loss by advecting heat upwards from the interior and conducting it through the lithosphere, they are probably not able to account for Venus' entire heat loss budget (Smrekar and Parmentier, 1996; Smrekar and Stofan, 1997). Another alternative to plate tectonics is stagnant-lid tectonics, in which the high yield strength of the lithosphere prevents the development of plates (Moresi and Solomatov, 1998; Reese *et al.*, 1999; Solomatov and Moresi, 2000). This scenario is distinct from simple conduction of heat through a lithosphere because the lithosphere does not thicken indefinitely. Specifically, the stagnant-lid model implies that the lithosphere acts to insulate the mantle, causing the mantle to heat up and the lithosphere to reach an equilibrium thickness; the lithosphere acts as an insulating lid that causes the mantle temperature to actually increase, possibly leading to enhanced volcanism and even resurfacing (Reese *et al.*, 1999). Vertical tectonics, in which deformation and volcanism are driven by uplift above mantle plumes and pressure-release melting, can operate within the context of a stagnant-lid convective regime.

2 Tectonic landforms and terrains

2.1 Introduction

With severely limited erosion, tectonically created landforms, terrains, and structures can remain pristine for geologically significant time spans. Thus, classifications of features by physiography are also tectonic classifications. Although there are minor differences, most past summaries of important landforms and terrains on Venus place these features into six or seven classes (e.g., Phillips and Hansen, 1994; Hansen *et al.*, 1997). For this chapter we have five classes: plains, volcanic rises, crustal plateaus and tesserae, coronae, and chasmata. Geophysical models relevant to explaining individual landforms and terrains are briefly discussed in the sections defining and describing these features. Global-scale geophysical and geological models are presented and discussed in a separate section at the end of this chapter.

2.2 Plains

By far the areally most important class of landform is plains. Depending on what is included as plains by various authors, the fraction of the planet's surface considered to be plains ranges between 40% and 80%, with most estimates nearer the high figure than the low one (Masursky *et al.*, 1980; Hansen *et al.*, 1997; Tanaka *et al.*, 1997; Banerdt *et al.*, 1997). The plains of Venus host abundant structural features visible in SAR images (Figures 3.2, 3.3). These include wrinkle ridges, radar-bright lineaments, polygonal patterns, grabens, ridge belts, fracture belts, and local complexes of nearly penetrative fractures, where "penetrative" implies that the fractures are so close together that little or no unfractured material can be resolved. There also are broad-scale ridges visible only in altimetry. All plains structures were divided into three categories by Banerdt *et al.* (1997); we have modified and simplified their scheme. Our categories are: (1) Distributed deformation, including wrinkle ridges, radar-bright lineations and patterns, and grabens; (2) concentrated deformation, including ridge belts and fracture belts; and (3) broad-scale, predominantly vertical deformation.

2.2.1 Distributed deformation

2.2.1.1 Wrinkle ridges Wrinkle ridges are long, narrow, sinuous features (Figures 3.2, 3) ranging in width from the limit of SAR resolution to about 1 km, and with

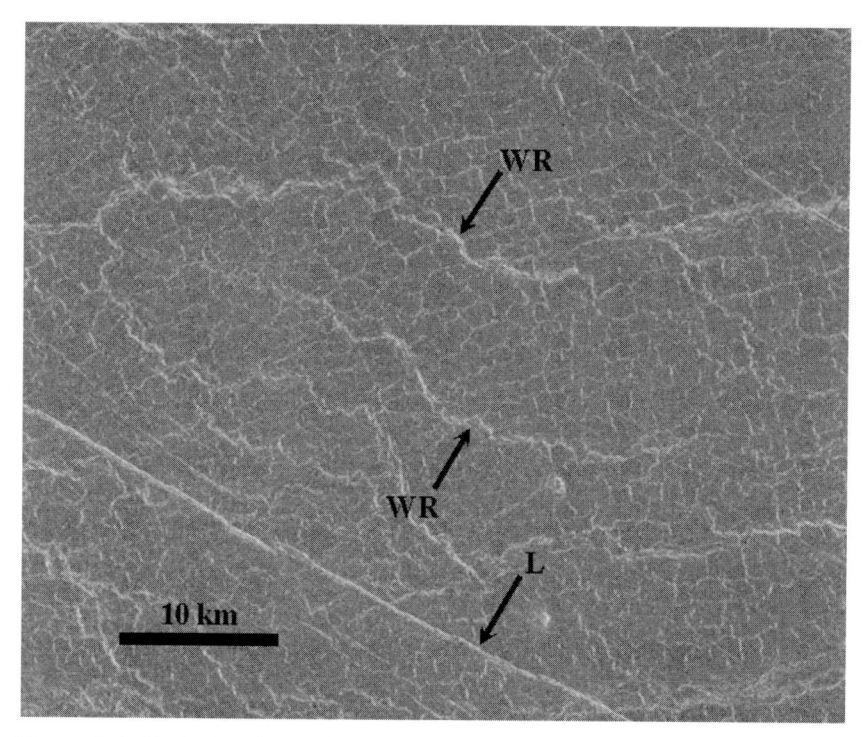

Figure 3.2. Typical polygonal terrain in Sedna Planitia, with cells averaging ~1–2 km across. WR = wrinkle ridges, L = radar bright lineament. Part of FMAP FL25N333. North up.

lengths up to several hundred km. They are brighter on Magellan SAR images than the plains materials that they deform, and this brightness contrast is present regardless of the trends of the ridges relative to radar look direction. From this it can be inferred that the brightness of wrinkle ridges is more related to wavelength-scale surface roughness than it is to topography. Most Venusian wrinkle ridges occur in sets of approximately evenly spaced, parallel ridges, and it is common for two or more of these sets to occur in the same area. Generally, however, one set is dominant and more through going than the others. Some of the larger Venusian wrinkle ridges exhibit the brightness contrast on SAR images expected of ridges (bright on radar-facing slopes, dark on slopes facing away from the radar), but most features interpreted as wrinkle ridges on Venus have insufficient relief to cause resolvable backslope darkening on Magellan SAR images. However, some wrinkle ridges have ponded flows derived from impact, implying positive relief. Wrinkle ridges on other bodies (Mars, Mercury, Moon) commonly have a complex morphology characterized by a gentle arch surmounted by a much narrower, sinuous ridge (Strom, 1972; Bryan, 1973; Maxwell *et al.*, 1975; Watters, 1988). With few exceptions, Venusian wrinkle ridges either do not include gentle arches, or else

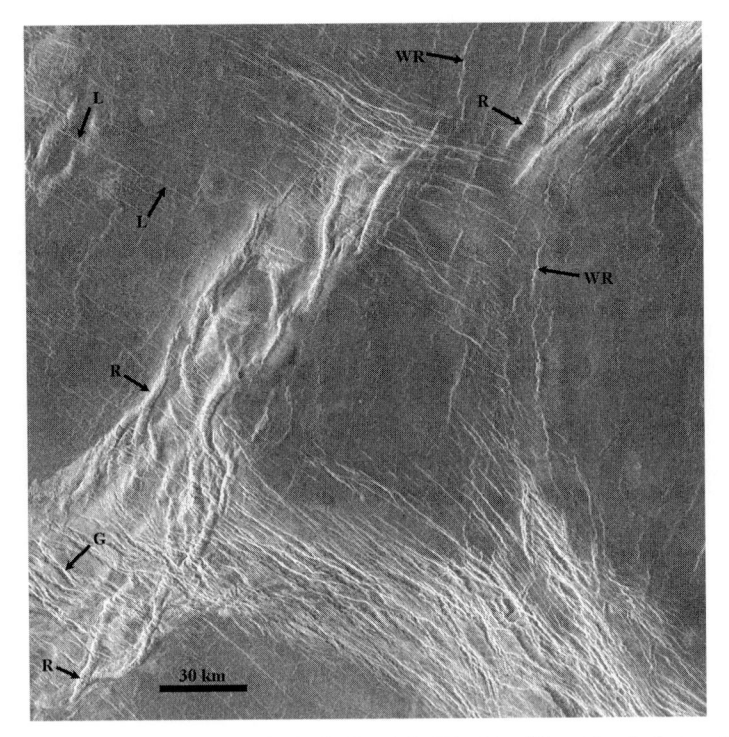

Figure 3.3. Ridge and fracture belts in Lavinia Planitia. The ridge belt trends NE, and is embayed by surrounding plains materials. The fracture belt trends NW, and the fractures cut the ridges of the ridge belt, and they also cut the plains materials that embay the ridge belt. G = graben; L = radar bright lineament; R = ridges of ridge belt; WR = wrinkle ridges. C1 MIDR 45S350, tile 4. North up.

these features are topographically too subdued to affect the radar return. Thus Venusian wrinkle ridges appear comparable to the simpler forms found on the other bodies.

Wrinkle ridges are extremely abundant on Venus. Tanaka *et al.* (1997) define six general types of plains material units, most of which are deformed by abundant wrinkle ridges. Wrinkle ridges are evidently not present (or not discernable) in tessera terrain, in intensely fractured terrains, and within mountain belts. They are relatively rare on lobate and digitate lava flows associated with young shield volcanoes and some coronae.

All recent models explain wrinkle ridges as due to shallow compressive stresses oriented approximately normal to the trend of the ridges. These stresses result in buckling (Watters, 1991) or thrust faulting, commonly with associated folding (Plescia and Golombek, 1986; Golombek *et al.*, 1991; Schultz, 2000). Wrinkle ridges generally are much smaller than the constituent ridges of ridge belts (Figure 3.3), and their trends do not seem to exhibit any consistent relationship to each other on a global scale; in places they are parallel, elsewhere they are not.

Almost all Venusian wrinkle ridges are superposed on the materials that make up the global plains, and thus the ridges must be younger than these plains materials. Within the plains, the abundance of wrinkle ridges locally decreases with decreasing relative age of specific plains units. Relative age relationships between impact craters and wrinkle ridges indicate that the ridges were not formed as a direct result of the emplacement of plains materials; rather, they are younger, superposed structures (McGill, 2004b).

Most plains regions of Venus may be divided into large domains, each characterized by a dominant through-going set of wrinkle ridges that maintains a roughly uniform trend or uniform curvature of trend over hundreds to thousands of km (Bilotti and Suppe, 1999). Most of these domains occur in areas that coincide with geoid lows (Bilotti and Suppe, 1999). Within these domains are more local wrinkle-ridge sets, generally trending at high angles to the trend of the dominant set. Commonly, it can be shown by intersection relationships that the local sets are younger than the dominant set (McGill, 1993).

Wrinkle ridges have been modeled as due to gravitational spreading on shallow slopes, an excellent example being the sets concentric to western and central Eistla Regio (e.g., Basilevsky, 1994; McGill, 1994), a model similar to that of Maxwell (1982) for wrinkle ridges in the Tharsis region of Mars. However, most wrinkle ridges have trends that are not consistent with a gravitational spreading model; indeed, dominant wrinkle-ridge sets trend at high angles to topographically elevated regions in places, such as near Aphrodite Terra (Bilotti and Suppe, 1999). Wrinkle ridges that are not readily explained as due to gravitational spreading seem to require the transmission of stresses in the shallow crust over hundreds to thousands of km. A possible mechanism for this was proposed by Solomon *et al.* (1999). Their model postulates a major increase in mean atmospheric temperature as a result of an inferred rapid emplacement of near-global plains materials (Bullock and Grinspoon, 1996). In time, the hotter atmosphere would cause heating of the rocks at and near the surface, and the associated thermal expansion would generate a near-global, shallow, compressive stress regime. Solomon *et al.* (1999) predict that these thermal stresses would be sufficient to produce the observed wrinkle ridges. The regional differences in wrinkle-ridge orientation would likely be due to the superposition of more local stresses varying in space and time. However, Anderson and Smrekar (1999), using a different model formulation, found that the strain resulting from thermal stresses due to climate change was approximately an order of magnitude less than that estimated for wrinkle ridges. Potentially, strain localization could increase the amount of strain to be consistent with the formation of wrinkle ridges via thermal contraction.

2.2.1.2 Lineaments and grabens Linear features of inferred extensional origin are abundant on the Venusian plains. Many of these show the brightening or darkening on SAR images that one would expect for slopes facing towards or away from the radar. Many of these slopes are paired, defining troughs that are inferred to be grabens (Figure 3.3). However, most radar-bright lineations are too narrow to define their boundaries because they are only one or two pixels across (Figures 3.2, 3). This limited pixel width also means that any relief associated with these lineations is not resolvable. Grabens clearly are extensional by direct analogy with similar features on Earth. The narrow lineations are inferred to be extensional as well, because many are parallel to larger structures that are resolvable as grabens or fault scarps, and because some increase in width along-trend into resolvable grabens. As for wrinkle ridges, the backscatter from radar-bright lineations is not related to their orientations relative to the radar look direction. Most likely, many or perhaps most narrow lineations are fractures with movement perpendicular to the fracture plane (mode 1 joints or joint zones) rather than faults.

Many lineament and graben complexes occur as radial or concentric sets associated with coronae or volcanic centers. Individual lineaments in the radial sets extend hundreds or even thousands of km from the source feature. These radial sets have been interpreted to overlie dikes derived from the central feature, analogous to examples on Earth (Anderson, 1951; Odé, 1957; Muller and Pollard, 1977; Grosfils and Head, 1994; Ernst *et al.*, 2001). At large distances from the central source, the radial sets commonly rotate into trends consistent with the regional maximum principal compression as defined by structures such as wrinkle ridges and radar-bright lineations. These sets also diverge around topographic lows (McKenzie *et al.*, 1992a), a pattern consistent with the behavior of compressive stress trajectories around holes in elastic plates.

Many other lineations have no apparent source. They commonly occur as closely spaced, parallel lineament sets up to hundreds of km long, or as suites of en-echelon lineaments a few km to a few tens of km long. It seems unlikely that these lineaments are due to the intrusion of blind dikes. However, they provide a means to infer the orientation of the least principal compressive stress, which should be approximately normal to the trends of the lineations. Locally, lineations define closely spaced, nearly penetrative fabrics. The close spacing of these fractures implies either that the fractured layer is very thin, possibly less than 1 km thick, or else comparable to the fracture spacing (Lachenbruch, 1961; Banerdt *et al.*, 1997).

2.2.1.3 Polygonal terrains Polygonal patterns of radar-bright lineations (Figure 3.2) occur in over 200 locations globally (Moreels and Smrekar, 2003). The polygons range in width from the resolution limit of the SAR up to about 10 km (Johnson and Sandwell, 1992). The larger polygons present a serious scale problem if they are tensile fractures formed in response to thermal stresses because of the approximate scaling of fracture spacing with fracture depth (Lachenbruch, 1961). Many of the polygons are rectangular, and thus could be due to a reorientation of principal stresses with time that produces two sets of lineations at high angles to each other.

Polygons have many different manifestations. In many places there are patterns of radar-bright lineations that are "cellular" rather than "polygonal." The lineations bounding the cells or polygons at these places are curved rather than straight, so that the resulting pattern resembles an array of irregular cells rather than a mesh of straight-sided polygons. At least some of the lineations defining cellular terrain may be contractional wrinkle ridges rather than extensional fractures. It also is possible for two intersecting sets of wrinkle ridges to produce a rectangular pattern that resembles polygonal fracturing.

Several models have been proposed to explain the formation of polygons. On Earth, polygons on the scale of up to decameters form in slowly cooling lava flows. The thickness of the flow and the required slow rate of cooling needed to produce polygons of the size seen on Venus make this mechanism improbable (Johnson and Sandwell, 1992; Anderson and Smrekar, 1999). Johnson and Sandwell (1992) proposed that the polygons are due to heating and cooling above a region of subsurface heating, consistent with the association of some polygons with shield fields and volcanic plains. Alternatively, polygons may result from thermal stresses due to cooling and heating cycles induced by climate change (Anderson and Smrekar, 1999). This hypothesis explains several characteristics of polygonal structures (Smrekar *et al.*, 2002). A range of sizes would occur as the thermal wave propagates deeper into the lithosphere. The compressional ridges seen in association with some polygons would result from either very large atmospheric temperature increases, or from stresses due to moderate temperature increases assisted by regional stresses. This model does not require a thin layer, and predicts strains that are comparable to estimates for polygons. It could also explain lineations over a very broad region in the presence of a regional stress field (Anderson and Smrekar, 1999). Additionally, the large-scale polygons (~10 km diameter) are predicted to occur as a result of stress in a thick ductile layer.

2.2.2 Concentrated deformation (deformation belts)

2.2.2.1 Ridge belt Ridge belts, or dorsa, are morphologically diverse. A few consist of a single, broad arch, but most ridge belts are long, relatively narrow elevated regions that incorporate an array of smaller ridges (Figure 3.3). Although ridge belts are widely scattered on the plains, there are two significant concentrations of these features: Atalanta Planitia/Vinmara Planitia in the northern hemisphere (Rosenberg and McGill, 2001), and Lavinia Planitia in the southern hemisphere (Ivanov and Head, 2001). In Lavinia Planitia, ridge belts trend approximately northeast and are crosscut by fracture belts trending mostly west–northwest (Figure 3.3). The Lavinia ridge belts are 10–20 km wide and up to a few hundred km long. In Atalanta/Vinmara Planitiae the belts are 20–230 km wide and 1000 to 2500 km long, and make up a complex anastomosing array with an overall northerly trend. Characteristically, these belts are spaced 300–400 km apart. Individual ridges within each belt average 15 km wide and 150 km long, and are spaced ∼25 km apart. In contrast to the ridge belts of Lavinia Planitia, the ridge belts of Atalanta/Vinmara Planitiae are composite; that is, they include both ridges and fractures, commonly with similar trends within a given belt. Ridge belts in areas other than these two concentrations generally have lengths and widths that lie between the characteristic dimensions in Lavinia, Atalanta, and Vinmara Planitiae. The individual ridges within ridge belts planet wide are mostly in the range 100–400 m high, with flank slopes of a few degrees to, rarely, as much as 30°.

Most ridge belts appear to be older than the plains materials that surround them. In both Atalanta/Vinmara and Lavinia Planitiae it can be demonstrated that at least some ridge belts deform a plains unit that is older than regional plains (Squyres *et al.*, 1992; McGill, 2003). One ridge belt in Atalanta Planitia (Laūma Dorsa) may be younger than the adjacent plains materials (Rosenberg and McGill, 2001). However, the relative ages of ridge belts and regional plains materials are basically ambiguous for many (probably most) ridge belts.

Ridge belts are generally interpreted as resulting from compressive stresses oriented normal to their trends. This inference is based on the arch-like morphology of almost all ridge belts, and on the morphology of the smaller ridges within the belts, which are inferred to be open anticlinal folds. The common tendency for ridge belts to be somewhat elevated also suggests modest associated crustal thickening that can be explained as a result of the same stresses (Solomon *et al.*, 1992; Squyres *et al.*, 1992). Sukhanov and Pronin (1989) correctly note that the volcanic centers associated with some ridge belts are difficult to explain in a compressional stress environment. However, an extensional origin for ridge belts cannot explain the most important characteristics of these belts: a broadly arched area containing sets of ridges.

2.2.2.2 Fracture belts Fracture belts, or lineae, are broad, elongated swells or arches that are characterized by numerous, roughly belt-parallel lineations, scarps, and grabens (Figure 3.3). In detail, the fractures and faults of these belts commonly occur as sets that intersect at low angles. Fracture belts typically range in size up to about 200 km wide and 1000 km long, and stand a few hundred meters to more than a km above adjacent plains. Some fracture belts grade lengthwise into chains of coronae; others simply terminate in broad fans. Fracture belts generally deform the materials of the plains in which they are located (Squyres *et al.*, 1992; Solomon *et al.*, 1992; McGill, 2004a) (Figure 3.3), although there is some disagreement about this (e.g., Basilevsky and Head, 2000; Ivanov and Head, 2001). Young digitate flows in Lavinia Planitia were clearly diverted by preexisting fracture belts and thus fracture belts appear to be older than the youngest volcanics in the same region.

The lineations (fractures?), scarps, and grabens making up fracture belts are all consistent with formation by extension normal to the belt trends. The presence of rhomboidal depressions bounded by faults on some fracture belts (Solomon *et al.*, 1991) suggests that many are actually due to transtension. On the other hand, the broad ridges characteristic of these belts suggest crustal thickening due to compression, as for ridge belts (Squyres *et al.*, 1992). If fracture belts are due to compression rather than extension, then the characteristic extensional structures must result from local tension due to uplift and bending (Solomon *et al.*, 1991) or from gravitational spreading (Phillips and Hansen, 1994). Fracturing due to bending stresses seems unlikely because the average slopes on the flanks of fracture belts are generally <1°, but this mechanism may be valid if the mechanical layer being arched is very thick.

2.2.3 Broad-scale vertical deformation

2.2.3.1 Stealth ridges In many places there are broad ridges resolvable in altimetry data but invisible on SAR images. The term "stealth" derives from the realization that many of these ridges define what have been called "stealth coronae" (Stofan *et al.*, 2001), which were so named because they are invisible on SAR images. Other stealth ridges bear no relationship to coronae. Commonly these ridges are hundreds of km long and tens of km wide.

2.2.3.2 Regional crustal movements At a still larger scale are uplifts of the surface that are discernable only because they affect regional slopes inferred from lava channels and digitate flows. It is impossible for the regional slope on which a flow or channel is present to reverse while emplacement is occurring; thus any reversals apparent in modern altimetry must be due to movements that occurred after flow ceased. Long lava channels are especially useful in defining these regional scale deformations, and some of these have as much as 2 km of relief between adjacent high and low points along the channels (Parker *et al.*, 1992; Komatsu and Baker, 1994; McLeod and Phillips, 1994). Two scales of deformation are defined, one with a wavelength of a few hundred km, the other with a wavelength of a few thousand km; the latter is based on only the two channels long enough to show this wavelength. The shorter wavelength is comparable to the characteristic spacing of ridge belts (Zuber, 1990).

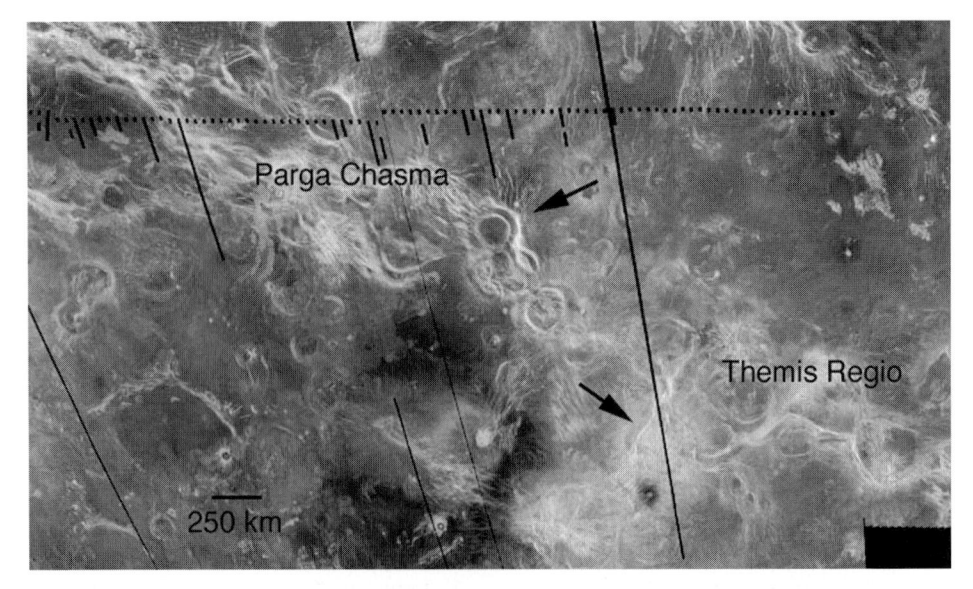

Figure 3.4. This Magellan SAR image covers a region in the southern hemisphere of Venus including Parga Chasma and Themis Regio. Themis Regio, in the lower right, is a corona-dominated volcanic rise. Parga Chasma intersects Themis Regio, and extends over 5000 km to the northwest. It is characterized by radar-bright fractures and faults and numerous coronae. The northern arrow points to Gertjon Corona, a 236 km wide corona with a double fracture annulus. The southern arrow points to Shulamite Corona, a 275 km plateau-shaped corona with extensive volcanism. The image is centered at approximately 36°S, 275° and is about 4000 km across. North up.

2.3 Volcanic rises

Ten volcanic rises have been identified on Venus, located predominantly in the equatorial region of the planet. Volcanic rises have dome-shaped topography, diameters of at least 1000 km, and associated large volcanoes. The ten rises identified to date are Beta, Atla, Bell, Western Eistla, Eastern Eistla, Central Eistla, Laufey, Imdr, Themis, and Dione Regiones (Figures 3.4, 3.5). Heights of the rises range from 0.5–2.5 km above the surrounding regions, with diameters of 1400–2500 km. All of the rises have associated volcanoes, and some are associated with coronae or rifts (chasmata) (Figures 3.4, 3.5). Volcanic rise volcanoes are among the largest volcanoes on Venus, with heights up to 4 km and flow aprons that extend for over 750 km. Minimum melt volumes estimated from volcano volumes are at least 10^4–10^6 km^3, similar to overall melt volumes at terrestrial hotspots (Stofan *et al.*, 1995). McGovern and Solomon (1998) modeled the growth of these volcanoes and found that large volcanoes form more readily when the elastic thickness is large (>30 km). Some volcanic rises show clear evidence for uplift, as well as volcanic construction and associated corona formation (Senske *et al.*, 1992; McGill, 1994; Stofan

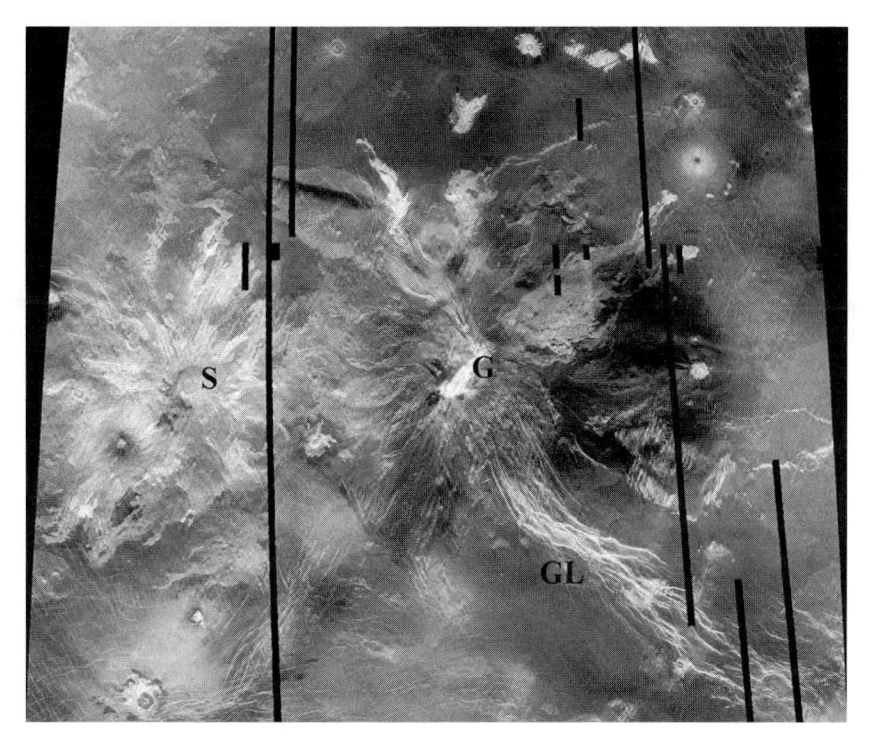

Figure 3.5. Magellan radar image of Western Eistla Regio in a sinusoidal projection covering 14°–30°N, 348°–8°E. The resolution is 106 pixels/degree. Black stripes occur where there are no data. The central part of this image covers the broad topographic rise that makes up Western Eistla. The two large volcanoes are Sif Mons (S) on the left and Gula Mons (G) on the right. Guor Linea (GL) is a rift system, ~700 km in length, which extends to the southeast from the main rise. North up.

et al., 1995; Brian *et al.*, 2004a). Volcanic rises on Venus are interpreted to form over mantle plumes because of their large diameters, dome-shaped topography, and volcanic edifices (e.g., Masursky *et al.*, 1980; Phillips and Malin, 1983; Campbell *et al.*, 1984; Senske *et al.*, 1992). Gravity data support the plume hypothesis, with several rises having very large compensation depths, which may indicate they are still active (Smrekar, 1994).

Rises have been classified into three categories based on their associated features: volcano-dominated, rift-dominated and corona-dominated (Stofan *et al.*, 1995). Five rises are volcano-dominated, three are corona-dominated, and two are rift-dominated. The classes do not indicate an age progression. The ten rises have been interpreted to be in the late to intermediate stages of evolution, based on their topography, morphology, and gravity signature (Stofan *et al.*, 1995; Smrekar *et al.*, 1997). Evidence for both uplift and subsidence at the rises also supports a plume origin. Wrinkle ridges concentric to rises suggest subsidence at Western Eistla

Regio (McGill, 1993; Bilotti and Suppe, 1999), Bell Regio, Themis Regio (Bilotti and Suppe, 1999), and Laufey Regio (Brian *et al.*, 2004a). Flows from a volcano on the northern flank of Western Eistla Regio trend uphill, demonstrating that the volcano formed prior to uplift of the rise (Stofan *et al.*, 1995). Uplift at Beta Regio is indicated by a lava channel that trends uphill and at Imdr Regio by the small amount of mappable volcanism that can be identified on the rise, implying that the rise topography is largely due to uplift.

The gravity and topography at large volcanic rises on Venus are consistent with upwelling mantle plumes (Kiefer and Hager, 1991, 1992; Smrekar and Parmentier, 1996). It has been suggested that the differences between the rises indicate variations in lithospheric structure, plume characteristics and regional tectonic environment (Stofan *et al.*, 1995; Smrekar *et al.*, 1997; McGill, 1998).

2.4 Tessera terrain and crustal plateaus

Tessera terrain was originally defined in Soviet Venera images (e.g., Basilevsky *et al.*, 1986). The name was chosen because many areas of this terrain resemble floor tiles. Most tessera terrain occurs in large crustal plateaus, such as Aphrodite and Ishtar Terra; and Alpha, Ovda, Phoebe, Tellus, and Thetis Regiones (Figure 3.6). In addition, there are numerous exposures of tessera terrain scattered around the planet as inliers within plains materials. We first discuss the general characteristics of tessera terrain and then the specifics of crustal plateaus.

2.4.1 Tessera terrain

Tessera terrain is characterized by high radar backscatter and at least two sets of structures at high angles to each other. Commonly, the structures in one or more of these sets are penetrative or nearly so. Tessera also is characterized by pronounced roughness at the centimeter to meter scale. The radar brightness is apparently due largely to the closely spaced structural elements and the high rms slopes, but locally there are small areas within tessera that are not so penetratively deformed, and these areas also are significantly brighter and rougher than plains materials. For almost all exposures, however, it is not possible to infer the nature of the material making up tessera because of the close spacing of younger structures. Many areas of tessera terrain contain pond-like areas of smooth, dark material (Figure 3.6), commonly referred to as "intratessera plains." These plains truncate tessera structures, and thus are younger than tessera materials. Some of these dark ponds are connected to the plains surrounding the tessera, and thus are simply areas flooded by these plains. Most intratessera plains are not continuous with the plains surrounding the tessera, and thus the age of these intratessera plains relative to the surrounding plains is indeterminate.

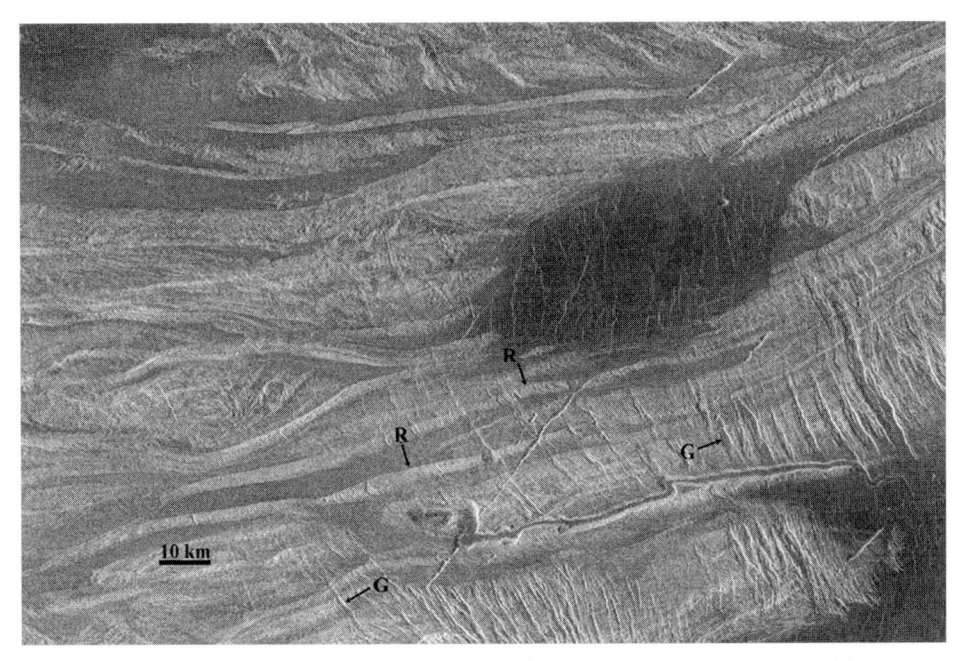

Figure 3.6. Tessera terrain in Ovda Regio. Broad ridges, assumed to be open folds, trend ENE. The ridges are cut by grabens trending NNW, and they are embayed by radar-dark material that fills topographic lows. The "eye-shaped" pattern in the western part of the image suggests that some of the deformation of this tessera terrain was ductile. G = grabens; R = broad ridges. Part of FMAP FL01N081. North up.

The structures deforming tessera terrain may be contractional ridges, extensional fractures or both (Figure 3.6), but in many places it is not possible to determine the nature of the deformation responsible for the linear trends observed. Large tessera areas, such as Aphrodite Terra or Alpha Regio, commonly consist of sets of presumably contractional ridges and troughs with 10–20 km spacing that are crosscut by extensional fractures or grabens. Some major areas of tessera also include long, narrow, steep-sided depressions referred to as "ribbon structures" (Hansen and Willis, 1996, 1998). They differ from grabens by being bound by vertical or nearly vertical fractures rather than the inward dipping faults that bound grabens. The relatively large scale structures present on crustal plateaus are rare or absent on tessera inliers in the plains, which are characterized by shorter and more closely spaced lineations.

Linear structures of tessera terrain are abruptly truncated at contacts with all other material units, and thus it appears that tessera is the locally oldest material unit. This has led to the assumption that tessera is approximately the same age globally (Basilevsky and Head, 2000). This very likely is an unwarranted assumption. Before the advent of radiometric dating, it was assumed that all high-grade metamorphic basement rocks on Earth are the same age (Precambrian), an assumption that is logically equivalent to assuming that all tessera are the same age. We now know that the ages of high-grade metamorphic rocks on Earth span almost the entire range of ages of preserved Earth rocks. The same could be true for tessera, but we have no way to determine "absolute" ages of rocks on Venus. Crater densities are of little use because the preserved record on Venus includes only about the last fourth of the age of the solar system, and because the dense atmosphere screens out bolides capable of creating craters smaller than about 2 km in diameter, depriving us of the large numbers of craters needed to separate material units by age. By combining crater counts for all tessera areas it can be inferred that the average age of tessera may be somewhat greater than the average age of plains (Gilmore *et al.*, 1997), but this observation does not constrain the possible range of tessera ages: it can be inferred to represent the age of tessera only if we assume a priori that all tessera areas expose materials of the same age. Furthermore, it is very likely that any age determined for tessera would be the age of deformation rather than the age of tessera material.

In addition to the two or more penetrative sets of lineations that define tessera terrain, many exposures also are cut by large grabens, most of which appear to be younger than the penetrative lineations. These grabens are flooded by surrounding plains materials, and thus are older than these plains. In some places, it is possible to determine relative ages of individual grabens by cross-cutting relationships with material that partially fills some grabens. Because these grabens have differing orientations, this implies that the principal stress orientations responsible for the grabens evolved during the time interval between the penetrative deformation of tessera materials and the emplacement of plains materials.

Not all tessera terrain exposures look exactly alike (Hansen and Willis, 1996), suggesting that there may have been a number of distinct material units that were deformed into the terrain we see today. It has been proposed that the isolated inliers of tessera represent local exposures of a widespread, possibly global basement (Ivanov and Head, 1996). If true, this basement could be stratigraphically very complex; that is, there is no need to assume that the basement is lithologically similar everywhere and of a common age. If this global basement hypothesis is correct, one would expect that the penetrative structures in inliers within relatively small areas would exhibit some coherence. They commonly do not; that is, there seems to be no continuity of lineation trends from one inlier to the next in some areas. This issue represents one of the most important unsolved problems concerning Venusian crustal evolution. Clearly, there must be something beneath the widespread plains materials on Venus; if not tessera, then what?

2.4.2 Crustal plateaus

Crustal plateaus are the most areally extensive highlands on Venus and account for the majority of tessera terrain. Alpha, Ovda (Figure 3.6), Phoebe, Thetis, and Tellus Regiones are classified as crustal plateaus. Ishtar Terra is also considered a crustal plateau, but has several unique characteristics, which we discuss below. Additionally, Western Ovda Regio may be a relaxed crustal plateau. Plateaus are roughly circular to ovoidal, approximately 1000–3000 km in diameter and stand 0.5–4 km above surrounding plains. The slopes of plateau perimeters are typically 3–5° (Bindschadler *et al.*, 1992). Another characteristic of several crustal plateaus is a discontinuous rim of higher topography along all or part of the plateau edge (Bindschadler *et al.*, 1992; Nunes *et al.*, 2004).

The gravity signatures of plateaus distinguish them from other highlands on Venus. Crustal plateaus have a shallow apparent depth of compensation, implying that they are isostatically compensated (leading to the designation "crustal plateau"), while most other highlands are more deeply compensated, implying dynamic compensation by mantle plumes (Smrekar and Phillips, 1991; Simons *et al.*, 1997). The isostatic state of compensation of crustal plateaus further implies that they are no longer active.

The origin of crustal plateaus is highly controversial. The leading candidates for a formation mechanism are mantle upwelling (Grimm and Phillips, 1991; Phillips *et al.*, 1991) and downwelling (Bindschadler and Parmentier, 1990; Lenardic *et al.*, 1995), two diametrically opposed processes. In upwelling models, the plateau is created above a laterally spreading mantle plume and maintained by crustal thickening generated by pressure release melting. Downwelling models create a plateau by thickening the crust above mantle downwelling in which the lower lithosphere eventually detaches. Models that use a stronger crust based on flow laws for dry diabase (Mackwell *et al.*, 1998) favor an upwelling origin for highland plateaus due to the long timescale required for crustal thickening to result from downwelling (Lenardic *et al.*, 1995; Kidder and Phillips, 1996). However, upwelling models of plateau highlands have not fully accounted for the pervasive compressional deformation. Both upwelling and downwelling models are more plausible if crustal plateaus formed at a time when the lithosphere on Venus was thinner than it is now. Upwelling models require that the plume generates a much greater volume of pressure-release melt than is inferred for the large, relatively young volcanic rises. Downwelling models need weak crust to allow deformation to occur within a reasonable timescale of $\sim 10^8$ yr (Lenardic *et al.*, 1995; Kidder and Phillips, 1996).

The surfaces of crustal plateaus consist of deformed tessera terrain, as described above. In some cases, the deformation pattern is either radial or circumferential, implying that it is associated with the formation of the plateau itself. These features, which include both radial extension and circumferential compression, have been used to try to distinguish between upwelling and downwelling models. Evidence for an early extensional phase of deformation in a thin brittle layer has been cited as support for the hypothesis of mantle upwelling (Hansen and Willis, 1998; Ghent and Hansen, 1999; Hansen et al., 2000). Conversely, others have interpreted the stratigraphic and strain relationships to indicate that compression occurred first, consistent with downwelling (Gilmore et al., 1998; Gilmore and Head, 2000; Marinangeli and Gilmore, 2000). In addition, there are many lineaments that are not easily associated with specific formation mechanisms.

The composition of crustal plateaus is also uncertain. Some have suggested that they are composed of more siliceous crust, similarly to terrestrial continents (Basilevsky et al., 1986). The ability of a planet to produce more evolved crust may be a function of the water content. Campbell and Taylor (1983) argue that the lack of water in the crust of Venus precludes the formation of granitic crust. However, Venus may have been wetter early in its history (Donahue et al., 1997), and thus it might still be possible to produce felsic igneous rocks comparable to the cores of Archean continents (Jull and Arkani-Hamed, 1995). Alternatively, crustal plateaus may simply be thicker blocks of basaltic crust (Bindschadler and Head, 1991; Solomon et al., 1992). The high elevations of many plateaus and their apparent formation during a period of higher heat flow argue for a relatively strong and thus low silica content crust (Smrekar and Solomon, 1992; Nunes et al., 2004).

2.4.2.1 Ishtar Terra Ishtar Terra resembles an Earth continent more than any other feature on Venus. It is a very complex feature that includes a large central plain (Lakshmi Planum) bordered by mountain belts and a massif (Freyja, Akna, Danu, and Maxwell Montes), or by scarps. Distal to the scarps and montes is a complex of terrains, mostly tessera, that includes Fortuna, Itzpapalotl, Atropos, Moira, and Clotho Tesserae (Kaula et al., 1992). The entire complex of terrains making up Ishtar Terra is bounded by $\sim$55–78° north latitude and $\sim$305–60° (through 0°) east longitude, an area comparable to that of Australia.

Lakshmi Planum is on an approximately Tibet-sized plateau that rises 3–4 km above the median planetary radius. Most of this plateau is surfaced by putative volcanic flows similar to those surfacing Venusian plains (Gaddis and Greeley, 1990). Several caldera-like depressions occur on Lakshmi Planum, including two large ones: Colette with a mean diameter of $\sim$150 km, and Sakajawea with a mean diameter of $\sim$235 km. These calderas may have been the source for many of the volcanic flows surfacing Lakshmi Planum, but it is not possible to rule out now-buried fissures as a source. Over most of Lakshmi Planum the plateau surface is nearly featureless, but the highest areas of the plateau expose tessera terrain that is inferred to be inliers of a basement beneath the flows that surface most of the plateau.

Lakshmi Planum is almost completely encircled by mountains: Danu Montes to the south, Akna Montes to the west, Freyja Montes to the north, and Maxwell Montes to the east. Between Danu Montes and Akna Montes the plateau is bordered by a steep scarp, Vesta Rupes. There also are scarps bounding the plateau between Danu and Maxwell Montes, and between Freyja and Maxwell Montes. Danu Montes consists of a long, narrow belt of ridges and valleys, inferred to be folds, that extends for >1000 km in an arc, concave towards the plateau, and with elevations up to ~1.5 km above the plateau surface. Akna and Freyja Montes are broader belts of ridges and valleys: Akna rises 2–3 km above the plateau, and Freyja rises 3–4 km above the plateau. Akna Montes defines a gentle arc convex to the plateau that extends for a distance of ~600 km along the plateau border. Freyja Montes defines an arc concave to the plateau that extends for a distance of ~900 km along the plateau border. Akna and Freyja Montes are morphologically similar to many of the ridge belts found on the plains of Venus; they differ mainly in being at a much higher elevation, and in being associated with a large plateau rather than lowland plains. Akna and Freyja Montes have length:width ratios generally <4, in contrast to ratios >10 for Earth mountain belts. This has been interpreted to mean that these mountain belts are not formed by plate tectonic processes (Hansen *et al.*, 1997).

The eastern border of Lakshmi Planum is Maxwell Montes, a massif that is ~850 km long in roughly NW–SE orientation, and ~700 km wide. The nearly equidimensional plan-view shape of Maxwell Montes is unlike the shapes of typical mountain belts on Earth, hence "mountain massif" is more appropriate than "mountain belt." The top of Maxwell Montes is the highest place on the planet, rising to an elevation of ~11 km above the median planetary radius. The north, west, and south margins of Maxwell are steep scarps, with slopes approaching 30° in places (Smrekar and Solomon, 1992). The interior structure of Maxwell Montes is dominated by ridges and valleys trending NW, approximately parallel to the length of the massif as a whole. Individual ridges vary in length from 150 to as much as 500 km; ridge spacing is widely variable, ranging from ~2 to as much as 25 km, with most ridges separated by 6–12 km (Keep and Hansen, 1994). In addition to the ridges and valleys, deformation features in Maxwell Montes include penetrative north-trending lineaments, and grabens trending northeast and northwest. The ridge topography on the relatively gentle east slope of Maxwell Montes grades gradually into the topography of Fortuna Tessera to the east.

As with crustal plateaus in general, models developed to explain the landforms and structures present in Ishtar Terra include either upwelling (Grimm and Phillips, 1991) or downwelling (Bindschadler *et al.*, 1992). At Ishtar, it has been proposed that the surrounding mountain belts formed via horizontal compression directed inward from all sides (Roberts and Head, 1990) or by deformation in a brittle upper crust that is detached from the lower crust and mantle (Keep and Hansen, 1994). Specifically, the ridges and troughs on Maxwell Montes have been explained as due to very shallow, brittle deformation of the upper crust which is inferred to be mechanically detached from a much-thickened lower crust forming a keel that supports the overall topography of Ishtar Terra (Keep and Hansen, 1994). This model interprets the ridge and trough topography of Maxwell Montes as ribbon terrain (Hansen and Willis, 1998), similar to the terrain found in Fortuna Tessera immediately east of Ishtar Terra. The implication is that the kilometer-scale deformation of Maxwell Montes is independent of the much larger-scale processes responsible for the elevation of this region. Gravitational collapse of the plateau margins has been proposed to account for the many extensional structures found in various parts of Maxwell and Danu Montes, which appear to be younger than the folds of the mountain belts (Smrekar and Solomon, 1992).

2.5 *Coronae*

Coronae are circular to elliptical features (Figure 3.4), typically surrounded by a ring of concentric ridges and/or fractures (Basilevsky *et al.*, 1986; Squyres *et al.*, 1992; Stofan *et al.*, 1997). They were first identified in Venera 15/16 images of Venus (Basilevsky *et al.*, 1986), although enigmatic large circular features had been identified previously in Earth-based radar data (Campbell and Burns, 1980). Coronae range in size from about 50–2600 km, with a mean of about 257 km (if the anomalously large 2600 km diameter Artemis is excluded) (Glaze *et al.*, 2002). Two types of coronae have been identified: 406 Type 1 coronae, defined as features having at least 180° of annular fractures; and 107 Type 2 coronae, with <180° of annular fractures and a topographic rim. Many Type 2 coronae completely lack annular fractures, and were identified solely on the basis of their topographic signature (Stofan *et al.*, 2001). Coronae have been classified into ten topographic groups, with about 50% of coronae topographically higher than the surrounding region, and 39% forming depressions (Smrekar and Stofan, 1997). Almost all coronae exhibit some evidence of volcanic activity, including small to intermediate volcanoes, and/or flows in their interiors and surrounding volcanic flow deposits. In addition to the annulus surrounding most coronae, coronae can also have radiating fractures, some possibly related to dikes.

The distribution of coronae is non-random, with a high concentration in the Beta-Atla-Themis region (Squyres *et al.*, 1993). Coronae are located on topographic rises (13%), along chasmata (68%), and as isolated features in the plains (18%) (Stofan *et al.*, 2001). One exception to the general association of coronae with extension and/or upwelling is a group of coronae and corona-like features in the Bereghinya Planitia (V-8) quadrangle that are located in plains cut by wrinkle ridges, some of which radiate from the coronae and corona-like features (McGill, 2004a). Coronae along chasmata have large amounts of associated volcanism (Roberts and Head, 1993), although coronae with little associated volcanism also occur along chasmata (Martin *et al.*, in review). Due to the similarity in size and topographic signature of Type 1 and Type 2 coronae, they have been interpreted as likely to have a similar mode of origin (Stofan *et al.*, 2001; Glaze *et al.*, 2002). Type 2 coronae are somewhat more likely to be found as isolated features in the plains (56%), with 43% along fracture belts and chasmata, and only 1% located on tesserae (Stofan *et al.*, 2001).

When coronae were first identified it was thought that they might be impact craters, or sites of volcanic and tectonic activity resulting from an impact. However, the distribution of coronae across the surface of Venus is non-random. Coronae tend to be concentrated along belts of fractures or chasmata. Also, the size distribution of coronae is not consistent with that of impact craters (Stofan *et al.*, 1992). Therefore, it was concluded that coronae must originate from endogenic processes. Another early suggestion was that the features might be similar to ring dikes, which they resemble. However, the arcuate ridges of terrestrial ring dikes are produced by differential erosion; erosional processes on Venus are negligible, as discussed earlier, due to the absence of water.

Most workers have concluded that coronae must originate from buoyant material rising from within the planet, with buoyancy resulting from thermal or compositional effects, or both (Stofan *et al.*, 1991; Janes *et al.*, 1992; Koch and Manga, 1996; Smrekar and Stofan, 1997; Hansen, 2003). The size range of coronae (50–2600 km diameter) indicates that the buoyant plumelets or diapirs must result from relatively shallow depths within the mantle of the planet. However, despite the general consensus on the origin of coronae, many questions remain. The features have an extremely broad range in size and vary greatly in morphology, from 2-km deep depressions to 4-km high plateaus. Other modes of origin for some coronae, such as sinking rather than rising mantle diapers (Hoogenboom and Houseman, 2005), cannot be ruled out, but the variations also could result from the nature of the upwelling (e.g., size, buoyancy, longevity), stage of evolution, and/or local controlling factors (lithospheric thickness, strength, etc.).

2.6 Chasmata

Chasmata are major rift systems (Figure 3.4) located primarily in the equatorial region of Venus; the terms "chasma" and "rift" are used interchangeably here (chasma means canyon). They are characterized by troughs hundreds to thousands of kilometers long, and relief of up to 7 km (Solomon et al., 1992). The troughs are overprinted by intense fracturing, which typically extends beyond the trough. Some of the fractures can be clearly identified as graben, but most are bright lineaments interpreted to be normal faults or narrow graben. There are five major chasmata: Devana, Ganis, Diana/Dali, Hecate and Parga (Figure 3.4); and a number of smaller rift systems, including Ix-Chel Chasma, Guor Linea (Figure 3.5), Juno Chasma, Kalaipahoa Linea, Kuanja Chasma, Morrigan Linea, and Virtus Linea. Devana, Guor, Virtus, and Ganis are associated with volcanic rises, while other chasmata such as Parga and Hecate cut across the plains of Venus. The major chasmata have deep, wide (~80 km) troughs and wide (~200 km) zones of faulting, while minor chasmata tend to have less deep troughs and are transitional to fracture belts. Phillips and Hansen (1994) divided chasmata into symmetric and asymmetric classes, with the latter being bounded on one side by a ridge that is as high as the trough is deep. Asymmetric chasmata are generally deeper than symmetric chasmata.

On the plains, chasmata typically have associated coronae, volcanoes, and volcanic flows (Hansen and Phillips, 1993; Hamilton and Stofan, 1996). Most of the coronae on chasma systems formed synchronously with extension, although some coronae predate and others postdate rifting (Hamilton and Stofan, 1996; Hoogenboom et al., 2005). No clear age progression of coronae along rifts has been observed (Hamilton and Stofan, 1996; Hansen and DeShon, 2002; Brian et al., 2004b). A few minor rift systems located near regions of tessera (Ix-Chel Chasma, Kuanja Chasma) lack coronae (Bleamaster and Hansen, 2004), and chasmata at volcanic rises also tend to have fewer coronae than those on the plains. Most of the trough systems appear to be younger than the plains units that surround them, but geologic units and relative ages are often difficult to determine owing to the intensity of the fracturing (Hansen and DeShon, 2002; Brian et al., 2004b).

Examples of features that have been cut by rift faults can be found along different chasma systems. The most dramatic example is the crater Somerville, which is split by Devana Chasma in Beta Regio (Solomon et al., 1992). The morphology, association with a "hotspot"-geologic setting, and estimated elastic thickness associated with Devana Chasma are similar to those of the rifts in the East African rise (Foster and Nimmo, 1996). Several volcanoes have been rifted apart along the Hecate Chasma rift (Stofan et al., 2005). Individual faults along the rifts have also been analyzed, giving an average slope of 36° (Conners, 1995; Conners and Suppe, 2001). This slope value suggests that most rift flanks are talus slopes, indicating that the effective cohesive strength of the crust at most fault surfaces is low (Conners and Suppe, 2001). Crustal extension at rifts estimated using the split crater Somerville as well as faults at Devana Chasma is small, on the order of 10–20 km, or 0.3 average strain (Solomon et al., 1992; Rathbun et al., 1999; Connors and Suppe, 2001).

The topography of many rift segments and corona outer rises can be fit with a model of an elastically bending plate. These studies produce a large range of elastic thickness estimates, ranging from 0 to over 100 km (Sandwell and Schubert, 1992; Johnson and Sandwell, 1994). Brown and Grimm (1996) also found a very low heat flow estimate, 4 K/km, at Artemis Chasma, equivalent to a large elastic thickness. Some of the large elastic thickness estimates may be inaccurate if plastic bending has occurred.

Chasmata at volcanic rises are likely related to uplift associated with an underlying plume. Models for the origin of chasma systems on the plains range from extension zones (Schaber, 1982) to crustal spreading (Head and Crumpler, 1987) to subduction zones (McKenzie *et al.*, 1992; Sandwell and Schubert, 1992). The general morphology of rift zones in the plains and their association with coronae and volcanism indicate that they are most likely to be extension zones accompanied by upwelling (Hansen and Phillips, 1993; Hamilton and Stofan, 1996). In addition, it has been suggested that the association of large flow fields and coronae with chasmata indicates decompression melting associated with regional-scale lithospheric extension (Magee and Head, 2001).

3 Models

3.1 Geological models

Early analysis of the Magellan data suggested a first-order similarity in the structural evolution of widely separated plains regions (Solomon *et al.*, 1992), an observation thoroughly documented by Basilevsky and Head (2000). Solomon *et al.* (1992) left open a key question: does this similarity imply a globally coherent history of material emplacement and deformation, or are we seeing a sequence that has been repeated at different times in different places? To a large extent, this question has polarized the discussion of tectonics and crustal evolution on Venus.

In a series of papers, Basilevsky and Head (1995, 1998, 2000) and Basilevsky *et al.* (1997) have proposed and defended the hypothesis that the similarity in sequence implies the existence of a globally correlated stratigraphic and tectonic history. Their model also infers a global evolution of volcanic styles. These authors explicitly address the question stated above (Basilevsky and Head, 2000), noting that there are three possible answers: (1) similar sequences of events occurred simultaneously everywhere, (2) similar sequences occurred at different times at different places, and (3) the situation is intermediate between these extremes. Guest and Stofan (1999) define option 1 as the "directional" model, and option 2 as the "non-directional" model. Basilevsky and Head (2000) conclude that the first option is correct, whereas Guest and Stofan (1999) conclude that the second option is more likely for the recent history of Venus, although there are elements in the geologic history of Venus suggesting a rough directionality. The authors of this chapter favor the third option.

The Earth provides some useful and thought-provoking examples of how a global interpretation of similar geologic relationships can lead one astray. As has been noted by others (e.g., Hansen and Willis, 1996), efforts to define a superposition-based global geology on Earth led to the inference that all highly metamorphosed "basement" rocks were old and globally correlative. The advent of radiometric dating destroyed this tempting simplification by demonstrating that metamorphic basement rocks can range in age from early Archean to Cenozoic; that is, over almost all of the preserved geologic history of the planet. Likewise, Archean greenstone belts commonly exhibit similar sequences of volcanic and sedimentary units globally, but again radiometric dating proves that these sequences are of different ages in different places. Basically, global similarities in sequence represent the repeated operation of processes that are favored by the tectonic style of the planet. These observations suggest that the directional model over-simplifies the crustal evolution of Venus, and we believe that it does, in fact, gloss over regional and local differences in the relative ages of some structures and materials. On the other hand, one must be careful not to assume that Venus has behaved like the Earth; that is, an otherwise Earth-like planet that lacks water and plate tectonics may well manifest volcanic and tectonic processes on a more nearly global scale than does Earth.

A specific example of the controversy surrounding the crustal evolution of Venus centers on a map unit referred to as "shield plains." Basilevsky and Head (2000) place this unit in their global stratigraphic and structural sequence as younger than fracture belts and older than "plains with wrinkle ridges." This stratigraphic position is used as partial evidence for a roughly monotonic change in volcanic styles from the formation of large clusters of small shield volcanoes, to voluminous plains-forming volcanism, and finally to the formation of large central volcanoes and associated flows. But there is abundant evidence that not all shield plains are older than the widespread flows making up the areally dominant plains of Venus (Guest and Stofan, 1999). Addington (2001) carried out a detailed study of the relative ages of shields, associated shield plains, and regional plains for 179 small shield clusters within seven U.S. Geological Survey 1:5 M scale quadrangles representing ~11% of the surface of Venus and selected to represent areas of both high and low volcanic feature concentration. Nearly half (47%) of these shield clusters yield ambiguous relative ages of shield plains and regional plains. Shield fields at least partially younger than regional plains make up ~42% of the population, whereas those at least partially older than regional plains make up only ~10% of the total population. Ivanov and Head (2004) carried out a similar study, but their results were that ~69% of shield fields predate regional plains, ~11% are the same age as regional plains, and ~8% are younger than regional plains, leaving only 12% considered to be ambiguous. This startling difference is at least partially due to the two studies using different criteria. It seems likely that some of Addington's ambiguous cases could be reinterpreted to be examples of shield plains older than regional plains based on some criteria employed by Ivanov and Head (2004). On the other hand, some of the criteria used by Ivanov and Head (2004) are probably not valid. For example, a spatial association of shield plains with an older unit may suggest an age older than regional plains, but does not require that this be true. Also, a difference in radar backscatter between shield plains and regional plains carries no obvious relative age information. Finally, the presence of wrinkle ridges on shield plains indicates that both regional plains and shield plains are older than the wrinkle ridges, but it implies nothing about the relative ages of regional and shield plains. It is possible that Addington overstated the number of ambiguous cases at the expense of examples of shield plains older than regional plains, and that Ivanov and Head overstated the number of shield plains older than regional plains because some of their criteria are not valid. The difference in the results of the two studies is thus likely less extreme than as published, but it is clear from both that shield plains are not all older than regional plains, and thus the monotonic evolution of volcanic styles that is embedded in the directional model cannot be rigorously true; that is, the truth probably lies somewhere between the strictly directional and strictly non-directional models.

The directional model also interprets all ridge belts to be older than regional plains (Basilevsky et al., 1997). For some ridge belts, the evidence for this age assignment is very robust, especially in Atalanta/Vinmara and Lavinia Planitiae, where ridge belts are very abundant and prominent (Squyres et al., 1992; Rosenberg and McGill, 2001; Ivanov and Head, 2001; McGill, 2003). However, one unusual ridge belt in Vinmara Planitia (Laūma Dorsa) is possibly younger than adjacent regional plains (Rosenberg and McGill, 2001). For most ridge belts, however, the evidence for age relative to regional plains is ambiguous. The cause of this common ambiguity is simple: although SAR images may suggest strongly that plains materials are embaying the ridges, the same appearance could be due to the ridges consisting of folded plains materials. The fold ridges will have different backscatter than the adjacent plains, even if they consist of folded plains materials, because the local incidence angle will be smaller than nominal on ridge flanks facing towards the radar, and larger than nominal on ridge flanks facing away from the radar. This expectation suggests that it may be possible to constrain the age of ridge-belt materials relative to adjacent plains materials using radar properties. As is true for more traditional geologic methods, the radar properties can determine if ridges are unlikely to be folded plains more robustly than the inverse; that is, the best one can do is determine that radar properties permit a ridge belt to consist of folded plains materials, but they can never prove that this is true. For most ridge belts, the radar properties are not consistent with ridges formed of folded regional plains materials (McGill and Campbell, 2006), and thus the tentative conclusion is that almost all ridge belts are older than regional plains. This conclusion is consistent with the directional model if we allow for rare exceptions, such as possibly Laūma Dorsa.

The directional model does not entirely separate structures from material units, resulting in some confusion and also some loss of kinematic information (e.g., Hansen, 2000). This is a particular problem for wrinkle ridges, which in the directional model are used to define a "plains with wrinkle ridges" stratigraphic unit. This implies that the wrinkle ridges are essentially the same age as the plains materials they cut, and are genetically related to plains emplacement. Very likely, this linkage of wrinkle-ridge formation to plains emplacement is responsible for citing superposition of wrinkle ridges on shield plains materials (discussed above) as evidence that the shield plains materials are the same age or older than regional plains materials. But there is abundant evidence that wrinkle ridges are younger structures superposed on plains and thus not directly related to plains emplacement. Much of this evidence derives from individual U.S. Geological Survey 1:5M scale quadrangle maps and from relative ages of impact craters and wrinkle ridges, as reviewed by McGill (2004b). As argued by McGill (2004b), it is likely that most wrinkle ridges did form in a relatively short interval of time on the order of 100 m.y. after emplacement of the regional plains. It is clear, however, that wrinkle-ridge formation continued after emplacement of younger units (e.g., Guest and Stofan, 1999; McGill, 2004b), an important aspect of Venusian crustal history that is lost by maintaining that all wrinkle ridges formed immediately following emplacement of regional plains and that no younger units are cut by wrinkle ridges (cf. Basilevsky *et al.*, 1997, Fig. 3.10). The global pattern of wrinkle ridges (McGill, 1993; Bilotti and Suppe, 1999) suggests that both spatial and temporal variations in wrinkle-ridge development are involved, a complexity that is not compatible with the extreme directional model.

The evidence presented here bearing on the directional vs. non-directional controversy is not exhaustive, but this evidence favors a Venus that exhibits Earth-like repetition of stratigraphic and structural sequences that differ in age from place to place, superposed on a broad global evolution of tectonic style that may reflect a general thickening of the lithosphere over the preserved geologic record.

3.2 Geophysical models

Geophysical models seek to address how Venus can be very much like Earth in its size and bulk density and thus in heat content, but dissimilar from Earth in terms of its tectonic style. Earth loses the majority of its heat through plate tectonics, and subduction in particular (e.g., Schubert, 1992). Some chasmata have been proposed to be retrograde subduction zones (Sandwell and Schubert, 1992; Schubert and Sandwell, 1995). As discussed above, this hypothesis is controversial, and the potential subduction zones, if they actually exist (cf. Hansen and Phillips, 1993), are not as extensive as those on Earth. Nor is simple advection of material and formation of such features as hot spots able to account for a very large part of Venus' heat budget (Smrekar and Parmentier, 1996; Schubert *et al.*, 1997). In addition to the question of how Venus loses its heat, a fundamental question is what process caused global resurfacing such that there are no large-scale areas of older terrain? An important aspect of the resurfacing history is that few craters are modified by volcanism or tectonism. There is not a strong constraint on the rate of resurfacing, as the implied rate is a function of the size of the area resurfaced in a given event (Phillips *et al.*, 1992). Resurfacing relatively small areas (hundreds of km^2) best fits the geologic observations and statistical constraints (Phillips *et al.*, 1992). Stofan *et al.* (2005) suggest that resurfacing is complex and occurs on a smaller scale than proposed by earlier workers.

Most hypotheses proposed to explain these fundamental differences between Venus and Earth focus on the consequences of the lack of water on Venus. The current low water content of the atmosphere and the high surface temperature imply that the crust is quite dry. This lack of water results in the crust of Venus being stronger than that on Earth, despite the higher temperature (Mackwell *et al.*, 1998). The mantle on Venus also appears to be dry. The primary evidence for a dry mantle is the apparent lack of an asthenosphere. The very large gravity anomalies over hotspots on Venus, as compared to those on Earth, indicate that the asthenosphere decouples the rising plume heads from the overlying lithosphere on Earth (Robinson and Parsons, 1988) but not on Venus (Kiefer and Hager, 1991; Smrekar and Phillips, 1991). The absence of an asthenosphere may also facilitate the formation of coronae on Venus, since the asthenosphere on Earth may shield the lithosphere from rising plume heads (Smrekar and Stofan, 1997).

The strength of the crust and lithosphere, resulting from a lack of water, may also be responsible for Venus having no plate tectonics. Numerous studies have explored the model of stagnant-lid convection for Venus and other planets (Moresi and Solomatov, 1998; Reese *et al.*, 1999; Solomatov and Moresi, 2000). These convection models simulate the effect of the lithosphere on the style of convection. When the yield strength of the lithosphere is sufficiently large, neither surface plates nor subduction can develop (Richards *et al.*, 2001). A dry crust and lithosphere and thus high yield strength on Venus are consistent with a stagnant-lid regime.

The absence of plate tectonics and associated subduction zones has been proposed to affect the overall planform of convection. Jellinek *et al.* (2002) proposed that large mantle plumes would not form on Venus in the absence of plate tectonics. They proposed that cold slabs are needed to create a sufficient temperature gradient at the core–mantle boundary to initiate plume formation. Under this hypothesis, the large hotspots on Venus must have formed earlier in Venus' history, prior to the establishment of a stagnant-lid regime. This contradicts the evidence from gravity data suggesting that many of the large hotspots are currently active.

Although indirect evidence suggests that the crust and mantle on Venus are presently dry, this may not always have been the case. The D/H ratio indicates that Venus once had at least a factor of 100 times more water than it does currently (e.g., Taylor *et al.*, 1997; Donahue *et al.*, 1997). The current amount of water in the atmosphere is equivalent to a global layer 2–10 cm thick (Yung and DeMore, 1999). Given the amount of water lost, Venus may once have had a volume of water equivalent to a global layer ~10 m thick. Exactly when Venus lost its water is not known. Modeling of hydrodynamic escape of hydrogen to space, following photolysis of water to hydrogen, indicates that water could have been lost via this process in 100–1000 m.y. (Kasting and Pollack, 1983). Thus, it is possible that granitic crust may have formed earlier in Venus' evolution.

The current lack of water on Venus may explain the absence of plate tectonics, but the question of global resurfacing remains. Numerous hypotheses have been proposed. Many have to do with cooling of the lithosphere, core or mantle. Cooling of the lithosphere results in a reduction in the rate of geologic activity over time (Solomon, 1993) and an increase in viscosity (Arkani-Hamed *et al.*, 1993; Arkani-Hamed, 1994; Grimm, 1994). Cooling of the mantle has also been proposed to increase lithospheric buoyancy and inhibit plate tectonics (Herrick, 1994). Although cooling of either the lithosphere or mantle would make it more difficult for volcanism and tectonic deformation to occur, the rate of cooling is probably inconsistent with the rate at which activity declined. Parmentier and Hess (1992) proposed that resurfacing occurred by foundering of the lithosphere caused by an increase in its chemical and thermal density due to cooling of the lithosphere and of a thick layer of mantle residuum. The formation of entirely new crust and lithosphere follows lithospheric foundering.

Another hypothesis is that stagnant-lid convection itself leads to a reduction in the amount of tectonic activity (Schubert *et al.*, 1997), or else to volcanic resurfacing of the planet. Because the stagnant lid reduces the amount of heat lost relative to convection with plate tectonics, the transition from plate tectonics to stagnant-lid convection causes the mantle temperature to increase. This increase in the temperature could result in much larger volumes of pressure release melting, thus resurfacing the planet (Reese *et al.*, 1999). Alternatively, a phase transition within the mantle could have triggered a global mantle avalanche from a two-layer mantle regime to whole mantle convection, allowing for complete resurfacing (Steinbach *et al.*, 1993; Steinbach and Yuen, 1994).

4 Conclusions

As is clear from the many uncertainties discussed in this chapter, Venus presents us with some difficult problems. From a geological point of view, the inability to resolve age differences with needed precision is perhaps the most serious obstacle to understanding the crustal history of Venus. This limitation stems both from the lack of impact craters less than about 2 km in diameter, and from the apparently extreme slowness of erosion. The practical effect of this is that similar stratigraphic and structural sequences could share a common age or else be separated in time by hundreds of millions of years; available data do not permit distinguishing between these possibilities.

The slowness of erosion results in the preservation of a bewildering array of structural features. We may well be looking at structures accumulated over hundreds of millions of years, a situation unlike that faced on Earth where structures are buried, destroyed, or overprinted by younger processes. Venus provides us with the potential to sort out long spans of structural history if we could reliably determine the kinematics. To date, this has not been done to any significant extent.

From a broader perspective, how does the apparent lack of plate tectonics and the extreme dryness of the crust affect the fundamental internal processes that determine Venus' tectonic style? Many models have been proposed for heat-loss mechanisms on Venus, but the internal workings of the planet remain largely unknown. The lack of time constraints on crustal history contributes to this unsatisfactory understanding of internal processes.

Another major issue is the apparent youth of preserved crustal materials and features. Although not well constrained, the probable age of the oldest terrains on Venus is of the order of 10^9 years. What was Venus like during the earlier 80% of solar system history? In particular, was it wetter and thus more conducive to the evolution of granite and related felsic rocks? What processes destroyed or buried all older materials? Are all materials from earlier times completely recycled, or are they present as a "basement" beneath the surface materials we see now? Is tessera terrain a remnant of this putative older basement? This last seems an unlikely model because tessera terrain does not preserve abundant large, old craters.

Many of these outstanding issues could be effectively addressed by future missions. The most desirable missions would include landers capable of collecting *in situ* seismic data to determine the internal structure of Venus; landers and rovers capable of collecting chemical and mineralogical data critical to inferring rock origin, data of special relevance to our need to understand the origin of tessera terrain; and sample-return landers that would provide rocks for analysis and radiometric dating on Earth. Other potentially useful missions could use balloons to collect high-resolution SAR or optical images, some of which could provide much improved topographic control. Also useful would be descent modules and landers capable of improving our understanding of the isotope chemistry of the atmosphere.

Venus is an important target for solar system exploration because it is our sister planet. We must understand how two very similar planets could evolve along such diverse paths.

Acknowledgments

Helpful reviews were provided by Francis Nimmo and Bruce Campbell.

References

Addington, E. A. (2001). A stratigraphic study of small volcano clusters on Venus. *Icarus*, **149**, 16–36.
Anderson, E. M. (1951). *The Dynamics of Faulting and Dyke Propagation with Applications to Britain*. Edinburgh: Oliver and Boyd.
Anderson, F. S. and Smrekar, S. E. (1999). Tectonic effects of climate change on Venus. *J. Geophys. Res.*, **104**, 30 743–30 756.
Arkani-Hamed, J. (1994). On the thermal evolution of Venus. *J. Geophys. Res.*, **99**, 2019–2033.
Arkani-Hamed, J., Schaber, G. G., and Strom, R. G. (1993). Constraints on the thermal evolution of Venus inferred from Magellan data. *J. Geophys. Res.*, **98**, 5309–5315.
Banerdt, W. B., McGill, G. E., and Zuber, M. T. (1997). Plains tectonics on Venus. In *Venus II*, ed. S. W. Gougher, D. M. Hunten and R. J. Phillips. Tucson, AZ: University of Arizona Press, pp. 901–930.
Basilevsky, A. T. (1994). Concentric wrinkle ridge pattern around Sif and Gula (abs.). *Lunar Planet. Sci. Conf. XXV*, 63–64.
Basilevsky, A. T. and Head, J. W. (1995). Global stratigraphy of Venus: Analysis of a random sample of thirty-six test areas. *Earth, Moon, and Planets*, **66**, 285–336.
Basilevsky, A. T. and Head, J. W. (1998). The geologic history of Venus: A stratigraphic view. *J. Geophys. Res.*, **103**, 8531–9544.
Basilevsky, A. T. and Head, J. W. (2000). Geologic units on Venus: Evidence for their global correlation. *Planet. Space Sci.*, **48**, 75–111.
Basilevsky, A. T., Pronin, A. A., Ronca, L. B., Kryuchkov, V. P., Sukhanov, A. L., and Markov, M. S. (1986). Styles of tectonic deformation on Venus: Analysis of Venera 15 and 16 data. *Proc. Lunar Planet. Sci. Conf. 16, J. Geophys. Res.*, **91**, D399–D411.

Basilevsky, A. T., Head, J. W., Schaber, G. G., and Strom, R. G. (1997). The resurfacing history of Venus. In *Venus II*, ed. S. W. Gougher, D. M. Hunten and R. J. Phillips. Tucson, AZ: University of Arizona Press, pp. 1047–1084.

Bilotti, F. and Suppe, J. (1999). The global distribution of wrinkle ridges on Venus. *Icarus*, **139**, 137–159.

Bindschadler, D. L. and Head, J. W. (1991). Tessera terrain, Venus: Characterization and models for origin and evolution. *J. Geophys. Res.*, **96**, 5889–5907.

Bindschadler, D. L. and Parmentier, E. M. (1990). Mantle flow tectonics: The influence of a ductile lower crust and implications for the formation of topographic uplands on Venus. *J. Geophys. Res.*, **95**, 21 329–21 344.

Bindschadler, D. L., Schubert, G., and Kaula, W. M. (1992). Coldspots and hotspots: Global tectonics and mantle dynamics of Venus. *J. Geophys. Res.*, **97**, 13 495–13 532.

Bleamaster, L. F. and Hansen, V. L. (2004). Effects of crustal heterogeneity on the morphology of chasmata, Venus. *J. Geophys. Res.*, **109**, doi:10.1029/2003JE002193.

Brian, A. W., Stofan, E. R., Guest, J. E., and Smrekar, S. E. (2004a). Laufey Regio: A newly discovered topographic rise on Venus. *J. Geophys. Res.*, **109**, doi:10.1029/2002JE002010.

Brian, A. W., Guest, J. E., and Stofan, E. R. (2004b). Geologic map of the Taussig Quadrangle (V39), Venus. U.S. Geol. Surv. Geol. Invest. Ser. SIM-2813.

Brown, C. D. and Grimm, R. E. (1996). Lithospheric rheology and flexure at Artemis Chasma, Venus. *J. Geophys. Res.*, **101**, 12 697–12 708.

Bryan, W. B. (1973). Wrinkle ridges as deformed surface crust on ponded mare lava. *Proc. Lunar Planet. Sci. Conf. 4*, 93–106.

Bullock, M. A. and Grinspoon, D. H. (1996). The stability of climate on Venus. *J. Geophys. Res.*, **101**, 7521–7530.

Campbell, B. A. (1999). Surface formation rates and impact crater densities on Venus. *J. Geophys. Res.*, **104**, 21 952–21 955.

Campbell, D. B. and Burns, B. A. (1980). Earth-based radar imagery of Venus. *J. Geophys. Res.*, **85**, 8271–8281.

Campbell, D. B., Head, J. W., Harmon, J. K., and Hine, A. A. (1984). Venus volcanism and rift formation in Beta Regio. *Science*, **226**, 167–170.

Campbell, I. H. and Taylor, S. R. (1983). No water, no granites – no oceans, no continents. *Geophys. Res. Lett.*, **10**, 1061–1064.

Connors, C. (1995). Determining heights and slopes of fault scarps and other surfaces on Venus using Magellan radar. *J. Geophys. Res.*, **100**, 14 361–14 382.

Conners, C. and Suppe, J. (2001). Constraints on magnitude of extension on Venus from slope measurements. *J. Geophys. Res.*, **106**, 3237–3260.

Donahue, T. M. and Russell, C. T. (1997). The Venus atmosphere and ionosphere and their interaction with the solar wind: An overview. In *Venus II*, ed. S. W. Bougher, D. M. Hunten and R. J. Phillips. Tucson, AZ: University of Arizona Press, pp. 3–31.

Donahue, T. M., Grinspoon, D. H., Hartle, R. E., and Hodges, R. R. Jr. (1997). Ion/neutral escape of hydrogen and deuterium: Evolution of water. In *Venus II*, ed. S. W. Bougher, D. M. Hunten and R. J. Phillips. Tucson, AZ: University of Arizona Press, pp. 385–414.

Ernst, R. E., Grosfils, E. B., and Mège, D. (2001). Giant dike swarms: Earth, Venus, and Mars. *Annu. Rev. Earth Planet. Sci.*, **29**, 489–534.

Foster, S. and Nimmo, F. (1996). Comparisons between rift systems of East Africa, Earth and Beta Regio, Venus. *Earth Planet. Sci. Lett.*, **143**, 183–195.

Gaddis, L. R. and Greeley, R. (1990). Volcanism in northwest Ishtar Terrs, Venus. *Icarus*, **87**, 327–338.

Ghent, R. and Hansen, V. (1999). Structural and kinematic analysis of eastern Ovda Regio, Venus: Implications for crustal plateau formation. *Icarus*, **139**, 116–136.

Gilmore, M. S. and Head, J. W. (2000). Sequential deformation of plains at the margin of Alpha Regio, Venus: Implications for tessera formation. *Meteorit. Planet. Sci.*, **35**, 667–687.

Gilmore, M. S., Ivanov, M. A., Head, J. W., III, and Basilevsky, A. T. (1997). Duration of tessera deformation on Venus. *J. Geophys. Res.*, **102**, 13 357–13 368.

Gilmore, M. S., Collins, G. C., Ivanov, M. A., Marinangeli, L., and Head, J. W. (1998). Style and sequence of extensional structures in tessera terrain, Venus. *J. Geophys. Res.*, **103**, 16 813–16 840.

Glaze, L. S., Stofan, E. R., Smrekar, S. E., and Baloga, S. M. (2002). Insights into corona formation through statistical analyses. *J. Geophys. Res.*, **107**, doi:10.1029/2002JE001904.

Golombek, M. P., Plescia, J. B., and Franklin, B. J. (1991). Faulting and folding in the formation of planetary wrinkle ridges. *Proc. Lunar Planet. Sci. Conf. 21*, 679–693.

Greeley, R., Bender, K. C., Saunders, R. S., Schubert, G., and Weitz, C. M. (1997). Aeolian processes and features on Venus. In *Venus II*, ed. S. W. Bougher, D. M. Hunten and R. J. Phillips. Tucson, AZ: University of Arizona Press, pp. 547–589.

Grimm, R. E. (1994). Recent deformation on Venus. *J. Geophys. Res.*, **99**, 23 163–23 171.

Grimm, R. E. and Phillips, R. J. (1991). Gravity anomalies, compensation mechanisms, and the geodynamics of western Ishtar Terra, Venus. *J. Geophys. Res.*, **96**, 8305–8324.

Grosfils, E. B. and Head, J. W. (1994). The global distribution of giant radiating dike swarms on Venus: Implications for the global stress state. *Geophys. Res. Lett.*, **21**, 701–704.

Guest, J. E. and Stofan, E. R. (1999). A new view of the stratigraphic history of Venus. *Icarus*, **139**, 55–66.

Hamilton, V. E. and Stofan, E. R. (1996). The geomorphology and evolution of Hecate Chasma, Venus. *Icarus*, **121**, 171–194.

Hansen, V. L. (2000). Geologic mapping of tectonic planets. *Earth Planet. Sci. Lett.*, **176**, 527–542.

Hansen, V. L. (2003). Venus diapirs: Thermal or compositional? *Geol. Soc. Am. Bull.*, **115**, 1040–1052.

Hansen, V. L. and Phillips, R. J. (1993). Tectonics and volcanism of eastern Aphrodite Terra, Venus: No subduction, no spreading. *Science*, **260**, 526–530.

Hansen, V. L. and DeShon, H. R. (2002). Geologic map of the Diana Chasma Quadrangle (V37), Venus. U.S. Geol. Surv. Geol. Invest. Ser. I-2752.

Hansen, V. L. and Willis, J. J. (1996). Structural analysis of a sampling of tesserae: Implications for Venus geodynamics. *Icarus*, **123**, 296–312.

Hansen, V. L. and Willis, J. J. (1998). Ribbon terrain formation, southwestern Fortuna Tessera, Venus: Implications for lithosphere evolution. *Icarus*, **132**, 321–343.

Hansen, V. L., Willis, J. J., and Banerdt, W. B. (1997). Tectonic overview and synthesis. In *Venus II*, ed. S. W. Gougher, D. M. Hunten and R. J. Phillips. Tucson, AZ: University of Arizona Press, pp. 797–844.

Hansen, V. L., Phillips, R. J., Willis, J. J., and Ghent, R. R. (2000). Structures in tessera terrain, Venus: Issues and answers. *J. Geophys. Res.*, **105**, 4135–4152.

Head, J. W. and Crumpler, L. S. (1987). Evidence for divergent plate boundary characteristics and crustal spreading on Venus. *Science*, **238**, 1380–1385.

Herrick, R. R. (1994). The resurfacing history of Venus. *Geology*, **22**, 703–706.

Hoogenboom, T. and Houseman, G. A. (2005). Rayleigh-Taylor instability as a mechanism for coronae formation on Venus. *Icarus*, in review.

Hoogenboom, T., Houseman, G. A., and Martin, P. (2005). Elastic thickness estimates for coronae associated with chasmata on Venus. *J. Geophys. Res.*, **110**, doi://10.1029.2004JE002394.

Ivanov, M. A. and Head, J. W. (1996). Tessera terrain on Venus: A survey of the global distribution, characteristics and relation to surrounding units from Magellan data. *J. Geophys. Res.*, **101**, 14 861–14 908.

Ivanov, M. A. and Head, J. W. III. (2001). Geologic map of the Lavinia Planitia Quadrangle (V-55), Venus. U.S. Geol. Surv. Geol. Invest. Ser., I-2684.

Ivanov, M. A. and Head, J. W. (2004). Stratigraphy of small shield volcanoes on Venus: Criteria for determining stratigraphic relationships and assessment of relative age and temporal abundance. *J. Geophys. Res.*, **109**, E10001, doi:10.1029/2004JE002252.

Janes, D. M, Squyres, S. W., Bindschadler, D. L., Baer, G., Schubert, G., Sharpton, V. L., and Stofan, E. R. (1992). Geophysical models for the formation and evolution of coronae on Venus. *J. Geophys. Res.*, **97**, 16 055–16 067.

Jellinek, A. M., Lenardic, A., and Manga, M. (2002). The influence of interior mantle temperature on the structure of plumes: Heads for Venus, tails for the Earth. *Geophys. Res. Lett.*, **29**, doi:1029/2001GL014624.

Johnson, C. L. and Sandwell, D. T. (1992). Joints in venusian lava flows. *J. Geophys. Res.*, **97**, 13 601–13 610.

Johnson, C. L. and Sandwell, D. T. (1994). Lithospheric flexure on Venus. *Geophys. J. Int.*, **119**, 627–647.

Jull, M. G. and Arkani-Hamed, J. (1995). The implications of basalt in the formation and evolution of mountains on Venus. *Phys. Earth Planet. Inter.*, **89**, 163–175.

Kasting, J. F. and Pollack, J. B. (1983). Loss of water from Venus: I. Hydrodynamic escape of hydrogen. *Icarus*, **53**, 479–508.

Kaula, W. M. and Phillips, R. J. (1981). Quantitative tests for plate tectonics on Venus. *Geophys. Res. Lett.*, **8**, 1187–1190.

Kaula, W. M., Bindschadler, D. L., Grimm, R. E., Hansen, V. L., Roberts, K. M., and Smrekar, S. E. (1992). Styles of deformation in Ishtar Terra and their implications. *J. Geophys. Res.*, **97**, 16 085–16 120.

Keep, M. and Hansen, V. L. (1994). Structural history of Maxwell Montes, Venus: Implications for Venusian mountain belt formation. *J. Geophys. Res.*, **99**, 26 015–26 028.

Kidder, J. G. and Phillips, R. J. (1996). Convection-driven subsolidus crustal thickening on Venus. *J. Geophys. Res.*, **101**, 23 181–23 194.

Kiefer, W. S. and Hager, B. H. (1991). A mantle plume model for the equatorial highlands of Venus. *J. Geophys. Res.*, **96**, 20 947–20 966.

Kiefer, W. S. and Hager, B. H. (1992). Geoid anomalies and dynamic topography from convection in cylindrical geometry: Applications to mantle plumes on Earth and Venus. *Geophys. J. Int.*, **108**, 198–214.

Koch, D. M. and Manga, M. (1996). Neutrally buoyant diapirs: A model for Venus coronae. *Geophys. Res. Lett.*, **23**, 225–228.

Komatsu, G. and Baker, V. R. (1994). Plains tectonism on Venus: Inferences from canali longitudinal profiles. *Icarus*, **110**, 275–286.

Lachenbruch, A. H. (1961). Depth and spacing of tension cracks. *J. Geophys. Res.*, **66**, 4273–4292.

Lenardic, W., Kaula, W. M., and Bindschadler, D. L. (1995). Some effects of a dry crustal flow law on numerical simulations of coupled crustal deformation and mantle convection on Venus. *J. Geophys. Res.*, **100**, 16 949–16 957.

Mackwell, S. J., Zimmerman, M. E., and Kohlstedt, D. L. (1998). High-temperature deformation of dry diabase with applications to tectonics on Venus. *J. Geophys. Res.*, **103**, 975–984.

Magee, K. P. and Head, J. W. (2001). Large flow fields on Venus: Implications for plumes, rift associations and resurfacing. *Geo. Soc. Am. Spec. Paper*, **352**, 81–101.

Marinangeli, L. and Gilmore, M. S. (2000). Geologic evolution of the Akna Montes-Atropos Tessera region, Venus. *J. Geophys. Res.*, **195**, 12 053–12 075.

Martin, P., Stofan, E. R., Glaze, L. S., and Smrekar, S. E. Coronae of Parga Chasma, Venus. *Icarus*, in review.

Masursky, H., Eliason, E., Ford, P. G., McGill, G. E., Pettengill, G. H., Schaber, G. G., and Schubert, G. (1980). Pioneer Venus radar results: Geology from images and altimetry. *J. Geophys. Res.*, **85**, 8232–8260.

Maxwell, T. A. (1982). Orientation and origin of ridges in the Lunae Palus – Coprates region of Mars. *Proc. Lunar Planet. Sci. Conf. 13*, A97-A108.

Maxwell, T. A., El Baz, F., and Ward, S. H.(1975). Distribution, morphology, and origin of ridges and arches in Mare Serenitatis. *Geol. Soc. Am. Bull.*, **86**, 1273–1278.

McGill, G. E. (1993). Wrinkle ridges, stress domains, and kinematics of Venusian plains. *Geophys. Res. Lett.*, **20**, 2407–2410.

McGill, G. E. (1994). Hotspot evolution and Venusian tectonic style. *J. Geophys. Res.*, **99**, 23 149–23 161.

McGill, G. E. (1998). Central Eistla Regio: Origin and relative age of topographic rise. *J. Geophys. Res.*, **103**, 5889–5896.

McGill, G. E. (2003). Kinematics of a linear deformation belt: The evolution of Pandrosos Dorsa, Venus (abs.). *Lunar Planet. Sci. Conf. XXXIV*, 1012. Houston, TX: Lunar and Planetary Institute (CD-ROM).

McGill, G. E. (2004a). Geologic map of the Bereghinya Planitia Quadrangle (V-8), Venus. U.S. Geol. Surv. Geol. Invest. Ser., I-2794.

McGill, G. E. (2004b). Tectonic and stratigraphic implications of the relative ages of Venusian plains and wrinkle ridges. *Icarus*, **172**, 603–612.

McGill, G. E. and Campbell, B. A. (2006). Radar properties as clues to relative ages of ridge belts and plains on Venus. *J. Geophys. Res.*, **111**, E12006, doi:10.1029/2006JE002705.

McGovern, P. J. and Solomon, S. C. (1998). Growth of large volcanoes on Venus: Mechanical models and implications for structural evolution. *J. Geophys. Res.*, **103**, 11 071–11 101.

McKenzie, D., Ford, P. G., Johnson, C., Parsons, B., Pettengill, G. H., Sandwell, D., Saunders, R. S., and Solomon, S. C. (1992). Features on Venus generated by plate boundary processes. *J. Geophys. Res.*, **97**, 13 533–13 544.

McKenzie, D., McKenzie, J. M., and Saunders, R. S. (1992). Dike emplacement on Venus and on Earth. *J. Geophys. Res.*, **97**, 15 977–15 990.

McKinnon, W. B., Zahnle, K. J., Ivanov, B. A., and Melosh, H. J. (1997). Cratering on Venus: Models and observations. In *Venus II*, ed. S. W. Gougher, D. M. Hunten and R. J. Phillips. Tucson, AZ: University of Arizona Press, pp. 969–1014.

McLeod, L. C. and Phillips, R. J. (1994). Venusian channel gradients as a guide to vertical tectonics (abs.). *Lunar Planet. Sci. Conf. XXV*, 885–886.

Moreels, P. and Smrekar, S. E. (2003). Identification of polygonal patterns on Venus using mathematical morphology. *IEEE Trans., Image Processing*, **12**, doi:10.1109/TIP.2003.814254.

Moresi, L.-N. and Solomatov, V. S. (1998). Mantle convection with a brittle lithosphere: Thoughts on the global tectonic style of the Earth and Venus. *Geophys. J.*, **133**, 669–682.

Morgan, P. and Phillips, R. J. (1983). Hot spot heat transfer: Its application to Venus and implications to Venus and Earth. *J. Geophys. Res.*, **88**, 8305–8317.

Muller, O. H. and Pollard, D. D. (1977). The state of stress near Spanish Peaks, Colorado, determined from a dike pattern. *Pure Appl. Geophys.*, **115**, 69–86.

Nunes, D. C., Phillips, R. J., Brown, C. D., and Dombard, A. J. (2004). Relaxation of compensated topography and the evolution of crustal plateaus on Venus. *J. Geophys. Res.*, **109**, E01006, doi:10.1029/2003JE002119.

Odé, H. (1957). Mechanical analysis of the dike pattern of the Spanish Peaks, Colorado. *Geo. Soc. Am. Bull.*, **68**, 567–576.

Parker, T. J., Komatsu, G., and Baker, V. R. (1992). Longitudinal topographic profiles of very long channels in Venusian plains regions (abs.). *Lunar Planet. Sci. Conf. XXIII*, 1035–1036.

Parmentier, E. M. and Hess, P. C. (1992). Chemical differentiation of a convecting planetary interior: Consequences for a one-plate planet. *Geophys. Res. Lett.*, **19**, 2015–2018.

Phillips, R. J. and Hansen, V. L. (1994). Tectonic and magmatic evolution of Venus. In *Annual Review of Earth and Planetary Sciences*, ed. G. W. Wetherill, assoc. ed. A. L. Albee and K. C. Burke. Palo Alto, CA: Annual Reviews, Inc., pp. 597–654.

Phillips, R. J. and Malin, M. C. (1983). The interior of Venus and tectonic implications. In *Venus*, ed. D. M. Hunten, L. Colin, T. M. Donahue and V. I. Moroz. Tucson, AZ: University of Arizona Press, pp. 159–214.

Phillips, R. J., Grimm, R. E., and Malin, M. C. (1991). Hotspot evolution and the global tectonics of Venus. *Science*, **252**, 651–658.

Phillips, R. J., Raubertas, R. F., Arvidson, R. E., Sarkar, I. C., Herrick, R. R., Izenberg, N., and Grimm, R. E., (1992). Impact craters and Venus resurfacing history. *J. Geophys. Res.*, **97**, 15 923–15 948.

Plescia, J. B. and Golombek, M. P. (1986). Origin of planetary wrinkle ridges based on study of terrestrial analogs. *Geo. Soc. Am. Bull.*, **97**, 1289–1299.

Rathbun, J. A., Janes, D. M., and Squyres, S. W. (1999). Formation of Beta Regio, Venus: Results from measuring strain. *J. Geophys. Res.*, **104**, 1917–1928.

Reese, C. C., Solomatov, V. S., and Moresi, L.-N. (1999). Non-Newtonian stagnant lid convection and magmatic resurfacing on Venus. *Icarus*, **139**, 67–80.

Richards, M. A., Yang, W.-S., Baumgardner, J. R., and Bunge, H.-P. (2001). Role of a low-viscosity zone in stabilizing plate tectonics: Implications for comparative terrestrial planetology. *Geochem., Geophys., Geosyst.*, 1040, doi:10.1029/2002GC000374.

Roberts, K. M. and Head, J. W., (1990). Western Ishtar Terra and Lakshmi Planum, Venus: Models of formation and evolution. *Geophys. Res. Lett.*, **17**, 1341–1344.

Roberts, K. M. and Head, J. W. (1993). Large-scale volcanism associated with coronae on Venus: Implications for formation and evolution. *Geophys. Res. Lett.*, **20**, 1111–1114.

Robinson, E. M. and Parsons, B. (1988). Effect of a shallow low-viscosity zone on the formation of midplate swells. *J. Geophys. Res.*, **93**, 3144–3156.

Rosenberg, E. and McGill, G. E. (2001). Geologic map of the Pandrosos Dorsa Quadrangle (V-5), Venus. U.S. Geol. Surv. Geol. Invest. Ser. I-2721.

Sandwell, D. T. and Schubert, G. (1992). Flexural ridges, trenches and outer rises around Venus coronae. *J. Geophys. Res.*, **97**, 16 069–16 084.

Schaber, G. (1982). Limited extension and volcanism along zones of lithospheric weakness. *Geophys. Res. Lett.*, **9**, 499–502.

Schaber, G. G., Strom, R. G., Moore, H. J., Soderblom, L. A., Kirk, R. L., Chadwick, D. J., Dawson, D. D., Gaddis, L. R., Boyce, J. M., and Russell, J. (1992). Geology and distribution of impact craters on Venus: What are they telling us? *J. Geophys. Res.*, **97**, 13 257–13 301.

Schubert, G. (1992). Numerical models of mantle convection. *Annu. Rev. Fluid Mech.*, **24**, 359–394.

Schubert, G. and Sandwell, D. T. (1995). A global survey of possible subduction sites on Venus. *Icarus*, **117**, 173–196.

Schubert, G., Solomatov, V. S., Tackley, P. J., and Turcotte, D. L. (1997). Mantle convection and the thermal evolution of Venus. In *Venus II*, ed. Brougher, S. W., Hunten, D. M. and Phillips, R. J. Tucson, AZ: University of Arizona Press, pp. 1245–1287.

Schultz, R. A., (2000). Localization of bedding-plane slip and backthrust faults above blind thrust faults: Keys to wrinkle ridge structure. *J. Geophys. Res.*, **105**, 12 035–12 052.

Sclater, J. G., Jaupart, C., and Galson, D. (1980). The heat flow through oceanic and continental crust and the heat loss of the Earth. *Rev. Geophys. Space Phys.*, **18**, 269–311.

Seif, A., Kirk, D. B., Young, R. E., Blanchard, R. C., Findlay, J. T., Kelly, G. M., and Sommer, S. C. (1980). Measurements of thermal structure and thermal contrasts in the atmosphere of Venus and related dynamical observations: Results from the four Pioneer Venus probes. *J. Geophys. Res.*, **85**, 7903–7933.

Senske, D. A., Schaber, G. G., and Stofan, E. R. (1992). Regional topographic rises on Venus: Geology of western Eistla Regio and comparisons to Beta Regio and Atla Regio. *J. Geophys. Res.*, **97**, 13 395–13 420.

Simons, M., Solomon, S. C., and Hager, B. H. (1997). Localization of gravity and topography: Constraints on the tectonics and mantle dynamics of Venus. *Geophys. J. Int.*, **131**, 24–44.

Smrekar, S. E. (1994). Evidence for active hotspots on Venus from analysis of Magellan gravity data. *Icarus*, **112**, 2–26.

Smrekar, S. E. and Parmentier, E. M. (1996). Interactions of mantle plumes with thermal and chemical boundary layers: Application to hotspots on Venus. *J. Geophys. Res.*, **101**, 5397–5410.

Smrekar, S. E. and Phillips, R. J. (1991). Venusian highlands: Geoid to topography ratios and their implications. *Earth Planet. Sci. Lett.*, **107**, 582–597.

Smrekar, S. E. and Solomon, S. C. (1992). Gravitational spreading of high terrain in Ishtar Terra, Venus. *J. Geophys. Res.*, **97**, 16 121–16 148.

Smrekar, S. E. and Stofan, E. R. (1997). Coupled upwelling and delamination: A new mechanism for coronae formation and heat loss on Venus. *Science*, **277**, 1289–1294.

Smrekar, S. E., Stofan, E. R., and Kiefer, W. S. (1997). Large volcanic rises on Venus. In *Venus II*, ed. Brougher, S. W., Hunten, D.M. and Phillips, R. J. Tucson, AZ: University of Arizona Press, pp. 845–878.

Smrekar, S. E., Moreels, P., and Franklin, B. J. (2002). Characterization and formation of polygonal fractures on Venus. *J. Geophys. Res.*, **107**, E11, 5098, doi:10.1029/2001JE001808.

Solomatov, V. S. and Moresi, L.-N. (2000). Scaling of time-dependent stagnant lid convection: Application to small-scale convection on Earth and other terrestrial planets. *J. Geophys. Res.*, **105**, 21 795–21 818.

Solomon, S. C. (1993). The geophysics of Venus. *Phys. Today*, **46**, 48–55.

Solomon, S. C., and Head, J. W. (1982). Mechanisms for lithospheric heat transport on Venus: Implications for tectonic style and volcanism. *J. Geophys. Res.*, **87**, 9236–9246.

Solomon, S. C., Head, J. W., Kaula, W. M., McKenzie, D., Parsons, B., Phillips, R. J., Schubert, G., and Talwani, M., (1991). Venus tectonics: Initial analysis from Magellan. *Science*, **252**, 297–312.

Solomon, S. C., Smrekar, S. E., Bindschadler, D. L., Grimm, R. E., Kaula, W. M., McGill, G. E., Phillips, R. J., Saunders, R. S., Schubert, G., Squyres, S. W., and Stofan, E. R. (1992). Venus tectonics: An overview of Magellan observations. *J. Geophys. Res.*, **97**, 13 199–13 255.

Solomon, S. C., Bullock, M. A., and Grinspoon, D. H. (1999). Climate change as a regulator of tectonics on Venus. *Science*, **286**, 87–90.

Squyres, S. W., Jankowski, D. G., Simons, M., Solomon, S. C., Hager, B. H., and McGill, G. E. (1992). Plains tectonism on Venus: The deformation belts of Lavinia Planitia. *J. Geophys. Res.*, **97**, 13 579–13 599.

Squyres, S. W., Janes, D. M., Baer, G., Bindschadler, D. L., Schubert, G., Sharpton, V. L., and Stofan, E. R. (1992). The morphology and evolution of coronae on Venus. *J. Geophys. Res.*, **97**, 13 611–13 634.

Squyres, S. W., Janes, D. M., Schubert, G., Bindschadler, D. L., Moersch, J. E., Turcotte, D. L., and Stofan, E. R. (1993). The spatial distribution of coronae and related features on Venus. *Geophys. Res. Lett.*, **20**, 2965–2968.

Steinbach, V. and Yuen, D. A. (1994). Effects of depth dependent properties on the thermal anomalies produced in flush instabilities from phase transitions. *Phys. Earth Planet. Inter.*, **86**, 165–183.

Steinbach, V., Yuen, D. A., and Zhao, W. L. (1993). Instabilities from phase transitions and the timescales of mantle thermal evolution. *Geophys. Res. Lett.*, **20**, 1119–1122.

Strom, R. G. (1972). Lunar mare ridges, rings and volcanic ring complexes. *Mod. Geol.*, **2**, 133–157.

Stofan, E. R., Bindschadler, D. L., Head, J. W., and Parmentier, E. M. (1991). Corona structures on Venus: Models of origin. *J. Geophys. Res.*, **96**, 20 933–20 946.

Stofan, E. R., Sharpton, V. L., Schubert, G., Baer, G., Bindschadler, D. L., Janes, D. M., and Squyres, S. W. (1992). Global distribution and characteristics of coronae and related features on Venus: Implications for the origin and relation to mantle processes. *J. Geophys. Res.*, **97**, 13 347–13 378.

Stofan, E. R., Smrekar, S. E., Bindschadler, D. L., and Senske, D. A. (1995). Large topographic rises on Venus: Implications for mantle upwellings. *J. Geophys. Res.*, **100**, 23 317–23 327.

Stofan, E. R., Bindschadler, D. L., Hamilton, V. E., Janes, D. M., and Smrekar, S. E. (1997). Coronae on Venus: Morphology and origin. In *Venus II*, ed. Brougher, S. W., Hunten, D. M. and Phillips, R. J. Tucson, AZ: University of Arizona Press, pp. 931–965.

Stofan, E. R., Tapper, S. W., Guest, J. E., Grinrod, P., and Smrekar, S. E. (2001). Preliminary analysis of an expanded corona database for Venus. *Geophys. Res. Lett.*, **28**, 4267–4270.

Stofan, E. R., Brian, A. W., and Guest, J. E. (2005). Resurfacing styles and rates on Venus: Assessment of 18 Venusian quadrangles. *Icarus*, **173**, 312–321.

Stofan, E. R., Guest, J. E., and Brian, A. W. Geologic map of the Hecate Chasma quadrangle (V-28), Venus. U.S. Geol. Surv. Geol. Invest. Ser., in review.

Sukhanov, W. L. and Pronin, A. A. (1989). Ridge belts on Venus as extensional features. *Proc. Lunar Planet. Sci. Conf.* 19, 335–348.

Tanaka, K. L., Senske, D. A., Price, M., and Kirk, R. L. (1997). Physiography, geomorphic/geologic mapping, and stratigraphy of Venus. In *Venus II*, ed. S. W. Brougher, D. M. Hunten and R. J. Phillips. Tucson, AZ: University of Arizona Press, pp. 667–694.

Taylor, F. W., Crisp, D., and Bezard, B. (1997). Near-infrared sounding of the lower atmosphere. In *Venus II*, ed. S. W. Brougher, D. M. Hunten and R. J. Phillips. Tucson, AZ: University of Arizona Press, pp. 325–351.

Watters, T. R. (1988). Wrinkle ridge assemblages on the terrestrial planets. *J. Geophys. Res.*, **93**, 10 236–10 254.

Watters, T. R. (1991). Origin of periodically spaced wrinkle ridges on the Tharsis Plateau of Mars. *J. Geophys. Res.*, **96**, 15 599–15 616.

Wood, J. A., and 9 others (1981). Geophysical and cosmochemical constraints on properties of mantles of the terrestrial planets. In *Basaltic Volcanism on the Terrestrial Planets*. New York: Pergamon Press, pp. 633–699.

Yung, D. L. and DeMore, W. B. (1999). *Photochemistry of Planetary Atmospheres*. Oxford: Oxford University Press.

Zuber, M. T. (1990). Ridge belts: Evidence for regional- and local-scale deformation on the surface of Venus. *Geophys. Res. Lett.*, **17**, 1369–1372.

4

Lunar tectonics

Thomas R. Watters
*Center for Earth and Planetary Studies, National Air and Space Museum,
Smithsonian Institution, Washington, DC*

and

Catherine L. Johnson
Earth and Ocean Sciences, University of British Columbia, Vancouver, Canada

Summary

Tectonic landforms on the Moon predominantly occur on the nearside, associated directly with the lunar maria. Basin-localized lunar tectonics is expressed by two landforms: wrinkle ridges, and linear and arcuate rilles or troughs. Wrinkle ridges are complex morphologic landforms found in mare basalts, interpreted to be contractional tectonic landforms formed by thrust faulting and folding. Linear and arcuate rilles are long, narrow troughs, interpreted to be graben formed by extension, deforming both mare basalts at basin margins and the highlands adjacent to the basins. In contrast to basin-localized tectonics, landforms of the nearside are the more broadly distributed lobate scarps. Lobate scarps on the Moon are relatively small-scale asymmetric landforms that are often segmented with lobate margins. These landforms are the surface expression of thrust faults and are the dominant tectonic feature on the lunar farside. Crater density ages indicate that crustal extension associated with lunar maria ceased at ∼3.6 Ga. Crustal shortening in the maria, however, continued to as recently as ∼1.2 Ga. The cessation of extension may have resulted from the superposition of compressional stresses from global contraction on flexural extensional stress due to loading from the mare basalts. The lobate scarps formed less than 1 Ga and appear to be among the youngest endogenic features on the Moon. The presence of young lobate scarp thrust faults supports late-stage compression of the lunar crust. Lunar seismic data provide insight into the lunar interior structure, and the spatial distribution and depth of some moonquakes suggests current tectonic activity on the Moon. Deep moonquakes tend to cluster around edges of, or beneath, the nearside

Planetary Tectonics, edited by Thomas R. Watters and Richard A. Schultz. Published by Cambridge University Press. Copyright © Cambridge University Press 2010.

basins. Although too deep to be directly correlated with the distribution of tectonic features, they may be related to long-lived, deep lateral thermal and/or compositional heterogeneities associated with either basin formation and/or mare basalt production. Shallow-depth moonquakes are also distributed around the nearside basins. Depths of these quakes are poorly constrained; although some arguments favor an upper mantle origin, it is plausible that at least some quakes occur within the lunar crust. Thus, some moonquakes may be associated with the observed lunar faults. Current lunar gravity models show that anomalies for the mare mascons have broad plateaus throughout the mare interior. The shoulders of the positive anomalies spatially correlate with basin-interior rings of wrinkle ridges and suggest a pan- rather than a bowl-shaped mare-fill geometry. Recent mascon stress models suggest that a more uniform thickness of mare basalts best fits the observed spatial distribution of the wrinkle ridges and linear and arcuate rilles. The contractional strain expressed by the relatively young lobate scarps is estimated to be ~0.008% over an area of roughly 10% of the lunar surface. If representative of the entire surface, this strain corresponds to a radius change of ~70 m, consistent with thermal history models that predict a small change in lunar radius in the last 3.8 Ga.

1 Introduction

The Moon has many characteristics that distinguish it from other moons and satellites in the solar system. This is particularly true in the context of planetary tectonics. Its greatest distinction is, of course, the fact that it is the only other body in the solar system that humans have explored from the surface. The Apollo 17 astronauts were the last to visit the Moon and the first to make a field examination of a fault scarp on another planetary body. The Lee-Lincoln scarp is not a very imposing structure in the landscape of the Apollo 17 landing site (Figure 4.1); the latter is dominated by the several kilometer-high mountains of the North and South Massif. However, when astronauts Eugene A. Cernan and Harrison H. Schmitt attempted to drive straight up the scarp face with the Lunar Roving Vehicle (LRV), the wheels began to slip, and they had to steer across-slope to reach the top, some 80 m above the lower floor of the Taurus-Littrow valley. The Lee-Lincoln scarp exemplifies an important class of lunar faults, lobate scarp thrust faults.

The dark maria are prominent features of the nearside, visible from Earth even with the naked eye because of their albedo contrast with the bright terrae or highlands. It is now well known that the maria are basalts that flooded topographic lows created by large impact events. The vast majority of the Moon's large-scale tectonic features are found in the basalt-filled impact basins and the adjacent highlands. The dominant tectonic landforms that resulted from this basin-localized

Figure 4.1. Tauras-Littrow valley near the Apollo 17 landing site. The Lee-Lincoln scarp is in the foreground, with the hills of Family Mountain in the background. The fault scarp offsets the floor of the Tauras-Littrow valley (Hasselblad Camera frame AS17-134-20443).

deformation on the Moon are wrinkle ridges, and linear and arcuate troughs or rilles (Plates 6A,B).

Wrinkle ridges are morphologically complex landforms that occur in mare basalts. In fact, they are often referred to as mare ridges because of their direct association with the maria (see Plates 6A,B) (see Head, 1976; Wilhelms, 1987). Their morphologic complexity is generally attributable to the superimposition of two landforms, a broad arch and a narrow ridge. Wrinkle ridges were first discovered and mapped using Earth-based telescopic observations of the lunar maria (Figure 4.2) (Gilbert, 1893; Fielder, 1961; Baldwin, 1963, 1965). Although interpreted to be anticlinal forms over a hundred years ago by Gilbert (1893), other investigators concluded that wrinkle ridges were volcanic features related to the emplacement of the mare basalts or the result of intrusion or extrusion of lava into fractures and zones of weakness following crustal extension related to basin-localized or global tectonic patterns (Fielder, 1961; Quaide, 1965; Whitaker, 1966; Tjia, 1970; Hartman and Wood, 1971; Colten *et al.*, 1972; Strom, 1972; Hodges, 1973; Scott,

Figure 4.2. Earth-based telescope view of Mare Serenitatis. Wrinkle ridges formed in the mare basalts are prominent landforms in low sun angle telescope observations. Rilles, long, narrow arcuate or linear troughs are also discernable (near the southwestern margin of Serenitatis). The image (Photo Number C1487) is from the Consolidated Lunar Atlas (Lunar and Planetary Laboratory, University of Arizona, 1967).

1973; Young *et al.*, 1973). Others concluded, as Gilbert had, that mare ridges are purely tectonic landforms (Baldwin, 1963, 1965; Bryan, 1973; Hodges, 1973; Howard and Muehlberger, 1973; Schaber, 1973a; Muehlberger, 1974; Maxwell *et al.*, 1975; Lucchitta, 1976, 1977; Maxwell, 1978; Sharpton and Head, 1981, 1982, 1988). Circular wrinkle ridges or wrinkle-ridge rings and basin-concentric ridge patterns (Plate 6A) were cited as evidence that subsidence of the mare basalts played an important role in their formation (Wilhelms and McCauley, 1971; Maxwell *et al.*, 1975; Brennan, 1976). Perhaps the most convincing support for a structural interpretation of mare ridges is the subsurface information provided by the Apollo Lunar Sounder Experiment (ALSE) data (Phillips *et al.*, 1973; Peeples *et al.*, 1978; Maxwell and Phillips, 1978). ALSE data over a mare ridge in southeastern Mare Serenitatis show evidence of an anticlinal rise in subsurface horizons and thinning of a mare unit (sequence of flows) on apparent structural relief (Maxwell, 1978). A compressional tectonic origin involving a combination of folding and thrust faulting is also supported by studies of terrestrial analogues (Plescia and Golombek, 1986; Watters, 1988). The tectonic interpretation of wrinkle ridges subsequently found on Mercury, Venus, and Mars is rooted in the early analysis of mare ridges. In the study of planetary wrinkle ridges, the volcanic versus tectonic debate has been replaced by general disagreement over the relative role of folding and faulting, the geometry and number of thrust faults, and whether faults are surface breaking or non-surface breaking (blind) (e.g., Golombek *et al.*, 1991; Watters, 1992; Schultz, 2000; Golombek *et al.*, 2001; Mueller and Golombek, 2004; Watters, 2004).

Rilles were also identified in Earth-based telescopic surveys of the Moon (Figure 4.2). They are long, narrow troughs that commonly exhibit three plan-view geometries: sinuous, arcuate, and linear. Because sinuous rilles are meandrous and confined to mare basalts they are interpreted to be volcanic in origin. Sinuous rilles are generally thought to be collapsed lava tubes or lava channels (Greeley, 1971; Schultz, 1976a; Masursky *et al.*, 1978; Wilhelms, 1987; Spudis *et al.*, 1988a). The relationship between sinuous rilles and mare ridges has been cited as evidence that some elements of wrinkle ridges are volcanic in origin (Greeley and Spudis, 1978). Linear and arcuate rilles are found in the highlands adjacent to, and in the margins of mare basins and generally have basin-concentric orientations (Plates 6A,B) (see McGill, 1971; Golombek, 1979; Wilhelms, 1987). In cross section, these rilles are flat-floored with steep walls. This cross-sectional geometry and the fact that the trough walls maintain roughly the same relief as they extend from mare basalts into basin rim or highlands material has led to the interpretation that rilles are formed by graben (Baldwin, 1963; Quaide, 1965; McGill, 1971; Golombek, 1979). Like terrestrial and other planetary graben, many lunar rilles are segmented with echelon steps indicating that the faults grow by segment linkage (see McGill, 2000). Less common than basin-concentric graben are basin-radial graben (Plate 6A). These graben are dominated by linear rille segments, many of which are radial or subradial to the Imbrium basin (Quaide, 1965; Wilhelms, 1987). Relatively small depressions that occur near or at intersections of linear rilles have been interpreted as possible sites of recent out-gassing from sources deep in the lunar interior (Schultz *et al.*, 2006).

It is clear that the mare ridges and the arcuate and linear rilles are indicative of a pattern of basin-localized contractional and extensional deformation (Plates 6A,B). Thus the types, locations, and relative timing of these two classes of tectonic features have been used to provide constraints on the tectonic evolution of the mare-filled nearside lunar basins (e.g., Phillips *et al.*, 1972; Melosh, 1978; Solomon and Head, 1979, 1980; Freed *et al.*, 2001). The combination of a mass concentration from the mare basalts (mascon) and an impact-induced thinned and weakened lithosphere is responsible for what is referred to as mascon tectonics, resulting in the observed spatial patterns of deformation.

Although lunar tectonics, in contrast to terrestrial planets like Mercury, Venus, and Mars, is largely basin-localized, lobate scarps is a class of tectonic landform on the Moon that is not directly associated with the mare basins. Analogous to tectonic landforms found on Mercury (Watters *et al.*, 1998, 2001, 2004; Watters and Nimmo, Chapter 2) and Mars (Watters, 1993, 2003; Watters and Robinson, 1999; Golombek and Phillips, Chapter 5), lunar lobate scarps are generally asymmetric landforms and are often lobate and segmented. The most significant contrast between planetary and lunar lobate scarps is scale. While lobate scarps on Mercury and Mars can have over a kilometer of relief, lunar scarps have a maximum relief of only tens of meters (Howard and Muehlberger, 1973; Lucchitta, 1976; Binder, 1982; Binder and Gunga, 1985). The lengths of the lunar scarps are proportionately smaller, reaching a maximum of only tens of kilometers (Binder and Gunga, 1985). Lunar lobate scarps, like their planetary counterparts, are interpreted to be the result of thrust faulting (Howard and Muehlberger, 1973; Lucchitta, 1976; Binder, 1982; Binder and Gunga, 1985). The evidence of offset is not as dramatic as in the case of large-scale lobate scarps on Mercury and Mars, however, the morphology and the linkage between individual segments of the lunar scarps supports the interpretation that they are the surface expression of shallow thrust faults. Although many of the lunar lobate scarps are found in the highlands (Plates 6A,B), there are some cases where scarps are associated with wrinkle ridges. The structures have been described as "mare–ridge, highland–scarp systems" (Lucchitta, 1976). The source of compressional stresses that formed the lobate scarps has been suggested to be thermal stresses from global cooling (Solomon and Chaiken, 1976; Solomon and Head, 1979; Binder and Lange, 1980). If so, the spatial distribution and scale of the lunar scarps has important implications for the Moon's thermal history and for constraining models for its origin.

The Moon is also unique in our solar system, because it is the only other body for which we have *in situ* seismic data. Data recorded by seismometers at Apollo sites 12, 14, 15 and 16, over the period 1969–1977, show that the Moon is seismically active, exhibiting present-day quakes in two depth ranges: shallow moonquakes occurring at depths of 100 km or less, and deep moonquakes occurring at depths of 700–1000 km. The deep moonquakes are numerous (many thousands recorded), and very early on during the Apollo era were observed to exhibit tidal periods. Repeatable or coherent seismograms have been interpreted as events that originate from the same location – thus, although many thousands of deep moonquakes have been identified, they appear to originate from only 100–200 source regions. Shallow moonquakes have been interpreted to be tectonic in origin, and so their locations with respect to tectonic features on the Moon are of interest. In addition to the location and timing of events, the seismic data provides information on the interior structure of the Moon to depths of about 1000 km.

The scale and complexity of lunar tectonics does not rival that on the terrestrial planets or even that on some of the icy moons of the outer planets. It is nonetheless important because the pre- and early-robotic and human investigation of lunar tectonic landforms is the foundation and touchstone for subsequent interpretation and analysis of crustal deformation on other bodies in the solar system. Our understanding of lunar tectonics is far from complete. To date, less than roughly 10% of the lunar surface has been imaged at high enough resolution and optimal illumination conditions to detect small-scale tectonic features such as the highland lobate scarps. With the international armada of new lunar robotic missions and the U.S. initiative to return humans to the Moon, the opportunity to fully explore lunar tectonics is close at hand.

In this chapter, we will describe the known tectonic landforms on the Moon, their spatial distribution, and the timing of their formation. The relationship between the tectonic features, the lunar maria, and the topography and the gravity field of mare basins is examined. Lunar seismicity and our current understanding of the Moon's interior structure are reviewed and the correlation between moonquakes and nearside tectonism is evaluated. Mascon stress models are considered in light of current topography and gravity data. The strain and contraction from lobate scarp thrust faults is estimated and implications for the thermal history models discussed. Finally, outstanding questions about the tectonic evolution of the Moon are identified and explored in the context of new data from upcoming missions.

2 Tectonic features of Moon

A full morphological description and characterization of lunar tectonic landforms requires high-resolution imaging as well as topographic data. Current topographic data sets for the Moon afford different resolution and accuracy, depending on the geographical region under study. We briefly review available topographic data sets, before turning to more detailed discussions of the tectonic features.

Early attempts to measure the topography of lunar features were made using telescope-based shadow measurements (Wu and Doyle, 1990). The Ranger spacecrafts and the Lunar Orbiters provided the first space-based stereo images used to derive lunar topographic maps. The highest quality Apollo-era topography was derived from the Apollo 15, 16, and 17 Metric (25–30 m resolution) and Panoramic (1–2 m resolution) cameras' stereo photography. These photographs were used to generate the Lunar Topographic Orthophotomap (LTO) series at scales of 1:250 000 down to 1:10 000 by means of analogue stereoplotters (Wu and Doyle, 1990). However, the Apollo spacecraft were confined to lunar equatorial orbits, limiting the coverage of the LTOs. Limited topographic profiles were obtained by Apollos 15, 16, and 17 from laser altimeters attached to the metric camera systems (e.g., Wollenhaupt et al., 1973), and by the Apollo 17 Lunar Sounder Experiment (Phillips et al., 1973; Peeples et al., 1978; Maxwell and Phillips, 1978; Sharpton, 1992).

Nearly global topography of the Moon was obtained by the laser ranging instrument (LIDAR) flown on the Clementine spacecraft. The LIDAR instrument collected topographic data between 75°S and 75°N latitude (see Nozette et al., 1994). Clementine's polar orbit provided altimetry data along north–south orbital tracks (roughly along lines of longitude) spaced by approximately 2.5° at the equator (~75 km) (Spudis et al., 1994; Zuber et al., 1994). The distribution of good returns is highly variable within an orbital track because of the influence of terrain roughness and solar phase angle (Spudis et al., 1994). The single-shot ranging precision of the LIDAR (vertical resolution) is estimated to be 40 m (Zuber *et al.*, 1994). Some of the highest along-track resolution was obtained over smooth mare surfaces, late in the mission when solar phase angles where larger (Smith *et al.*, 1997).

Earth-based radar interferometry has been used to determine the topography of the polar regions where no Clementine LIDAR data was collected (Margot *et al.*, 1998, 1999). Using digital stereo methods, Clementine ultraviolet-visible (UVVIS) nadir and off-nadir stereo images were used to produce Digital Elevation Models (DEMs) for the polar regions beyond that covered by Earth-based radar and globally (Cook *et al.*, 2000; U.S. Geological Survey, 2002; Rosiek *et al.*, 2007). The nadir-pointing Clementine UVVIS images are not ideal because the stereo angle between adjacent UVVIS images is generally weak (3° to 5°) and the lighting geometry was not optimized for morphologic studies. Stereo images from the Galileo spacecraft flyby in 1992 have also been used to generate a DEM of the lunar north polar region (Schenk and Bussey, 2004). At present, the highest resolution stereo images of the lunar surface are from the Apollo Metric and Panoramic cameras.

2.1 Wrinkle ridges

Wrinkle ridges are landforms that have been found in volcanic plains on Mercury, Venus, and Mars, and analogous structures occur in continental flood basalts on Earth (Plescia and Golombek, 1986; Watters, 1988, 1992). Mare wrinkle ridges are the most common and probably the best described of those observed on the terrestrial planets, and are found in nearly all lunar maria (Strom, 1972; Bryan, 1973; Maxwell *et al.*, 1975; Head, 1976; Plescia and Golombek, 1986; Watters, 1988; Golombek *et al.*, 1991). Mare ridges typically occur both radial to and concentric with the centers of mare basins (Bryan, 1973; Maxwell *et al.*, 1975) (Plates 6A,B, Figure 4.3). The association between wrinkle ridges, maria and mare basalts has led some workers to suggest a genetic relation between the basalts and the origin of the structures (see Strom, 1972; Bryan, 1973; Watters, 1988).

Mare wrinkle ridges and their planetary counterparts are morphologically complex structures that may be composed of a number of superimposed landforms, often consisting of a composite of a broad arch and narrow, asymmetric ridges (Figures 4.3, 4.4). These morphologic elements may also occur independently of one another. Arches are broad, gently sloping, curvilinear topographic rises that are commonly asymmetric in profile and often only distinguishable in low sun angle images (Figure 4.4) (Strom, 1972; Bryan, 1973; Maxwell *et al.*, 1975). Ridges are long, relatively narrow, segmented features that are commonly strongly asymmetric in cross section. The sense of the asymmetry may change either along strike or from one ridge segment to the next, and segments often occur in en-echelon

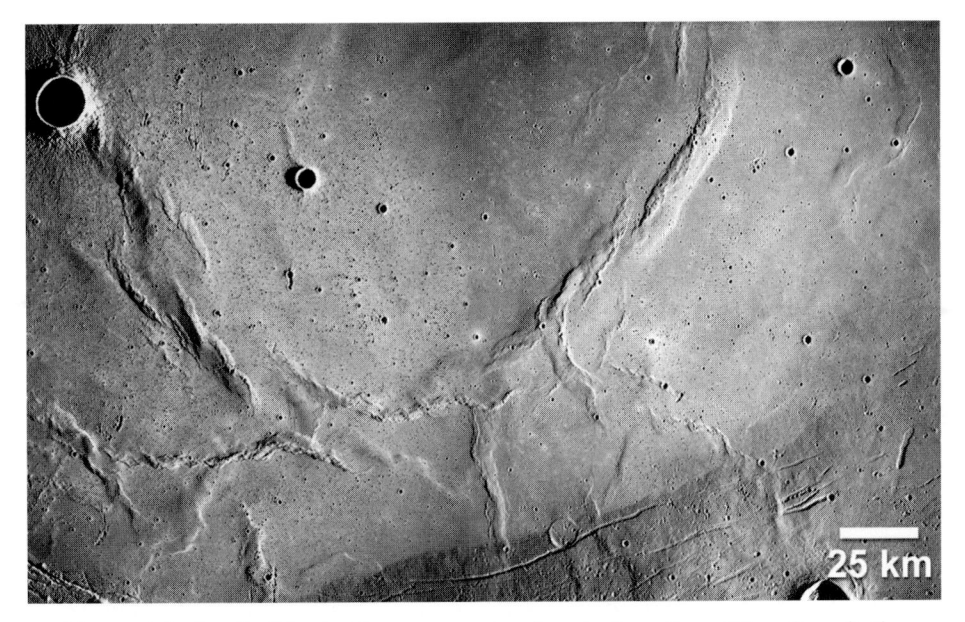

Figure 4.3. Apollo Metric Camera mosaic of part of southern Mare Serenitatis. Wrinkle ridges, also described as mare ridges, are morphologically complex features often composed of a number of superimposed landforms. Both basin-concentric and basin-radial wrinkle ridges occur in Mare Serenitatis. Arcuate rilles or troughs occur near the basin margin. The mosaic was generated using Metric camera frames AS17-450 and AS17-454.

Figure 4.4. Mare ridge in Oceanus Procellarum near the northwestern edge of the Aristarchus Plateau (lower left). Wrinkle ridges often consist of broad, low-relief arches and superposed, narrow ridges (Apollo 15 Metric Camera AS15-2487).

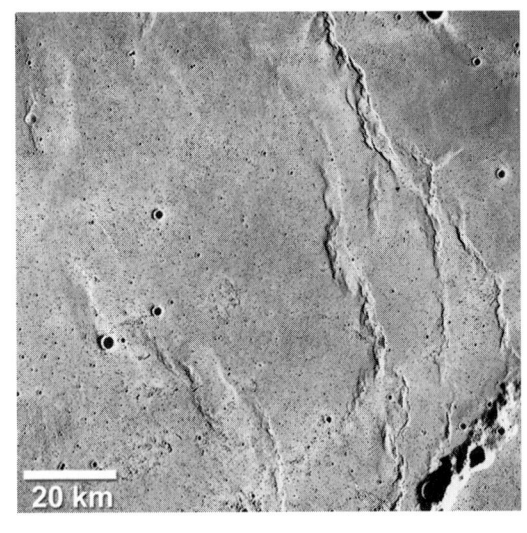

arrangements (see Tjia, 1970). In wrinkle-ridge assemblages in mare basins, the ridge is often superposed on the arch (Strom, 1972; Bryan, 1973; Maxwell et al., 1975).

In addition to this ridge–arch association, there is evidence of smaller secondary ridges that occur on or near larger primary ridges. These second-order mare ridges

130   *Planetary Tectonics*

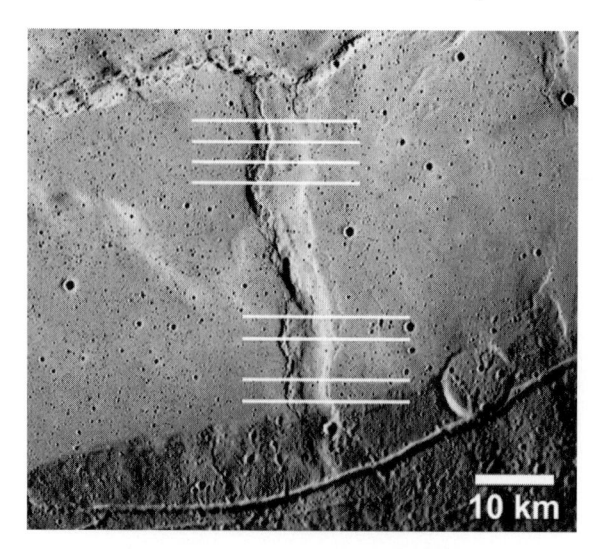

Figure 4.5. The morphology of wrinkle ridge Dorsum Nicol in Mare Serenitatis. The complex morphology of this prominent wrinkle ridge is the result of several superposed landforms (Metric Camera frames AS-17-453).

are sharp, narrow prominences that are very similar in morphology to the first-order ridges and commonly flank or cap the larger ridges (Watters, 1988). Even smaller third-order ridges can be found that flank or commonly cap larger ridges. These small-scale ridges can only be easily resolved in Apollo Metric Camera and Panoramic Camera images (see Scott, 1973; Watters, 1988).

2.1.1 Topography of wrinkle ridges

The best available topography for lunar wrinkle ridges is from Lunar Topographic Orthophotomaps. The morphology and dimensions of a number of wrinkle ridges in Mare Serenitatis, Mare Imbrium, and Mare Procellarum have been described using these data (Watters, 1988; Golombek *et al.*, 1991). High-resolution Apollo Metric and Panoramic Camera images and stereo derived topography illustrate the distinct morphology elements of these landforms (Figure 4.5); the broad rise or arch and the superposed ridge. Using the LTOs, the maximum relief of 12 wrinkle-ridge segments was measured (Table 4.1). The maximum relief of the measured ridges varies from about 50 to 410 m (mean ∼253 m), in agreement with previous surveys (Watters, 1988; Golombek *et al.*, 1991). Topographic profiles across Dorsum Nicol in Mare Serenitatis show that the vergent side of the ridge changes along strike (Figure 4.6). The vergence may change either along strike or from one ridge segment to the next. Changes in vergence is a common characteristic of wrinkle ridges on the Moon, Mercury, and Mars and terrestrial analogues such as the anticlinal ridges of the Columbia Plateau in the northwestern United States (see Reidel, 1984; Watters, 1988, 1991, 1992, 2004).

Clementine LIDAR profiles provide excellent long-wavelength topography of mare surfaces (Smith *et al.*, 1997). In rare cases, the along-track spatial resolution

Table 4.1 *Dimensions of wrinkle ridges on the Moon*

Index	Latitude	Longitude	Maximum Relief (m)	Length (km)	D $\theta = 30°$ (m)
Dorsum Nicol	18.5°N	22.8°E	240	49	480
Dorsum Lister S	20°N	23.3°E	408	79	816
Dorsum Lister N	21.2°N	24.8°E	358	61	716
Dorsum Lister W	19.5°N	19.5°E	262	49	524
Dorsum Zirkel	30°N	25.5°W	272	145	544
Dorsum Buckland E	18.3°N	19.5°W	207	80	414
Dorsum Buckland Mid	18.8°N	17.5°W	223	24	446
Dorsum Buckland W	19.3°N	16°E	238	35	476
Dorsa Ewing	29.2°N	24.5°W	309	110	618
Dorsa Rubey	9.8°S	42.3°W	48	50	96
Dorsa Smirnov	25°N	25.5°E	300	68	600
Dorsum Von Cotta	25.5°N	12°E	166	54	332

Relief was determined using Lunar Topographic Orthophotomaps (LTOs).

was high enough to resolve the cross-sectional topography of wrinkle ridges in Mare Serenitatis. LIDAR data across a segment of Dorsum Buckland (located at ~17°E) indicates the relief of the ridge is ~190 m (Figure 4.7). The maximum relief of this segment of Buckland (mid-segment, Table 4.1) determined from an LTO is ~220 m, in good agreement with the LIDAR-based measurement. The same LIDAR profile shows the long-wavelength topography of the mare surface in Serenitatis and indicates that this segment of Dorsum Buckland is imposed on a gentle regional slope (Figure 4.8). It also indicates that the center of Mare Serenitatis is higher than the margins along this transect (Watters and Konopliv, 2001). This is the case elsewhere in Mare Serenitatis. Generally, the lowest elevations of the mare surface are outside the prominent mare ridge ring in the interior (Watters and Konopliv, 2001) (Figure 4.8, Plate 7). The elevation difference between the center and the lowest elevation of the mare surface is up to 400 m, located on its southeast margin (Watters and Konopliv, 2001). A central topographic high in Mare Serenitatis is also evident in east–west profiles across the central basin, as revealed in Apollo Laser Altimeter data (see Wollenhaupt *et al.*, 1973, Fig. 33–24a) and Apollo Lunar Sounder Experiment (ALSE) data (see Sharpton, 1992, Fig. 2).

2.1.2 Elevation offsets across wrinkle ridges

It has been observed that topographic data for some mare ridges from LTOs and Apollo Lunar Sounder Experiment (ALSE) data exhibit elevation offsets from one side of a ridge to the other (Maxwell *et al.*, 1975; Lucchitta, 1976; Golombek

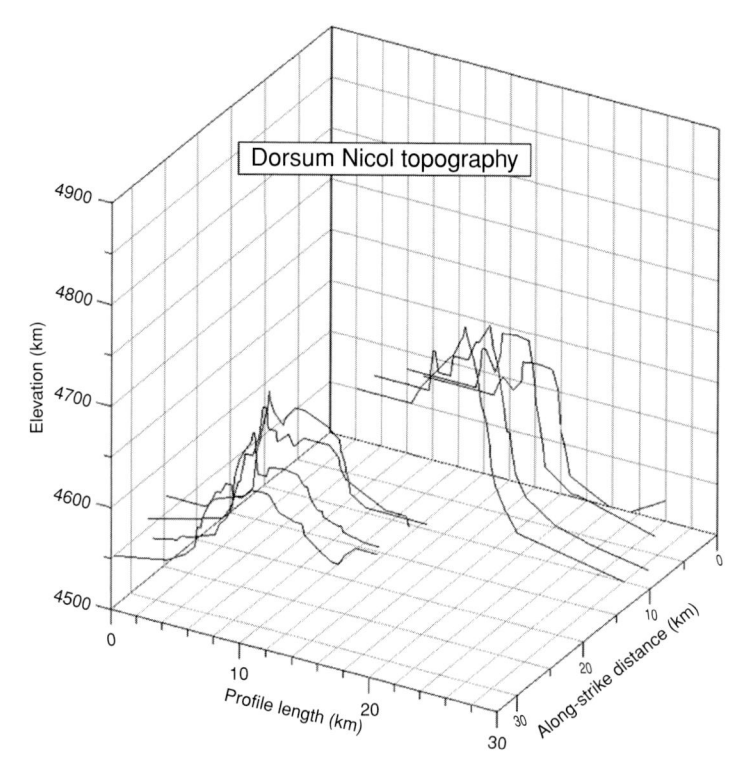

Figure 4.6. Topographic profile constructed from two 1:50 000 LTOs (42C4S1 and 42C4S2) show the major morphologic elements of wrinkle ridges. The profiles also show a reversal in the vergent side of the ridge from the northern section (upper four profiles) to the southern section (lower four profiles). The first profile (southernmost) is at the origin of the along strike distance. Profile locations are shown in Figure 4.5. Elevations are relative to an arbitrary zero vertical datum of 1 730 000 m. Vertical exaggeration is 75:1.

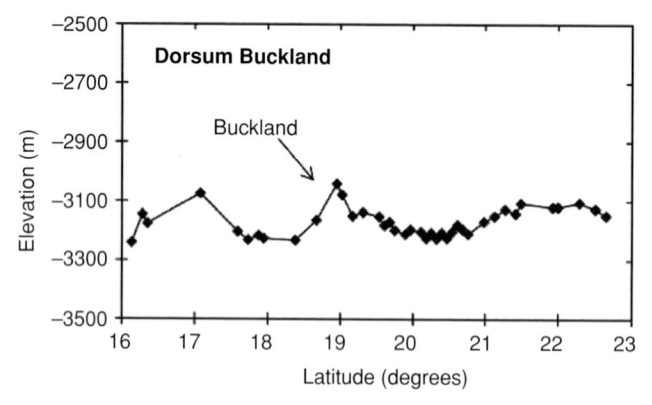

Figure 4.7. Clementine LIDAR profile located at approximately 17.25°E longitude crossing Dorsum Buckland in Mare Serenitatis. The LIDAR data were extracted from the dataset of Smith *et al.* (1997). Elevations are in meters above an ellipsoid of radius 1738 km at the equator with a flattening of 1/3234.93 corresponding to the flattening of the geoid. The vertical exaggeration is ~135:1.

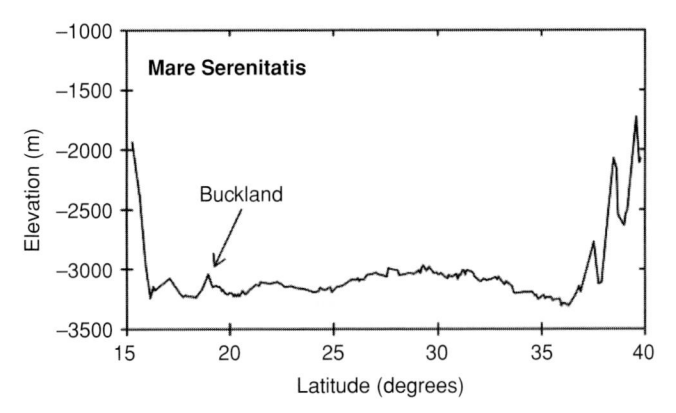

Figure 4.8. Clementine LIDAR profiles located at approximately 17.25°E longitude crossing Mare Serenitatis. The LIDAR data were extracted from the dataset of Smith *et al.* (1997). Elevations are in meters above an ellipsoid of radius 1738 km at the equator with a flattening of 1/3234.93 corresponding to the flattening of the geoid. The vertical exaggeration is ~194:1.

Figure 4.9. Dorsa Aldrovandi wrinkle ridge system near the eastern margin of Mare Serenitatis. Ridge segments are associated with a significant topographic offset of the mare surface with the lower elevations consistently on the interior side of the ridge. Extensional troughs (Fossae Pavlova) trend parallel to subparallel to Dorsa Aldrovandi (Metric camera frame AS17-939).

et al., 1991). The most prominent elevation offset is found on a wrinkle ridge near the eastern margin of Mare Serenitatis. Dorsa Aldrovandi is about 130 km long and segments of the ridge have offsets of as much as ~300 m, with the lower mare surface consistently on the basin-interior side of the ridge (Figure 4.9). One segment of Aldrovandi, referred to as the Littrow ridge, has linear fissures or troughs that run along the ridge crest (Figure 4.10) (Howard and Muehlberger, 1973; Maxwell *et al.*, 1975). The troughs are interpreted to be evidence of significant layer–parallel extension associated with the formation of the ridge over a high-angle reverse or

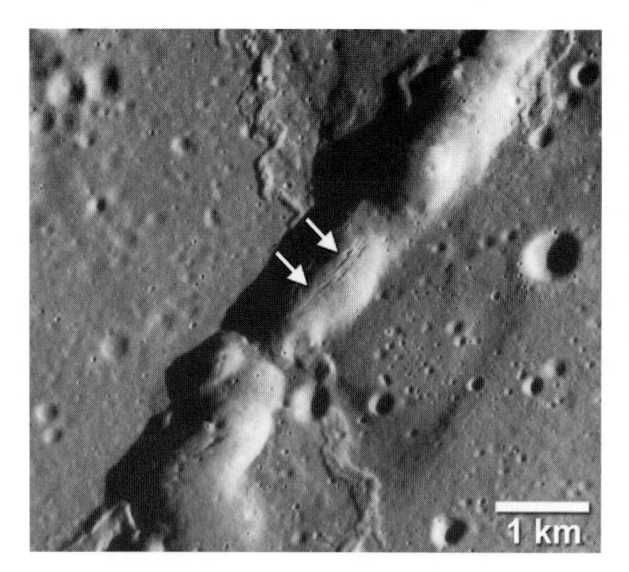

Figure 4.10. The Littrow ridge in the Taurus-Littrow region near the eastern margin of Mare Serenitatis. The lineations along the crest of the ridge (see arrows) may be evidence of layer–parallel extension resulting from folding and thrust faulting of the mare basalts (Panoramic camera frame AS17-2313).

thrust fault (Watters, 1988). A network of extensional troughs (Fossae Pavlova) also flanks Dorsa Aldrovandi on the margin side of the ridge system. These troughs trend parallel to subparallel to Aldrovandi (Figure 4.9), a rare pattern in lunar mare but common in the Caloris basin of Mercury (Watters and Nimmo, Chapter 2).

Elevation offsets across other mare ridges are not as large compared to Dorsa Aldrovandi (see Golombek *et al.*, 1991). It has been suggested that elevation offsets across mare ridges, as in the case of Dorsa Aldrovandi, are an indication of deeply rooted thrust faults (Golombek *et al.*, 1991). These deeply rooted thrust faults separate crustal material into structural blocks resulting in elevation steps of the mare surface. An alternative explanation for the elevation offsets across some of the mare ridges is that they are an artifact of regional slope, and the apparent offset is due to the short-wavelength topography of the ridge superposed on the long-wavelength topography of the mare surface (see Sharpton, 1992; Watters and Robinson, 1997). In Mare Serenitatis, for example, the lowest elevations in the basin generally lie outside the prominent ring of mare ridges in the interior (Plates 6A,B, 7). Thus, the elevation of the mare surface on the margin side appears to be lower than the interior side. The exception to this regional trend is Dorsa Aldrovandi (see above).

2.1.3 Subsurface structure at wrinkle ridges

The first application of electromagnetic (EM) sounding to planetary exploration was the Apollo 17 Lunar Sounder Experiment (ALSE). The radar sounder revealed subsurface reflecting horizons in the basalts of Mare Serenitatis and Mare Crisium (Phillips *et al.*, 1973; Peeples *et al.*, 1978; Maxwell and Phillips, 1978). Echoes from the 5 MHz frequency of ALSE indicated two subsurface reflectors at depths of 0.9

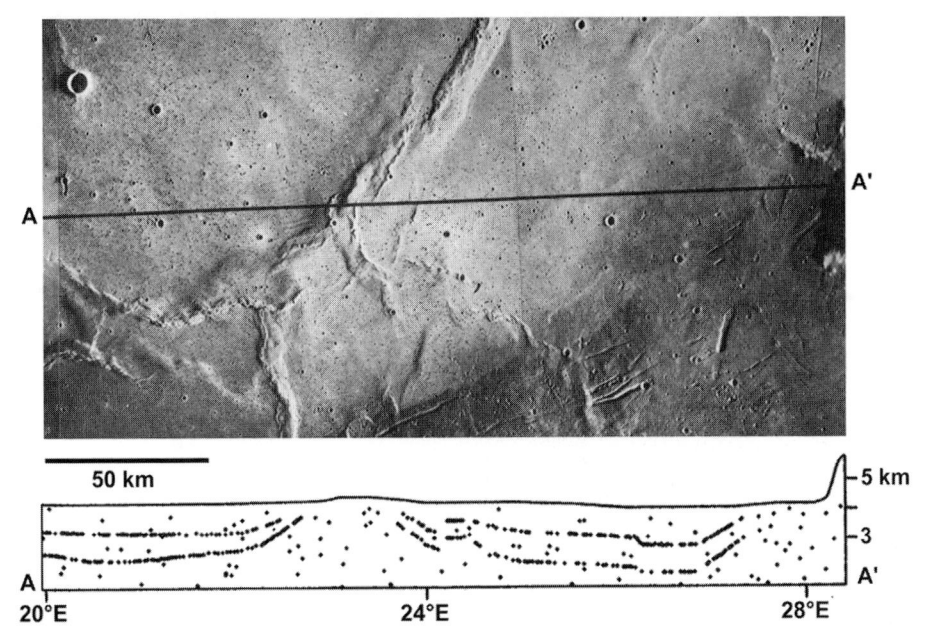

Figure 4.11. Apollo Lunar Sounder Experiment (ALSE) radar returns over part of Mare Serenitatis and a segment of the mare ridge Dorsa Lister. Echoes from the 5 MHz frequency of ALSE indicate two subsurface reflectors at depths of 0.9 and 1.6 km below the surface of Mare Serenitatis. Figure taken from Peeples *et al.*, 1978, Plate 1.

and 1.6 km below the surface of Mare Serenitatis (Figure 4.11) and at 1.4 km below the surface of Mare Crisium (Peeples *et al.*, 1978; Maxwell and Phillips, 1978). They were interpreted to be deep-lying density inversions consisting of regolith or pyroclastic deposits (Peeples *et al.*, 1978). The thicknesses of the interbeds were estimated to be on the order of several meters. Radar sounder data obtained by Kaguya also shows evidence of subsurface reflectors in Mare Serenitatis (Ono *et al.*, 2009). The ALSE data over Dorsa Lister shows that the subsurface radar reflectors dip away from the center of the ridge (Figure 4.11) (see Maxwell, 1978). The data also shows evidence of thinning of a mare unit (sequence of flows) on apparent structural relief (Maxwell, 1978). It is likely that the mare basalt thins on structural relief due to faulting and buckling over a pre-mare topographic prominence in the basin floor (see Maxwell, 1978; Sharpton and Head, 1982; Watters, 1988). Kaguya sounder data also indicate that subsurface reflectors beneath wrinkle ridges curve upward suggesting folded basalt layers (Ono *et al.*, 2009). The ring of mare ridges in Mare Serenitatis and other mare basins strongly suggests that wrinkle ridge thrust faults were localized by pre-mare topography, particularly interior basin rings (Maxwell *et al.*, 1975). These patterns of mare ridges are the only basis for identifying the location of the inner rings of Mare Serenitatis and other mare

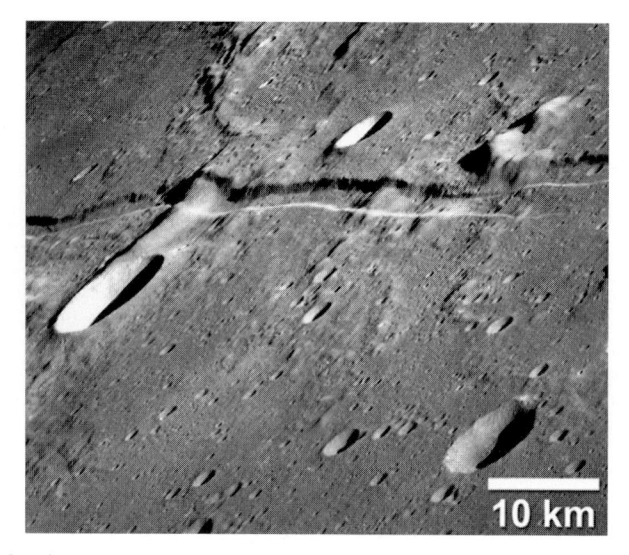

Figure 4.12. Linear graben of Rima Ariadaeus (6.5°N, 12.7°E). The troughs have a symmetric cross-sectional geometry, flat floors and steeply dipping walls. The trough walls maintain roughly the same relief in both mare basalts and highlands material suggesting that the bounding normal faults of the graben have about the same dip (Apollo Hasselblad Camera frame AS10-4645).

basins (Wilhelms, 1987). This suggests that subsurface discontinuities play a major role in localizing mare ridges. Impact craters buried by mare basalts also influence the formation of wrinkle ridges. Arcuate segments of the Dorsa Aldrovandi that include the Littrow ridge clearly indicate the influence of a buried impact crater in the basement of Mare Serenitatis (Figure 4.9). Wrinkle-ridge rings formed over shallow buried impact craters are common in ridged plains on both Mercury and Mars (Watters, 1993; Watters and Robinson, 1997; Watters and Nimmo, Chapter 2).

2.2 Lunar graben

Analogous landforms to linear and arcuate rilles or troughs, first identified in Earth-based telescopic observations, are now known to occur on Mercury, Venus, and Mars, many of the icy satellites of the outer planets, and even some small solar system objects. Their highly symmetric cross-sectional geometry and characteristic flat floors and steep inward dipping walls (Figure 4.12) have led to the nearly unanimous interpretation that they are graben formed by crustal extension (Baldwin, 1963; Quaide, 1965; McGill, 1971; Lucchitta and Watkins, 1978). The spatial correlation between linear and arcuate rilles and the nearside maria is striking (Plate 6A). Equally striking is the absence of extensional troughs on the farside (Scott *et al.*, 1977), outside of Mare Orientale (Plate 6B).

Figure 4.13. Arcuate graben system of Rimae Hippalus (23.5°S, 29°W). The graben are regularly spaced over a ~50-km wide zone east of Mare Humorum (Lunar Orbiter frame IV-132-H1).

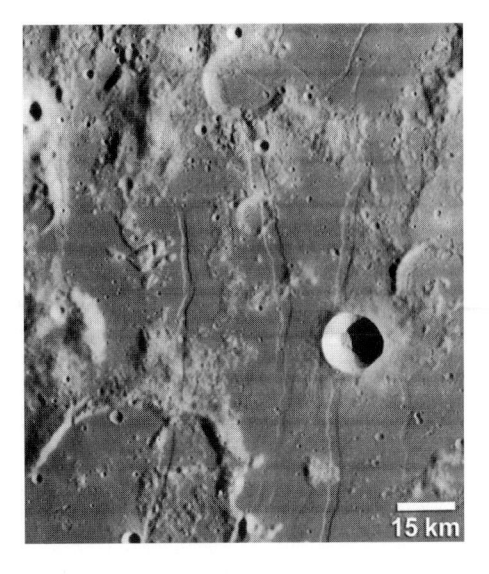

The strongly symmetric cross-sectional geometry and the lack of offset of the trough walls across the structure (i.e., the walls maintain roughly the same relief as they extend from mare basalts into highlands material) indicate that the bounding antithetic normal faults of the graben have about the same dip (Figures 4.12, 4.13). This class of graben and analogue structures is often described as simple graben (see Golombek, 1979; Golombek and McGill, 1983). Complex graben by contrast exhibit crosscut and offset floors and walls, reflecting multiple episodes of extension. These structures are common in broad zones of crustal extension and rifting, not found on the Moon. Zones of basin-localized extension on the Moon are more distributed in nature, and graben are often regularly spaced. The most striking example is the regularly spaced, arcuate graben of Rimae Hippalus, just east of Mare Humorum (Figure 4.13).

Kinematic models for simple graben differ on the geometry of the faults as they converge at depth. One kinematic model has the two faults intersecting in a mechanical discontinuity (Golombek, 1979; Golombek and McGill, 1983). Failure initiates in the discontinuity and conjugate faults develop. In the conjugate fault model, the fault plane dip and the width of the graben expressed at the surface may be used to infer the depth of the mechanical discontinuity (Golombek, 1979; Golombek and McGill, 1983). On the Moon the mechanical discontinuity is assumed to be the megaregolith (Golombek, 1979; Golombek and McGill, 1983). An alternative kinematic model involves the development of a master fault that triggers the formation of the secondary antithetic fault (Melosh and Williams, 1989). In this model the initial fault is the major factor controlling graben width. Finite element modeling suggests that the presence of mechanical discontinuities has only a small effect on the width of the graben (Melosh and Williams, 1989). The master fault

138   *Planetary Tectonics*

Table 4.2 *Dimensions of lunar arcuate and linear rilles or graben*

Index	Latitude	Longitude	Maximum Depth (m)	Maximum Width (km)	Length (km)	D $\theta = 60°$ (m)
Brackett	17.5°N	23.2°E	230	1.71	93	266
Tetrazzini	26°N	0.6°W	140	5.0	50	162
Patricia	24.8°N	0.5°E	100	1.4	10	115
Bradley	24°N	1°W	400	3.25	127	462
Alphonsus	13°S	1.8°W	120	0.8	51	139
Littrow	22.2°N	29.4°E	280	2.25	34	323

Depth and dimensions were determined using Lunar Topographic Orthophotomaps (LTOs).

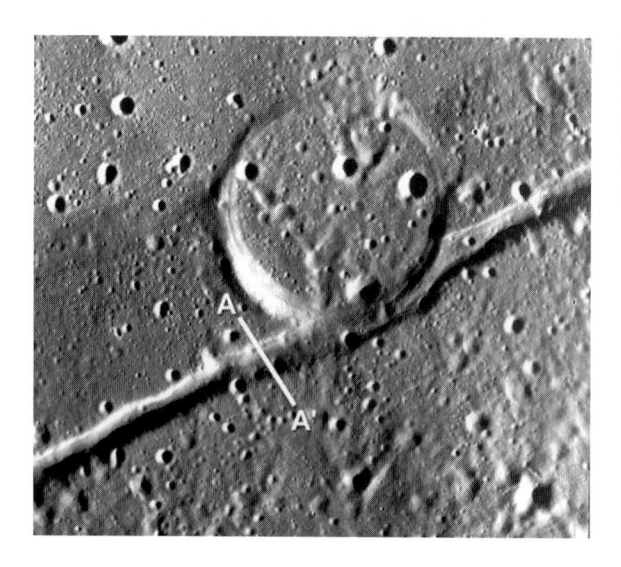

Figure 4.14. Brackett graben of Fossae Plinius (17.8°N, 23.5°E). Brackett is an arcuate graben on the southern margin of Mare Serenitatis (Panoramic Camera frame AS17-2331).

kinematic model is also supported by field studies of the graben in Canyonlands National Park (Moore and Schultz, 1999).

2.2.1 Topography of lunar graben

As with mare ridges, the best available topography for lunar graben is from Lunar Topographic Orthophotomaps. However, only a few lunar graben are covered by LTOs with sufficient resolution. The maximum relief of six lunar graben segments was measured (Table 4.2). The measured graben have a maximum relief that varies from about 100 to 400 m (mean ~210 m). The maximum width of these lunar graben ranges from about 800 m to 5 km (mean ~2.4 km). Profiles across the graben confirm that they are characterized by relatively steep walls and flat floors. A topographic profile across a prominent arcuate graben in southern Mare Serenitatis (a graben of Fossae Plinius) shows that it is relatively narrow (~1.7 km in width)

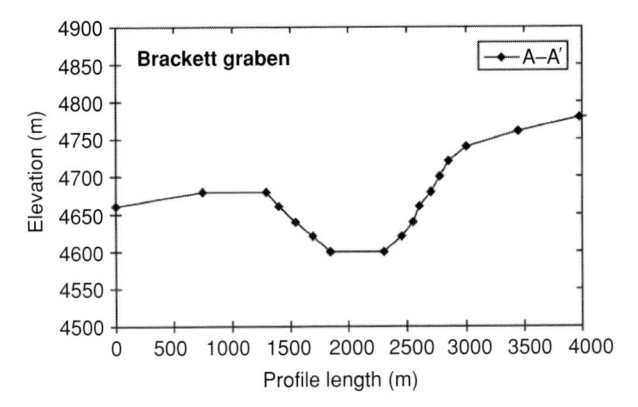

Figure 4.15. Topographic profile across Brackett indicates that the southern wall of the graben is offset relative to the northern wall. The topographic profile was constructed from a 1:50 000 LTO (42C4S1). Profile location is shown in Figure 4.14. Elevations are relative to an arbitrary zero vertical datum of 1 730 000 m.

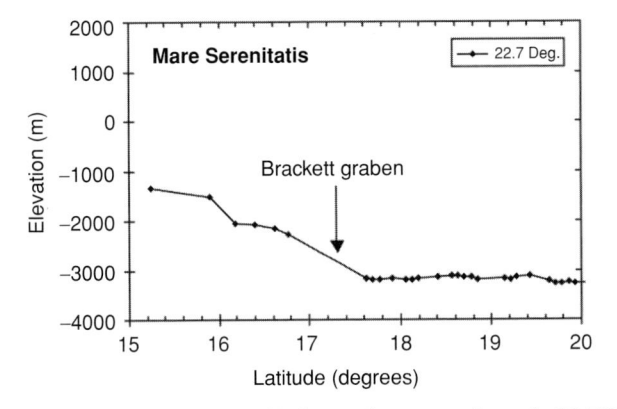

Figure 4.16. Clementine LIDAR profile located at approximately 22.7°E longitude crossing Brackett graben in Mare Serenitatis. The approximate location of Brackett graben is shown by the arrow. The LIDAR data were extracted from the dataset of Smith *et al.* (1997). Elevations are in meters above an ellipsoid of radius 1738 km at the equator with a flattening of 1/3234.93 corresponding to the flattening of the geoid. The vertical exaggeration is ~16:1.

with a flat floor (Figure 4.14). The graben appears to crosscut the southern end of Dorsum Nicol (Figure 4.5). The graben walls have relatively shallow maximum slopes ranging from ~8° to 12°. The topography also shows an offset across the graben of 60 m (Figure 4.15). This indicates that the graben is formed on a sloping mare surface, increasing in elevation to the south. Clementine LIDAR data show a rapid increase in elevation of the mare surface near the margins of Mare Serenitatis where arcuate graben of Fossae Plinius are located (see Plate 7, Figure 4.16). A topographic offset also occurs across an arcuate graben of Fossae Littrow, located

140   *Planetary Tectonics*

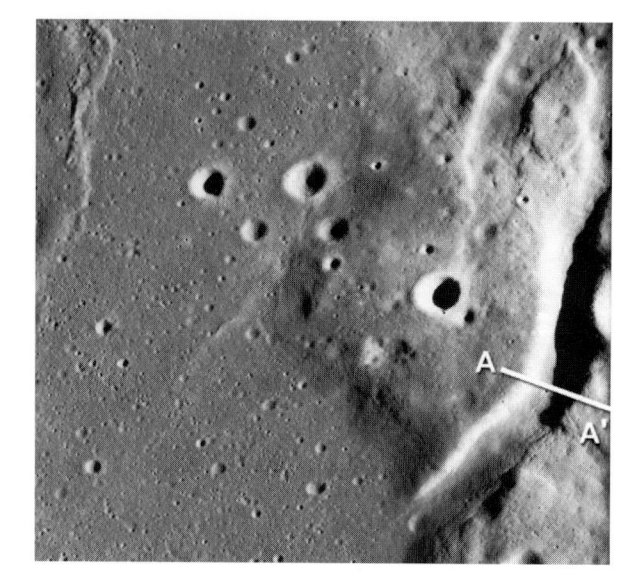

Figure 4.17. The Littrow graben of Fossae Littrow (22°N, 29.3°E) on the eastern margin of Mare Serenitatis. The Littrow graben abruptly terminates in the mare suggesting that it predates the emplacement of mare basalts that flooded the eastern margin of the basin (Panoramic Camera frame AS17-2313).

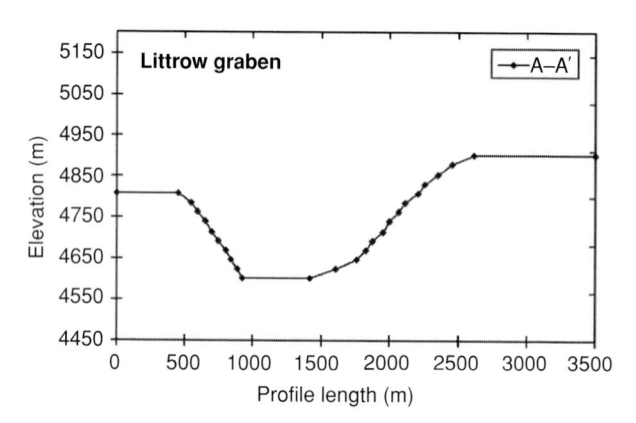

Figure 4.18. Topographic profile across the Littrow graben indicates an offset of the eastern wall relative to the western wall. The topographic profile was constructed from a 1:50 000 LTO (42C2S1). Profile location is shown in Figure 4.17. Elevations are relative to an arbitrary zero vertical datum of 1 730 000 m.

on the eastern margin of Mare Serenitatis (Figure 4.17). The Littrow graben has a maximum measured relief of ~280 m, and a maximum width of ~2.3 km. The walls of the Littrow graben are steeper than the Plinius graben slopes, ranging from ~16° to 23°, and the elevation offset across the structure is 80 m (Figure 4.18). Of the lunar graben examined (Table 4.2), most that are circumferential to mare basins exhibit elevation offsets (e.g., Plinius, Littrow, Tetrazzini). The longest and deepest of the graben examined is Fossa Bradley (Table 4.2). Bradley is a linear graben, circumferential to Mare Imbrium and exhibits no significant elevation offset.

2.2.2 Crater floor graben

Graben and fractures in the floors of impact craters are common lunar tectonic structures (Schultz, 1976b; Wilhelms, 1987). Often these graben form a rough polygonal pattern. Examples are the graben of Fossae Alphonsus that occur in the floor material of the Alphonsus impact crater located near the edge of Mare Nubium (Carr, 1969; McCauley, 1969). This 115 km diameter highland crater was the site of impact of Ranger 9 (see McCauley, 1969). Topographic profiles across a graben of Fossae Alphonsus, the narrowest of the graben examined, indicate no elevation offset (Table 4.2). It is generally agreed that uplift of the crater floor is the source of the stresses that form the graben and fractures. The fractured floor material is often at a higher elevation relative to the rims than the floors of craters without fractures (Pike, 1971; Wilhelms, 1987). The cause of the uplift, however, is not completely clear. Uplift and subsequent faulting of the crater floor material may have resulted from viscous relaxation (Hall *et al.*, 1981). Alternatively, uplift may result from igneous intrusions since many of the floor-fractured craters occur near maria (Brennan, 1975; Schultz, 1976b; Wichman and Schultz, 1995). Dombard and Gillis (2001) used finite element analysis to model elastoviscoplastic deformation of a reasonable analogue to lunar crustal materials and concluded that topographic relaxation cannot account for the majority of floor-fractured craters (diameters less than ∼60 km). Thus, they favor igneous intrusion over relaxation.

2.2.3 Rupes Recta normal fault

The vast majority of the extensional landforms on the Moon are troughs, clearly reflecting a set of graben-forming antithetic normal faults. There are, however, rare cases of extension involving a single normal fault expressed as a scarp. The best example is Rupes Recta, commonly called the "Straight Wall." Located in Mare Nubium, it casts a prominent shadow on the mare surface during the lunar sunrise, making it a favorite among amateur astronomers. Rupes Recta is generally linear over much of its ∼112 km length (Figure 4.19). The fault cuts mare basalts that partially flooded a pre-Nectarian crater on the southeastern edge of Nubium. Wrinkle ridges appear to outline the western half of this ancient crater; the buried rim likely localized the mare ridges (Plate 6A). The most curvilinear section of the scarp is near its southern terminus. This is also where the Rupes Recta normal fault extends from the Nubium mare basalts into an embayed inlier of highlands material (Figure 4.19). Also remarkable is the lack of segmentation of the fault. There is only one clear fault segment; the southernmost segment where the fault is the most curvilinear. Other segment boundaries may be marked by several slump blocks along the scarp wall. The elevation of the Rupes Recta has been estimated from shadow measurement to be a maximum of ∼300 m. Clementine LIDAR data across

142   *Planetary Tectonics*

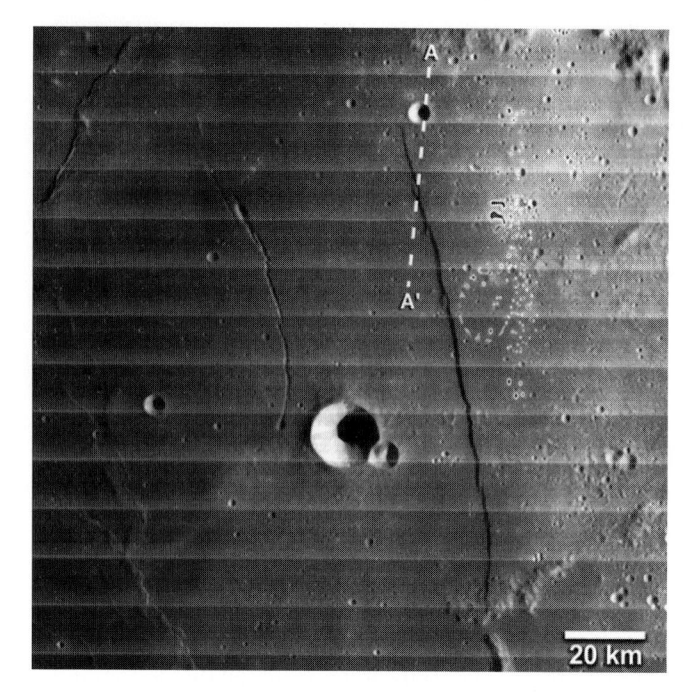

Figure 4.19. Rupes Recta located in southeastern Mare Nubium. The scarp, commonly known as the "Straight Wall," is the surface expression of a normal fault (mosaic of Lunar Orbiter frames IV-133-H1, H2).

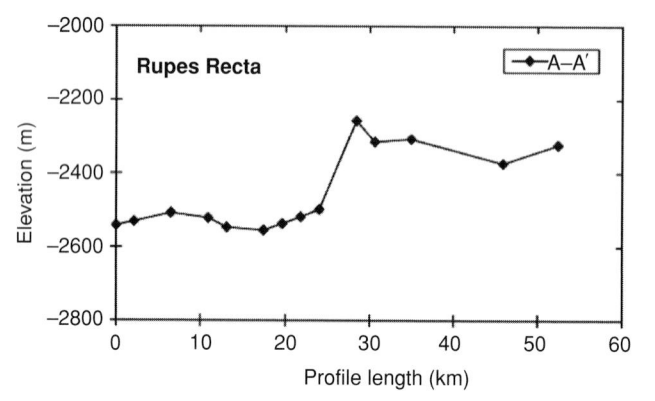

Figure 4.20. Clementine LIDAR profile crossing Rupes Recta. The LIDAR data were extracted from the dataset of Smith *et al.* (1997). Elevations are in meters above an ellipsoid of radius 1738 km at the equator with a flattening of 1/3234.93 corresponding to the flattening of the geoid. Profile location is shown in Figure 4.19. The vertical exaggeration is 48:1.

the northern section of Rupes Recta indicates a maximum relief of ∼240 m at this location (Figure 4.20). The Clementine orbit track crosses Rupes Recta at an acute angle (Figure 4.19) and traverses one of the slump blocks along the wall. Thus, the slope at this location cannot be accurately measured from the LIDAR profile. However, with a measured relief of ∼240 m and a scarp face width of ∼600 m (measured from a Clementine 750 nm mosaic), the maximum slope at this location

Figure 4.21. Lobate scarp in the farside highlands (6.8°N, 129.7°E). This series of en-echelon stepping lobate scarps, also referred to as Morozov scarp for a nearby crater (Binder and Gunga, 1985), has an estimated maximum relief of ~20 m and a combined length of ~10 km. The scarp cuts a small impact crater suggesting it is the surface expression of a thrust fault (Panoramic Camera frame AS16-4970).

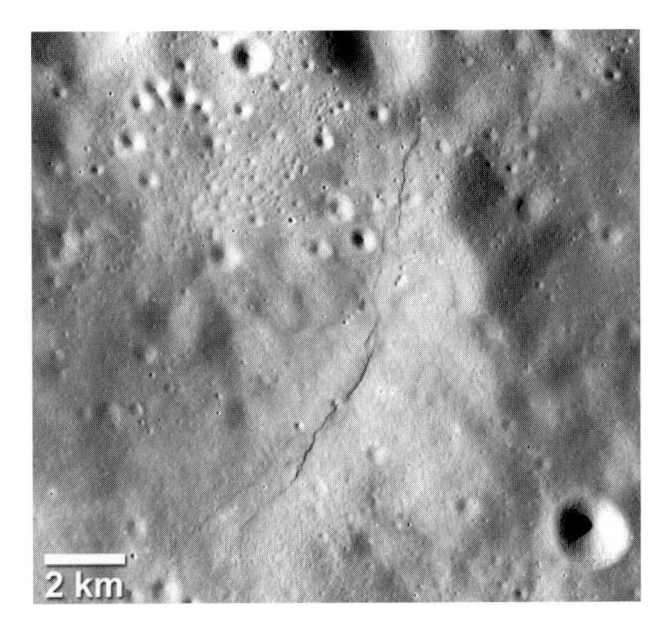

is >20°. This is in the upper range of slopes measured on lunar graben walls (see Section 2.2.1). The LIDAR profile also shows that the scarp is flanked by a rise (Figure 4.20). The presence of the rise is subtly expressed in high-resolution Lunar Orbiter images of Rupes Recta. Flanking topographic rises are also associated with graben in the Caloris basin on Mercury (see Watters and Nimmo, Chapter 2).

The Rupes Recta normal fault appears to be relatively young. Because the fault offsets the floor of Mare Nubium, it clearly postdates the emplacement of the mare basalts. A young age is also suggested by the fact that the fault cuts and offsets the rim walls of two small impact craters (Figure 4.19). The larger of the two craters is ~2 km in diameter. No craters of this diameter or larger are superposed on the fault scarp.

2.3 Lunar scarps

Unlike the other major tectonic features of the Moon, the lunar lobate scarps were not detected in Earth-based telescopic surveys. This is because lunar scarps are generally small-scale structures compared to many mare ridges and rilles. In fact, most lunar scarps are only easily resolved in the highest resolution images of the lunar surface, acquired by the Apollo Panoramic Cameras (Mattingly *et al.*, 1972; Schultz, 1976a; Masursky *et al.*, 1978). In a survey of the lunar highlands using Apollo 15, 16 and some 17 Panoramic Camera images, Binder and Gunga (1985) found a total of 71 scarps. They grouped these features into three broad morphologic classes: linear, arcuate, and irregular. Many of the scarps consist of a series of

Figure 4.22. Lobate scarp segment in series that cut Mandel'shtam crater in the farside highlands (6.9°N, 161.5°E). The scarp is lobate with a relatively steeply sloping scarp face and a gently sloping back scarp (Panoramic camera frame AS16-4150).

en-echelon stepping segments (Figure 4.21). Other scarps occur in clusters that cover relatively small areas, and the scarps are closely spaced and exhibit parallel to subparallel orientations (Binder and Gunga, 1985, Fig. 7). Lunar scarps are often lobate, one-sided structures with a relatively steeply sloping scarp face and a gently sloping back scarp (Figure 4.22). The lobate nature of many lunar scarps led the Apollo 16 astronauts to describe them as having the appearance of flow fronts (Mattingly et al., 1972). They also noted reversals in the direction of the scarp face along strike. Reversals in vergence of en-echelon stepping scarps is a common characteristic of planetary lobate scarp thrust faults (Watters and Nimmo, Chapter 2; Golombek and Phillips, Chapter 5). Many of the identified lobate scarps occur in the lunar highlands (Plates 6A,B) and are thus often referred to as "highland scarps". However, some lobate scarps are found in mare basalts, specifically in Mare Nectaris, Mare Cognitum, and Mare Serenitatis (Plate 6A).

2.3.1 Topography of lunar scarps

The resolution of nearly all the currently available topographic data for the Moon is too low to characterize lunar lobate scarps. Only topography derived from Panoramic Camera stereo pairs has sufficient spatial and vertical resolution. Such high resolution topographic maps were generated for the Apollo landing sites.

Figure 4.23. Lee-Lincoln scarps in the Tauras-Littrow valley. The Lee-Lincoln thrust fault scarp cuts across the mare basalt floor of the Tauras-Littrow valley and extends up onto the highlands of North Massif. This fault scarp was examined and traversed by the Apollo 17 astronauts Eugene A. Cernan and Harrison H. Schmitt. The "X" marks the approximate location of the Apollo 17 landing site (Panoramic camera frame AS17-2309).

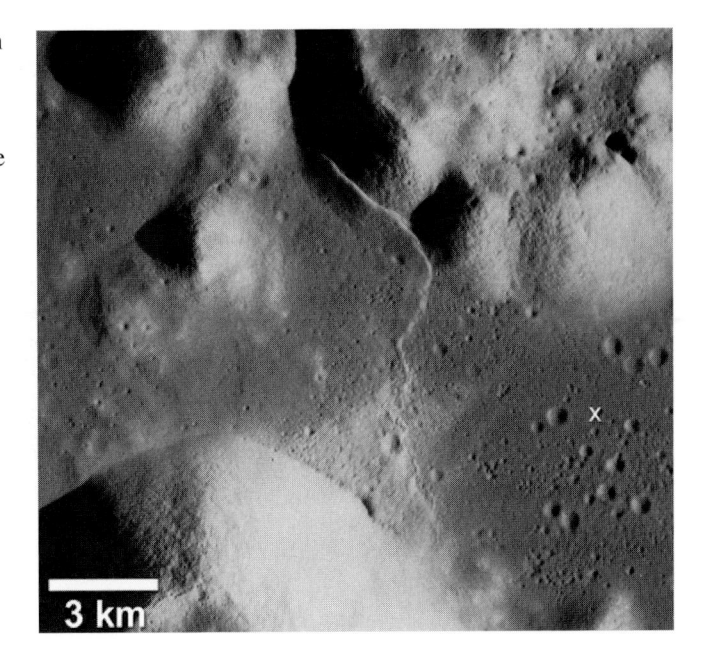

The Apollo 17 landing site is located in the Tauras-Littrow Valley on the southeastern margin of Mare Seneratitis. The Lee-Lincoln scarp cuts the floor of the Tauras-Littrow Valley, trending roughly north–south between the South Massif and North Massif (Schmitt and Cernan, 1973; Scott, 1973). The scarp extends into the highlands of the North Massif, cutting upslope, but abruptly changes trend to the northwest, cutting along slope (Figure 4.23). There are two special scale topographic maps of the Tauras-Littrow Valley, a 1:25 000 scale LTO and a 1:50 000 scale U.S. Geologic Survey map (USGS, 1972). Robinson and Jolliff (2002) generated a DEM of the Tauras-Littrow Valley using the USGS map (Plate 8). The Lee-Lincoln scarp has the greatest relief ($\sim$80 m) in the Tauras-Littrow Valley where the thrust fault offsets the mare basalts that make up the floor of the valley (Plate 8). The relief of the scarp in the highlands is markedly less than the maximum reached in the valley. The largest measured slope on the scarp face is $\sim$27° and occurs in the valley on the southern section of the scarp, near the South Massif. Just before the Apollo 17 astronauts started up the slopes of the southern section of the Lee-Lincoln scarp in the LRV, Harrison Schmitt described the scarp as very rolling and relatively smooth with no exposed outcrops. The maximum slope of the scarp face near where Cernan and Schmitt traversed the scarp is $\sim$20°.

In an effort to estimate the relief of other lunar lobate scarps, shadow measurements were made on scarp segments near the craters Madler A and Morozov. Portions of Panoramic Camera images (second generation positive film transparencies) were scanned at high resolution and shadows were measured from the digital

Table 4.3 *Dimensions of lobate scarps on the Moon*

Index	Latitude	Longitude	Maximum Relief (m)	Length (km)	D $\theta = 30°$, (m)
Madler S1	10.8°S	31.8°E	53	1.75	106
Madler S2	10.8°S	31.75°E	25	1.58	50
Madler S3	10.7°S	31.5°E	54	7.77	108
Madler S4	10.6°S	31.45°E	38	4.94	76
Morozov S1	6.76°N	129.68°E	17	5.72	34
Morozov S2	6.85°N	129.73°E	7	0.72	14
Morozov S3	6.95°N	129.75°E	6	1.25	12
Morozov S4	6.99°N	129.75°E	10	1.78	20
Lee-Lincoln	20.27°N	30.56°E	80	13.86	160

Relief was determined using shadow measurements made on Apollo Panoramic camera images with the exception of Lee-Lincoln. Relief on the Lee-Lincoln scarp was determined from a digital elevation model (DEM) of the Taurus-Littrow Valley (see Plate 7).

images. The estimated maximum relief of the scarps measured varies from 6 ± 2 to 54 ± 2 m ($n = 8$) (Table 4.3). The mean maximum relief of all the scarps measured (including the Lee-Lincoln scarp) is $\sim$32 m ($n = 9$).

2.4 Wrinkle ridge – lobate scarp transitions

Mare ridges and lobate scarps have distinct differences in morphology. As described above, wrinkle ridges are complex structures that may have multiple superposed landforms. Lobate scarps, by contrast, are relatively simple structures morphologically, usually consisting of a relatively steep scarp face and a gently sloping back scarp. Both structures, however, are to some degree surface manifestations of thrust faults and are thus related tectonic landforms. This relationship is clearly demonstrated by wrinkle ridge – lobate scarp transitions. A prominent example is found on the eastern edge of Oceanus Procellarum where a mare ridge extends to the Montes Riphaeus, which separates Procellarum and Mare Cognitum (Masursky *et al.*, 1978). In the highlands of Montes Riphaeus ($\sim$8°S, 28°W) the lobate scarp consists of both linear and lobate segments and has a rare subsidiary, flanking scarp (Figure 4.24). Although rare, multiple, flanking (or piggybacking) scarps are also found on Mercury (Watters and Nimmo, Chapter 2). The Montes Riphaeus scarp cuts directly across the highlands into Mare Cognitum. In Cognitum, the lobate scarp is expressed by a series of linear segments that appear to be localized at the contact between the mare and highlands (Figure 4.24) (see Masursky *et al.*, 1978). Some wrinkle ridge – scarp transitions are not located at the margins of mare basins. In the highlands south of Mare Australe ($\sim$53°S, 104°E), a lobate scarp transitions

Figure 4.24. Wrinkle ridge – lobate scarp transition in the area of Montes Riphaeus between Mare Cognitium and Oceanus Procellarum. The lobate scarp cuts the highlands of Montes Riphaeus and has two distinct terraces in the central montes that may reflect imbricate thrust faulting (Panoramic Camera frame AS16-5452).

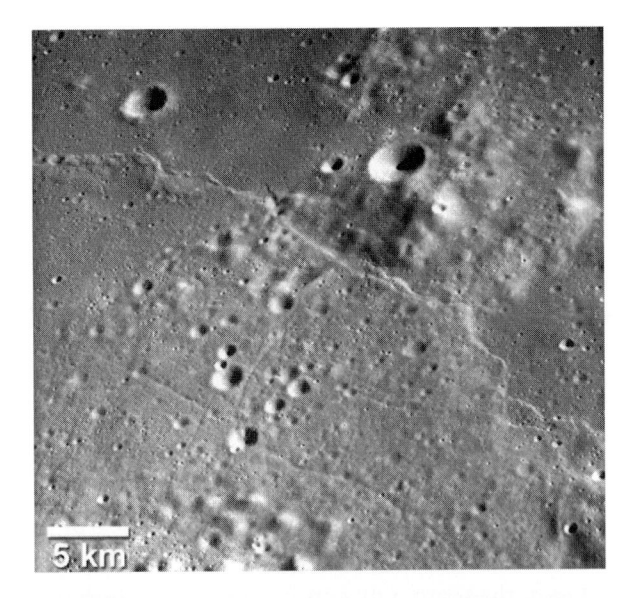

Figure 4.25. Lobate scarp in the highlands south of Mare Humorum. The northeast trending lobate scarp (left-pointing black arrows) extends across a small patch of mare basalts in the valley where it transitions into a wrinkle ridge (white arrow). The wrinkle ridge crosses the mare basalts to the highlands where the structure transitions back to a lobate scarp (right-pointing black arrow) and cuts across and along slope (Lunar Orbiter frame IV-136-H3).

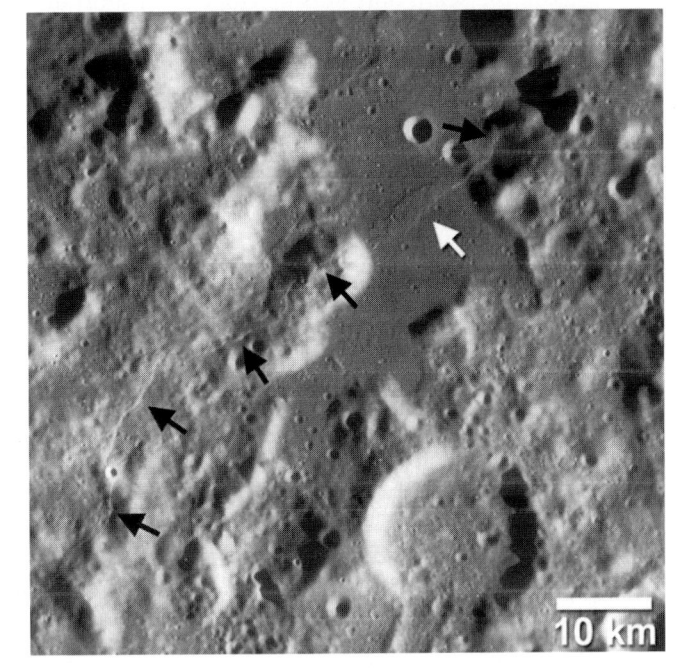

to a mare ridge as the structure extends into the mare-filled Kugler crater (Schultz, 1976a; Raitala, 1984). A very similar ridge – scarp transition involving an impact crater filled with volcanic plains occurs on Mars in the area just northeast of the Herschel basin (Watters, 1993, Fig. 4b). One of the most unusual wrinkle ridge – lobate scarp transitions is found in the highlands south of Mare Humorum. Here a lobate scarp transitions into a wrinkle ridge where the structure crosses a small valley filled with mare basalts (Figure 4.25). The wrinkle ridge then transitions back to a lobate scarp in the highlands to the north of the mare basalts, where the scarp cuts across and along slope.

There are other cases of wrinkle ridge – lobate scarp transitions on the Moon where the scarps parallel mare–highland contacts. One is associated with Dorsa Aldrovandi, located on the eastern margin of Mare Serenitatis (Figure 4.9) (see Howard and Muehlberger, 1973). As described previously (see subsection 2.1.2), segments of Dorsa Aldrovandi (like the Littrow ridge) have large elevation offsets from the basin-interior side to the margin side of the ridge segments. The northernmost segment of the ridge reaches the highlands at the basin margin. The structure, however, does not terminate at the contacts between the mare basalts and the highlands. The structure extends into the highlands as a lobate scarp, cutting along slope (Howard and Muehlberger, 1973, Figs. 31–33). This is also the case for a transition on the western margin of Mare Serenitatis, referred to by Lucchitta (1976) as the West Serenitatis scarp. As described in subsection 2.3.1, where the Lee-Lincoln scarp extends from the Taurus-Littrow valley into the North Massif, the scarp only cuts up slope for a short distance before its trend changes to the northwest, where it cuts along slope (Figure 4.23). Although the Lee-Lincoln scarp has been described as a wrinkle ridge – lobate scarp transition (Howard and Muehlberger, 1973; Lucchitta, 1976; Watters, 1988), only the southernmost segment (near South Massif) has some characteristics of wrinkle-ridge morphology. Much of the segment of the scarp in the mare basalts of the valley floor has the morphology of a lobate scarp (Figure 4.23). The fact that in some cases lobate scarp thrust faults do not cut across highland massifs but roughly parallel the contacts between the mare basalts and the highlands is an important constraint on the kinematics of their formation. Howard and Muehlberger (1973) suggested that the lobate scarp thrust faults are rooted in a decollement, formed at the base of the highlands regolith.

The change in deformation style between the mare basalts and the highlands material may be the result of a contrast in the mechanical properties of the materials (Watters, 1988), specifically the presence or absence of layering (Watters, 1991). Photogeologic evidence and radar sounder data (see subsection 2.1.4) suggest the mare basalts that flooded the nearside basins consist of a sequence of flows separated by interbeds. The highlands materials by contrast are generally mechanically isotropic (i.e. lacking layering). The presence of layering, however, may not result in a sufficient contrast in the mechanical properties. The strength of a multilayer is determined by the mechanical nature of the contacts between the layers (Johnson, 1980). A multilayer will have a greater tendency to deform by folding and faulting when the contacts between layers have low yield strengths (Johnson, 1980). For example, an upward propagating blind thrust fault will induce folding of a multilayer if the layer contacts are weak (Nino et al., 1998). If on the other hand the contacts are strong, faulting is favored and the fault is expected to propagate through the sequence and become surface-breaking (Roering *et al.*, 1997; Nino *et al.*, 1998). Since lobate scarps appear to form over surface-breaking thrust faults, if there is layering in the lunar highlands the contacts must be strong and resist slip. In mare basalt sequences, slip between layers probably occurred in interbeds with low shear strengths (see Watters, 1991).

2.5 Displacement–length relationships of lunar tectonic features

It has been shown that the maximum displacement D on a fault scales with the planimetric length L of the fault (e.g., Walsh and Watterson, 1988; Cowie and Scholz, 1992a,b; Gillespie *et al.*, 1992; Cartwright *et al.*, 1995). The scaling relationship applies to normal, strike-slip, and thrust faults found in a wide variety of terrestrial tectonic settings and over eight orders of magnitude in length scale (Cowie and Scholz, 1992b). Planetary faults also exhibit this scaling relationship (Schultz and Fori, 1996; Schultz, 1997, 1999; Schultz *et al.*, Chapter 10; Watters and Robinson, 1999; Watters *et al.*, 1998, 2000; Watters, 2003). The displacement, D, is related to fault length, L, by $D = cL^n$, where c is a constant related to material properties and n is estimated to be >1 (Walsh and Watterson, 1988; Marrett and Allmendinger, 1991; Gillespie *et al.*, 1992). Studies of terrestrial fault populations formed in uniform rock types suggest the relationship $D = \gamma L$ is linear, where the ratio γ is a constant determined by rock type and tectonic setting (for $n = 1$, $\gamma = c$) (Dawers *et al.*, 1993; Clark and Cox, 1996; Hardacre and Cowie, 2003). Cowie and Scholz (1992a) suggest that γ reflects the mechanical properties of the rock and the regional stresses. The values of γ for the fault populations analyzed by Cowie and Scholz (1992b) range from 10^0 and 10^{-3}. Fault segmentation, uncertainties in fault dip and depth or aspect ratio, interaction with other faults, and ambiguities in determining the maximum value of D along the scarp trace are possible sources of scatter in the D–L data (Cartwright *et al.*, 1995, 1996; Dawers and Anders, 1995; Wojtal, 1996; Schultz, 1999).

Assuming the maximum relief h of the wrinkle ridge is a function of the total slip on the underlying thrust fault and the dip of the fault plane θ, the displacement can be estimated by $D = h/\sin\theta$ (see Wojtal, 1996; Watters *et al.*, 2000). This approach assumes that the bulk of the horizontal shortening is due to displacement on thrust faults and does not account for the component of shortening due to folding (see Watters, 1988; Golombek *et al.*, 1991). The displacement on mare ridge thrust faults is estimated to range from 100 to 820 m with an average of $\sim$510 m ($n = 12$), assuming fault plane dips of 30° (Table 4.1). A linear fit to the D–L data for the mare ridges yields a value of $\gamma \cong 6.5 \times 10^{-3}$ for $\theta = 30°$. This is consistent with estimates of γ for Mercury wrinkle ridges ($\sim 3.5 \times 10^{-3}$ for $\theta = 30°$) (Watters and

Nimmo, Chapter 2). Estimates of γ for analogue structures on the Columbia Plateau (Watters, 1988, 1992) are $2.9 \pm 1.0 \times 10^{-2}$ larger (Mege and Reidel, 2001), likely due to the much larger accumulated shortening on the anticlinal ridges (Reidel, 1984).

The D–L relationship for lunar graben can be determined from the available topography (Table 4.2) if the fault plane dips are known. The slopes of the measured graben walls do not exceed 25° (see preceding section), which is much lower than likely fault plane dips on the bounding normal faults (>45°). The low slopes of the bounding walls are not unique to lunar graben. The walls of graben in the Caloris basin on Mercury also have low slopes (Watters and Nimmo, Chapter 2), and are likely the result of mass wasting related to impact processes. Thus, the fault plane dips of the bounding normal faults were probably much higher than the current slopes of the graben walls.

It has been observed that the width of lunar graben increases with increasing elevation (McGill, 1971; Golombek, 1979). This relationship can be used to estimate the fault plane dip θ using a two point method

$$\tan\theta = 2\Delta E/(W_u - W_l), \tag{4.1}$$

where ΔE is the change in elevation and W_u and W_l are the widths of the graben at the high- and low-elevation points, respectively (McGill, 1971; Golombek, 1979). It is assumed that the bounding faults have roughly equal (and constant along their length) dips and that the faults intersect at a constant depth. A variation of the two point method uses the slope of a linear least-squares fit to a plot of width versus elevation along the lengths of lunar graben to estimate the average dips of the bounding faults (McGill, 1971; Golombek, 1979). McGill (1971) and Golombek (1979) examined a total of 19 lunar graben and reported fault plane dips with a mean of 61°. The large variation in average dip yielded by the linear regression method led to the conclusion that the bounding faults do not have a constant depth of intersection along their lengths (Golombek, 1979).

Assuming the fault plane dips of the bounding faults of lunar graben are $\sim 60°$, the estimated displacement on the normal faults ranges from 0.12 to 0.46 km with an average of ~ 0.24 km ($n = 6$) (Table 4.2). A linear fit to the D–L data for the graben yields a value of $\gamma \cong 3.6 \times 10^{-3}$ ($\theta = 60°$). The range of displacements and γ for the lunar graben are in good agreement with those determined for graben in the Caloris basin on Mercury ($\gamma \cong 2.2 \times 10^{-3}$ for $\theta = 60°$) (Watters et al., 2005; Watters and Nimmo, Chapter 2).

An accurate estimate of the D–L ratio for lobate scarp thrust faults on the Moon is more challenging because of their scale. Using estimates of the maximum relief obtained from the available topography and shadow measurements, and

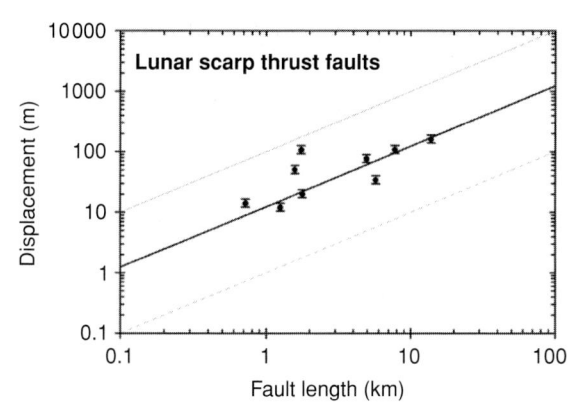

Figure 4.26. Log–log plot of maximum displacement as a function of fault length for nine lunar lobate scarp segments. The ratio γ ($\sim 1.2 \times 10^{-2}$ using estimates of D based on fault plane dips $\theta = 30°$) for the thrust faults (Table 4.3) were obtained by a linear fit to the D–L data with the intercept set to the origin (Cowie and Scholz, 1992b).

measurements of fault segment lengths, the displacement on lunar scarp thrust faults ranges from 12 to 160 m, assuming fault plane dips $\theta = 30°$ (Table 4.3) with a mean of ~ 64 m ($n = 9$). Lengths of the measured scarp segments range from ~ 0.72 to 13.9 km with a mean of ~ 4.4 km. The value of γ, determined from a linear fit to the D–L data for the lunar lobate scarps, for a range in θ of 25° to 35° is $\sim 1.5 \times 10^{-2}$ to $\sim 1.1 \times 10^{-2}$ and $\gamma \cong 1.2 \times 10^{-2}$ for $\theta = 30°$ (Figure 4.26). This range of γ is larger than estimates of γ for larger scale lobate scarp populations on Mercury and Mars ($\sim 6.9 \times 10^{-3}$ and 6.2×10^{-3} respectively for $\theta = 30°$) (Watters, 2003; Watters and Nimmo, Chapter 2), but less than the γ for terrestrial thrust faults ($\sim 8.0 \times 10^{-2}$) (Watters et al., 2000). Thus, the estimated γ for lunar lobate scarps is intermediate between that of lobate scarp thrust faults on Mercury and Mars and terrestrial thrust faults.

A likely explanation for the higher value of γ for terrestrial thrust faults compared to those on the Moon, Mercury and Mars is the friction on the faults (see Watters and Nimmo, Chapter 2; Schultz et al., Chapter 10). The presence of water on Earth lubricates faults, effectively reducing the residual frictional stress σ_f on the faults. The absence of water is thus expected to result in higher values of σ_f on the Moon and Mercury and stronger faults. The similar values of γ for thrust faults on Mercury and Mars suggest the first-order mechanical properties of intercrater plains on the two planets are similar (Watters et al., 2000) and also suggest that water was not abundant in the Martian cratered highlands at the time the thrust faults formed.

3 Timing of wrinkle-ridge, graben, and lobate scarp formation

The age of wrinkle-ridge and graben formation is critical to understanding the evolution of tectonic stresses on the Moon. The age of these tectonic features can be constrained by determining the age of the geologic units they deform. Unlike other planetary surfaces, the ages of geologic units based on crater counts, particularly some mare basalts, can be calibrated in absolute age by radiometric dating of lunar rock samples returned in the Apollo and Luna missions (see Wilhelms, 1987; Neukum and Ivanov, 1994; Stöffler and Ryder, 2001; Hiesinger *et al.*, 2000, 2003).

In an effort to characterize the superposition relationship of linear and arcuate rilles on the nearside, Lucchitta and Watkins (1978) determined that 64% of the combined length of the graben (17 000 km) occurs in highlands, highland plains, and Imbrian crater material, and 35% occurs in mare units (see Plates 9A,B). The large cumulative length of graben superposed on mare units and highland plains material indicates that graben postdate the formation of the nearside impact basins and the emplacement of the mare basalts. This is supported by the strong spatial and azimuthal correlation between linear and arcuate graben and the mare basins (Plate 6A), and the absence of graben on the farside outside of Orientale basin material (Plate 9B). There is no evidence that the observed graben formed before the end of the period of heavy bombardment (Lucchitta and Watkins, 1978). This does not preclude the possibility that extension of the lunar crust occurred before or during the period of heavy bombardment, since no record of these faults would be expected to be preserved. Examination of the superposition relationships of graben that occur in mare basalts suggests that they are restricted to relatively old mare units surrounding Serenitatis, Tranquillitatis and Humorum (Boyce, 1976; Lucchitta and Watkins, 1978) (Plate 9A). Crater density ages of the mare units indicated that no mare basalts younger than $\sim 3.6 \pm 0.2\,\text{Ga}$ are cut by graben (Boyce, 1976; Lucchitta and Watkins, 1978). This is consistent with more recent crater density age estimates of the nearside mare basalts (Hiesinger *et al.*, 2000, 2003). Thus, crustal extension associated with lunar maria appears to have ceased at $\sim 3.6\,\text{Ga}$.

Crustal shortening responsible for wrinkle-ridge formation in the mare basalts was much longer lived. Recent crater density age estimates by Hiesinger *et al.* (2000, 2003) suggest that basalt volcanism in nearside mare initiated $\sim 4\,\text{Ga}$. The largest pulse of mare volcanism occurred in the Late Imbrian period that ended about 3.2 Ga (see Tanaka *et al.*, Chapter 8, Fig. 8.1). Lesser volumes of mare basalt were emplaced during the Eratosthenian and Copernican periods (Hiesinger *et al.*, 2000, 2003). The youngest mare basalt units occur in Oceanus Procellarum, embaying the southern portions of the Aristarchus plateau. The basalts are estimated to be $\sim 1.2\,\text{Ga}$ old (Hiesinger *et al.*, 2003). Wrinkle ridges deform these relatively young mare basalts as well as the oldest mare units on the nearside in Mare Serenitatis, Mare Tranquillitatis, and Mare Nubium (Plate 9A). Thus, crustal shortening associated with lunar mare occurred as recently as ~1.2 Ga.

The timing of wrinkle-ridge formation can be further constrained by the relationship between structural relief and emplacement of the mare basalts. Evidence of ponded flows on mare ridges suggests that structural relief developed simultaneously with and following the emplacement of the mare basalts (Bryan, 1973; Schaber, 1973a,b; Schaber et al., 1976). The greatest relief on the wrinkle ridges appears to have developed after the emplacement of the mare basalts ceased. A similar relationship between emplacement history and structural relief has been observed with anticlinal ridges of the Columbia Plateau (Watters, 1988) (see Section 2.1). Flows of the Columbia River Basalts Plateau are embayed by, and thin or pinch out on the flanks of the anticlines. Buckling was simultaneous with the emplacement of the oldest basalt flows and structural relief from cumulative displacement on thrust faults, and increased fold amplification was repeatedly buried by younger basalt flows (Reidel, 1984). Subsequent to the emplacement of the youngest flow, structural and topographic relief developed together. Thus, as is the case of the Columbia anticlinal ridges, deformation on mare ridges likely increases with depth and age. This suggests a long history of contractional deformation of mare basalts that is closely correlated spatially and temporally with the volcanic flooding of the maria.

The early termination of mascon-related extension and the subsequent dominance of compression in the maria until ~1.2 Ga suggest a compressional stress bias in the lunar lithosphere after ~3.6 Ga. The cessation of extension has been suggested to result from the superposition of compressional stresses from global contraction on flexural extensional stress due to loading by the mare basalts (Solomon and Head, 1979, 1980). This may have marked a stage in the Moon's thermal history around 3.6 Ga where interior cooling resulted in a shift from net expansion to net contraction (Solomon and Head, 1980).

Among the youngest lunar tectonic landforms are the lobate scarps. In fact, they may be some of the youngest endogenic features on the Moon. Schultz (1976a) first noted that the scarps were expressions of very young thrust faults. The age of the lobate scarps has been estimated by Binder and Gunga (1985) to be less than 1 Ga old. A young age for the scarps is supported by their relatively undegraded appearance and the absence of superimposed impact craters (see Figures 4.21, 4.22). Young lobate scarp thrust faults have been cited as evidence to support a late-stage compressional stress bias in the lunar lithosphere. Their young appearance raises the intriguing possibility that some of the lobate scarp thrust faults may be currently active.

4 Lunar seismicity

The Apollo seismic "network" comprised four stations at Apollo sites 12, 14, 15, and 16, forming a roughly triangular array with stations 12 and 14 at one corner, and stations 15 and 16 each about 1100 km away. Each station included a 3-component long-period seismometer and a single, vertical-component short-period seismometer (Lammlein *et al.*, 1974). The passive seismic experiment recorded data from 1969 until 1977, and 12558 events were documented in the lunar event catalogue (Nakamura *et al.*, 1981). Recorded events exhibit different signal characteristics, and about 3000 of the catalogued events were classified as being from four types of sources: artificial impacts (8 events), meteoroid impacts (~1700 events), shallow moonquakes (28 events), and deep moonquakes (1360 events noted in the catalogue of Nakamura *et al.*, 1981). In addition, the thermal signature of the lunar sunrise and sunset is evidenced in the records (see, for example, Fig. 3 of Bulow *et al.*, 2005). Event identification and classification was originally made by eye, using hard copy print-outs and overlays of the seismograms recorded at Apollo stations 12, 14, 15, and 16.

Lunar seismicity provides information on the current state of the Moon's interior. Naturally occurring moonquakes indicate that there are locations or regions that are undergoing brittle failure in response to imposed stresses. Understanding the sources of the imposed stresses, and why certain regions can undergo failure, is important to establishing the present-day tectonic state of the Moon. Insofar as it is possible to estimate stress drops from the Apollo-era seismograms, these provide constraints on the energy release associated with moonquakes. Body-wave arrival times (i.e., P- and S-wave arrivals) recorded at multiple stations for a given moonquake allow the location of the quake to be determined, and simultaneously provide information (albeit limited) on the seismic velocity structure of the lunar interior. Seismic velocities in turn constrain elastic properties, which are controlled by temperature and composition.

There have been numerous investigations of lunar seismicity; recent reviews can be found in Wieczorek *et al.* (2006) and Lognonné and Johnson (2007). In this chapter we focus on the aspects of moonquakes that may, directly or indirectly, shed light on present and past lunar tectonics. We begin by examining the temporal and spatial occurrence of shallow and deep moonquakes, along with inferences regarding quake magnitudes and stress drops.

Interpretation of the Apollo seismic data is difficult for two main reasons. The first is the nature of the seismic station array – there are only three geographically distinct stations, and sensitivity to events is limited to an aperture extending beyond the array by approximately ~1100 km. Clearly then, recorded events will be biased to those occurring on the nearside, and this is evidenced in the data. The second difficulty is the low signal-to-noise of events seen in the seismograms

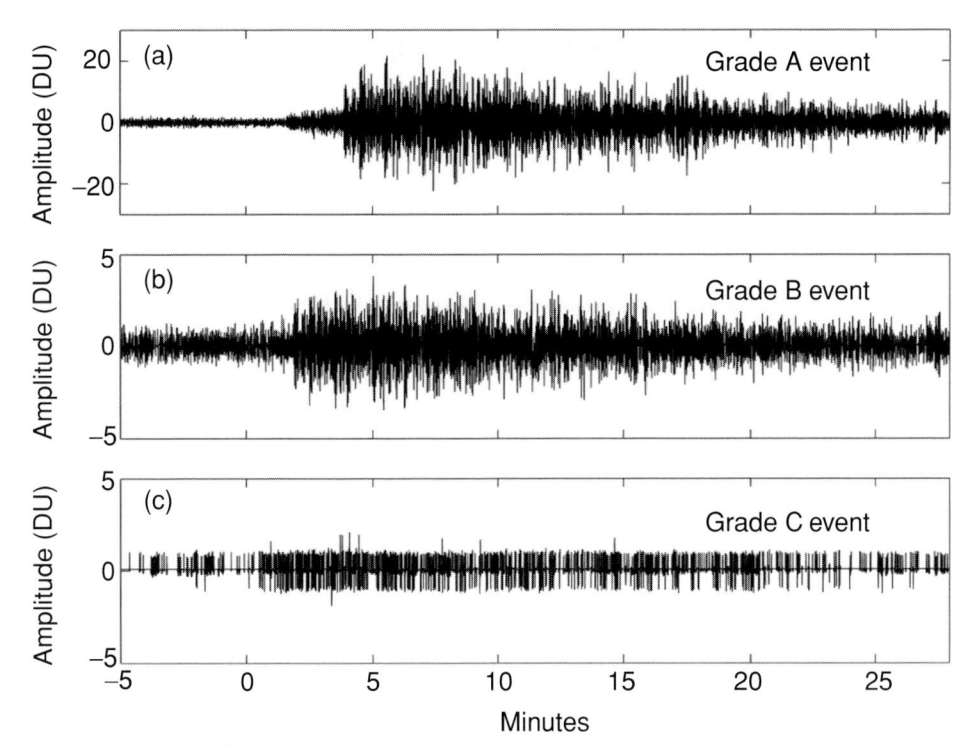

Figure 4.27. Typical data quality, as measured by visual inspection of the seismograms, of events in the lunar catalogue (Nakamura et al., 1981). Waveform amplitudes are in digital units (DU). Approximately 2% of events have quality comparable to that in (a), 8% comparable to that in (b), and 57% comparable to that in (c). The remaining 33% are of poorer quality than that in (c). Figure taken from Bulow et al. (2005).

(Figure 4.27), and this is the combined result of several factors: (a) the limited dynamic range offered by the 10-bit instruments used, (b) the small magnitudes of most lunar events, and (c) long coda (i.e., seismic energy arriving long after the event itself) resulting from scattering in the lunar regolith. Consequently, identification of seismic phases in the seismograms is restricted to first arrivals of P and S waves; ideally these should be observed on as many stations as possible for successful determination of the moonquake location. This is often not the case, contributing to considerable uncertainty in the discussion of geographical variations in lunar seismicity.

4.1 Deep moonquakes

Among the first inferences from observations of the lunar seismic data was that some moonquakes tend to occur with tidal periods (Ewing et al., 1971). Investigations revealed this population of quakes to have waveform characteristics that enabled them to be identified as distinct from other seismic events, such as meteorite impacts (Lammlein, 1973; Lammlein *et al.*, 1974; Lammlein, 1977). Furthermore, it was found that subsets of the population exhibit similar waveforms – originally observed by overlaying seismogram traces, and now quantifiable through digital waveform cross-correlation (Nakamura, 2003, 2005; Bulow *et al.*, 2005, 2007). These observations, together with P- and S-wave arrival times, suggested that the quakes originate from deep (∼800 to 1000 km) source regions that undergo repeated failure, giving rise to sets of moonquakes with similar waveforms and periodic occurrence times (Lammlein *et al.*, 1974). Despite the difficulties encountered in analyzing lunar seismic data, this first-order inference has stood the test of time.

Deep moonquakes have typical Richter magnitudes ∼1, with the largest events having a magnitude ∼3 and associated stress drops of about 10 kPa (Goins *et al.*, 1981). Such moonquakes are detectable due to the low seismic noise level on the Moon, specifically the absence of micro-seismic noise (Lognonné and Johnson, 2007). The waveform repeatability of deep moonquakes from a given source region has been of great help in both identification and classification of these events and in locating them. Typical seismograms for individual quakes from a given deep cluster are shown in Figure 4.28 – the low signal-to-noise ratio is apparent. However, seismograms for individual events from a given deep source region have coherent waveforms and so can be stacked, to enhance the signal-to-noise ratio (Figure 4.28). This has allowed a significant number of events to be added to the original 1360 deep moonquake population; the new additions are events that were either (a) previously identified but not recognizable as deep moonquakes (i.e., were listed as unclassified events in the original catalogue) or (b) not previously identified in the lunar seismograms (i.e., not in the original catalogue, even as an unclassified event). Nakamura (2003) used digital waveform cross-correlation, stacking, and single-link cluster analyses to classify many of the previously 9128 unclassified events in the lunar catalogue as deep moonquakes. His study has expanded the number of known deep moonquakes in the original catalogue to 7245, and has established that these events originate from at least 160 deep source regions. In a subsequent study, Nakamura (2005) established the locations of deep source regions with sufficient travel time picks – 98 can be located on the nearside, and 8 may be farside source regions. Bulow *et al.* (2005, 2007) used waveform coherence and stacking to produce a "target" waveform for a known deep source region, and then cross-correlated this target waveform with the entire continuous time series to search for previously unidentified events from that source region (see Bulow *et al.* (2005) for details of the approach). This has resulted in an average increase in the number of events at eight nearside deep source regions of 30%, with one region yielding a 50% increase in the number of events. All of these eight clusters contain at least 140 events, and the study of Bulow *et al.* (2007) contributes an additional 503 events to the 7245 events of Nakamura (2005). The most-studied

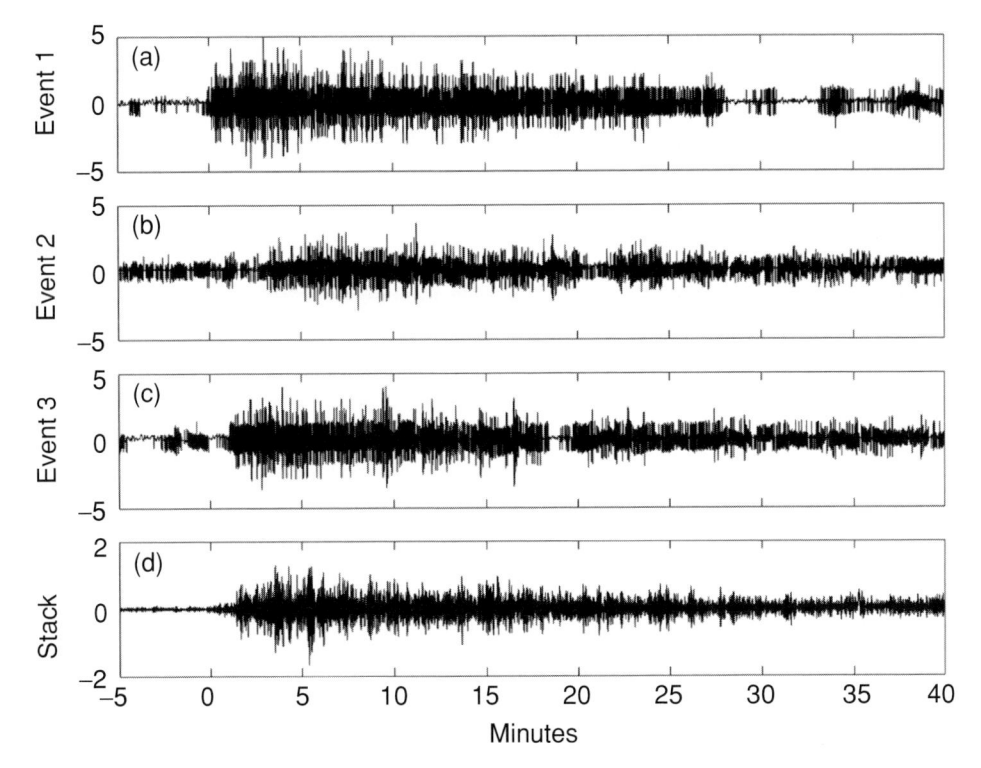

Figure 4.28. Waveform repeatability of deep moonquake source regions. Figures (a)–(c) denote typical seismograms for single events from source region A1. Waveform amplitudes are in digital units (DU). Figure (d) is a stack of waveforms meeting a given coherence criterion – the increased signal-to-noise ratio is evidenced, in particular, via the decreased noise level immediately preceding the event onset. Figure taken from (Bulow *et al.*, 2005)

deep moonquake source region is A1, as it contains the largest number (443) of individual events.

Deep source regions – also referred to in the lunar literature as deep event groups or clusters – can be located using P- and S-wave arrival times. The internal seismic velocity structure must be either assumed, or simultaneously inverted for, in moonquake location studies. As a result of the limited number of seismic stations, reliable location estimates require both P and S arrival times to be measured on all the seismic stations or on stations 15, 16 and at least one of stations 12 and 14. In practice this is usually not the case, and even where they are available, the travel time picks tend to have large uncertainties due to the noisy records. This results in trade-offs between moonquake locations and seismic velocity structure. Two approaches have been taken: in the first, sufficient travel time picks are required to be able to estimate moonquake location and seismic velocity simultaneously – the most recent such study is that of Gagnepain-Beyneix *et al.* (2006) for which 23 deep moonquakes have sufficient arrival time picks to allow simultaneous location and velocity inversion. In the second approach, a seismic velocity model is assumed and the deep source region located using that model. This approach was adopted in the study of Nakamura (2005) in which he used a velocity model from Nakamura (1983) to locate the previously-mentioned 98 nearside clusters. The effects of different models on uncertainties in source region location are illustrated in Figure 4.5 of Bulow *et al.* (2007). Importantly, the absence of arrival picks from particular stations leads to very large uncertainties (e.g., over 20° in latitude) in deep moonquake locations.

Plate 10 shows the locations of nearside deep clusters from both the study of Nakamura (2005) and Gagnepain-Beyneix *et al.* (2006). For the Nakamura (2005) study we consider only those deep clusters for which depth estimates were determined, and for which uncertainties in the latitude and longitude of the deep cluster are less than 10°. This results in retaining 50 of his 98 nearside clusters; 20 deep source regions are common to this dataset and that of Gagnepain-Beyneix *et al.* (2006), and we distinguish these in Plate 10. The additional source regions contributed by the study of Nakamura (2005) provide improved geographical sampling compared with the Gagnepain-Beyneix *et al.* (2006) dataset alone. It is clear that the deep source regions are not distributed randomly even within some nominal region of sensitivity about the centroid of the Apollo seismic network. It has been noted (Lammlein, 1977) that the deep source regions are located in an approximately NE–SW band. Plate 10 shows qualitative agreement with this statement, but that in particular, few deep clusters are located below the nearside highlands. We discuss the geographical distribution of deep moonquake clusters further in Section 6.

4.2 Shallow moonquakes

Over the 8 years the Apollo passive seismic experiment collected data, 28 events were detected with signal characteristics that distinguish them from both meteorite impacts and deep moonquakes (Nakamura, 1977, 1980; Nakamura *et al.*, 1979). These events were originally designated as "high frequency teleseismic" (HFT) events, on the basis of the frequency content of their waveforms, and hence their stronger appearance on the short-period versus long-period instruments. Given the small number of events, the occurrence statistics of these events cannot be robustly assessed; however, unlike the deep moonquakes, there is no clear correlation with particular tidal phases (Nakamura, 1980).

Following the termination of the passive seismic experiment, there was some debate as to whether these events occur at the lunar surface, within the crust, or within the upper mantle; initially they were thought to be surface events. All 28 events occurred outside the seismic network; P- and S-wave arrival times are thus rather insensitive to the depth of the foci, and depths in the range 0 to 200 km (and even above the surface!) are permitted by the travel times alone. Nakamura et al. (1979) compared the observed variation in amplitude of the short period seismograms with epicentral distance with theoretical amplitude variations based on an earlier lunar seismic model (LM-761, Nakamura et al., 1976). The observed amplitude variation is the most convincing evidence to date of a sub-crustal origin for the shallow moonquakes – 100 km depth was suggested (Nakamura et al., 1979). This average depth is based on the amplitude–distance variation of the complete population, and was set in part by a lunar interior model, LM-761, that included a 55 km thick crust. In the light of recent revisions to estimates of the seismically determined crustal thickness (see Section 5), an average depth less than 100 km for the shallow events may be permitted.

As with the deep moonquake population, shallow moonquakes can be located either by assuming or co-estimating an internal seismic velocity model. The study of Gagnepain-Beyneix et al. (2006) finds eight shallow moonquakes to have sufficient travel time arrivals to allow the latter approach – inspection of Table 4.1 of their paper shows 2 of the events to have best-fit depths at the surface, 3 to have best-fit depths in the crust and 3 to have best-fit depths in the mantle, respectively. The studies of Nakamura et al. (1979) and Nakamura (1980) located 26 of the 28 events using the velocity model mentioned above (Plate 10). Because of the small number of shallow moonquakes we examine all 26 events, with the caveat that there may be large uncertainties in many of these locations. In general, shallow moonquakes are not located near the centers of the nearside basins, and, as has been noted previously (Nakamura et al., 1979), appear to show some correlation with the edges of impact basins. The small number of events, together with uncertainties in their locations, renders statistical tests of preferred locations tenuous at best. We return to discussion of shallow moonquake locations in Section 6.

Nakamura et al. (1979) estimated magnitudes for all 28 shallow moonquakes by measuring the amplitude of the envelope function for the short-period seismograms. Their magnitude estimates correspond to Richter magnitudes in the range ~1.5 to just under 5.0. Using displacement spectra, Binder and Oberst (1985) and Oberst (1987) estimated seismic moments and stress drops. The results indicate that most shallow quakes have stress drops of a few MPa, but that the three largest events show body wave magnitudes greater than 5.5 and stress drops of over 100 MPa.

5 Internal structure of the Moon

The mechanical properties of the outer part of the Moon determine its tectonic deformation, and these properties are intimately related to the thermal and compositional evolution of the body. The main compositional layers in the Moon are a small core, a silicate mantle, and a less dense silicate crust. Thermally, the Moon is comprised of an outer thermal boundary layer or lithosphere across which heat is transported via conduction. Unlike larger terrestrial planetary bodies, it is unclear whether below this there is a region in which convective motions are presently occurring. The part of the lithosphere that can deform elastically is usually referred to as an effective elastic thickness, T_e (we shall refer to this simply as elastic thickness for brevity).

While we are primarily interested in lunar interior structure at depths shallower than ∼1000 km, we briefly review what is understood about the deep lunar interior because it is linked to the present and past shallower structure through the Moon's thermal history. Gross geophysical constraints on lunar interior structure include the moment of inertia factor (0.3932 ± 0.0002; Konopliv et al., 1998), the mean mass, and the k_2 Love number. These, together with magnetic sounding measurements (Hood et al., 1999) suggest a conducting core with a radius of 340 ± 90 km. Moreover Williams et al. (2001) have shown that the rotation of the Moon is influenced by a dissipation source, which has been interpreted as the signature of a liquid core. Recent interior-structure models (Lognonné and Johnson, 2007) indicate that a wide range of acceptable core models in the range of 1%–2% lunar mass fit the data. Core densities less than 6000 kg/m^3 are preferred, consistent with a core containing some light element(s), in turn allowing a currently liquid state (Lognonné et al., 2003; Gagnepain-Beyneix et al., 2006; Khan et al., 2006). To date there is no seismic constraint on lunar core size or physical state (see review in Lognonné and Johnson, 2007), though renewed attempts are being made to search for core phases in the Apollo seismic data (Bulow and Lognonné, 2007).

Several approaches can be used to probe the thermal and mechanical structure of the outer part of the Moon. Seismic data provide information on the present-day elastic properties of the lunar interior, and hence indirectly offer constraints on current compositional and thermal structure (Section 5.1). Gravity and topography data together allow investigation of models for support or compensation of topography. Typically the quantities of interest in such studies are estimates of crustal and/or elastic thickness, although models are non-unique and there are additional trade-offs with the assumed density structure. Studies of elastic thickness have mainly focused on the nearside basins that exhibit large positive gravity anomalies. The results are rather inconclusive and, as alternative constraints are provided by the locations of tectonic features, we discuss elastic thickness in the context of integrated tectonic studies of the nearside basins (Section 6). Here we focus on the constraints on global crustal structure provided by gravity and topography data (Section 5.2) and compare these results with those provided by the seismic data.

5.1 Seismological constraints

From a practical perspective, the limited source–receiver geometries and the noisy seismograms mean that only averaged 1-D seismic velocity profiles can be established. Inspection of Plate 10 shows that these velocity–depth profiles will mainly reflect seismic velocities beneath the lunar maria. Of particular interest are discontinuities or sharp velocity gradients corresponding to major compositional boundaries.

By analogy with Earth, the lunar seismological crust–mantle interface is defined by a velocity discontinuity or steep gradient (>0.1 km s^{-1} km^{-1}), below which the P-wave velocity should attain a value of at least 7.6 km s^{-1} (Plate 11). Velocity structure at crustal depths is mainly constrained by P- and S-wave arrival times due to impacts (see discussion in Lognonné and Johnson, 2007) and variations among existing models reflect the limited seismic ray path coverage at depths less than 60 km. In terrestrial studies, crustal thickness can be obtained by studies of converted seismic phases; however, the latter are not readily observed on lunar seismograms and only phases from the Apollo 12 station have been successfully investigated (Vinnik *et al.*, 2001; Gagnepain-Beyneix *et al.*, 2006).

While there are uncertainties in crustal thickness estimates, three main conclusions can be drawn from the seismological data. First, the averaged crustal thickness reflects primarily crustal structure at the Apollo 12 and 14 sites. Second, recent investigations indicate that a 30 to 45-km thick crust at the Apollo 12/14 sites (Khan and Mosegaard, 2002; Lognonné *et al.*, 2003; Gagnepain-Beyneix *et al.*, 2006) is preferred over previous estimates of 58 km (Nakamura, 1983; and see earlier work by Toksöz, 1974). Third, these more recent 1-D crustal thickness estimates, and a recent attempt to map geographical variations in crustal thickness using the Apollo seismic data (Chenet *et al.*, 2006), are consistent with those based on analyses of gravity and topography data (see Wieczorek *et al.*, 2006 and Section 5.2).

It is not possible to resolve whether the correct model for the seismological crustal structure is one of distinct constant-velocity layers, or gradually increasing velocity with depth. Perhaps the most succinct summary of the seismic data is that it is broadly consistent with a crustal structure in which there are two major compositional layers – an upper, very feldspathic, and lower, moderately feldspathic, crust – consistent with inferences from gravity and topography data (Wieczorek *et al.*, 2006). Seismologically, higher velocities are associated with the noritic lower crust. Some suggestions of a mid-crustal reflector have been made (e.g., the 20 km discontinuity of Khan and Mosegaard, 2002), but this is not observed in all models. Furthermore, a near-surface regolith and fractured layer results in scattering of seismic energy and significantly reduced seismic velocities in the upper ~1 km (see review in Wieczorek *et al.*, 2006).

P- and S-wave arrival times from moonquakes provide some information on seismic velocity structure down to ~1000 km depth – i.e., the depth of the deep moonquake source regions. A review of mantle seismic velocity structure is beyond the scope of this chapter and can be found in Lognonné and Johnson (2007). In general, P-wave velocities in the lunar mantle are around 8 km s^{-1} and S-wave velocities are around 4.5 km s^{-1}. Of interest here are the predictions of thermal models based on these seismic velocity profiles for temperatures in the deep moonquake source regions and we return to this in Section 6.2.

5.2 Constraints from gravity and topography

Global measurements of gravity and topography data allow constraints to be placed on crustal and lithospheric structure. A thorough review of this topic can be found in Wieczorek *et al.* (2006) and we summarize only the main points here. Lunar topography data sets were reviewed in Section 2. For analyses involving both gravity and topography data, the spherical harmonic topography model derived from the Clementine data (GLTM2c, Smith *et al.*, 1997) is the most useful, and contains information to spherical harmonic degree and order 90, equivalent to a maximum spatial resolution of ~120 km. Gravity field measurements were obtained during the Lunar Prospector mission; the most recent gravity field model is that of Konopliv *et al.* (2001) which contains information to spherical harmonic degree and order 150, equivalent to a maximum spatial resolution of 72 km (Plates 12A,B).

Nearside free-air gravity anomalies (Plate 12A) are typically close to zero over the highland regions, indicating isostatic compensation of topography. Large positive gravity anomalies are seen over several nearside basins, in particular Imbrium, Serenitatis, Crisium, Nectaris and Humorum. These "mascons" were first observed in Lunar Orbiter tracking data (Muller and Sjogren, 1968), and have been mapped in more detail by Lunar Prospector. Peak anomalies are over 300 m Cal. The farside gravity field is poorly determined because of the lack of tracking data away from the limbs on the farside.

Three different approaches can be used to estimate lunar crustal thickness from gravity and topography; one of these techniques also allows estimation of elastic thickness. For topography compensated solely by variations in crustal thickness (commonly referred to as Airy compensation), the ratio of geoid height to topography over a region is proportional to the average crustal thickness. This approach was pioneered by Ockendon and Turcotte (1977) in cartesian coordinates (i.e., the region of interest must be small enough for planetary curvature to be negligible) and adapted for use in spherical coordinates by Wieczorek and Phillips (1998). Regions where Airy isostasy clearly does not hold (e.g., mascons) can easily be excluded from analyses. Using this approach Wieczorek *et al.* (2006) infer a globally averaged crustal thickness of 49 ± 15 km. Crustal thicknesses at Apollo sites 12 and 14 are in the range 16–56 km, depending on the crustal and mantle densities assumed, and are consistent with the more recent seismologically-based inferences of crustal thickness (Khan and Mosegaard, 2002; Lognonné *et al.*, 2003; Gagnepain-Beyneix *et al.*, 2006).

Spectral admittance techniques examine the relationship between gravity and topography in the wavenumber domain. The advantage over the geoid–topography ratio approach is that if topography is supported by more than one mechanism (e.g., a combination of crustal thickening and lithospheric flexure), then it may be possible to identify these different contributions by their spectral signature, and furthermore to estimate both crustal and elastic thickness for a given region. The drawback is that it is less easy to investigate localized regions using spectral approaches, although work by Wieczorek and Simons (2005) make this problem more tractable (for a full discussion, see Wieczorek *et al.*, 2006, and references therein). Localized spectral admittance modeling suggests mean crustal thicknesses beneath the lunar basins in the range 35–70 km (Aoshima and Namiki, 2001), and the two basins that likely satisfy a pre-mare isostatic assumption (the density deficit due to excavation of the basin was balanced by an associated positive density contrast at depth due to uplift of the crust–mantle boundary) (Wieczorek and Phillips, 1998) both yield a mean crustal thickness of 50 km.

Global crustal thickness models have also been constructed using lunar gravity and topography data (Neumann *et al.*, 1996; Wieczorek and Phillips, 1998; Wieczorek *et al.*, 2006). The contribution to the gravity field due to the surface topography is subtracted from the free-air gravity and the resulting gravity anomaly (the Bouguer anomaly) inverted for relief on the crust–mantle boundary. Previous studies have investigated both single- and dual-layer crustal structure: all models show substantial crustal thinning beneath major basins, and several of the mascon basins exhibit super-isostatic thinning (i.e., uplift of the crust–mantle boundary to a depth shallower than that needed for hydrostatic equilibrium beneath an excavated basin). In two-layer crustal models, the upper crust is completely removed at major basins, and the observed gravity anomalies can only be matched by including further thinning of the lower crust and the contribution of a mare basaltic load (see Wieczorek *et al.*, 2006).

6. Basin-localized tectonics and seismicity

The temporal and spatial patterns of deformation in the nearside maria, characterized by basin-concentric and basin-radial wrinkle ridges and arcuate and linear rilles, provide important, albeit incomplete, information on the tectonic history of the lunar basins. The large positive gravity anomalies over several of the nearside basins indicate uncompensated positive density anomalies that likely result from a combination of loading by mare basalts, intrusion, and a possible super-isostatic pre-mare basin state (see Section 5). The mare basins exhibit low elevations and generally flat topography (see Section 2 and Plates 6A,B); in the case of Mare Serenitatis, a long-wavelength, low-amplitude rise is observed, with the highest topography in the center of the basin. Taken together, these observations have been used to both probe the sub-mare mechanical structure of the lunar lithosphere, and to develop loading models that can explain the observed tectonism.

6.1 Lithospheric structure beneath mare basins

The classic loading model of a mass concentration due to the mare basalts on an impact-thinned and impact-weakened lithosphere results in compression in the interior of the basin and extension near the margins (Phillips *et al.*, 1972; Melosh, 1978; Solomon and Head, 1979, 1980; Freed *et al.*, 2001). Early mascon tectonic models involved axisymmetric loading of a thin elastic plate and these predict basin-interior compression (see Section 6.2) and concentric normal faulting near the basin margins. For a given basin, the radial distance to the concentric rilles and to the wrinkle ridges can be used to estimate elastic lithospheric thickness, T_e, at the time of loading. This approach was adopted in a series of studies (Comer *et al.*, 1979; Solomon and Head, 1979, 1980), and combined with age information for the two classes of tectonic features to investigate spatial and temporal variations in T_e. Solomon and Head (1980) examined eight basins: Imbrium, Serenitatis, Crisium, Humorum, Grimaldi, Orientale, Smythii, and Nectaris, and concluded that T_e at the time of rille formation (i.e., ages > 3.6 Ga) was generally less than about 75 km, but showed considerable variations among the different basins, ranging from less than 25 km at Grimaldi basin to up to 75 km at Imbrium and Crisium. T_e estimates at the time of graben formation (i.e., younger ages) are greater and generally at least 100 km, implying cooling and thickening of the lithosphere over time. In addition, T_e at the time of wrinkle-ridge formation shows less between-basin variability (Solomon and Head, 1980).

As mentioned briefly in Section 5, gravity and topography data can be used to estimate T_e. Estimates derived in this way will not necessarily correspond to T_e at the time of loading, since the present-day gravity signature of the mare basins is being matched. Such an approach has been attempted for the lunar basins, though with limited success. Arkani-Hamed (1998) found T_e in the range 30 to 50 km beneath six nearside basins including Serenitatis, and T_e of 20 km beneath Orientale; however, Aoshima and Namiki (2001) concluded that for Orientale and Serenitatis, T_e is not well constrained. A study by Crosby and McKenzie (2005) indicates that T_e at Imbrium must be greater than 25 km. It is not surprising that gravity and topography studies have returned a variety of estimates for T_e, since such studies require a-priori assumptions about the vertical density structure, and this is not straightforward to establish given the likely complex loading history of the mare basins.

6.2 Predictions for tectonic deformation

The mascon loading model described above predicts stresses that would result in the following pattern of faulting with increasing radial distance from the basin center: basin-interior radial thrust faults (the maximum principal stress σ_1 is horizontal, minimum principal stress $\sigma_3 =$ vertical stress σ_v), a zone of strike-slip faulting (σ_1 is horizontal, $\sigma_2 = \sigma_v$), and finally concentric normal faulting near the basin margins (σ_1 is vertical, $\sigma_1 = \sigma_v$) (Melosh, 1978; Freed *et al.*, 2001). Two issues are immediately apparent: the absence of documented strike-slip faults in mare basins, and the dominance of concentric rather than radial thrust faults.

The zone of strike-slip faults in mare mascons is predicted by Anderson's criterion (Anderson, 1951). It results from the transition of σ_v from the minimum to the maximum principal stress (see Schultz *et al.*, Chapter 10). The absence of evidence of strike-slip faulting in the mare mascons has been described as the "strike-slip faulting paradox" (Freed *et al.*, 2001; Wieczorek *et al.*, 2006). A number of factors may act to suppress the formation of lunar strike-slip faults. These include the preexisting state of stress in the lunar lithosphere (Solomon and Head, 1980), the depth of fault nucleation (Golombek, 1985), and the load history and evolution of stress (Schultz and Zuber, 1994). Freed *et al.* (2001) showed that the predictions of a large strike-slip zone and the dominance of radial thrust faults are the result of the use of elastic plate models and the Anderson faulting criterion. They employed a viscoelastic model and showed that the inclusion of lunar curvature and the use of the fault criterion of Simpson (1997) (that accounts for stress transitions), significantly reduce the zone of predicted strike-slip faulting and permit a wider range of faulting styles than the classic mascon model. In addition, strike-slip faulting is further inhibited by larger radii loads, loading of a thin lithosphere (e.g., in the Freed *et al.* (2001) study, a 25 km versus 100 km lithosphere is considered), an initially super-isostatic stress state, and the inclusion of heterogeneities in crustal strength.

Another paradox in the mascon tectonic model arises from the discrepancy between the predicted and observed pattern of wrinkle ridges in the maria (see Wieczorek *et al.*, 2006). As mentioned above, the mascon model predicts basin-interior, radial oriented thrust faulting. The dominant pattern of wrinkle ridges in the interiors of many of the mascons, however, is concentric (see Plate 6A).

Many of the radial oriented wrinkle ridges occur near the basin margins, closer to the location of concentric graben. In Mare Serenitatis, basin-radial oriented wrinkle ridges are crosscut by prominent arcuate graben of Fossae Plinius (see Figure 4.3). Freed *et al.* (2001) concluded that both concentric and radial mare ridges are expected if a global compressive stress bias and the loading history from the mare basalts are accounted for. Two other factors explored by Freed *et al.* (2001) may also influence the pattern of wrinkle ridges. One is the geometry of the load and the other is the pre-mare compensation state of the basin (see Wieczorek *et al.*, 2006). A radial pattern of mare ridges is predicted if the basin was initially sub-isostatic, and rebound occurred after emplacement of the mare basalts (Freed *et al.*, 2001, Fig. 12). A concentric pattern of mare ridges is favored if the basin was initially super-isostatic, and subsidence occurred after the emplacement of the mare basalts.

The geometry of the load from the mare basalts also has a strong influence on the expected pattern of mare ridges (Freed *et al.*, 2001, Fig. 10). A radial pattern of mare ridges is predicted if the mare basalts are thicker in the center of the basin than near its margin. A concentric pattern of ridges is expected if the mare basalts are more uniform in thickness. It is generally concluded that the mare basalts have the greatest thickness in the centers of the basins, based on a decrease in the number of pre-mare craters and ALSE subsurface profiles (see Baldwin, 1970; Wilhelms, 1987). Insight into the geometry of the load from the mare basalts may be obtained from an analysis of the long-wavelength topography of the basins. As discussed in Section 2.1.1, Clementine LIDAR data over Mare Serenitatis show a topographic high in the mare surface located in the interior of the basin. In contrast, Mare Crisium and Mare Humorum have generally flat interior floors. Mare Imbrium also has a flat floor, but it has a pronounced regional slope. The mare surface at the north rim of Imbrium may be as much as 1000 m higher than at the south rim. Unlike Mare Serenitatis, there is no obvious spatial correlation between interior rings of wrinkle ridges and elevation changes in the Mare Imbrium, Mare Crisium, or Mare Humorum.

As discussed in Section 5, gravity data collected by the Lunar Orbiters, the Apollo missions, Clementine, and the Lunar Prospector mission, indicate that Mare Imbrium, Mare Serenitatis, Mare Crisium, Mare Humorum, and Mare Nectaris have strong positive gravity anomalies ringed by negative anomalies (Zuber *et al.*, 1994, Fig. 2b; Lemoine *et al.*, 1997, Plate 1; Konopliv *et al.*, 1998, Fig. 1a). Positive anomalies are attributed to the uncompensated mare basalts (Phillips *et al.*, 1972) and the presence of a dense mantle "plug" formed by post-impact mantle rebound (Wise and Yates, 1970; Solomon and Head, 1979). The surrounding negative anomaly suggests subsurface mass deficiencies that result from crustal thickening related to basin formation or modification (Lemoine *et al.*, 1997). Future high-resolution gravity data, e.g., from the Gravity Recovery and Anterior

Laboratory or GRAIL mission, are needed to evaluate possible contributions to the negative anomalies from ringing due to the finite spherical harmonic expansion.

Pre-Lunar Prospector mission gravity models often show lunar mascons with "bull's eye" patterns, with the high located roughly in the center of the mare (e.g., Lemoine *et al.*, 1997, GLGM-2 free-air model). This pattern would suggest an axially symmetric, Gaussian-shaped accumulation, with the greatest thickness of the basalts near the basin center if the dominant contribution to the gravity anomaly is from the basalt fill (Phillips *et al.*, 1972). The most recent lunar gravity models that incorporate Lunar Prospector data (Konopliv *et al.*, 1998, 2001; Konopliv and Yuan, 1999), however, show that anomalies for the mare mascons have broad plateaus throughout the mare interior. The spatial correlation between the edges of the positive anomalies and basin-interior wrinkle-ridge rings in Mare Imbrium, Mare Serenitatis, Mare Crisium, and Mare Humorum is striking (Plate 12A). This supports the interpretation that positive anomalies are dominated by the near-surface mare fill (Phillips *et al.*, 1972; Konopliv *et al.*, 1998, 2001; Watters and Konopliv, 2001), since contributions to the gravity anomalies from topography on the crust–mantle boundary will be more smoothly varying. This suggests that the basalt sequence in the interior of the basin may be generally uniform in thickness and thin rapidly at the margins. The modeling of Freed *et al.* (2001) suggests that a more uniform thickness of mare basalts will result in a basin-interior concentric pattern of mare ridges with radial mare ridges near the margin, consistent with the observed distribution. Interestingly, Mare Nubium has no associated mascon, but is dominated by radial wrinkle ridges. Similarly there is a group of radially oriented wrinkle ridges at the western edge of Mare Tranquillitatis, with no associated free-air gravity anomaly. Finally, the fact that some concentric graben crosscut radial oriented mare ridges, as is the case in Mare Serenitatis, suggests some mare have a more complex stress history, where zones dominated by compression and extension have changed over time.

Not all the Moon's wrinkle ridges and graben are associated with mascons. The most significant occurrence of nearside tectonic features outside a lunar mascon is in Oceanus Procellarum (see Plates 6A and 12A). Although the youngest mare basalt units, estimated to be ~1.2 Ga old (Plate 9A) (Hiesinger *et al.*, 2003), occur in Oceanus Procellarum, it lacks a strong positive anomaly and the spatial correlation of the anomaly with the wrinkle ridges indicative of mascons. Also, many graben are radial to Procellarum (Plate 6A). Only hints of weak to moderate positive and negative anomalies are resolved in the current gravity model (Plate 12A). The origin of Procellarum remains uncertain. The hypothesis that Oceanus Procellarum is a giant impact basin (see Wilhelms, 1987) is still debated (see Hiesinger and Head, 2006). Current geophysical evidence suggests overprinting by known basins, but provides no definite indication of a giant pre-Nectarian basin (Neumann *et al.*, 1996). An alternative hypothesis is that Procellarum is connected to the Imbrium event, possibly the outermost ring structure of Imbrium (Spudis *et al.*, 1988b; Spudis, 1993) having formed by down-dropping on bounding normal faults (Cooper *et al.*, 1994).

6.3 Basin-localized seismicity

The deep moonquake source regions generally occur around the edges of, or beneath, the major nearside basins (Plate 10). Furthermore, with the exception of two source regions in Serenitatis and one in Crisium, deep moonquakes generally occur exterior to the concentric wrinkle ridges and hence exterior to large positive gravity anomalies. Deep moonquake depths are of course too deep to indicate any direct correlation between their geographical distribution and the distribution of surface tectonic features. However, their spatial distribution suggests that the influence of long-lived deep lateral thermal and/or compositional heterogeneities is associated with either basin formation and/or mare basalt production.

Very little is known about deep moonquake source regions. A routine procedure in analyzing earthquake data is the investigation of focal mechanisms; however, this has not been possible for deep moonquakes because of the limited number of receivers and the paucity of deep moonquakes that occurred inside the seismic array. Some attempts have been made to extract information about putative fault plane orientations by examining whether elastic tidal stresses resolved onto a given plane can be used to predict moonquake occurrence times. However, such studies have not reached consistent conclusions because of differing assumptions regarding the role of shear versus normal stresses, a-priori background stress fields (e.g., due to global thermal contraction), and the lack of removal of the long-term tidal stress (Lammlein, 1977; Toksöz *et al.*, 1977; Cheng and Toksöz, 1978; Minshull and Goulty, 1988). In addition, most previous work that has investigated the relationship between tidal stress and moonquake occurrence has focused on the A1 source region because of the large number of A1 events. Comparison of results among existing studies is further hampered by differences in the choice of coordinates systems used (Cheng and Toksöz, 1978; Minshull and Goulty, 1988). A recent study (Weber *et al.*, 2009) has examined clusters with more than 30 events (51 in total), and investigated plane orientations that minimize the variance in stresses at moonquake occurrence times for a Coulomb failure criterion (i.e., failure can result from a combination of normal and shear stresses). Moonquake occurrence times at some of the clusters examined appear to be quite well predicted by failure on a plane, however, at other clusters no particular plane orientation appears to be preferred. Importantly, nearby clusters do not seem to respond similarly to tidal forcing, suggesting that the brittle failure is controlled by physical properties that vary on spatial scales much shorter than inter-cluster distances.

This latter inference is supported by earlier and recent investigations of the possible spatial extent of deep moonquake source regions. Investigations of the A1 source region suggest that the hypocenters of individual events are confined within a radius of about 1 km (Nakamura, 1977, 2007), and 12 other clusters appear to have a similarly small spatial extent (Nakamura, 2007).

An unresolved, and major, issue is why brittle failure occurs at depths of 700 to 1000 km. Recent thermal models based on seismic data suggest temperatures of 1100 to 1400 °C (Lognonné et al., 2003), and although those based on electromagnetic data are lower (Khan et al., 2007), all estimates suggest a regime in which ductile, rather than brittle, failure should occur.

As with the deep moonquakes, many outstanding questions surround the cause and locations of shallow moonquakes. As mentioned earlier, shallow moonquakes have been suggested to occur at the edges of basins (see Plate 10), plausibly related to lateral heterogeneities in mechanical structure. If these events are in fact upper mantle events, as suggested by Nakamura et al. (1979), then any connection to surface tectonic features is unlikely. If however, some of these events do occur within the crust, then any association with tectonic features would be of interest.

As reviewed earlier, the majority of basin-localized tectonism mainly ceased ~1.2 Ga, and the youngest class of tectonic features, lobate scarps, are not basin-localized. However, current seismicity does appear to be largely basin-localized, offering the tantalizing suggestion that lateral heterogeneities in temperature, and/or composition, and/or mechanical structure resulting from basin and mare-basalt formation have persisted to the present day.

7 Global strain from young lobate scarps

The relatively small-scale lobate scarps may be the youngest tectonic features on the Moon. The source of compressional stresses that formed the young lobate scarps may have been thermal stresses from interior cooling. Thermal models for a magma ocean (Solomon and Chaiken, 1976; Solomon and Head, 1979) and an initially total molten Moon (Binder and Lange, 1980) both predict a late-stage compressional stress bias in the lunar lithosphere. The true distribution of thrust faults in the lunar highlands cannot be evaluated at present because most of the scarps can only be identified in Apollo Panoramic camera images of the lunar equatorial zone, and in some high-resolution Lunar Orbiter III and V images. Of the ~20% of the lunar surface imaged by the Apollo Panoramic Cameras, only about 10% was imaged with the optimal lighting geometry for detecting the lobate scarps. This makes an accurate estimate of the strain expressed by the lobate scarps difficult. However, the contractional strain in the area covered by high-resolution Panoramic Camera images can be estimated using the *D–L* relationship of the lobate scarp thrust faults (see Section 2.5). If the *D–L* relationship of a fault population is known, the strain can be calculated using fault lengths alone (Scholz and Cowie, 1990; Cowie et al., 1993). The strain for faults where the fault length $L \geq$ the maximum depth of faulting is given by

$$\varepsilon = \frac{\cos(\theta)}{A} \sum_{k=1}^{n} D_k L_k, \quad (4.2)$$

where θ is the fault plane dip, A is the survey area, n is the total number of faults, and $D = \gamma L$ (Cowie et al., 1993). Using the value of γ of $\sim 1.2 \times 10^{-2}$ for $\theta = 30°$ (see Section 2.5) and the total length of the lobate scarps measured in this study, the strain is estimated to be $\sim 0.008\%$ for a study area covering about 10% of the lunar surface. For comparison, the strain reflected by the lobate scarps on Mercury is $\sim 0.06\%$ (for $\theta = 30°$) (Watters and Nimmo, Chapter 2; Watters et al., 2009). Thus, the contractional strain from the young lobate scarps on the Moon appears to be nearly an order of magnitude less than the strain from the larger scale thrust faults on Mercury.

If the lobate scarps are the result of thermal stresses from global cooling, then the surface strain can be used to estimate the corresponding radius change of the Moon. The pre-deformation lunar radius R_u is related to the contractional strain by

$$R_u = \left[\frac{R_d^2}{\varepsilon + 1}\right]^{0.5}, \quad (4.3)$$

where the radius change is expressed by $\Delta R = R_d - R_u$ where R_d is the post-deformation (current) radius (Watters and Nimmo, Chapter 2). Assuming the $\sim 10\%$ of the lunar surface covered by Panoramic Camera images is representative of the entire surface, a strain ε of $\sim 0.008\%$ corresponds to a ΔR of ~ 70 m. The isotropic stresses σ resulting from global contraction (Melosh and McKinnon, 1988) are given by

$$\sigma = 2\mu \frac{(1+\nu)}{(1-\nu)} \frac{\Delta R}{R_d}, \quad (4.4)$$

where μ is the shear modulus, and ν is the Poisson's ratio. The stress corresponding to the relatively small estimated radius change is <10 MPa.

Global contraction is predicted on the basis of thermal evolution models, but distributed large-scale lobate scarp thrust faults like those on Mercury are not found on the Moon. In an effort to reconcile this mismatch, Pritchard and Stevenson (2000) concluded that the near-surface regolith might be able to accommodate more contractional strain than intact rock before failing, or the Moon may have experienced a non-monotonic thermal evolution, or that contractional strain was accommodation by numerous small faults (Weisberg and Hager, 1998). Thus, relating surface strain to internal processes may not be straightforward and the lack of large-scale thrust faults may not strongly constrain the change in lunar volume (see Shearer *et al.*, 2006). However, the small-scale lobate scarp thrust faults, possibly rooted in the regolith (or megaregolith), may be an expression of late-stage global contraction. If this is the case, they could provide important insight into the recent compressional stress state of the lunar lithosphere.

Thermal history models based on the formation of a magma ocean (Solomon and Chaiken, 1976; Solomon and Head, 1979) and an initially totally molten Moon (Binder and Lange, 1980) both predict a late-stage compressional stress bias in the lunar lithosphere. The initially totally molten model predicts stresses of up to 350 MPa (Binder and Lange, 1980; Binder and Gunga, 1985). Such a high state of global compressional stress might be expected to result in thrust faults comparable to large-scale lobate scarp thrust faults on Mercury and Mars. The magnitude of compressional stresses predicted by the magma ocean thermal history model is less than 100 MPa after the end of the period of heavy bombardment (Solomon and Chaiken, 1976; Solomon and Head, 1979). Although the estimated strain from the lunar lobate scarps must be considered a lower limit in light of the possibility that significant contractional strain may go without expression (Pritchard and Stevenson, 2000), they support thermal history models predicting a small change in lunar radius of about ±1 km in the last 3.8 Ga (Solomon and Chaiken, 1976; Solomon and Head, 1979, 1980).

8 Conclusions and outstanding questions

The majority of the Moon's large-scale tectonic landforms occur on the nearside, in and around the maria. This basin-localized tectonics is reflected in the wrinkle ridges and linear and arcuate rilles or troughs. Wrinkle ridges are the expression of crustal shortening in the basin interiors, while linear and arcuate rilles express crustal extension at basin margins and in the adjacent highlands. The spatial distribution of contractional and extensional tectonic features is the result of lithospheric flexure due to mascon loading. Basin-localized crustal extension ceased ~3.6 Ga while crustal shortening continued to as recently as ~1.2 Ga. Crustal extension may have ceased when compressional stresses from global contraction were superimposed on flexural extensional stress due to mascon loading.

Mascon gravity anomalies exhibit broad plateaus throughout mare interiors. In some basins, interior wrinkle-ridge rings correlate spatially with the edges of the positive anomalies, suggesting that the mare basalt fill results in a disk-like load.

Stress models of lunar mascons suggest that the spatial distribution of the wrinkle ridges and rilles are best fit by a uniform thickness of mare basalts.

Exterior to the nearside maria, lobate scarps are the dominant lunar tectonic feature. These small-scale thrust faults are less than 1 Ga and are likely among the youngest endogenic features on the Moon. Young lobate scarp thrust faults support late-stage compression of the lunar crust. The lobate scarps express a relatively small contractional strain ($\sim$0.008%), corresponding to a radius change of $\sim$70 m. This is consistent with thermal history models that predict a small change in lunar radius in the last 3.8 Ga.

The Apollo seismic data provide information on the crustal and upper mantle velocity structure of the Moon, from which interior thermal profiles have been constructed. Naturally occurring lunar seismicity comprises two distinct classes of events: (1) a few (28) shallow moonquakes that occur at depths of 100 km or less and can result in stress drops of over 100 MPa, and (2) thousands of small stress drop moonquakes that originate from distinct source regions at 700 to 1000 km depth; a given source region exhibits moonquakes with tidal periods and repeatable waveforms. The deep moonquake source regions cluster around the edges of nearside basins or beneath them. Limited information regarding these source regions is available – the failure volumes appear to be small (on the order of 1 km^3), and it is unclear whether the canonical view of failure on a plane describes these quakes. The spatial distribution of deep moonquakes suggests that they may be related to long-lived, deep, lateral thermal and/or compositional heterogeneities associated with either basin formation and/or mare basalt production. Shallow moonquakes are mostly distributed around the edges of major basins. The spatial distribution of some moonquakes and their approximate correlation with tectonic features suggests that there may be current tectonic activity on the Moon.

Many outstanding questions remain to be addressed using data returned by recent lunar missions like KAGUYA (SELENE), Chang'e-1 and Chandrayaan-1 and the current Lunar Reconnaissance Orbiter mission. Critical to understanding the tectonic history of the Moon is determining the global distribution and ages of the lobate scarp thrust faults. This will require global high-resolution imaging (2 m/pixel or better). Elucidating the origin of tectonic stresses in nearside maria without associated mascons is also important in more fully understanding lunar tectonism. Related to this question is determining the origin of Oceanus Procellarum. The non-uniqueness inherent to the problem of interpreting gravity anomalies over the lunar basins could be significantly helped by better constraints on the depths of mare basalts. The radar sounder on KAGUYA has already extended our knowledge of subsurface interfaces beneath the maria gained from the ALSE experiment. Additional knowledge regarding the depths of mare basaltic fill is clearly an important constraint in tectonic loading models.

Continuous, broad-band, seismic measurements made over a period of at least 6 years by a surface geophysical network, such as the proposed International Lunar Network (ILN) are essential to further our understanding of the lunar interior. Source mechanisms for both deep and shallow moonquakes are needed, and the underlying cause of deep moonquakes is one of the outstanding problems in lunar geophysics. Essential to understanding connections between current seismicity and current or past tectonism are accurate moonquake locations for both deep and shallow events. These, in turn, are needed for improved determination of interior structure. Clearly a major gap in our understanding of lunar tectonism results from the very limited knowledge of crustal structure. For example, more detailed crustal imaging will aid in understanding the depth extent and origin of faulting.

The prospect of returning humans to the Moon has the potential to continue and expand on the wealth of lunar science made possible by the Apollo missions. Our ability to unravel the mysteries of the Moon's origin and geologic evolution may ultimately depend on a program of investigation that combines robotic missions with human exploration.

Acknowledgments

We wish to thank Paul Spudis and Walter Kiefer for their thorough and thoughtful reviews that greatly improved the chapter. This work was supported by the National Aeronautics and Space Administration under Grants issued through the Office of the Planetary Geology and Geophysics Program.

References

Anderson, E. M. (1951). *The Dynamics of Faulting and Dyke Formation, with Applications to Britain*. Edinburgh, UK: Oliver & Boyd.

Aoshima, C. and Namiki, N. (2001). Structures beneath lunar basins: Estimates of Moho and elastic thickness from local analysis of gravity and topography (abs.). *Lunar Planet. Sci. Conf. XXXII*, 1561–1562. Houston, TX: Lunar and Planetary Institute (CD-ROM).

Archinal, B. A., Rosiek, M. R., Kirk, R. L., and Redding, B. L. (2006). The Unified Lunar Control Network 2005, USGS Open File Report 2006–1367. http://pubs.usgs.gov/of/2006/1367/.

Arkani-Hamed, J. (1998). The lunar mascons revisited. *J. Geophys. Res.*, **103**, 3709–3739.

Baldwin, R. B. (1963). *The Measure of the Moon*. Chicago, IL: University of Chicago Press.

Baldwin, R. B. (1965). *A Fundamental Survey of the Moon*. New York, NY: McGraw-Hill.

Baldwin, R. B. (1970). A new method of determining the depth of the lava in lunar maria. *Astron. Soc. Pacific Publ.* **82**, 857–864.

Binder, A. B. (1982). Post-Imbrian global lunar tectonism: Evidence for an initially totally molten Moon. *Earth, Moon, and Planets*, **26**, 117–133.

Binder, A. B. and Gunga, H.-C. (1985). Young thrust-fault scarps in the highlands: Evidence for an initially totally molten Moon. *Icarus*, **63**, 421–441.

Binder, A. B. and Lange, M. A. (1980). On the thermal history, thermal state, and related tectonism of a Moon of fission origin. *The Moon*, **17**, 29–45.

Binder, A. B. and Oberst, J. (1985). High stress shallow moonquakes: Evidence for an initially totally molten Moon. *Earth Planet. Sci. Lett.*, **74**, 149–154.

Boyce, J. M. (1976). Ages of flow units in the lunar nearside maria based on Lunar Orbiter IV photographs. *Proc. Lunar Planet. Sci. Conf.* 7, 2717–2728.

Brennan, W. J. (1975). Modification of premare impact craters by volcanism and tectonism. *The Moon*, **12**, 449–461.

Brennan, W. J. (1976). Multiple ring structures and the problem of correlation between lunar basins. *Proc. Lunar. Planet. Sci. Conf.* 7, 2833–2843.

Bryan, W. B. (1973). Wrinkle-ridges as deformed surface crust on ponded mare lava. *Geochim. Cosmochim. Acta*, **1**, (Suppl.), 93–106.

Bulow, R. and Lognonné, P. (2007). Lunar internal structure from reflected and converted core phases (abs.). *Eos Trans. AGU*, P51B-0483.

Bulow, R., Johnson, C. L., and Shearer, P. (2005). New events discovered in the lunar apollo seismic data. *J. Geophys. Res.*, **110**, E10003, doi:10.1029/2005JE002414.

Bulow, R., Johnson, C. L., Bills, B., and Shearer, P. (2007). Temporal and spatial properties of some deep moonquake clusters. *J. Geophys. Res.*, **112**, E09003, doi:10.1029/2006JE002847.

Carr, M. H. (1969). Geologic map of the Alphonsus region of the Moon. U.S. Geol. Surv. Map 1–599 (RLC-14), scale 1:250 000.

Cartwright, J. A., Trudgill, B. D., and Mansfield, C. S. (1995). Fault growth by segment linkage: An explanation for scatter in maximum displacement and trace length data from the Canyonlands grabens of SE Utah. *J. Struct. Geol.*, **17**, 1319–1326.

Cartwright, J. A., Mansfield, C. S., and Trudgill, B. D. (1996). The growth of normal faults by segment linkage. *Geol. Soc. Am. Spec. Publ.*, **99**, 163–177.

Chenet, H., Lognonné, P., Wieczorek, M., and Mizutani, H. (2006). Lateral variations of lunar crustal thickness from Apollo seismic dataset. *Earth Planet. Sci. Lett.*, **243**, 1–14.

Cheng, C. H. and Toksöz, M. N. (1978). Tidal stresses in the Moon. *J. Geophys. Res.*, **83**, 845–853.

Clark, R. and Cox, S. (1996). A modern regression approach to determining fault displacement–length scaling relationships. *J. Struct. Geol.*, **18**, 147–154.

Colton, G. W., Howard, K. A., and Moore, H. J. (1972). Mare ridges and arches in southern Oceanus Procellarum, Photogeology. *Apollo 16 Prel. Sci. Rep., NASA Spec. Publ.*, **SP-315**, 29–90 – 29–93.

Comer, R. P., Solomon, S. C., and Head, J. W. (1979). Elastic lithospheric thickness on the Moon from mare tectonic features: a formal inversion. *Proc. Lunar Planet. Sci. Conf.* 10, 2441–2463.

Cook, A. C., Watters, T. R., Robinson, M. S., Spudis, P. D., and Bussey, D. B. J. (2000). Lunar polar topography derived from Clementine stereoimages. *J. Geophys. Res.*, **105**, 12 023–12 033.

Cooper, B. L., Carter, J. L., and Sapp, C. A. (1994). New evidence for graben origin of Oceanus Procellarum from lunar sounder optical imagery. *J. Geophys. Res.*, **99**, 3799–3812.

Cowie, P. A. and Scholz, C. H. (1992a). Physical explanation for the displacement–length relationship of faults using a post-yield fracture-mechanics model. *J. Struct. Geol.*, **14**, 1133–1148.

Cowie, P. A. and Scholz, C. H. (1992b). Displacement–length scaling relationship for faults data synthesis and discussion. *J. Struct. Geol.*, **14**, 1149–1156.

Cowie, P. A., Scholz, C. H., Edwards, M., and Malinverno, A. (1993). Fault strain and seismic coupling on mid-ocean ridges. *J. Geophys. Res.*, **98**, 17 911–17 920.

Crosby, A. and McKenzie, D. (2005). Measurements of elastic thickness under ancient lunar terrain. *Icarus*, **173**, 100–107.

Dawers, N. H. and Anders, M. H. (1995). Displacement–length scaling and fault linkage, *J. Struct. Geol.*, **17**, 607–614.

Dawers, N. H., Anders, M. H., and Scholz, C. H. (1993). Growth of normal faults: displacement–length scaling. *Geology*, **21**, 1107–1110.

Dombard, A. J. and Gillis, J. J. (2001). Testing the viability of topographic relaxation as a mechanism for the formation of lunar floor-fractured craters. *J. Geophys. Res.*, **106**, 27 901–27 910.

Ewing, M., Latham, G., Press, F., Sutton, G., Dorman, J., Nakamura, Y., Meissner, R., Duennebier, F., and Kovach, R. (1971). Seismology of the Moon and implications on internal structure, origin and evolution (abs.). In *Highlights of Astronomy*, ed. C. De Jager. Dordrecht: Reidel, pp. 155–172.

Fielder, G. (1961). *Structure of the Moon's Surface*. New York, NY: Pergamon.

Freed, A. M., Melosh, H. J., and Solomon, S. C. (2001). Tectonics of mascon loading: Resolution of the strike-slip faulting paradox. *J. Geophys. Res.*, **106**, 20 603–20 620.

Gagnepain-Beyneix, J., Lognonné, P., Chenet, H., and Spohn, T. (2006). Seismic models of the Moon and constraints on temperature and mineralogy. *Phys. Earth Planet. Inter.*, **159**, 140–166.

Gilbert, G. K. (1893). The Moon's face, a study of the origin of its features. *Philos. Soc. Washington Bull.*, **12**, 241–292.

Gillespie, P. A., Walsh, J. J., and Watterson, J. (1992). Limitations of dimension and displacement data from single faults and the consequences for data analysis and interpretation. *J. Struct. Geol.*, **14**, 1157–1172.

Goins, N. R., Dainty, A. M., and Toksöz, M. N. (1981). Seismic energy release of the Moon. *J. Geophys. Res.*, **86**, 378–388.

Golombek, M. P. (1979). Structural analysis of lunar grabens and the shallow crustal structure of the Moon. *J. Geophys. Res.*, **84**, 4657–4666.

Golombek, M. P. (1985). Fault type predictions from stress distributions on planetary surfaces: Importance of fault initiation depth. *J. Geophys. Res.*, **90**, 3065–3074.

Golombek, M. P. and McGill, G. E. (1983). Grabens, basin tectonics, and the maximum total expansion of the Moon. *J. Geophys. Res.*, **88**, 3563–3578.

Golombek, M. P., Plescia, J. B., and Franklin, B. J. (1991). Faulting and folding in the formation of planetary wrinkle ridges. *Proc. Lunar Planet Sci. Conf. 21*, 679–693.

Golombek, M. P., Anderson, F. S., and Zuber, M. T. (2001). Martian wrinkle ridge topography: Evidence for subsurface faults from MOLA. *J. Geophys. Res.*, **106**, 23 811–23 821.

Greeley, R. (1971). Lunar Hadley Rille: Considerations of its origin. *Science*, **172**, 722–725.

Greeley, R. and Spudis, P. D. (1978). Mare volcanism in the Herigonius region of the Moon. *Proc. Lunar Planet. Sci. Conf. 9*, 3333–3349.

Hall, J. L., Solomon, S. C., and Head, J. W. (1981). Lunar floor-fractured craters: Evidence for viscous relaxation of crater topography. *J. Geophys. Res.*, **86**, 9537–9552.

Hardacre, K. M. and Cowie, P. A. (2003). Controls on strain localization in a two-dimensional elastoplastic layer: Insights into size-frequency scaling of extensional fault populations. *J. Geophys. Res.*, **108**, 2529.

Hartmann, W. K. and Wood, C. A. (1971). Moon: Origin and evolution of multiring basins. *The Moon*, **3**, 3–78.
Head, J. W. (1976). Lunar volcanism in space and time. *Rev. Geophys. Space Phys.*, **14**, 265–300.
Hiesinger, H. and Head, J. W. (2006). New views of lunar geoscience: An introduction and overview. *Rev. Mineral. Geochem.*, **60**, 1–81.
Hiesinger, H., Jaumann, R., Neukam, G., and Head, J. W. (2000). Ages of mare basalts on the lunar nearside. *J. Geophys. Res.*, **105**, 29 239–29 276.
Hiesinger H., Head III, J. W., Wolf, U., Jaumann, R., and Neukum, G. (2003). Ages and stratigraphy of mare basalts in Oceanus Procellarum, Mare Nubium, Mare Cognitum, and Mare Insularum. *J. Geophys. Res.*, **108** (E7), 5065, doi:10.1029/2002JE001985.
Hodges, C. A. (1973). Mare ridges and lava lakes. *Apollo 17 Prel. Sci. Rep., NASA Spec. Publ.*, **SP-330**, 31–12 – 31–21.
Hood, L. L., Mitchell, D. L., Lin, R. P., Acuna, M. H., and Binder, A. B. (1999). Initial measurements of the lunar induced magnetic dipole moment using Lunar Prospector magnetometer data. *Geophys. Res. Lett.*, **26**, 2327–2330.
Howard, K. A. and Muehlberger, W. R. (1973). Lunar thrust faults in the Taurus−Littrow region. *Apollo 17 Prel. Sci. Rep., NASA Spec. Publ.*, **SP-330**, 31–32 – 31–25.
Johnson, A. M. (1980). Folding and faulting of strain-hardening sedimentary rocks. *Tectonophysics*, **62**, 251–278.
Khan, A. and Mosegaard, K. (2002). An inquiry into the lunar interior – A non linear inversion of the Apollo seismic data. *J. Geophy. Res.*, **107**, doi:10.1029/2001JE001658.
Khan, A., Maclennan, J., Taylor, S. R., and Connolly, J. A. D. (2006). Are the Earth and the Moon compositionally alike? Inferences on lunar composition and implications for lunar origin and evolution from geophysical modeling. *J. Geophys. Res.*, **111**, E05005, doi:10.1029/2005JE002608.
Khan, A., Connolly, J. A. D., Olsen, N., and Mosegaard, K. (2007). Constraining the composition and thermal state of the moon from an inversion of electromagnetic lunar day-side transfer functions (abs.). *Lunar Planet. Sci. Conf. XXXVIII*, 1086. Houston, TX: Lunar and Planetary Institute (CD-ROM).
Konopliv, A. S. and Yuan, D. N. (1999). Lunar Prospector 100th degree gravity model development. *Proc. Lunar Planet. Sci. Conf. 30*, 1067–1068.
Konopliv, A. S., Binder, A. B., Hood, L. L., Kucinskas, A. B., Sjogren, W. L., and Williams, J. G. (1998). Improved gravity field of the Moon from Lunar Prospector. *Science*, **281**, 1476–1480.
Konopliv, A. S., Asmar, S. W., Carranza, E., Sjogren, W. L., and Yuan, D. N. (2001). Recent gravity models as a result of the Lunar Prospector mission. *Icarus*, **150**, 1–18.
Lammlein, D. R. (1973). *Lunar seismicity, structure and tectonics*. PhD dissertation, Columbia University, New York.
Lammlein, D. (1977). Lunar seismicity and tectonics. *Phys. Earth Planet. Inter.*, **14**, 224–273.
Lammlein, D., Latham, G. V., Dorman, J., Nakamura, Y., and Ewing, M. (1974). Lunar seismicity, structure and tectonics. *Rev. Geophys. Space Phys.*, **12**, 1–21.
Lemoine, F. G. R., Smith, D. E., Zuber, M. T., Neumann, G. A., and Rowlands, D. D. (1997). A 70th degree lunar gravity model (GLGM-2) from Clementine and other tracking data. *J. Geophys. Res.*, **102**, 16 339–16 359.
Lognonné, P. and Johnson, C. L. (2007). Planetary seismology. In *Treatise on Geophysics*, Vol. 10, Ch. 4., ed. G. Schubert. New York: Elsevier.

Lognonné, P., Gagnepain-Beyneix, J., and Chenet, H. (2003). A new seismic model of the Moon: Implication in terms of structure, formation and evolution. *Earth Plan. Sci. Lett.*, **6637**, 1–18.

Lucchitta, B. K. (1976). Mare ridges and related highland scarps: Results of vertical tectonism. *Geochim. Cosmochim. Acta*, **3**, (Suppl.), 2761–2782.

Lucchitta, B. K. (1977). Topography, structure, and mare ridges in southern Mare Imbrium and northern Oceanus Procellarum. *Proc. Lunar Sci. Conf. 8, Geochim. Cosmochim. Acta*, **3**, (Suppl.), 2691–2703.

Lucchitta, B. K. and Watkins, J. A. (1978). Age of graben systems on the Moon. *Proc. Lunar Planet. Sci. Conf. 9, Geochim. Cosmochim. Acta*, **3**, (Suppl.), 3459–3472.

Margot, J. L., Campbell, D. B., Jurgens, R. F., Slade, M. A., and Stacy, N. J. (1998). The topography of the lunar polar regions from Earth-based radar interferometry (abs.). *Lunar Planet. Sci. Conf. 29*, 1845. Houston, TX: Lunar and Planetary Institute (CD-ROM).

Margot, J. L., Campbell, D. B., Jurgens, R. F., and Slade, M. A. (1999). Topography of the lunar poles from radar interferometry: a survey of cold trap locations. *Science*, **284**, 1658, doi:10.1126/science.284.5420.1658.

Marrett, R. and Allmendinger, R. W. (1991). Estimates of strain due to brittle faulting: sampling of fault populations. *J. Struct. Geol.*, **13**, 735–738.

Masursky, H., Colton, G. W., and El-Baz, F. (1978). Apollo over the Moon: A view from orbit. *NASA Spec. Publ.*, **SP-362**.

Mattingly, T. K., El-Baz, F., and Laidley, R. A. (1972). Observations and impressions from lunar orbit. *Apollo 16 Prel. Sci. Rep.*, 28–1 – 28–16.

Maxwell, T. A. (1978). Origin of multi-ring basin ridge systems: An upper limit to elastic deformation based on a finite-element model. (Proc. Lunar Planet. Sci. Conf. 9), *Geochim. Cosmochim. Acta*, **3**, (Suppl.), 3541–3559.

Maxwell, T. A. and Phillips, R. J. (1978). Stratigraphic correlation of the radar-detected subsurface interface in Mare Crisium. *Geophys. Res. Lett.*, **5**, 811–814.

Maxwell, T. A., El-Baz, F., and Ward, S. W. (1975). Distribution, morphology, and origin of ridges and arches in Mare Serenitatis. *Geol. Soc. Am. Bull.*, **86**, 1273–1278.

McCauley, J. F. (1969). Geologic map of the Alphonsus region Ga region of the Moon. U.S. Geol. Surv. Map 1–586, scale 1:50 000.

McGill, G. E. (1971). Attitude of fractures bounding straight and arcuate lunar rilles. *Icarus*, **14**, 53–58.

McGill, G. E. (2000). Fault growth by segment linkage: an explanation for scatter in maximum displacement and trace length data from the Canyonlands grabens of SE Utah: Discussion. *J. Struct. Geol.*, **22**, 135–140.

Mege, D. and Reidel, S. P. (2001). A method for estimating 2D wrinkle ridge strain from application of fault displacement scaling to the Yakima folds, Washington. *Geophys. Res. Lett.*, **28**, 3545–3548.

Melosh, H. J. (1978). The tectonics of mascon loading. *Proc. Lunar Planet. Sci. Conf. 9*, 3513–3525.

Melosh, H. J. and McKinnon, W. B. (1988). The tectonics of Mercury. In *Mercury*, ed. F. Vilas, C. R. Chapman and M. S. Matthews. Tucson, AZ: University of Arizona Press, pp. 374–400.

Melosh, H. J. and Williams, C. A., Jr. (1989). Mechanics of graben formation in crustal rocks: A finite element analysis. *J. Geophys. Res.*, **94**, 13 961–13 973.

Minshull, T. A. and Goulty, N. R. (1988). The influence of tidal stresses on deep moonquake activity. *Phys. Earth Planet. Inter.*, **52**, 41–55.

Moore, J. M. and Schultz, R. A. (1999). Processes of faulting in jointed rocks of Canyonlands National Park. *Geol. Soc. Am. Bull.*, **111**, 808–822.

Muehlberger, W. R. (1974). Structural history of southeastern Mare Serenitatis and adjacent highlands. *Proc. Lunar Sci. Conf. 5, Geochim. Cosmochim. Acta*, **1**, (Suppl.), 101–110.

Mueller, K. and Golombek, M. P. (2004). Compressional structures on Mars. *Annu. Rev. Earth Planet. Sci.*, doi:10.1146/annurev.earth.1132.101802.120553.

Muller, P. M. and Sjogren, W. L. (1968). Masons: lunar mass concentrations. *Science*, **161**, 680–684.

Nakamura, Y. (1977). HFT events: shallow moonquakes? *Phys. Earth Planet. Inter.*, **14**, 217–223.

Nakamura, Y. (1980). Shallow moonquakes: How they compare with earthquakes. *Proc. Lunar Planet. Sci. Conf. 11*, 1847–1853.

Nakamura, Y. (1983). Seismic velocity structure of the lunar mantle. *J. Geophys. Res.*, **88**, 677–686.

Nakamura, Y. (2003). New identification of deep moonquakes in the Apollo lunar seismic data. *Phys. Earth Planet. Inter.*, **139**, 197–205.

Nakamura, Y. (2005). Farside deep moonquakes and deep interior of the Moon. *J. Geophys. Res.*, **110**, E01001, doi:10.1029/2004JE002332.

Nakamura, Y. (2007). Within-nest hypocenter distribution and waveform polarization of deep moonquakes and their possible implications (abs.). *Lunar Planet. Sci. Conf. XXXVIII*, 1160. Houston, TX: Lunar and Planetary Institute (CD-ROM).

Nakamura, Y., Duennebier, F. K., Latham, G. V., and Dorman, H. J. (1976). Structure of the lunar mantle. *J. Geophys. Res.*, **81**, 4818–4824.

Nakamura, Y., Latham, G. V., Dorman, H. J., Ibrahim, A.-B. K., Koyama, J., and Horvath, P. (1979). Shallow moonquakes: Depth, distribution and implications as to the present state of the lunar interior. *Proc. Lunar Sci. Conf. 10*, 2299–2309.

Nakamura, Y., Latham, G. V., Dorman, J., and Harris, J. (1981). Passive seismic experiment long-period event catalog: Final version. *Technical Report* No. 18, Galveston Geophysics Laboratory, University of Texas at Austin.

Neukum, G. and Ivanov, B. A. (1994). Crater size distributions and impact probabilities on Earth from lunar, terrestrial-planet, and asteroid cratering data, hazards due to comets and asteroids. In *Space Science Series*, ed. T. Gehrels, M. S. Matthews and A. Schumann. Tucson, AZ: University of Arizona Press, 359.

Neumann, G. A., Zuber, M. T., Smith, D. E., and Lemoine, F. G. (1996). The lunar crust: Global structure and signature of major basins. *J. Geophys. Res.*, **101**, 16 843–16 863.

Nino, P., Philip, H., and Chery, J. (1998). The role of bed-parallel slip in the formation of blind thrust faults. *J. Struct. Geol.*, **20**, 503–516.

Nozette, S., Rustan, P., Pleasance, L. P., Kordas, J. F., Lewis, I. T., Park, H. S., Priest, R. E., Horan, D. M., Regeon, P., Lichtenberg, C. L., Shoemaker, E. M., Eliason, E. M., McEwen, A. S., Robinson, M. S., Spudis, P. D., Acton, C. H., Buratti, B. J., Duxbury, T. C., Baker, D. N., Jakosky, B. M., Blamont, J. E., Corson, M. P., Resnick, J. H., Rollins, C. J., Davies, M. E., Lucey, P. G., Malaret, E., Massie, M. A., Pieters, C. M., Reisse, R. A., Simpson, R. A., Smith, D. E., Sorenson, T. C., Vorder Breugge, R. W., and Zuber, M. T. (1994). The Clementine mission to the Moon: Scientific overview. *Science*, **266**, 1835–1839.

Oberst, J. (1987). Unusually high stress drops associated with shallow moonquakes. *J. Geophys. Res.*, **92**, 1397–1405.

Ockendon, J. R. and Turcotte, D. L. (1977). On the gravitational potential and field anomalies due to thin mass layers. *Geophys. J.*, **48**, 479–492.

Ono, T., Kumamoto, A., Nakagawa, H., Yamaguchi, Y., Oshigami, S., Yamaji, A., Kobayashi, T., Kasahara, Y., and Oya, H. (2009). Lunar radar sounder observations of subsurface layers under the nearside maria of the Moon. *Science*, **323**, 909–912.

Peeples, W. J., Sill, W. R., May, T. W., Ward, S. H., Philips, R. J., Jordan, R. L., Abbott, E. A., and Killpack, T. J. (1978). Orbital radar evidence for Lunar subsurface layering in Maria Serenitatis and Crisium. *J. Geophys. Res.*, **83**, 3459–3468.

Phillips, R. J., Conel, J. E., Abbott, E. A., Sjogren, W. L., and Morton, J. B. (1972). Mascons: Progress toward a unique solution for mass distribution. *J. Geophys. Res.*, **77**, 7106–7114.

Phillips, R. J., Adams, G. F., Brown, W. E., Jr., Eggleton, R. E., Jackson, P., Jordan, R., Peeples, W. J., Porcello, L. J., Ryu, J., Schaber, G., Sill, W. R., Thompson, T. W., Ward, S. H., and Zelenka, J. S. (1973). The Apollo 17 Lunar Sounder (Proc. Lunar Science Conf. 4). *Geochim. Cosmochim. Acta*, **3**, (Suppl. 4), 2821–2831.

Pike, R. J. (1971). Genetic implications of the shapes of Martian and lunar craters. *Icarus*, **15**, 384–395.

Plescia, J. B. and Golombek, M. P. (1986). Origin of planetary wrinkle ridges based on the study of terrestrial analogs. *Geol. Soc. Am. Bull.*, **97**, 1289–1299.

Pritchard, M. E. and Stevenson, D. J. (2000). Thermal aspects of a lunar origin by giant impact. In *Origin of the Earth and Moon*, ed. R. Canup, and K. Righter. Tucson, AZ: University of Arizona Press, pp. 179–196.

Quaide, W. L. (1965). Rilles, ridges, and domes: Clues to maria history. *Icarus*, **4**, 374.

Raitala, J. (1984). Terra scarps indicating youngest terra faults on the Moon. *Earth, Moon, and Planets*, **31**, 63–74.

Reidel, S. P. (1984). The Saddle Mountains: The evolution of an anticline in the Yakima fold belt. *Am. Jour. Sci.*, **284**, 942–978.

Robinson, M. S. and Joliff, B. L. (2002). Apollo 17 landing site: Topography, photometric corrections, and heterogeneity of the surrounding highland massifs. *J. Geophys. Res. Planets*, **107**, 20–1, E11, 5110, doi:10.1029/2001JE001614.

Roering, J. J., Cooke, M. L., and Pollard, D. D. (1997). Why blind thrust faults do not propagate to the Earth's surface: Numerical modeling of coseismic deformation associated with thrust-related anticlines. *J. Geophys. Res.*, **102**, 12 901–12 912.

Rosiek, M. R., Cook, A. C., Robinson, M. S., Watters, T. R., Archinal, B. A., Kirk, R. L., and Barrett, J. M. (2007). A revised planet-wide digital elevation model of the moon (abs.), *Lunar Planet. Sci. Conf. XXXVIII*, 2297. Houston, TX: Lunar and Planetary Institute (CD-ROM).

Schaber, G. G. (1973a). Lava flows in Mare Imbrium: Geologic evaluation from Apollo orbital photography. *Proc. Lunar Sci. Conf. 4*, 73–92.

Schaber, G. G. (1973b). Eratosthenian volcanism in Mare Imbrium: Sources of youngest lava flows. *Apollo 17 Prel. Sci. Rep., NASA Spec. Publ.*, **SP-330**, 31–22 – 31–25.

Schaber, G. G., Boyce, J. M., and Moore, H. J. (1976). The scarcity of mappable flow lobes on the lunar maria: Unique morphology of the Imbrium flows. *Proc. Lunar Sci. Conf. 7*, 2783–2800.

Schenk, P. M., and Bussey, B. J. (2004). Galileo stereo topography of the lunar north polar region. *Geophys. Res. Lett.*, **31**, L23701.

Schmitt, H. H. and Cernan, E. A. (1973). Geological investigation of the Apollo 17 landing site. *Apollo 17 Prel. Sci. Rep., NASA Spec. Publ.*, **SP-330**, 5–1 – 5–21.

Scholz, C. H. and Cowie, P. A. (1990). Determination of geologic strain from fault slip data. *Nature*, **346**, 837–839.

Schultz, P. H. (1976a). *Moon Morphology: Interpretations Based on Lunar Orbiter Photography*. Austin, TX: University of Texas Press.

Schultz, P. H. (1976b). Floor-fractured lunar craters. *The Moon*, **15**, 241–273.

Schultz, P. H., Staid, M. I., and Pieters, C. M. (2006). Lunar activity from recent gas release. *Nature*, **444**, 184–186, doi:10.1038/nature05303.

Schultz, R. A. (1997). Displacement–length scaling for terrestrial and Martian faults: Implications for Valles Marineris and shallow planetary grabens. *J. Geophys. Res.*, **102**, 12 009–12 015.

Schultz, R. A. (1999). Understanding the process of faulting: Selected challenges and opportunities at the edge of the 21st century. *J. Struct. Geol.*, **21**, 985–993.

Schultz, R. A. (2000). Localizaton of bedding-plane slip and backthrust faults above blind thrust faults: Keys to wrinkle ridge structure. *J. Geophys. Res.*, **105**, 12 035–12 052.

Schultz, R. A. and Fori, A. N. (1996). Fault-length statistics and implications of graben sets at Candor Mensa, Mars. *J. Struct. Geol.*, **18**, 272–383.

Schultz, R. A. and Zuber, M. A. (1994). Observations, models, and mechanisms of failure of surface rocks surrounding planetary surface loads. *J. Geophys. Res.*, **99**, 14 691–14 702.

Scott, D. H. (1973). Small structures of the Taurus-Littrow region. *Apollo 17 Prel. Sci. Rep., NASA Spec. Publ.*, **SP-330**, pp. 31–25 – 31–29.

Scott, D. H., Diaz, J. M., and Watkins, J. A. (1977). Lunar farside tectonics and volcanism. *Proc. Lunar Sci. Conf. 8*, 1119–1130.

Sharpton, V. L. (1992). Apollo 17: One giant step toward understanding the tectonic evolution of the Moon: Geology of the Apollo 17 landing site. *LPI Tech. Rep. 92–09, Part 1*, 50–53.

Sharpton, V. L. and Head, J. W. (1981). The origin of mare ridges: Evidence from basalt stratigraphy and substructure in Mare Serenitatis (abs.). *Lunar Planet. Sci. Conf. XII*, 961–963.

Sharpton, V. L. and Head, J. W. (1982). Stratigraphy and structural evolution of southern Mare Serenitatis: A reinterpretation based on Apollo Lunar Sounder Experiment data. *J. Geophys. Res.*, **87**, 10 983–10 998.

Sharpton, V. L. and Head, J. W. (1988). Lunar mare ridges: Analysis of ridge-crater intersection and implications for the tectonic origin of mare ridges. *Proc. Lunar Sci. Conf. 18*, 307–317.

Shearer, C. K., Hess, P. C., Wieczorek, M. A., Pritchard, M. E., Permentier, E. M., Borg, L. E., Longhi, J., Elkins-Tanton, L. T., Neal, C. R., Antonenko, I., Canup, R. M., Halliday, A. N., Grove, T. L., Hager, B. H., Less, D.-C., and Wiechert, U. (2006). Thermal and magmatic evolution of the Moon. *Rev. Mineral. Geochem.*, **60**, 365–518.

Simpson, R. W. (1997). Quantifying Anderson's fault types. *J. Geophys. Res.*, **102**, 17 909–17 920.

Smith, D. E., Zuber, M. T., Neumann, G. A., and Lemoine, F. G. (1997). Topography of the Moon from the Clementine LIDAR. *J. Geophys. Res.*, **102**, 1591–1611.

Solomon, S. C. and Chaiken, J. (1976). Thermal expansion and thermal stress in the Moon and terrestrial planets: Clues to early thermal history. *Proc. Lunar Sci. Conf. 7*, 3229–3243.

Solomon, S. C. and Head, J. W. (1979). Vertical movement in mare basins: Relation to mare emplacement, basin tectonics and lunar thermal history. *J. Geophys. Res.*, **84**, 1667–1682.

Solomon, S. C. and Head, J. W. (1980). Lunar mascon basins: Lava filling, tectonics, and evolution of the lithosphere. *Rev. Geophys. Space Phys.*, **18**, 107–141.

Spudis, P. D. (1993). *The Geology of Multi-Ring Impact Basins: The Moon and Other Planets*. Cambridge: Cambridge University Press.

Spudis, P. D., Swann, G. A., and Greeley, R. (1988a). The formation of Hadley Rille and implications for the geology of the Apollo 15 region. *Proc. Lunar Sci. Conf. 18*, 243–254.

Spudis, P. D., Hawke, B. R., and Lucey, P. G. (1988b). Materials and formation of the Imbrium Basin. *Proc. Lunar Sci. Conf. 18*, 155–168.

Spudis, P. D., Reisse, R. A., and Gillis, J. J. (1994). Ancient multiring basins on the Moon revealed by Clementine laser altimetry. *Science*, **266**, 1848–1851.

Stöffler, D. and Ryder, G. (2001). Stratigraphy and isotope ages of lunar geologic units: Chronological standard for the inner Solar System. *Space Sci. Rev.*, **96**, 9–54.

Strom, R. G. (1972). Lunar mare ridges, rings, and volcanic ring complexes. *Mod. Geol.*, **2**, 133–157.

Tjia, H. D. (1970). Lunar wrinkle ridges indicative of strike-slip faulting. *Geol. Soc. Am. Bull.*, **81**, 3095–3100.

Toksöz, M. N. (1974). Geophysical data and the interior of the Moon. *Annu. Rev. Earth Planet. Sci.*, **2**, 151.

Toksöz, M. N., Goins, N. R., and Cheng, C. H. (1977). Moonquakes: Mechanisms and relations to tidal stresses. *Science*, **196**, 979–981.

U. S. Geological Survey (USGS) (1972). Preliminary topographic map of part of the Littrow region of the Moon, scale 1:50 000. Flagstaff, AZ.

U. S. Geological Survey (USGS) (2002). Color-coded topography and shaded relief of the lunar north and south hemispheres, U.S. Geol. Surv. Geol. Invest. Ser., 2769.

Vinnik, L., Chenet, H., Gagnepain-Beyneix, J., and Lognonné, P. (2001). First seismic receiver functions on the Moon. *Geophys. Res. Lett.*, **28**, 3031–3034.

Walsh, J. and Watterson, J. (1988). Analysis of the relationship between displacements and dimensions of faults. *J. Struct. Geol.*, **10**, 239–247.

Watters, T. R. (1988). Wrinkle ridge assemblages on the terrestrial planets. *J. Geophys. Res.*, **93**, 10 236–10 254.

Watters, T. R. (1991). Origin of periodically spaced wrinkle ridges on the Tharsis plateau of Mars. *J. Geophys. Res.*, **96**, 15 599–15 616.

Watters, T. R. (1992). A system of tectonic features common to Earth, Mars, and Venus. *Geology*, **20**, 609–612.

Watters, T. R. (1993). Compressional tectonism on Mars. *J. Geophys. Res.*, **98**, 17 049–17 060.

Watters, T. R. (2003). Thrust faulting along the dichotomy boundary in the eastern hemisphere of Mars. *J. Geophys. Res.*, **108**, 5055, doi:10.1029/2002JE001934.

Watters, T. R. (2004). Elastic dislocation modeling of wrinkle ridges on Mars. *Icarus*, **171**, 284–294.

Watters, T. R. and Konopliv, A. S. (2001). The topography and gravity of Mare Serenitatis: Implications for subsidence of the mare surface. *Planet. Space Sci.*, **49**, 743–748.

Watters, T. R. and Robinson, M. S. (1997). Radar and photoclinometric studies of wrinkle ridges on Mars. *J. Geophys. Res.*, **102**, 10 889–10 904.

Watters, T. R. and Robinson, M. S. (1999). Lobate scarps and the Martian crustal dichotomy. *J. Geophys. Res.*, **104**, 18 981–18 900.

Watters, T. R., Robinson, M. S., and Cook, A. C. (1998). Topography of lobate scarps on Mercury: New constraints on the planet's contraction. *Geology*, **26**, 991–994.

Watters, T. R., Robinson, M. S., and Schultz, R. A. (2000). Displacement–length relations of thrust faults associated with lobate scarps on Mercury and Mars: Comparison with terrestrial faults. *Geophys. Res. Lett.*, **27**, 3659–3662.

Watters, T. R., Robinson, M. S., and Cook, A. C. (2001). Large-scale lobate scarps in the southern hemisphere of Mercury. *Planet. Space Sci.*, **49**, 1523–1530.

Watters, T. R., Robinson, M. S., Bina, C. R., and Spudis, P. D. (2004). Thrust faults and the global contraction of Mercury. *Geophys. Res. Lett.*, **31**, L04701, doi:10.1029/2003GL019171.

Watters, T. R., Nimmo, F., and Robinson, M. S. (2005). Extensional troughs in the Caloris Basin of Mercury: Evidence of lateral crustal flow. *Geology*, **33**, 669–672.

Watters, T. R., Solomon, S. C., Robinson, M. S., Head, J. W., André, S. L., Hauck, S. A., and Murchie, S. L. (2009). The tectonics of Mercury: The view after MESSENGER'S first flyby. *Earth Planet. Sci. Lett.*, doi:10.1016/j.epsl.2009.01.025.

Weber (formerly Bulow), R., Bills, B. G., and Johnson, C. L. (2009). Constraints on deep moonquake focal mechanisms through analyses of tidal stress. *J. Geophys. Res. Planets*, **114**, E05001, doi:10.1029/2008JE003286.

Whitaker, E. A. (1966). The surface of the Moon. In *The Nature of the Lunar Surface: Proceedings of the 1965 IAU-NASA Symposium*, ed. W. N. Hess, D. H. Menzel and J. A. O'Keefe. Baltimore, MD: Johns Hopkins Press, pp. 79–98.

Wichman, R. W. and Schultz, P. H. (1995). Floor-fractured craters in Mare Smythii and west of Oceanus Procellarum: Implications of crater modification by viscous relaxation and igneous intrusion models. *J. Geophys. Res.*, **100**, 21 201–21 218.

Wieczorek, M. A. and Phillips, R. J. (1998). Potential anomalies on a sphere: Applications to the thickness of the lunar crust. *J. Geophys. Res.*, **103**, 1715–1724.

Wieczorek, M. A. and Simons, F. J. (2005). Localized spectral analysis on the sphere. *Geophys. J. Int.*, **162**, 655–675.

Wieczorek, M. A., Jolliff, B. L., Khan, A., Pritchard, M. E., Weiss, B. P., Williams, J. G., Hood, L. L., Righter, K., Neal, C. R., Shearer, C. K., McCallum, I. S., Tompkins, S., Hawke, B. R., Peterson, C., Gillis, J. J., and Bussey, B. (2006). The constitution and structure of the lunar interior. *Rev. Mineral. Geochem.*, **60**, 221–364.

Wilhelms, D. E. (1987). *The Geologic History of the Moon*. Washington, DC: U.S. Government Printing Office.

Wilhelms, D. E. and McCauley, J. F. (1971). Geologic map of the near side of the Moon. USGS Map I-703, scale 1:5 000 000.

Williams, J. G., Boggs, D. H., Yoder, C. F., Ratcliff, J. T., Todd, J., and Dickey, J. O. (2001). Lunar rotational dissipation in solid body and molten core. *J. Geophys. Res.*, **106**, 27 933–27 968.

Wise, D. U. and Yates, M. T. (1970). Mascons as structural relief on a lunar "Moho". *J. Geophys. Res.*, **75**, 261–268.

Weisberg, O. and Hager, B. H. (1998). Global lunar contraction with subdued surface topography (abs.): *Origin of the Earth and Moon*, LPI contribution no. 957. Houston, TX: Lunar and Planetary Institute, p. 54.

Wojtal, S. F. (1996). Changes in fault displacement populations correlated to linkage between faults. *J. Struct. Geol.*, **18**, 265–279.

Wollenhaupt, W. R., Sjogren, W. L., Lingenfelter, R. E., Schubert, G., and Kaula, W. M. (1973). Apollo 17 Laser Altimeter. *Apollo 17 Prel. Sci. Rep., NASA Spec. Publ.*, **SP-330**, 33–41 – 33–44.

Wu, S. S. C. and Doyle, F. J. (1990). Topographic mapping. In *Planetary Mapping*, ed. R. Greeley and R. M. Batson. Cambridge: Cambridge University Press, pp. 169–207.

Young, R. A., Brennan, W. J., Wolfe, R. W., and Nichols, P. J. (1973). Volcanism in the lunar maria. *Apollo 17 Prel. Sci. Rep., NASA Spec. Publ.*, **SP-330**, 31–1 – 31–11.

Zuber, M. T., Smith, D. E., Lemoine, F. G., and Neumann, G. A. (1994). The shape and internal structure of the Moon from the Clementine Mission. *Science*, **266**, 1839–1843.

5

Mars tectonics

Matthew P. Golombek

Jet Propulsion Laboratory, California Institute of Technology, Pasadena

and

Roger J. Phillips

Planetary Science Directorate, Southwest Research Institute, Boulder

Summary

Mars is a key intermediate-sized terrestrial planet that has maintained tectonic (and overall geologic) activity throughout its history, and preserved a record in rocks and terrains exposed at the surface. Among the earliest recorded major geologic events was lowering of the northern plains, relative to the southern highlands, possibly by a giant, oblique impact (or endogenic process) that left an elliptical basin with a thinned crust. Sitting on the edge of this global crustal dichotomy is Tharsis, an enormous elevated volcanic and tectonic bulge that rises ~10 km above the datum. It is topped by four giant shield volcanoes, and is surrounded by radial extensional grabens and rifts and concentric compressional wrinkle ridges that together deform the entire western hemisphere and northern plains. Deformation in the eastern hemisphere is more localized in and around large impact basins and volcanic provinces. Extensional structures are dominantly narrow grabens (several kilometers wide) that individually record of order 100 m extension, although larger (100 km wide), deeper rifts are also present. Compressional structures are dominated by wrinkle ridges, interpreted to be folds overlying blind thrust faults that individually record shortening of order 100 m, although larger compressional ridges and lobate scarps (thrust fault scarps) have also been identified. Strike-slip faults are relatively rare and typically form in association with wrinkle ridges or grabens. Mapping of extensional structures in deformed regions that have rich stratigraphies shows that structures on Mars formed during five main stages: (1) Noachian deformation concentrated in Claritis Fossae, Thaumasia, Sirenum, Syria Planum, Ceraunius, and Tempe Terra; (2) Late Noachian – Early Hesperian deformation concentrated

Planetary Tectonics, edited by Thomas R. Watters and Richard A. Schultz. Published by Cambridge University Press. Copyright © Cambridge University Press 2010.

in Valles Marineris and Thaumasia; (3) Early Hesperian deformation concentrated in Noctis Labyrinthus, Tempe Terra, Valles Marineris, and Thaumasia; (4) Late Hesperian – Early Amazonian deformation concentrated in Alba Patera, Tempe Terra, and around Olympus Mons and the Tharsis Montes; and (5) Middle–Late Amazonian deformation concentrated around the large shield volcanoes. About half of the extensional structures formed during the Noachian (>3.8 Ga), indicating that tectonic activity peaked early and generally decreased with time. Wrinkle-ridge formation peaked in the Hesperian (both around Tharsis and in the eastern hemisphere, far from Tharsis), suggesting an overprint and modulation by global compressional cooling stresses. Lithospheric deformation models resulting from elastic-shell loading show that loading over the scale of Tharsis (large relative to the radius of the planet) is dominated by membrane stresses, and produces the concentric extensional stresses around the periphery and the radial compressional stresses closer in that are needed to explain the radial grabens and rifts and concentric wrinkle ridges. Because elastic-shell models based on present-day gravity and topography can explain the observed distribution and strain of radial (over half of which are Noachian) and concentric tectonic features, the basic lithospheric structure of the province has probably changed little since the Noachian and elastic support of the Tharsis load by a thickening lithosphere has been the dominant geodynamical mechanism. The origin of Tharsis required a positively buoyant mantle region accompanied by voluminous partial melting, of which a core–mantle plume is one possibility. This enormous volcanic load produced a moat around it, which shows up most dramatically as a negative gravity ring, and an antipodal bulge that contributes to the first-order shape and gravity field of the planet. If the load is composed of basaltic magmatic products as suggested by fine layers within Valles Marineris, water released with the magma would be equivalent to a global layer up to 100 m thick, which might have enabled the early warm and wet Martian climate indicated by valley networks, degraded craters and terrains, sulfates, and phyllosilicates found by rovers and orbiters exploring Mars. Additional geophysical constraints on Tharsis include the observation that remanent magnetic anomalies occur high on the Tharsis rise and likely require uplift of the Tharsis crust early in its evolution.

1 Introduction

Because of its intermediate size, Mars is a key planet among our solar system's collection of terrestrial planets. It is large enough to have had geologic and tectonic activity throughout geologic time, but not so active as to have totally destroyed all record of its early geologic history. Because most geologic and tectonic processes are driven by the internal heat engine of a planet or satellite, large planets such as the Earth and Venus have had active and prolonged geological histories due to their greater volume to surface area ratio (proportional to radiogenic heat flux) than the small planetary bodies such as the Moon and Mercury (Kaula, 1975). Of our assortment of terrestrial planets, the Earth and Venus have been so active that most of their early histories have been destroyed by ongoing vigorous geologic activity. In contrast, Mercury and the Moon have largely cratered surfaces with little geologic activity after the first ∼2 Ga. Mars appears to be just the right size to have maintained geologic and tectonic activity throughout time, and to have preserved the record in rocks and terrains exposed at the surface for study.

On Earth, it is now accepted that the plate tectonic system has directly modulated and maintained clement conditions and liquid oceans for most of geologic time and thus has been important for the evolution of life (Ward and Brownlee, 2000, 2002). Plate tectonics has maintained a constant freeboard or relative level of exposed continents and oceans throughout the Phanerozoic (Wise, 1974) by liberating, through subduction, volatiles trapped in weathered rocks, and has controlled the level of greenhouse gases in the atmosphere. Without this process, volatiles can be trapped in weathering products and lost from the atmosphere and hydrosphere, resulting in a dry, waterless climate.

On Mars, it is therefore perhaps not surprising that there is growing evidence that its tectonic evolution has been critically important to its climatic history (Phillips *et al.*, 2001). During the early history of Mars (Plate 14) – the oldest geologic period is the Noachian, estimated by crater density to be >3.6 Ga (Hartmann and Neukum, 2001; Hartmann, 2005; Tanaka and Hartmann, 2008) – there is abundant evidence for wetter and possibly warmer conditions in valley networks, dry lake beds, and high erosion rates preserved in degraded craters and terrain (Carr, 1996; Craddock and Howard, 2002; Howard *et al.*, 2005; Irwin *et al.*, 2005). Further, surface exploration by the Mars Exploration Rovers (MER) has found sulfates in Meridiani Planum that have been interpreted as "dirty" evaporites (Squyres *et al.*, 2004) (likely fed by groundwater (Andrews-Hanna *et al.*, 2007)), and the aqueous alteration of rocks in the Columbia Hills (Squyres *et al.*, 2006) that also indicate wetter conditions in the Late Noachian. The imaging spectrometers OMEGA on Mars Express and CRISM on the Mars Reconnaissance Orbiter have revealed the widespread occurrence of phyllosilicates in Noachian terrain (Bibring *et al.*, 2006; Mustard *et al.*, 2008), with clear evidence in places that this material has been excavated from the crust by impacts (Mustard *et al.*, 2008). Clearly water was a pervasive commodity during the Noachian period on Mars. In contrast, the current climate is dry and desiccating, with extremely low erosion rates consistent with eolian activity, and little or no erosion by liquid water in the Hesperian (the intermediate geologic period on Mars, estimated to be 2.6–3.6 Ga) and Amazonian – the youngest geologic period, estimated to be <2.6 Ga (Hartmann and Neukum, 2001; Hartmann, 2005; Tanaka and Hartmann, 2008) (Arvidson *et al.*, 1979; Golombek and Bridges, 2000; Golombek *et al.*, 2006a,b; Grant *et al.*, 2004) (Plate 14).

Evidence for early warmer and wetter conditions have generally fueled speculation that the early environment on Mars may have been conducive to the formation of life at a time when life started on Earth. If Mars was indeed wetter early in its history, what caused this brief more clement period and what led to its change to the current cold and dry conditions that appear to have been in place since the Hesperian? The answer to this question may be locked in the development and rise of the Tharsis province on Mars, the largest and longest lived tectonic entity in the solar system.

Any discussion of Mars tectonics centers on Tharsis and the highland–lowland dichotomy (aka global dichotomy). Tharsis is an enormous elevated region of the planet, covered by huge volcanoes and surrounded by a system of generally radial extensional tectonic features (including the huge Valles Marineris), and generally concentric compressional tectonic features that imprint the entire western hemisphere of the planet (Banerdt et al., 1992) and most of the northern plains (Withers and Neumann, 2001; Head et al., 2002). The elevated region of Tharsis covers one quarter of the planet ($\sim 3.0 \times 10^7$ km^2 in areal extent) and rises to 10 km above the Mars Orbiter Laser Altimeter (MOLA) defined geoid near Syria Planum, with young (Amazonian) shield volcanoes rising to over 20 km elevation (Smith et al., 2001). Tharsis appears to sit on the highland–lowland boundary that separates the northern plains that are less cratered, and therefore have a younger surface age, and are several kilometers lower than the southern highlands that are more heavily cratered and higher in elevation. Located at the edges of Tharsis and the highland–lowland boundary are the catastrophic outflow channels that funneled huge volumes of water into the northern plains, intermediate in Mars history (Hesperian and Early Amazonian) (Baker et al., 1992). To first order then, some process or processes lowered and resurfaced about a third of the planet to create the northern plains early in its history and another (or perhaps related) process created a huge topographic rise, topped by young volcanics that produced tectonic features that deform, to various degrees, most of the planet. The resultant major questions in Mars tectonics involve understanding the timing and causes of these two primary tectonic events.

Our understanding of the tectonics of Mars has been fueled by data returned from spacecraft that have visited the Red Planet, and is experiencing a renaissance with the data returned from the Mars Global Surveyor (MGS), Mars Odyssey, Mars Express, and Mars Reconnaissance Orbiter (MRO) spacecraft. Prior to the data returned by these spacecraft, our interpretations of structural features and tectonics have been based on Viking topography and gravity, and global imaging at $\sim$200 m/pixel scale. The imaging data allowed the mapping of regions and areas that defined the timing of structural events within the established stratigraphic framework of Mars (Tanaka, 1986). However, the poor topography and accurate knowledge of only the lower degree gravity and topographic fields severely inhibited the realistic modeling of local and regional deformational processes. MGS has returned exquisite global topographic data from MOLA (Smith *et al.*, 1999b; Smith *et al.*, 2001) and gravity data accurate to spherical harmonic degree and order 60 (Smith *et al.*, 1999a; Lemoine *et al.*, 2001; Yuan *et al.*, 2001), which have ushered in a renewed era of much more constrained modeling of deformational processes. The addition of tracking data from MRO has yielded significant improvements in the spherical harmonic gravity field and derived crustal thickness models (Zuber *et al.*, 2007; Neumann *et al.*, 2008). Similar advances in geologic and tectonic mapping and characterization have been realized from the high-resolution Mars Orbiter Camera (MOC on MGS), High Resolution Imaging Science Experiment (HiRISE on MRO), High Resolution Stereo Camera (HRSC on Mars Express), and Thermal Emission Imaging System (THEMIS on Odyssey) images (Golombek, 2003).

In this chapter on Mars tectonics (and associated geophysics), we begin with a description of the global geology, topography and gravity of Mars as a basis for understanding the tectonic framework of the planet. We then discuss the geometry and kinematics of structural features on Mars (extensional, compressional and strike-slip). The tectonic history of these features, including their distribution in structural terrains, timing and orientation, are next discussed for the western and eastern hemispheres of Mars. These sections then provide the basic framework to discuss deformational mechanisms at a global, regional and local scale. We will refer to, but not repeat, information in the last major review of Mars tectonics, a summary of knowledge based on the Viking data (Banerdt *et al.*, 1992), and will focus on new information and advances since that publication.

2 Global geology, topography and gravity

2.1 Physiography

To first order, the major physiographic units of Mars are also distinguishable in its geology, topography and gravity (Plates 13 and 15). The three major units: the southern highlands, northern plains, and the Tharsis plateau have been known since Mariner 9 acquired imaging data in the early 1970s. The southern highlands are elevated, heavily cratered terrain that includes large impact basins (Argyre and Hellas) that date back to the end of heavy bombardment. High crater densities and mapping results place the southern highlands as ancient Noachian units most affected by the high cratering rates early in solar system history. The northern plains, separated from the highlands by the global dichotomy boundary and the Tharsis rise, are several kilometers lower in elevation and have much lower crater densities, suggesting Late Hesperian and Amazonian surface ages that are much younger than the southern highlands (Tanaka *et al.*, 2003, 2005). The physiographic boundary, where not obscured by the Tharsis rise, is in places a smooth transition from cratered terrain to volcanic/sedimentary plains, but in other places is more abrupt, showing the effects of both tectonic and erosional processes (McGill and Dimitriou, 1990; Smith *et al.*, 1999b; Smrekar *et al.*, 2004).

Remnants of large craters are evident in the northern plains (e.g., Isidis and Utopia) and careful inspection of MOLA topography revealed a large number of subtle circular depressions (termed "quasi-circular depressions," QCDs) that have been reasonably interpreted as shallowly buried impact craters and basins (Frey *et al.*, 2002; Frey, 2006a,b). The size–frequency distributions of these circular depressions argue strongly that the underlying surface dates from before the earliest Noachian (i.e., pre-Noachian) (Nimmo and Tanaka, 2005; Frey, 2006a,b) and that the total fill in the northern plans is mostly thin (a few kilometers), suggesting that the northern plains have been low throughout virtually all of Martian history recorded in surface or near-surface terrains.

The enormous plateau of the Tharsis rise is ringed by ridged plains, interpreted as deformed Hesperian flood basalts, and capped by Amazonian lavas and shield volcanoes. The five largest volcanoes are the three central Tharsis Montes (from south to north: Arsia, Pavonis and Ascreaeus), the tallest, Olympus Mons to the northwest, and Alba Patera to the north. To the east, south and southwest of Tharsis are Noachian cratered highlands and within the plateau are portions of highlands crust surrounded by younger Amazonian volcanics (Plescia and Saunders, 1982; Scott and Tanaka, 1986). To the northeast, north and northwest of Tharsis are the lower and younger northern plains covered by Hesperian through Amazonian lava flows, sediments and outflow channel deposits. The highest portion of Tharsis, aside from the five giant volcanoes, is Syria Planum, which is bounded on the west by Claritas Fossae, a high-standing ridge of heavily fractured Noachian terrain that extends south to the Thaumasia region, and to the north by Noctis Labyrinthus, a huge system of intersecting troughs. From the eastern end of Noctis Labyrinthus open the giant canyons of Valles Marineris that cleave the Hesperian plains for 2000 km to the east before opening into a series of chaos-filled depressions. Some of the Hesperian giant catastrophic outflow channels that begin in the Noachian highlands and drain into the northern plains appear to originate in these chaos-filled depressions.

The eastern hemisphere of Mars is dominated by heavily cratered terrain to the south, and across the dichotomy boundary lie the northern plains that have been imprinted by ancient impact basins (Isidis and Utopia). Just north of the boundary is the Elysium volcanic construct with grabens, surrounding wrinkle ridges and young Amazonian volcanics and channels. Elsewhere in the highlands are

Hesperian volcanics and the Hellas and Argyre impact basins. The volcanic areas are characterized by wrinkle ridges, and the dichotomy boundary by lobate scarps in the highlands and extensional structures towards the lowlands. Wrinkle ridges related to Tharsis loading are found throughout the lowlands of both hemispheres (Head *et al.*, 2002; Tanaka *et al.*, 2003).

2.2 Shape of Mars and crustal structure

The shape of Mars, consisting of the long-wavelength components of the topography, is to first order determined (1) by its global spin oblateness, (2) by a pole-to-pole slope that renders the northern hemisphere ~5 km lower than the southern hemisphere (Smith *et al.*, 1999b) and that is locally expressed by the highland–lowland dichotomy, and (3) by the topography of the Tharsis rise and any global-scale deformation induced by its attendant mass load (Phillips *et al.*, 2001). The voluminous extrusive and intrusive magmatic deposits of Tharsis have loaded the lithosphere, leading to pervasive fracturing and folding over vast regions of the planet (Solomon and Head, 1982; Banerdt *et al.*, 1992), and are responsible specifically for the radial extensional tectonic and concentric compressional tectonic features described earlier. Indeed, global-scale tectonics has been largely the result of the planetary response to Tharsis, plus the changing thermal state of the lithosphere. The Tharsis load has affected the shape of the planet by deforming the lithosphere as a thin spherical membrane, inducing the circumferential Tharsis flexural trough (surrounding moat) and contributing to high-standing topography in parts of Arabia Terra (the "Arabia bulge") (Phillips *et al.*, 2001) (Plate 16). This global-scale deformation appears to control the azimuth of many Late Noachian valley networks, implying that most of the Tharsis mass was in place by the end of the Noachian period. This is consistent with the observation that over half of Tharsis-related extensional tectonic features are Noachian in age (Anderson *et al.*, 2001). The Tharsis trough (surrounding moat) contains the vast majority of outflow channels on the planet (Phillips *et al.*, 2001).

Bouguer gravity anomalies correlated with topography provided the first hard evidence that Mars has a distinct crust (Phillips *et al.*, 1973), though such early analyses provided no information on mean crustal thickness or global crustal structure. Recent work has greatly improved our knowledge of crustal structure (Zuber *et al.*, 2000; Zuber, 2001; Neumann *et al.*, 2004, 2008). Geophysically, the mean crustal thickness has a lower bound of ~50 km from the requirement that model Moho relief does not lead to negative crustal thickness anywhere on the planet (Zuber *et al.*, 2000; Neumann *et al.*, 2008). Using both geochemical and geophysical constraints, Wieczorek and Zuber (2004) find a mean crustal thickness of 50 ± 12 km. An upper bound on mean crustal thickness of ~100 km results from the constraint that Moho relief must be preserved against ductile collapse over billions of years (Nimmo and Stevenson, 2001; Zuber *et al.*, 2000). Gravity/topography admittance estimates at the hemispheric dichotomy boundary yield an upper bound on crustal thickness of 75 km in that region (Nimmo, 2002).

Gravity and topography data show that crustal thickness is distinctly bimodal, with modal thicknesses of 32 and 58 km in the northern and southern hemispheres, respectively (Neumann *et al.*, 2004). Regionally, the thickest crust is associated with the Tharsis volcanic construct, and the thinnest with major impact basins, such as Hellas, Argyre, and Isidis. Gravity and topography are both high on the Tharsis plateau and crustal thickness estimates here exceed 80 km at the higher elevations (Zuber *et al.*, 2000; Zuber, 2001; Neumann *et al.*, 2004, 2008).

The origin of the variation in crustal structure (the crustal dichotomy: the crustal expression of the global dichotomy) between the two hemispheres has been controversial. Given that the northern hemisphere crust beneath the Hesperian and Amazonian cover dates to the pre-Noachian (Frey *et al.*, 2002; Frey, 2006a,b), global-scale crustal thickness variations could be a feature of primordial crustal fractionation, and the north–south crustal dichotomy must certainly predate the ancient impact basins Utopia and Hellas, whose Moho relief is preserved (Neumann *et al.*, 2004). Global crustal thickness variations could also be due to a variety of other causes, including lower crustal flow induced by degree-1 convection (Wise *et al.*, 1979; Zhong and Zuber, 2001; Roberts and Zhong, 2006) and very early plate tectonics (Sleep, 1994). The low elevation of the northern hemisphere could also be the result of one or more giant impacts (Wilhems and Squyres, 1984). By stripping away the cover of the Tharsis rise, Andrews-Hanna *et al.* (2008a) have shown that the global dichotomy boundary and lowlands have an elliptical shape, suggesting formation of the northern lowlands by a large oblique impact. Marinova *et al.* (2008) performed numerical simulations of single mega-impacts and found that for plausible impact energies, low impact velocities, and, in particular, impact angles in the 30–60° range, elliptical boundary shapes and basins similar to the shape found by Andrews-Hanna *et al.* (2008a) are achieved. For these models, the melt distribution stays within the impact basin and does not cover and obscure the dichotomy boundary.

2.3 Gravity field and lithospheric structure

Spectrally, the Tharsis rise and its global response largely account for (aside from the spin oblateness) the low-degree ($l \leq 10$) spherical harmonic gravity field and geoid of Mars. Spatially, the long-wavelength geoid of Mars (apart from its oblateness) is dominated by a strong anomaly (∼1000 m) centered on the Tharsis rise, a lesser high (∼500 m) over Arabia Terra (antipodal to Tharsis), and a negative anomaly (∼−500 m) surrounding Tharsis (Smith *et al.*, 1999b) (Plate 15). The last two features correspond to the antipodal Arabia bulge and the circumferential Tharsis trough discussed above (Plate 16). Shorter wavelength anomalies correspond to specific physiographic features (Lemoine *et al.*, 2001), including the Valles Marineris, large shield volcanoes, and impact basins (notably Isidis and Utopia). The gravity anomalies are strong enough for most of these features to indicate a non-isostatic state and flexural support by the lithosphere (Zuber *et al.*, 2000; McGovern *et al.*, 2002, 2004b). The Utopia basin is unique in that it seems to support flexurally approximately 10 km (or more) of fill that has a large igneous component (Searls *et al.*, 2006). Both the long- and short-wavelength gravity anomalies imply significant stress in the lithosphere; much of the tectonic fabric of the planet can be tied to these stresses, with the long-wavelength component, in turn, directly related to Tharsis.

The high-fidelity topographic information acquired by MOLA on the MGS mission, as well as the superb gravity field information derived from tracking the MGS, Mars Odyssey, and MRO spacecraft, have enabled a comprehensive survey of effective elastic lithosphere thicknesses (T_e) and from these estimates, the heat flow (Zuber *et al.*, 2000; McGovern *et al.*, 2002, 2004b; McKenzie *et al.*, 2002; Belleguic *et al.*, 2005; Milbury *et al.*, 2007; Grott and Breuer, 2008; Phillips *et al.*, 2008). Selected elastic thickness estimates are given in Table 5.1. Heat flow values are obtained from a yield-strength-envelope (Brace and Kohlstedt, 1980) analysis that equates the bending moment of an elastic lithosphere to that of a temperature-gradient-dependent elastic-plastic lithosphere (McNutt, 1984). These analyses show in general that with decreasing age, lithospheric loads were deforming increasingly thicker mechanical lithospheres. The implication is that heat flow declined rapidly during the Noachian, and, subsequently, more slowly; this is of course entirely consistent with both the exponential decrease in parent isotopes of heat-producing elements and the secular cooling of a terrestrial planet (e.g., Hauck and Phillips, 2002).

Most determinations of T_e have been derived in the spectral (wavelength) domain using gravity/topography admittance and correlation estimates (McGovern *et al.*, 2002, 2004b; Belleguic *et al.*, 2005). In the spatial domain, one or both of the gravity and topography data types have been used to estimate elastic thickness by comparing model stresses to fault locations circumferential to shield volcanoes and impact basins (Comer *et al.*, 1985; Solomon and Head, 1990), and by analyzing linear fault deformation and rift flank uplift (Grott *et al.*, 2005, 2007a; Kronberg *et al.*, 2007). Phillips *et al.* (2008) used the lack of deflection of the substrate beneath the icy north polar cap load to derive a lower bound on present-day elastic thickness of 300 km. This implies that Mars has sub-chondritic heat sources, or that heat flow is regionally variable; for example, excess heat flow might be expected

Table 5.1. *Elastic lithosphere thicknesses through time on Mars*

Feature	Surface Age*	T_e, km
Olympus Mons[1,2]	A	>70, 93 ± 40
Ascraeus Mons[1,2]	A	2–80, 105 ± 40
Pavonis Mons[1]	A	<100
Arsia Mons[1]	A	>20
North Polar deposits[3]	A	>300
Alba Patera[1,2]	A–H	38–65, 66 ± 20
Elysium Rise[1,2,4]	A–H	15–45, 56 ± 20, 25–29
Solis Planum[1]	H	24–37
Hellas west rim[1]	H–N	<20
Coracis Fossae[5]	H–N	10–13
Hellas basin[1]	N	<13
Noachis Terra[1]	N	<12
Northeastern Terra Cimmeria[1]	N	<12
Northeastern Arabia Terra[1]	N	<16
Southern Thaumasia[6]	N	21–35
Acheron Fossae[7]	N	9–11

Selected elastic thickness (T_e) estimates from [1]McGovern *et al.* (2004b), [2]Belleguic *et al.* (2005), [3]Phillips *et al.* (2008), [4]McKenzie *et al.* (2002), [5]Grott *et al.* (2005), [6]Grott *et al.* (2007a), [7]Kronberg *et al.* (2007).
*A = Amazonian, H = Hesperian, N = Noachian.

in the Tharsis province, and, therefore, less than the planetary average elsewhere. Alternatively, in response to the polar load, the lithosphere might be in a transient state controlled by the viscosity of the mantle; however, this also implies subchondritic planetary heat sources.

2.4 Core, magnetic field, and true polar wander

Two-way ranging and Doppler tracking of the Mars Pathfinder lander yielded the first assumption-free estimate of the polar moment of inertia. This was done by estimating the precession rate of the rotational pole, which incorporated knowledge of the position of the Viking landers and derived pole 20 years earlier. The estimated polar moment of inertia of 0.3662 ± 0.0017 constrains the central metallic core to be between 1300 and 2000 km in radius (Folkner *et al.*, 1997), though a reanalysis of the data lowers the moment of inertia to 0.3650 ± 0.0012 (Yoder *et al.*, 2003). The estimate of the potential Love number k_2 of 0.153 ± 0.017 from MGS tracking constrains, under mild assumptions, the core radius to lie between 1520 and 1840 km, although these values would decrease somewhat if there is presently partial melt in the mantle (Yoder *et al.*, 2003). The value of the Love number also implies that the outer part of the core is liquid, consistent with the generation of a magnetic field early in Mars' history.

One of most startling discoveries of the MGS mission was that magnetic field mapping revealed the existence of strong remanent magnetic anomalies of various strengths, geometries, and locales (Acuña *et al.*, 1999). The Terra Cimmeria region (in the Noachian southern highlands) (Plate 13) contains the strongest anomalies, and early interpretations of apparent alternating positive and negative quasi-linear zones of anomalies invoked a hypothesis for a plate-tectonics origin (Connerney *et al.*, 1999), although the impression of linearity was aided by the choice of map projection. Other, weaker anomalies are scattered through the highlands, in the Tharsis region, and in the northern lowlands. The magnetic field information from MGS has gone through a number of data inversions that have improved the quality of the solutions (e.g., Purucker *et al.*, 2000; Arkani-Hamed, 2001; Langlais *et al.*, 2004; Hood *et al.*, 2005). Mitchell *et al.* (2007) used electron reflectometry data from MGS to produce a nearly global map of the magnetic field that resolved anomalies about seven times weaker than determinations from the MGS magnetometer. Numerous small anomalies are found in the northern lowlands, suggesting that the northern crust was at one time more magnetized than at present. In particular, the anomalies ring, but do not occur in, the Utopia basin, supporting the view that this feature was formed subsequent to the dynamo die-off, as proposed by Frey (2006b). Additionally, the area immediately to the north of the Tharsis rise is devoid of magnetic anomalies, suggesting a thermal demagnetization connection to this massive volcanic province (Mitchell *et al.*, 2007).

Collectively, the magnetic field anomalies require that Mars had a global-scale internal field early in its history, but the origin of the spatial distribution of the anomalies and the timing of the internal dynamo remain as unsolved problems (Solomon *et al.*, 2005). Although other interpretations are possible, the most likely reason for the absence of magnetic field anomalies over the Hellas and Argyre basins is that the dynamo had ceased to exist by the time these basins had formed. Magnetization lost by the high interior temperatures and shock associated with the impact would not have been reacquired. Tighter bounds on dynamo extinction in the pre-Noachian are estimated by the crater ages of large impact basins (some are QCDs) that are spatially associated with varying strengths of crustal magnetic fields (Frey, 2006b).

Of particular interest to the tectonics of Mars is the presence of magnetic anomalies high on the Tharsis rise (Johnson and Phillips, 2005). Either these anomalies were acquired by cooling Tharsis magmas when the dynamo was active, or they are in older basement that was uplifted during the formation of Tharsis. If the magnetic field ceased by the time Hellas/Argyre formed, then the former hypothesis requires that the Tharsis origin dates back to the earliest Noachian. These anomalies on

Tharsis are not found in the highest parts of the rise or in the vicinity of the large shield volcanoes (Tharsis Montes), suggesting thermal demagnetization has been an important process (see also Mitchell et al., 2007). Jellineck et al. (2008) used the presence and scale of the Tharsis demagnetized region as a constraint on geodynamical properties of the regional interior, deriving a Noachian elastic thickness range of 29–40 km, excess mantle temperature of 205–240 °C, and a plume heat flux of 60–100 mWm^{-2}.

The plethora of magnetic field models from the MGS mission data has led to estimates of magnetic paleopole locations by determining the vector magnetization direction of isolated remanent magnetic anomalies and assuming that the global magnetic field was dominated by the dipole term. Such an exercise should be approached cautiously because of the non-uniqueness of potential field inversions, unknown causative body geometry, and the likely poor resolution of distinct sources at MGS altitudes (Langlais et al., 2004; Biswas and Ravat, 2005). Estimated paleopole locations have been all over the map, with the most consistent results showing paleopole locales in the vicinity of the Tharsis rise (Arkani-Hamed and Boutin, 2004; Hood et al., 2005; Quesnel et al., 2007). If the magnetic pole was always close to the rotational pole, then magnetic paleopoles that do not coincide, more or less, with the present spin axis imply that Mars has undergone true polar wander (Melosh, 1980; Willemann, 1984; Zuber and Smith, 1997; Sprenke et al., 2005; Matsuyama et al., 2006; Roberts and Zhong, 2007). If this is so, then the most likely suspect is the growth of Tharsis, which if not already on the equator, would migrate there via true polar wander to maintain rotational stability of the planet.

True polar wander has been a contentious topic for Mars. The geologic evidence for true polar wander is ambiguous at best (Banerdt et al., 1992, and references therein). The long-wavelength structures in the proposed shorelines of putative northern lowlands oceans (itself a contentious topic and beyond the scope of this review) are consistent with the deformation expected from a post-Tharsis loading true polar wander event, though the source of the driving mass is unclear (Perron et al., 2007). However, on Mars, true polar wander sets up large stress fields (Melosh, 1980) and well-defined tectonic patterns from moving the lithosphere over the equatorial (spin) bulge, which have not been observed (Grimm and Solomon, 1986). These tectonic features likely should have been preserved at least on post-Tharsis Mars, and possibly even from the proposed Tharsis true polar wander event itself.

3 Tectonic features

3.1 Extensional structures

Extensional structures identified on Mars include very narrow cracks or joints to fairly narrow linear negative relief structures (grabens that are a few kilometers wide and of order hundred meters deep) to enormous troughs and rifts that are up to 100 km wide and several kilometers deep (see review by Banerdt *et al.*, 1992 and references therein). Most interpretations of the grabens and rifts suggest they are bounded by steeply dipping normal faults whose strike is generally perpendicular to the least compressive principal stress. Large troughs and rifts (Plates 17 and 25) are likely bounded by relatively high-angle normal faults that cut through the entire brittle lithosphere (Frey, 1979; Schultz, 1991, 1995; Mege and Masson, 1996b; Anderson and Grimm, 1998; Hauber and Kronberg, 2001, 2005; Wilkins and Schultz, 2003; Grott *et al.*, 2007b).

Viking images of narrow grabens show fairly fresh looking fault scarps and flat floors suggesting simple structures. Photoclinometry on Viking images and MOLA data show the scarps overall have extremely low slopes (~9°) (Davis *et al.*, 1995; Golombek *et al.*, 1996; Harrington *et al.*, 1999) and high-resolution MOC images show evidence for significant erosion and deposition within the structures, which could conceal a more complex subsurface structure (Figure 5.1a and b). Bounding faults that dip inward at ~60° would intersect at depths of several kilometers (Tanaka and Golombek, 1989; Davis and Golombek, 1990), which could represent a mechanical discontinuity (Tanaka *et al.*, 1991; Thomas and Allemand, 1993; Davis *et al.*, 1995; Borraccini *et al.*, 2006). The faults could extend below their intersection depths in hourglass structures or one fault could terminate against a master fault that extends to greater depth (Schultz *et al.*, 2007). Alternatively, dikes could underlie the normal faults and accommodate the strain at greater depth (Schultz *et al.*, 2004; Goudy and Schultz, 2005). The association of narrow grabens and tension cracks with volcanic flows, pit chains and outflow channels is well documented for certain areas (Tanaka and Golombek, 1989; Tanaka and Chapman, 1990; Burr *et al.*, 2002a,b; Hanna and Phillips, 2006), although not always in the same area as most pit crater chains do not show associated volcanics (Ferrill *et al.*, 2004; Wyrick *et al.*, 2004). Subsequent work has argued that most, if not all grabens are the surface manifestations of subsurface dikes (Mege and Masson, 1996a; Ernst *et al.*, 2001; Scott and Wilson, 2002; Scott *et al.*, 2002; Wilson and Head, 2002) and that massive dikes could underlie Valles Marineris troughs and be responsible for the associated catastrophic outflow channels (McKenzie and Nimmo, 1999). Mege *et al.* (2003) have further argued that multiple dikes or dike swarms underlie each graben based on terrestrial analogues and experimental extension and deflation experiments. Analysis of the occurrence, size, and characteristics of pit crater chains associated with grabens on Mars argues that most are not volcanogenic, however, but are rather related to dilational normal faulting (Ferrill *et al.*, 2004; Wyrick *et al.*, 2004). Regardless of what happens to the faults beneath narrow grabens on Mars, their relief suggests they accommodate small amounts of extension (order 100 m or less) and strain (locally <10%).

196 *Planetary Tectonics*

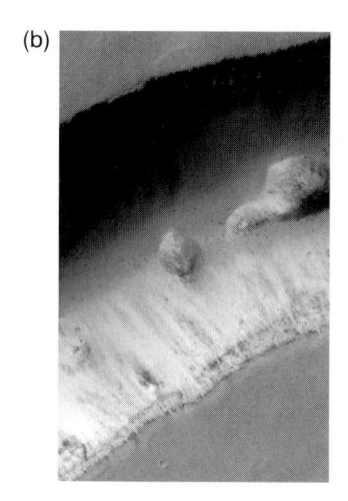

(a)

(b)

Figure 5.1. High-resolution (1.4 m/pixel) MOC narrow-angle images of grabens on Mars showing highly modified appearance. (a) 3-km wide graben in Memnonia Fossae (MOC image M02-02352 is 2.9 km wide; north is roughly up). (b) 2-km wide graben in Sirenum Fossae (MOC image M02-04357 is 1.4 km wide; north is roughly up). Both grabens have steep upper walls ($\sim30°$) with exposures of rock and lower walls with no rocks apparent, suggesting deposition of fine-grained material. Both floors are littered with boulders ~1.5 to 15 m in diameter that likely eroded from the upper walls. Deposition of sediment is apparent by eolian bedforms in (a), suggesting deposition of eolian material and downward streaks in (b). Although the floor of (a) is reasonably flat, the floor of (b) includes mounds of material that could be wall rock and or interior hanging wall remnants of a complex interior structure. The highly modified floor and walls make inferences about the geometry of the structure difficult and allow a complexly faulted downthrown block that is covered up by sediment.

3.2 Compressional structures

Wrinkle ridges are the most common compressional structures on Mars and are composed of linear to arcuate asymmetric ridges hundreds of meters high, up to a few tens of kilometers wide, and can be hundreds of kilometers long (Plate 18). They are commonly evenly spaced and typically have a small crenulation or surface wrinkle and a broad rise adjacent or as part of the feature. Initial work on lunar wrinkle ridges suggested origins relating to both volcanic and tectonic processes, but more recent work consistently argues that these features are compressional structures that involve thrust faulting and folding (e.g., Plescia and Golombek, 1986; Mueller and Golombek, 2004). Detailed analysis and modeling of MOLA topography and comparison to compressional structures on Earth has shown that many aspects of wrinkle ridges are consistent with surface folding overlying blind thrusts at depth (e.g., Schultz, 2000; Golombek *et al.*, 2001; Vidal *et al.*, 2003; Mueller and Golombek, 2004). In this interpretation, slip on the underlying thrust fault is accommodated by asymmetric flexural slip folding in strong, but weakly bonded, layered materials (volcanics or sedimentary rocks) near the surface and thus wrinkle ridges represent fault-propagation folds (e.g., Suppe and Medwedeff, 1990) or trishear fault-propagation folds (Erslev, 1991) (Figure 5.2). The asymmetric fold shape with steep forelimbs and shallow backlimbs of these structures are consistent with thrust faults that dip beneath the shallow backlimb. Smaller crenulations are likely high-level backthrusts that nucleate at weak layers in the upper crust or by flexural slip folds that facilitate bending of layered materials (Schultz, 2000; Okubo and Schultz, 2003, 2004; Mueller and Golombek, 2004). A variety of kinematic and dynamic models show that the relief of the ridge (hundreds of meters) is a measure of the shortening, and that wrinkle ridges accommodate very small amounts of shortening (order 100 m) and strain (of order 0.1% or less) (Watters, 1988, 2004; Golombek *et al.*, 1991; Plescia, 1991a, 1993; Schultz, 2000; Golombek *et al.*, 2001).

Lobate scarps (Figure 5.3) and other large ridges also have been identified on Mars (Chicarro *et al.*, 1985; Watters, 1993, 2003b; Schultz and Tanaka, 1994; Watters and Robinson, 1997; Watters *et al.*, 2000; Anguita *et al.*, 2001, 2006; Grott *et al.*, 2007a). These structures are larger (hundreds of meters to several kilometers high) and have been inferred to represent lithospheric scale thrust faults in heavily cratered Noachian terrain (Schultz and Tanaka, 1994). Displacement–length ratios for the lobate scarps (Watters, 2003b) are much greater (10 times) than wrinkle ridges (Tate *et al.*, 2002b; Mueller and Golombek, 2004), but still about an order of magnitude smaller than for typical thrust faults on Earth (Watters and Robinson, 1997; Watters *et al.*, 2000; Watters, 2003b). The morphology of these ridges is characterized by fairly simple linear to arcuate (lobate), asymmetric

198   *Planetary Tectonics*

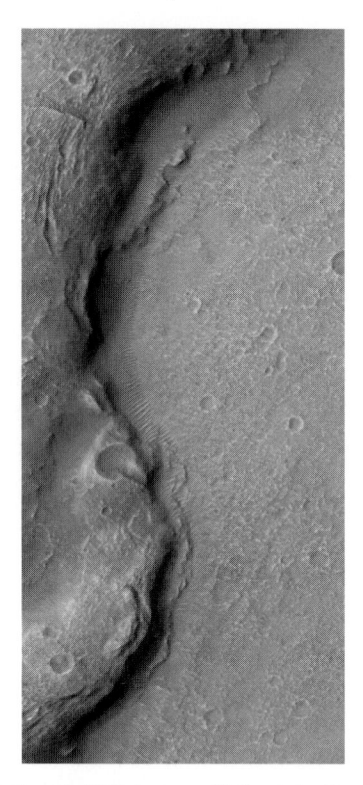

Figure 5.2. High-resolution MOC image (5.6 m/pixel) showing well-preserved detailed structure of wrinkle ridge. Note surface fold with narrow (25–100 m wide) extensional structures along the crest, likely generated in response to bending stresses over the fold (e.g., Mueller and Golombek, 2004; Plescia and Golombek, 1986). Eolian bedforms deposited at the base of the structure partially cover the crenulations. Similar surface texture and preservation of fine scale structures suggest that total modification of the structure has been relatively modest. MOC image E05-01229 is 2.9 km wide, north is roughly up.

scarps (Figure 5.3), suggesting a simple subsurface thrust fault that actually breaks the surface (or in which folding of surface layers is less important) (Vidal et al., 2005). In certain cases, wrinkle ridges change along-strike to lobate scarps, coincident with a change from smooth layered materials to highlands material (Watters, 1993). Based on the relief of lobate scarps, shortening is estimated to be hundreds of meters to kilometers, which is substantially greater than wrinkle ridges.

Although most workers agree that lobate scarps and other large ridges likely represent thrust faults that extend through most, if not all of the brittle lithosphere (Schultz and Tanaka, 1994), there is no such agreement on the depth of penetration of faults beneath wrinkle ridges. Thin-skinned deformation has been argued to explain the regular spacing (Plate 18) of wrinkle ridges (buckle folds in a thin

Figure 5.3. Viking mosaic of lobate scarp Amenthes Rupes, interpreted to be a thrust fault that breaks the surface of Mars. This large lobate scarp with over 1 km of relief and a shortened impact crater suggests several kilometers of shortening on a shallowly, northeastward dipping thrust fault (Vidal *et al.*, 2005). Mars Digital Image Mosaic is ~500 km wide centered at 111°E, 2.5°N; north is up.

strong layer) and the abrupt change in wrinkle-ridge trend around the rims of some craters (suggesting a shallow fault and/or décollemont) (Watters, 1991, 2004; Allemand and Thomas, 1995; Mangold *et al.*, 1998). Thick-skinned deformation in which subsurface faults extend through most or all of the lithosphere is based on observed elevation offsets and stair step topography across some adjacent wrinkle ridges, suggesting they are underlain by stacked thrust faults and the mechanical difficulty in preventing the strain observed at the surface from being accommodated by faulting at deeper levels (Zuber and Aist, 1990; Golombek *et al.*, 1991, 2001; Zuber, 1995; Montesi and Zuber, 2003; Mueller and Golombek, 2004).

3.3 Strike-slip faults

Strike-slip faults are uncommon on Mars (and other single plate planets) (Golombek, 1985; Tanaka *et al.*, 1991; Banerdt *et al.*, 1992) and typically are found as accommodation structures associated with wrinkle ridges and compressional push ups (Schultz, 1989; Watters, 1992; Anguita *et al.*, 2001, 2006) or associated with radial graben and concentric wrinkle ridges to the west of Tharsis (Okubo and Schultz, 2006; Andrews-Hanna *et al.*, 2008b). Ancient strike-slip faults of lithospheric proportion also have been proposed (Forsythe and Zimbelman, 1988).

4 Tectonic history, orientation and distribution of structures

The broad pattern of deformation in the western hemisphere is dominated by the giant radiating pattern of extensional structures and the sweeping family of compressional ridges and wrinkle ridges associated with Tharsis (Wise *et al.*, 1979; Anderson *et al.*, 2001). Structures around Tharsis are not uniformly distributed, but are found concentrated in particular regions separated by areas with different types of structures or no apparent deformation. Detailed mapping in these deformed regions, which have rich stratigraphies (Scott and Tanaka, 1986; Tanaka and Davis, 1988; Tanaka, 1990; Scott and Dohm, 1990a,b; Witbeck *et al.*, 1991; Morris and Tanaka, 1994; Dohm and Tanaka, 1999; Dohm *et al.*, 2001), has revealed a complex deformational history that has spanned most of the preserved geologic record on Mars. This detailed mapping has been pieced together at a global scale (Plate 24 in Tanaka *et al.*, Chapter 8) to provide a record of the deformation through time on Mars (Anderson *et al.*, 2001, 2004, 2008; Tanaka *et al.*, Chapter 8). We will first briefly describe the deformed regions and then the global tectonics.

4.1 Alba Patera and Ceraunius Fossae

North of the Tharsis Montes and their associated Amazonian volcanic cover are heavily faulted Noachian terrains of Ceraunius Fossae, and to the north, the Amazonian volcano Alba Patera (Plate 13). Ceraunius Fossae is composed of closely spaced narrow grabens and fault scarps that formed mostly during the Late Noachian and Early Hesperian (Tanaka, 1990). Deformation towards the north in the younger Hesperian and Amazonian volcanics appears to splay around Alba Patera, extending into the northern plains, with most extensional structures on the eastern side (Scott and Wilson, 2002; Cailleau *et al.*, 2003, 2005; Ivanov and Head, 2006). Grabens here are wider than most (2–8 km) with complex fault scarps and shallow, relatively flat floors a few hundred meters deep (Plescia, 1991a). Most of these structures formed in the Early Amazonian, although deformation continued into the Middle to Late Amazonian concentrated on regularly spaced, wider and deeper complex grabens with pit chains along their axes (Tanaka, 1990). A transect across the southern part of Alba at 35°N, in which the depth of grabens and fault scarps was measured from shadows and photoclinometry, yielded a total extension of about 8 km for assumed normal fault dips of 60° (Plescia, 1991a). Strain estimates vary from ~5% across individual structures to ~0.5% across the entire ~1500 km transect.

Measurement of the displacement across Ceraunius Fossae, just south of the above transect, in stacked MOLA profiles yields much larger extensions of 36 km and 42 km (Borraccini *et al.*, 2005), which correspond to about 2% strain across the entire region. The factor of five difference in extension between the two estimates that cross similar concentric sectors may be related to the older age of Ceraunius Fossae basement (Noachian), and so may be recording more extension compared with the Amazonian transect through Alba.

4.2 Structures associated with volcanoes

The large volcanoes of Tharsis, Olympus Mons, Arsia Mons, Pavonis Mons, Ascraeus Mons and Alba Patera, also have locally associated structures (Crumpler and Aubele, 1978; McGovern and Solomon, 1993). All have central calderas likely related to shallow magma chambers (Mouginis-Mark, 1981; Zuber and Mouginis-Mark, 1992; Crumpler *et al.*, 1996), most have fissure and flank rift structures, often in preferred directions (Crumpler and Aubele, 1978), and many have peripheral circumferential normal faults and grabens that have been related to local lithosphere flexure due to loading (Comer *et al.*, 1985; McGovern and Solomon, 1993), edifice spreading (McGovern and Solomon, 1993), and dike or sill intrusion (Scott *et al.*, 2002; Cailleau *et al.*, 2003, 2005; Ivanov and Head, 2006). Many of the grabens and flank structures have pits along their floors and end in sinuous rilles and channels, suggesting that they are underlain by volcanic dikes (Tanaka and Golombek, 1989; Ernst *et al.*, 2001; Wilson and Head, 2002; Mege *et al.*, 2003). Calderas are bounded by large normal faults that have accommodated kilometers of slip down to caldera floors that can have smaller parallel graben and radial wrinkle ridges. Compressional deformation on the higher flanks of the volcanoes has been suggested (Thomas *et al.*, 1990; McGovern and Solomon, 1993), as has folding of the flanks from the weight of the volcanic mass (Borgia *et al.*, 1990). Disrupted aureole deposits (Harris, 1977) and a large scarp around Olympus Mons have been interpreted as due to landslide deposits from gravity sliding headed by listric normal faults (Lopes *et al.*, 1980, 1982; Francis and Wadge, 1983; Tanaka, 1985; McGovern *et al.*, 2004a).

4.3 Tempe Terra

To the east of Alba Patera and northeast of (and in line with) the Tharsis Montes is Tempe Terra (Plate 13), a plateau of Noachian through Hesperian terrain that borders the northern plains to the north–northeast. Tempe has a complex faulting history with a peak in northeast trending grabens and normal faults in the Middle–Late Noachian to Early Hesperian and in the Late Hesperian (Scott and Dohm, 1990a,b; Moore, 2001; Fernández and Anguita, 2007). Most of the structures in Tempe Terra are fairly narrow grabens, but large rifts of Tempe Fossae (Plate 25) are analogous to large continental rifts on Earth (Hauber and Kronberg, 2001). Northwest striking wrinkle ridges are found on Hesperian ridged plains that appear to extend north from Lunae Planum. Extension across Tempe Terra has been estimated from shadow and fault scarp widths and deformed craters (Golombek *et al.*, 1996) to be about 22 km across two ∼1000 km traverses (regional strain of ∼2%). MOLA elevation profiles indicate lower scarp slopes across the rifts (Harrington *et al.*, 1999) consistent with measurements by Hauber and Kronberg (2001) of the large rifts and indicate the total extension across Tempe Terra is probably slightly lower (roughly 20 km).

4.4 *Lunae and Solis Plana*

South of Tempe Terra and to the east and southeast of the Tharsis Montes are the classic wrinkle ridge provinces of Lunae and Solis Plana (Plate 13). Wrinkle ridges on Tempe, Lunae and Solis define a sweeping roughly concentric system to Tharsis (Wise *et al.*, 1979; Maxwell, 1982; Watters and Maxwell, 1986; Anderson *et al.*, 2001). Ridges are larger and more widely spaced on Solis Planum than Lunae Planum (Watters, 1991; Golombek *et al.*, 2001). These areas are mapped as Hesperian ridged plains (Scott and Tanaka, 1986), interpreted to be underlain by layers of flood basalts totaling several kilometers thick (e.g., Watters, 1991). MOLA elevations indicate some ridges in Solis Planum accommodate adjacent stair step offsets in elevation, suggesting a package of stacked west-dipping underlying thrust faults that root in the ductile lower crust and are analogous to small foreland basement uplifts on Earth (Golombek *et al.*, 2001; Montesi and Zuber, 2003). Estimates of radial shortening across Lunae Planum (Plate 18) suggest very low regional strain of 0.1–0.3% (Golombek *et al.*, 1991; Plescia, 1991b; Tate *et al.*, 2002a,b). The compressional deformation that formed the wrinkle-ridges appears coeval for all of the wrinkle-ridge provinces and is constrained globally to have been dominantly in the Hesperian (Scott and Dohm, 1990b; Mangold *et al.*, 2000), although some formed in the Late Noachian and Early Amazonian (Chapter 8; Tanaka *et al.*, 1991, 2003, 2005).

4.5 *Valles Marineris and Noctis Labyrinthus*

Separating Lunae and Solis Plana is the great rift of Valles Marineris (Plate 13), which is composed of troughs that are up to 10 km deep, tens to hundreds of kilometers wide, and many hundreds to a thousand kilometers long. The formation of Valles Marineris began in the Late Noachian–Early Hesperian with faulting, broad basin formation and subsequent major trough formation (Witbeck *et al.*, 1991; Lucchitta *et al.*, 1992; Peulvast and Masson, 1993a,b; Schultz, 1998). Younger deposits and landslides within the troughs are Amazonian in age, and collapse of Noctis Labyrinthus troughs occurred in the Late Hesperian–Early Amazonian (Tanaka and Davis, 1988). The origin of Valles Marineris as a large rift is based on the identification of large fault scarps along the base of the chasmata, mapped downdropped blocks, structural models, and orientation radial to Tharsis (Lucchitta *et al.*, 1992; Peulvast and Masson, 1993a,b; Anderson and Grimm, 1998; Schultz, 1998; Schultz and Lin, 2001; Wilkins and Schultz, 2003). Extension across the central portion of Vallis Marineris from relief and structural offsets (that account for deposition within the troughs) varies between 10–60 km (20–30 km for 60° dipping faults) (Schultz, 1995; Mege and Masson, 1996b). Strain varies from ~20% across the individual chasmata to about 0.6% if measured across the entire shadow region (without extensional deformation) to the next extensional fault terrains (Tempe to Thaumasia) radial to Tharsis.

4.6 *Claritas Fossae, Thaumasia, and Sirenum*

South of the Tharsis Montes and west of Syria Planum is a high standing ridge of highly fractured terrain known as Claritas Fossae, which extends farther south into Thaumasia (Plate 13). This ridge of Noachian terrain continues in an arc south and to the east, where it joins the Coprates rise, south of Coprates Chasma. It is on-lapped by Hesperian plains to the north and records a well-preserved sequence of structural events (Dohm and Tanaka, 1999; Dohm *et al.*, 2001). The southern margin of the ridge itself is likely a thrust associated with Tharsis (Courtillot *et al.*, 1975; Masson, 1977; Wise *et al.*, 1979) and appears related to a series of Noachian lithospheric buckles and thrusts that are roughly concentric to Tharsis and extend in an arc from Coprates through Thaumasia to Terra Sirenum and Memnonia (Schultz and Tanaka, 1994; Anguita *et al.*, 2001, 2006; Grott *et al.*, 2007a). Claritas Fossae is dominated by generally north striking normal faults and grabens that continue south into Thaumasia. The Thaumasia rift (Hauber and Kronberg, 2005) is part of Claritas Fossae and is roughly 100 km wide and several kilometers deep (Plate 17). Similar size rifts (Coracis Fossae) are found in the Thaumasia, highlands to the southeast (Grott *et al.*, 2007b). A distinct fanning set of grabens radiates to the south, and southeast into Thaumasia, and to the southwest and west in Sirenum (Borraccini *et al.*, 2007).

Detailed mapping has established multiple faulting episodes (Dohm and Tanaka, 1999; Dohm *et al.*, 2001) with first Claritas faulting occurring in the Early to Middle

Noachian with a series of east–west (Tharsis-circumferential) trending fault scarps (Tanaka and Davis, 1988), which may record the earliest structural uplift of Tharsis (Phillips *et al.*, 1990; Johnson and Phillips, 2005). Subsequent faulting events occurred in the Late Noachian–Early Hesperian (NE trending structures, including Coracis Fossae), with the fanning set forming in the Early Hesperian (Anderson *et al.*, 2001). Estimates of extension across the Thaumasia rift in Claritas and both Coracis Fossae rifts in Thaumasia to the southeast from MOLA topography are 1–4 km (Hauber and Kronberg, 2005) and 2–2.5 km (Grott *et al.*, 2007b), respectively, assuming 60° dipping normal faults. Measurements of fault scarp widths and shadows across grabens of Thaumasia and southern Claritas, and Sirenum yield extension estimates of about 11 km and 3 km, respectively (Golombek *et al.*, 1994; Golombek *et al.*, 1997). If the rift extensions in Claritas and Coracis Fossa are added, the total extension across the Thaumasia region could be as high as 18 km assuming displacement across 60° dipping normal faults.

4.7 Western hemisphere tectonic history

Placing the mapping results of the geologic and tectonic activity within these deformed regions into the global stratigraphic framework of Mars (Tanaka, 1986) has revealed a complex structural history involving five main stages of tectonic activity in the western hemisphere, with distinct centers of tectonic activity through time (Anderson *et al.*, 2001). Careful testing of the geometry of all structures in the western hemisphere shows statistically significant radial (extensional) and concentric (compressional) orientations about the highest standing portion of Tharsis (excluding the volcanoes) near Syria Planum (Anderson *et al.*, 2001). More than half of the structures mapped on Mars are Noachian in age (stage 1; >3.8 Ga), concentrated in exposures of Noachian age crust in Tempe Terra, Ceraunius Fossae, Syria Planum, Claritas Fossae, Thaumasia, and Sirenum (see Plate 24). About half of these grabens are radial to a center of activity in Claritas Fossae, the ancient Noachian ridge west of Syria. By Late Noachian–Early Hesperian (stage 2), activity was concentrated in Thaumasia and Valles Marineris. Early (to Late) Hesperian (stage 3) included the development of concentric wrinkle ridges concentrated along the edge of the topographic rise, in the northern plains and in volcanic provinces in the eastern hemisphere. Normal faulting also occurred north of Alba, in Tempe Terra, in Ulysses Fossae, in Syria Planum and Valles Marineris, and in Claritas Fossae and Thaumasia. Stage 3 structures are generally radial (grabens) or concentric (wrinkle ridges) to Syria Planum. Stage 4 activity during the Late Hesperian–Early Amazonian was concentrated in and around Alba Patera and Middle to Late Amazonian activity (stage 5) was concentrated in and around the Tharsis Montes volcanoes.

Given that about half of all extensional structures preserved in the western hemisphere formed before the Late Noachian, Tharsis tectonic activity peaked early and decreased with time. The observation that Noachian through Amazonian radial grabens and concentric wrinkle ridges formed, centered at slightly different locations (local centers of volcanic and tectonic activity) within the present-day highest standing terrain of Tharsis, indicates that the basic lithospheric structure of the western hemisphere has changed little since the Middle Noachian (Anderson et al., 2001). This result is in agreement with geophysical results discussed below.

4.8 Eastern hemisphere

Elysium is a smaller volcanic province than Tharsis (Mouginis-Mark et al., 1984) with some surrounding structures and a small, elevated plateau with a thickened crust (Neumann et al., 2004). Elysium Mons (Plate 13), the largest shield volcano in the province, is similar to the giant Tharsis shield volcanoes, with a central caldera and peripheral concentric grabens and normal faults attributed to loading (Comer et al., 1985). Cerberus Fossae, southeast of Elysium Mons, trend east–southeast and appear radial to Tharsis (Hall et al., 1986) and north- to northeast-trending wrinkle ridges in Elysium and Arcadia Planitae may be concentric to Tharsis. Cerberus Rupes have been reactivated repeatedly and are the source of very young volcanism and fluids forming the channels and volcanic plains of Athabasca and Marte Valles (Burr et al., 2002a,b; Jaeger et al., 2007). Other extensional grabens (Plate 24) appear radial to Elysium (Anderson et al., 2008). Wrinkle ridge formation in Arcadia formed on a thin ridged plains unit in the Late Hesperian (Tanaka et al., 2005), coeval with other ridged plains, and the compressional strain has been estimated to be $\sim$0.06% across a $\sim$1000 km transect (Plescia, 1993).

Hesperia Planum, Syrtis Major Planum (Plate 13) and other plains (possibly volcanic) regions in Arabia Terra also have populations of wrinkle ridges in the eastern hemisphere (Anderson et al., 2008). Wrinkle ridges in Hesperia Planum form a reticulate pattern with northeast and northwest trending sets (Raitala, 1988; Goudy and Schultz, 2005). Wrinkle ridges in Syrtis Major show radial and concentric patterns (Hiesinger and Head, 2004). Circular rings of ridges are also found, suggesting their trends are guided by subsurface shallowly buried craters (Raitala and Kauhanen, 1989; Allemand and Thomas, 1995). Both the ridges and plains formed in the Hesperian (Scott and Dohm, 1990b; Watters, 1993; Mangold et al., 2000) (Plate 24).

Structures in the eastern hemisphere are also associated with the large basins: Isidis, Utopia and Hellas (Plate 13). Isidis has concentric grabens still preserved in Noachian crust to the northwest and southeast, and Hellas has extensional structures preserved to the west at distances that have been related to mascon loading (Comer et al., 1985; Wichman and Schultz, 1989). The structures are Late

Noachian–Early Hesperian in age (Wichman and Schultz, 1989; Tanaka and Leonard, 1995; Leonard and Tanaka, 2001). Concentric grabens have not been recognized at Utopia. Ridges within the Isidis and Utopia basins, recently discovered in MOLA data (similar to those in the northern plains, see next section), are numerous, showing a complex radial and concentric pattern (Thomson and Head, 2001; Head *et al*., 2002; Tanaka *et al*., 2003), suggesting that stresses related to basin subsidence may be important (although finite-element elastic modeling of Utopia basin does not support this view (Searls and Phillips, 2007)). Fewer ridges are present in Hellas, although they tend to be radial. The ridges have been interpreted as wrinkle ridges formed from compressional deformation and are Early Hesperian in age, although reactivation may have occurred in Early and Middle Amazonian (Tanaka *et al*., 2005).

The dichotomy boundary in the eastern hemisphere is characterized by scarps and troughs aligned with the boundary that were proposed to represent Late Noachian–Early Hesperian normal faulting that accommodated the lowering of the northern plains (Maxwell and McGill, 1988; McGill and Dimitriou, 1990; Smrekar *et al*., 2004). As discussed earlier, it is very likely that the northern lowlands crust dates from the pre-Noachian (Frey *et al*., 2002; Nimmo and Tanaka, 2005; Frey, 2006a,b). Therefore, the scarps and troughs likely represent a later southward retreat of the dichotomy boundary, which originally may have been sharpened by faulting (Guest and Smrekar, 2005), but then may have been eroded by fluvial processes during a period of "wet" climate that occurred during the Noachian–Hesperian transition (Howard *et al*., 2005; Irwin *et al*., 2005).

Lobate scarps trend parallel to the boundary (Watters, 2003a) and formed in the Late Noachian, roughly coeval with the Noachian ridge belt south of Tharsis (Schultz and Tanaka, 1994). The topography of adjacent highlands in the Amenthes region rises to a crest up to 1 km above its surroundings over a distance of 200 km away from the margin, suggestive of a passive margin (Frey *et al*., 1998) or flexed elastic plate (Watters, 2003a). Bending stresses in a simple, broken flexed plate, subject to a vertical load and bending moment at the broken end, would be extensional over the crest where the compressional lobate scarps are found, suggesting that erosional and/or global stresses may be important in the formation of the lobate scarps (Watters, 2003a,b). Alternatively, a faulted plate model with finite extension and specified fault dip will generate compressional stresses in the footwall and extensional stresses in the hanging wall (Weissel and Karner, 1989). This is consistent with observations and may obviate the need for global compressional stresses for the formation of these lobate scarps; footwall compressional stresses in this case would be amplified by erosion of the footwall block. Alternatively, Nimmo (2005) has shown that lower crustal flow leads to differential thickness changes north and south of the dichotomy boundary, correctly predicting faulting styles.

4.9 Northern plains

Not seen in generally poor, high latitude Viking images, an extensive system of ridges in the northern plains was discovered in the MOLA topographic data (Withers and Neumann, 2001). The height, width, length, organized pattern and spacing of the ridges resemble wrinkle ridges found on the classic ridged plains. In addition, the ridges continue in the northen plains in a large concentric arc between Tempe Terra and Arcadia Planitia (Plates 13 and 24), suggesting they are produced by radial compressive stresses from Tharsis (Withers and Neumann, 2001). In general, ridges on the northern plains appear to have lower relief and a more rounded appearance (less crenulated), suggesting partial burial (Head *et al.*, 2002) and/or formation in layered materials (sediments) of different rheology (Tanaka *et al.*, 2003, 2005). The wider spacing of the ridges in the northern plains has been attributed to burial (Head *et al.*, 2002) or to a thinner crust resulting in a greater brittle–ductile transition depth, which controls fault spacing (Montesi and Zuber, 2003). Constraints on the timing of formation appear consistent with the global period of wrinkle ridge compression in the Hesperian (Head *et al.*, 2002; Tanaka *et al.*, 2003) and perhaps continued regional compression in the northern plains into the Amazonian (Tanaka *et al.*, 2005).

5 Tharsis geodynamical models and comparisons to tectonics

5.1 Models for the origin of Tharsis

Two general models have been proposed to explain the gravity field and topography associated with Tharsis that bear on the origin of this planetary-scale feature: elastic spherical shell loading (Banerdt *et al.*, 1982, 1992) and plume uplift with the attendant deformation of the lithospheric shell. Elastic spherical shell loading models propose that whatever the origin of Tharsis, by the end of the Noachian, mechanically Tharsis was a massive load supported largely by membrane stresses in a rigid lithosphere. Further, this mass distribution has changed little since that time. Plume models posit that the long-wavelength topography of Tharsis is presently supported to a significant fraction by a plume, and that plume-related radial subsurface dikes are responsible for the radial graben system (Mege and Masson, 1996a; McKenzie and Nimmo, 1999; Ernst *et al.*, 2001; Wilson and Head, 2002). Such models can be consistent with the observed degree two and three (and higher) topography and gravity fields (Kiefer *et al.*, 1996), but lack the overwhelming success in predicting the style and distribution of stress and strain predicted by elastic shell loading models (see below).

The challenge to plume models for Mars is to generate but a single plume to create Tharsis (and possibly a second plume to create Elysium). In a series of papers, Harder (1998, 2000) and Harder and Christensen (1996) show that if the mantle endothermic spinel–perovskite phase transition exists just above the Martian core, then only a few plumes are generated in the mantle (with time, only one), and that about half of the geoid at Tharsis might be explained by such a process. We note that the time necessary to form one or two plumes in these models is extraordinarily long, exceeding the age of the planet. Breuer *et al.* (1998) showed that that the exothermic olivine to β-spinel and β-spinel to γ-spinel phase transitions in the mantle tend to create a single stable area full of plumes, again a mechanism for Tharsis genesis. However, the model time necessary for this process to occur is about 1.5 Ga, far exceeding the bulk of Tharsis emplacement (Anderson *et al.*, 2001; Phillips *et al.*, 2001). Additionally, using a viscoelastic formulation for mantle flow, Zhong (2002) showed that plume structures likely account for less than 10% of the present geoid. Further, simultaneous consideration of both plume buoyancy and surface load to calculate the geoid-to-topography ratio indicates that a plume presently contributes less than 25% to the topography of Tharsis (Zhong and Roberts, 2003). Additional detailed analyses support the view that Tharsis is presently dominated by surface (and lithosphere) loading, as opposed to buoyant uplift from beneath the lithosphere (Lowry and Zhong, 2003; Roberts and Zhong, 2004). Redmond and King (2004) showed that it is possible in the present day for a plume model to contribute only very modestly to the Tharsis geoid and topography, while still producing a small amount of partial melt (as suggested by recent volcanism (Hartmann *et al.*, 1999)).

These results do not rule out the possibility that Tharsis had, at least in part, a plume origin (Spohn *et al.*, 2001). Most considerations are of a deep mantle plume (e.g., Schubert *et al.*, 1992), although Reese *et al.* (2004) suggest that the putative Tharsis plume could be thermochemical in origin, induced by an enormous impact. In support of a plume origin, we note that the Early–Middle Noachian circumferential graben in the Claritas Fossae region are consistent with buoyant uplift, not surface loading (Phillips *et al.*, 1990). However, buoyant uplift does not require a plume per se, and all models proposed thus far for Tharsis require such uplift because of the unavoidability of partial melting in the upper mantle and the intrusion of magmatic products in the crust (Finnerty *et al.*, 1988; Johnson and Phillips, 2005). Solomon and Head (1982) proposed that Tharsis formed over a region of warm mantle, resulting in substantial partial melting and a weak, thin lithosphere that focused the development of Tharsis. There are a number of possible mechanisms for creating a regionally warmer mantle. For example, Wenzel *et al.* (2004) show that the increased insulating effects of a thicker southern hemisphere crust (relative to a thinner northern one) (Neumann *et al.*, 2004) lead to long-lived upwellings beneath the southern highlands that may be a source for Tharsis volcanism. It is also possible, for a thin lithosphere, that the upper mantle was heated by a flushing of the hot lower mantle via cold downwelling plumes from a convective upper boundary layer (van Thienen *et al.*, 2006).

As we discuss below, models that treat Tharsis as a load on an elastic spherical shell have been very successful in predicting, quantitatively and spatially, the tectonic features associated with Tharsis. Further, there is a strong indication that this type of lithospheric loading has been operative since the end of the Noachian, and that by that time whatever dynamic process had been involved with the origin and early evolution of Tharsis had declined to a minor role. Thus we next discuss briefly the deformation styles on an elastic spherical shell.

5.2 Models for deformation on an elastic spherical shell

Mass loads on an elastic spherical shell will be supported by bending stresses, but more importantly by a resistance to changes in the radius of curvature of the shell if the horizontal scale of the load is of the order of or exceeds the planetary radius (Turcotte *et al.*, 1981). Isostatic equilibrium will not equate to mass balance, even when bending stresses are negligible, because of partial to essentially full support of loads by membrane stresses.

Plate 19 shows an isostatic response function for Mars, and indicates that loads that are spatially represented by spherical harmonic degrees less than about eight are supported largely by membrane stresses. The load represented by the massive Tharsis rise falls into this category (in fact most of the power lies in $l = 2$ to 4), which indicates that the stress distribution is controlled by membrane forces.

Plate 20 illustrates the fundamental difference in stress distribution for bending and membrane stresses. For features small relative to planetary radius, a positive mass load at the surface creates an annular zone of radial extensional stress that would form circumferential normal faults. One-dimensionally, this is commonly observed on Earth at the flexural bulge that defines the subduction zone outer rise. On the Moon, this is manifested as circumferential graben surrounding impact basins with mass loads (mascons) (Solomon and Head, 1980; Golombek and McGill, 1983; Freed *et al.*, 2001), and on Mars as circumferential graben surrounding large shield volcanoes, as well as the Isidis mascon basin (Comer *et al.*, 1985). In contrast, when the diameter of the positive mass load approaches or exceeds the radius of the planet, the annular zone is subject to concentric extensional stresses, which can lead to radial normal faults (Banerdt *et al.*, 1992), as observed emanating from the Tharsis region and discussed above.

The most rigorous model of deformation of a thin elastic spherical shell involves both vertical and horizontal loads (Vlasov, 1964) and has been adopted for planetary-scale geodynamical problems by Banerdt (1986). The horizontal load potential arises from horizontal gradients in lithostatic stress acting on vertical planes, assuming uniaxial strain, as well as lithostatic stress acting on inclined horizontal planes. This term was missing from earlier thin-shell derivations (e.g., Turcotte *et al.*, 1981), but incorporated in early thick-shell loading models (Phillips and Lambeck, 1980; Banerdt *et al.*, 1982) for Tharsis. The solution of Banerdt (1986) is constrained to satisfy the geoid everywhere, and global surface and crust–mantle boundary topography (both referenced to the geoid) contribute to the load, in addition to net mass accommodated by deformation. Stress and strain are calculated from spherical coordinate system spatial gradients in deformation. A second approach modifies the model above by zeroing-out topography outside of Tharsis and proposes that the long-wavelength surface here is a deformed membrane (Phillips *et al.*, 2001, 2004). The model has been successful in describing the shape of Mars, as described earlier, although the results are not necessarily unique. Finally, Dimitrova *et al.* (2006) approached this problem by solving the three-dimensional force balance equations for a thin lithosphere and crustal model constrained by the gravity field (Neumann *et al.*, 2004) and topography, with the solution producing a global minimum in the second stress invariant.

5.3 *Tharsis-induced stress and strain from elastic shell models*

Previous lithospheric deformation models employing elastic-shell loading and based on Viking era gravity and topography (through degree and order eight), summarized in Banerdt *et al.* (1992), indicated that two different stress conditions were required to explain the tectonic features around Tharsis (Banerdt *et al.*, 1982, 1992). Isostatic conditions lead to concentric extensional stresses on the topographic rise and radial compressional stresses around the edge that could account for the radial grabens on the rise and the concentric wrinkle ridges around the edge. A global compressive stress field may have modulated the formation of the concentric wrinkle ridges at the time when their formation peaked on Mars (Tanaka *et al.*, 1991). However, the isostatic model could not explain all of the Tharsis-related tectonic features, as flexural (bending plus membrane) loading stresses were needed to develop the concentric extensional stresses needed to produce the radial grabens beyond the edge of the topographic rise.

Lithospheric deformation models based on the Mars Global Surveyor gravity and MOLA topography simplified the stress states required to explain most of the tectonic features around Tharsis (Banerdt and Golombek, 2000) (Plates 21 and 22). Flexural loading stresses based on present-day gravity and topography appear to explain successfully the location and orientation of most tectonic features. Specifically, the orientation of extensional grabens and rifts in Memnonia, Sirenum, Thaumasia, southern Claritas, Valles Marineris, Tempe and Mareotis Fossae are all accounted for by the flexural loading stresses, and the distribution agrees with modeled areas with high strain. In addition, the orientation of compressional wrinkle ridges in Lunae, Solis, and Syria Plana, Sirenum, and Acidalia are explained by flexural loading stresses, and their distribution corresponds to highly strained areas in the model. Only extensional structures around Alba (which appears related to local volcanotectonic processes (Zuber *et al.*, 2000; Cailleau *et al.*, 2003, 2005)) and local structures around the Tharsis Montes and in Noctis Labyrinthus are not explained by the model. Early Noachian circumferential extensional structures in the Noctis Labyrinthus region (Phillips *et al.*, 1990; Johnson and Phillips, 2005) can be explained with a thin elastic shell model and a net upward load (Banerdt *et al.*, 1992).

In addition to explaining the orientation and distribution of most tectonic features around Tharsis, the extensional hoop strain predicted by the flexural loading model (0.2–0.4%) (Plate 21) agrees with the estimated extension and averaged strain across deformed regions around Tharsis. Estimates of extension across Tempe Terra (Golombek *et al.*, 1996; Harrington *et al.*, 1999; Hauber and Kronberg, 2001), Alba Patera (Plescia, 1991a) and Ceraunius Fossae (Borraccini *et al.*, 2005), the central portion of Vallis Marineris (Schultz, 1995; Mege and Masson, 1996b), Sirenum and Thaumasia (Golombek *et al.*, 1994, 1997; Hauber and Kronberg, 2005; Grott *et al.*, 2007b) are reported in Table 5.2. Tempe, Alba, Sirenum and Thaumasia each occupy about an eighth section of Tharsis (roughly 2000 km across); Valles Marineris and Olympus, between Alba and Sirenum (which has negligible extension), each occupy a quarter section (roughly 4000 km across). As a result, extensional strain varies significantly from 0 to 1–2% across different 2000-km and 4000-km wide regions around Tharsis (Table 5.2), even though strain across individual structures with smaller widths can be as high as 5–20%. Taken together these estimates suggest a total circumferential extension of order ~70 km, with an uncertainty of about ±50 km (from possible variations in fault dip). For a 2500 km radius (or circumference of 15 700 km), these results suggest a total circumferential extensional strain of 0.4% ± 0.3% now expressed at the surface (not accounting for buried structures), which matches the model result. Estimates of the measurable shortening across wrinkle ridges in Lunae and Solis Plana and Arcadia Planitia radial to Tharsis suggest comparably low regional strains of 0.1–0.3% (Golombek *et al.*, 1991, 2001; Plescia, 1991b, 1993). Concentric Late Noachian/Early Hesperian thrust structures south of Tharsis (Schultz and Tanaka, 1994) also suggest bulk shortening of 2–4 km and strain of 0.1–0.3% over radial transects of ~1500 km in Sirenum. Both of these estimates of radial compressional strain agree with model results of 0.2–0.3% in Lunae and Solis Plana and 0.1–0.2% in Sirenum (Banerdt and Golombek, 2000) (Plate 22).

The observations that flexural (bending plus membrane) loading lithospheric stress models based on present-day gravity and topography can explain the observed

Table 5.2. *Hoop extension and strain around Tharsis*

Tharsis Province	Hoop extension ± uncertainty (km)*	Strain (%)**	Length (km)***
Alba	8 ± 5	0.4	2000
Ceraunius Fossae	40 ± 24	2	2000
Tempe Terra	20 ± 15	1	2000
Valles Marineris	20–30 ± 15–20	0.6	4000
Thaumasia (total)	18 ± 11	1	2000
Claritas Fossae	4 ± 2		
Thaumasia Fossae	11 ± 8	0.5	2000
Coracis Fossae	2 ± 1		
Sirenum	3 ± 2	0.1	2000
Olympus	0	0	4000

*Extension concentric to the center of Tharsis at a radial distance of ~2500 km as estimated from references in the text. The quoted formal uncertainty is from the possible variations in fault dip (Golombek *et al.*, 1996). Strain **estimated from the extension over length ***of pie-shaped sector that province occupies.

distribution and strain of radial and compressional structures, coupled with the mapping results that suggest over half of the radial structures formed by the Middle Noachian, argue strongly that the basic lithospheric structure of the province has changed little since the Middle Noachian. The lithospheric deformation models require the load to be huge (of the scale of Tharsis, large relative to the radius of the planet), and the mapping requires the load that caused the flexure to have been in place by the Middle Noachian (>3.8 Ga). Subsequent deformation events around Tharsis appear consistent with the same overall lithospheric structure, with relatively minor changes in the centers of deformation (e.g., Anderson *et al.*, 2001). Fine layers within Valles Marineris revealed by the Mars Orbiter Camera have been interpreted as being lava flows that are Late Noachian or older (McEwen *et al.*, 1999). These volcanics are therefore likely part of the load that caused the flexure around Tharsis.

The emplacement of the Tharsis load ($\sim 3 \times 10^8$ km^3) (Phillips *et al.*, 2001) may have contributed to wetter and likely warmer periods during the Noachian. If the load is composed largely of volcanics, as suggested by fine layers within Valles Marineris (McEwen *et al.*, 1999), water released with the magma would be equivalent to a global layer up to 100 m thick, which could have had a significant impact on the Martian climate (Phillips *et al.*, 2001). Impact processes may have also contributed to short-lived clement environments (e.g., Segura *et al.*, 2002).

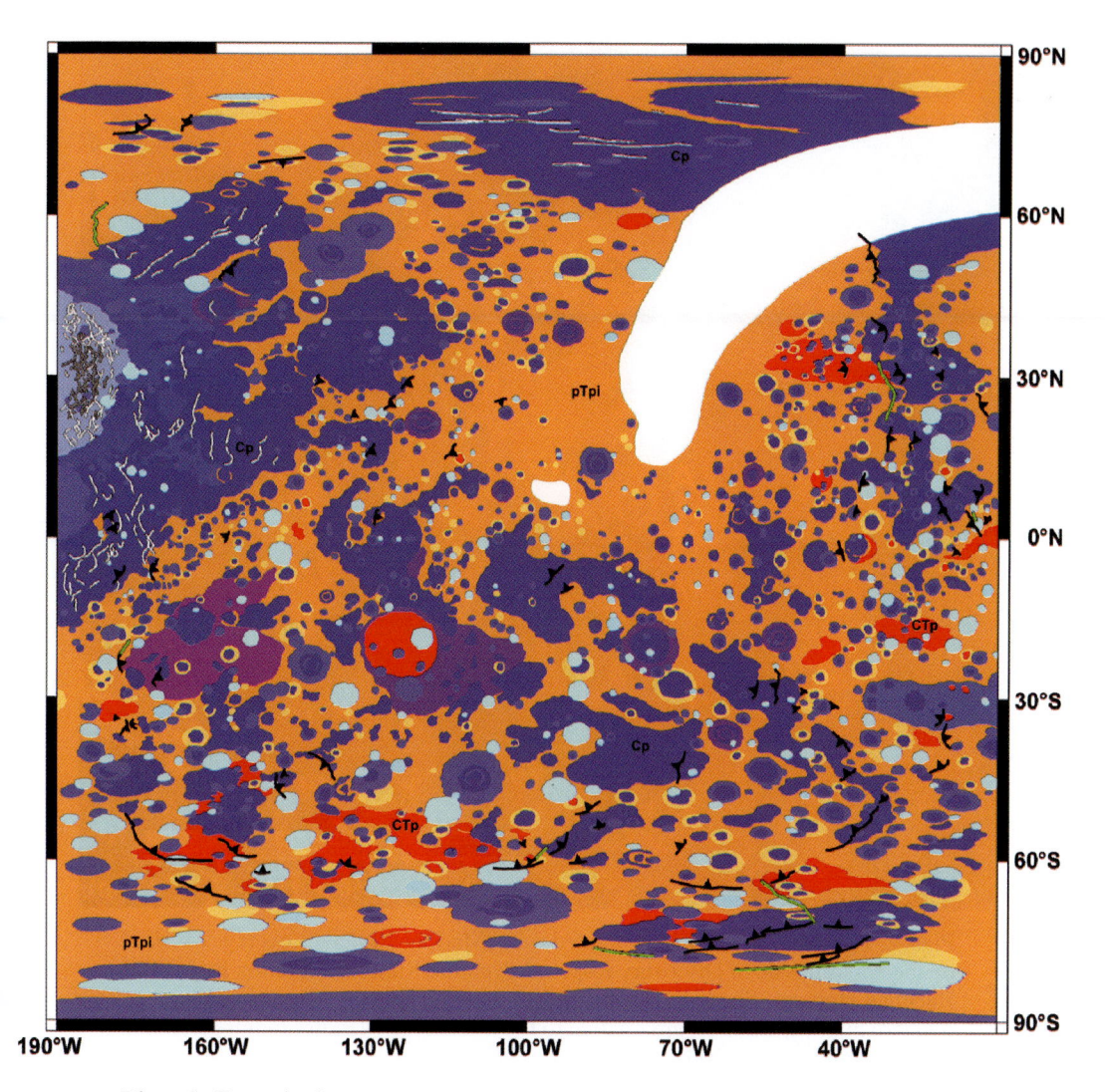

Plate 1. Tectonic features on Mercury imaged by Mariner 10. Digitized segments of lobate scarps (black), wrinkle ridges (white), high-relief ridges (green), and troughs (gray) are overlaid on the geologic map of Mercury (Spudis and Guest, 1988). The major geologic units are intercrater plains material pTpi (tan), Calorian–Tolstojan plains material CTp (red), and Calorian plains material Cp (blue). The inferred dip directions of thrust faults underlying lobate scarps are indicated by black triangles.

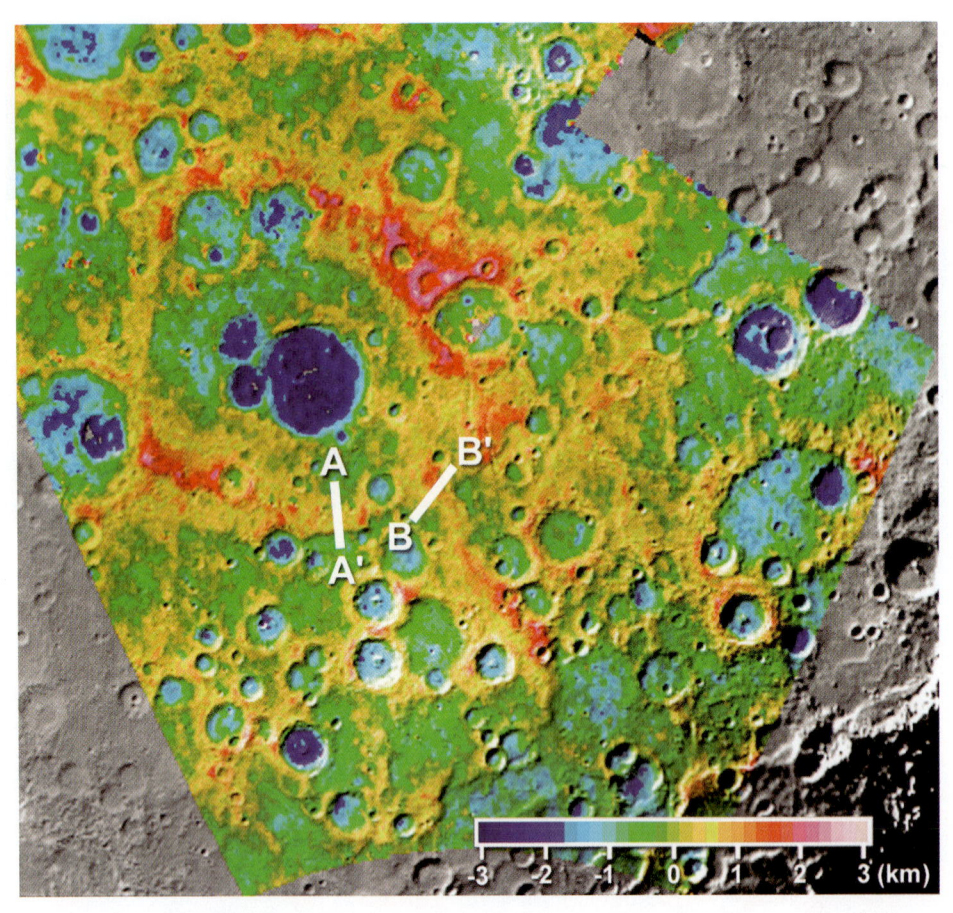

Plate 2. Regional-scale DEM generated using Mariner 10 stereo coverage overlaid on the photomosaic of Discovery Rupes, Adventure Rupes, Resolution Rupes and Rabelais Dorsum. Shades of cyan to dark blue are lows, and shades of red to pink are highs. Elevations are relative to the 2439.0 km Mercury radius reference sphere (Watters *et al.*, 2001).

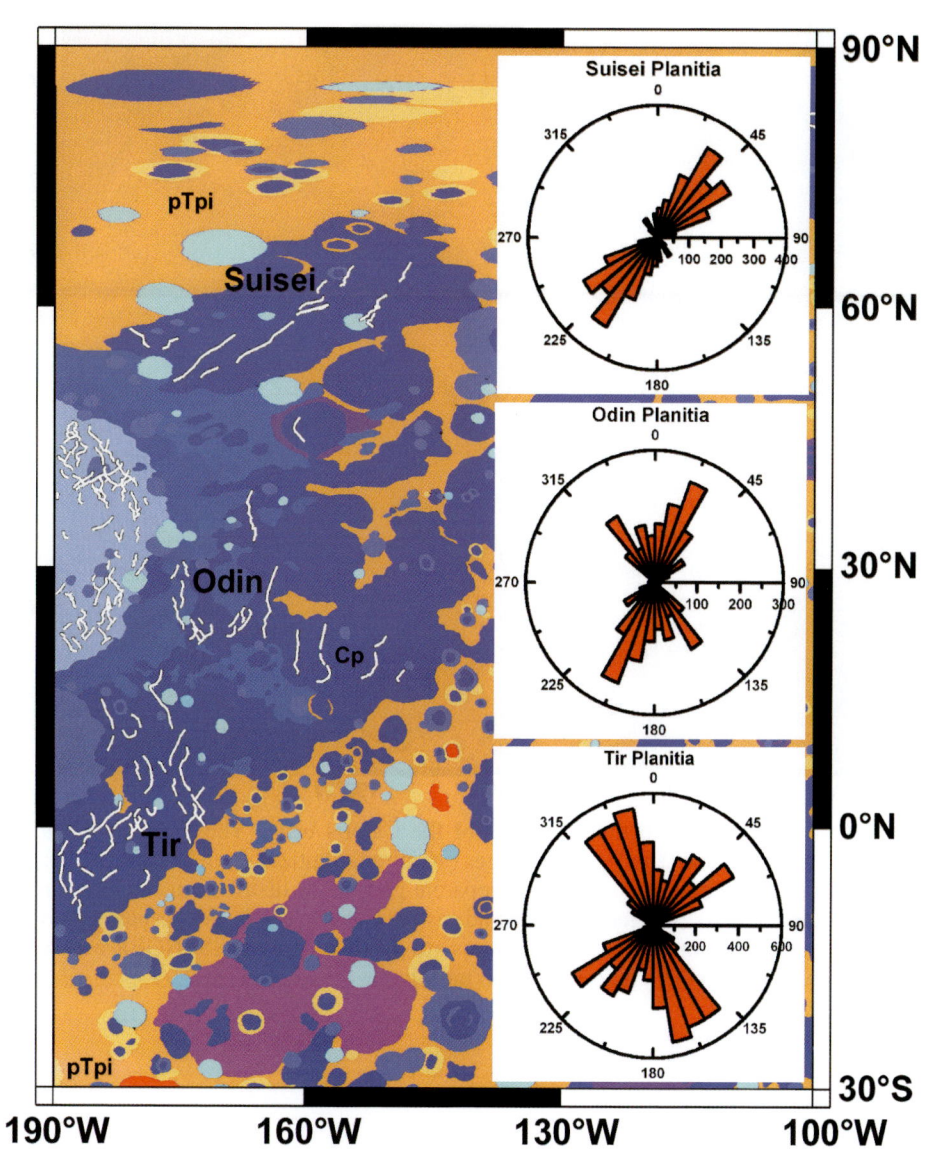

Plate 3. Rose diagrams showing the distribution of orientations of wrinkle ridges in Tir, Odin, and Suisei Planitiae. The rose diagrams are overlaid on the geologic map showing the location of mapped wrinkle ridges. Orientations are length weighted by dividing segments of constant orientation into 1 km intervals. The horizontal axes are in units of kilometers.

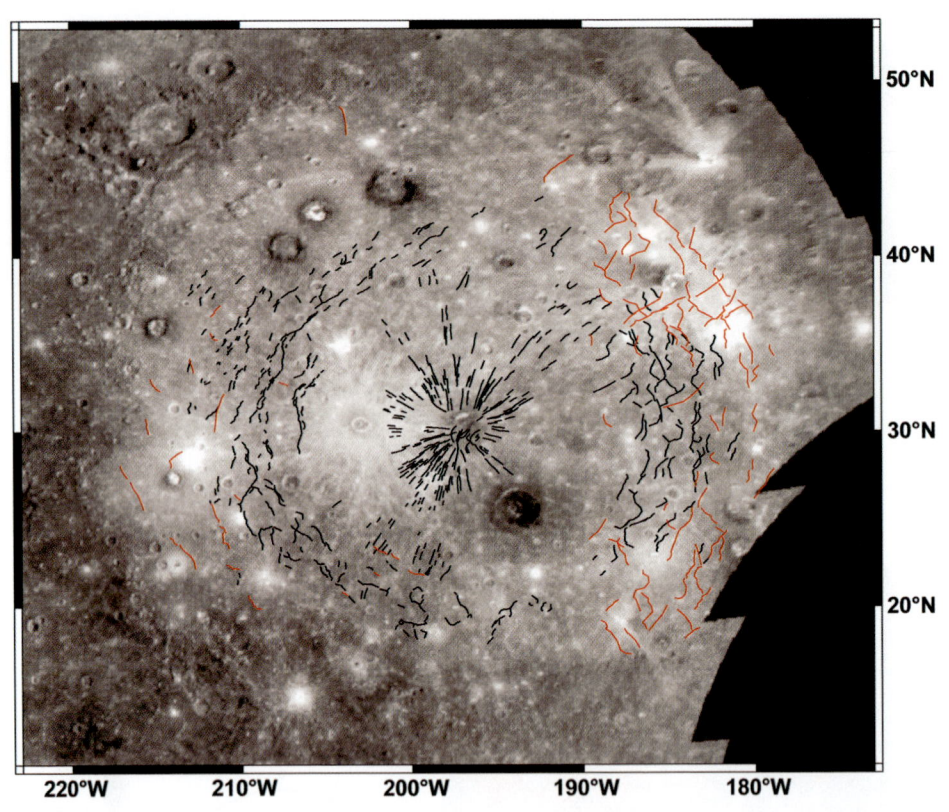

Plate 4. Map of the tectonic features of the Caloris basin. The map combines graben (560 in black) and wrinkle ridges (96 in red) digitized from MESSENGER and Mariner 10 image mosaics. The apparent gaps in the distribution and number of tectonic features between western and eastern Caloris may be due, in part, to the poor lighting geometry. Tectonic features are overlain on a MESSENGER MDIS Narrow Angle Camera high-resolution mosaic.

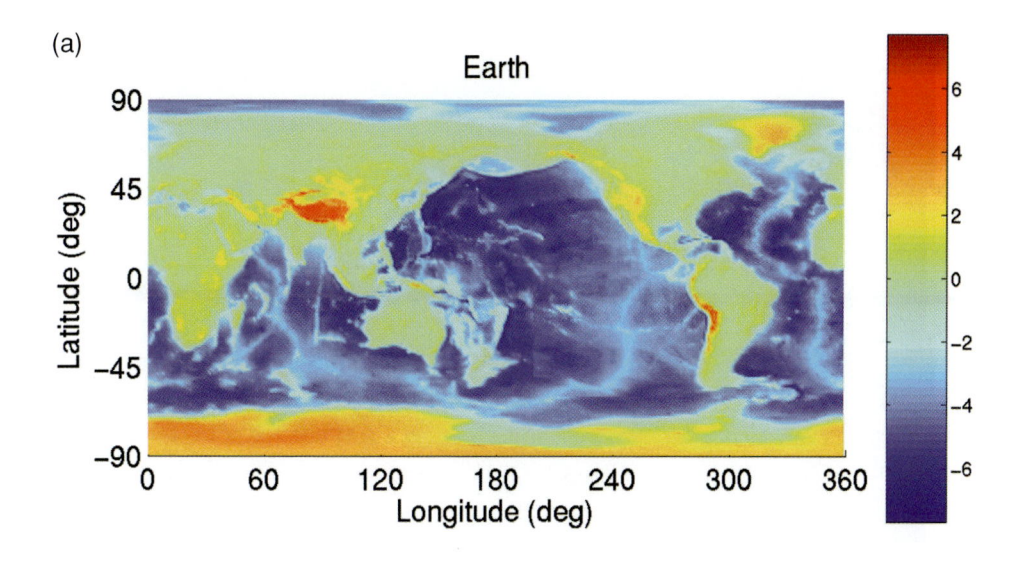

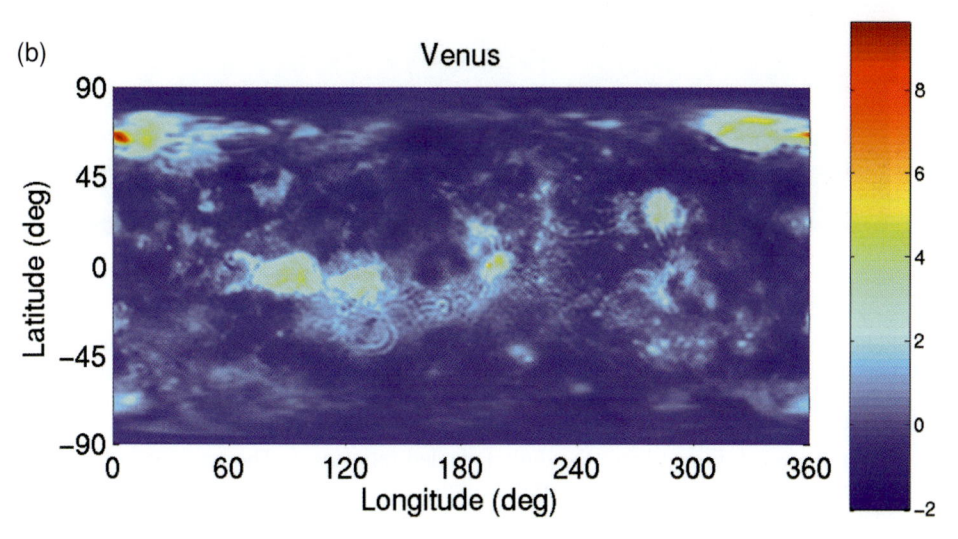

Plate 5. Topographic comparison of Venus and Earth. (a) Earth topography with "transparent" oceans; (b) Venus topography. Note prominent elevations along plate boundaries on Earth and absence of comparable plate-tectonic signature on Venus.

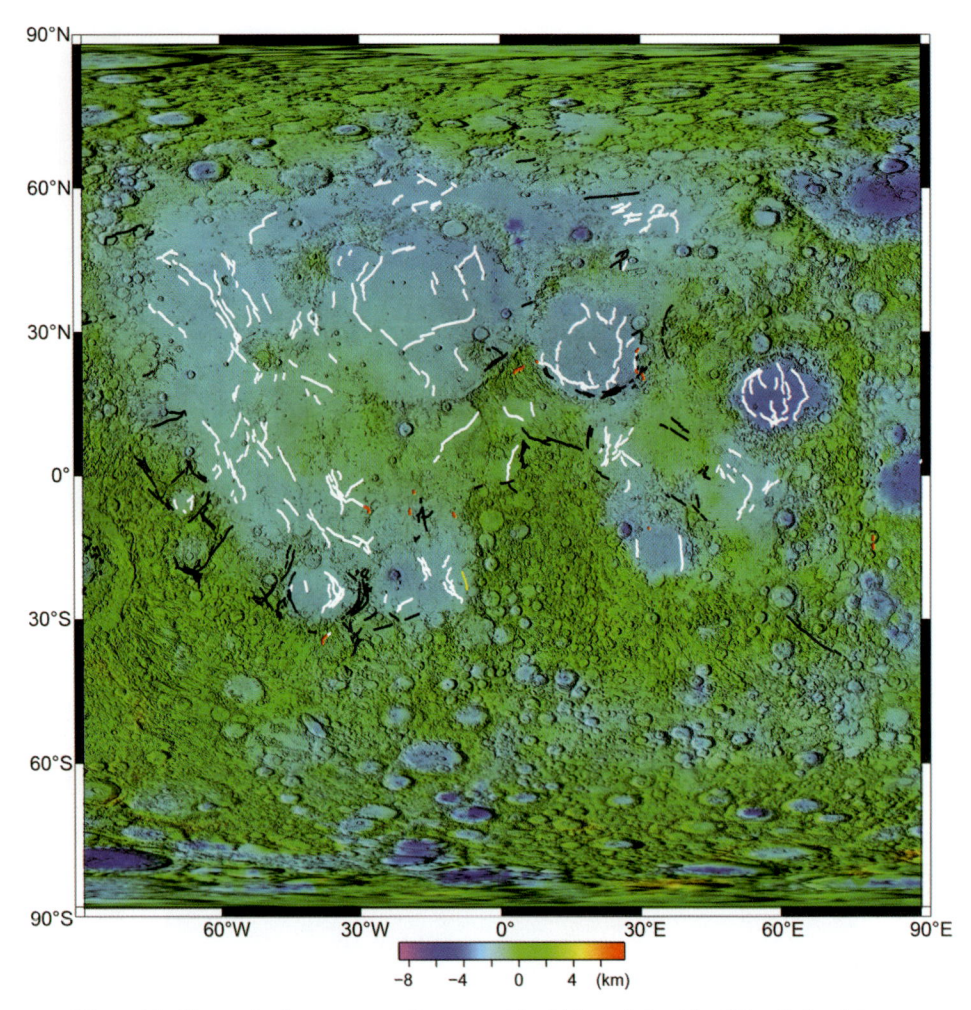

Plate 6A. Tectonic features on the nearside (A) and farside (B) of the Moon. Digitized segments of wrinkle ridges or mare ridges (white), linear and arcuate rilles (black), Rupes Recta (yellow), and lobate scarps (red) are overlaid on a shaded relief map digitally merged with color coded topographic data. The tectonic features were digitized from the digital shaded relief map. The locations of the tectonic features were checked using available Earth-based image surveys of the nearside, Lunar Orbiter images, and Apollo Metric and Panoramic camera images. The locations of features not shown on or below the limits of resolution of the shaded relief map were approximated. The topographic data is the global ULCN 2005 topographic model (Archinal *et al.*, 2006). Elevations are referenced to a sphere of 1737.4 km.

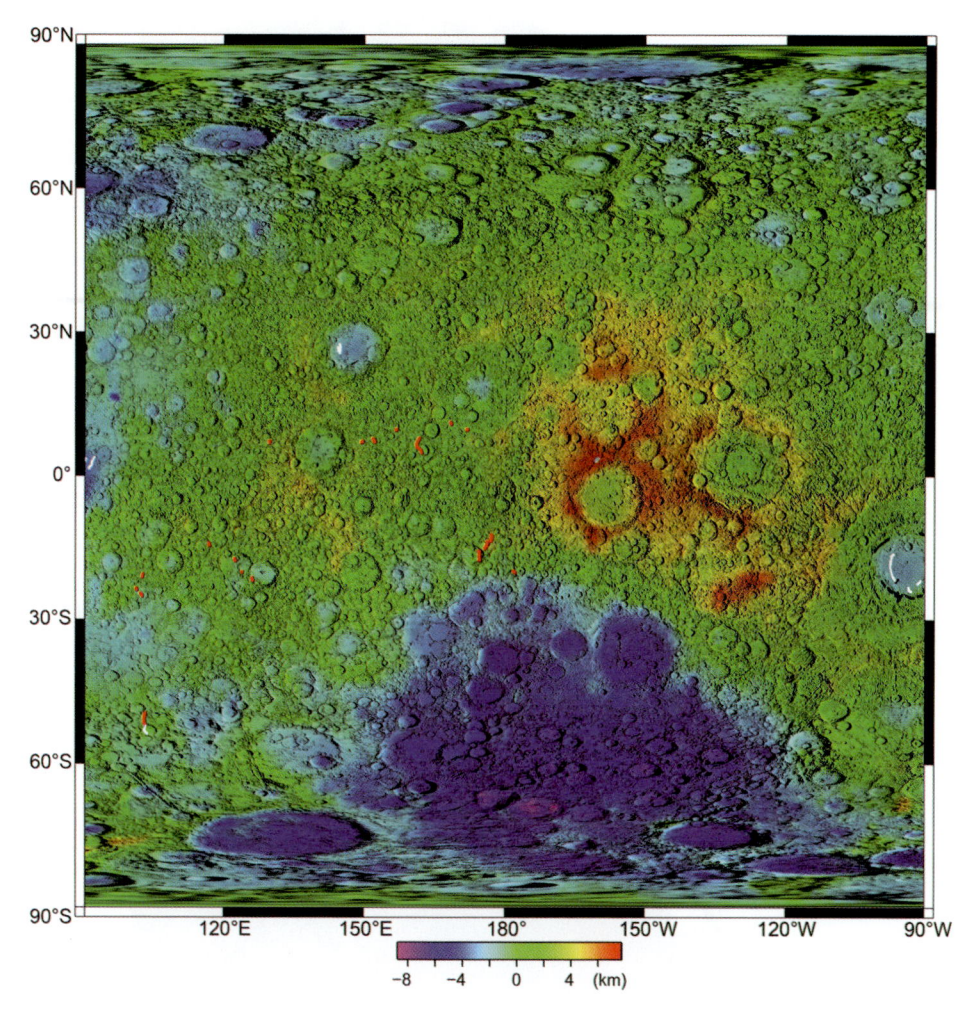

Plate 6B.

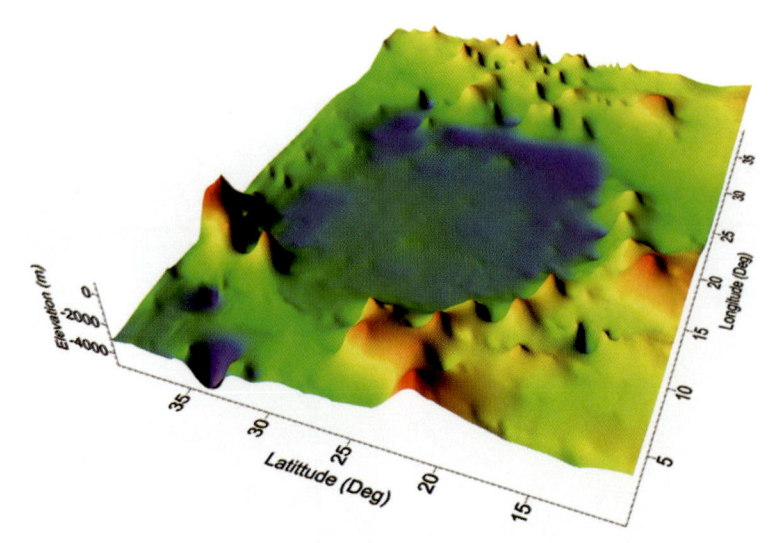

Plate 7. Topography of Mare Serenitatis. This perspective view shows that the interior of much of the Mare surface is higher than the margins. The break in slope from the interior to the margin generally corresponds to the location of the prominent wrinkle-ridge ring. The DEM was generated using Clementine LIDAR data (Smith *et al.*, 1997). Elevations are in meters above an ellipsoid of radius 1738 km at the equator with a flattening of 1/3234.93 corresponding to the flattening of the geoid.

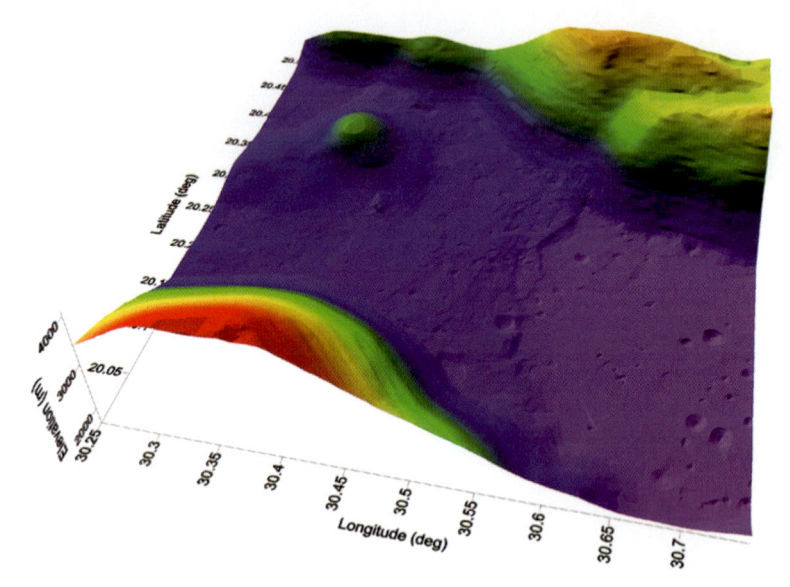

Plate 8. Digital elevation model of the Tauras-Littrow valley. The Lee-Lincoln scarp offsets the floor of the Tauras-Littrow valley by as much as 80 m. The DEM was generated by Robinson and Jolliff (2002) from a 1:50 000 scale U.S. Geologic Survey map (USGS, 1972). Elevations are relative to an arbitrary zero vertical datum.

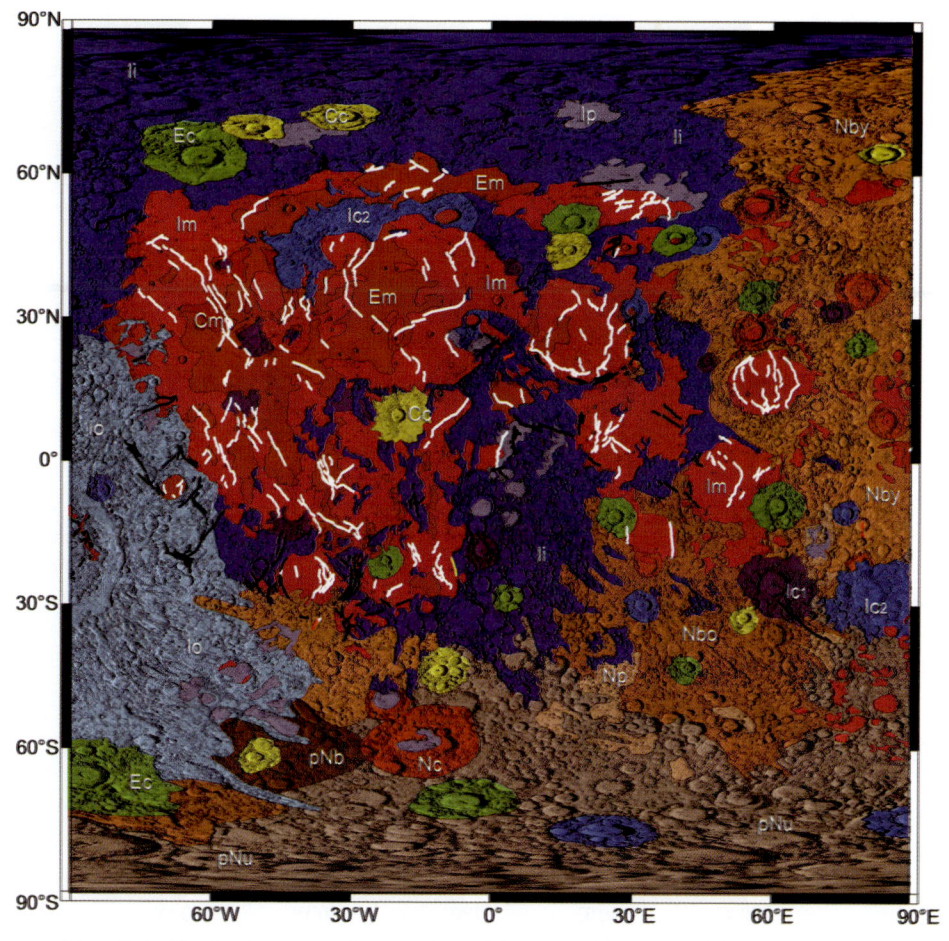

Plate 9A. Geologic map and tectonics of the nearside (A) and farside (B) of the Moon. Digitized segments of wrinkle ridges or mare ridges (white), linear and arcuate rilles (black), lobate scarps (red), and Rupes Recta (yellow) are overlaid on the geologic map of the Moon by Wilhelms (1987). The major geologic units are pre-Nectarian undivided material (pNu) and basin material (pNb), Nectarian older basin material (Nbo) and younger basin material (Nby), and crater (Nc) and plains material (Np), lower Imbrian-basin material (Ii), crater material (Ic1), Orientale-basin material (Io), and plains material (Ip), upper Imbrian mare basalts (Im), and crater material (Ic2), Eratosthenian mare basalts (Em) and crater material (Ec), and Copernican mare basalt (Cm) and crater material (Cc). For absolute ages of the lunar geologic epochs see Tanaka *et al.* (Chapter 8, Fig. 8.1).

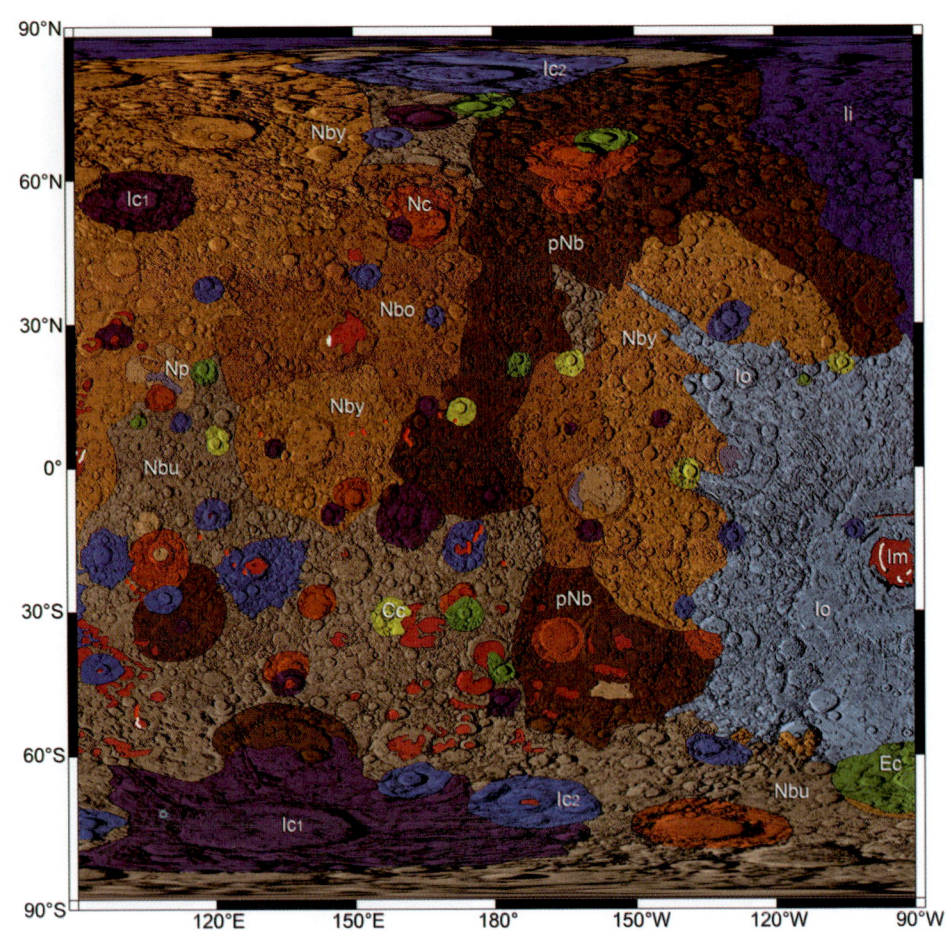

Plate 9B.

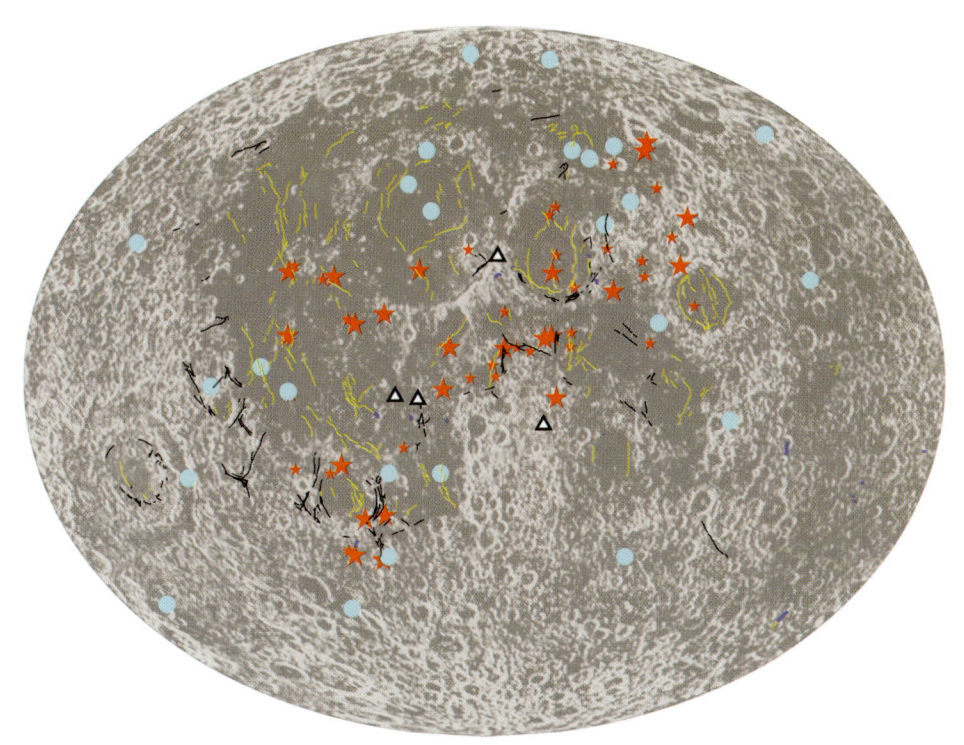

Plate 10. Shaded relief map of the lunar nearside (see Plate 5A for details) with mapped wrinkle ridges (yellow), graben (black), and lobate scarps (blue). Apollo stations 12, 14, 15, and 16 are denoted by the white triangles. Deep moonquake source region locations (red stars): 23 events from Gagnepain-Beyneix *et al.* (2006) along with the 50 events of Nakamura (2005) that have less than ±10° uncertainty in latitude/longitude and for which depth estimates were available. The large symbols denote the 20 clusters common to both Gagnepain-Beyneix *et al.* (2006) and Nakamura (2005), and for these events the locations in Gagnepain-Beyneix *et al.* (2006) were plotted. Small symbols are events occurring in only one of the two data sets. Turquoise circles denote the 26 shallow moonquakes located in Nakamura (1980), assuming a depth of 100 km.

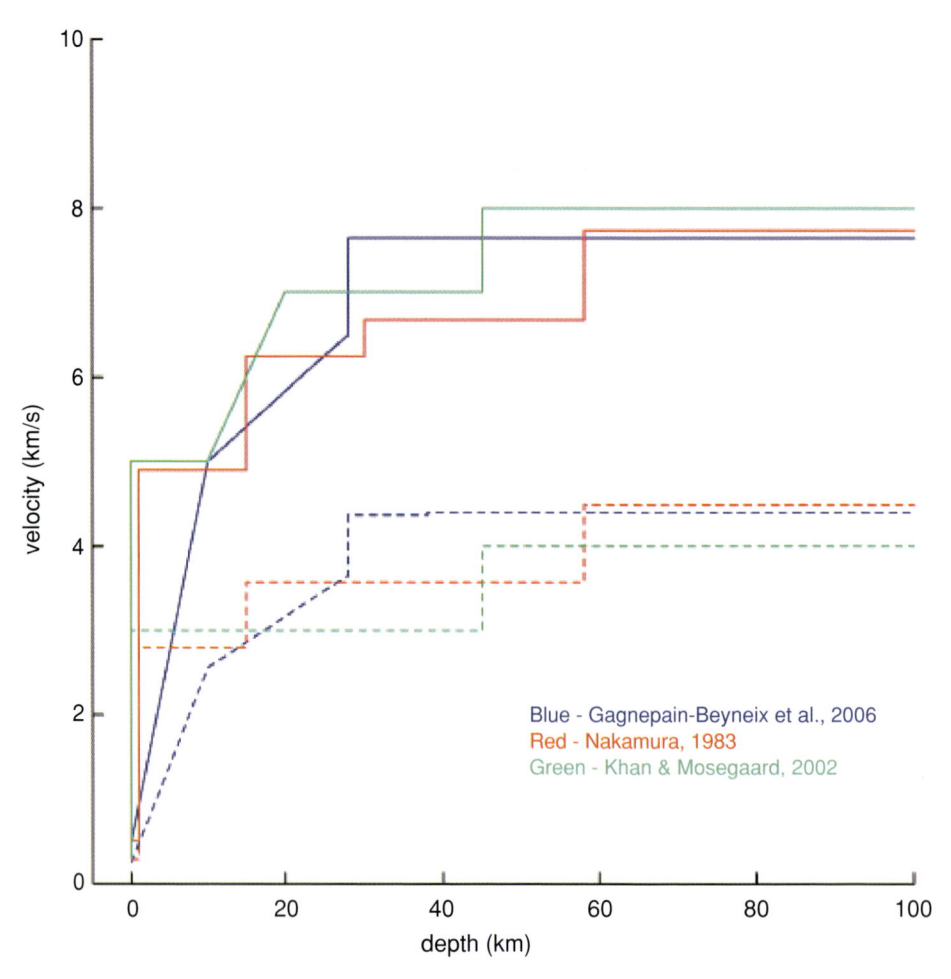

Plate 11. Seismic velocity structure of the lunar crust for three different lunar interior models: P-wave (solid) and S-wave (dashed) velocities for the three models are shown in blue (Gagnepain-Beyneix *et al.*, 2006), red (Nakamura, 1983), and green (Kahn and Mosegaard, 2002). Differences in crustal thickness among the models are seen: from 58 km (Nakamura, 1983) to 45 km (Kahn and Mosegaard, 2002) to 30 km (Gagnepain-Beyneix *et al.*, 2006).

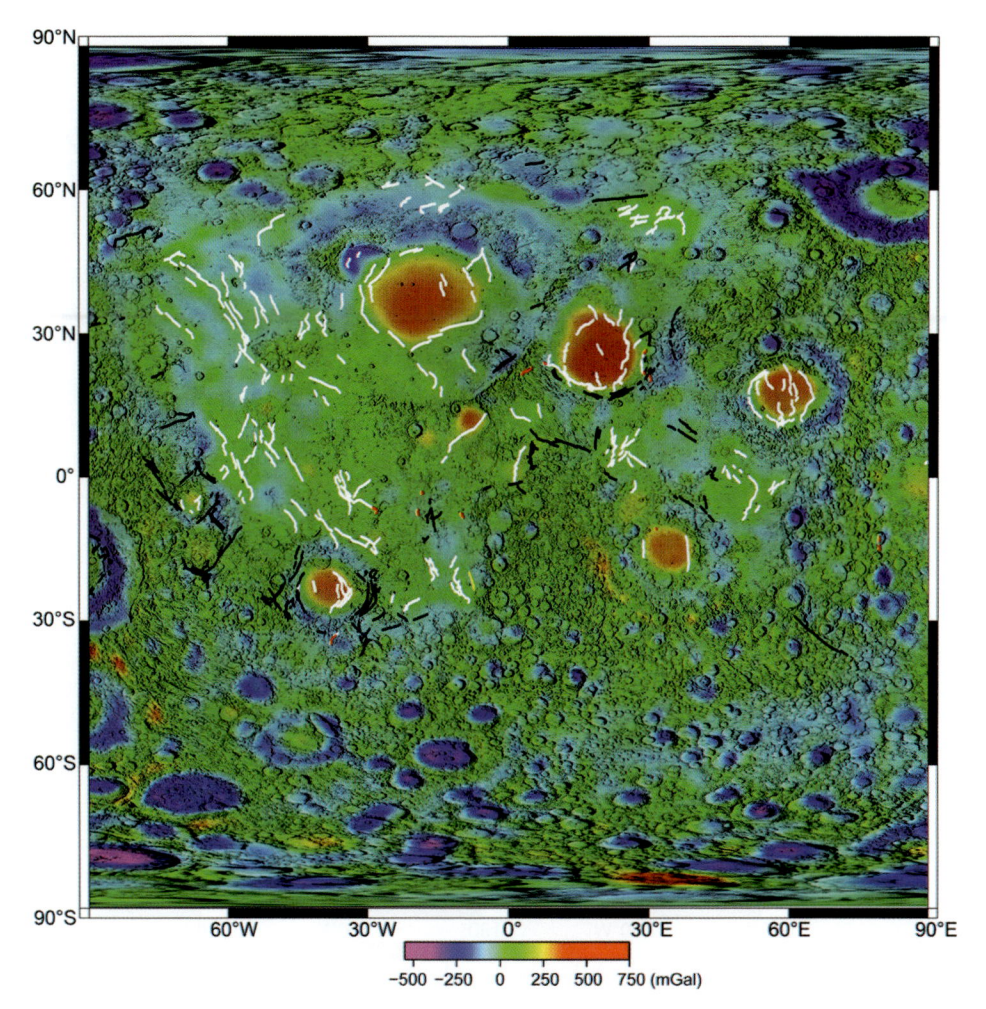

Plate 12A. Gravity and tectonics of the nearside (A) and farside (B) of the Moon. Free-air gravity derived from the Lunar Prospector LP150Q gravity model provided by Alex Konopliv, Jet Propulsion Laboratory (see Konopliv *et al.*, 2001). The grid was generated by truncating LP150Q at degree and order 140. Gravity is overlaid on the shaded relief map with digitized segments of wrinkle ridges or mare ridges (white), rilles or graben (black), Rupes Recta (yellow), and lobate scarps (red).

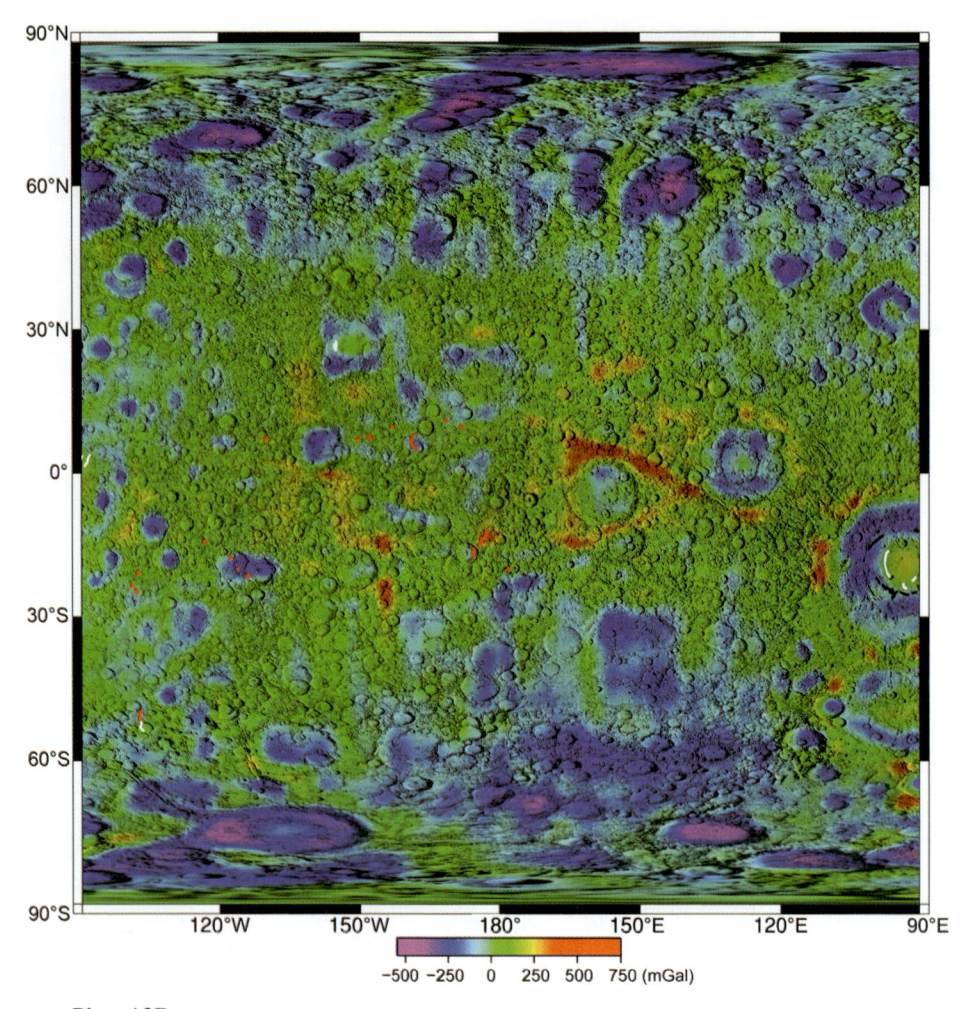

Plate 12B.

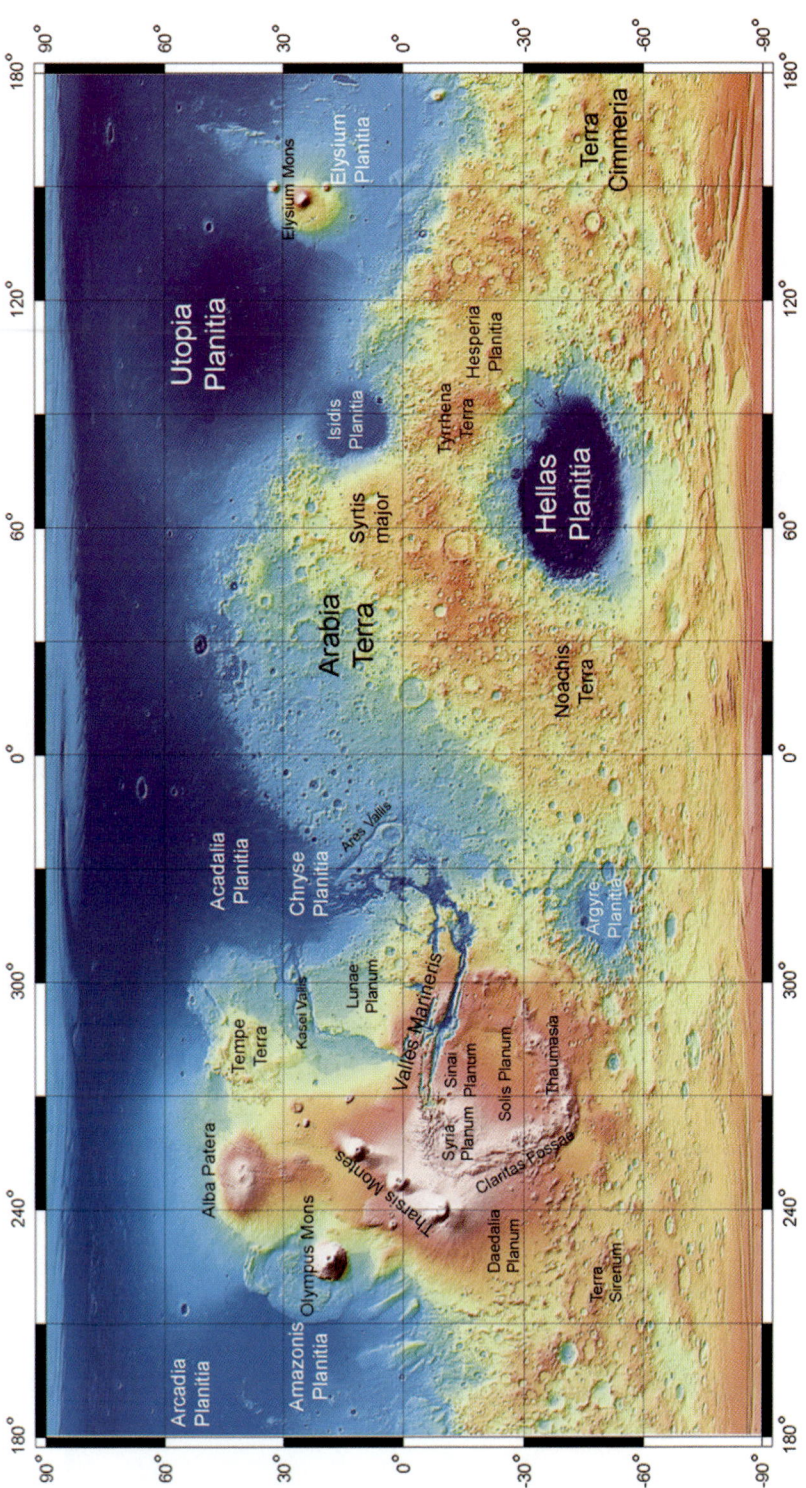

Plate 13. Topographic map and place names for Mars. The elevation range extends from ~−8 km (blue) to ~+8 km (light pink in Thaumasia and Tharsis rise regions). The Tharsis Montes and Olympus Mons extend to higher elevations. Simple cylindrical projection of MOLA topography.

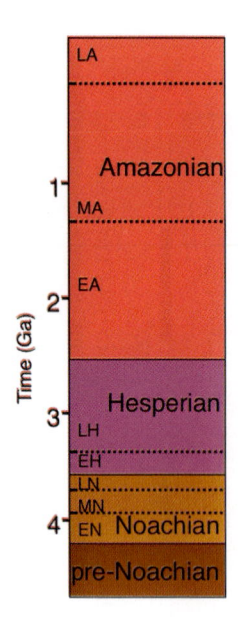

Plate 14. Geologic time scale of Mars based on major geologic/stratigraphic units exposed at the surface (Tanaka, 1986). Absolute time scale is based on correlating crater densities on Mars with radiometrically dated surfaces on the Moon (Hartmann and Neukum, 2001 and references therein; Hartmann, 2005; Tanaka and Hartmann, 2008). Within each era, epochs are shown as Early (EA, EH, EN) and Late (LA, LH, LN), Amazonian, Hesperian and Noachian, respectively. Middle Amazonian (MA) and Noachian (MN) epochs also shown.

Plate 15. Crustal thickness model (Neumann et al., 2008) based on MRO spherical harmonic degree 95 gravity model 95a (A. S. Konopliv). Mean crustal thickness is 45 km and crustal density 2900 kg m^{-3}, with density adjustments made for the Tharsis volcanoes, Elysium Mons, and the polar caps.

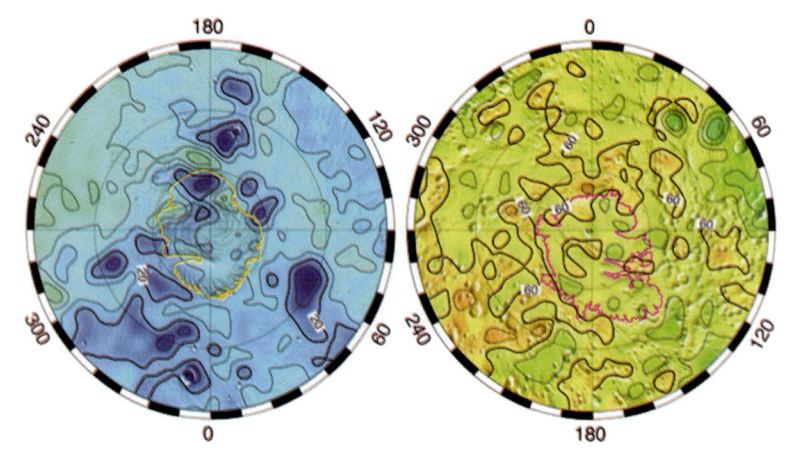

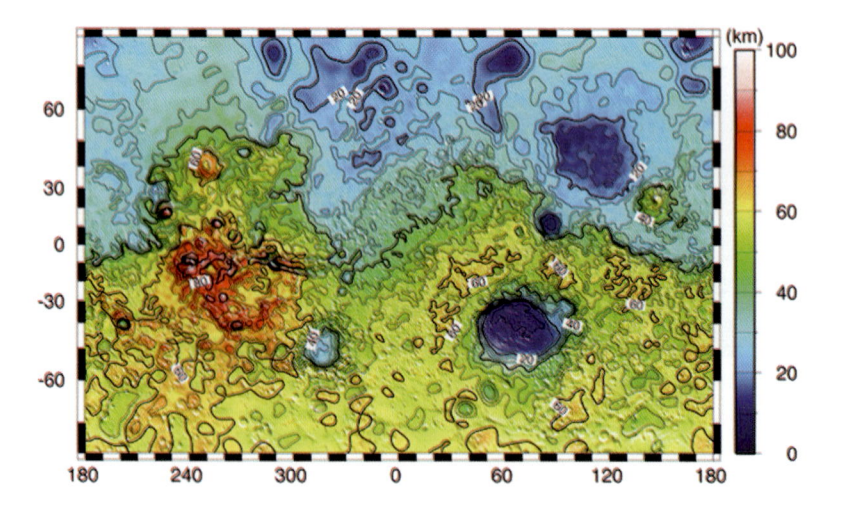

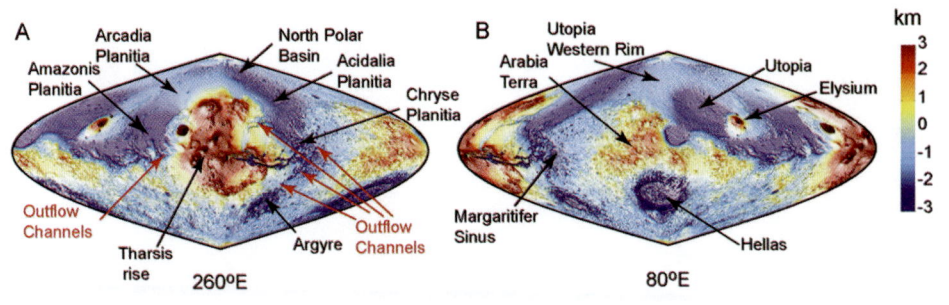

Plate 16. Observed Martian topography for hemisphere centered on Tharsis (A) and the opposite hemisphere (B). The pole-to-pole slope (the J_1 term of a spherical harmonic expansion) has been removed to emphasize other global features. A trough or moat surrounds much of Tharsis and contains most of the planet's outflow channels. To the southwest of Tharsis the trough may be obscured by both Tharsis volcanics and crustal folding. In the other hemisphere, Arabia Terra is the site of a topographic bulge. The trough and bulge are predicted by a model that localizes the Tharsis load on an elastic lithospheric shell (Phillips et al., 2001). Sinusoidal projection.

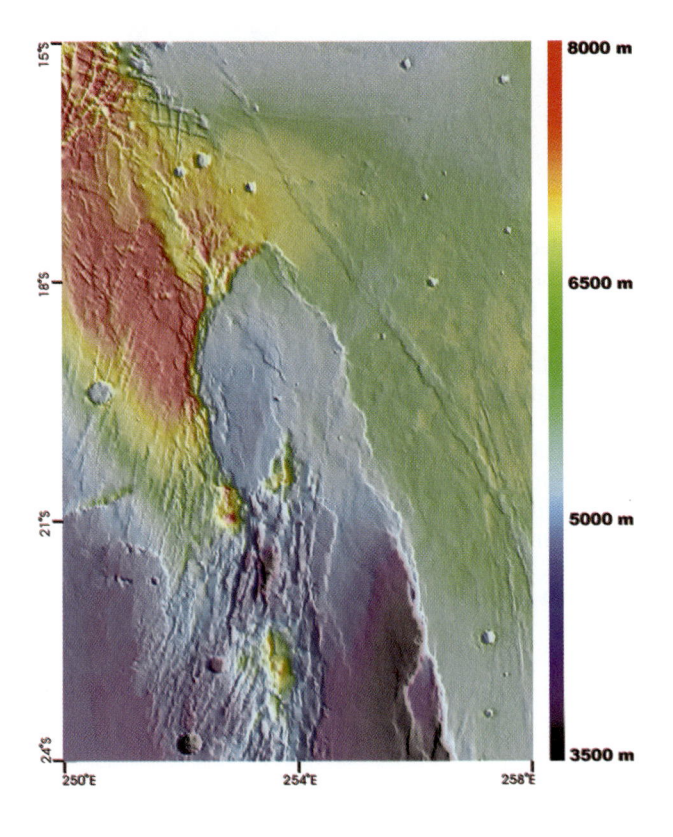

Plate 17. MOLA topographic, shaded relief map of the Thaumasia rift, analogous to a continental rift on Earth, in Claritas Fossae, Mars. The rift is 100–200 km wide with 1–2 km of relief. Note segmented nature of the rift, with the floor north of the intrarift horst tilted into the master fault on the west and the southern floor tilted into the master fault on the east. The north–northeast trending fault set is mapped as Early Hesperian in age, compared with the major rifting, which occurred in the Early Amazonian (Tanaka and Davis, 1988). The older fault set was clearly reactivated during rifting. The 360-km wide map is derived from the 0.5 km gridded MOLA elevations (Smith et al., 2001), illuminated from the west.

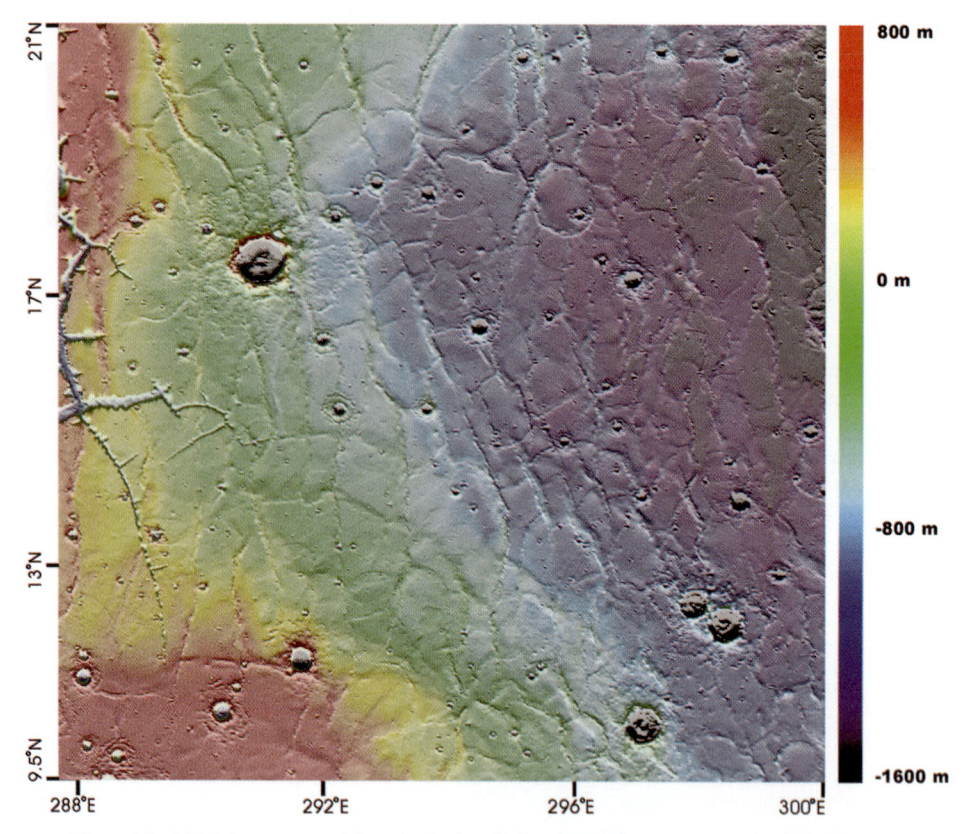

Plate 18. MOLA topographic, shaded relief map of Lunae Planum showing the classic ridged plains on Mars. Wrinkle ridges show up prominently as positive relief hills with sharper wrinkles or crenulations. The primary regularly spaced set of ridges trends generally north to north–northwest and a secondary set trends generally east–west. Note how the plains surface generally decreases in elevation to the east across wrinkle ridges (red to yellow, yellow to green, green to blue, etc.), suggesting they accommodate elevation offsets via deep-seated faulting. Subtle arches between and adjacent to many ridges can also be seen that are likely structurally related to the wrinkle ridges, but could not be seen in visible images. Ridge rings, interpreted as wrinkle ridges that follow the rims of shallowly buried craters, can be seen at 10°N, 296°E and at 18°N, 296°E. MOLA data are from the 0.5 km gridded product (Smith *et al.*, 2001), illuminated from the south. The map is 720 km wide; each degree is about 60 km.

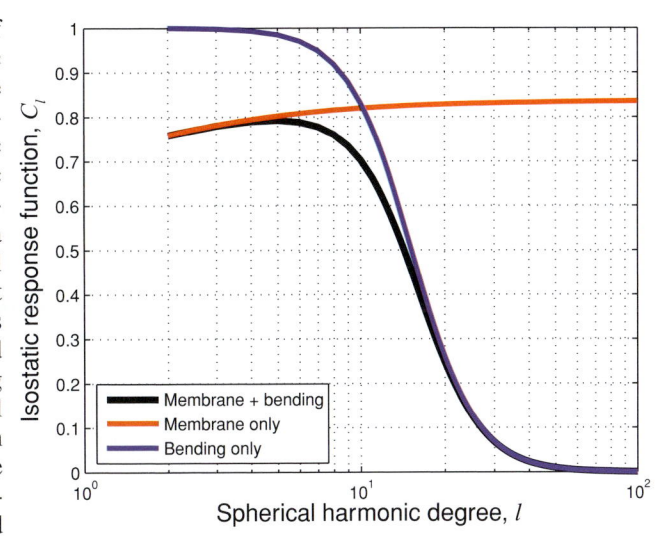

Plate 19. The "degree of compensation" or isostatic response function (Turcotte *et al.*, 1981), C_l, as a function of spherical harmonic degree, l, for a topographic load on a planetary lithosphere represented by a thin elastic spherical shell with a density contrast at its base. Perfect mass balance yields $C_l = 1$ and is possible at very long wavelengths (low spherical harmonic degree) when only bending forces are considered. The complete support of the load ("Membrane + Bending") is shown, as well as the responses obtained by isolating the membrane contribution and the bending contribution. For l less than about 10, the load is supported largely by membrane deformation. The elastic shell has a thickness, Te, of 100 km and a Young's modulus, E, of 5×10^{10} Pa. Crust and mantle densities are 2900 kg m^{-3} and 3500 kg m^{-3}, respectively.

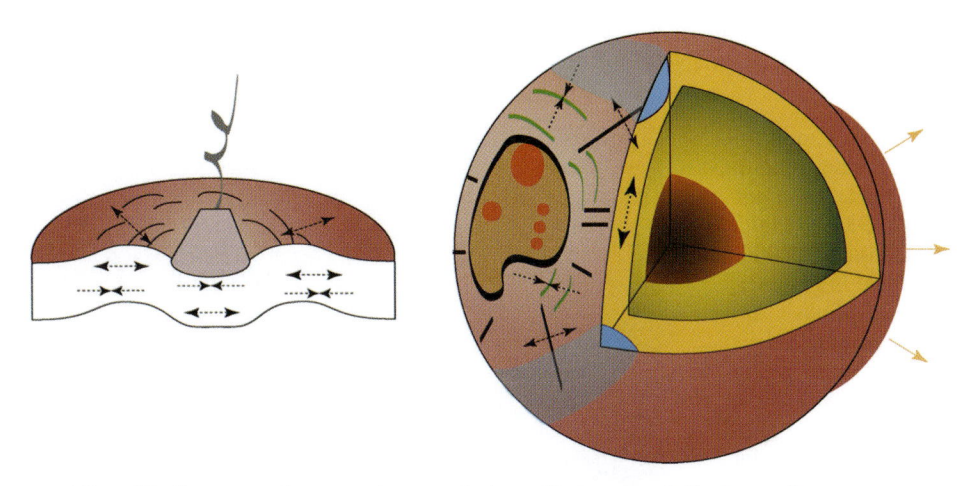

Plate 20. Stress modes on sphere and plate. Cartoon on left shows flexure on a flat elastic lithospheric plate with circumferential grabens resulting from radial extensional stresses in the flexural bulge surrounding the load. Cartoon on right indicates that for a large-scale load on a spherical elastic shell, the stress directions are switched relative to the flat plate due to the effects of membrane forces. A large load, relative to the radius of the planet, such as Tharsis, induces a trough or moat (gray) and an antipodal bulge (yellow arrows). This appears to explain the first-order gravity and topography of Mars. Outward black arrowheads indicate extensional stresses; inward black arrowheads indicate compressional stresses. Black lines and curves are graben; green curves are wrinkle ridges.

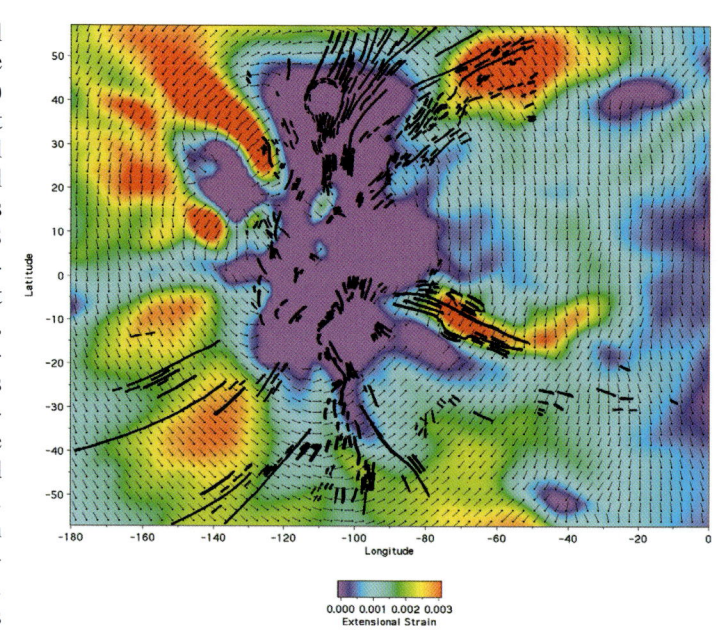

Plate 21. Extensional strain magnitude (color background) and direction (short lines) from model of elastic spherical shell loading as constrained by MGS gravity and topography data (Banerdt and Golombek, 2000). Major extensional structures are shown as heavier lines and are from Scott and Tanaka (1986). Results are shown for Tharsis (western) hemisphere. Model calculations use $T_e = 100$ km, crustal thickness = 50 km, crustal density = 2900 kg m^{-3} and mantle density = 3500 kg m^{-3}. Note structures are generally perpendicular to the extension direction, and the preponderance of structures in areas with high strain, such as Valles Marineris to the east–southeast, Tempe Terra to the northeast, Thaumasia to the south, and Sirenum to the southwest.

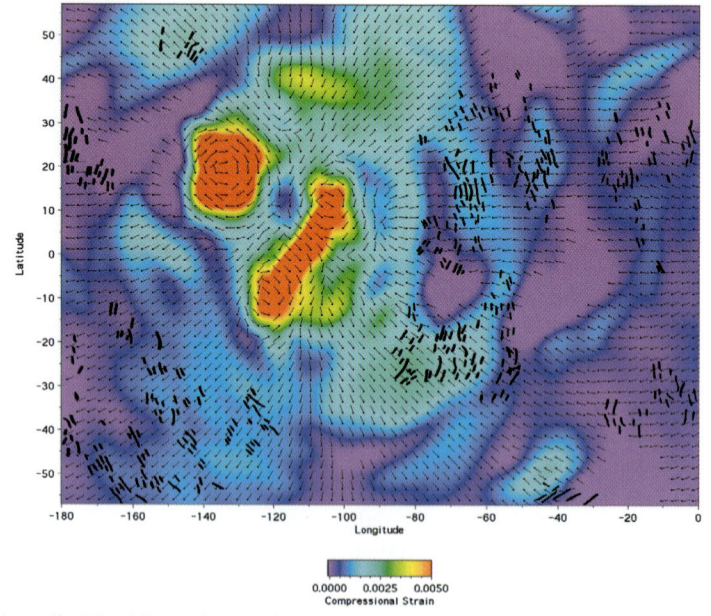

Plate 22. Compressional strain magnitude and direction from model (Banerdt and Golombek, 2000) described in Plate 21. Heavier lines are major wrinkle ridges from Scott and Tanaka (1986). Note structures are generally perpendicular to the compression direction, and the preponderance of structures in areas with relatively high strain, such as Lunae Planum to the east, Solis Planum to the southeast, Sirenum to the southwest, and Arcadia Planitia to the northwest.

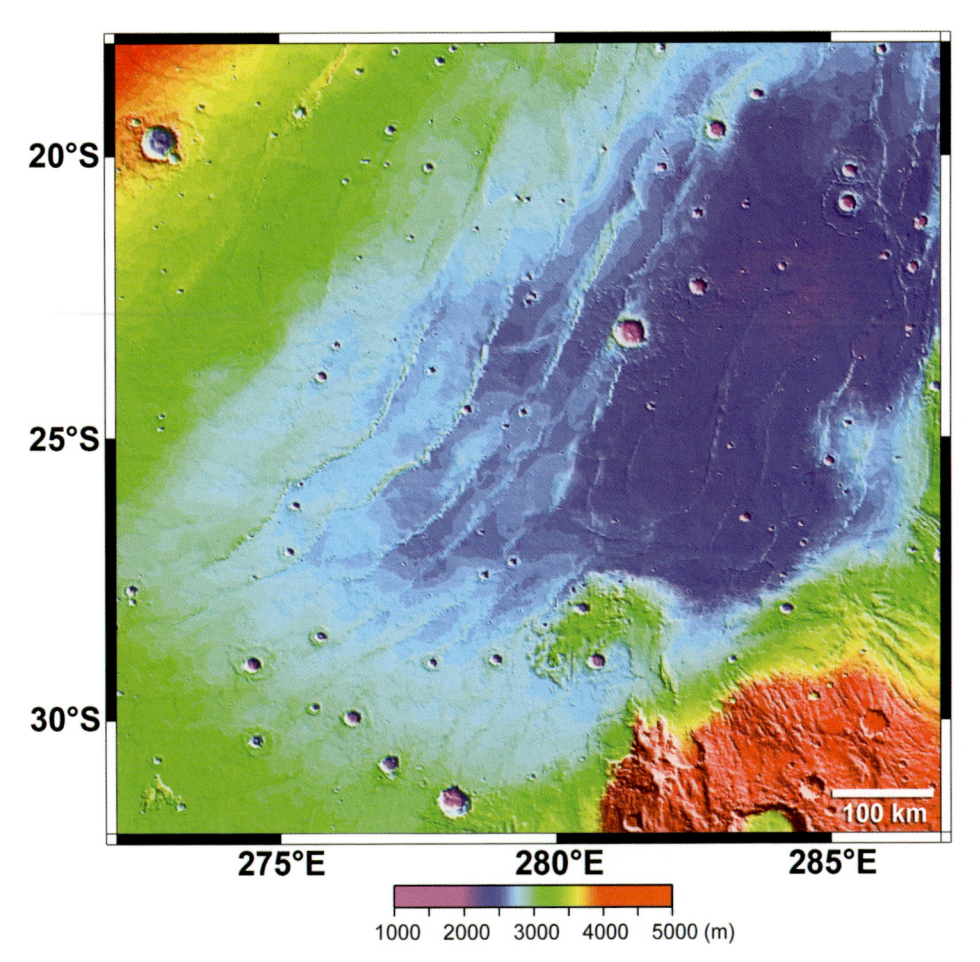

Plate 23. Ridged plains of Solis Planum, Mars. The color-coded shaded relief map generated using MOLA topographic data shows wrinkle ridges formed in ridged plains volcanic material partially filling a regional depression (see Watters, 2004). Narrow grabens and a possible rift cut highland areas at lower right. Topographic data are from the MOLA gridded 1/64° per pixel resolution model; artificial illumination from the left.

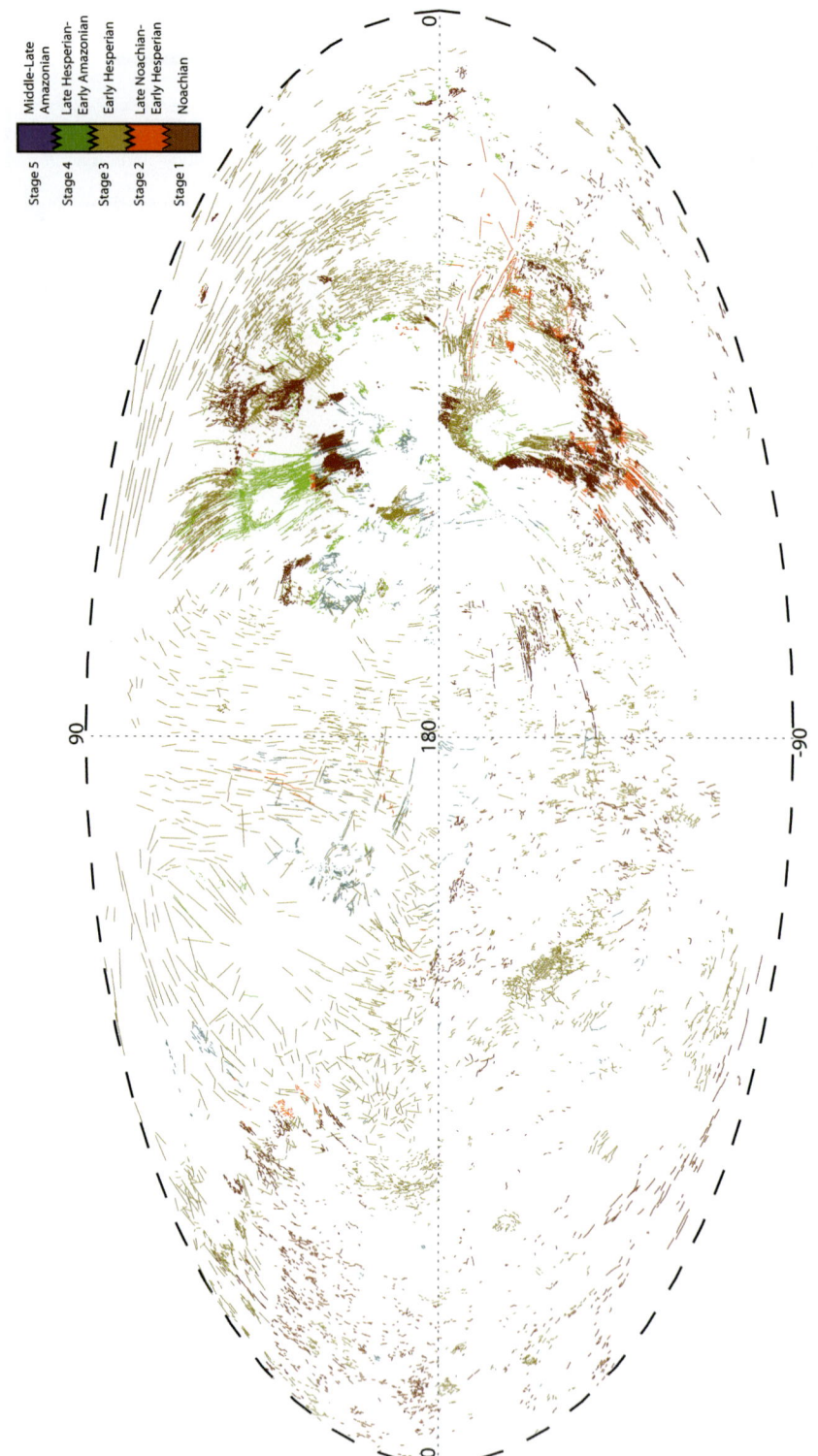

Plate 24. Map of tectonic structural features on Mars. Color coding reflects different ages of faults (grabens and wrinkle ridges) that relate to apparent major episodes of activity (adapted from Anderson *et al.*, 2001).

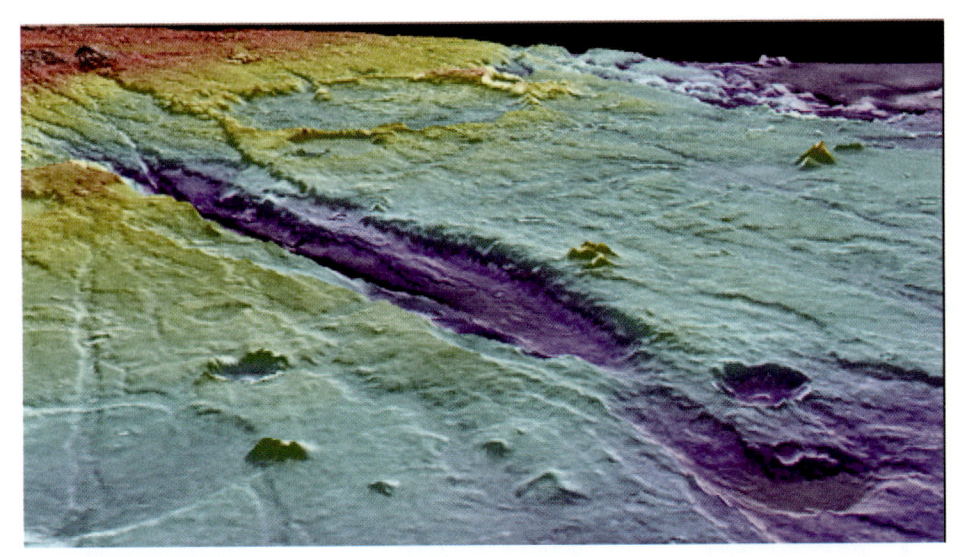

Plate 25. Oblique view of Tempe rift on Mars, looking southeast, created from MOLA DEM (Wilkins *et al.*, 2002; Okubo *et al.*, 2004). The main rift graben is ~550 km long by ~60 km wide and shows footwall uplift; many shallow grabens may represent an earlier stage of faulting in this part of northern Tharsis (e.g., Polit, 2005).

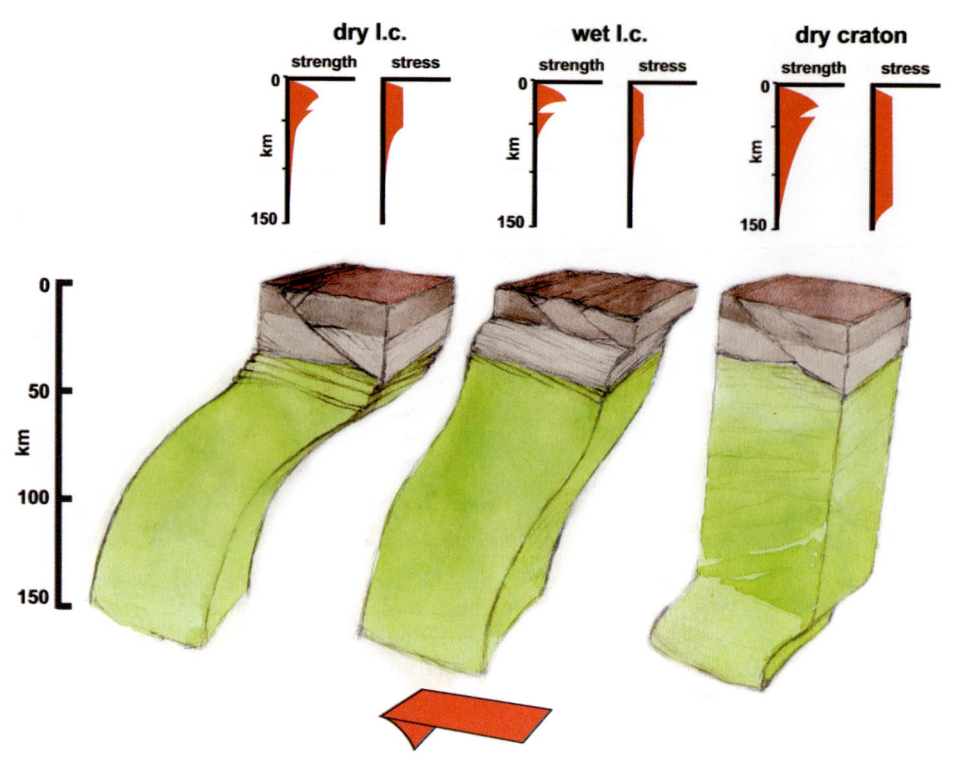

Plate 26. Cartoons illustrating generalized behavior of three regions in the Earth when subjected to shearing to the left from below. Plots showing the strength envelope and stress distribution for each cartoon are shown above. In the left cartoon, dry crust deforming predominantly by frictional sliding on faults overlies a mantle characterized by dislocation glide at shallow depths and dislocation creep below; such a cartoon may provide a reasonable approximation to oceanic lithosphere. The crustal thickness in this cartoon is exaggerated for clarity. In the center cartoon, dry upper crust overlies a weaker lower crust that deforms plastically, presumably due to water weakening; the mantle lithosphere deforms as in the left cartoon. Note that there is a weak lower crustal zone separating stiffer upper crustal and upper mantle regions (the jelly sandwich model). This cartoon might be appropriate to a back-arc setting where water released from the downgoing slab may weaken the lower crust. The right cartoon shows a possible scenario for cratonic lithosphere, where the crust overlies a stable, presumably dry, continental root. A stiff crust deforming predominantly by frictional sliding at shallow depths and localized plastic flow at greater depths overlies a stiff dry mantle lithosphere, which is underlain by wet convecting asthenosphere at depths greater than 100 km. Drawing courtesy of B. K. Holtzman.

6 Models and tectonic comparisons: other global-scale features

6.1 Models for global isotropic stress and strain

Investigations on other terrestrial planets show that regional to global tectonics can be the result of the superposition of stresses from regional loads and stresses from global contraction of the cooling planet. On Venus, regional control of otherwise isotropic compressional strain appears to have oriented ridges surrounding Aphrodite Terra (Sandwell *et al.*, 1997), and global contraction has been suggested to modulate the formation of wrinkle ridges in and around mascon basins on the Moon (Solomon and Head, 1980). Mercury has long been recognized for its surface manifestation (thrust faults) of global contraction, presumably due to cooling (Watters and Nimmo, 2009), and imaging from the MESSENGER spacecraft has increased (and will continue to increase as more data are acquired) the estimated magnitude of the areal density of thrust faults and thus estimates of total contractional strain (Solomon *et al.*, 2008).

Mars appears to be no different than these other planets. Pervasive wrinkle ridge occurrence on volcanic plains of Early Hesperian age strongly suggests a global contractional event of Early to Late Hesperian. Stresses from Tharsis have global influences (Phillips *et al.*, 2001) and this is an obvious source of deformation. Furthermore, most of the wrinkle ridges in the western hemisphere (including those in northern plains (Head *et al.*, 2002)) are circumferential to Tharsis, implying a causal relationship (Wise *et al.*, 1979; Watters, 1993; Anderson *et al.*, 2001). Tharsis-circumferential wrinkle ridges are also present in the northern lowlands of the eastern hemisphere (Head *et al.*, 2002). However, significant populations of wrinkle ridges are also found in the eastern hemisphere (e.g., Hesperia and Syrtis Major Plana) and northern basins (e.g., Utopia and Isidis) that have little obvious relationship with Tharsis, and yet most wrinkle ridges appear to have formed in the time period spanning the Early Hesperian to the Early Amazonian (Tanaka *et al.*, 1991, 2003, 2005; Mangold *et al.*, 2000). These globally-pervasive wrinkle ridges may have resulted from a globally isotropic contractional strain event during this time interval that was modulated by Tharsis stresses in the western hemisphere and northern plains, and by Tharsis and other loads, such as Utopia in the eastern hemisphere. It is possible that due to growing global contractional stresses, the onset of pervasive wrinkle-ridge formation marks a reorientation of principal stresses to form thrust faults rather than extensional or strike-slip faults induced by Tharsis and other loads (e.g., Okubo and Schultz, 2006; Andrews-Hanna *et al.*, 2008b).

Secular cooling is the obvious mechanism for producing isotropic global contraction on Mars (Schubert *et al.*, 1992; Mangold *et al.*, 2000). Modeling by Andrews-Hanna *et al.* (2008b) has shown that the isotropic compressive stresses required to explain the transition from strike-slip faulting to thrust faulting in circum-Tharsis regions produces the observed amount of compressional strain of about 0.1% (e.g., Golombek *et al.*, 1991, 2001). They show that such strain magnitudes also result from contractional strain produced by thermal models, with the addition of loading effects from Tharsis.

6.2 Models for origin of the global dichotomy

The origin of the planetary dichotomy that separates the smooth northern lowlands from the heavily cratered southern highlands has remained an enigma since first discussed over thirty years ago (Hartmann, 1973). Hypotheses for its genesis can be broadly divided into endogenic and exogenic. Exogenic theories have it that one (Wilhems and Squyres, 1984) or more (Frey and Schultz, 1988) large impacts were responsible for the lowering of the crustal surface in the northern hemisphere. Endogenic theories involve several types of processes for thinning the crust dynamically from within (e.g., Wise *et al.*, 1979), and include the possibility of plate tectonics (Sleep, 1994). In the last five years, improved sophistication in dynamic modeling, as well as new data from missions, have provided fresh constraints on the origin of the dichotomy.

The most important findings bearing on the origin of the dichotomy come from MGS data: (1) The underlying crust of the northern lowlands is nearly as old (or as old, given uncertainties) as the crust of the southern highlands (Frey *et al.*, 2002; Frey, 2006a,b), implying that their crustal ages are essentially the same. (2) The northern and southern crusts have distinctive thicknesses (Zuber *et al.*, 2000; Zuber, 2001; Neumann *et al.*, 2004, 2008). (3) As mentioned earlier, the northern lowlands crust contains widespread but modest magnetic anomalies (Mitchell *et al.*, 2007). This implies that this crust formed, (a) during the waning stages of the dynamo, (b) lacked large igneous intrusions to acquire stronger remanent magnetization, or (c) was affected by demagnetization processes associated with water. Regarding (c), it is possible that magnetization in the northern lowlands is more robust than observed, but the dominant wavelengths of the anomalies are below the resolution of the MGS magnetic field or electron reflectometry data sets. Selective removal of magnetic anomalies by hydrothermal alteration has been proposed as a means of shifting the power in the magnetic field spectrum to shorter wavelengths (Solomon *et al.*, 2005).

If the northern crust is pre-Noachian (Nimmo and Tanaka, 2005; Frey, 2006a), then it could be a feature of primary crustal fractionation, possibly associated with core formation, which took place immediately after planetary accretion (Lee and Halliday, 1997). Variations in crustal thickness associated with the Utopia basin in the northern lowlands are mostly preserved, suggesting that any crustal thinning must have occurred before the basin-forming impact (Neumann *et al.*, 2004, 2008).

As noted above, a very early origin for the crustal dichotomy raises the question as to why there are so few magnetic anomalies in the northern lowlands, and challenges the primary crustal fractionation hypothesis. A multiple-impact origin (Frey and Schultz, 1988) for the crustal dichotomy requires an unlikely preferential bombardment in what is now the northern hemisphere. Multiple impacts also provide no simple way to remove ejecta from the northern hemisphere (McGill and Squyres, 1991), and thus it is difficult to see how multiple, large impacts could strip the northern hemisphere crust nearly uniformly (Zuber *et al.*, 2000; Zuber, 2001; Neumann *et al.*, 2004) to a depth of ~3 km.

As discussed earlier, the arguments for a single impact origin for the northern lowlands (Wilhems and Squyres, 1984) have been considerably strengthened by the work of Andrews-Hanna *et al.* (2008a) on the shape of the northern lowlands and by Marinova *et al.* (2008) on numerical modeling of a single impact. Nimmo *et al.* (2008) have provided a potential constraint on the timing of the mega-impact in that their numerical simulations show that the lowlands crust is likely derived from a deep, depleted mantle source, which may be akin to shergottite meteorites, which likely formed ~100 Myr after planet formation. In this scenario, the dynamo was waning by this time in order to produce weaker anomalies in the lowlands crust. Overall, the impact hypothesis for the origin of the crustal dichotomy appears to suffer fewer problems than others that have been proposed. It occurs very early in Mars' history, provides a ready explanation for the elliptical shape of the lowlands and thinned crust, and fits in conceptually with large impacts having an important role in the early formation and evolution of the terrestrial planets (aka the giant impact hypothesis for the origin of the Earth's moon).

The relation of the formation of the northern lowlands to the formation of Tharsis is another issue that has been considered qualitatively (Wise *et al.*, 1979) and quantitatively (Wenzel *et al.*, 2004). Endogenic theories for the formation of the crustal dichotomy, whether Tharsis is involved or not, are usually concerned with the generation of spherical harmonic degree $l = 1$ mantle convection (i.e., single regions of upgoing and downgoing flow). Zhong and Zuber (2001) showed that degree-1 mantle convection is possible in Mars with a central core if the upper mantle viscosity is ~100 times smaller than that of the lower mantle. The onset of degree-1 mantle convection is delayed in their model runs, which may be in conflict with the pre-Noachian age of both the northern and southern crusts; more recent work with degree-1 convection in layered viscosity models shows that timescales as short as 100 Myr can be achieved (Roberts and Zhong, 2006).

The idea that the northern hemisphere crust is a relic "oceanic" plate from a plate tectonics era was advanced by Sleep (1994), based mainly on the low elevation and presumed (at the time of publication) thinner crust in the north. Subsequently, the thinner (and approximately constant thickness) crust was confirmed (Zuber *et al.*, 2000). Under this hypothesis, the dichotomy boundary is made up of the three classic types of oceanic plate margins. However, geological predictions of this model appear to fail (Pruis and Tanaka, 1995; Sleep and Tanaka, 1995). Further, the recognition that the northern highlands crust is pre-Noachian in age (Frey *et al.*, 2002; Frey, 2006a) with younger circular basins (QCDs and Utopia) unaffected by plate tectonic processes pushes the episode to very early in Martian history, and thus this hypothesis is difficult to test. It may be worth noting that the areal growth of the southern highlands crust might have shut down a very early (no longer recognizable) episode of plate tectonics when the fractional planetary coverage exceeded 50% (Lenardic *et al.*, 2004), a scenario seemingly at odds with a primordial origin for the crustal dichotomy.

The dichotomy boundary separates the two distinct crustal provinces of Mars, except in northwestern Arabia Terra where low-standing highlands terrain appears to belong to the northern province (Zuber *et al.*, 2000), or is a transitional form between the two provinces (Zuber, 2001; Neumann *et al.*, 2004, 2008), and where Tharsis volcanics have obscured the boundary. Subsequent erosional processes and possible deformation in the Late Noachian/Early Hesperian (McGill and Dimitriou, 1990; Smrekar *et al.*, 2004) suggest ongoing activity at the dichotomy boundary. Once the basic elevation difference of the crustal provinces was created, processes could have operated subsequently to create a sharp boundary in places (Guest and Smrekar, 2005). Mechanisms proposed include faulting in response to flexural stresses caused by the differences in crustal thickness and the resulting broad-scale topography (Watters, 2003a,b), and faulting due to relaxation of the dichotomy boundary by lower crustal flow (Guest and Smrekar, 2005). As noted above, mass wasting and fluvial erosion could have modified the steep gradients caused by faulting, as well as the crustal stress fields.

7 Concluding remarks

Our knowledge of Mars tectonics has improved dramatically since publication of the last major review based on Viking data (Banerdt *et al.*, 1992). Mars Global Surveyor has provided our first high-definition view of the global topography, geoid and gravity fields, which have allowed definition of global crustal thickness and substantially improved modeling of global and regional geodynamical and geological processes. In addition, geologic mapping of tectonized regions has defined the major phases of deformation within the global stratigraphic framework (Tanaka, 1986), and investigations of the geometry and kinematics of observed structures have yielded broad estimates of shortening and extension across these regions. As a result, higher fidelity geodynamical models can be constructed and then tested and constrained much better than was possible before.

This work shows that some process or processes in the pre-Noachian thinned the crust and lowered the surface by several kilometers in the northern hemisphere of Mars. This thinning could have been the result of primordial crustal fractionation, degree-1 convection processes, or a single mega-impact. The latter hypothesis has been considerably strengthened by the recognition that the shape of the northern lowlands with Tharsis removed match a large elliptical basin consistent with a large oblique impact early in Mars' history. Subsequently, these northern lowlands (roughly a third of the planet) were resurfaced (thinly, 1–2 km) over a period of time extending into the Hesperian (and, modestly, into the Amazonian).

The global shape of Mars (outside of the spin oblateness) is dominated by the pole-to-pole slope and the enormous Tharsis rise (load) and the global deformation it produced (surrounding flexural moat and antipodal bulge). Lithospheric membrane/flexural loading models based on present-day global gravity and topography appear capable of explaining the orientation, distribution and strain found in the giant radiating extensional grabens and concentric compressional ridges that cover the entire western hemisphere and northern plains.

Geologic mapping shows that deformation (radiating grabens) around Tharsis peaked in the Noachian and decreased with relatively minor changes in centers of deformation through time, arguing that the basic lithospheric structure of Mars has changed little since the Middle Noachian. Structures exposed in the eastern hemisphere record local loading around basins (including those in the northern plains) and volcanoes, compressional lobate ridges and extensional faults associated with the dichotomy boundary, and far-field effects of Tharsis. Wrinkle ridge formation in volcanic provinces of the eastern hemisphere, in the northern plains, and around Tharsis appears to have peaked globally in the Hesperian, likely modulated by global contraction.

The outpouring of volcanics responsible for the development and growth of Tharsis appear capable of producing the likely early warmer and wetter conditions that formed the valley networks, dry lake beds, and high erosion rates preserved in degraded craters and terrain, the exposures of Noachian phyllosilicates mapped by OMEGA and CRISM, and the wet conditions indicated by the Meridiani Planum sulfate evaporites and aqueous altered rocks of the Columbia Hills discovered by the Mars Exploration Rovers.

Acknowledgments

MPG was supported by grants from the Planetary Geology and Geophysics Program of the National Aeronautics and Space Administration at the Jet Propulsion Laboratory, California Institute of Technology, under a contract with NASA. RJP was supported by NASA grants from the Mars Data Analysis and Planetary Geology and Geophysics Programs. The authors appreciate constructive comments from K. Tanaka, P. McGovern and G. McGill and Figures from Steven A. Hauck II (Plate 13), Kris Larsen (Plates 17 and 18) and Bruce Banerdt (Plates 21 and 22).

References

Acuña, M. H., Connerney, J. E. P., Ness, N. F., Lin, R. P., Mitchell, D., Carlson, C. W., McFadden, J., Anderson, K. A., Rème, H., Mazelle, C., Vignes, D., Wasilewski, P., and Cloutier, P. (1999). Global distribution of crustal magnetization discovered by the Mars Global Surveyor MAG/ER experiment. *Science*, **284**, 790–793.

Allemand, P. and Thomas, P. G. (1995). Localization of Martian ridges by impact craters: Mechanical and chronological implications. *J. Geophys. Res.*, **100**, 3251–3262.

Anderson, S. and Grimm, R. E. (1998). Rift processes at the Valles Marineris, Mars: Constraints from gravity on necking and rate-dependent strength evolution. *J. Geophys. Res.*, **103**, 11 113–11 124.

Anderson, R. C., Dohm, J. M., Golombek, M. P., Haldemann, A. F. C., Franklin, B. J., Tanaka, K. L., Lias, J., and Peer, B. (2001). Primary centers and secondary concentrations of tectonic activity through time in the western hemisphere of Mars. *J. Geophys. Res.*, **106**, 20 563–20 585.

Anderson, R. C., Dohm, J. M., Haldemann, A. F. C., Hare, T. M., and Baker, V. R. (2004). Tectonic histories between Alba Patera and Syria Planum, Mars. *Icarus*, **171**, 31–38.

Anderson, R. C., Dohm, J. M., Haldemann, A. F. C., Pounders, E., Golombek, M., and Castano, A. (2008). Centers of tectonic activity in the eastern hemisphere of Mars. *Icarus*, **195**, 537–546.

Andrews-Hanna, J. C., Phillips, R. J., and Zuber, M. T. (2007). Meridiani Planum and the global hydrology of Mars. *Nature*, **446**, 163–166.

Andrews-Hanna, J. C., Zuber, M. T., and Banerdt, W. B. (2008a). The Borealis basin and the origin of the Martian crustal dichotomy. *Nature*, **453**, 1212–1215.

Andrews-Hanna, J. C., Zuber, M. T., and Hauck I, S. A. (2008b). Strike-slip faults on Mars: Observations and implications for global tectonics and geodynamics. *J. Geophys. Res.*, **113**, doi:10.1029/2007JE002980.

Anguita, F., Farelo, A. F., Lopez, V., Mas, C., Munoz-Espadas, M. J., Marquez, A., and Ruiz, J. (2001). Tharsis dome, Mars: New evidence for Noachian-Hesperian thick-skin and Amazonian thin-skin tectonics. *J. Geophys. Res.*, **106**, 7577–7589.

Anguita, F., Fernández, C., Cordero, G., Carrasquilla, S., Anguita, J., Núñez, A., Rodríguez, S., and García, J. (2006). Evidences for a Noachian–Hesperian orogeny in Mars. *Icarus*, **185**, 331–357.

Arkani-Hamed, J. (2001). A 50-degree spherical harmonic model of the magnetic field of Mars. *J. Geophys. Res.*, **106**, 23 197–23 208.

Arkani-Hamed, J. and Boutin, D. (2004). Paleomagnetic poles of Mars: Revisited. *J. Geophys. Res.*, **109**, doi:10.1029/2003JE002229; see also correction doi:10.1029/2004JE002278.

Arvidson, R., Guiness, E., and Lee, S. (1979). Differential aeolian redistribution rates on Mars. *Nature*, **278**, 533–535.

Baker, V. R., Carr, M. H., Gulick, V. C., Willemann, C. R., and Marley, M. S. (1992). Channels and valley networks. In *Mars*, ed. H. H. Kieffer, B. M. Jakosky, C. W. Snyder and M. S. Matthews. Tucson, AZ: University of Arizona Press, pp. 493–522.

Banerdt, W. B. (1986). Support of long-wavelength loads on Venus and implications for internal structure. *J. Geophys. Res.*, **91**, 403–419.

Banerdt, W. B. and Golombek, M. P. (2000). Tectonics of the Tharsis region of Mars: Insights from MGS topography and gravity (abs.). *Lunar Planet. Sci. Conf. XXXI*, 2038. Houston, TX: Lunar and Planetary Institute (CD-ROM).

Banerdt, W. B., Phillips, R. J., Sleep, N. H., and Saunders, R. S. (1982). Thick shell tectonics on one-plate planets: Applications to Mars. *J. Geophys. Res.*, **87**, 9723–9733.

Banerdt, W. B., Golombek, M. P., and Tanaka, K. L. (1992). Stress and tectonics on Mars. In *Mars*, ed. H. H. Kieffer, B. M. Jakosky, C. W. Snyder and M. S. Matthews. Tucson, AZ: University of Arizona Press, pp. 249–297.

Belleguic, V., Lognonné, P., and Wieczorek, M. (2005). Constraints on the Martian lithosphere from gravity and topography data. *J. Geophys. Res.*, **110**, doi:10.1029/2005JE002437.

Bibring, J.-P., Langevin, Y., Mustard, J., Poulet, F., Arvidson, R., Gendrin, A., Gondet, B., Mangold, N., Pinet, P., Forget, F., and the-Omega-team (2006). Global mineralogical and aqueous Mars history derived from OMEGA/Mars Express data. *Science*, **312**, 400–404.

Biswas, S. and Ravat, D. (2005). Why meaningful paleopoles can't be determined without special assumptions from Mars Global Surveyor data? (abs.) *Lunar Planet. Sci. Conf. XXXVI*, 2192. Houston, TX: Lunar and Planetary Institute (CD-ROM).

Borgia, A., Burr, J., Montero, W., Morales, L. D., and Alvarado, G. E. (1990). Fault propagation folds induced by gravitational failure and slumping of the volcanic edifices. *J. Geophys. Res.*, **95**, 14 357–14 382.

Borraccini, F., Lanci, L., Wezel, F. C., and Baioni, D. (2005). Crustal extension in the Ceraunius Fossae, Northern Tharsis Region, Mars. *J. Geophys. Res.*, **110**, doi:10.1029/2004JE002373.

Borraccini, F., Lanci, L., and Wezel, F. C. (2006). Does a detachment level exist beneath the Ceraunius Fossae? Insights from graben mapping and lost-area balancing analysis. *Planet. Space Sci.*, **54**, 701–709.

Borraccini, F., Di Achille, G., Ori, G. G., and Wezel, F. C. (2007). Tectonic evolution of the eastern margin of the Thaumasia Plateau (Mars) as inferred from detailed structural mapping and analysis. *J. Geophys. Res.*, **112**, doi:10.1029/2006JE002866.

Brace, W. F. and Kohlstedt, D. L. (1980). Limits on lithospheric stress imposed by laboratory experiments. *J. Geophys. Res.*, **85**, 6248–6252.

Breuer, D., Yuen, D. A., Spohn, T., and Zhang, S. (1998). Three dimensional models of Martian mantle convection with phase transitions. *Geophys. Res. Lett.*, **25**, 229–232.

Burr, D. M., Grier, J. A., Keszthelyi, L. P., and McEwen, A. S. (2002a). Repeated aqueous flooding from the Cerberus Fossae: Evidence for very recently extant, deep groundwater on Mars. *Icarus*, **159**, 53–73.

Burr, D. M., McEwen, A. S., and Sakimoto, S. E. H. (2002b). Recent aqueous floods from the Cerberus Fossae. *Geophys. Res. Lett.*, **29**, doi:10.1029/2001GL013345.

Cailleau, B., Walter, T. R., Janle, P., and Hauber, E. (2003). Modeling volcanic deformation in a regional stress field: Implications for the formation of graben structures on Alba Patera, Mars. *J. Geophys. Res.*, **108**, doi:10.1029/2003JE002135.

Cailleau, B., Walter, T. R., Janle, P., and Hauber, E. (2005). Unveiling the origin of radial grabens on Alba Patera volcano by finite element modeling. *Icarus*, **176**, 44–56.

Carr, M. H. (1996). *Water on Mars*. New York: Oxford University Press.

Chicarro, A., Schultz, P. H., and Masson, P. (1985). Global and regional ridge patterns on Mars. *Icarus*, **63**, 153–174.

Comer, R. P., Solomon, S. C., and Head, J. W. (1985). Mars: Thickness of the lithosphere from the tectonic response to volcanic loads. *Rev. Geophys.*, **23**, 61–92.

Connerney, J. E. P., Acuña, M. H., Wasilewski, P., Ness, N. F., Rème, H., Mazelle, C., Vignes, D., Lin, R. P., Mitchell, D., and Cloutier, P. (1999). Magnetic lineations in the ancient crust of Mars. *Science*, **284**, 794–798.

Courtillot, V. E., Allegre, C. J., and Mattauer, M. (1975). On the existence of lateral relative motions on Mars. *Earth Planet. Sci. Lett.*, **25**, 279–285.

Craddock, R. A. and Howard, A. D. (2002). The case for rainfall on a warm, wet early Mars. *J. Geophys. Res.*, **107**, doi:5110.1029/2001JE001505.

Crumpler, L. S. and Aubele, J. C. (1978). Structural evolution of Arsia Mons, Pavonis Mons, and Ascreus Mons: Tharsis region of Mars. *Icarus*, **34**, 496–541.

Crumpler, L. S., Head, J. W., and Aubele, J. C. (1996). Calderas on Mars: Characteristics, structure, and associated flank deformation. In *Volcano Instability on the Earth and Other Planets. Geol. Soc. Am. Spec. Publ. No. 110*, ed. W. J. Mcguire, A. P. Jones and J. Neuberg, pp. 307–348.

Davis, P. A. and Golombek, M. P. (1990). Discontinuities in the shallow Martian crust at Lunae, Syria, and Sinai Plana. *J. Geophys. Res.*, **95**, 14 231–14 248.

Davis, P. A., Tanaka, K. L., and Golombek, M. P. (1995). Topography of closed depressions, scarps, and grabens in the north Tharsis region of Mars: Implications for shallow crustal discontinuities and graben formation. *Icarus*, **114**, 403–422.

Dimitrova, L. L., Holt, W. E., Haines, A. J., and Schultz, R. A. (2006). Toward understanding the history and mechanisms of Martian faulting: The contribution of gravitational potential energy. *Geophys. Res. Lett.*, **33**, L08202, doi:10.1029/2005GL025307.

Dohm, J. M. and Tanaka, K. L. (1999). Geology of the Thaumasia region, Mars: Plateau development, valley origins, and magmatic evolution. *Planet. Space Sci.*, **47**, 411–431.

Dohm, J. M., Tanaka, K. L., and Hare, T. M. (2001). Geologic, paleotectonic, and paleoerosional maps of the Thaumasia Region, Mars. U.S. Geol. Surv. Geol. Invest. Ser., Map I-2650.

Ernst, R. E., Grosfils, E. B., and Mege, D. (2001). Giant dike swarms: Earth, Venus and Mars. *Annu. Rev. Earth Planet. Sci.*, **29**, 489–534.

Erslev, E. A. (1991). Trishear fault-propagation folding. *Geology*, **19**, 617–620.

Fernández, C. and Anguita, F. (2007). Oblique rifting at Tempe Fossae, Mars. *J. Geophys. Res.*, **112**, doi:10.1029/2007JE002889.

Ferrill, D. A., Wyrick, D. Y., Morris, A. P., Sims, D. W., and Franklin, N. M. (2004). Dilational fault slip and pit chain formation on Mars. *GSA Today*, **14**, 4–12.

Finnerty, A. A., Phillips, R. J., and Banerdt, W. B. (1988). Igneous processes and the closed system evolution of the Tharsis region of Mars. *J. Geophys. Res.*, **93**, 10 225–10 235.

Folkner, W. M., Yoder, C. F., Yuan, D. N., Standish, E. M., and Preston, R. A. (1997). Interior structure and seasonal mass redistribution of Mars from radio tracking of Mars Pathfinder. *Science*, **278**, 1749–1752.

Forsythe, R. D. and Zimbelman, J. R. (1988). Is the Gordii Dorsum escarpment on Mars an exhumed transcurrent fault? *Nature*, **336**, 143–146.

Francis, P. W. and Wadge, G. (1983). The Olympus Mons aureole: Formation by gravitational spreading. *J. Geophys. Res.*, **88**, 8333–8344.

Freed, A. M., Melosh, H. J., and Solomon, S. C. (2001). Tectonics of mascon loading: Resolution of the strike-slip faulting paradox. *J. Geophys. Res.*, **106**, 20 603–20 620.

Frey, H. (1979). Martian canyons and African rifts: Structural comparisons and implications. *Icarus*, **37**, 142–155.
Frey, H. V. (2006a). Impact constraints on the age and origin of the lowlands of Mars. *Geophys. Res. Lett.*, **33**, doi:10.1029/2005GL024484.
Frey, H. V. (2006b). Impact constraints on, and a chronology for, major events in early Mars history. *J. Geophys. Res.*, **111**, doi:10.1029/2005JE002449.
Frey, H. V. and Schultz, R. A. (1988). Large impact basins and the mega-impact origin for the crustal dichotomy on Mars. *Geophys. Res. Lett.*, **15**, 229–232.
Frey, H., Sakimoto, S. E., and Roark, J. (1998). The MOLA topographic signature at the crustal dichotomy boundary zone on Mars. *Geophys. Res. Lett.*, **25**, 4409–4412.
Frey, H. V., Roark, J. H., Shockey, K. M., Frey, E. L., and Sakimoto, S. E. H. (2002). Ancient lowlands on Mars. *Geophys. Res. Lett.*, **29**, doi:10.1029/2001GL013832.
Golombek, M. P. (1985). Fault type predictions from stress distributions on planetary surfaces: Importance of fault initiation depth. *J. Geophys. Res.*, **90**, 3065–3074.
Golombek, M. P. (2003). The surface of Mars: Not just dust and rocks. *Science*, **300**, 2043–2044.
Golombek, M. P. and Bridges, N. T. (2000). Erosion rates on Mars and implications for climate change: Constraints from the Pathfinder landing site. *J. Geophys. Res.*, **105**, 1841–1853.
Golombek, M. P. and McGill, G. E. (1983). Grabens, basin tectonics, and the maximum total expansion of the Moon. *J. Geophys. Res.*, **88**, 3563–3578.
Golombek, M. P., Plescia, J. B., and Franklin, B. J. (1991). Faulting and folding in the formation of planetary wrinkle ridges. *Proc. Lunar Planet. Sci. Conf. 21*, 679–693.
Golombek, M. P., Tanaka, K. L., Chadwick, D. J., Franklin, B. J., and Davis, P. A. (1994). Extension across Tempe Terra and Sirenum provinces on Mars from measurements of fault scarp widths (abs.). *Lunar Planet. Sci. Conf. XXV*. Houston, TX: Lunar and Planetary Institute, 443–444.
Golombek, M. P., Tanaka, K. L., and Franklin, B. J. (1996). Extension across Tempe Terra, Mars, from measurements of fault scarp widths and deformed craters. *J. Geophys. Res.*, **101**, 26 119–26 130.
Golombek, M. P., Franklin, B. J., Tanaka, K. L., and Dohm, J. M. (1997). Extension through time across Thaumasia (abs.). *Lunar Planet. Sci. Conf. XXVIII*. Houston, TX: Lunar and Planetary Institute, 431–432.
Golombek, M. P., Anderson, F. S., and Zuber, M. T. (2001). Martian wrinkle ridge topography: Evidence for subsurface faults from MOLA. *J. Geophys. Res.*, **106**, 23 811–23 821.
Golombek, M. P., Crumpler, L. S., Grant, J. A., Greeley, R., Cabrol, N. A., Parker, T. J., Rice Jr., J. W., Ward, J. G., Arvidson, R. E., Moersch, J. E., Fergason, R. L., Christensen, P. R., Castaño, A., Castaño, R., Haldemann, A. F. C., Li, R., Bell III, J. F., and Squyres, S. W. (2006a). Geology of the Gusev cratered plains from the Spirit rover transverse. *J. Geophys. Res.*, **111**, doi:10.1029/2005JE002503.
Golombek, M. P., Grant, J. A., Crumpler, L. S., Greeley, R., Arvidson, R. E., Bell III, J. F., Weitz, C. M., Sullivan, R., Christensen, P. R., Soderblom, L. A., and Squyres, S. W. (2006b). Erosion rates at the Mars Exploration Rover landing sites and long-term climate change on Mars. *J. Geophys. Res.*, **111**, doi:10.1029/2006JE002754.
Goudy, C. L. and Schultz, R. A. (2005). Dike intrusions beneath grabens south of Arsia Mons, Mars. *Geophys. Res. Lett.*, **32**, doi:10.1029/2004GL021977.
Grant, J. A., Arvidson, R., Bell III, J. F., Cabrol, N. A., Carr, M. H., Christensen, P., Crumpler, L. S., Des Marais, D. J., Ehlmann, B. L., Farmer, J., Golombek, M., Grant, F. D., Greeley, R., Herkenhoff, K., Li, R., McSween, H. Y., Ming, D. W., Moersch, J.,

Rice Jr., J. W., Ruff, S., Richter, L., Squyres, S., Sullivan, R., and Weitz, C. (2004). Surficial deposits at Gusev crater along Spirit rover traverses. *Science*, **305**, 807–810.

Grimm, R. E. and Solomon, S. C. (1986). Tectonic tests of proposed polar wander paths for Mars and the Moon. *Icarus*, **65**, 110–121.

Grott, M. and Breuer, D. (2008). The evolution of the Martian elastic lithosphere and implications for crustal and mantle rheology. *Icarus*, **193**, 503–515.

Grott, M., Hauber, E., Werner, S. C., Kronberg, P., and Neukum, G. (2005). High heat flux on ancient Mars: Evidence from rift flank uplift at Coracis Fossae. *Geophys. Res. Lett.*, **32**, doi:10.1029/2005GL023894.

Grott, M., Hauber, E., Werner, S. C., Kronberg, P., and Neukum, G. (2007a). Mechanical modeling of thrust faults in the Thaumasia region, Mars, and implications for the Noachian heat flux. *Icarus*, **186**, 517–526.

Grott, M., Kronberg, P., Hauber, E., and Cailleau, B. (2007b). Formation of the double rift system in the Thaumasia Highlands, Mars. *J. Geophys. Res.*, **112**, doi:10.1029/2006JE002800.

Guest, A. and Smrekar, S. E. (2005). Relaxation of the Martian dichotomy boundary: Faulting in the Ismenius Region and constraints on the early evolution of Mars. *J. Geophys. Res.*, **110**, E12S25, doi:10.1029/2005JE002504.

Hall, J. L., Solomon, S. C., and Head, J. W. (1986). Elysium Region, and Mars: Tests of lithospheric loading models for the formation of tectonic features. *J. Geophys. Res.*, **91**, 11 377–11 392.

Hanna, J. C. and Phillips, R. J. (2006). Tectonic pressurization of aquifers in the formation of Mangala and Athabasca Valles, Mars. *J. Geophys. Res.*, **111**, doi:10.1029/2005JE002546.

Harder, H. (1998). Phase transitions and the three-dimensional planform of thermal convection in the Martian mantle. *J. Geophys. Res.*, **103**, 16 775–16 797.

Harder, H. (2000). Mantle convection and the dynamic geoid of Mars. *Geophys. Res. Lett.*, **27**, 301–304.

Harder, H. and Christensen, U. (1996). A one-plume model of Martian mantle convection. *Nature*, **380**, 507–509.

Harrington, B. W., Phillips, R. J., and Golombek, M. P. (1999). Extension across Tempe Terra, Mars from MOLA topographic measurements (abs.). *The Fifth International Conference on Mars*, 6130. Houston, TX: Lunar and Planetary Institute (CD-ROM).

Harris, S. A. (1977). The aureole of Olympus Mons. *J. Geophys. Res.*, **82**, 3099–3107.

Hartmann, W. K. (1973). Martian surface and crust: Review and synthesis. *Icarus*, **19**, 550–575.

Hartmann, W. K. (2005). Martian cratering 8: Isochron refinement and the chronology of Mars. *Icarus*, **174**, 294–320, doi:10.1016/j.icarus.2004.11.023.

Hartmann, W. K. and Neukum, G. (2001). Cratering chronology and evolution of Mars. *Space Sci. Rev.*, **96**, 165–194.

Hartmann, W. K., Malin, M. C., McEwen, A. S., Carr, M. H., Soderblom, L., Thomas, P., Danielson, E., James, P., and Veverka, J. (1999). Evidence for recent volcanism on Mars from crater counts. *Nature*, **397**, 586–589.

Hauber, E. and Kronberg, P. (2001). Tempe Fossae, Mars: A Planetary analog to a terrestrial continental rift? *J. Geophys. Res.*, **106**, 20 587–20 602.

Hauber, E. and Kronberg, P. (2005). The large Thaumasia graben on Mars: Is it a rift? *J. Geophys. Res.*, **110**, doi:10.1029/2005JE002407.

Hauck II, S. A. and Phillips, R. J. (2002). Thermal and crustal evolution of Mars. *J. Geophys. Res.*, **107**, doi:10.1029/2001JE001801.

Head, J. W., Kreslavsky, M. A., and Pratt, S. (2002). Northern lowlands of Mars: Evidence for widespread volcanic flooding and tectonic deformation in the Hesperian period. *J. Geophys. Res.*, **107**, doi:10.1029/2000JE001445.

Hiesinger, H. and Head, J. W. (2004). The Syrtis Major volcanic province, Mars: Synthesis from Mars Global Surveyor data. *J. Geophys. Res.*, **109**, doi:10.1029/2003JE002143.

Hood, L. L., Young, C. N., Richmond, N. C., and Harrison, K. P. (2005). Modeling of major Martian magnetic anomalies: Further evidence for polar reorientations during the Noachian. *Icarus*, **177**, 144–173.

Howard, A. D., Moore, J. M., and Irwin III, R. P. (2005). An intense terminal epoch of widespread fluvial activity on early Mars: 1. Valley network incision and associated deposits. *J. Geophys. Res.*, **110**, doi:10.1029/2005JE002459.

Irwin III, R. P., Howard, A. D., Craddock, R. A., and Moore, J. M. (2005). An intense terminal epoch of widespread fluvial activity on early Mars: 2. Increased runoff and paleolake development. *J. Geophys. Res.*, **110**, doi:10.1029/2005JE002460.

Ivanov, M. A. and Head, J. W. (2006). Alba Patera, Mars: Topography, structure, and evolution of a unique late Hesperian – early Amazonian shield volcano. *J. Geophys. Res.*, **111**, doi:10.1029/2005JE002469.

Jaeger, W. L., Keszthelyi, L. P., Mcewen, A. S., Dundas, C. M., and Russell, P. S. (2007). Athabasca Valles, Mars: A lava-draped channel system. *Science*, **317**, 1709–1711.

Jellinek, A. M., Johnson, C. L., and Schubert, G. (2008). Constraints on the elastic thickness, heatflow and melt production at early Tharsis from topography and magnetic field observations. *J. Geophys. Res.*, **113**, E09004, doi:10.1029/2007JE003005.

Johnson, C. L. and Phillips, R. J. (2005). Evolution of the Tharsis region of Mars: Insights from magnetic field observations. *Earth Planet. Sci. Lett.*, **230**, 241–254.

Kaula, W. M. (1975). The seven ages of a planet. *Icarus*, **26**, 1–15.

Kiefer, W. S., Bills, B. G., and Nerem, R. S. (1996). An inversion of gravity and topography for mantle and crustal structure on Mars. *J. Geophys. Res.*, **101**, 9239–9252.

Kronberg, P., Hauber, E., Grott, M., Werner, S. C., Schäfer, T., Gwinner, K., Giese, B., Masson, P., and Neukum, G. (2007). Acheron Fossae, Mars: Tectonic rifting, volcanism, and implications for lithospheric thickness. *J. Geophys. Res.*, **112**, doi:10.1029/2006JE002780.

Langlais, B., Purucker, M. E., and Mandea, M. (2004). Crustal magnetic field of Mars. *J. Geophys. Res.*, **109**, doi:10.1029/2003JE002048.

Lee, D.-C. and Halliday, A. N. (1997). Core formation on Mars and differentiated asteroids. *Nature*, **388**, 854–857.

Lemoine, F. G., Smith, D. E., Rowlands, D. D., Zuber, M. T., Neumann, G. A., Chinn, D. S., and Pavlis, D. E. (2001). An improved solution of the gravity field of Mars (GMM-2B) from Mars Global Surveyor. *J. Geophys. Res.*, **106**, 23 359–23 376.

Lenardic, A., Nimmo, F., and Moresi, L. (2004). Growth of the hemispheric dichotomy and the cessation of plate tectonics on Mars. *J. Geophys. Res.*, **109**, doi:10.1029/2003JE002172.

Leonard, G. J. and Tanaka, K. L. (2001). Geologic map of the Hellas region of Mars. U.S. Geol. Surv. Geol. Invest. Ser., Map I-2694.

Lopes, R. M., Guest, J. E., and Wilson, C. J. N. (1980). Origin of the Olympus Mons aureole and perimeter scarp. *Moon and Planets*, **22**, 221–234.

Lopes, R. M., Guest, J. E., Hiller, K., and Neukum, G. (1982). Further evidence for a mass movement origin for the Olympus Mons aureole. *J. Geophys. Res.*, **87**, 9917–9928.

Lowry, A. R. and Zhong, S. (2003). Surface versus internal loading of the Tharsis rise. *J. Geophys. Res.*, **108**, doi:10.1029/2003JE002111.

Lucchitta, B. K., McEwen, A. S., Clow, C. D., Geissler, R. B., Singer, R. B., Schultz, R. A., and Squyres, S. W. (1992). The canyon system on Mars. In *Mars*, ed. H. H. Kieffer, B. M. Jakosky, C. W. Snyder and M. S. Matthews. Tucson, AZ: University of Arizona Press.

Mangold, N., Allemand, P., and Thomas, P. G. (1998). Wrinkle ridges of Mars: Structural analysis and evidence for shallow deformation controlled by ice-rich decollements. *Planet. Space Sci.*, **46**, 345–356.

Mangold, N., Allemand, P., Thomas, P. G., and Vidal, G. (2000). Chronology of compressional deformation on Mars: Evidence for a single and global origin. *Planet. Space Sci.*, **48**, 1201–1211.

Marinova, M. M., Aharonson, O., and Asphaug, E. (2008). Mega-impact formation of the Mars hemispheric dichotomy. *Nature*, **453**, 1216–1219.

Masson, P. (1977). Structure pattern analysis of the Noctis Labyrinthus-Valles Marineris regions of Mars. *Icarus*, **30**, 49–62.

Matsuyama, I., Mitrovica, J. X., Manga, M., Perron, J. T., and Richards, M. A. (2006). Rotational stability of dynamic planets with elastic lithospheres. *J. Geophys. Res.*, **111**, doi:10.1029/2005JE002447.

Maxwell, T. A. (1982). Orientation and origin of ridges in the Lunae Palus-Coprates region of Mars. *Proc. Lunar Planet. Sci. Conf. 13 J. Geophys. Res.*, **87**, A97-A108.

Maxwell, T. A. and McGill, G. E. (1988). Ages of fracturing and resurfacing in the Amenthes region, Mars. *Proc. Lunar Planet. Sci. Conf. 18*, 701–711.

McEwen, A. S., Malin, M. C., Carr, M. H., and Hartmann, W. K. (1999). Voluminous volcanism on early Mars revealed in Valles Marineris. *Nature*, **397**, 584–586.

McGill, G. E. and Dimitriou, A. M. (1990). Origin of the Martian global dichotomy by crustal thinning in the Late Noachian or Early Hesperian. *J. Geophys. Res.*, **95**, 2573–2759.

McGill, G. E. and Squyres, S. W. (1991). Origin of the Martian crustal dichotomy: Evaluating hypotheses. *Icarus*, **93**, 386–393.

McGovern, P. J. and Solomon, S. C. (1993). State of stress, faulting and eruption characteristics of large volcanoes on Mars. *J. Geophys. Res.*, **98**, 23 553–23 579.

McGovern, P. J., Solomon, S. C., Smith, D. E., Zuber, M. T., Simons, M., Wieczorek, M. A., Phillips, R. J., Neumann, G. A., Aharonson, O., and Head, J. W. (2002). Localized gravity/topography admittance and correlation spectra on Mars: Implications for regional and global evolution. *J. Geophys. Res.*, **107**, doi:10.1029/2002JE001854.

McGovern, P. J., Smith, J. R., Morgan, J. K., and Bulmer, M. H. (2004a). Olympus Mons aureole deposits: New evidence for a flank failure origin. *J. Geophys. Res.*, **109**, doi:10.1029/2004JE002258.

McGovern, P. J., Solomon, S. C., Smith, D. E., Zuber, M. T., Simons, M., Wieczorek, M. A., Phillips, R. J., Neumann, G. A., Aharonson, O., and Head, J. W. (2004b). Correction to "Localized gravity/topography admittance and correlation spectra on Mars: Implications for regional and global evolution." *J. Geophys. Res.*, **109**, doi:10.1029/2004JE002286.

McKenzie, D. and Nimmo, F. (1999). The generation of Martian floods by the melting of ground ice above dykes. *Nature*, **397**, 231–233.

McKenzie, D., Barnett, D. N., and Yuan, D.-N. (2002). The relationship between Martian gravity and topography. *Earth Planet. Sci. Lett.*, **195**, 1–16.

McNutt, M. K. (1984). Lithospheric flexure and thermal anomalies. *J. Geophys. Res.*, **89**, 11 180–11 194.

Mege, D. and Masson, P. (1996a). A plume tectonics model for the Tharsis Province, Mars. *Planet. Space Sci.*, **44**, 1499–1546.

Mege, D. and Masson, P. (1996b). Amounts of crustal stretching in Valles Marineris, Mars. *Planet. Space Sci.*, **44**, 749–781.

Mege, D., Cook, A. C., Garel, E., Lagabrielle, Y., and Cormier, M. H. (2003). Volcanic rifting at Martian grabens. *J. Geophys. Res.*, **108**, doi:10.1029/2002JE001852.

Melosh, H. J. (1980). Tectonic patterns on a reoriented planet: Mars. *Icarus*, **44**, 745–751.

Milbury, C. A. E., Smrekar, S. E., Raymond, C. A., and Schubert, G. (2007). Lithospheric structure in the eastern region of Mars' dichotomy boundary. *Planet. Space Sci.*, **55**, 280–288.

Mitchell, D. L., Lillis, R. J., Lin, R. P., Connerney, J. E. P., and Acuña, M. H. (2007). A global map of Mars' crustal magnetic field based on electron reflectometry. *J. Geophys. Res.*, **112**, doi:10.1029/2005JE002564.

Montesi, L. G. J. and Zuber, M. T. (2003). Clues to the lithospheric structure of Mars from wrinkle ridge sets and localization instability. *J. Geophys. Res.*, **108**, doi:10.1029/2002JE001974.

Moore, H. J. (2001). Geologic map of the Tempe-Mareotis region of Mars. U.S. Geol. Surv. Geol. Invest. Ser., Map I-2727.

Morris, E. C. and Tanaka, K. L. (1994). Geologic maps of the Olympus Mons region of Mars. U.S. Geol. Surv. Misc. Invest. Ser., Map I-2327-B.

Mouginis-Mark, P. J. (1981). Late-stage summit activity of Martian shield volcanoes. *Proc. Lunar Planet. Sci. Conf. 12*, 1431–1447.

Mouginis-Mark, P. J., Wilson, L., Head, J. W., Brown, S. H., Hall, J. L., and Sullivan, K. (1984). Elysium Planitia, Mars: Regional geology, volcanology, and evidence for volcano-ground interactions. *Earth, Moon and Planets*, **30**, 149–173.

Mueller, K. and Golombek, M. P. (2004). Compressional structures on Mars. *Annu. Rev. Earth Planet. Sci.*, **32**, 435–464, doi:10.1146/annurev.earth.32.101802.120553.

Mustard, J. F., Murchie, S. L., Pelkey, S. M., Ehlmann, B. L., Milliken, R. E., Grant, J. A., Bibring, J.-P., Poulet, F., Bishop, J., Noe Dobrea, E., Roach, L., Seelos, F., Arvidson, R. E., Wiseman, S., Green, R., Hash, C., Humm, D., Malaret, E., McGovern, J. A., Seelos, K., Clancy, T., Clark, R., Des Marais, D., Izenberg, N., Knudson, A., Langevin, Y., Martin, T., McGuire, P., Morris, R., Robinson, M., Roush, T., Smith, M., Swayze, G., Taylor, H., Titus, T., and Wolff, M. (2008). Hydrated silicate minerals on Mars observed by the Mars Reconnaissance Orbiter CRISM instrument. *Nature*, **454**, 305–309.

Neumann, G. A., Zuber, M. T., Wieczorek, M. A., McGovern, P. J., Lemoine, F. G., and Smith, D. E. (2004). Crustal structure of Mars from gravity and topography. *J. Geophys. Res.*, **109**, E08002, doi:10.1029/2004JE002262.

Neumann, G. A., Lemoine, F. G., Smith, D. E., and Zuber, M. T. (2008). Marscrust3-a crustal thickness inversion from recent MRO gravity solutions (abs.). *Lunar Planet. Sci. Conf. XXXIX*, 2167. Houston, TX: Lunar and Planetary Institute (CD-ROM).

Nimmo, F. (2002). Admittance estimates of mean crustal thickness and density at the Martian hemispheric dichotomy. *J. Geophys. Res.*, **107**, doi:10.1029/2000JE001488.

Nimmo, F. (2005). Tectonic consequences of Martian dichotomy modification by lower-crustal flow and erosion. *Geology*, **33**, 533–536.

Nimmo, F. and Stevenson, D. J. (2001). Estimates of Martian crustal thickness from viscous relaxation of topography. *J. Geophys. Res.*, **106**, 5085–5098.

Nimmo, F. and Tanaka, K. L. (2005). Early crustal evolution of Mars. *Annu. Rev. Earth Planet. Sci.*, **33**, 133–161.

Nimmo, F., Hart, S. D., Korycansky, D. G., and Agnor, C. B. (2008). Implications of an impact origin for the Martian hemispheric dichotomy. *Nature*, **453**, 1220–1223.

Okubo, C. H. and Schultz, R. A. (2003). Thrust fault vergence directions on Mars: A foundation for investigating global-scale Tharsis-driven tectonics. *Geophys. Res. Lett.*, **30**, 2154, doi:10.1029/2003GL018664.

Okubo, C. H. and Schultz, R. A. (2004). Mechanical stratigraphy in the western equatorial region of Mars based on thrust fault-related fold topography and implications for near-surface volatile reservoirs. *GSA Bulletin*, **116**, 594–605, doi:10.1130/B25361.1.

Okubo, C. H. and Schultz, R. A. (2006). Variability in Early Amazonian Tharsis stress state based on wrinkle ridges and strike-slip faulting. *J. Struct. Geol.*, **28**, 2169–2181.

Perron, J. T., Mitrovica, J. X., Manga, M., Matsuyama, I., and Richards, M. A. (2007). Evidence for an ancient Martian ocean in the topography of deformed shorelines. *Nature*, **447**, 840–843.

Peulvast, J. P. and Masson, P. L. (1993a). Erosion and tectonics in central Valles Marineris (Mars): A new morpho-structural model. *Earth, Moon and Planets*, **61**, 191–217.

Peulvast, J. P. and Masson, P. L. (1993b). Melas Chasma: Morphology and tectonic patterns in central Valles Marineris (Mars). *Earth, Moon and Planets*, **61**, 219–248.

Phillips, R. J. and Lambeck, K. (1980). Gravity fields of the terrestrial planets: Long-wavelength anomalies and tectonics. *Rev. Geophys. Space Phys.*, **18**, 27–76.

Phillips, R. J., Saunders, R. S., and Conel, J. E. (1973). Mars: Crustal structure inferred from Bouguer gravity anomalies. *J. Geophys. Res.*, **78**, 4815–4820.

Phillips, R. J., Sleep, N. H., and Banerdt, W. B. (1990). Permanent uplift in magmatic systems with application to the Tharsis region of Mars. *J. Geophys. Res.*, **95**, 5089–5100.

Phillips, R. J., Zuber, M. T., Solomon, S. C., Golombek, M. P., Jakosky, B. M., Banerdt, W. B., Smith, D. E., Williams, R. M. E., Hynek, B. M., Aharonson, O., and Hauck I, S. A. (2001). Ancient geodynamics and global-scale hydrology on Mars. *Science*, **291**, 2587–2591.

Phillips, R. J., Johnson, C. L., and Dombard, A. J. (2004). Localized Tharsis loading on Mars: Testing the membrane surface hypothesis (abs.). *Lunar Planet Sci. Conf. XXXV*, 1427. Houston, TX: Lunar and Planetary Institute (CD-ROM).

Phillips, R. J., Zuber, M. T., Smrekar, S. E., Mellon, M. T., Head, J. W., Tanaka, K. L., Putzig, N. E., Milkovich, S. M., Campbell, B. A., Plaut, J. J., Safaeinili, A., Seu, R., Biccari, D., Carter, L. M., Picardi, G., Orosei, R., Mohit, P. S., Heggy, E., Zurek, R. W., Egan, A., Giacomoni, E., Russo, F., Cutigni, M., Pettinelli, E., Holt, J. W., Leuschen, C. J., and Marinangeli, L. (2008). Mars north polar deposits: Stratigraphy, age and geodynamical response. *Science*, **230**, 1182–1185.

Plescia, J. B. (1991a). Graben and extension in northern Tharsis, Mars. *J. Geophys. Res.*, **96**, 18 883–18 895.

Plescia, J. B. (1991b). Wrinkle ridges in Lunae Planum, Mars: Implications for shortening and strain. *Geophys. Res. Lett.*, **18**, 913–916.

Plescia, J. B. (1993). Wrinkle ridges of Arcadia Planitia, Mars. *J. Geophys. Res.*, **98**, 15 049–15 059.

Plescia, J. B. and Golombek, M. P. (1986). Origin of planetary wrinkle ridges based on the study of terrestrial analogs. *Geol. Soc. Am. Bull.*, **97**, 1289–1299.

Plescia, J. B. and Saunders, R. S. (1982). Tectonic history of the Tharsis Region, Mars. *J. Geophys. Res.*, **87**, 9775–9791.

Pruis, M. J. and Tanaka, K. L. (1995). The Martian northern plains did not result from plate tectonics (abs.). *Lunar Planet. Sci. Conf. XXVI*. Houston, TX: Lunar and Planetary Institute, 1147–1148.

Purucker, M., Ravat, D., Frey, H., Voorhies, C., Sabaka, T., and Acuña, M. (2000). An altitude-normalized magnetic map of Mars and its interpretation. *Geophys. Res. Lett.*, **27**, 2449–2452.

Quesnel, Y., Langlais, B., and Sotin, C. (2007). Local inversion of magnetic anomalies: Implication for Mars' crustal evolution. *Planet. Space Sci.*, **55**, 258–269.

Raitala, J. (1988). Superposed ridges of the Hesperia Planum Area on Mars. *Earth, Moon and Planets*, **40**, 71–99.

Raitala, J. and Kauhanen, K. (1989). Tectonics of Syrtis Major Planum on Mars. *Earth, Moon and Planets*, **46**, 243–260.

Redmond, H. L. and King, S. D. (2004). A numerical study of a mantle plume beneath the Tharsis Rise: Reconciling dynamic uplift and lithospheric support models. *J. Geophys. Res.*, **109**, doi:10.1029/2003JE002228.

Reese, C. C., Solomatov, V. S., Baumgardner, J. R., Stegman, D. R., and Vezolainen, A. V. (2004). Magmatic evolution of impact-induced Martian mantle plumes and the origin of Tharsis. *J. Geophys. Res.*, **109**, doi:10.1029/2003JE002222.

Roberts, J. H. and Zhong, S. (2004). Plume-induced topography and geoid anomalies and their implications for the Tharsis rise on Mars. *J. Geophys. Res.*, **109**, doi:10.1029/2003JE002226.

Roberts, J. H. and Zhong, S. (2006). Degree-1 convection in the Martian mantle and the origin of the hemispheric dichotomy. *J. Geophys. Res.*, **111**, doi:10.1029/2005JE002668.

Roberts, J. H. and Zhong, S. (2007). The cause for the north–south orientation of the crustal dichotomy and the equatorial location of Tharsis on Mars. *Icarus*, **190**, 24–31.

Sandwell, D. T., Johnson, C. L., Bilotti, F., and Suppe, J. (1997). Driving forces for limited tectonics on Venus. *Icarus*, **129**, 232–244.

Schubert, G., Solomon, S. C., Turcotte, D. L., Drake, M. J., and Sleep, N. H. (1992). Origin and thermal evolution of Mars. In *Mars*, ed. H. H. Kieffer, B. M. Jakosky, C. W. Snyder and M. S. Matthews. Tucson, AZ: University of Arizona Press, pp. 147–183.

Schultz, R. A. (1989). Strike-slip faulting of ridged plains near Valles Marineris, Mars. *Nature*, **341**, 424–428.

Schultz, R. A. (1991). Structural development of Coprates Chasma and western Ophir Planum. *J. Geophys. Res.*, **96**, 22 777–22 792.

Schultz, R. A. (1995). Gradients in extension and strain at Valles Marineris. *Planet. Space Sci.*, **43**, 1561–1566.

Schultz, R. A. (1998). Multiple-process origin of Valles Marineris basins and troughs. *Planet. Space Sci.*, **46**, 827–834.

Schultz, R. A. (2000). Localization of bedding-plane slip and backthrust faults above blind thrust faults: Keys to wrinkle ridge structure. *J. Geophys. Res.*, **105**, 12 035–12 052.

Schultz, R. A. and Lin, J. (2001). Three-dimensional normal faulting models of Valles Marineris, Mars, and geodynamic implications. *J. Geophys. Res.*, **106**, 16 549–16 566.

Schultz, R. A. and Tanaka, K. L. (1994). Lithospheric-scale buckling and thrust structures on Mars: The Coprates rise and south Tharsis ridge belt. *J. Geophys. Res.*, **99**, 8371–8385.

Schultz, R. A., Okubo, C. H., Goudy, C. L., and Wilkins, S. J. (2004). Igneous dikes on Mars revealed by Mars Orbiter Laser Altimeter topography. *Geology*, **32**, 889–892.

Schultz, R. A., Moore, J. M., Grosfils, E. B., Tanaka, K. L., and Mège, D. (2007). The Canyonlands model for planetary grabens: Revised physical basis and implications. In *The Geology of Mars: Evidence from Earth-Based Analogs*, ed. M. Chapman. Cambridge: Cambridge University Press, pp. 371–399.

Scott, D. H. and Dohm, J. M. (1990a). Chronology and global distribution of fault and ridge systems on Mars. *Proc. Lunar Planet. Sci. Conf. 20*. Houston, TX: Lunar and Planetary Institute, 487–501.

Scott, D. H. and Dohm, J. M. (1990b). Faults and ridges: Historical development in Tempe Terra and Ulysses Patera regions of Mars. *Proc. Lunar Planet. Sci. Conf. 20*. Houston, TX: Lunar and Planetary Institute, 503–513.

Scott, D. H. and Tanaka, K. L. (1986). Geologic map of the western equatorial region of Mars. U.S. Geol. Surv. Map I-1802-A.

Scott, E. D. and Wilson, L. (2002). Plinian eruptions and passive collapse events as mechanisms of formation for Martian pit chain craters. *J. Geophys. Res.*, **107**, doi:10.1029/2000JE001432.

Scott, E. D., Wilson, L., and Head III, J. W. (2002). Emplacement of giant radial dikes in the northern Tharsis region of Mars. *J. Geophys. Res.*, doi:10.1029/2000JE001431.

Searls, M. L. and Phillips, R. J. (2007). Tectonics of Utopia Basin, Mars: Results from finite element loading models (abs.). *Lunar Planet. Sci. Conf. XXXVIII*, 1965. Houston, TX: Lunar and Planetary Institute (CD-ROM).

Searls, M. L., Banerdt, W. B., and Phillips, R. J. (2006). Utopia and Hellas basins, Mars: Twins separated at birth. *J. Geophys. Res.*, **111**, doi:10.1029/2005JE002666.

Segura, T. L., Toon, O. B., Colaprete, A., and Zahnle, K. (2002). Environmental effects of large impacts on Mars. *Science*, **298**, 1977–1980.

Sleep, N. H. (1994). Martian plate tectonics. *J. Geophys. Res.*, **99**, 5639–5655.

Sleep, N. H. and Tanaka, K. L. (1995). Point-counterpoint: Did Mars have plate tectonics? *Mercury*, **24**, 10–11.

Smith, D. E., Sjogren, W. L., Tyler, G. L., Balmino, G., Lemoine, F. G., and Konopliv, A. S. (1999a). The gravity field of Mars: Results from Mars Global Surveyor. *Science*, **286**, 94–97.

Smith, D. E., Zuber, M. T., Solomon, S. C., Phillips, R. J., Head, J. W., Garvin, J. B., Banerdt, W. B., Muhleman, D. O., Pettingill, G. H., Neumann, G. A., Lemoine, F. G., Abshire, J. B., Aharonson, O., Brown, C. D., Hauck II, S. A., Ivanov, A. B., McGovern, P. J., Zwally, H. J., and Duxbury, T. C. (1999b). The global topography of Mars and implications for surface evolution. *Science*, **284**, 1495–1503.

Smith, D. E., Zuber, M. T., Frey, H. V., Garvin, J. B., Head, J. W., Muhleman, D. O., Pettingill, G. H., Phillips, R. J., Solomon, S. C., Zwally, H. J., Banerdt, W. B., Duxbury, T. C., Golombek, M. P., Lemoine, F. G., Neumann, G. A., Rowlands, D. D., Aharonson, O., Ford, P. G., Ivanov, A. B., Johnson, C. L., McGovern, P. J., Abshire, J. B., Afzal, R. S., and Sun, X. (2001). Mars Orbiter Laser Altimeter (MOLA): Experiment summary after the first year of global mapping of Mars. *J. Geophys. Res.*, **106**, 23 689–23 722.

Smrekar, S. E., McGill, G. E., Raymond, C. A., and Dimitriou, A. M. (2004). Geologic evolution of the Martian dichotomy in the Ismenius area of Mars and implications for plains magnetization. *J. Geophys. Res.*, **109**, doi:10.1029/2004JE002260.

Solomon, S. C. and Head, J. W. (1980). Lunar mascon basins: Lava filling, tectonics, and evolution of the lithosphere. *Rev. Geophys.*, **18**, 107–141.

Solomon, S. C. and Head, J. W. (1982). Evolution of the Tharsis province of Mars: The importance of heterogeneous lithospheric thickness and volcanic construction. *J. Geophys. Res.*, **87**, 9755–9774.

Solomon, S. C. and Head, J. W. (1990). Heterogeneities in the thickness of the elastic lithosphere of Mars: Constraints on heat flow and internal dynamics. *J. Geophys. Res.*, **95**, 11 073–11 083.

Solomon, S. C., Aharonson, O., Aurnou, J. M., Banerdt, W. B., Carr, M. H., Dombard, A. J., Frey, H. V., Golombek, M. P., Hauck II, S. A., Head III, J. W., Jakosky, B. M., Johnson, C. L., McGovern, P. J., Neumann, G. A., Phillips, R. J., Smith, D. E., and Zuber, M. T. (2005). New perspectives on ancient Mars. *Science*, **307**, 1214–1220.

Solomon, S. C., McNutt Jr., R. L., Watters, T. R., Lawrence, D. J., Feldman, W. C., Head, J. W., Krimigis, S. M., Murchie, S. L., Phillips, R. J., Slavin, J. A., and Zuber, M. T. (2008). Return to Mercury: A global perspective on MESSENGER's first Mercury flyby. *Science*, **321**, 59–62.

Spohn, T., Acuña, M. H., Breuer, D., Golombek, M. P., Greeley, R., Halliday, A., Hauber, E., Jaumann, R., and Sohl, F. (2001). Geophysical constraints on the evolution of Mars. *Space Sci. Rev.*, **96**, 231–262.

Sprenke, K. F., Baker, L. L., and Williams, A. F. (2005). Polar wander on Mars: Evidence in the geoid. *Icarus*, **174**, 486–489.

Squyres, S. W., Grotzinger, J. P., Arvidson, R. E., Bell III, J. F., Calvin, W., Christensen, P. R., Clark, B. C., Crisp, J. A., Farrand, W. H., Herkenhoff, K. E., and Johnson, J. R. (2004). In-situ evidence for an ancient aqueous environment at Meridiani Planum, Mars. *Science*, **306**, 1709–1714.

Squyres, S. W., Arvidson, R. E., Blaney, D. L., Clark, B. C., Crumpler, L., Farrand, W. H., Gorevan, S., Herkenhoff, K. E., Hurowitz, J., Kusack, A., McSween, H. Y., Ming, D. W., Morris, R. V., Ruff, S. W., Wang, A., and Yen, A. (2006). Rocks of the Columbia Hills. *J. Geophys. Res.*, **111**, doi:10.1029/2005JE002562.

Suppe, J. and Medwedeff, D. A. (1990). Geometry and kinematics of fault-propagation folding. *Eclogae Geo. Helv.*, **83**, 409–454.

Tanaka, K. L. (1985). Ice-lubricated gravity spreading of the Olympus Mons aureole deposits. *Icarus*, **62**, 191–206.

Tanaka, K. L. (1986). The stratigraphy of Mars. *Proc. Lunar and Planet. Sci. Conf. 17. J. Geophys. Res.*, **91**, E139–E158.

Tanaka, K. L. (1990). Tectonic history of the Alba Patera-Ceraunius Fossae region of Mars. *Proc. Lunar Planet. Sci. Conf. 20*. Houston, TX: Lunar and Planetary Institute, 515–523.

Tanaka, K. L. and Chapman, M. G. (1990). The relation of catastrophic flooding of Mangala Valles, Mars, to faulting of Menmonia Fossae and Tharsis volcanism. *J. Geophys. Res.*, **95**, 14 315–14 323.

Tanaka, K. L. and Davis, P. A. (1988). Tectonic history of the Syria Planum province of Mars. *J. Geophys. Res.*, **93**, 14 893–14 917.

Tanaka, K. L. and Golombek, M. P. (1989). Martian tension fractures and the formation of grabens and collapse features at Valles Marineris. *Proc. Lunar Planet. Sci. Conf. 19*, 383–396.

Tanaka, K. L. and Hartmann, W. K. (2008). The planetary timescale. *The Concise Geologic Time Scale*, ed. J. G. Ogg, G. M. Ogg and F. M. Gradstein. New York: Cambridge University Press, pp. 13–22.

Tanaka, K. L. and Leonard, G. J. (1995). Geology and landscape evolution of the Hellas region of Mars. *J. Geophys. Res.*, **100**, 5407–5432.

Tanaka, K. L., Golombek, M. P., and Banerdt, W. B. (1991). Reconciliation of stress and structural histories of the Tharsis region of Mars. *J. Geophys. Res.*, **96**, 15 617–15 633.

Tanaka, K. L., Skinner Jr., J. A., Hare, T. M., Joyal, T., and Wenker, A. (2003). Resurfacing history of the northern plains of Mars based on geologic mapping of Mars Global Surveyor data. *J. Geophys. Res.*, **108**, doi:10.1029/2002JE001908.

Tanaka, K. L., Skinner Jr., J. A., and Hare, T. M. (2005). Geologic map of the northern plains of Mars. U.S. Geol. Surv., Map SIM-2888.

Tate, A., Mueller, K., and Golombek, M. P. (2002a). Kinematics and structural inversion of wrinkle ridges on Lunae and Solis Planum: Implications for the early history of Tharsis (abs.). *Lunar Planet. Sci. Conf. XXXIII*, 1828. Houston, TX: Lunar and Planetary Institute (CD-ROM).

Tate, A., Mueller, K. J., and Golombek, M. P. (2002b). Geometry and kinematics of wrinkle ridges on Lunae and Solis Planum: Implications for fault/fold growth history (abs.). *Lunar Planet. Sci. Conf. XXXIII*, 1836. Houston, TX: Lunar and Planetary Institute (CD-ROM).

Thomas, P. G. and Allemand, P. (1993). Quantitative analysis of the extensional tectonics of the Tharsis bulge, Mars: Geodynamic implications. *J. Geophys. Res.*, **98**, 13 097–13 108.

Thomas, P. J., Squyres, S. W., and Carr, M. H. (1990). Flank tectonics of Martian volcanoes. *J. Geophys. Res.*, **95**, 14 345–14 355.

Thomson, B. J. and Head, J. W. (2001). Utopia basin, Mars: Characterization of topography and morphology and assessment of the origin and evolution of basin internal structure. *J. Geophys. Res.*, **106**, 23 209–23 230.

Turcotte, D. L., Willemann, R. J., Haxby, W. F., and Norberry, J. (1981). Role of membrane stress in the support of planetary topography. *J. Geophys. Res.*, **86**, 3951–3959.

van Thienen, P., Rivoldini, A., Van Hoolst, T., and Lognonné, P. (2006). A top-down origin for Martian mantle plumes. *Icarus*, **185**, 197–210.

Vidal, A., Mueller, K., and Golombek, M. P. (2003). Axial surface mapping of wrinkle ridges on Solis Planum, Mars from MOLA topography: Constraints on subsurface blind thrust geometry (abs.). *Lunar Planet. Sci. Conf. XXXIV*, 1125. Houston, TX: Lunar and Planetary Institute (CD-ROM).

Vidal, A., Mueller, K. M., and Golombek, M. P. (2005). Geometry of thrust faults beneath Amenthes Rupes, Mars (abs.). *Lunar Planet. Sci. Conf. XXXVI*, 2333. Houston, TX: Lunar and Planetary Institute (CD-ROM).

Vlasov, V. Z. (1964). *General Theory of Shells and Its Applications in Engineering.* NASA Tech. Trans. T T F-99.

Ward, P. D. and Brownlee, D. (2000). *Rare Earth.* New York: Springer-Verlag.

Ward, P. D. and Brownlee, D. (2002). *The Life and Death of Planet Earth.* New York: H. Holt.

Watters, T. R. (1988). Wrinkle ridge assemblages on the terrestrial planets. *J. Geophys. Res.*, **93**, 10 236–10 254.

Watters, T. R. (1991). Origin of periodically spaced wrinkle ridges on the Tharsis Plateau of Mars. *J. Geophys. Res.*, **96**, 15 599–15 616.

Watters, T. R. (1992). System of tectonic features common to Earth, Mars, and Venus. *Geology*, **20**, 609–612.

Watters, T. R. (1993). Compressional tectonism on Mars. *J. Geophys. Res.*, **98**, 17 049–17 060.

Watters, T. R. (2003a). Lithospheric flexure and the origin of the dichotomy boundary on Mars. *Geology*, **31**, 271–274.

Watters, T. R. (2003b). Thrust faults along the dichotomy boundary in the eastern hemisphere of Mars. *J. Geophys. Res.*, **108**, doi:10.1029/2002JE001934.

Watters, T. R. (2004). Elastic dislocation modeling of wrinkle ridges on Mars. *Icarus*, **171**, 284–294.

Watters, T. R. and Maxwell, T. A. (1986). Orientation, relative age, and extent of the Tharsis plateau ridge system. *J. Geophys. Res.*, **91**, 8113–8125.

Watters, T. R. and Robinson, M. S. (1997). Radar and photoclinometric studies of wrinkle ridges on Mars. *J. Geophys. Res.*, **102**, 10 889–10 903.

Watters, T. R., Schultz, R. A., and Robinson, M. S. (2000). Displacement–length relations of thrust faults associated with lobate scarps on Mercury and Mars: Comparison with terrestrial faults. *Geophys. Res. Lett.*, **27**, 3659–3662.

Weissel, J. K. and Karner, G. D. (1989). Flexural uplift of rift flanks due to mechanical unloading of the lithophere during extension. *J. Geophys. Res.*, **94**, 13 919–13 950.

Wenzel, M. J., Manga, M., and Jellinek, A. M. (2004). Tharsis as a consequence of Mars' dichotomy and layered mantle. *Geophys. Res. Lett.*, **31**, doi:10.1029/2003GL019306.

Wichman, R. W. and Schultz, P. H. (1989). Sequence and mechanisms of deformation around the Hellas and Isidis impact basins on Mars. *J. Geophys. Res.*, **94**, 17 333–17 357.

Wieczorek, M. A. and Zuber, M. T. (2004). Thickness of the Martian crust: Improved constraints from geoid-to-topography ratios. *J. Geophys. Res.*, **109**, doi:01010.01029/02003JE002153.

Wilhems, D. E. and Squyres, S. W. (1984). The Martian hemispheric dichotomy may be due to a giant impact. *Nature*, **309**, 138–140.

Wilkins, S. J. and Schultz, R. A. (2003). Cross faults in extensional settings: Stress triggering, displacements localization, and implications for the origin of blunt troughs at Valles Marineris. *J. Geophys. Res.*, **108**, doi:10.1029/2002JE001968.

Willemann, R. J. (1984). Reorientation of planets with elastic lithospheres. *Icarus*, **60**, 701–709.

Wilson, L. and Head III, J. W. (2002). Tharsis-radial graben systems as the surface manifestation of plume-related dike intrusion complexes: Models and implications. *J. Geophys. Res.*, **107**, doi:10.1029/2001JE001593.

Wise, D. U. (1974). Continental margins, freeboard and the volume of continents and oceans through time. In *The Geology of Continental Margins*, ed. C. A. Burk and C. L. Drake. New York: Springer Verlag, pp. 45–58.

Wise, D. U., Golombek, M. P., and McGill, G. E. (1979). Tharsis province of Mars: Geologic sequence, geometry, and a deformation mechanism. *Icarus*, **38**, 456–472.

Witbeck, N. E., Tanaka, K. L., and Scott, D. H. (1991). Geologic map of the Valles Marineris region, Mars (East Half And West Half): 1:2M. U.S. Geol. Surv. Misc. Invest. Ser., Map I-2010.

Withers, P. and Neumann, G. A. (2001). Enigmatic northern plains of Mars. *Nature*, **410**, 651.

Wyrick, D., Ferrill, D. A., Morris, A. P., Colton, S. L., and Sims, D. W. (2004). Distribution, morphology, and origins of Martian pit crater chains. *J. Geophys. Res.*, **109**, doi:10.1029/2004JE002240.

Yoder, C. F., Konopliv, A. S., Yuan, D. N., Standish, E. M., and Folkner, W. M. (2003). Fluid core size of Mars from detection of the solar tide. *Science*, **300**, 299–303.

Yuan, D. N., Sjogren, W. L., Konopliv, A. S., and Kucinskas, A. B. (2001). Gravity field of Mars: A 75th degree and order model. *J. Geophys. Res.*, **106**, 23 377–23 401.

Zhong, S. (2002). Effects of lithosphere on the long-wavelength gravity anomalies and their implications for the formation of the Tharsis rise on Mars. *J. Geophys. Res.*, **107**, doi:10.1029/2001JE001589.

Zhong, S. and Roberts, J. H. (2003). On the support of the Tharsis rise on Mars. *Earth Planet. Sci. Lett.*, **214**, 1–9.

Zhong, S. and Zuber, M. T. (2001). Degree-1 mantle convection and the crustal dichotomy on Mars. *Earth Planet. Sci. Lett.*, **189**, 75–84.

Zuber, M. T. (1995). Wrinkle ridges, reverse faulting and the depth penetration of lithospheric strain in Lunae Planum, Mars. *Icarus*, **114**, 80–92.

Zuber, M. T. (2001). The crust and mantle of Mars. *Nature*, **412**, 220–227.

Zuber, M. T. and Aist, L. L. (1990). The shallow structure of the Martian lithosphere in the vicinity of the ridged plains. *J. Geophys. Res.*, **95**, 14 215–14 230.

Zuber, M. T. and Mouginis-Mark, P. J. (1992). Caldera subsidence and magma chamber depth of Olympus Mons volcano, Mars. *J. Geophys. Res.*, **97**, 18 295–18 307.

Zuber, M. T. and Smith, D. E. (1997). Mars without Tharsis. *J. Geophys. Res.*, **102**, 28 673–28 685.

Zuber, M. T., Solomon, S. C., Phillips, R. J., Smith, D. E., Tyler, G. L., Aharonson, O., Balmino, G., Banerdt, W. B., Head, J. W., Johnson, C. L., Lemoine, F. G., McGovern, P. J., Neumann, G. A., Rowlands, D. D., and Zhong, S. (2000). Internal structure and early thermal evolution of Mars from Mars Global Surveyor topography and gravity. *Science*, **287**, 1788–1793.

Zuber, M. T., Phillips, R. J., Andrews-Hanna, J. C., Asmar, S. W., Konopliv, A. S., Lemoine, F. G., Plaut, J. J., Smith, D. E., and Smrekar, S. E. (2007). Density of Mars' south polar layered deposits. *Science*, **317**, 1718–1719.

6

Tectonics of small bodies

Peter C. Thomas
Center for Radiophysics and Space Research, Cornell University, Ithaca

and

Louise M. Prockter
Applied Physics Laboratory, Laurel

Summary

Solar system bodies smaller than ~200 km mean radius have little internal heat energy to drive tectonics typical of the terrestrial environment. Short-lived high stresses from impacts or long-term, low stresses are the primary shapers of these bodies. This chapter provides an overview of the basic features and processes that can be regarded as small-body tectonics.

1 Introduction: types of small bodies, their properties, and environments

Small bodies of the solar system are here taken to be those too small for gravitationally driven viscous relaxation to have determined their shapes. This definition restricts consideration to objects less than about 150 km radius (Johnson and McGetchin, 1973; Thomas, 1989). Within this definition are some dozens of satellites of planets, and thousands of asteroids, cometary nuclei, and Centaur and Kuiper-Edgeworth belt objects (Binzel *et al.*, 2003). As of early 2006, spacecraft have visited small satellites, asteroids, and four cometary nuclei (Figure 6.1). Resolved information on these objects is dominated by the NEAR mission that orbited and then landed on 433 Eros, by images of the Martian satellites, Phobos and Deimos, and by images of comet Tempel 1 (A'Hearn *et al.*, 2005). Radar images of near-Earth objects are beginning to show some details of asteroid shapes and surface features (Hudson *et al.*, 2003). Meteorites provide small samples of asteroids, though only in the case of asteroid Vesta (larger than the size range considered here) are there positive connections of meteorite samples to a specific object (Binzel *et al.*, 1993; Keil, 2002).

Figure 6.1. A sampler of small solar system objects. Top row, left to right: Phobos, Eros, Ida. Second row: Deimos and Gaspra. Bottom row: comet Borrelly and asteroid Mathilde.

Interest in these small bodies stems from a desire to fit them in the sequence of formation and modification of all solar system bodies. Large planets undoubtedly formed from accumulations of smaller components, and subsequent continued impacts of small objects have perhaps modified the chemistry and surfaces of planets (Bottke *et al.*, 2002; Stevenson, 2004). On Earth, impacts are associated with features as diverse as ore-body initiation to global biologic extinctions (Melosh, 1989). Study of small bodies is advanced enough that interpreting their history and internal makeup are required to evaluate their past and future effects on planets. For instance, are most asteroids loose accumulations of fragments, or are they monolithic? Even with spacecraft visits, this question has been difficult to answer positively, and many indirect methods, such as orbital dynamics of remnants of catastrophic collisions, are used to infer asteroid structures (Michel *et al.*, 2003).

2 Small bodies: characteristics

2.1 Asteroids

While most asteroids orbit the Sun between Mars and Jupiter, gravitational perturbations of planets place some into more eccentric orbits that cross those of the inner planets, with collisions or even ejection from the solar system limiting their lifetimes (Murray and Dermott, 2003). Most asteroids fall into our categorization of small objects; tens of thousands are between 1 and 150 km in diameter. They encompass a variety of compositions as indicated by meteorite samples and spectroscopy. Compositions include iron-nickel metals, olivine, pyroxene, plagioclase and other silicates, clays and organics, and some organics and possibly ices and other oxides (Gaffey *et al.*, 2002). The asteroid belt is compositionally zoned, reflecting probable different condensation compositions and temperatures as a function of distance from the Sun (Gradie and Tedesco, 1982; Grimm and McSween, 1993). After formation, asteroids suffered orbital perturbations, collisions, and different amounts of internal heating, and in some instances, differentiation (Bell *et al.*, 1989).

2.2 Comets

Comets are distinguished by having significant volatile components. They have been perturbed into orbits that bring them to the inner solar system from reservoirs tens of AU distant (Kuiper belt) or from much farther out (Oort cloud) (Levinson *et al.*, 2001). Their lifetimes in the inner solar system are geologically brief, chiefly due to volatile loss, but also due to impacts into planets and the Sun. Volatile compositions include H_2O, CO_2, CO, and other "icy" compounds. Silicates and organics are part of the non-volatile component. Spacecraft imaging, and behavior observed from the ground, show that the great majority of comet surfaces are composed of dark (albedo of a few percent), non-volatile materials. Only small fractions of comet surfaces are actively sublimating due to solar radiation (A'Hearn *et al.*, 1992; Britt *et al.*, 2004; Brownlee *et al.*, 2004; A'Hearn *et al.*, 2005). Comet nuclei are generally a few km across; a 20 km length is a very large comet.

2.3 Small satellites

Small satellites may be composed of materials formed with the parent planet or objects captured into orbit around the planet. The compositions of most small satellites are poorly known, spectral and color comparisons being the best guides, such as for the Martian satellites (Rivkin *et al.*, 2002). The best data are infrared spectroscopy from the Cassini spacecraft of the Saturnian moon Phoebe (Clark *et al.*, 2005), which show the surface to be very compositionally diverse, including bound water, trapped CO_2, organics, nitriles, cyanide compounds, iron minerals, and probable phyllosilicates. In general, small satellites close to the primary in circular orbits are thought to have formed with the planet; those in distant, elliptical, and inclined orbits are likely to have been captured into orbit and may be asteroidal or cometary bodies (Burns, 1978; Gladman *et al.*, 2001).

3 Stress environments of small bodies

On the Earth and other terrestrial planets, most tectonics are driven by stresses of internal origin, in most cases tied, directly or indirectly, to thermally driven convection. The small size of the objects we are considering limits any thermal convective activity to very early in their history ($<10^8$ y), when exotic heat sources, such as short-lived radionuclides (Grimm and McSween, 1993) or inductive heating (Sonnett *et al.*, 1968), may have been effective in some objects. The presence of iron-nickel meteorites and other mineral compositions requiring melting and differentiation shows that at least some bodies did experience periods of thermal activity with likely accompanying stresses. Our opportunity to detect structures resulting from such activity is limited by the subsequent impact processing of these bodies that has, in many instances, apparently stripped silicate mantles from metallic cores, and disaggregated mantle materials and large-scale structures (Binzel *et al.*, 2002).

Thermal evolution of cometary bodies is even less well known, and is probably very different from that of asteroids. Their initial large distances from the Sun and high volatile content reduce possible internal heating, although parts of the surface that have lost volatiles can become quite warm near perihelion (Soderblom *et al.*, 2004). Additionally, for many of the volatiles (H_2O, CO_2, CO) the low internal pressures mean that transition of phases may involve solid–gas rather than solid–liquid changes. In combination with the minimal buoyancy forces possible, convection and resulting structures are unlikely. As will be noted below, though, comets may be so weak that very small stresses associated with volatilization may be effective modifiers of their structures.

3.1 Impact environment

Collisions among small bodies, asteroids, comets, and small satellites, provide the highest stresses suffered by these objects. Velocities of objects in different solar orbits can vary tremendously, but relative velocities of 5–10 km/s are common in the asteroid belt and inner solar system. The cratering event can be divided into three stages: contact and compression, excavation, and modification (Gault *et al.*, 1968; Melosh, 1989). Upon contact with the target both are compressed, generating shock waves that propagate into both objects. Pressures developed during high velocity impacts commonly reach hundreds of GPa, far above yield strengths for target and projectile. This stage lasts only as long as it takes the shock wave to traverse the projectile, perhaps milliseconds for an object tens of meters across. During the excavation stage, the shock wave expands into the target; it and the following rarefaction put target material in motion, which lasts much longer than the compression stage. After excavation, the "transient" crater is modified by gravity, with material slumping into the crater to define a final profile. Structurally, the importance of this process is the application of high stresses, and high strain rates, for a volume of the target somewhat larger than the final crater. Some material is melted and vaporized, fractures can extend several crater radii out, and some materials are very highly compressed and otherwise strained. The resulting crater's depth is approximately half due to excavation of material and half due to compression or other flow within the target body (Melosh, 1989).

The ultimate impact outcome is dispersal of the target object in many pieces that exceed the escape velocity of the object's mass. All but the largest asteroids, and most of the small bodies that we are considering, are probably the result of impact fragmentation and are themselves subsequently cratered. Crater populations approximate size–number distributions expected of products of collisional fragmentation (Chapman and McKinnon, 1986); thus, there is a mutual grinding down of the small objects over geologic time. The fragmentary origins of many asteroids are also inferred from observations of families of asteroids in similar orbits, apparently formed by breakup of a parent object (Michel *et al.*, 2003). The irregular shapes of many asteroids also suggest fragmentation origins.

3.2 Tidal stresses

Stresses may be induced in small objects by the gradient of gravitational acceleration due to another, usually larger, object (Soter and Harris, 1977; Weidenschilling,

1979). For asteroids, these stresses are usually imposed during the occasional close encounters with much larger planets (Bottke et al., 1999), but can be induced by close encounters with other asteroids.

For satellites, tidal forces are always present, and may change over time due to evolution of the orbit toward or away from the planet. The period of the satellite's orbit is determined by the mass of the system (planet and satellites) and the size of the satellite's orbit. Yet parts of the satellite that are closer or farther from the planet are forced to have the same orbital period as the satellite's center of mass. This imposes a slight tensional stress along the vector to the planet. The majority of satellites are locked into a spin period equal to their orbital period, thus these stresses along the object, although weak, can act for long periods of time. They also may provide a stress field that can combine with other stresses. The magnitude of these stresses is quite small; for instance, modeling by Dobrovolskis (1982) of the satellite Phobos (27 km long) yields tidally imposed deviatoric stresses of $\sim 10^{-2}$ MPa. The central pressure in Phobos is $\sim 7 \times 10^{-2}$ MPa.

For asteroids and comets, close encounters with planets can provide short periods of tidal stress; these are hypothesized to affect the distribution of loose material on some asteroids, and even their shapes (Bottke et al., 1999). Some comets are known to break apart under tidal stresses, most notably Shoemaker-Levy 9 (Asphaug and Benz, 1996). In the cometary examples, tidal stresses can be less than 10 Pa and still effect disruption of km-sized objects. This behavior has been used to infer extremely low tensile strengths for comets, and even a distribution of sizes of agglomerated particles or fracture spacing within the object (Asphaug and Benz, 1996).

3.3 Thermal stresses

Thermal stresses within asteroids may arise from temperature changes. Those arising from short period changes, such as rotation or seasons, will likely penetrate only a few meters at most and be taken up in the regolith. Long-term changes in average distance from the Sun might generate global temperature gradients that could induce a few MPa stress on Eros (Dombard and Freed, 2002). Some grooves, if compressional in origin, match predicted orientations, yet most likely would be tensional in origin. Thermal effects are most likely expressed as downslope creep of loose material (Mantz et al., 2004) rather than as large-scale fracturing or tectonics. Thermal effects on the lunar regolith have been studied by microseismicity (Duennebier, 1976), and have been suggested to apply to small objects such as Deimos (Thomas, 1998) and Eros (Mantz et al., 2003).

On comets, the effect of variable solar insolation is largely taken up by variations in evolution of volatiles, and sometimes strong heating of the very dark, non-volatile crusts that develop (Soderblom et al., 2004.). Significant fracturing and

Table 6.1. *Spacecraft remote sensing data of small bodies*

Object	Primary	Mean Radius (km)	Mission	Pixel Scale Range
Satellites:				
Phobos	Mars	11	Viking 1976–1980	6–100+ m
Deimos	Mars	6	Viking 1976–1980	3–100+ m
Janus	Saturn	86	Voyager 1980–1981	3–6 km
Epimetheus	Saturn	55	Voyager 1980–1981	1.6–6 km
Phoebe	Saturn	106	Cassini 2004	13–2000 m
Proteus	Neptune	140	Voyager 1989	1.6 km
Asteroids:				
Gaspra		6.1	Galileo 1991	64–500 m
Ida		15.7	Galileo 1993	31–500 m
Mathilde		26.5	NEAR 1997	160–1200 m
Eros		8.3	NEAR 2000–2001	0.1–100 m
Itokawa		0.2	Hyabusa 2005	0.02–10 m
Comets:				
Halley		6	Giotto, Vega 1986	150 m
Borrelly		3	DS-1 2001	60 m
Wild 2		2	Stardust 2004	15 m
Tempel 1		3	Deep Impact	0.1–1000 m

even breakup may develop due to stresses associated with elevated vapor pressures. However, observed breakup of comets has been difficult to associate with onsets of heating.

4 Observing structures in small bodies: methods and limitations

The nature of asteroid and cometary exploration so far limits us to remote detection of structures and tectonic features. Remote sensing of small bodies relevant to study of tectonic features consists of visible and infrared imaging, radar imaging, and one instance of laser altimetry. Table 6.1 briefly summarizes the types and quality of data available for some of the well-resolved objects of concern here.

The most fundamental limitation on remote sensing of satellite and asteroid structures is the presence of regolith. Here we use "regolith" to be any of the loose material, of whatever origin or particle size, on the surface. Virtually all regolith on these objects is probably the result of impact fragmentation. Despite the low gravity on objects only a few km across (typically of order 10^{-3} that on the Earth), cratering does leave significant fractions of the excavated material as distributed debris, apparently including fine grained components. On parts of Eros, regolith is estimated to be over tens of meters deep (Robinson *et al.*, 2002), and possibly over 100 m in places on the Martian satellites (Horstmann and Melosh, 1989; Thomas, 1998). This covering has much the same camouflage effect that soil, alluvium, or other surficial debris does on seeing terrestrial features. However, as with the Earth, the covering can respond to the topography and motion of the more solid substrate, and thus structures of some sort have been mapped on most of the small bodies that have been imaged at close range. Interpretation of this response is one of the major hurdles in interpreting the structural patterns on these objects (Section 7 below).

5 Accretional and precursor body structures

Original structures from accretion of asteroids might be present if the bodies have not differentiated or been mechanically altered after formation. The vast majority of asteroids (<200 km diameter) are regarded as in some way fragmentary remains of larger precursor objects. A substantial fraction of these may have been differentiated in whole or in part (Grimm and McSween, 1993) and would not be expected to show obvious remnants of accretion. Some asteroid classes, such as C and D, may be thermally less processed (McSween *et al.*, 2002), so there might be the possibility of remaining accretionary structures. However, from the standpoint of remote sensing, structures induced later by impact fragmentation might be very similar to original accretionary ones. Thus, indications of layering or the like have to be evaluated from the standpoint of when in the history of the object they might have been formed.

5.1 Layers in small bodies

Any small body that is a fragment of a larger precursor body might expose layering formed in that body by differentiation, accretion, or reassembly of pieces after earlier fragmentations. Accretion is usually thought to involve a range of particle sizes, including those of a significant fraction of the "final" object size, so well-organized, global layering is probably not accretionary, or from reassembly of pieces after fragmentation. Layering from differentiation might not be perfectly smooth and would probably tend to have a radius of curvature appropriate to the depth within the parent body. Expression of parent body layers in fragments we observe today could be effectively planar to noticeably curved.

There are only three instances yet observed where layers or planar structures extend for a large fraction of an asteroid's length: Mathilde, Eros, and Ida. All of these present some interpretive complications.

Figure 6.2. The surface of Mathilde. The brighter linear feature is visible in many images, and shows no topographic expression at 160 m/pixel. It is unlike features seen on other small bodies, and might represent exposure of a layer or crater sampling of a layer through a regolith of tens of meters or less in depth. NEAR Multispectral Imager image 42826484.

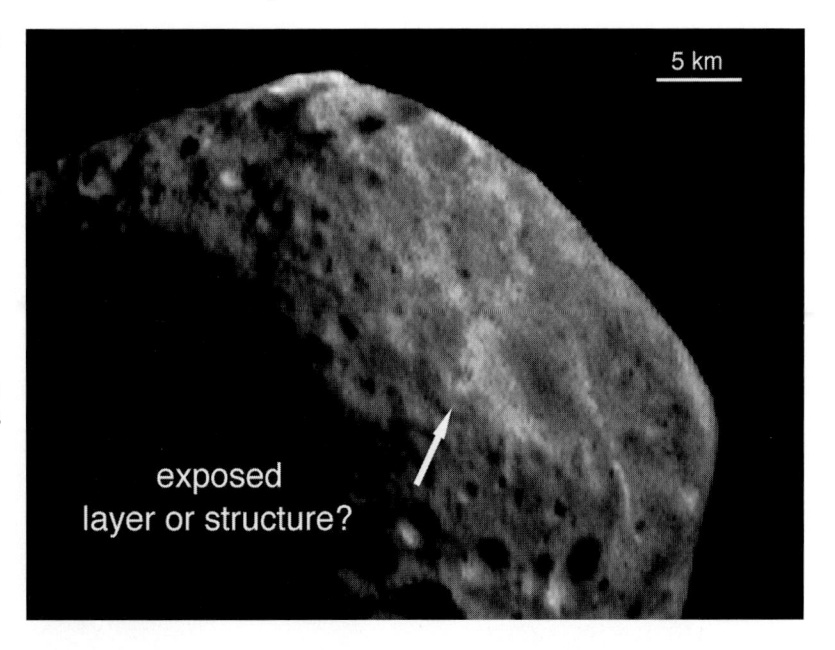

Mathilde (Figure 6.2), with a mean radius of ~26 km, was imaged at approximately 250 m/pixel (Veverka *et al.*, 1997). Mathilde's low density of approximately 1.3 g cm^{-3} in combination with the likely composition of the object, suggests a porosity of well over 30%, and possibly as much as 50% (Veverka *et al.*, 1999). The images of Mathilde showed a bright marking, about 1 km wide, extending about 20 km along the surface (Thomas *et al.*, 1999). It appears to be an albedo feature, interrupted on scales of 200–400 m by small craters or other topography. As subtle as it is, it is the only real candidate among the asteroid images for something that might be called an "exposed" layer in the body of the object (as opposed to exposed regolith layers; these exist on Phobos and possibly Eros). The bright marking might be the locus of excavation by small craters through a regolith into an underlying feature. Craters 200 m in diameter, individually invisible at this resolution, could excavate material from beneath 20–30 m of regolith, and leave a surficial trace of different properties. If this marking indeed is from a layer, it suggests a structure in something originally much larger than the present size of Mathilde.

Eros was thoroughly mapped by the NEAR-Shoemaker spacecraft; nearly all the surface was imaged at better than 4 m/pixel, and parts were imaged at less than 1 m/pixel (Veverka *et al.*, 2001). This detailed imaging allows for definitive mapping of the presence or absence of different types of surface features. Evidence of loose material on the surface is ubiquitous, as shown by both morphologic and albedo features (Robinson *et al.*, 2002). Early images showed suggestions of aligned features, and efforts were made to establish global patterns. Only one global

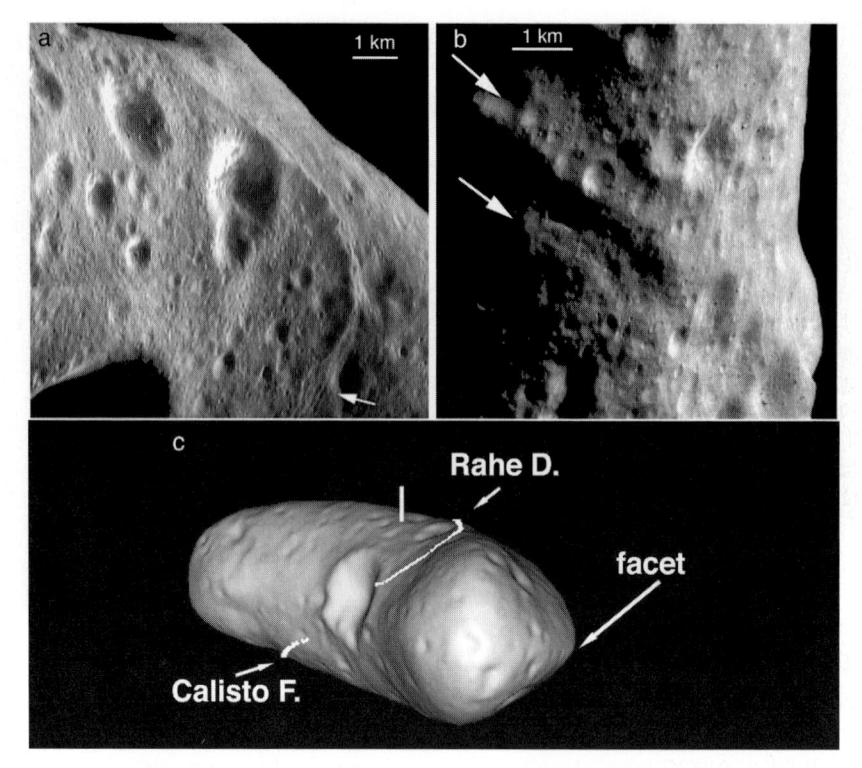

Figure 6.3. Global structures of Eros. (a) Portion of Rahe Dorsum, showing the narrow, sinuous nature of this ridge near crater Himeros. NEAR images 129901617–129901897. (b) Calisto Fossae, a double trough–ridge feature with many superposed craters that is coplanar with Rahe Dorsum. Image 130230097. (c) Relation of Rahe Dorsum, Calisto Fossae, and a nearly planar facet on Eros. Image-derived shape model with mapped locations of features.

pattern (Thomas et al., 2002) emerged from the mapping of surface features and asteroid shape (Figure 6.3). Two structural features, a ridge (Rahe Dorsum) and a ridge/trough system (Calisto Fossae) can individually be well fit by planes, and these are coplanar. This alignment of the largest linear features on the object strongly suggests a structural link 22 km from Calisto Fossae to the farthest part of Rahe Dorsum (Thomas et al., 2002). Furthermore, one of the facets of the shape of Eros (Figure 6.3) is nearly parallel (within $\sim 7°$) to the orientation of this plane. The detailed morphology of Rahe Dorsum and other structural expressions on Eros will be discussed below. Although these features indicate a planar structure, and possibly two parallel structural trends, there are three different morphologic expressions that probably formed at different times. The ridges and troughs in Calisto Fossae are very heavily cratered (Prockter et al., 2002; Thomas et al., 2002), and suggest exploitation of weaknesses by the cratering process. The feature and the surface

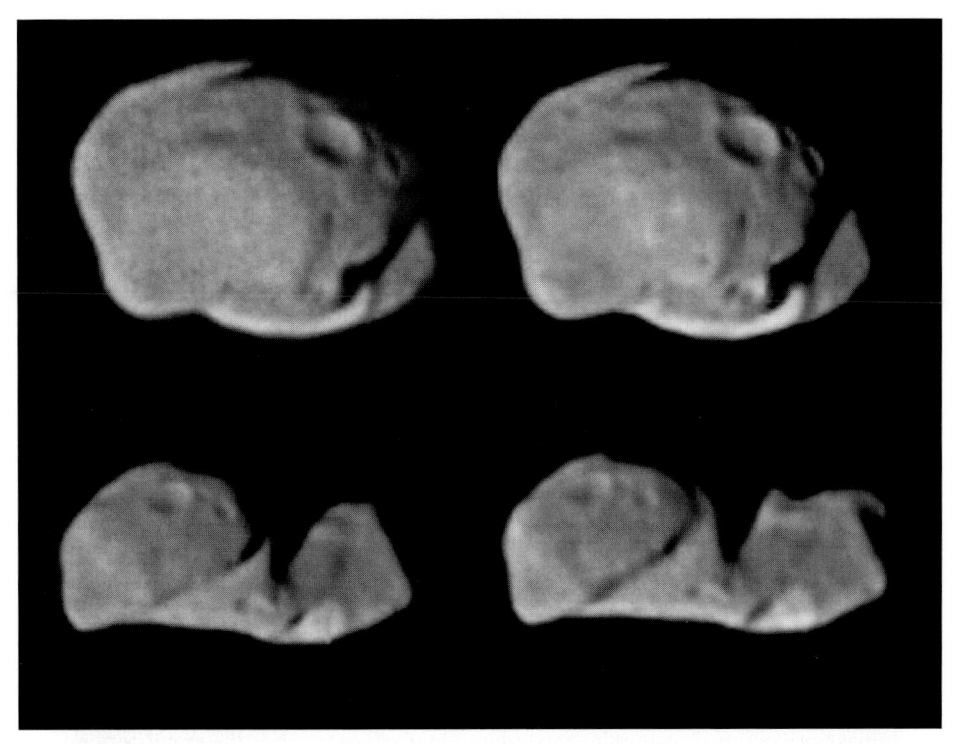

Figure 6.4. Stereo views of planar topographic boundary on Ida. Top view looks along the long axis of the asteroid, Vienna Regio is the concavity in the end facing out. Galileo images S0202557100 and S0202557300. Bottom pair shows view from the right relative to the top pair, images S0202553700 and S0202554800. The distinct crater in the upper part is 5.2 km across.

expression are very old. Rahe Dorsum has crisper topography, is much narrower, and does include a change in morphology along its length (see Section 7 below), and it is by no means certain that all sections are of equivalent age. This feature suggests displacement along a fault rather than the morphologic exploitation by cratering. The parallel (but not coplanar) facet also may suggest exploitation of weaknesses by spallation. This facet's surface may be very old as well. Thus, there is a likely near-planar fabric to Eros, exploited in at least two different ways, probably at different times. If it represents inherited structure, it might indicate a substantially larger precursor object.

Asteroid Ida, imaged by the Galileo spacecraft in 1993, has a mean radius of 15.7 km, a maximum length of 56 km, and a somewhat bent profile when viewed along its spin pole. In terms of global structure, it has a ridge, Townsend Dorsum, which traverses 40 km and about 150° of longitude (Figure 6.4). The crest of this ridge is well fit by a plane (within about a km over its length). The ridge is imaged at low resolution (~500 m/pxl); one end of it just reaches an area of very much higher

244   *Planetary Tectonics*

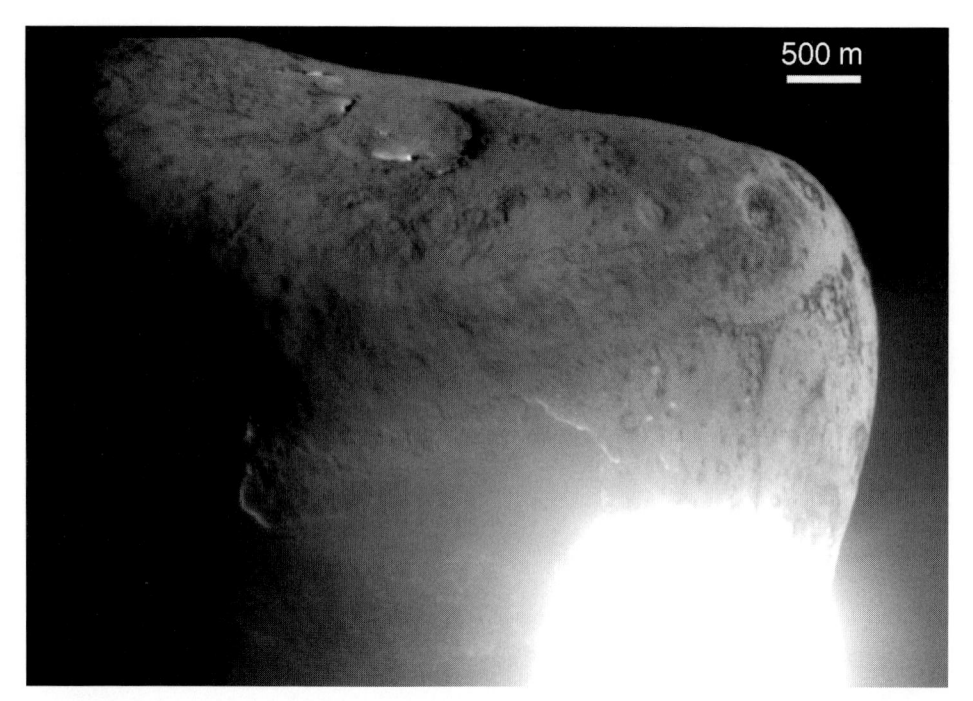

Figure 6.5. Comet Tempel 1. Portion of image 173728459 obtained by the Deep Impact spacecraft showing the expanding cloud of ejecta from the crater formed by the impactor spacecraft. Some of the layered structure is visible in the middle, running left to slightly upper right. A smooth deposit of material is visible just to the upper left of the ejecta cloud.

resolution imaging (31 m/pxl), but this area displays no discernable continuation of the feature. The ridge, especially the morphology shown in Figure 6.4, might suggest displacement along a fault (Sullivan *et al.*, 1996; Thomas *et al.*, 1996). However, the projection of the ridge into the area of high-resolution data shows no sign of fault topography. Thus, it might be an older feature, such as a compositional discontinuity, that has been etched by cratering and spallation.

Comet Tempel 1 shows evidence of two kinds of layers. One may be recent deposition of material, but others, expressed as bands of different albedo, may reflect erosional exposure of deeper layers perhaps related to accretionary structures (Figure 6.5) (A'Hearn *et al.*, 2005).

5.2 Binary objects

Perhaps the ultimate asteroid structural features are binary asteroids: two objects in orbit about each other. Satellites of asteroids were long suspected to exist (Van Flandern *et al.*, 1979; Weidenschilling *et al.*, 1989), but after many false indications, started to be discovered in the 1990s, beginning with Dactyl orbiting Ida (Belton *et al.*, 1996). Satellites of asteroids were expected to result from catastrophic and near-catastrophic impacts (Melosh and Stansberry, 1991). Objects such as Dactyl are very small compared to the primary asteroid, but other objects more closely match the connotation of binary objects, that is, pairs of bodies of somewhat comparable sizes. Binary main belt asteroids are being found, and many of the Kuiper belt objects appear to be binary (Goldreich *et al.*, 2002). The near-Earth asteroid population of objects may be ~1/6th binary (Margot *et al.*, 2002). Binaries have also been detected among the Trojan asteroids (Merline *et al.*, 2001) orbiting ahead and behind of Jupiter. One of the early hints of a noticeable population of small binary objects was the number of binary impact craters on the Moon, Mars, and other objects. These were thought to form by tidal separation of objects orbiting close to one another just before impact (Aggerwal and Oberbeck, 1974).

5.3 Differentiation

Some types of asteroids appear to represent differentiated objects, and meteoritic evidence shows many small bodies did indeed suffer enough melting to differentiate. Those likely to have suffered melting are concentrated in the inner asteroid belt, where it is thought accretion occurred more rapidly, allowing short-lived radionuclides, especially ^{26}Al, to heat them (McSween *et al.*, 2002). Alternative suggestions of heating by electromagnetic induction remain difficult to model (Sonnett *et al.*, 1968). However, the clear evidence of melting in the form of metallic and various silicate-rich achondritic meteorites shows that some differentiation did occur.

Melting and differentiation can generate many types of structures, including global shell structures and intrusive bodies of various shapes. All of these may have attendant mechanical and thermal modifications on their boundaries, faulting, and fracturing, and volcanic forms at the surface. Metallic cores often retain structures developed in the cooling metal. Yet, it is not clear that any of these structures have been detected by spacecraft. While Gaspra may be differentiated, even metal-rich, the only possible structural relation to differentiation yet seen might be its overall shape, which is somewhat rhomboidal (Thomas *et al.*, 1996). Eros, best approximated by an LL chondrite meteorite composition, appears undifferentiated (McCoy *et al.*, 2001).

6 Interior structure from impacts

6.1 "Rubble piles"

The expectation of catastrophic collisions having occurred for even 100 km diameter objects suggests that most asteroids record features of severe fracturing or global fragmentation (Britt *et al.*, 2002; Richardson *et al.*, 2002). A crucial aspect is the expectation that fragmentation yields particles with a range of sizes and velocities relative to the center of mass, which means that many fragments may reaccumulate under the influence of gravity.

Objects that are gravitationally bound collections of fragments have been termed "rubble piles." Recent work by Richardson *et al.* (2002) has attempted to classify the possible states of fragmentation of asteroidal bodies. Their classification uses porosity and relative tensile strength, which is the ratio of the tensile strength of the object to the tensile strength of the components. Objects with high porosity and low strength are likely to be gravitational *aggregates*; high strength and low porosity are *monolithic* forms. Objects that have tensile strength reduced by faults or cracks yet retain original structures are termed *fractured bodies*. *Shattered bodies* applies to those objects with a fractionally large reduction in tensile strength. *Rubble piles* are those that have been shattered and reassembled. A slightly different classification has been presented by Britt *et al.* (2002) that divides asteroids into three general groups: those that are essentially monolithic, heavily fractured ones with macroporosities of ∼20%, and loosely consolidated ones with porosities over 30%. Structurally, the primary factors are whether any individual fragments are rotated from their "original" positions, and whether porosity is due to opening of fractures or to irregularities of packed material that has moved relative to each other.

Although there is a growing sense that a large fraction of 0.1–100 km sized asteroids are loosely bound aggregates, the evidence until 2005 was in most measure circumstantial. Asteroid spin periods derived from lightcurves have been used to infer a lack of tensile strength for objects over 200 m in size (Pravec and Harris, 2000). Asteroids, and some small satellites, appear to have large porosities on the basis of mean density and their probable compositions. Mathilde, with a density of ∼1.3 g cm^{-3}, is estimated to have a porosity of up to 50% (Veverka *et al.*, 1999). Other asteroids may also have high porosities (Merline *et al.*, 2002; Britt and Consolmagno, 2001; Britt *et al.*, 2002). Two icy satellites of Saturn, Epimetheus and Janus, have compositions thought to be largely water ice, and densities less than 0.7 g cm^{-3} (Nicholson *et al.*, 1992), suggesting porosities of ∼30%. Craters with diameters equal to or even greater than the object's mean radius are common on small objects (Thomas, 1999). Hydrocode modeling of these impacts predicts that they can form only in fractured or weak targets (Asphaug *et al.*, 1998). Shapes of asteroids that display gravitational slopes well below an angle of repose have also been suggested to imply loose materials (Ostro *et al.*, 2000). Objects with a somewhat S-shaped profile viewed along the spin axis might be consistent with reassembly after tidal breakup (Bottke *et al.*, 1999) and thus may indicate a loose constitution. More circumstantial evidence for loose aggregations comes from

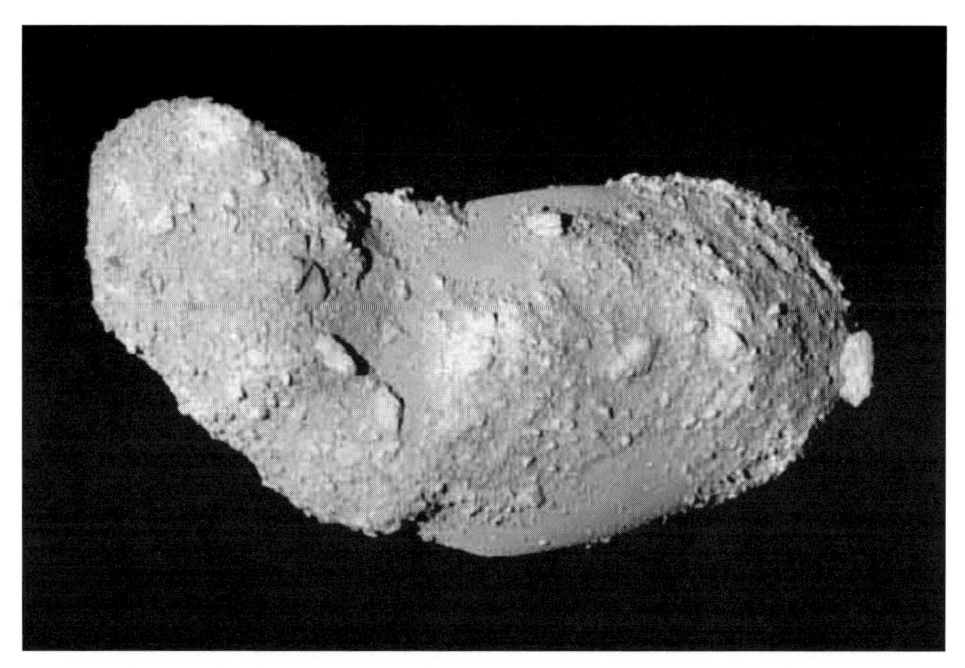

Figure 6.6. Asteroid Itokawa from the Hyabusa spacecraft. Object is about 500 m across in this view. Courtesy of JAXA.

doublet craters (Aggarwal and Oberbeck, 1974) and asteroid families (Michel *et al.*, 2003), where breakup of cohesionless objects rather than of monolithic ones is predicted by the modeling of current fragment orbital distributions.

Most close-up views of asteroids do not show collections of particles or the significant lumpiness expected of loose aggregates of particles. Most can be described as moderately smooth surfaces modified by craters and grooves (Figure 6.1). Some of the radar data may be the best direct evidence of contact binaries (Benner *et al.*, 2005). However, the radar images do not yet show multiple lumpiness; rather, the second-order features appear to be impact craters. Some asteroids that have a constriction in their shape, such as Ida, might be thought to be two pieces accumulated together with smaller sized particles filling in (Thomas *et al.*, 1996). However, in the instance of Ida, there is high-resolution imaging in the constricted area, and it shows no evidence of any disturbance of loose material that relatively fine-grained fill between two large pieces might show as a result of impacts on one or the other of the large pieces.

After years of searching and anticipation, the Hyabusa mission to asteroid Itokawa returned images of what is almost certainly a rubble pile (Figure 6.6). This object, only 0.16 km in mean radius, has exactly the appearance one would expect of a loose collection of fragments (Fujiwara *et al.*, 2006). Because this one is so obvious, other objects thought to be rubble piles but showing much different surfaces may indeed be more monolithic.

6.2 Porosity

Porosities of asteroids and small satellites can be estimated using their mean density, estimated composition, and grain densities of analogue meteoritic materials (see Britt *et al.*, 2002). Mean density calculations require good measures of volume and mass. Masses for several asteroids have been determined by observations of the periods and sizes of orbits of natural satellites (Merline *et al.*, 2002). Mean densities of asteroids range from slightly above $1.0\,\mathrm{g\,cm^{-3}}$ to $3.4\,\mathrm{g\,cm^{-3}}$ (Hilton, 2002). Modeled total porosities range from 10–70%. The total porosity (Britt *et al.*, 2002; Wilkison *et al.*, 2002) includes microporosity that has little structural effect and macroporosity that is mostly caused by, and affects, impact and other structural processes. Several asteroids, including Eros, have estimated porosities of less than 30%. This is approximately the value at which materials are expected to lose coherency (Britt *et al.*, 2002). They are thought to be heavily fractured, but not to have been dispersed and reassembled in catastrophic collisions.

6.3 Center of mass – center of figure offsets: significance for structure

The relative positions of center of mass and center of figure, or more generally, the predicted and observed gravity fields, may help constrain interior models of small asteroids. Only for Eros do we have the gravity and shape data that allow any meaningful comparison. Although different techniques yield different locations of the center of figure, the offset of center of mass and center of figure for Eros is small: between 10 and 50 m (Thomas *et al.*, 2002; Konopliv *et al.*, 2002). The direction of offset does not fit any simple layer model based on the global planar structures (Thomas *et al.*, 2002), and can be matched by fairly small variations in mean density that might be due to modest changes in porosity. Certainly no great variation in composition through the asteroid is indicated, and a uniform composition is consistent with the globally uniform spectral and color data (Bell *et al.*, 2002).

For Ida we do not have a gravity map, but the coincidence of the spin pole with the maximum moment of inertia of the shape (Thomas *et al.*, 1996), assuming homogeneous density distribution, suggests there are no major density variations within the object. As noted above, there is a strong suggestion of a major structural feature in Ida, but it does not separate areas of different density, apparently.

7 Morphology of surface expressions of structures

The surface expressions of structural features on small bodies can be largely classified as grooves, troughs, ridges, and modifications of crater shapes. They occur over a wide range of scales. The majority appear to represent surface disturbances in regolith overlying fractures in a competent substrate, but some ridges and troughs are large enough to be visible at a global scale, and probably result from significant tectonic events.

7.1 Grooves

Grooves appear to be a common feature of small bodies, as they have been observed on Gaspra, Ida, Eros and Phobos. The grooves on these bodies share a range of surprisingly similar morphological types, from linear, straight-walled depressions, to rows of coalesced pits, but nearly all are less than 100 m in depth and are concave up (Figure 6.7). Images of Phobos and Eros are of sufficient resolution to discriminate rows of discrete pits, and floors of grooves that may be smooth or hummocky. On all of these bodies, grooves are present in patterns that crosscut or otherwise intersect with each other.

The primary distinguishing feature of grooves among these small bodies is scale. Phobos' grooves are many kilometers long, and at least one can be traced for a distance of over 30 km. They are typically 100–200 m wide and 10–20 m deep. They are grouped in four sets, each of which has parallel members (Thomas *et al.*, 1979). Ida's grooves are less than 4 km in length, with the majority less than 100 m wide. They are estimated to be less than 50 meters in depth (Sullivan *et al.*, 1996). Ida's grooves are concentrated in three regions; however, they are not obviously related to one crater or structural grain (Sullivan *et al.*, 1996). Grooves on Gaspra are generally 100–200 m wide, 0.8 to 2.5 km long, and 10–20 m deep. Most occur in two groups of nearly parallel members, the most prominent of which roughly parallels the long axis of the body. There is no obvious connection to specific impact features (Veverka *et al.*, 1994). The majority of grooves on Eros are up to 2 km long (although most are less than 800 m), with typical widths of 75–100 m. Depths are in the range of a few meters for smaller grooves, and tens of meters for the largest structures (Prockter *et al.*, 2002). Eros has grooves distributed unevenly across its surface. The highest concentration of parallel grooves is within the concavity of the largest impact feature, Himeros. There are no obvious concentrations of grooves around individual impact features. The occurrence of grooves and the relation of some to tidal stresses is dealt with in detail in Morrison *et al.* (2009).

7.2 Troughs

We distinguish troughs from grooves by their greater widths and depths, and by the high density of superposed craters and accompanying subdued morphology.

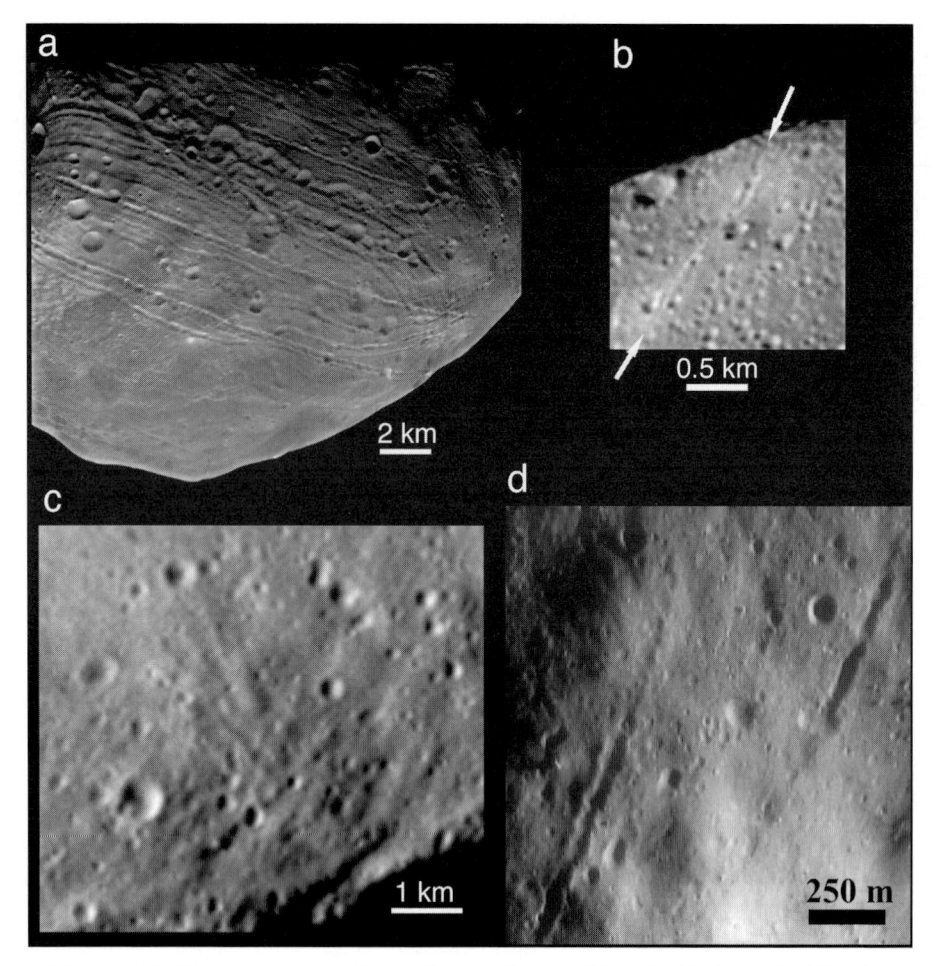

Figure 6.7. Grooves on small satellites and asteroids. (a) Phobos, (b) Ida, (c) Gaspra, (d) Eros.

The largest structural features on Eros are part of the Calisto Fossae, two parallel adjacent troughs separated by a narrow ridge (Figure 6.3b). The troughs are several km long, range from 500–700 m wide, and are up to 120 m deep. They appear to be very old, with hummocky, degraded interiors and several large impact craters superimposed upon them. They have relatively flat floors, to within ±20 m (Cheng et al., 2002), and many subparallel fractures are present both within the floors and adjacent to the troughs. The Calisto Fossae troughs have a morphology consistent with graben, and they may have formed by extension early in Eros' visible history.

7.3 Ridges

Elongate, positive relief forms that are not part of crater rims are found on Eros, and a few examples are found on Gaspra. Close-packed grooves have intervening

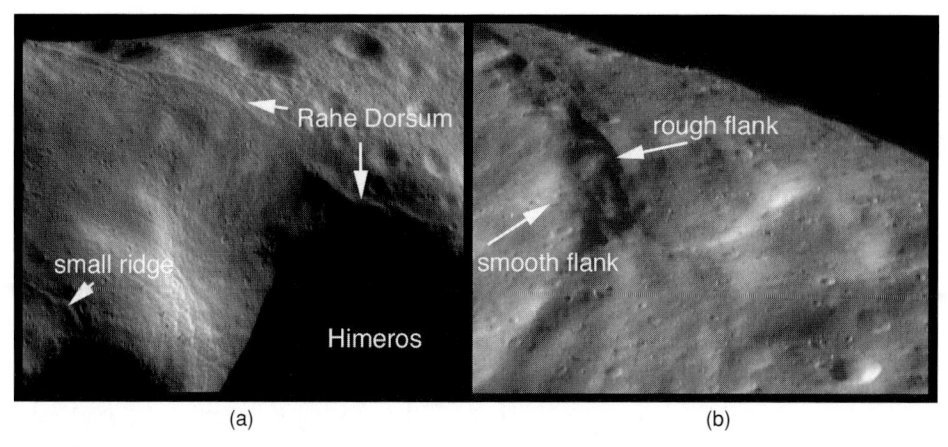

Figure 6.8. (a) Segment of Rahe Dorsum, trending from the crater Himeros across the northern hemisphere. A small subparallel ridge is visible on the opposite side of Himeros. (b) In this segment, one flank of Rahe Dorsum is relatively smooth and has a low slope, while the opposite flank is much steeper and rougher textured, a morphology indicative of a thrust fault. MSI image 133972060; image width ~2.3 km. The limb of the asteroid is at the upper right.

crests that can appear to be ridges, but which we do not include in this discussion. The term Dorsum is also applied to a form on Ida, such as Townsend Dorsum, but as shown in Figure 6.4, this feature probably reflects a differential response to fragmentation rather than being a strain-related form.

One of the most prominent features on Eros is Rahe Dorsum (IAU: Hinks Dorsum), an 18-km long arrangement of segments of somewhat varying morphology (Figures 6.3 and 6.8). This path curves around the central "waist" of the asteroid and comes close to the north pole. Rahe Dorsum is comprised of a string of en echelon, asymmetric segments, with a variety of morphological characteristics. All segments are generally less than 300 m wide and a few tens of meters in height. A notable exception is a scarp at the Himeros end of the ridge, which exhibits a near vertical drop of 120 m into an adjacent trough. Bright material at the foot of this scarp is interpreted to be material that has moved downslope, by analogy to materials seen in other crater walls and talus observed around other scarps (Figure 6.9). Some segments are asymmetric in cross section: the more northerly flank has a relatively gentle slope and smooth surface, while the other flank is steeper and significantly rougher textured. The best example of this morphology occurs near Eros' north pole, where the ridge has the characteristics of a thrust fault, in which the steeper flank corresponds to the hanging wall face (Figure 6.8) (Prockter et al., 2002). NEAR Laser Rangefinder profiles across the ridge are consistent with this interpretation and further suggest a succession of fault blocks (Cheng et al., 2002). At its Himeros terminus, Rahe Dorsum intersects a series of shallow troughs, while at the Psyche terminus, it appears to continue as a series of closely spaced

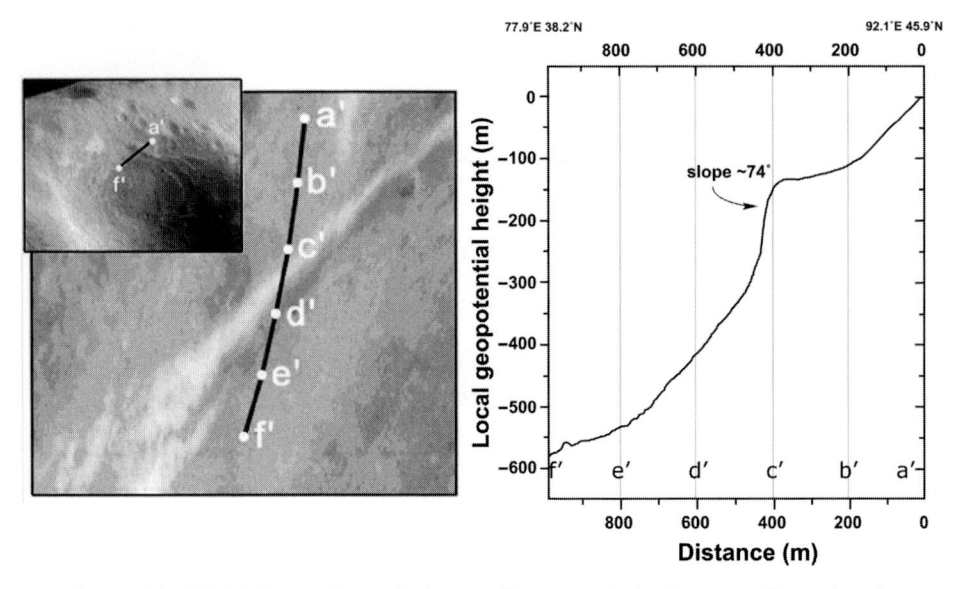

Figure 6.9. NEAR Laser Rangefinder profile across Rahe Dorsum. Note that the profile is not perpendicular to the ridge, which is therefore steeper than the slope shown here (adapted from the NEAR Image of the Day website).

fractures, which themselves intersect with at least two other subparallel fracture sets.

Although there are some impact craters on Rahe Dorsum, these are generally small, indicating that it may have formed in the latter stages of the asteroid's evolution, after any large crater formation. Superposition relationships between the ridge and the large craters Himeros and Psyche imply that it postdates these features. However, Rahe Dorsum is itself crosscut by a prominent groove, indicating that tectonic activity persisted subsequent to its formation (Prockter et al., 2002). The presence of fine-scale material on and at the foot of the ridge flanks suggests that it has undergone relatively recent mass wasting, possibly in response to seismic shaking from impacts elsewhere on the asteroid.

Watters and Robinson (2003) have modeled the topographic profiles of Rahe Dorsum and found it consistent with a thrust fault of less than 240 m depth, dipping ~35–40°, with several tens of meters of slip. The modeling suggests a material shear strength of 1–10 MPa.

At least two smaller ridges are present in the same vicinity as Rahe Dorsum, but neither are longer than a few km. One of these is subparallel to Rahe Dorsum, but is on the opposite rim of Himeros (Figure 6.8), and may have formed in response to the same stresses that caused Rahe Dorsum to be thrust above the surface.

7.4 Crater shape modification

Some craters on Eros have "squared off" outlines, a form that is controlled by joints, faults, or other fabric within the target rock (Prockter *et al.*, 2002). This type of structural control is seen in Arizona's Barringer meteor crater, the outline of which is the result of two sets of orthogonal vertical joints in the underlying sedimentary rock (Shoemaker, 1963). The final crater shape was influenced by planes of weakness that were exploited by the cratering flow. The sides of the crater are oriented at approximately 45° to the joints as a result of the way excavation flow exploits fracture weaknesses. Although crater side orientations have not been mapped for Eros, locally several craters can have similar orientations of straight segments of crater walls, suggesting consistent fracture/joint orientations over a couple of kilometers extent.

8 Implications of grooves for structure and material properties

Several hypotheses have been proposed for the formation of grooves on small bodies. An early suggestion was that they are chains of secondary craters (Veverka and Duxbury, 1977), but this is not supported by observations of their morphology or their patterns. Most grooves lack raised rims, are linear in nature, and have uniform sizing and spacing of pits. These characteristics are difficult to match with expectations of secondary crater morphologies. Their patterns also do not match expectations from rolling ejecta (Thomas *et al.*, 2000).

The presence of blocks on the surface of Ida, Phobos, and Eros are indicative of significant regolith retention on those bodies (Lee *et al.*, 1996; Thomas *et al.*, 2000, 2002). It is likely that the majority of grooves represent the disturbance of this regolith over deep, steeply dipping fractures or fracture zones (Thomas *et al.*, 1979). Two models have been suggested for how this would occur. The first involves the ejection of material from fractures, probably through gas venting from the subsurface. This could occur if the body contains sufficient volatiles, which would be driven off when heated by a large impact event (Thomas *et al.*, 1979). The resulting grooves would have raised rims and shallow cross-profiles from redeposition of material back into the groove, or slumping of the groove wall itself. However, it is not certain that sufficient amounts of bound water are present in the bodies in question, and few of the grooves actually have raised rims. If gas venting has occurred, it is likely to have been only a minor factor in groove development.

A more probable model for groove formation involves the drainage of regolith into fractures. Experimental modeling of this process suggests a series of steps occur in the formation of grooves (Horstman and Melosh, 1989). These are: (a) the formation of a shallow trough over a fracture; (b) some local deepening, leading to early pits within the trough; (c) the existing pits become better developed, and new pits form; (4) neighboring pits meet, marking the beginning of groove formation, and (5) the pits coalesce, forming a prominent groove. Comparison of the model results to grooves on small bodies shows remarkable similarities. It has been noted that the fractures are probably deep, since individual grooves retain a planar pattern across vertical changes in topography of over 1 km (Thomas *et al.*, 1979).

If grooves do form by this method, their tendency to have a preferred width could result from a uniform spacing or depth of fractures, or a uniform depth of regolith. The pits have a characteristic spacing, which corresponds to the thickness of the material layer in which they form, if it is dry, unconsolidated, and non-cohesive. Perhaps surprisingly, this spacing appears to be uninfluenced by grain shape, size, rounding, angle of repose, or regolith bulk density (Horstman and Melosh, 1989), allowing a crude estimate of the regolith depth to be made for small bodies. For Phobos, pit spacing suggests the regolith is up to 100 m in depth (Horstman and Melosh, 1989), while for Eros it is at least 30 m (Prockter *et al.*, 2002). Both of these values are supported by other estimates of regolith thickness, based on morphology and topography (Robinson *et al.*, 2002).

9 Patterns of linear features and inferred structures

If grooves represent disturbances in regolith overlying fractures, the question remains of how the underlying fractures form. The geometric pattern of grooves or ridges in relation to likely stress orientations is the primary tool for inferring their origins, as our data are all from remote sensing, although the morphology of grooves and ridges might help in determining displacements or associations.

9.1 Phobos grooves: impact and/or tidal stresses?

The grooves on Phobos have a strong morphologic association with crater Stickney, which is ~10 km in diameter (over 40% of the mean diameter of Phobos), while their orientations are closely tied to the principal axes of the satellite (Thomas *et al.*, 1979). Phobos' grooves are widest and deepest near Stickney, and are virtually absent from the region of the satellite roughly opposite this crater. Laboratory simulations of impact fracturing indicated a relation to the crater Stickney (Fujiwara, 1991). The grooves are also consistent with forming at about the same time as Stickney, based on crater counts and the relative crispness of the topography of the crater and the grooves. Some grooves cut the rim of Stickney and indicate, whatever the process for their formation, it was not exactly coincident with excavation of Stickney; this is still consistent with drainage of material into fractures.

Figure 6.10. Projection of prominent grooves on Phobos, looking from 0°N, 270°W, along the direction of orbital motion.

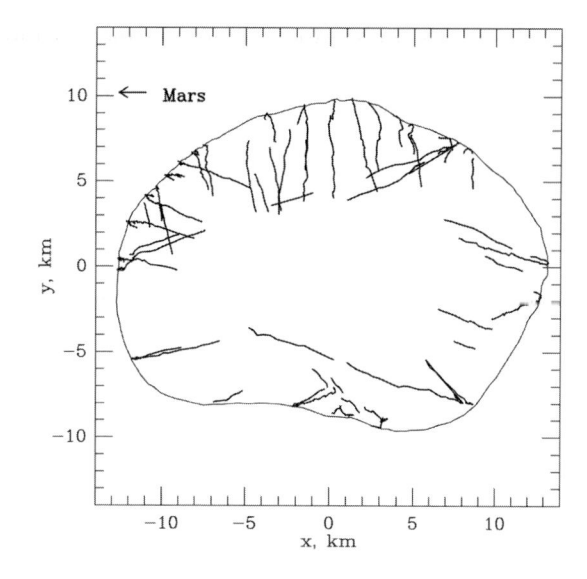

However, the geographic pattern of the grooves suggests an important role for stresses unrelated to Stickney (Figure 6.10). A large fraction of the Phobos grooves are grouped into sets of parallel members, with paths that follow the intersection of planes with the surface. Normals to these planes lie very close to the plane defined by the longest and shortest axes of the satellite. Because the satellite is in synchronous rotation about Mars, tidal stresses are essentially fixed with respect to the body of the satellite. These forces are tensile along the long axis of the satellite, and the resulting stress fields are consistent with fracturing in planes similar to the groove-defined ones (Aggarwal and Oberbeck, 1974; Dobrovolskis, 1982).

This combination of associations has complicated interpretations of the grooves. Burns (1978) proposed that the grooves on Phobos could have formed by fracturing that occurred when Phobos was captured by Mars. The morphologic association with the large crater led to suggestions that the grooves are related to Stickney's formation (Veverka and Duxbury, 1977; Thomas et al., 1979). Soter and Harris (1977) proposed the grooves formed through tidal stressing of the satellite by Mars as the satellite evolved inward toward Mars. Weidenschilling (1979) made calculations that demonstrated the sequence of development of tidal-induced fractures could vary considerably with different shapes. He suggested that a large impact such as Stickney could have broken the satellite's synchronous rotation, and stresses associated with recovery to synchronous spin might have caused the groove fracture pattern. One aspect of tidal stresses does not fit well with the apparent ages of the grooves. The grooves have a modest population of superposed craters. Tidal stresses increase as a^{-3}, where a is the satellite's semimajor axis (Soter and Harris, 1977). Thus, the stresses would increase progressively as the satellite orbit shrinks. As a result, tidal fracturing by itself would be expected to be recent, or increasingly prominent, whereas the grooves are not the most recent features on the satellite (Soter and Harris, 1977).

Modeling of the Stickney impact by Asphaug and Melosh (1993) predicted that an event of this magnitude could fracture the interior of Phobos on a scale of several kilometers, a value consistent with the extent of the grooves. The predicted fragment sizes were consistent with groove spacing, which, with the attendant assumptions, would suggest that Phobos was coherent prior to the impact. However, the strong association of groove patterns with expected tidal effects suggests that whatever the role of impact stresses, the tidal field helped define the principal stress orientations responsible for groove fractures. Given the magnitude of the impact stresses compared to the weak tidal ones, tidally driven fracturing in a weak Phobos may have been exploited by the impact, rather than formed entirely new. While it may appear counterintuitive that a weak object would not show dominant effects of a very large impact, findings from the porous asteroid Mathilde may be relevant. A weak, fractured target can absorb much of the impact energy by inefficient transmission of the seismic energy and by a large fraction of the crater being produced by compression rather than excavation (Housen *et al.*, 1999). Damage outside the visible crater is minimized. Thus, a Phobos weak enough to be fractured by tides could still suffer a large impact that could widen the fractures, but may not drastically rearrange the object.

9.2 Eros grooves: no single pattern

The Eros grooves (Figure 6.11) provide a strong contrast to the global organization of the ones on Phobos (Figure 6.10). Eros' grooves are not obviously related to specific craters, nor do they cluster in orientations related to the overall shape of the object. There is a global pattern only in the sense noted earlier that Rahe Dorsum and Calisto Fossae may reflect a global planar fabric. The grooves do maintain consistent morphologic and orientation patterns over areas of several km, such as between 300–330°W and 0–30°N, and 20–90°W, 60–30°S (Figure 6.11). These patterns suggest mechanical continuity over a few km. Additionally, the lack of obvious relation to any individual craters suggests that many fractures might result from seismic focusing or reactivation from later impact events.

9.3 Ida grooves: antipodal effects?

Ida displays a modest collection of grooves, none of which are obviously concentrated about a particular crater. One set is essentially opposite the largest distinct concavity on Ida, a 15-km wide feature, Vienna Regio (Figure 6.4). Using hydrocode techniques, Asphaug *et al.* (1996) showed that stress waves from the

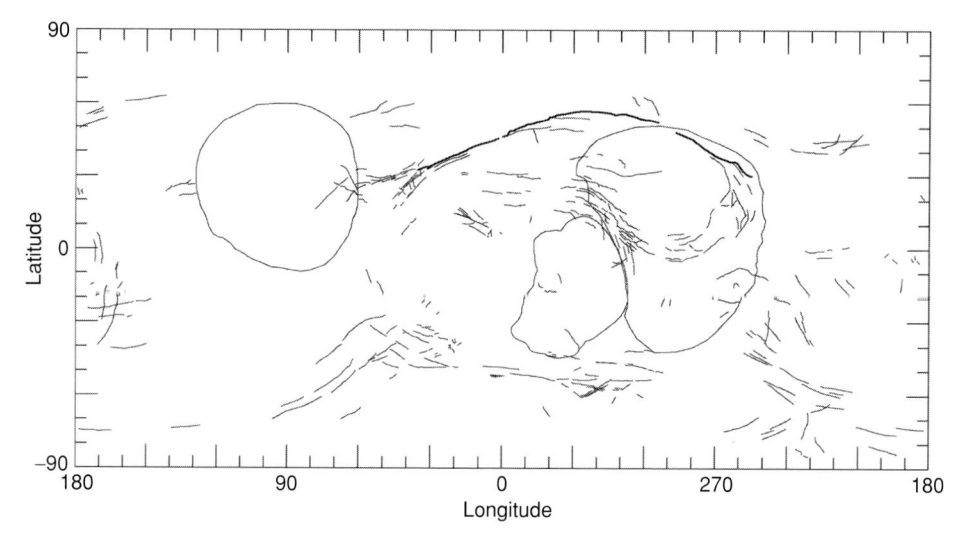

Figure 6.11. Map of grooves and ridges on Eros; simple cylindrical projection, which greatly expands the relative area of regions near longitude 90° and 270° and at the poles. Outlines of the three largest craters, from left to right, Psyche, Shoemaker, and Himeros, are shown for reference. Most features plotted are grooves, but ridges, such as the heavy line marking Rahe Dorsum, are included.

Vienna Regio impact on Ida could have disrupted the surface at the opposite end of the asteroid to form the Pola Regio grooves. The apparent freshness of the grooves implies that although they may have been originally formed by the Vienna impact, a somewhat degraded crater, they could have been subsequently reactivated by a later impact, probably crater Azzurra, which is about 10 km across and located about 15 km from the grooves. If fractures underlying grooves are formed through the focusing of seismic waves, then the body on which they appear must be sufficiently mechanically coupled to transmit seismic waves. This interpretation implies that such objects are not rubble piles, although they may have a measure of fracturing in their interiors.

9.4 Gaspra structures

Grooves on Gaspra (Veverka *et al.*, 1994) are more globally organized (within the approximately 25% of the surface that is well imaged) than those on Eros (Figure 6.12). Most occur in what may be either two groups of slightly different orientations, or as part of a globally curving pattern. Thomas *et al.* (1994) showed that the asteroid also had two nearly flat facets that imply a structural fabric extending through much of Gaspra. The effect on its shape presumably arises from fracturing and spallation during very large impacts. Gaspra also shows a few ridges

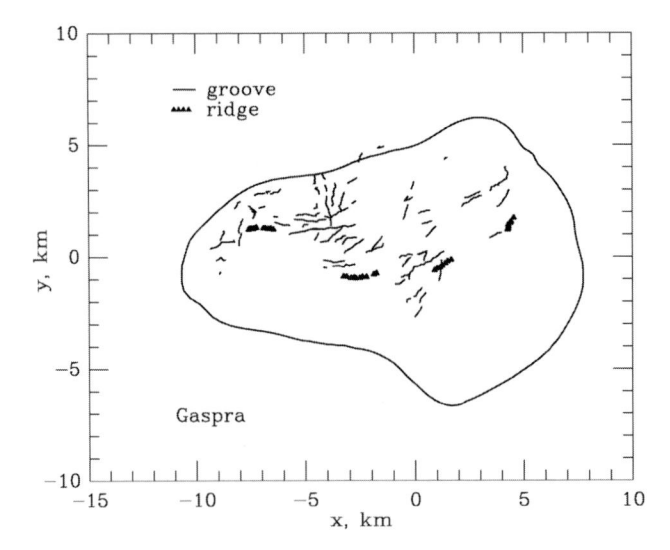

Figure 6.12. Gaspra groove and ridge patterns. Projection of 3-D positions from 90°N, equatorial profile for reference. Pattern covers the well-imaged area. Ridge at the right (x = 4.5, y = 1.2) deforms the interior of a crater.

that may be part of a curving pattern. Most notable among these is a ridge that deformed a small crater near the larger end of the object, and was the first indication of compressive failure on a small object (Thomas *et al.*, 1994).

10 Overview and outstanding questions

Objects that have no internally driven tectonics display a remarkable variety of forms, generated primarily by impacts and tidal stresses. Different compositions, sizes, and orbital histories may explain much of this variety.

Our knowledge of the tectonics and structures of small bodies is demonstrably crude at the moment. The limitations of a few examples of remote sensing of objects covered by loose debris in interpreting interior features are all too clear. Yet our experience is sufficient to pose questions in ways that future missions may provide more definitive answers. These interior views are still important in evaluating the assembly and evolution of solar system objects.

The clear evidence of many styles of fracturing and a wide variety of porosities in these objects should allow design of seismic and radar sounding probes of the interiors. Better estimates of regolith cover can help position active and passive interior investigations. More successful associations of meteorites with specific asteroids will help place the structures in more specific contexts.

Acknowledgments

We thank Karla Consroe and Brian Carcich for technical support, and the reviewers for significant help. The work necessarily includes results of long-term discussions with numerous colleagues.

Glossary

Asteroid: One of the small bodies orbiting the Sun, chiefly between the orbits of Mars and Jupiter. The largest, Ceres, is less than 1000 km across; the total mass of the asteroids (many thousands) is much less than that of the Earth's Moon.
Astronomical unit (AU): mean distance of the Earth from the Sun: 149 598 000 km.
Comet: solar system body, visible from Earth chiefly by its release of volatiles and gas from solar heating. The cometary nucleus is the actual solid body, usually less than 20 km across. Their orbits range from nearly parabolic to nearly circular.
Gravitational slopes: angle between a surface normal and the local gravitational acceleration. For small, rapidly rotating objects this may be non-intuitive, as the rotational accelerations can change the net gravity significantly from a radius vector.
Regolith: loose, particulate material on the surface of a planetary body. Usually thought of as impact generated, but may apply to any unconsolidated deposit.

References

Aggarwal, H. R. and Oberbeck, V. R. (1974). Roche limit of a solid body. *Astrophys. J.*, **191**, 577–588.
A'Hearn, M. F., Millis, R. L., Schleicher, D. G., Osip, D. J., and Birch, P. V. (1992). The ensemble properties of comets: Results from narrow band photometry of 85 comets, 1976–1992. *Icarus*, **118**, 223–270.
A'Hearn, M. F. *et al.* (2005). Deep impact: Excavating comet Tempel 1. *Science*, **310**, 258–264.
Asphaug, E. and Benz, W. (1996). Size, density and structure of comet Shoemaker-Levy 9 inferred from the physics of tidal breakup. *Icarus*, **121**, 225–248.
Asphaug, E. and Melosh, H. J. (1993). The Stickney impact of Phobos: A dynamical model. *Icarus*, **101**, 144–164.
Asphaug, E. *et al.* (1996). Mechanical and geological effects of impact cratering on Ida. *Icarus*, **120**, 158–184.
Asphaug, E., Ostro, S. J., Hudson, R. S., Scheeres, D. J., and Benz, W. (1998). Disruption of kilometre-sized asteroids by energetic collision. *Nature*, **393**, 437–440.
Bell, J. F., Davis, D. R., Hartmann, W. K., and Gaffey, M. J. (1989). Asteroids: The big picture. In *Asteroids II*, ed. R. P. Binzel, T. Gehrels and M. S. Mathews. Tucson, AZ: University of Arizona Press, pp. 921–945.
Bell III, J. F. *et al.* (2002). Near-IR reflectance spectroscopy of 433 Eros from the NIS instrument on the NEAR mission 1: Low phase angle observations. *Icarus*, **155**, 119–144.
Belton, M. J. S. *et al.* (1996). The discovery and orbit of 1993 (243) 1 Dactyl. *Icarus*, **120**, 185–199.
Benner, L. *et al* (2005). Radar images of near-Earth asteroid 2005 CR37 (abs.). Division of Planetary Sciences, 37, 15.02, 639.
Binzel, R. P. and Xu, S. (1993). Chips off of asteroid 4 Vesta: Evidence for the parent body of achondritic meteorites. *Science*, **260**, 186–191.
Binzel, R. P., Lupshko, D. F, Di Martino, M., Whiteley, R. J., and Hahn, G. J. (2002). Physical properties of near-Earth objects. In *Asteroids III*, ed. W. Bottke, A. Cellino, P. Paolicci and R. Binzel. Tucson, AZ: University of Arizona Press, pp. 255–271.

Binzel, R. P. et al. (2003). Interiors of small bodies: Foundations and perspectives. *Planet. Space Sci.*, **51**, 443–454.
Bottke, W. F. Jr., Richardson, D. C., Michel, P., and Love, S. G. (1999). 1620 Geographos and 433 Eros: Shaped by planetary tides? *Astron. J.*, **117**, 921–1928.
Bottke, W. F. Jr., Cellino, A., Paolichi, P., and Binzel, R. P. (2002). An overview of the asteroids: The Asteroids III perspective. In *Asteroids III*, ed. W. F. Bottke, Jr., A. Cellino, P. Paolichi and R. P. Binzel. Tucson, AZ: University of Arizona Press, pp. 3–15.
Britt, D. T. and Consolmagno, G. J. (2001). Modeling the structure of high-porosity asteroids. *Icarus*, **152**, 134–139.
Britt, D. T., Yeomans, D., Housen, K., and Consolmagno, C. (2002). Asteroid density, porosity, and structure. In *Asteroids III*, ed. W. Bottke, A. Cellino, P. Paolicci and R. Binzel, Tucson, AZ: University of Arizona Press, pp. 485–499.
Britt, D. T. et al. (2004). The morphology and surface processes of Comet 19/P Borrelly. *Icarus*, **167**, 45–53.
Brownlee, D. E. et al. (2004). Surface of young Jupiter family comet 81/P Wild 2: View from the Stardust spacecraft. *Science*, **304**, 1764–1769.
Burns, J. A. (1978). The dynamical evolution and origin of the Martian moons. *Vistas Astron.*, **22**, 193–210.
Chapman, C. R. and McKinnon, W. (1986). Cratering of planetary satellites. In *Satellites*, ed. J. Burns and M. Matthews. Tucson, AZ: University of Arizona Press, pp. 492–580.
Chapman, C. R. et al. (1995). Discovery and physical properties of Dactyl, a satellite of asteroid 243 Ida. *Nature*, **374**, 783–785.
Clark, R. N. et al. (2005). Compositional maps of Saturn's moon Phoebe from imaging spectroscopy. *Nature*, **435**, 66–69.
Cheng, A. F. et al. (2002). Small-scale topography of 433 Eros from laser altimetry and imaging. *Icarus*, **155**, 51–74.
Dobrovolskis, A. (1982). Internal stresses in Phobos and other triaxial bodies. *Icarus*, **52**, 136–148.
Dombard, A. J. and Freed, A. M. (2002). Thermally induced lineations on the asteroid Eros: Evidence of orbit transfer. *Geophys. Res. Lett.*, **29**, doi 10.1029/202GL015181.
Duennebier, F. (1976). Thermal movement of the regolith. *Proc. Lunar Sci. Conf. 7*, 1073–1086.
Fujiwara, A. (1991). Stickney-forming impact on Phobos: Crater shape and induced stress distribution. *Icarus*, **89**, 384–391.
Fujiwara, A. and 21 colleagues (2006). The Rubble-pile asteroid Itokawa as observed by Hayabusa. *Science*, **312**, 1330–1334.
Gaffey, M. J., Cloutis, E. A., Kelley, M. S., and Keil, K. S. (2002). Mineralogy of asteroids. In *Asteroids III*, ed. W. Bottke, A. Cellino, P. Paolicci and R. Binzel. Tucson, AZ: University of Arizona Press, pp. 183–204.
Gault, D. E., Quaide, W. L., and Oberbeck, V. R. (1968). Impact cratering mechanics and structures in shock metamorphism of natural materials. In *Shock Metamorphism of Natural Materials*, ed. B. M. French and N. M. Short. Baltimore MD: Mono Book Co., pp. 87–99.
Gladman, B. et al. (2001). Discovery of 12 satellites of Saturn exhibiting orbital clustering. *Nature*, **412**, 163–166.
Goldreich, P., Lithwick, Y., and Sari, R. (2002). Formation of Kuiper-belt binaries by dynamical friction and three-body encounters. *Nature*, **420**, 643–646.

Gradie, J. C. and Tedesco, E. F. (1982). Compositional structure of the asteroid belt. *Science*, **216**, 1405–1407.

Grimm, R. E. and McSween, H. Y. (1993). Heliocentric zoning of the asteroid belt by aluminum-26 heating. *Science*, **259**, 653–655.

Hilton, J. L. (2002). Asteroid masses and densities. In *Asteroids III*, ed. W. Bottke, A. Cellino, P. Paolicci and R. Binzel. Tucson, AZ: University of Arizona Press, pp. 103–112.

Housen, K. R., Holsapple, K. A., and Voss, M. E. (1999). Compaction as the origin of the unusual craters on asteroid Mathilde. *Nature*, **402**, 155–157.

Horstman, K. C., and Melosh, H. J. (1989). Pits in cohesionless materials: Implications for the surface of Phobos. *Icarus*, **94**, 12 433–12 441.

Hudson, R. S., Ostro, S. J., and Scheeres, D. J. (2003). High resolution model of asteroid 4179 Toutatis. *Icarus*, **161**, 346–355.

Johnson, T. V. and McGetchin, T. R. (1973). Topography on satellite surfaces and the shape of asteroids. *Icarus*, **39**, 317–351.

Keil, K. (2002). Geological history of asteroid 4 Vesta: The "smallest terrestrial planet". In *Asteroids III*, ed. W. Bottke, A. Cellino, P. Paolicci and R. Binzel. Tucson, AZ: University of Arizona Press, pp. 573–584.

Konopliv, A. S. et al. (2002). A global solution for the gravity field, rotation, landmarks, and ephemeris of Eros. *Icarus*, **160**, 289–299.

Lee, P. et al. (1996). Ejecta blocks on 243 Ida and on other asteroids. *Icarus*, **120**, 87–105.

Levinson, H. F., Dones, L., and Duncan, M. J. (2001). The origin of Halley-type comets: Probing the Oort cloud. *Astron. J.*, **121**, 2253–2267.

Mantz, A., Sullivan, R., and Veverka, J. (2004). Regolith transport in craters on Eros. *Icarus*, **167**, 197–203.

Margot, J.-L. et al. (2002). Binary asteroids in the near-Earth object population. *Science*, **296**, 1445–1448.

McCoy, T. J. et al. (2001). The composition of 433 Eros: A mineralogical-chemical synthesis. *Meteorit. Planet. Sci.*, **36**, 1661–1672.

McSween, Jr., H. Y., Ghosh, A., Grimm, R. E., Wilson, L., and Young, A. D. (2002). Thermal evolution models of asteroids. In *Asteroids III*, ed. W. Bottke, A. Cellino, P. Paolicci and R. Binzel. Tucson, AZ: University of Arizona Press, pp. 559–571.

Melosh, H. J. (1989). *Impact Cratering*. New York: Oxford University Press.

Melosh, H. J. and Stansberry, J. A. (1991). Doublet craters and the tidal disruption of binary asteroids. *Icarus*, **94**, 171–179.

Merline, W. J., and 12 colleagues (2001). S/2001 (617) 1. International Astronomical Union Circular 7741, 2.

Merline, W. J. et al. (2002). Asteroids do have satellites. In *Asteroids III*, ed. W. F. Bottke Jr., A. Cellino, P. Paolicci and R. Binzel. Tucson, AZ: University of Arizona Press, pp. 289–312.

Michel, P., Benz, W., and Richardson, D. C. (2003). Disruption of fragmented parent bodies as the origin of asteroid families. *Nature*, **421**, 608–611.

Morrison, S., Thomas, P. C., Tiscareno, M. S., Burns, J. A., and Veverka, J. (2009). Grooves on small Saturnian satellites and other objects: Characteristics and significance, *Icarus*, doi:10.1016/j.icarus.2009.06.003.

Murray, C. D. and Dermott, S. F. (2000). *Solar System Dynamics*. Cambridge: Cambridge University Press.

Nicholson, P. D., Hamilton, D. P., Matthews, K., and Yoder, C. F. (1992). New observations of Saturn's coorbital satellites. *Icarus*, **100**, 464–484.

Ostro, S. J. et al. (2000). Radar observations of asteroid 216 Kleopatra. *Science*, **288**, 836–839.
Pravec, P. and Harris, A. W. (2000). Fast and slow rotation of asteroids. *Icarus*, **148**, 12–20.
Prockter, L. et al. (2002). Surface expressions of structural features on Eros. *Icarus*, **155**, 75–93.
Richardson, D. C., Leinhardt, Z. M., Melosh, H. J., Bottke Jr., W. F., and Asphaug, E. (2002). Gravitational aggregates: Evidence and evolution. In *Asteroids III*, ed. W. Bottke, A. Cellino, P. Paolicci and R. Binzel. Tucson, AZ: University of Arizona Press, pp. 501–551.
Rivkin, A. S., Brown, R. H., Trilling, D. E., Bell, J. F., and Plassman, J. H. (2002). Near-infrared spectrophotometry of Phobos and Deimos. *Icarus*, **156**, 64–75.
Robinson, M. S., Thomas, P. C., Veverka, J., Murchie, S. L., and Wilcox, B. B. (2002). The geology of 433 Eros. *Meteorit. Planet. Sci.*, **37**, 1651–1684.
Shoemaker, E. M. (1963). Impact mechanics at Meteor Crater, Arizona. In *The Moon, Meteorites, and Comets*, The Solar System, **4**, ed. B. M. Middlehurst and G. P. Kuiper. Chicago: University of Chicago Press, pp. 301–336.
Soderblom, L. A. et al. (2004). Short-wavelength infrared (1.3–2.6 m) observations of the nucleus of Comet 19P/Borrelly. *Icarus*, **167**, 100–112.
Sonnett, C. P., Colburn, D. S., and Schwartz, K. (1968). Electrical heating of meteorite parent bodies and planets by dynamo induction from a premain sequence T Tauri "solar wind." *Nature*, **219**, 924–926.
Soter, S. and Harris, A. (1977). Are striations on Phobos evidence for tidal stress? *Nature*, **268**, 421–422.
Stevenson, D. J. (2004). Planetary diversity. *Phys. Today*, **57**, 43–48.
Sullivan, R. et al. (1996). Geology of 243 Ida. *Icarus*, **120**, 119–139.
Thomas, P. (1989). The shapes of small satellites. *Icarus*, **77**, 248–277.
Thomas, P. C. et al. (1994). The shape of Gaspra. *Icarus*, **107**, 23–36.
Thomas, P. C. (1998). Ejecta emplacement on the Martian satellites. *Icarus*, **131**, 78–106.
Thomas, P. C. (1999). Large craters on small objects: Occurrence, morphology, and effects. *Icarus*, **142**, 89–96.
Thomas, P., Veverka, J., Bloom, A., and Duxbury, T. (1979). Grooves on Phobos: Their distribution, morphology, and possible origin, *J. Geophys. Res.*, **84**, 8457–8477.
Thomas, P. C., Veverka, J., Morrison, D., Davies, M., and Johnson, T. V. (1982). Saturn's small satellites: Voyager imaging results. *J. Geophys. Res.*, **88**, 8743–8754.
Thomas, P. C. et al. (1996). The shape of Ida. *Icarus*, **120**, 20–32.
Thomas, P. C. et al. (2000). Phobos: Regolith and ejecta blocks investigated with Mars Orbiter camera images. *J. Geophys. Res.*, **105**, 15 091–15 106.
Thomas, P. C. et al. (2002). Eros: Shape, topography, and slope processes. *Icarus*, **155**, 18–37.
Van Flandern, T. C., Tedesco, E. F., and Binzel, R. P. (1979). Satellites of asteroids. In *Asteroids*, ed. T. Gehrels. Tucson, AZ: University of Arizona Press pp. 443–465.
Veverka, J. and Duxbury, T. (1977). Viking observations of Phobos and Deimos: Preliminary results. *J. Geophys. Res.*, **82**, 4213–4223.
Veverka, J. et al. (1994). Discovery of grooves on Gaspra. *Icarus*, **107**, 72–83.
Veverka, J. et al. (1997). NEAR's flyby of 253 Mathilde: Images of a C asteroid. *Science*, **278**, 2109–2114.

Veverka, J. et al. (1999). NEAR encounter with asteroid 253 Mathilde: Overview. *Icarus*, **140**, 3–16.

Veverka, J. et al. (2001). Imaging of small-scale features on 433 Eros from NEAR: Evidence for a complex regolith. *Science*, **292**, 484–488.

Watters, T. R. and Robinson, M. S. (2003). Boundary element modeling of the Rahe Dorsum thrust fault on asteroid 433 Eros (abs.). *Lunar Planet. Sci. Conf. XXXIV*, 1928. Houston, TX: Lunar and Planetary Institute (CD-ROM).

Weidenschilling, S. J. (1979). A possible origin for the grooves on Phobos. *Nature*, **282**, 697–698.

Weidenschilling, S. J., Paolicchi, P., and Zappala, V. (1989). Do asteroids have satellites? In *Asteroids II*, ed. R. P. Binzel, T. Gehrels, and M. S. Mathews. Tucson, AZ: University of Arizona Press, pp. 643–660.

Wilkison, S. L. et al. (2002). An estimate of Eros's porosity and implications for internal structure. *Icarus*, **155**, 94–103.

7

Tectonics of the outer planet satellites

Geoffrey C. Collins
Wheaton College, Norton

William B. McKinnon
Washington University, Saint Louis

Jeffrey M. Moore
NASA Ames Research Center, Moffett Field

Francis Nimmo
University of California, Santa Cruz

Robert T. Pappalardo
Jet Propulsion Laboratory, California Institute of Technology, Pasadena

Louise M. Prockter
Applied Physics Laboratory, Laurel

and

Paul M. Schenk
Lunar and Planetary Institute, Houston

Summary

Tectonic features on the satellites of the outer planets range from the familiar, such as clearly recognizable graben on many satellites, to the bizarre, such as the ubiquitous double ridges on Europa, the twisting sets of ridges on Triton, or the isolated giant mountains rising from Io's surface. All of the large and middle-sized outer planet satellites except Io are dominated by water ice near their surfaces. Though ice is a brittle material at the cold temperatures found in the outer solar system, the amount of energy it takes to bring it close to its melting point is lower than for a rocky body. Therefore, some unique features of icy satellite tectonics may be influenced by a near-surface ductile layer beneath the brittle surface material, and several of the icy satellites may possess subsurface oceans. Sources of stress to drive

Planetary Tectonics, edited by Thomas R. Watters and Richard A. Schultz. Published by Cambridge University Press. Copyright © Cambridge University Press 2010.

tectonism are commonly dominated by the tides that deform these satellites as they orbit their primary giant planets. On several satellites, the observed tectonic features may be the result of changes in their tidal figures, or motions of their solid surfaces with respect to their tidal figures. Other driving mechanisms for tectonics include volume changes due to ice or water phase changes in the interior, thermoelastic stress, deformation of the surface above rising diapirs of warm ice, and motion of subsurface material toward large impact basins as they fill in and relax. Most satellites exhibit evidence for extensional deformation, and some exhibit strike-slip faulting, whereas contractional tectonism appears to be rare. Io's surface is unique, exhibiting huge isolated mountains that may be blocks of crust tilting and foundering into the rapidly emptying interior as the surface is constantly buried by deposits from hyperactive volcanoes. Of the satellites, diminutive Enceladus is spectacularly active; its south polar terrain is a site of young tectonism, copious heat flow, and tall plumes venting into space. Europa's surface is pervasively tectonized, covered with a diverse array of exotic and incompletely understood tectonic features. The paucity of impact craters on Europa suggests that its tectonic activity is ongoing. Geysers on Triton show that some degree of current activity, while tectonic features that cross sparsely cratered terrain indicate that it may also be tectonically active. Ganymede and Miranda both exhibit ancient terrains that have been pulled apart by normal faulting. On Ganymede these faults form a global network, while they are confined to regional provinces on Miranda. Ariel, Dione, Tethys, Rhea, and Titania all have systems of normal faults cutting across their surfaces, though the rifting is less pronounced than it is on Ganymede and Miranda. Iapetus exhibits a giant equatorial ridge that has defied simple explanation. The rest of the large and middle-sized satellites show very little evidence for tectonic features on their surfaces, though the exploration of Titan's surface has just begun.

1 Introduction

The four new worlds orbiting Jupiter, as reported in Galileo's *Siderius Nuncius* (Galilei, 1610), forever changed humankind's worldview by demonstrating that Jupiter, like Earth, is a center of celestial motion, in strong support of the Copernican model of the heavens (Copernicus, 1543). Four centuries later, the consequences of this seemingly simple motion of satellites about their primary planet are still being understood. On the worlds now known as the Galilean satellites of Jupiter, as well as the satellites of the other giant planets, Saturn, Uranus, and Neptune, the interactions of tides with satellite interiors has played a large part in determining geological activity. Tides raised by the giant planets power volcanoes on Io, maintain a liquid water ocean beneath Europa's ice, drive plumes on Enceladus, and probably played a large role in fracturing the surfaces of Ganymede, Dione, and other satellites.

We begin this chapter by examining the behavior of ice lithospheres, contrasting the rheology of ice to the more familiar rock behavior on the inner planets. We then summarize the global and local stress mechanisms that can affect outer planet satellites, as linked to their geophysics and geodynamics. After this background, we examine the tectonics on groups of outer planet satellites, starting with Io, the only large rocky outer planet satellite, then proceeding from currently active icy satellites, to satellites that formerly had tectonic activity, to satellites with very little evidence of any activity. We rely on insights gained from analysis of data from the twin Voyager spacecraft, which flew through all of the giant planet systems between 1979 and 1989; from the Galileo spacecraft, which orbited Jupiter from 1995 to 2003; and from the Cassini mission, which arrived in orbit around Saturn in 2004. The chapters that comprise Burns and Matthews (1986), Morrison (1982), Gehrels and Matthews (1984), Bergstralh *et al.* (1991), and Cruikshank (1995) offer excellent reviews of Voyager-based understanding of the satellites. Comprehensive Galileo-based syntheses of Jupiter's satellites are provided by several chapters of Bagenal *et al.* (2004), and the understanding of Saturn's satellites from Cassini is currently evolving as the spacecraft mission proceeds.

2 Rheology of ice

2.1 Introduction

The observable consequences of tectonic stresses depend mainly on the response of the material being stressed, i.e., its rheology. Hence, understanding the rheology of ice and rock is of fundamental importance to interpreting the tectonics of outer solar system satellites. In this section, we focus on the rheology of water ice, because its behavior is less well known to terrestrial geologists and it has some important differences to the behavior of rock. In particular, we discuss the ways in which ice may respond to imposed stresses, and the consequences of these different response mechanisms.

It is well known that a material's rheological properties depend on its homologous temperature, that is the ratio of its absolute temperature to the melting temperature (Frost and Ashby, 1982). On Earth, ice is rarely at temperatures lower than about 80% of the melting point (a homologous temperature of 0.8). On the icy satellites, typical surface temperatures correspond to a homologous temperature of about 0.4, similar to that of rocks at the surface of the Earth. Thus, one would expect the behavior of ice at the surface of the icy satellites to mimic that of rocks on Earth, and this is exactly what is observed; as discussed later in this chapter, many tectonic features observed on icy satellites have counterparts on the Earth.

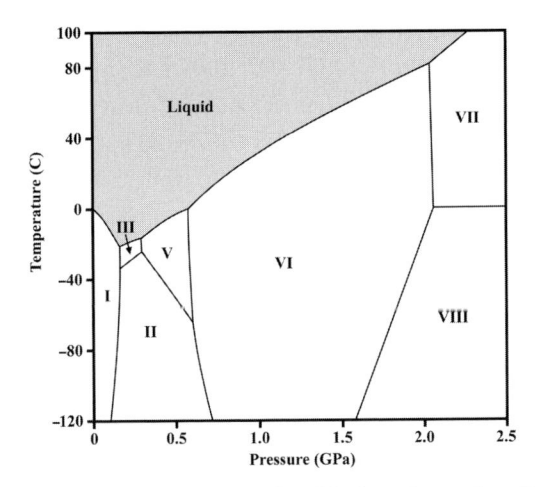

Figure 7.1. Phase diagram of water ice. Ice I is less dense than liquid water, but the other solid ice phases are more dense than the liquid.

As we also discuss further below, water ice has several important differences when compared to silicate materials. First of all, solid ice is less dense than its molten equivalent (water). This makes the surface eruption of water difficult, and means that any density-driven overturn (e.g. solid state convection or subduction) would have to take place entirely within a floating ice shell. In contrast to most terrestrial contexts, a modest thermal gradient can allow ice to reach a high homologous temperature at relatively shallow depths (several km). Ice close to its melting point flows more readily than silicates. Thus, viscous flow timescales are much shorter in icy bodies than in their silicate equivalents (Section 2.4), and a ductile layer may be more shallow, with more influence on the surface, than in a typical terrestrial context. In addition to being more ductile, ice is intrinsically weaker than rock in that it is less rigid (lower Young's modulus, Section 2.2), and undergoes brittle tensile and compressive failure at lower stresses (Weeks and Cox, 1984).

As with silicate materials, ice transforms to denser phases at elevated pressures (Figure 7.1). These higher pressure ice phases have a higher density than liquid water, rather than possessing the peculiar density and melting behavior of low-pressure ice I. The rheological properties of high-pressure ice phases are not as well known as the properties of ice I. However, because they are likely only found at depths of several hundred kilometers within the largest icy satellites (Titan, Ganymede, Callisto, and perhaps Rhea and Triton), they are usually less directly relevant to near-surface tectonic processes.

Just as with silicate materials, ice under stress can respond in one of three idealized ways. At low stresses and strains, the ice will deform in an *elastic* (recoverable) manner, but at strains greater than roughly 10^{-4}, the ice will undergo irrecoverable deformation. At low temperatures and relatively high strain rates, this deformation will be accomplished by *brittle* failure. At higher temperatures and lower strain rates, the result will be *ductile* behavior or *creep*. A closer approximation to reality is to say that materials behave in an *elastoviscoplastic* manner, combining elements of elastic and ductile behavior, as well as brittle or "plastic" failure.

2.2 Elastic deformation

At low stresses and strains, ice will deform elastically, and the relationship between stress σ and strain ε depends on the Young's modulus E of the material as $\sigma = E\varepsilon$ (for uniaxial deformation). Measurements of Young's modulus in small laboratory specimens of ice are straightforward and yield a value near 9 GPa (Gammon *et al.*, 1983). However, the effective Young's modulus of large bodies of deformed ice is less obvious (e.g., Nimmo, 2004a). Observations of ice shelf response to tides on Earth (Vaughan, 1995) give an effective $E \sim 0.9$ GPa, an order of magnitude smaller than the laboratory values. This discrepancy is most likely due to the fact that a large fraction of the ice shelf thickness is not responding in a purely elastic fashion (e.g., Schmeltz *et al.*, 2002). On icy satellites, porosity and/or fracturing in the near-surface may result in a reduction in the effective value of E (Nimmo and Schenk, 2006).

2.3 Brittle deformation

At shallow depths, where the overburden pressure is low, ice can undergo tensile failure (for a review of ice fracture, see Schulson, 2006). The tensile strength of ice at temperatures above $-50\,°$C has been studied for glaciological applications, but preliminary experiments of tensile failure of polycrystalline ice at much colder temperatures relevant to icy satellite surfaces (Sklar *et al.*, 2008) show that tensile strength rises with decreasing temperature, up to a few MPa. At greater depths, the overburden pressures are such that failure occurs by shear motion. For silicate materials, shear failure on preexisting faults occurs when the shear stresses exceed some fraction, typically 0.85, of the normal stresses (e.g., Turcotte and Schubert, 2002). This behavior is largely independent of composition and is known as Byerlee's rule. Cold ice in the laboratory obeys a Byerlee-like rule at low sliding velocities and stresses, with a coefficient of friction $\mu_f \approx 0.55$ (Beeman *et al.*, 1988). At higher sliding velocities, the behavior becomes more complex (Rist, 1997), with ice friction dependent on temperature at sufficiently high sliding velocities (Petrenko and Whitworth, 1999; Maeno and Arakawa, 2004). Brittle deformation is expected to dominate on icy satellites at shallow depths where normal stresses are small and temperatures are low.

2.4 Ductile deformation

At sufficiently high temperatures, ice responds to applied stress by deforming in a ductile fashion, also known as creep. The response is complicated by the fact that individual ice crystals can deform in several different ways: by diffusion of defects within grain interiors, by sliding of grain boundaries, and by dislocation creep (Goldsby and Kohlstedt, 2001; Durham et al., 2001). Which mechanism dominates depends on the specific stress and temperature conditions, but each individual mechanism can be described by the following generalized equation:

$$\dot{\varepsilon} = A\sigma^n d^{-p} \exp\left(-\frac{Q + PV}{RT}\right). \tag{7.1}$$

Here $\dot{\varepsilon}$ is the resulting strain rate of the deforming ice, σ is the differential applied stress, A, n and p are experimentally determined constants, d is the grain size, Q and V are the experimentally determined activation energy and volume, respectively, R is the gas constant, and P and T are pressure and temperature, respectively. Here the strain rate and stress variables are scalar representations of the appropriate tensors. For icy satellites, the PV contribution is generally small enough to be ignored, and strain rates increase with increasing temperature and stress and decreasing grain size, as expected. Because several different deformation mechanisms can operate together, the total strain rate is a function of the individual strain rates (see Equation 3 in Goldsby and Kohlstedt, 2001).

For a given ice grain size, the applied stress and temperature control the deformation mechanism (see Figure 1 in Barr and Pappalardo, 2005). At low stresses and high temperatures, diffusion creep is expected to dominate and is predicted to result in Newtonian flow (that is, $n = 1$, a linear relationship between stress and strain rate) with a grain-size dependence ($p = 2$). At higher stresses and lower temperatures, the dominant creep regimes are basal slip and grain boundary sliding, which result in non-Newtonian behavior ($n \sim 2$) and grain-size dependence. At even higher stresses, strongly non-Newtonian dislocation creep ($n = 4$) dominates. Deformation rates are enhanced within about 20 K of the melting temperature (Goldsby and Kohlstedt, 2001), presumably because of the presence of thin films of water along grain boundaries (e.g., De La Chapelle et al., 1999).

Because stresses and to some extent strain rates within the warm interiors of icy satellites are expected to be low, the most relevant deformation mechanism within the warm icy interior is probably diffusion creep (McKinnon, 2006; Moore, 2006), while other mechanisms may be important at higher stresses (for instance, near the surface). Diffusion creep has the advantage of resulting in Newtonian behavior, but the disadvantage that the viscosity ($\eta \approx \sigma/\dot{\varepsilon}$) is dependent on the grain size. Ice grain size evolution is poorly constrained, because it depends both on the presence of secondary (pinning) phases and because of dynamic recrystallization processes (e.g., Barr and McKinnon, 2007). Given the uncertainties, it is often acceptable to assume for modeling purposes that ice has a Newtonian viscosity near its melting temperature in the range 10^{13}–10^{15} Pa s (e.g., Pappalardo et al., 1998a). However, more complex models taking into account the non-Newtonian and viscoelastic behavior of ice have also been developed (e.g., Dombard and McKinnon, 2006a), defining an effective viscosity assuming an effective composite strain rate.

Although grain size is a major unknown for describing the ductile deformation of ice in outer planet satellites, other effects can also be important. The presence of even small amounts of fluid significantly enhances creep rates (e.g., De La Chapelle et al., 1999). On the other hand, the presence of rigid impurities (e.g., silicates) at levels greater than 10 percent serves to increase the viscosity (Friedson and Stevenson, 1983; Durham et al., 1992). Finally, some higher pressure phases of ice, clathrates incorporating other chemical species such as methane, and ice with hydrated sulfate salts, all tend to have much higher viscosities than pure water ice at the same P,T conditions (Durham et al., 1998, 2005; McCarthy et al., 2007), whereas ammonia–water ice mixtures and other ice phases can be weaker (Durham et al., 1993; Durham and Stern, 2001).

2.5 Viscoelastic behavior

In reality, materials do not exhibit entirely elastic or entirely viscous behavior. Rather, they exhibit elastic-like behavior if the timescale over which deformation occurs is very short compared to a characteristic deformation timescale of the material (proportional to η/E), known as the Maxwell time τ_M (Turcotte and Schubert, 2002). Conversely, if the deformation timescale is long compared to the Maxwell time, the material behaves in a viscous fashion. Such compound materials are termed viscoelastic. The principal importance of viscoelastic materials with reference to icy satellites is that the amount of tidal heating in an ice layer may be controlled by the viscoelastic properties of the ice, in particular its viscosity structure (e.g., Ross and Schubert, 1986).

2.6 Application to icy satellites

Temperatures near the surface of icy satellites are sufficiently cold, and overburden pressures sufficiently small, that tectonic stresses are likely to result in brittle deformation. However, at greater depths, temperatures will increase, allowing ductile deformation to dominate. At any particular depth, the mechanism with the lowest yield stress (ductile/elastic/brittle) dominates. For example, in pure extension, brittle failure at the surface transitions to ductile flow at depth. As another example, if

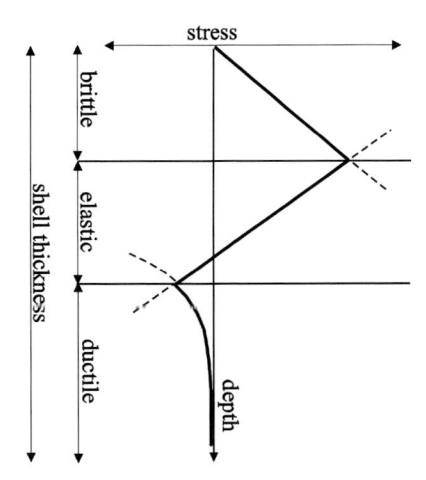

Figure 7.2. Schematic stress profile within generic icy satellite shell. This example shows how the mechanical layers interact in the case of bending the shell, with opposite stresses at the top and bottom. Near the surface the ice is cold and brittle and stress increases proportionately with overburden pressure. The elastic bending stresses decrease towards the midpoint of the shell and lead to an elastic "core"; the slope of the elastic stress curve depends on the local curvature of the shell and the Young's modulus (e.g., Turcotte and Schubert, 2002). At greater depths, the ice is sufficiently warm that the shell deforms in a ductile fashion. The first moment of the stress profile about the midpoint controls the effective elastic thickness T_e of the ice shell as a whole (e.g., Watts, 2001). Note that T_e is dominated by the brittle and elastic portions of the shell (see Nimmo and Pappalardo, 2004).

the principal source of stress is bending (e.g., due to fault motion), then near the mid-plane of the bending shell the stresses are sufficiently low that deformation is accommodated elastically. Thus, one would expect an ice layer deformed in bending to consist of three regions (Figure 7.2): a brittle near-surface layer; an elastic "core"; and a ductile base (e.g., Watts, 2001). The upper and lower halves of the shell experience opposite stresses, hence the change in sign.

The thickness of the near-surface brittle layer in Figure 7.2 depends mainly on the temperature gradient, and to a lesser extent on the degree of curvature (bending). Thus, if the brittle layer thickness can be constrained, for example by observations of fault spacing (Jackson, 1989), then the temperature structure can be deduced for a given tectonic interpretation (e.g., Golombek and Banerdt, 1986).

It has long been recognized that the brittle/elastic/ductile structure encountered in real silicate or icy bodies can be more simply modeled as a purely elastic layer overlying an inviscid interior (e.g., McNutt, 1984). The thickness of this modeled elastic layer, termed the effective elastic thickness T_e, may in some cases be deduced from observations such as surface topography when a load is bending the elastic layer (e.g., Watts, 2001). Furthermore, the value of T_e thus deduced may be related back to the more realistic stress profile shown in Figure 7.2 if the topographic curvature is known (McNutt, 1984; Watts, 2001). Because the stress profile depends primarily on temperature gradient, determination of the effective elastic thickness of the lithosphere can be used along with strength envelopes such as that shown in Figure 7.2 to determine the thermal structure of ice shells (e.g., Nimmo and Pappalardo, 2004). For a conductive ice layer, knowing the thermal structure in turn allows the layer thickness to be deduced.

Many of the equations used in the rest of this chapter implicitly make the simplifying assumption that tectonic behavior on icy satellites is adequately modeled using the elastic lithosphere model. We emphasize here that T_e is the thickness of the elastic layer which deforms in a manner equivalent to the more complicated strength envelope likely to be present in real satellites. The utility of T_e is that it can be measured directly, and that it may be used to infer the real strength envelope and thus the thermal gradient. The drawback of this approach is that even cold ice topography may viscously relax over geologic time (e.g., Dombard and McKinnon, 2006a).

2.7 Comparison with silicate behavior

Much of the above analysis can be equally applied to silicate materials, such as the lithosphere of Io. However, there are some important differences. Silicate materials have higher Young's moduli ($\sim$100 GPa) and near their melting point have viscosities $\sim 10^4$–10^8 times higher than ice, implying much longer Maxwell timescales. Silicate materials have much higher melting temperatures ($\sim$950 to 1500 K, or higher, depending on pressure; Best, 2003) than ices, so a given homologous temperature is achieved at much greater depth within a rocky body than an icy body for a given thermal gradient. The low relative density of silicate melts means that the melting products are much more likely to be erupted to the surface than is water on an icy body. Silicate melts are typically more viscous than water, which means melt drainage is slower and thus that partially molten regions are likely to persist for longer in silicate systems than in ices.

3 Global and local stress mechanisms

The tectonic features observed on a satellite provide clues to its geological and orbital history. In order to understand the origin of the strain represented by landforms on outer planet satellites, we must understand the possible range of stress mechanisms that may have operated on the satellites. Here we review processes likely to be most relevant, grouping them into global and local mechanisms. A seminal review of this material is provided by Squyres and Croft (1986). As before, our focus will be on icy satellites; where appropriate we will also mention mechanisms relevant to silicate bodies.

3.1 Background

3.1.1 Satellite figures

The global stresses experienced by satellites commonly relate to changes in their shape, so we present a brief discussion of satellite figures here. Satellite figures depart from a spherical shape primarily due to rotation and tides. The relative magnitude of the two effects is given by $\omega^2 a_s^3/3GM_p$ (Murray and Dermott, 1999), where $\omega = 2\pi/P_s$ is the rotational frequency of the satellite, a_s is the semimajor axis of its orbit, G is the gravitational constant, and M_p is the mass of the primary planet. For satellites in synchronous rotation (P_s = orbital period = rotational period), the tidal and rotational potentials that characterize the satellite shape are in the ratio 3:1. Thus the tidal bulge of a synchronously rotating satellite is larger than the rotational bulge along the equator.

The equilibrium shape of a satellite is the shape it would assume if it is in hydrostatic equilibrium, that is, if its shape approximates that of a fluid that is incapable of supporting global-scale forces by elastic (or dynamic) stresses. In this case, the satellite will have an ellipsoidal shape with three unequal axes a (the tidal axis, oriented along a line from the center of the satellite to the center of the primary planet), b (oriented tangent to the satellite's orbit), and c (the spin axis of the satellite). For an equilibrium satellite shape in synchronous orbit, the ratio $(a-c)/(b-c) = 4$ (with a small correction if the satellite is spinning rapidly), while the magnitudes of $(a-c)$ and $(b-c)$ depend on the rotation rate and internal density distribution (Murray and Dermott, 1999). If the rotation rate or magnitude of the tide changes, then so too will the shape of the satellite, generating global-scale patterns of stress.

3.1.2 Rigidity and effective elastic thickness

Outer planet satellites tend to have low surface temperatures, and as a result the near-surface region, hereafter referred to as the lithosphere, will be cold and rigid. As discussed above (Section 2.6), it is a convenient simplification to describe this near-surface region as a fully elastic layer overlying an inviscid interior. The thickness of this layer, T_e, is termed the effective elastic thickness of the lithosphere, and depends on the strength envelope (e.g., Figure 7.2) and thus the thermal gradient within the lithosphere. The lithosphere is able to support loads of some spatial scale over geological timescales, and is also the region in which permanent tectonic deformation (such as faulting) is recorded. The elastic thickness of the lithosphere T_e controls the characteristic bending (or flexural) wavelength α at which deformation occurs in response to loading, and is given by (e.g., Turcotte and Schubert, 2002)

$$\alpha^4 = \frac{1}{3}\frac{ET_e^3}{(1-\nu^2)\rho g}, \tag{7.2}$$

where E is the Young's modulus, ν is Poisson's ratio, ρ the density of subsurface material being displaced (ductile ice or water), and g is gravity.

Many of the expressions given below assume, usually implicitly, that the satellite's figure is well described by hydrostatic equilibrium. This approximation is generally a good one for the large satellites. However, a thick lithosphere (either near-surface, or in a rocky core, if present) may invalidate the hydrostatic assumption, as is relevant to Ganymede (where non-hydrostatic mass anomalies have been inferred within the ice shell or core; Palguta et al., 2006) and potentially Callisto (McKinnon, 1997), and particularly for the mid-sized icy satellites of Saturn and Uranus (e.g., Rhea; Mackenzie et al., 2008). Long-wavelength loads are more readily supported on smaller planetary bodies due to the influence of membrane stresses (Turcotte et al., 1981). The most serious consequence of this result is that first-order interior structure models making use of the hydrostatic assumption may be incorrect (e.g., McKinnon, 1997). Increased rigidity will also reduce the amplitude of tidal and rotational deformation, as described below. A satellite with significant rigidity may also retain its shape from an epoch when the tidal and rotational characteristics were different from their present-day values (e.g., Iapetus), complicating present-day analysis.

3.1.3 Tides

With the significant exception of Hyperion at Saturn, all the regular satellites of the outer solar system are in (or very close to) synchronous rotation, showing the same face toward the primary planet. This is because tidal torques exerted by the primary on each satellite act to quickly tidally evolve any initially rapidly spinning satellites into this configuration (Peale, 1977; Murray and Dermott, 1999). Tides not only influence the orbital and rotational evolution of the satellites, but are a major source of stress and heat (e.g., Peale, 2003). Thus, one key manner in which satellite tectonics and geophysics differ from those of the terrestrial planets is in the influence of tides.

A satellite in orbit around its primary will experience a tidal bulge due to the difference in gravitational attraction of the primary from the near to the far side of the satellite. If a synchronously rotating satellite's orbital eccentricity is zero, the tidal bulge will then be at a fixed geographical point, and of constant amplitude. The maximum amplitude H of this static (or permanent) tidal bulge is given by (e.g., Murray and Dermott, 1999)

$$H = h_2 R_s \left(\frac{M_P}{M_s}\right) \left(\frac{R_s}{a_s}\right)^3, \quad (7.3)$$

where R_s is the satellite radius, a_s is the semimajor axis, M_p and M_s are the masses of the primary and satellite, respectively, and h_2 is a Love number that describes the radial response of the satellite to a gravitational potential. The Love number h_2 has a value of 5/2 for an idealized incompressible fluid body with uniform density ρ, but this is reduced as the rigidity (or shear modulus) μ of the satellite increases. For a homogeneous satellite, the reduction is a factor of $1 + \frac{19}{2}\tilde{\mu}$, where $\tilde{\mu} = \mu/\rho g R_s$ is a dimensionless measure of the importance of the rigidity (Murray and Dermott, 1999). A thin ice shell decoupled from the underlying material by an ocean has a low global rigidity, so a satellite with a thin floating ice shell can have a tidal response that approaches that of a fluid, provided that the entire interior can respond as a fluid on a sufficiently short timescale (e.g., Moore and Schubert, 2003).

For a synchronous satellite with non-zero orbital eccentricity, the magnitude of the tide varies as the satellite moves closer and farther from the primary planet, generating a time-varying radial tide. Moreover, the position of the sub-planet point will oscillate in longitude over the course of an orbit, and the amplitude and longitudinal position of the tidal bulge will thus vary in time, generating a librational tide. The combined result of the radial and librational tides is a time-varying diurnal tide with an amplitude given by $3eH$, where e is the eccentricity (e.g., Greenberg et al., 2002). Table 7.1 gives nominal permanent and diurnal tidal bulge amplitudes for the satellites of the outer solar system. If the obliquity (the angle between the spin pole and the normal vector to the satellite's orbital plane) of a satellite is non-zero, then the sub-planet point will librate in latitude as well as longitude. Such obliquity librations are potentially an additional source of tidal stress (Bills, 2005).

A time-varying diurnal tide leads to time-varying stresses within the satellite. The lithosphere can flex elastically from the diurnal tide, but if the viscoelastic interior has a Maxwell time (Section 2.5) comparable to the orbital period, then the tidal energy can be dissipated as heat within the satellite (e.g., Peale, 2003). Although the diurnal tidal strains are generally small (10^{-5} or less), the periods are short, and thus the rate of tidal heating can be considerable, as at Io (Peale et al., 1979). Possible additional sources of tidal dissipation include obliquity variation (Bills, 2005), forced libration (Wisdom, 2004), or a large impact that triggers transient chaotic rotation (Marcialis and Greenberg, 1987).

Because the orbital speed of a satellite in an eccentric orbit varies while its rotation rate stays constant, the diurnal tidal component will both lead and lag the sub-planet point over the course of one orbit, generating a net non-zero torque (Goldreich, 1966; Greenberg and Weidenschilling, 1984). If this torque is not opposed by a sufficiently large permanent mass asymmetry, it can produce a rotation rate slightly faster than synchronous, so that the geographic location of the sub-planet point very slowly changes over time. As shown below, stresses due to both diurnal tides and nonsynchronous rotation can have important tectonic effects.

Table 7.1. *Properties of the outer planet satellites*

	R (km)	M_s (10^{20} kg)	ρ (kg m^{-3})	a_s (10^3 km)	P (days)	e	H (m)	$3eH$ (m)	Tidal (GW)	Rad (GW)	Stress (kPa)
Io	1821	893	3530	422	1.77	.0041	7780	96	93000	310	158
Europa	1565	480	2990	671	3.55	.0101	1964	59	8100	130	110
Ganymede	2634	1482	1940	1070	7.15	.0015	1259	5.7	73	190	6.5
Callisto	2403	1076	1850	1883	16.7	.007	220	4.6	15	130	5.7
Mimas	198	0.37	1150	186	0.942	.0202	9180	556	780	0.0098	8400
Enceladus	252	1.08	1610	238	1.37	.0045	3940	53	20	0.094	630
Tethys	533	6.15	970	295	1.89	0	7270	0	0	0.016	0
Dione	562	11	1480	377	2.74	.0022	2410	16	8.6	0.77	85
Rhea	764	23	1230	527	4.52	.001	1440	4.3	0.67	0.85	17
Titan	2575	1346	1880	1222	15.95	.0292	254	22	450	170	26
Hyperion	185*			1481	21.3	.1042					
Iapetus	736	18	1080	3561	79.3	.0283	5	0.4	2.6×10^{-4}	0.31	1.7
Phoebe				12952	551R	.163					
Miranda	240*	0.66	1200	130	1.41	.0027	4960	40	5.5	0.022	500
Ariel	581*	13.5	1670	191	2.52	.0034	2616	27	37	1.3	140
Umbriel	585	11.7	1400	266	4.14	.005	1151	17	6.7	0.70	87
Titania	789	35.3	1710	436	8.71	.0022	287	1.9	0.14	3.5	7.2
Oberon	761	30.1	1630	583	13.46	.0008	122	0.3	1.8×10^{-3}	2.7	1.2
Triton	1353	214	2060	355	5.88R	0	897	0	0	31	0
Charon	606	15.2	1630	19.6	6.387	0	385	0	0	3.2	0

Data from Yoder (1995) except for Saturn satellite radii and densities, which are from Thomas *et al.* (2007) and Triton and Charon data from Perron *et al.* (2006). H is the permanent tidal bulge assuming $h_2 = 2.5$. $3eH$ is the approximate magnitude of the diurnal tidal bulge; note that this is likely an overestimate for most satellites because the rigidity of their ice shells or cores will reduce h_2. "Tidal" is the tidal heat production assuming a homogeneous body with the Love number $k_2 = 1.5$ (again, a likely overestimate for small satellites) and dissipation quality factor $Q = 100$. "Rad" is the present-day radiogenic heat production assuming a chondritic rate of 3.5×10^{-12} W kg^{-1} for the rocky portions of the bodies. "Stress" is the approximate diurnal tidal stress given by EeH/R where E is Young's modulus (assumed 9 GPa); see Section 3.2.1. The actual tidal stress varies according to location on the surface and position in the orbit. Asterisks denote bodies that are significantly non-spherical, and R denotes retrograde rotation.

Figure 7.3. The time-averaged strain rate as a function of location for diurnal tidal flexing, normalized to a value of unity at the poles, modified from Ojakangas and Stevenson (1989a). Note that the strain rate is highest at the poles and lowest on the tidal axis.

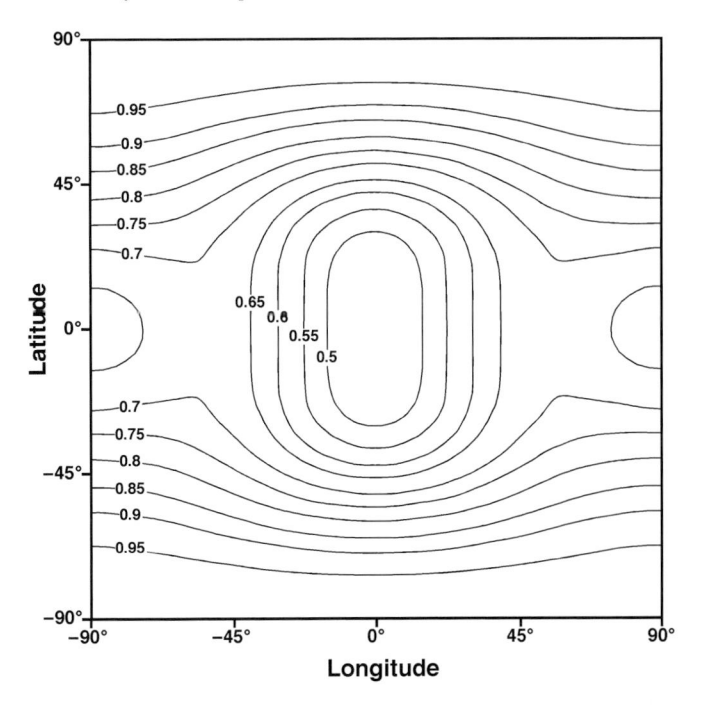

3.2 Global stress mechanisms

Most of the stresses discussed in this section arise from changes in the satellite's figure (subsection 3.1.1). Matsuyama and Nimmo (2008) give detailed expressions for the various kinds of stress-raising mechanisms enumerated here.

3.2.1 Diurnal tides

Subsection 3.1.3 discussed the important role of tides for a satellite in an elliptical orbit, as a function of its rigidity and semimajor axis. The maximum elastic stress generated by the diurnal tide is $\sim 3EeH/R_s$, where E is the Young's modulus of ice, and H is the static tidal bulge amplitude (Equation 7.2); the diurnal tidal stresses are thus a strong function of distance from the primary planet and the total rigidity of the satellite (Table 7.1). Analytical expressions for the different components of the diurnal tidal stress tensor are complicated (Tobie *et al.*, 2005), even in the case of a thin shell (Ojakangas and Stevenson, 1989a). Figure 7.3 shows a map of the average strain rate for a thin shell, demonstrating that strain rates and stresses are, on average, largest at the poles, and minimized at the sub- and anti-primary points. Note that the pattern is quite different for a satellite in which the shell is not thin (Segatz *et al.*, 1988).

Diurnal stresses change with orbital position (Greenberg *et al.*, 1998; Hoppa *et al.*, 1999a). On the equator, the magnitudes and signs of the horizontal principal stresses change as the satellite progresses through an orbit, alternating between compression and tension. Off the equator, the principal stress magnitudes and signs change, and the principal stress orientations also rotate through a diurnal cycle, counterclockwise in the northern hemisphere and clockwise in the southern. Consideration of the resultant out-of-phase variation in normal and shear stresses implies that contraction, extension, and (at high latitudes) opposite-sign horizontal principal stresses that promote strike-slip motions are all possible as a satellite orbits (Hoppa et al., 1999a). This motion can potentially lead to strike-slip displacement along faults, with a predicted preferred sense of left-lateral motion in the northern hemisphere and right-lateral motion in the southern hemisphere (Hoppa et al., 1999a), with the tendency for strike-slip motion increasing with higher latitude. For strike-slip displacement to occur, the fault friction μ_f, overburden stress ($\rho g z$), and tidal normal stress (σ_n) must be sufficiently low so that (according to the Coulomb criterion),

$$|\tau_s| > \mu_f(\rho g z + \sigma_n), \qquad (7.4)$$

where τ_s is the tidal shear stress acting on the fault, and z is depth (e.g., Smith-Konter and Pappalardo, 2008).

3.2.2 Nonsynchronous rotation

As noted above, synchronously rotating satellites in elliptical orbits experience torques that may lead to slow nonsynchronous rotation (NSR) unless permanent mass asymmetries exist that resist the applied torques (Greenberg and Weidenschilling, 1984). In the case of floating ice shells maintained by tidal dissipation, there may be no equilibrium shell configuration that satisfies both thermal and rotational constraints, a situation that also can induce nonsynchronous rotation (Ojakangas and Stevenson, 1989a). Because satellites in synchronous rotation experience more impacts on the leading hemisphere than the trailing hemisphere, NSR might be detected by examining the longitudinal distribution of impact features (e.g., Zahnle et al., 2001).

If NSR occurs, the reorientation of the surface with respect to the tidal axis leads to surface stresses (Helfenstein and Parmentier, 1985; Leith and McKinnon, 1996; Greenberg et al., 1998). The principal horizontal stresses are both tensile in regions spanning 90° of longitude and reaching to ~±40° latitude, centered 45° westward of the tidal axis, as these regions are being stretched as they rotate up onto the tidal bulge. Similarly, the principal horizontal stresses are both compressive in similar zones centered 45° eastward of the tidal axis, as these regions are rotating off of the tidal axis. Principal stresses with opposite signs, consistent with strike-slip motion, exist elsewhere across the satellite (Figure 7.4a). Stresses from NSR have been modeled as increasing with the angular amount of rotation, with the maximum

Figure 7.4. Horizontal principal stresses on the surface of a satellite arising from changes in the satellite's figure. The +/+ symbol denotes areas where both principal stresses are compressive (which could lead to thrust faulting), −/− denotes areas where both principal stresses are tensile (which could lead to normal faulting), and +/− denotes areas with one compressive and one tensile principal stress (which could lead to strike-slip faulting). The magnitude of the stress, and thus whether faults are actually formed, depends on the magnitude of change. (a) Instantaneous nonsynchronous rotation of the ice shell to the east with respect to the tidal axis (at 0° and 180° longitude). (b) Polar wander of the ice shell by 90° with respect to the original rotation axis (solid circle). (c) Despinning. (d) Orbital recession of a tidally locked satellite.

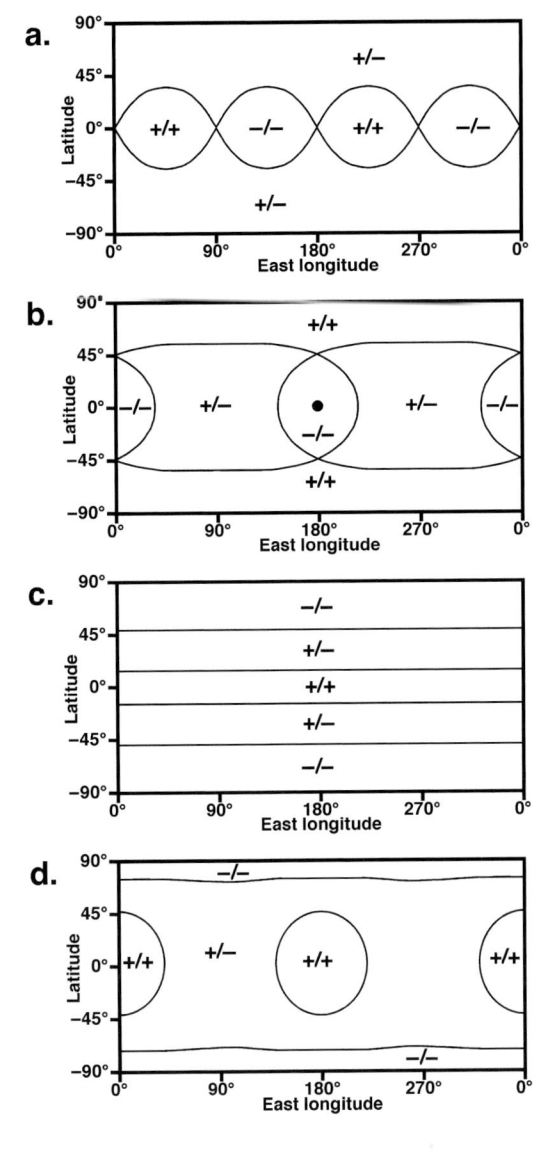

tensile stresses described by

$$\sigma = 6f\mu \left(\frac{1+\nu}{5+\nu}\right) \sin\theta = 3fE\frac{\sin\theta}{5+\nu}, \qquad (7.5)$$

(Leith and McKinnon, 1996) where μ is the shear modulus, θ is the reorientation angle, and f is the satellite flattening in the tidal direction. For a hydrostatic body, this flattening is given by

$$f = \frac{3}{2}\frac{R_s^3 \omega^2}{GM_s} h_2, \qquad (7.6)$$

(Leith and McKinnon, 1996). The flattening will be reduced if the satellite has significant rigidity (see subsection 3.1.2). The stresses due to nonsynchronous rotation will exceed the diurnal tidal stresses for $\theta > e$, or for a longitudinal reorientation of more than a few tenths of a degree for typical eccentricities. However, for likely nonsynchronous rotation rates, the strain rates generated will be much smaller than the strain rates caused by diurnal tides (Ojakangas and Stevenson, 1989a; Nimmo et al., 2007a).

Wahr et al. (2009) consider the combined NSR and diurnal stresses in a viscoelastic ice shell based on the gravitational potential fields of a rotating satellite and its primary planet, especially as relevant to Europa. The NSR stress affecting the ice shell is parameterized by rotation period P_{NSR} relative to the Maxwell time τ_M of the ice, $\Delta = P_{NSR}/2\pi\tau_M$. For an ice shell with a sufficiently short rotation period or high viscosity ($\Delta \ll 1$), the ice behaves elastically on the NSR period and NSR stresses are large, overwhelming diurnal stresses; however, for an ice shell with a sufficiently long rotation period or low viscosity ($\Delta \gg 1$), the ice behaves viscously so NSR stresses relax away and are small, permitting diurnal stresses to dominate. When $\Delta \approx 1/e$ for a satellite with orbital eccentricity e, NSR and diurnal stresses are of comparable magnitude, combining to influence the satellite's tectonics.

3.2.3 Polar wander

In certain circumstances, the surface of a satellite may reorient relative to its axis of rotation (Willemann, 1984; Matsuyama and Nimmo, 2007). This process of polar wander is conceptually very similar to NSR, since it also involves motion of the surface relative to the satellite's axes (Melosh, 1980a; Leith and McKinnon, 1996), but in addition to potentially moving through the tidal axis, the surface also moves through the polar axis, changing the stress pattern significantly. In general, the reorientation direction occurs roughly perpendicular to the tidal axis, because such paths are energetically favored (Matsuyama and Nimmo, 2007).

Floating ice shells maintained by tidal dissipation may undergo 90° reorientations about the fixed tidal axis due to increased ice thickness at the poles (Ojakangas and Stevenson, 1989b). Long-wavelength density anomalies may lead to smaller amounts of reorientation (Janes and Melosh, 1988; Nimmo and Pappalardo, 2006), as may volatile redistribution (Rubincam, 2003). Finally, large impacts may cause reorientation directly (e.g., Chapman and McKinnon, 1986) or due to the mass asymmetry from the creation of a new impact basin (e.g., Melosh, 1975; Murchie and Head, 1986; Nimmo and Matsuyama, 2007).

Polar wander induces tensile principal stresses (encouraging tension fractures or normal faulting) in the quadrant leading the reorientation direction because the region originally located along the spin axis must lengthen, and compressional stresses (encouraging thrust faulting or folding) in the trailing quadrant because that region moves toward the shorter spin axis (Leith and McKinnon, 1996; Figure 7.4b). The maximum tensile stress that develops is given by Equation (7.5), with the value of the flattening f depending on whether any reorientation of the tidal axis has occurred. If no such reorientation has occurred, the value of f is simply one-third of the value given in Equation (7.6); in the more general case, the stresses will be larger and may be obtained from expressions given in Matsuyama and Nimmo (2008).

3.2.4 Despinning

Satellites that are initially rotating at a rate faster than synchronous will reduce their rotation rate, or despin, to that of synchronous rotation in timescales generally very short compared to the age of the solar system (e.g., Murray and Dermott, 1999). A reduction in spin rate means that the satellite will be less rotationally flattened; the resulting shape change gives rise to stresses – compressive at the equator as the equatorial bulge collapses and tensile at the poles as the poles elongate (Figure 7.4c). The maximum differential stress caused by despinning for a thin elastic shell in hydrostatic equilibrium is given by

$$\sigma = 2\left(\omega_1^2 - \omega_2^2\right) \frac{R_s^3}{GM_s} \mu \left(\frac{1+\nu}{5+\nu}\right) h_2, \qquad (7.7)$$

(Melosh, 1977) where ω_1 and ω_2 are the initial and final angular rotation velocities. Note that if the satellite's rigidity is large enough to reduce the rotational flattening (an effect included in h_2), the stresses will be reduced by a factor of $1 + \frac{19}{2}\tilde{\mu}$ for a homogeneous satellite (see Melosh, 1977; Squyres and Croft, 1986). Note also that spin-up is possible in principle in some circumstances (e.g., if the satellite undergoes differentiation, and if spin-up overcomes the strong despinning torque noted above); in this case the signs of all the stresses would be reversed.

3.2.5 Orbital recession and decay

Tidal dissipation within the primary planet causes satellites in prograde orbits outside synchronous altitude to recede, or spiral outward (as the Earth's Moon is doing), and satellites in retrograde orbits or orbits under synchronous altitude to decay, or spiral inward. This will change the amplitude of the tidal bulge, and if the satellite is rotating synchronously, it will also change the spin rate. Most major satellites (with the notable exception of Triton) are in prograde orbits, and so generally undergo tidal recession. Recession causes a reduction in rotation rate, lengthening the c axis, and a decrease in the static tide, shortening the a axis. This combination of despinning and tidal bulge reduction causes a stress pattern in which a region around the sub-planet point experiences compressive stress, mid-latitudes experience horizontal shear stress, and the poles undergo tension (Figure 7.4d;

Melosh, 1980b; Helfenstein and Parmentier, 1983). The maximum principal stress difference is twice the maximum stress given by Equation (7.7) (Melosh, 1980b). A satellite undergoing orbital decay experiences exactly the opposite changes in static tide and spin rate, and so experiences the same pattern of stress, but with the signs reversed.

Internal differentiation of a satellite may or may not cause significant global volume change (discussed in the next section), but since it concentrates mass toward the center of the body, it decreases the moment of interia. This will reduce h_2, and thus reduce both the tidal and rotational distortions. As long as the satellite remains in synchronous rotation, the surface stresses are the same as for orbital recession (Collins et al., 1999).

3.2.6 Volume change

A large number of different mechanisms can lead to volume changes within a satellite, and thus extensional or contractional features on the surface (Squyres and Croft, 1986; Kirk and Stevenson, 1987; Mueller and McKinnon, 1988). Such volume changes will generate isotropic stress fields on the surface. In many cases, the most important effect is that ice at high pressures (roughly >0.2 GPa) is considerably more dense than ice at low pressures. Thus when a satellite undergoes internal differentiation, high pressure ice in the interior is displaced by silicates (which are less compressible), leading to an overall increase in volume. The concomitant increase in surface area can reach several percent in the case of Ganymede and Callisto (Squyres, 1980; Mueller and McKinnon, 1988). Smaller amounts of expansion will occur as the satellite warms (e.g., through radioactive decay or tidal heating) and ice transforms from the high- to the low-pressure phase (Ellsworth and Schubert, 1983). These effects will be reversed if the satellite cools. Warming may also lead to silicate dehydration (Finnerty et al., 1981; Squyres and Croft, 1986), which in turn leads to expansion, unless dehydration occurs at pressures in excess of ~2 GPa (e.g., Dobson et al., 2002). In the latter case the situation is more complicated: dehydration itself results in volume contraction, but eventual migration of released water to the satellite's outer layers may lead to overall expansion.

Another potential source of expansion or contraction is the large density contrast between ice I and water. As water freezes to ice I, large tensile surface stresses can result (Cassen et al., 1979; Nimmo, 2004b). This effect is most important for small icy satellites and satellites with thin water layers, since oceans freezing in icy satellites with very thick water layers also produce higher density ice phases at the bottom. The volume change of water to high-pressure ice phases is opposite to that due to the freezing of ice I, and the combination of the two reduces the net volume change (cf. Squyres, 1980; Showman et al., 1997).

Solidification of initially molten iron cores in silicate bodies can lead to substantial contraction, as in the case of Mercury (e.g., Melosh and McKinnon, 1988), but is likely to be a minor effect in the case of icy satellites (Ganymede and Titan are possible exceptions).

The isotropic surface stresses σ resulting from global expansion or contraction (McKinnon, 1982; Melosh and McKinnon, 1988) are given by

$$\sigma = 2\mu \frac{(1+\nu)}{(1-\nu)} \frac{\Delta R}{R_s} = \frac{E}{1-\nu} \frac{\Delta R}{R_s}, \quad (7.8)$$

where ΔR is the radial contraction or expansion. A fractional change in radius of 0.1% gives rise to stresses of order 10 MPa. A closely related stress mechanism, which appears to be confined to Io, is burial of the surface by more recently erupted material (subsection 4.1.3; Schenk and Bulmer, 1998). This radially inward motion results in isotropic compressive stresses within the buried layers, as described by Equation (7.8).

Similar expressions arise for thermal expansion or contraction, where in this case the radial change is the product of the globally averaged thermal expansivity and the temperature change (Ellsworth and Schubert, 1983; Zuber and Parmentier, 1984; Hillier and Squyres, 1991; Showman *et al.*, 1997). A global temperature change of 100 K gives rise to stresses of order 30 MPa for a volume thermal expansivity (α_V) of 1×10^{-4} K^{-1}. Thickening ice shells result in stresses due to both volumetric and thermal effects, but the former dominate (Nimmo, 2004b). Direct thermoelastic stresses will arise in cooling or warming lithospheres with or without underlying global volume change, since the lithospheres are laterally confined (Turcotte, 1983; Hillier and Squyres, 1991). Maximum stresses are $E\alpha_V \Delta T/3(1-\nu) \sim 3 \times (\Delta T/10$ K) MPa for water ice near 150 K, but decline for colder ice since α_V is strongly temperature dependent (Petrenko and Whitworth, 1999).

3.3 Local stress mechanisms

Satellites are quite likely to have non-axisymmetric structures, in which case some of the above global mechanisms may lead to local deformation. For instance, local zones of weakness can result in enhanced tidal dissipation and deformation (e.g., Sotin *et al.*, 2002), as may be occurring at the south pole of Enceladus. However, there are also stress-generating mechanisms that are intrinsically local in character, which will be itemized here.

3.3.1 Convection

Thermal convection is a potentially important process on icy satellites (e.g., McKinnon, 1999), since it will substantially alter the thermal evolution of such bodies, including the potential for subsurface oceans, and may also cause local deformation (e.g., Nimmo and Manga, 2002). For large icy satellites, layers of high- and low-pressure ice may convect separately (e.g., McKinnon, 1998). Whether convection occurs depends on the ice shell thickness, local gravity, and most importantly, the viscosity of the ice. As outlined in Section 2.4, the viscosity is both temperature- and grain-size dependent, and the biggest uncertainty in assessing whether convection occurs is due to uncertainties in the grain size (Barr and Pappalardo, 2005). The characteristic stresses generated by convection on the overlying lithosphere depend on the temperature difference available to drive convection; for strongly temperature-dependent viscosity (Solomatov and Moresi, 2000) the stresses are given by

$$\sigma \approx 0.03 \alpha_v \rho g \Delta T_{rh} \delta_{rh}, \qquad (7.9)$$

where ΔT_{rh} and δ_{rh} are the rheologically controlled temperature drop and boundary layer thickness. The latter controls the length scale of convective features, and depends on the vigor of convection; for ice it is typically of order a few km (see Solomatov and Moresi, 2000). The temperature contrast ΔT_{rh} for Newtonian rheologies is given by

$$\Delta T_{rh} \approx \frac{2.4 R T_i^2}{Q}, \qquad (7.10)$$

where R is the gas constant, T_i is the interior temperature, and Q is the activation energy. Assuming $Q = 60$ kJ/mol and $T_i = 250$ K, we have $\Delta T_{rh} \approx 20$ K. The resulting stresses, from Equation (7.9), are of order 1 kPa for a boundary layer thickness of 10 km and a gravity of $1\,\text{m s}^{-2}$. These stresses are generally small compared to the stresses arising due to other mechanisms. Furthermore, even though larger stresses than given by Equation (7.9) are generated near the surface in a thick, stagnant lid (Solomatov, 1995; Solomatov and Moresi, 2000), the lid is cold and immobile, which further reduces surface deformation (e.g., Nimmo and Manga, 2002). In general, therefore, one might not expect convection in an ice shell to generate identifiable surface features.

Convective stresses on silicate bodies tend to be larger, because the higher operating temperatures overwhelm the effect of the larger silicate activation energy in Equation (7.10), and the rheological length scales are typically greater. The kilometer high long-wavelength convective uplifts of the kind seen on Earth or Venus are typical of silicate convection, but not of that in ice shells.

Under certain circumstances, convection can lead to larger stresses. For instance, satellite differentiation may lead to a deep ice layer with a high potential temperature, which drives large-scale convective overturn and generates large tensile stresses (Kirk and Stevenson, 1987). If the near-surface ice is sufficiently weak to undergo yielding, then larger surface deformations and stresses may result (Showman and Han, 2005; Han and Showman, 2008; Barr, 2008). Compositional convection may also give rise to large stresses, because density contrasts driven by compositional variations can be much larger than those driven by thermal differences. It is possible that thermal convection, or other local heating mechanisms, can generate compositional contrasts by preferentially melting salt-rich, dense phases (Nimmo et al., 2003a; Pappalardo and Barr, 2004). However, unlike thermal convection, composition-driven overturn will only happen once, unless there is some mechanism continuously generating new compositional contrasts.

3.3.2 Lateral pressure gradients

For a floating ice shell, if isostatically compensated lateral shell thickness variations exist, then lateral pressure gradients occur. The force per unit length (e.g., Buck, 1991) is given by

$$F \approx g \Delta \rho t_c \Delta t_c, \qquad (7.11)$$

where $\Delta \rho$ is the density contrast between shell and underlying ocean and t_c and Δt_c are the mean shell thickness and the thickness variation, respectively. If these forces are distributed uniformly across the entire crust, the stress is simply F/t_c. To produce a 0.1 km variation in elevation requires a Δt_c of about 1 km if $\Delta \rho = 100 \text{ kg m}^{-3}$. The resulting stress distributed across the entire ice shell is 10 kPa for $g = 1 \text{ ms}^{-2}$.

One consequence of these pressure gradients for a floating ice shell is that lateral flow of the low-viscosity ice at the base of the shell will occur, removing shell thickness contrasts (Ojakangas and Stevenson, 1989a; Nimmo, 2004c). The resulting vertical motions of the ice shell may result in the generation of surface stresses. Lateral flow in the shell is also important in determining how ice shells respond to tensile stresses, and in particular whether wide or narrow rifts develop (Nimmo, 2004d).

3.3.3 Flexure

If an ice shell has long-term rigidity, then loads emplaced on it will be partially supported by the strength of the lithosphere. The deflection of the lithosphere in response to the load will in turn result in stresses. For a lithosphere with an effective elastic thickness T_e, when the load wavelength λ is much less than the satellite radius, the maximum deflection of the ice shell w_0 in response to a sinusoidal load that generates a final topography h_0 is given by (Turcotte and Schubert, 2002)

$$w_0 = h_0 \frac{\rho}{\Delta \rho} \left(1 + \frac{k^4 E T_e^3}{12(1-v^2)g\Delta \rho}\right)^{-1}, \qquad (7.12)$$

where $\Delta\rho$ is the density contrast between the shell and the water beneath and k is the wavenumber ($=2\pi/\lambda$). For long-wavelength loads, or if the elastic thickness is zero, then this result simplifies to the usual isostatic case. Larger elastic thicknesses result in reduced deflections.

For this same load, the maximum stresses experienced are given by (Turcotte and Schubert, 2002)

$$\sigma = \frac{E}{(1-\nu^2)} \frac{T_e}{2} k^2 w_0. \quad (7.13)$$

In the simplified elastic model, bending stresses at any location relative to the point of greatest deflection are maximized at the surface and base of the elastic layer, respectively, but with opposite signs.

3.3.4 Impacts

Impacts obviously generate transient, large local stresses. Sufficiently large impacts may induce lateral flow, generating tectonic features significantly outside the original impact area (McKinnon and Melosh, 1980). In some cases, focusing of impact-generated waves may in principle also result in tectonic features being generated at the impact antipode (Bruesch and Asphaug, 2004; Moore et al., 2004a). Over the longer term, impact sites may act as zones of weakness that affect the spatial distribution of other tectonic features (cf. McKenzie et al., 1992). Impact basins may also generate stresses through secondary mechanisms such as polar wander (subsection 3.2.3) or lateral flow in the ice shell (subsection 3.3.2).

4 Io

Io is the only outer planet satellite with a rocky (i.e., water-ice free) surface, and it is also the most geologically active of the satellites. Io and the Earth are rivals for the title of most geologically active body in the solar system, but the two bodies could not be more different in how they exhibit their dynamic personalities. It has become the paradigm that planets release excess internal heat through volcanism and tectonism. On Earth, the main mechanism of heat loss is by the creation of new lithospheric plates at mid-ocean ridges. These oceanic plates cool conductively and thicken as they move laterally toward subduction zones, leading to the horizontal and vertical conveyor belt of crustal recycling known as plate tectonics.

Io is very similar to the Earth's moon in size and density, and is inferred to have a dominantly silicate composition (with significant differences such as abundant sulfur). Thus, when Voyager discovered Io's intense global volcanism in 1979 (Peale et al., 1979; Smith et al., 1979), the lack of long linear mountain chains was also immediately obvious and a source of some consternation. How could such an

Figure 7.5. Volcanic calderas (or paterae) are the most common tectonic feature on Io. They are probably associated with volcanic collapse, although it is not yet clear that the mechanism on Io is related to that on Earth. This example, Shamshu Patera, is nestled adjacent to an eroded mountain. To the south lies Shamshu Mons, a faulted mountain at least 3 km high. North is to the top, and the illumination is from the left.

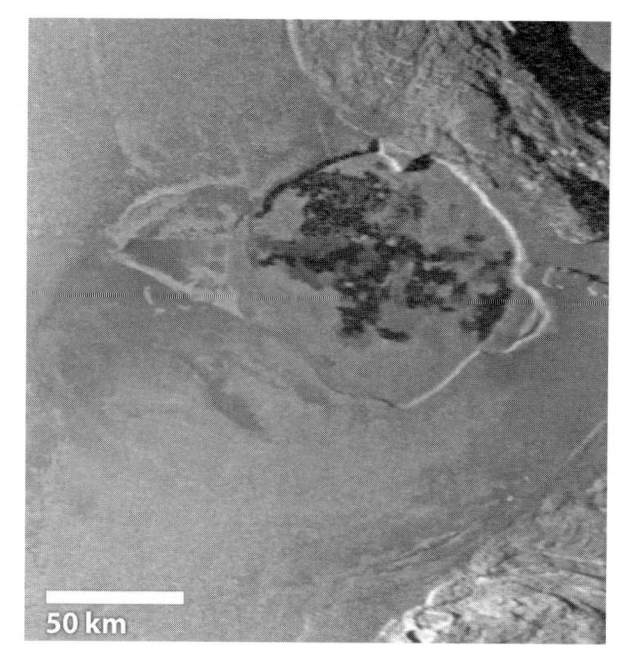

active volcanic world lack the organized tectonic network observed on Earth? If Io's interior is so hot and active, why is there no plate tectonics? It seems that Io has its own unique form of vertical crustal recycling.

4.1 Tectonic features on Io

Io may be dominated by volcanism, but it certainly has tectonic features. The most common tectonic features are volcanic calderas (e.g., Radebaugh *et al.*, 2001), referred to as paterae on Io. Calderas, fault-bounded quasi-circular depressions, form as collapse features over magma chambers and are usually associated with extensive lava flows. There are over 500 volcanic centers on Io, many of which are associated with calderas of some type (Figure 7.5). In addition, there are a few curvilinear fractures on Io, most likely volcanic fissures, flanked by short lava flows a few tens of kilometers long. These lava flows form symmetric butterfly-like patterns on the surface, again indicating fissure eruptions of lava. Calderas and fissures are directly associated with volcanism and are relatively well understood in that context. In the rest of this section we will focus on tectonic features that are not obviously of direct volcanic origin.

4.1.1 Fractures

Narrow curvilinear features are scattered across Io's volcanic plains. These features are dwarfed by more prominent neighboring volcanoes and mountains, and so they

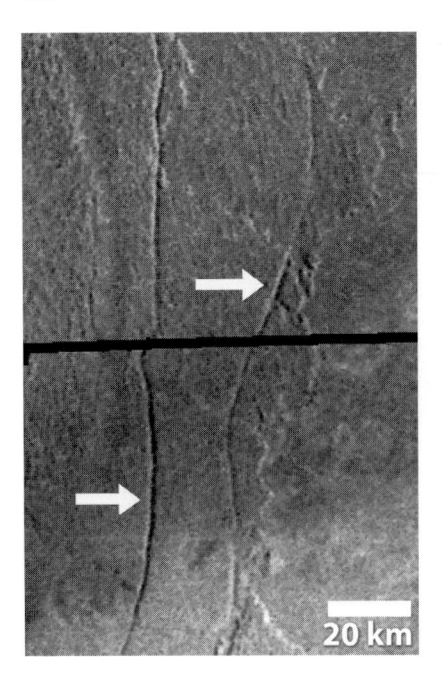

Figure 7.6. Examples of graben-like features on the surface of Io (marked by arrows). These features are typically a few kilometers wide. None of these features were observed at resolution better than a few hundred meters per pixel. Although they are likely to be extensional in origin, their relationship to local or global stress fields is unknown. Note how the graben-like features cross the cuspate scarp of an elevated plain (either an erosional scarp or a lava flow front). Illumination is from the right, and the dark stripe is a data gap.

have been largely ignored in the literature. Typically 1–3 kilometers across, some of these features can stretch for over 500 kilometers in length (Figure 7.6). Their morphology is consistent with simple extensional fractures or graben.

The formation of these fractures may be related to global tidal stresses. Io differs from most other worlds in the intense daily tidal deformation of its surface. Due to its close proximity to giant Jupiter and its orbital resonance with neighboring moons Europa and Ganymede, Io's orbit is eccentric and its solid surface experiences daily tides of up to ∼0.1 km (Table 7.1), leading to repetitive surface strains of 10^{-4} or greater. These tides flex and stress the lithosphere and can cause it to fracture (as also occurs extensively on neighboring Europa – see Section 5.1). However, no correlation has yet been found between fracture orientation and the predicted stresses resulting from tides. The situation can be confused if the features formed at different times or if the stress pattern shifts due to nonsynchronous rotation of the lithosphere (Milazzo et al., 2001).

Alternatively, curvilinear or concentric extensional fractures could be related to local loading of planetary lithospheres. On Io, this could be the result of construction of volcanic edifices or global convection patterns forming localized sites of upwelling and downwelling (e.g., Tackley et al., 2001). However, constructional volcanic edifices are quite rare on Io (Schenk et al., 2004a) and convective stresses on Io are likely to be quite small (Kirchoff and McKinnon, 2009). On Io, constant global resurfacing by lavas and deposition from volcanic plumes can locally erase tectonic patterns of this sort, in part or entirely. In the future, more sophisticated mapping and analysis will be required to explain these fractures, including correlations with topographic and gravity information.

4.1.2 Ridges

Sets of small-scale ridges are observed in certain high-resolution Galileo images of Io. They are short, typically a few kilometers in length, closely spaced (~1 km apart), and of modest amplitude (probably less than 100 m in height). What is remarkable is their consistent orientation within each set across several tens of kilometers in extent. Bart *et al.* (2004) find that ridge orientations are not inconsistent with dominant stress directions during the diurnal tidal cycle, but these orientations vary during the cycle, especially for locations away from the equator, and ridge directions are not uniformly consistent with either a compressional or tensional tidal stress origin. Bart *et al.* (2004) list specific mechanisms that might account for the ridges: compressional folding, infill of open tension fractures, and differential compaction of a porous surface layer. Whatever the mechanism, it is clear that these features are young even by Io standards, since they are unburied, and they involve deformation of very shallow units or layers. Understanding their global distribution and their relation to other features on Io will require better and more consistent image coverage.

4.1.3 Mountains

The most prominent tectonic features on Io are the ~150 mountains scattered across the surface (e.g., Schenk *et al.*, 2001a; Turtle *et al.*, 2001). These peaks average around 6 kilometers high, but the highest, Boosaule Montes, towers 17 kilometers above the flat volcanic plains surrounding it (Schenk *et al.*, 2001a). Most mountains on Io are a few hundred kilometers across or less. There are several curious aspects to these mountains. Much like the numerous volcanic centers, each mountain keeps a respectable distance from its neighbors. Mountains rarely line up in close chains or clusters. Mass wasting of mountain faces is also evident. Several cases have been identified whereby large portions of mountains have failed by landslide, including one of the largest known landslides in the solar system (e.g., Schenk and Bulmer, 1998). More frequently, evidence of downslope creep of material is observed, as mountains apparently are slowly deformed by gravitational collapse (Turtle *et al.*, 2001). Some mountains are merely isolated promontories only a few kilometers across. Given Io's prodigious volcanic outpourings, it would not be surprising if we are observing mountains in various stages of burial.

With the exception of four known and easily recognizable conical shield volcanoes (Moore *et al.*, 1986; Schenk *et al.*, 2004a), no evidence has been found for

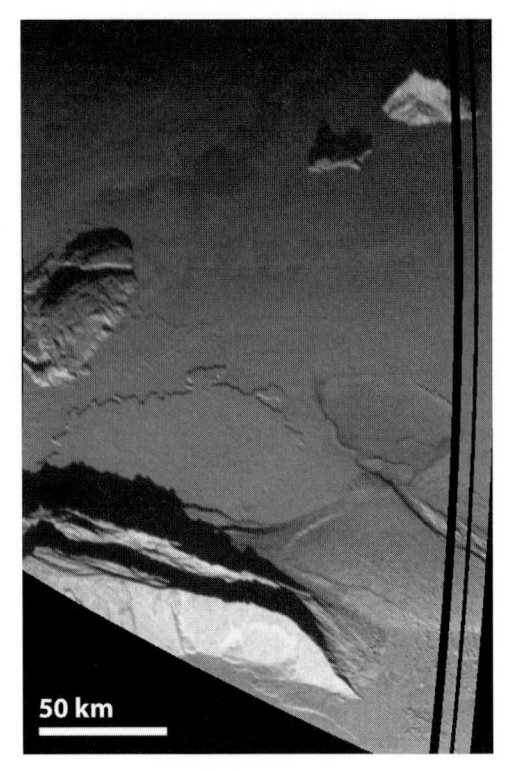

Figure 7.7. Oblique Galileo view showing two of the major mountain types seen on Io. In the foreground (bottom) is Mongibello Mons, a lineated mountain consisting of two major linear ridges 6 to 8 km high. The linearity could reflect bedding planes, fault planes, or another unknown process. Eroded or etched plains are visible just above the ridged mountain. In the background (top) is a smaller sharp-edged single mountain peak. North is to the left, and the illumination is from the bottom. The dark curves are data gaps.

volcanic flow or caldera formation on the slopes or tops of any Ionian mountains. Thus, Io's mountains must form by tectonic deformation and not by volcanic construction. This is consistent with the variety of mountain morphologies observed (Schenk *et al.*, 2001a; Turtle *et al.*, 2001), including flat-topped mesas, small narrow peaks, craggy massifs, rounded "eroded" massifs, elongate ridges, and asymmetric "flatirons" (Figure 7.7). These last morphologies are consistent with fault formation and are a key to understanding the origin of mountains on Io.

Flatiron mountains on Earth, such as those along the front range of the Rocky Mountains in Colorado, are associated with the tilting of thick, erosion-resistant layers by some combination of folding and thrust faulting, when two large masses of crust are forced together or upward by horizontal compression. Such flatirons on the Earth are fluvially dissected into flat triangular pieces, but on Io they appear as solid flat walls tilting up from the surrounding plains. It is difficult to envision how extension would uplift lithospheric blocks 10 or more kilometers above the surrounding plains, so the apparent presence of large tilted blocks of crust on Io suggests that compression may indeed be driving the formation of Io's mountains (Schenk and Bulmer, 1998).

Understanding the source of a global compressive stress field requires an understanding of how Io's global volcanism is processed. Io's volcanism is so intense that an estimated >1 mm to ~1 cm global layer of new lava is deposited on the surface each year (Johnson and Soderblom, 1982; McEwen et al., 2004). Volcanic eruptions are local phenomena, so this estimate is a global average calculated over millions of years, but the global effect of this volcanism is that the entire surface is continually being renewed over millennia (hence the lack of impact craters). It is clear from global mapping (e.g., Crown et al., 1992) that Io does not have lateral tectonics like the Earth: there are no subduction zones or spreading ridges. As a result, these new volcanic deposits must go somewhere, and on Io the only place to go is down. The new deposits force the older cooled lavas downward into the interior, and the result is a remarkably cool lithosphere of potentially great thickness (>25 km), as the recycling time of the crust (thickness divided by burial rate) is faster than the thermal equilibration time of the crust, for all but very thin crusts (O'Reilly and Davies, 1981; McKinnon et al., 2001). It is this rigid lithosphere that supports the large surface loads implied by the mountains (subsection 3.1.2).

The cool lithosphere was recognized shortly after the Voyager discoveries. A more subtle and elegant consequence of Io's regime of "heat piping" and vertical recycling became clear much later. The volcanic layers (i.e., units of similar age) that were once at the surface no longer fit the decreasing surface area as they descend into the interior, due to the simple geometric reality that the surface area of a sphere decreases as it shrinks in radius. The unavoidable result is that these older buried layers must be subject to increasing lateral compression as burial continues. This model can be thought of as an onion that grows from the outside rather than from the inside. Rocks under compression can fail by ductile shortening or by brittle failure. Lower portions of Io's crust may sometimes break under the strain, thrusting a large intact block upward, locally relieving some of the stress and forming a mountain on the surface (Schenk and Bulmer, 1998). Thus, mountain building on Io can be thought of as a direct consequence of Io's high global volcanic resurfacing rates.

A further corollary to the vertical recycling model described above suggests that fluctuations in Io's volcanic resurfacing rate cause variations in the heat flow at the base of the lithosphere (McKinnon et al., 2001; Kirchoff and McKinnon, 2009). These heat flow "pulses" might then increase stresses in the lower crust, triggering mountain formation. That is, the close proximity of a deep, cool lithosphere, whose temperature gradient is suppressed by downward advection, to a subjacent hot, tidally heated asthenosphere, is potentially thermally unstable. The compressive thermoelastic strains that could result, for a $\Delta T \sim 1000$ K, are of order 1% (subsection 3.2.6), comparable to those which result directly from crustal subsidence, z/R, when z is tens of kilometers. In both cases, the resulting stresses easily exceed Byerlee's rule in compression (Jaeger *et al.*, 2003; Kirchoff and McKinnon, 2009). The unsolved parameter is the importance of the thermal stresses relative to the subsidence stresses that most assuredly occur on Io.

The lack of a global thrust fault network or pattern can also be understood in the vertical tectonics scenario. If the global subsidence rate is more or less uniform across the surface, as suggested by the apparently uniform distribution of volcanic centers, then compressive stresses will also be nearly uniform within the lithosphere (see the next section for exceptions). Uniform compressive stresses in a thick lithosphere would likely be relieved locally, chiefly through thrust faulting triggered at local discontinuities or other failure points such as volcanic intrusions, preexisting faults, compositional boundaries, or areas of high thermoelastic stress, and mountain formation will occur in globally distributed "random" locations over time. Moreover, both the subsidence stress produced by downward vertical crustal advection, and compressive thermal stresses at depth that would develop during a local slowdown of volcanic advection of heat, would generate substantial bending moments, which would be released by mountain formation. Kirchoff and McKinnon (2009) suggest that the spacing of Ionian mountains and their relative isolation from each other are flexurally controlled.

4.2 Global distribution of mountains and volcanoes

Mapping of the distribution of mountains and volcanic centers shows that they are not exactly uniform, and there is a subtle but significant global pattern. There are two large areas on the surface where mountains appear to be significantly more numerous than the global average (Schenk *et al.*, 2001a). These areas are not sharply defined, but rather are broad zones of increased mountain formation on opposite sides of the globe from each other, displaced about 75° west of the current sub- and anti-Jovian regions. A similar antipodal pattern emerges for the global distribution of volcanic centers (as distinct from individual flows that may emanate from specific centers) (Schenk *et al.*, 2001a; Radebaugh *et al.*, 2001). However, the distribution of mountains and volcanic centers are significantly anticorrelated: the areas where mountains are less frequent are the areas where volcanic centers are more numerous (McKinnon *et al.*, 2001; Kirchoff, 2006). This seems counterintuitive if we believe the subsidence tectonics model described above. Areas with more volcanism should experience more subsidence and hence more mountain building. Io is not so simple, apparently, and several factors could explain this anomaly. If internal convection is active on Io (Tackley *et al.*, 2001) in a globally symmetric fashion, stress fields over rising and descending mantle plumes could potentially modify the compressive stresses due to subsidence, by enhancing volcanism and suppressing mountain building. Similarly, if nonsynchronous rotation occurs on Io (as proposed for Europa, see subsections 3.2.2 and Section 5.1), then this would impose an additional globally symmetric stress field within Io and could inhibit (or encourage) mountain formation in antipodal areas. Higher rates of volcanism could lead to thicker crust in those areas, although this should lead to more instead of less mountain formation.

The stresses generated by high Rayleigh number convection, as in Tackley (2001), should be quite modest in comparison with subsidence stresses (Kirchoff and McKinnon, 2009) because internal viscosities are small, and are thus unlikely to influence mountain formation. On the other hand, nonsynchronous rotation stresses could reach levels as great as 220 MPa (Equation 7.5) for $E = 65$ MPa, and $h_2 = 2.3$ (Schubert *et al.*, 2004). Such stresses would certainly be important for Io's lithosphere (especially its upper portion), but a large amount of nonsynchronous rotation, approaching 90°, would be necessary to match the actual positions of mountain and volcano concentrations, which would preferentially form in compression and tension, respectively. Strong mountain and volcanic feature orientation patterns would also be predicted if nonsynchronous rotation was an important source of stress, but such patterns have not yet been reported.

An alternative view of the anticorrelated mountain and volcanic feature distribution patterns invokes not spatially variable stress sources, but temporally varying ones. For example, regions that experience a slowdown or cessation of volcanism, and thus a slowdown in vertical crustal advection, should experience a buildup of compressive thermal stresses in the lower crust and possibly crustal thinning (McKinnon *et al.*, 2001; Kirchoff and McKinnon, 2009). This would lead to a period of enhanced mountain formation, which could last until the volcanic "heat pipe" network was able to reestablish itself (which could in turn depend on the creation or activation of mountain-forming faults). In the context of the presently observed degree-2 patterns of volcanism and mountain building, slow nonsynchronous rotation is a possible cause, in that the present regions of volcanic concentration are centered on the sub- and anti-Jovian points, which is where tidal heating should be maximized for an asthenosphere-bearing Io (e.g., Tackley, 2001). Regions 90° away would be *former* regions of enhanced volcanism, now diminished, that are now undergoing a period of enhanced orogenesis.

Volcanism and mountain formation may be intimately related on a local scale as well. Several instances have been identified in which volcanic paterae occur along the outer margins of mountains, in some cases cutting into the basal scarps (Figure 7.7). In a few instances, both ends of a mountain are abutted by volcanic centers (Turtle *et al.*, 2001; Schenk *et al.*, 2001a; Jaeger *et al.*, 2003). Normally, volcanism should be difficult under the compressional lithospheric stress regime described above, as strong compression would inhibit dike propagation, but local faulting could relieve these stresses enough to permit magma ascent. On the other hand, localized heating from a magma body could form a lithospheric discontinuity which could trigger faulting and stress release. The physical link between mountains and the formation of eruptive centers is not well understood, however. There are numerous counterexamples of isolated mountains that have formed hundreds of kilometers from the nearest volcano, which is remarkable given the ubiquitous presence of volcanic centers on Io.

After initially confounding us with their mere presence, it is now apparent that Io's isolated but prominent mountains are intimately linked to its high heat flow and volcanic resurfacing rates. Most of the mountains have probably formed due to global inward subsidence of the volcanic surface and resulting compressional stress in the lower crust. Following uplift, additional volcanism may occur along faults formed by the mountains, and erosive processes such as downslope creep, landslides, and volcanic burial will begin to degrade mountain landforms, and ultimately return them from whence they came.

5 Active icy satellites

In this section, we discuss three ice-covered satellites that stand out from the others, in terms of having very young surfaces and even displaying current activity on their surfaces.

5.1 Europa

Europa, the smallest of Jupiter's Galilean satellites, is a dynamic body so covered with overlapping linear ridges that it has often been compared in appearance to a giant ball of twine. From analysis of data from the Voyager and Galileo spacecraft, a comprehensive picture is emerging of how Europa's global stress mechanisms, interior structure, and surface geology are inherently linked. Tectonism provides this link, and therefore an understanding of Europa's tectonics is crucial for understanding the great unresolved questions of where tidal heat is deposited in Europa and the thickness of the ice shell. In this section we will address the tectonics of specific Europan landform types (fractures, ridges, bands, folds, domes and lenticulae, chaos, and large impacts), the global tectonic patterns and their links to stress mechanisms (diurnal stressing and nonsynchronous rotation), and the question of

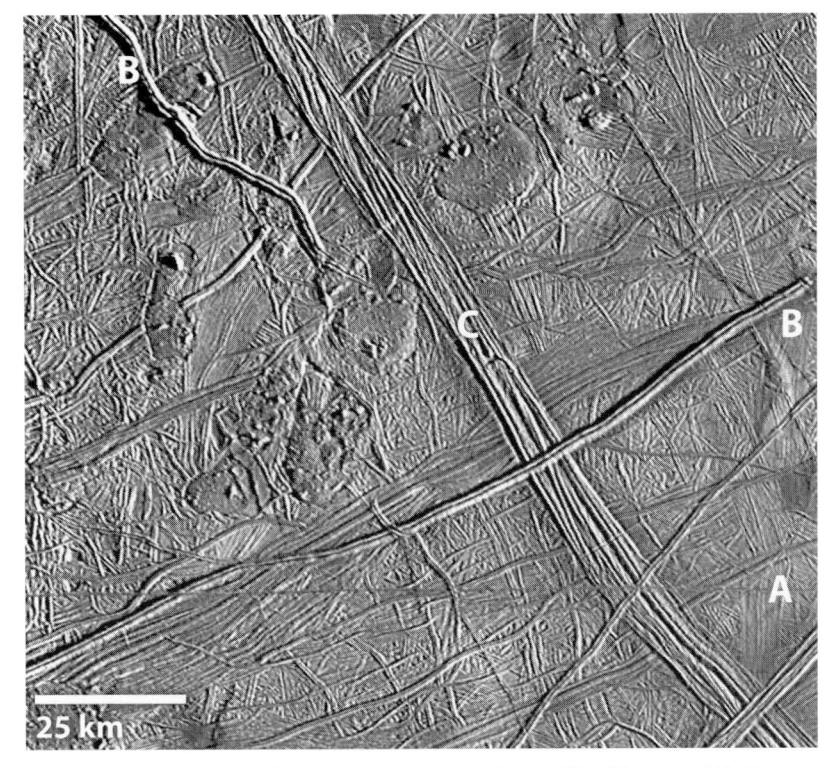

Figure 7.8. Typical tectonic features on Jupiter's satellite Europa. (A) denotes a trough with no flanking ridges, trending east–west; (B) denotes two large examples of the ubiquitous double ridges; (C) denotes a ridge complex. North is to the top, and the illumination is from the right.

active tectonism (see Greeley *et al.*, 2004, for a more general discussion of Europa surface geology).

5.1.1 Tectonics of Europa's landforms

5.1.1.1 Isolated troughs The simplest of Europa's landforms are individual linear to curvilinear troughs (Figure 7.8a). They are generally ~100–300 m wide and V-shaped in cross section, and they can be several to hundreds of kilometers long. Their rims can be relatively level, or raised with respect to the surrounding terrain. Interactions between propagating troughs are consistent with linear elastic fracture mechanics models of perturbed stress fields around fracture tips (Kattenhorn and Marshall, 2006). All of these characteristics suggest an origin as tension fractures.

The widths of visible troughs imply that they have widened subsequent to the original fracturing. The surface width w and depth z of a fracture can be related through the Young's modulus E as $w \sim \rho g z^2 / E$ (Nur, 1982). For $E \sim 10^9$ Pa (reduced from the solid-state value; Nimmo *et al.*, 2003b), crevasse depths of

100 m, 1 km, and 10 km could produce surface widths ∼1 cm, 1 m, and 100 m, respectively. A significant portion of the width of isolated troughs may be due to mass wasting of debris from the sides of the trough and/or due to tectonic movement along the trough, e.g., from upbowing or strike-slip movement (discussed below).

5.1.1.2 Normal faults Normal faults are ubiquitous in extensional regions on Earth (e.g., Jackson, 1989), but have proved difficult to detect on Europa owing to limited imaging and topographic coverage. Imbricate normal faults have been inferred within Europa's bands (Figueredo and Greeley, 2000; Prockter *et al.*, 2002; Kattenhorn, 2002), as discussed in subsection 5.1.1.4. Two prominent normal faults have been described by Nimmo and Schenk (2006), with vertical offsets of several hundred meters and flexural flanks implying elastic thicknesses in the range 0.15–1.2 km. The maximum displacement/length ratio is ∼0.02, comparable to values on silicate bodies (e.g., Watters *et al.*, 2000). The driving stresses implied by the existence of these faults are several MPa, and the derived near-surface shear modulus is apparently lower than that of intact ice, perhaps as a result of porosity and/or fracturing (Section 2.2).

5.1.1.3 Ridges Ridges are Europa's most ubiquitous landform (Figure 7.8), yet their origin remains poorly understood (Pappalardo *et al.*, 1999; Greeley *et al.*, 2000, 2004). They most commonly take the form of a double ridge, i.e., a ridge pair with a medial trough (Figure 7.8b). Understanding their mode of formation is important to the satellite's geophysical state, including the presence and distribution of liquid water. Suggested classification schemes (Greenberg *et al.*, 1998; Head *et al.*, 1999; Figueredo and Greeley, 2000) indicate a morphological transition from isolated troughs, to double ridges, to wider ridge complexes that commonly show a series of subparallel component ridges (Figure 7.8c). This morphological progression suggests an evolutionary sequence in which isolated troughs evolve into double ridges, then into more complex ridge morphologies. Some double ridges instead transition into wider bands, apparently as they are pulled apart along their axes (see subsection 5.1.1.4).

Double ridges have average widths of a few hundred meters, with prominent ridges being ∼2 km wide. Double ridges are characterized by a continuous axial trough that is V-shaped in cross section, and not as deep as its flanking ridges are tall. Ridge slopes are near the angle of repose (Kadel *et al.*, 1998). Preexisting topography is sometimes partially recognized up the outer flanks of ridges, with the most prominent topography extending to near the top of the outer flanks (Head *et al.*, 1999). Mass wasting is prevalent along ridge flanks, with the debris apparently draping over preexisting terrain (Sullivan *et al.*, 1999).

Ridge complexes are commonly ∼5–10 km wide, consisting of multiple subparallel lineations and ridges, which can interweave or merge along their trends (e.g., Figueredo and Greeley, 2000). Ridge complexes may be transitional between narrower double ridges and wider bands (subsection 5.1.1.4).

Many double ridges and ridge complexes show evidence for strike-slip motion along them, a characteristic not shared by isolated troughs (Hoppa *et al.*, 1999a). Reconstruction of preexisting lineaments provides evidence for extension across some ridges (Tufts *et al.*, 2000) and minor contraction across others (Patterson *et al.*, 2006; Bader and Kattenhorn, 2007).

Some ridges are flanked by topographic depressions and/or fine-scale fractures (Tufts *et al.*, 2000; Billings and Kattenhorn, 2005; Hurford *et al.*, 2005). The downwarping adjacent to the ridge suggests loading of the lithosphere either from above (due to the weight of the overburden) or from below (due to withdrawal of subsurface material). With the assumption that ridges have loaded the surface, Billings and Kattenhorn (2005) use the distance to ridge-flanking cracks to estimate an effective elastic thickness of $\sim$0.2 to 3 km, and Hurford *et al.* (2005) fit photoclinometric profiles to derive an average elastic lithospheric thickness of $\sim$0.2 km (results which are sensitive to the assumed value of E). Such thin elastic lithospheres near cracks and ridges may not be representative of Europa's ice shell as a whole, however.

At low solar incidence angle, it is apparent that some ridges have diffuse dark material infilling the flanking topographic depressions. The dark flanks that define these "triple bands" may have been created by ballistic emplacement of dark material entrained in gas-driven cryovolcanic eruptions (e.g., Crawford and Stevenson, 1988); alternatively, the flanks may be thin dark lag deposits due to sublimation of surface frosts and local concentration of refractory materials adjacent to a subsurface heat source associated with the ridge formation (Fagents *et al.*, 2000).

Several models have been proposed for the origin of Europa's ridges (cf. Pappalardo *et al.*, 1999; Greeley *et al.*, 2004), and these models have various implications for the presence and distribution of liquid water at the time of ridge formation. Some models invoke a shallow subsurface ocean; some rely on the action of warm mobile ice with perhaps an ocean at depth; some imply that liquid water exists in the shallow subsurface on at least an intermittent basis.

Tidal squeezing. Greenberg *et al.* (1998) propose that fractures penetrate completely through Europa's ice shell, and open and close in response to changing diurnal stress, allowing water and icy debris to be pumped toward the surface with each tidal cycle to build ridges. In this model, opposing lithospheric blocks pull apart along $\sim$1 m wide cracks that penetrate through the entire ice shell to an ocean below. As they pull apart, water rises up into the cracks hydrostatically. Europa's diurnal stress cycle soon reverses the strain direction and closes the crack again, driving material to the surface. This "pumping" mechanism every 3.5 days is envisioned to pile up enough ice and slush on the surface adjacent to the crack to form double ridges. Because this model explicitly assumes that the entire ice shell of Europa is penetrated by cracks, it envisions a very thin ice shell, one which more readily allows cracks to penetrate completely through the ice (Golombek and Banerdt, 1990; Leith and McKinnon, 1996). While Crawford and Stevenson (1988) discuss the difficulty of cracking from the base of the ice shell upward through warm ductile ice at the base of the ice shell, Lee *et al.* (2005) suggest that fractures formed at the top of an ice lithosphere can more easily penetrate downward through the entire plate because stress is concentrated at their tips. However, a complication for the tidal squeezing model is that water in narrow cracks is expected to freeze faster than the tidal cycle (Nimmo and Gaidos, 2002).

Linear volcanism. Kadel *et al.* (1998) propose that double ridges are linear volcanic constructs, built of debris associated with gas-driven fissure eruptions. Volcanic models suggest that volatiles such as CO_2 or SO_2 are capable of driving eruptions, overcoming the negative buoyancy of water relative to ice (Fagents *et al.*, 2000). Like the tidal pumping model, the volcanic model is challenged in presuming open conduits extend from a subsurface ocean to the surface. It is possible that shallow melt chambers feed conduits instead, or that volatiles have driven pinched off water-filled cracks toward the surface (Crawford and Stevenson, 1988). However, this model also has difficulty accounting for the great linearity and continuity of Europa's ridges, as terrestrial volcanic ridges tend to pinch and swell, due to eruption and coalescence of material into discrete eruption centers.

Dike intrusion. Melosh and Turtle (2004) propose that ridges form by intrusion of melt water into a shallow vertical crack to build a double ridge. In this model, melt intrudes into the shallow subsurface within dikes and subsequently freezes, causing outward and upward plastic deformation of the near-surface to create a ridge.

Compression. Sullivan *et al.* (1998) propose that ridges are contractional structures, deformed along plate boundaries. Reconstruction of preexisting structures suggests that compression is a viable model for some ridges (Sarid *et al.*, 2002; Patterson *et al.*, 2006; Bader and Kattenhorn, 2007). Contractional strain at ridges can help to compensate for the large degree of extensional strain represented by Europa's bands.

Linear diapirism. Head *et al.* (1999) propose that double ridges form in response to cracking and consequent diapiric rise of tabular walls of warm ice, which intrude and uplift the surface to form ridges. This model suggests that cracks penetrate down to a subsurface ductile ice layer, rather than through the ice shell. Warm subsurface ice moves buoyantly into the fracture, aided by tidal heating concentrated along the fracture (Stevenson, 1996; Gaidos and Nimmo, 2000; Nimmo and Gaidos, 2002). The process is envisioned as analogous to the rise of tabular "salt walls" that rise along extensional fractures on Earth (e.g., Jenyon, 1986) and can cause intrusive uplift. In some cases the trend of the preexisting topography appears to be deflected at a ridge as the ridge flank is encountered, consistent with the idea that ridge flanks may have formed by upwarping of the preexisting surface rather than by volcanic construction. As with the caveat regarding the linear volcanism model, the symmetry, uniformity, continuity, and great lengths of ridges on Europa are unmatched by any linear diapir on the Earth.

Shear heating. Gaidos and Nimmo (2000) and Nimmo and Gaidos (2002) suggest that diurnally induced strike-slip motion along fractures creates frictional and viscous heating. If the velocity of motion along a fracture is great enough (~10 cm per tidal cycle), shear heating can be sufficient to trigger upwelling of warm ice or compression along the weakened zone to form a ridge (Han and Showman, 2008). Shear heating may provide a more uniform mechanism for buoyancy generation that may explain some differences between ridges on Europa and terrestrial linear diapirs. Partial melting might occur along the ridge axis, with downward drainage of melt contributing to formation of the axial depression. On the other hand, neither shear heating nor linear diapirism would explain flexural troughs or tension fractures that flank large ridges.

Volumetric deformation. Aydin (2006) notes the similarity in morphology and merging relationships of Europa's ridges to some compaction and dilation bands in terrestrial rocks, localized zones of volumetric strain that occur in high porosity materials such as sandstone. Moreover, the multiple sets of ridges that comprise Europa's ridge complexes have a strong resemblance to shear bands in terrestrial rocks. Europa's ridges and bands are several orders of magnitude larger than the terrestrial analogues formed by volumetric deformation. The potential mechanisms for concentrating strain in large structures along discrete boundaries on Europa need to be understood in order to evaluate the viability of this model.

The origin of ridges, their relationships to isolated troughs and to bands, and their connectedness to the subsurface remain key open issues in our understanding of Europa's tectonics, but the shear heating model is the most quantitatively developed, and in some ways the most testable model (see also Section 5.2).

5.1.1.4 Bands Bands are polygonal areas of smoother terrain with sharp boundaries. Bands, like other features on Europa, appear to brighten with age; and the youngest bands are commonly of lower albedo than their surroundings, while older bands show little or no albedo contrast. Opposing sides of bands on Europa can be reconstructed with few gaps, restoring structures that were split and displaced as the bands opened along fractures (Schenk and McKinnon, 1989; Pappalardo and Sullivan, 1996; Sullivan et al., 1998) (Figure 7.9). Reconstruction of bands implies that Europa's surface layer has behaved in a brittle manner, separating and translating atop a low-viscosity subsurface material, with the region of separation being infilled with relatively dark, mobile material (Schenk and McKinnon, 1989;

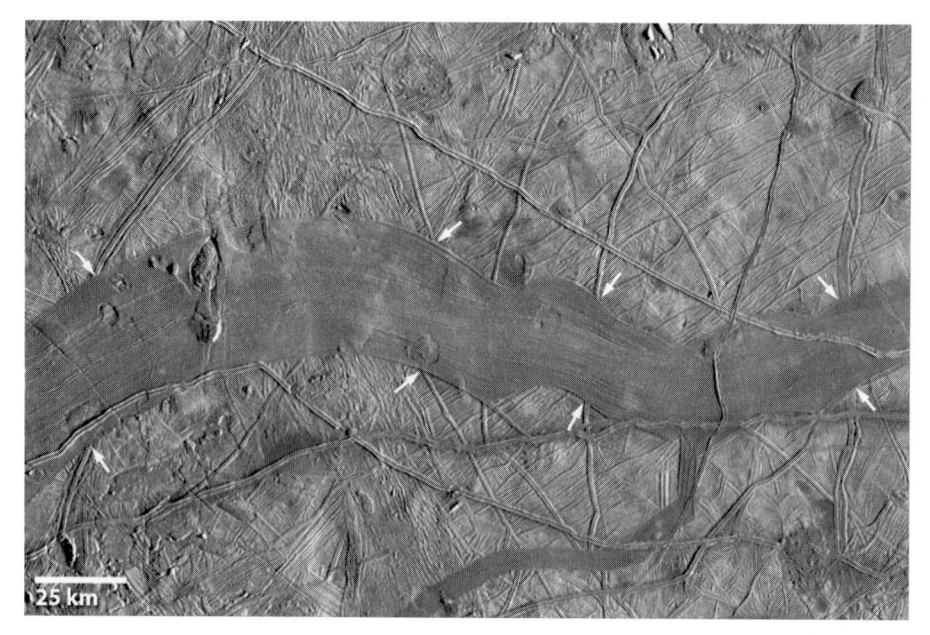

Figure 7.9. Prominent band on Europa. Arrows point to some of the preexisting ridges cut by this band. Preexisting features reconstruct almost perfectly if the band is closed, demonstrating that the bands form by crustal spreading. North is to the top, and the illumination is from the right.

Golombek and Banerdt, 1990). Thus, bands offer compelling evidence for warm, mobile material in the shallow Europan subsurface at the time of their formation.

It has generally been inferred that bands have formed in response to tension (Schenk and McKinnon, 1989; Golombek and Banerdt, 1990), with the bands near the anti-Jovian point perhaps forming under nearly isotropic tension (Pieri, 1981), consistent with current-day nonsynchronous rotation stresses west of the anti-Jovian point (see subsection 3.2.2). However, some bands, notably Astypalaea Linea, show a significant strike-slip component, suggesting oblique opening (Tufts *et al.*, 1999). Structural relationships within the anomalous bright band Agenor Linea suggest that it formed by right-lateral strike-slip motion (Prockter *et al.*, 2000a), while an analogous bright band Crick Linea on the sub-Jovian hemisphere may have formed with a component of compression (Greenberg, 2004). Wedge-shaped bands at the anti-Jovian point also show evidence of shear, aligned with the opening direction (Schenk and McKinnon, 1989); such shear is not consistent with isotropic tension or with present-day nonsynchronous rotation. Motion along strike-slip faults can generate "wing cracks" splaying from the region of the fault tip where the surface is being extended. Some bands on Europa appear to have originated as wing cracks associated with larger strike-slip features (Schulson, 2002; Kattenhorn, 2004; Kattenhorn and Marshall, 2006).

Regional-scale Galileo images of pull-apart bands show an overall bilateral symmetry (Sullivan *et al.*, 1998; Prockter *et al.*, 2002). Band margins are generally sharp, and reconstruction suggests that some have opened along preexisting ridges, which may have served as zones of weakness (Prockter *et al.*, 2002), although it is not known whether there is any further relationship between bands and ridges. A narrow central trough is common along bands and is remarkably linear and uniform in width along the length of each band. A hummocky textured zone commonly occurs to either side of this trough. Toward the margins of some bands are regularly spaced subparallel ridges and troughs, which are probably domino-style tilted normal fault blocks of the hummocky unit (Figueredo and Greeley, 2000; Prockter *et al.*, 2000a).

The units and characteristics of Europan pull-apart bands are analogous to those in terrestrial oceanic-spreading environments (e.g., Macdonald, 1982), suggesting that a spreading analogue may be appropriate (Sullivan *et al.*, 1998; Prockter *et al.*, 2002). The axial trough observed in many bands may be the site of plate separation, the flanking hummocky material may represent cryovolcanic material emplaced symmetrically on either side of the band axis, and the subparallel ridges and troughs are likely analogous to the abyssal hills observed along terrestrial mid-ocean ridges. Examples of contemporaneous three-band junctions have been found, with analogy made to ridge-ridge-ridge triple junctions, suggesting further similarities to terrestrial oceanic spreading processes (Head, 2000; Patterson and Head, 2003). The location and spacing of the ridges and troughs were used by Stempel *et al.* (2005) to obtain local strain rates of 10^{-15}–10^{-12} s^{-1} and local stresses of 0.4–2 MPa. These strain rates and stresses are consistent with theoretical models of the formation of narrow band-like rifts (Nimmo, 2004d).

An alternate model for band formation is based on the formation of leads in terrestrial sea ice (Pappalardo and Coon, 1996; Greeley *et al.*, 1998). Greenberg *et al.* (1998) and Tufts *et al.* (2000) consider that cyclical tension and compression due to Europa's diurnal tidal flexing might create bands through a ratcheting process. In this view, cracks open during the tensile phase of the diurnal cycle, allowing water to rise and freeze. These cracks are unable to close completely during the compressional phase due to the addition of the new material; hence, the band widens with time as new material is added. As discussed above in the context of the Greenberg *et al.* (1998) ridge formation model, this model relies on complete cracking through of Europa's ice shell to the depth of liquid water below. Experiments with wax analogue models show that cyclic strain on an opening rift zone can form band-like features in a thin brittle layer on top of a ductile substrate (Manga and Sinton, 2004). On a deeper level, one may expect that cyclic ratcheting is the native mechanism of crustal extension on Europa, as opposed to quasi-steady mid-ocean ridge spreading driven by distant subduction on the Earth.

Imaging of several bands indicates that they commonly stand topographically higher than the surrounding ridged plains (Malin and Pieri, 1986; Pappalardo and Sullivan, 1996; Giese et al., 1999; Prockter et al., 2000a; Tufts et al., 2000). This is consistent with emplacement of thermally or compositionally buoyant material, such as ice that is warm and/or clean relative to the cold and/or saltier surrounding lithospheric material (cf. Nimmo et al., 2003a).

5.1.1.5 Folds The means by which Europa accommodates the extensional strain from band formation is not well understood. Prockter and Pappalardo (2000) analyzed high-resolution Galileo images of Europa and presented evidence for regional-scale folds in several locations on Europa, the strongest of which is found in the region of Astypalaea Linea. High-resolution Galileo images reveal subtle shading variations suggestive of folds across Astypalaea Linea roughly perpendicular to its trend, and with a fold wavelength of ~25 km. Strong corroborative evidence for the fold interpretation comes from small-scale structures along the inferred anticline and syncline axes: discrete sets of small-scale fractures (troughs) occur along the crests of the regional-scale anticlines, while small-scale closely spaced single ridge crests occur within the inferred synclinal lows, trending parallel to the fold axes. They are inferred to be contractional structures (folds and/or thrust blocks) formed within regional-scale synclines.

The Astypalaea folds can be used to constrain the character and thickness of the lithosphere at the time of deformation, and the nature of the stresses that likely formed them. Folds may have formed by means of a compressional instability of a frictionally controlled brittle ice lithosphere overlying a ductile asthenosphere, in which ice strength decreases with depth (Herrick and Stevenson, 1990). If the local thermal gradient is very high ($\sim$100 K km^{-1}) and the brittle lithosphere is correspondingly thin ($\sim$2.5 km), compressional instability can be achieved with approximately 10 MPa of compressional stress, which is a relatively high amount of stress to achieve on Europa (Dombard and McKinnon, 2006b). The high stress derived may be a clue that brittle ice on Europa's surface is weaker in compression than currently thought.

Other more tentative examples of regional-scale folds have been identified, in the gray band Libya Linea and in the Manannán region (Prockter and Pappalardo, 2000) and in the satellite's northern leading hemisphere (Figueredo and Greeley, 2000). Some sets of rounded ridges in the ridged plains may represent small-scale folds (Patel et al., 1999a), but the mechanism of creating such small-wavelength fold structures is unclear. Overall, these folds can accommodate only small amounts of strain. As noted above, possible convergent bands have been identified (Greenberg et al., 2002; Sarid et al., 2002; Greenberg, 2004; Kattenholm and Marshall, 2006;

Patterson *et al.*, 2006), and may help to accommodate the strain from extensional bands.

5.1.2 Nonsynchronous rotation of Europa's ice shell

The smooth, presumably floating, ice shell of Europa is unlikely to have large mass asymmetries, and thus may rotate nonsynchronously. A lower limit of 10^4 years has been derived for the period of any ongoing nonsynchronous rotation, based on comparison of terminator views of the same features in Voyager 2 and Galileo images obtained 17 years apart (Hoppa *et al.*, 1999c). For their estimated shell thickness t_c of ~15–25 km (from tidal heating calculations), Ojakangas and Stevenson (1989a) predicted a nonsynchronous rotation time of ~10 Myr, consistent with the lower limit of Hoppa *et al.* (1999c). This timescale goes as $\sqrt{t_c}$; thus, a 10-km thick shell could rotate in 2.5 Myr.

Helfenstein and Parmentier (1985) first predicted the stress pattern that should result from nonsynchronous rotation, based on an eastward shift of Europa's surface relative to its fixed tidal axes (see subsection 3.2.2). Voyager global-scale lineaments were compared to this pattern by McEwen (1986) and Leith and McKinnon (1996). These workers concluded that the best match of the nonsynchronous stress pattern to Europa's global-scale lineaments occurred by considering a westward longitudinal shift in the locations of surface features relative to the fixed tidal axes. If the longitude of surface features is shifted westward (or equivalently, the tidal axes shifted eastward) to "back up" nonsynchronous rotation by ~25°, then lineament orientations achieve a best fit in being approximately perpendicular to the least compressive (greatest tensile) stress direction, as expected if the lineaments originated as tension fractures. The implication of this best fit is that Europa's major lineaments may have formed over a range of ~50° of nonsynchronous rotation. Stresses generated by nonsynchronous rotation can be significant. Maximum stresses of ~0.14 MPa can be achieved per degree of rotation (Equation 7.5); thus, accumulated tensile stress can exceed the tensile strength of cold laboratory ice upon ~12° of nonsynchronous rotation (Leith and McKinnon, 1996). However, the proper value of tensile strength to use for ice at the surface of Europa is uncertain due to a paucity of relevant laboratory data (see Section 2.3) and the problems inherent in scaling-up laboratory results to describe the bulk properties of Europa's lithosphere.

Galileo color images support and strengthen this argument, as imaging at near-infrared wavelengths can discriminate older lineaments that were invisible to Voyager (Clark *et al.*, 1998; Geissler *et al.*, 1998a). Geissler *et al.* (1998b) categorized lineament age based on color characteristics, and found that lineament orientations have progressively rotated clockwise over time, implying that stress orientation rotated similarly. This rotation sense is just as predicted by eastward migration of the surface relative to fixed tidal axes due to nonsynchronous rotation. Nonsynchronous rotation is not necessarily the formational stress mechanism, however, as the orientations of the most recent lineaments mapped by Geissler et al. (1998b) are better fit by diurnal stressing. It is plausible that diurnal stressing may create some cracks, while nonsynchronous rotation opens those cracks into wider ridges and bands.

Higher resolution Galileo imaging shows that Europa's ridged plains are overprinted by ridges and ridge sets of various orientations. Crosscutting relationships inferred from these higher resolution images have been cited as evidence for at least one full rotation of Europa's ice shell (Geissler et al., 1999; Figueredo and Greeley, 2000; Kattenhorn, 2002). Others have argued from stratigraphic relationships that few structures have formed over each rotation of the ice shell, implying that the surface records several shell rotations (Sarid et al., 2004, 2005), or even hundreds or thousands of shell rotations (Hoppa et al., 2001). Stratigraphic analysis of cycloidal cracks (subsection 5.1.3), which are very sensitive to the ice shell orientation, shows almost two complete rotations are required (Groenleer and Kattenhorn, 2008).

The equatorial region of isotropic tension west of the anti-Jovian point, predicted by nonsynchronous rotation, correlates to the zone of pull-apart bands originally recognized in Voyager imaging (Helfenstein and Parmentier, 1980; Pieri, 1981; Lucchitta and Soderblom, 1982; Schenk and McKinnon, 1989), and recognized from Galileo imaging to extend westward to ~250° longitude (Sullivan et al., 1998). A similar extensional region is predicted west of the sub-Jovian point, but is not observed in Galileo hemispheric-scale imaging, perhaps because cracks formed in this region did not open into bands (Hoppa et al., 2000). Zones of contractional tectonics are also predicted, centered 90° in longitude away from the extensional zones (Figure 7.4a). Evidence is mounting that shear failure may occur in these zones, as suggested by the "X"-patterned orientations of structures within and just east of these regions (Spaun et al., 2003; Stempel and Pappalardo, 2002).

More complex (but uncertain) stress sources are implied by the findings that the anti-Jovian extensional zone is centered ~15° south of the equator (implying that polar wander may have occurred; cf. subsection 3.2.3), and that dark and wedge-shaped band opening directions within have preferred orientations (Schenk and McKinnon, 1989; Sullivan et al., 1998). More evidence for polar wander comes from a survey of strike-slip offsets on Europa, showing that the dividing line between the expected north–south hemispheric dichotomy (see subsection 3.2.1) is tilted ~30° relative to the equator (Sarid et al., 2002). Mysterious troughs on Europa that exactly follow antipodal small circles centered near the equator are also offset as if several degrees of polar wander have occurred, and may themselves be the result of an episode of much greater (~90°) polar wander (Schenk et al., 2008).

Figure 7.10. Cycloidal features on Europa. These cycloids appear as single ridges, double ridges, or ridge complexes. The illumination is from the left.

5.1.3 Diurnal tidal variations

Europa is also subject to diurnal tidal variations that impose a daily rotating stress field on the surface, which may change the orientation at which cracks open relative to a larger stress field (e.g., from nonsynchronous rotation), and can lead to ratcheting of strike-slip faults (cf. subsection 3.2.1). Galileo images show many convincing examples of strike-slip offsets on Europa. Hoppa *et al.* (1999a) note that there is a preferred sense of strike-slip motion in each hemisphere, with a propensity for right-lateral strike-slip faults in the southern hemisphere, and left-lateral in the northern.

Even more definitive evidence for the role of diurnal stress variations is provided by the elegant explanation they provide for the previously mysterious patterns of cycloid ridges (flexus) and other cycloidal structures (Hoppa *et al.*, 1999b) (Figure 7.10). If a fracture propagates across Europa's surface at an appropriate speed (about $3\,\mathrm{km\,h^{-1}}$), the stress orientation rotates during a fraction of the Europan day such that the propagating fracture traces out the curvature of a single cycloidal arc. Tensile cracks typically propagate faster than this, but the cracks may propagate incrementally, slowing the overall rate (Lee *et al.*, 2005). The diurnal stress then drops below the critical value for fracture propagation until the following orbit, when tensile stress again increases, reinitiating the fracture propagation and thus generating the next cycloidal arc. The propagation of each succeeding crack may be aided by tailcracks that form during the strike-slip sliding that follows the tension in the diurnal cycle (Kattenhorn and Marshall, 2006; Groenleer and Kattenhorn, 2008).

This model explains several important observable aspects of cycloidal structures. First, the arcs of an individual cycloidal chain always show a consistent direction of convexity, while different chains can have opposite convexity directions. In the diurnal cracking model, convexity direction simply depends of the fracture propagation direction relative to the sense of stress rotation. Some cycloidal features transition into linear features, accounted for in the model as fractures propagate into regions in which the fracture can propagate for only a small fraction of the diurnal cycle. Similarly, the overall curvature of a cycloidal chain reflects the regional change in stress orientations from the latitude and longitude regime in which the fracture initiated, and into which it propagates.

The shape of cycloidal chains on Europa can be closely matched if tensile failure occurs at a stress of about 25 kPa, if propagation speed drops in proportion to tensile stress (producing a good match to arc skewness), and propagation halts when stress drops below 15 kPa. This suggests that Europa's uppermost lithosphere has a strength of only ~25 kPa. To what depth might these cycloidal fractures penetrate, and are they expected to transition into normal faults at depth? Leith and McKinnon (1996) suggest modeling fractures as crevasses, which can extend at least to a depth $z = \pi\sigma/2\rho g$. For an applied tensile stress $\sigma = 40$ kPa, the corresponding fracture depth is about 50 m or deeper if the brittle shell is thin (Lee *et al.*, 2005). Thus, diurnal stressing of a weak Europan lithosphere (tensile strength $\sigma_0 \sim 25$ kPa) may produce tensile fractures ~50 m to a few hundred meters deep.

5.1.4 Is Europa currently active?

The observed number of impact craters on Europa can be used to estimate the satellite's age if accurate estimates of the impactor flux can be made. By modeling the dynamics of small solar system bodies, Zahnle *et al.* (1998) and Levison *et al.* (2000) conclude that Jupiter family comets are the most common impactors onto the Galilean satellites in the present epoch. These authors model a current formation rate of one crater >20 km diameter each 3.2 Myr. The number of observed large craters on Europa implies a surface age of ~60 Myr, with a factor of three uncertainty (Zahnle *et al.*, 2003; Schenk *et al.*, 2004b).

An independent method for constraining the age of Europa's surface comes from estimates of ice sputtering, the ejection of ice particles due to the flux onto Europa of high-energy particles corotating with Jupiter's magnetic field. Based on Galileo Energetic Particle Detector (EPD) measurements, Ip *et al.* (2000) and Cooper *et al.* (2001) each have estimated H_2O sputtering rates, with estimates varying from 1.6 to 56 cm Myr^{-1}. From high-resolution images, Europa's stratigraphically oldest units (the ridged plains) display topography on vertical scales of tens of meters. If sputtering rates have been essentially constant over time, then 10 m of topography would be erased in $\sim 2 \times 10^7$ to 6×10^8 yr, so the oldest regions of Europa must be much less than a billion years old.

If Europa's average surface age is only ~50 Myr old, then it seems likely that Europa continues to be tectonically active today. One indication of relatively recent geological activity comes from the photometric properties of some ridge flanks, which suggest immature materials compared to other terrains (Helfenstein *et al.*, 1998), but the aging rate is unknown. As yet there is no certain evidence for current activity on the satellite. Comparison of Voyager and Galileo images obtained at similar resolution and lighting geometries shows no apparent changes, constraining the average surface age to ≥30 Myr (Phillips *et al.*, 2000). Searches for plumes and plume deposits have also been fruitless (Phillips *et al.*, 2000), though the opportunities to search for such activity during the Galileo mission were limited.

5.2. Enceladus

Enceladus is one of the most intriguing icy satellites. Despite its small size (252 km radius), Voyager 2 images revealed that portions of its surface are extensively tectonically deformed (Smith *et al.*, 1982; Squyres *et al.*, 1983; Kargel and Pozio, 1996). More recently, Cassini images of the south pole revealed active plumes of water vapor and ice crystals (Porco *et al.*, 2006) emanating from localized thermal anomalies (Spencer *et al.*, 2006) associated with tectonic features referred to as "tiger stripes." Because Cassini images are still being acquired and interpreted, all conclusions presented here are necessarily preliminary, and tectonics on Enceladus is likely to remain a field of active investigation for some time to come.

The internal structure of Enceladus probably consists of an ice shell 100 km or so thick overlying a silicate core (e.g., Schubert *et al.*, 2007). Because of the tectonic deformation and current activity observed at the surface, there is likely a subsurface ocean that decouples the ice shell from the silicate interior and increases the tidal deformation. Based on shape data, Enceladus probably is not in global hydrostatic equilibrium (Thomas *et al.*, 2007), which makes determination of its moment of inertia and interior structure very difficult.

Tidal stresses are likely an important source of deformation on Enceladus. If the shell of Enceladus responded in a fluid fashion, then the diurnal stresses and strain rates would be factors of 6 and 15 larger than Europa, respectively (Nimmo *et al.*, 2007b; Table 7.1). In practice, the ice shell of Enceladus is likely to be sufficiently thick that the deformation is reduced (Ross and Schubert, 1989); nonetheless, the stresses generated can be significant. A rough estimate of their magnitude is $250\,h_2$ kPa, where the Love number h_2 gives the response of the surface to tides and is 2.5 for a homogeneous fluid body (Nimmo *et al.*, 2007b).

Voyager 2 encountered Enceladus during northern summer, and thus observations were concentrated on the northern hemisphere. Cassini observations to date have been complementary to the Voyager observations, as they were obtained during southern summer and mainly focused on the south polar region. Most of the satellite (except for a swath centered on roughly 80°W) has now been imaged well enough to carry out global tectonic mapping. Broadly speaking, the satellite may be divided into three terrains: older cratered terrain, tectonically disrupted terrain, and the south polar terrain. Cratered terrains (Figure 7.11) occupy a broad band encircling the satellite along the 0° and 180° longitude lines (over the north pole, through the sub- and antisaturnian points). The craters in the older terrain are in some cases superimposed on sets of ancient, even older, subdued linear structures, and are often cut by younger fractures. In high-resolution images, many of the recent fractures in the cratered terrain are observed to be chains of pits (Michaud *et al.*, 2008), which are most likely formed by regolith drainage over dilational normal faults (e.g., Wyrick *et al.*, 2004) with a small amount of extensional strain.

Centered on the leading and trailing hemispheres (90° and 270°) are younger, tectonically disrupted terrains roughly 90° wide at the equator. This terrain is characterized by densely spaced sets of subparallel ridges and troughs (Figure 7.11), reminiscent of grooved terrain on Ganymede (subsection 6.1.2), and is probably the result of normal faulting. This terrain also harbors some prominent rounded, branching ridges that rise well above the surrounding terrain, the origin of which is not well understood.

Although the north polar region is apparently undeformed cratered terrain, the south polar region (south of 55°S) is complexly tectonized and essentially uncratered (Figure 7.12; Porco *et al.*, 2006). The margins of this region are scalloped, and cusps that extend northward into the surrounding regions contain arcuate scarps that are likely fold-and-thrust belts (Helfenstein *et al.*, 2006a). North–south trending extensional-tectonic structures stretch towards the equator from the ends of the cusps. Within the south polar terrain, the most prominent features are a set of subparallel, ridge-flanked troughs up to 0.5 km deep, 2 km wide and ∼130 km

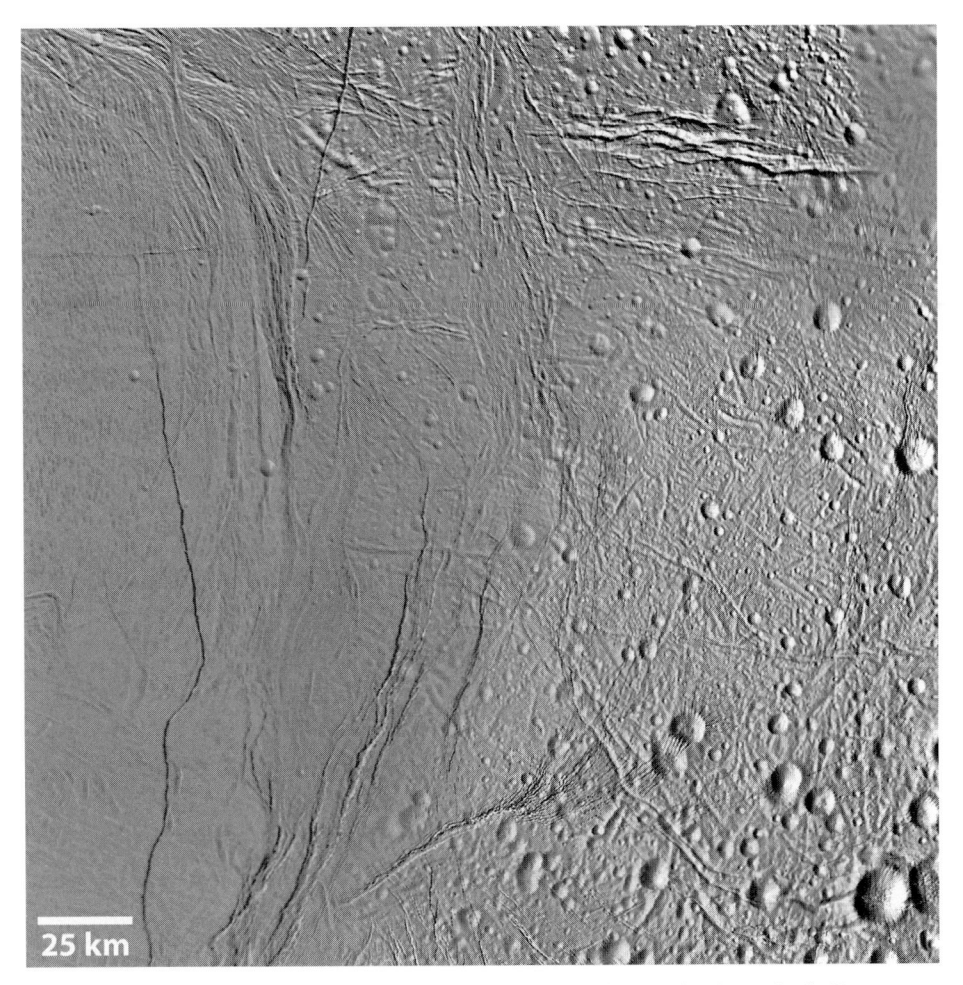

Figure 7.11. Cratered terrain (right) cut by a swath of tectonized terrain (left) near the equator of Saturn's satellite Enceladus. North is to the top, and the illumination is from the left.

long, informally termed "tiger stripes" (Porco *et al.*, 2006). These troughs trend at about 45° from the tidal axis of Enceladus' figure, and have a spacing of about 35 km.

A broad range of crater densities confirms that Enceladus has had a long and at least intermittently active tectonic history (Smith *et al.*, 1982; Kargel and Pozio, 1996; Porco *et al.*, 2006). The most heavily cratered regions have impact crater densities, suitably scaled, that are comparable to those of the lunar highlands (Porco *et al.*, 2006) and suggest that some terrains have survived mostly intact for billions of years. On the other hand, some of the tectonically deformed regions possess very few craters, indicating that these terrains were deformed within the last few

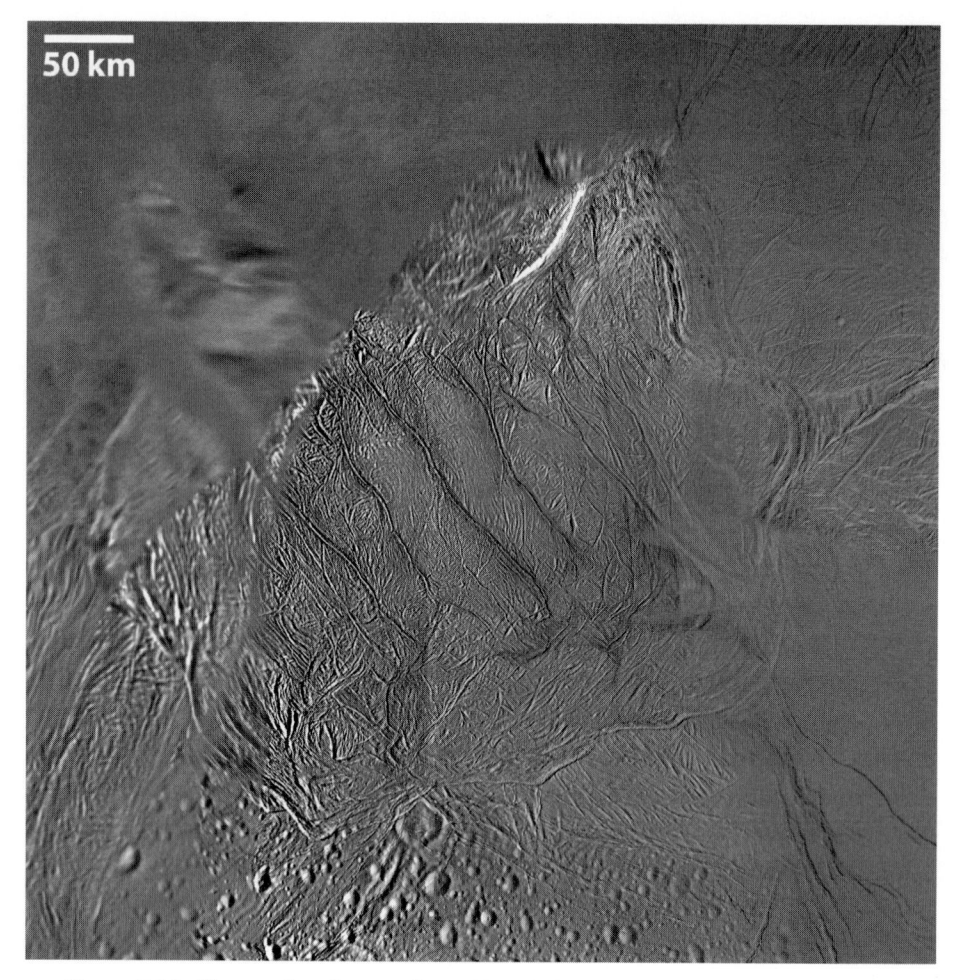

Figure 7.12. The south polar terrain on Enceladus forms a complexly tectonized area bounded by a band of cuspate ridges. Bands of extensional faults tend to radiate toward the equator from the outermost projections of the cuspate boundary. The "tiger stripes" are the dark ridge pairs in the center of the image. Polar stereographic map projection, centered on the south pole. The blurry area in the upper left quadrant has not been imaged at high resolution at the time of this writing.

tens of millions of years, and the south polar region is devoid of craters >1 km diameter, indicating a surface age of ~ 1 Myr (Porco et al., 2006). In some cases, the observed craters are much shallower than would be expected, indicating that relaxation has occurred, presumably as a result of relatively high subsurface heat fluxes (Passey, 1983; Schenk and Moore, 1995; Porco et al., 2006).

Because of its intense geological activity and low surface gravity, Enceladus is topographically relatively rough. Limb profiles can in some cases be correlated with geological features such as Samarkand Sulci (one of the equatorial bands of deformed terrain), and reveal local relief of up to 1 km vertical over ~100 km distance (Kargel and Pozio, 1996; cf. Bland et al., 2007). Certain trough flanks are suggestive of flank uplift, implying flexural parameters of roughly 30 km and indicating elastic thicknesses of about 4 km (Kargel and Pozio, 1996). The south polar region is depressed relative to the surrounding terrain by ~0.5 km (Thomas et al., 2007), possibly as a result of melting of ice by a subsurface heat source, creating a trapped sea beneath the south pole (Collins and Goodman, 2007).

Our understanding of the geological evolution of Enceladus, and the mechanisms responsible, is still evolving. The wide range of surface ages suggests patchy resurfacing, perhaps in a somewhat analogous manner to Ganymede (Bland et al., 2007). Based on Voyager 2 images, cryovolcanism was favored as the principal resurfacing mechanism (Squyres et al., 1983), but Cassini images suggest that this resurfacing is primarily tectonic (Porco et al., 2006; Barr, 2008). In addition, fallout from the plumes may cause mantling of surface features, and localized cryovolcanism has been suggested.

Most Cassini-based work to date has focused on the evolution of the south polar region. The orientation of the tiger stripes is consistent with their original formation as tension cracks, because they are oriented perpendicular to the maximum present-day diurnal stresses (Nimmo et al., 2007b). The tiger stripes may currently be undergoing strike-slip motion, possibly explaining the existence and timing of the plume eruptions (Nimmo et al., 2007b; Hurford et al., 2007; Smith-Konter and Pappalardo, 2008). No evidence has yet been reported for geological strike-slip offsets, but the deformation is expected to be cyclic. The existence of the unusually high heat flow must ultimately be due to tidal heating in Enceladus' south pole region, and may be tied to either the presence of a subsurface warm diapir (Nimmo and Pappalardo, 2006) or to melting and subsidence associated with heating in the silicate core or in the ice shell (Collins and Goodman, 2007; Tobie et al., 2008). Both of these mechanisms are likely to produce gravity anomalies leading to poleward reorientation of the region, thus explaining its current polar location (Nimmo and Pappalardo, 2006), although tidal heating should theoretically be maximized at the poles in any case. They also lead to predictable local tectonic stresses: extensional for a rising diapir, and compressional for a subsiding surface (see Section 6.2 for a similar situation on Miranda).

Poleward reorientation in turn generates its own set of global stresses (e.g., Melosh, 1980a; Leith and McKinnon, 1996; subsection 3.2.3). For synchronous satellites, reorientation is complicated by the triaxial shape of the body, with the result that reorientation tends to happen most readily around the tidal axis (e.g., Matsuyama and Nimmo, 2007). The resulting stresses can be calculated (Leith and McKinnon, 1996) and compared with existing tectonic features. Whether such reorientation stresses can explain the tectonic features observed is currently a topic of investigation. The identification of potential "tiger stripe analogues" in near-equatorial regions on Enceladus (Helfenstein *et al.*, 2006b) is a potential argument for reorientation, but this is also a topic of continuing research.

Another potentially important source of stress arises from an ice shell thickening above a subsurface ocean (Nimmo, 2004b). Because ice expands as it freezes, the surface moves radially outwards, leading to isotropic extension. Thus, the predominance of extensional features on Enceladus may be a function of its ice shell having thickened with time. Ice shell freezing can also lead to pressurization of the underlying ocean, and potentially cryovolcanism (Manga and Wang, 2007).

In summary, Enceladus is a geologically active, heavily deformed satellite that has a visible geological history stretching from billions of years ago to the present day. Our understanding is currently at a crude level. With further Cassini flybys planned and intensive data analysis only now beginning, the unraveling of this fascinating body's tectonic history has just begun.

5.3 Triton

Triton is the only large satellite of Neptune, and has a mass ~40% greater than Pluto's. As Voyager 2 flew by in 1989, slightly more than half the surface was imaged and ~20% of that was obtained at resolutions of 1 km/pixel or better, sufficient to discriminate geological features at regional scales. These images revealed a young surface with relatively few impact craters and a wealth of geological landforms. Some are interpreted to be cryovolcanic in origin, including flow lobes and pyroclastic sheets, and some are likely to have a tectonic origin, such as ridges and troughs. The geological age of the portion of Triton's surface imaged by Voyager is of the order 100 Myr (Stern and McKinnon, 2000; Schenk and Zahnle, 2007), implying that it follows Io, Europa, and portions of Enceladus in its level of geological activity. The dominant "bedrock" material is thought to be water ice or ammonia hydrate ice (Croft *et al.*, 1995; Cruikshank *et al.*, 2000), although a number of other exotic ices have been observed (e.g., Na, CO_2; Quirico *et al.*, 1999). Tidal heating may have sustained warm interior temperatures for upwards of a billion years, and if ammonia is present in the icy mantle, a subsurface liquid ocean may still persist today (Hussmann *et al.*, 2006).

Triton is in a retrograde and highly inclined (157°) orbit around Neptune, suggesting that it did not originate in its current position but is instead a captured object. Several models for Triton's capture have been proposed, including aerodynamic drag in a protosatellite disk (McKinnon and Leith, 1995), capture by collision with an existing satellite (Goldreich *et al.*, 1989; Benner and McKinnon, 1995), or exchange capture between a binary system and Neptune, in which one

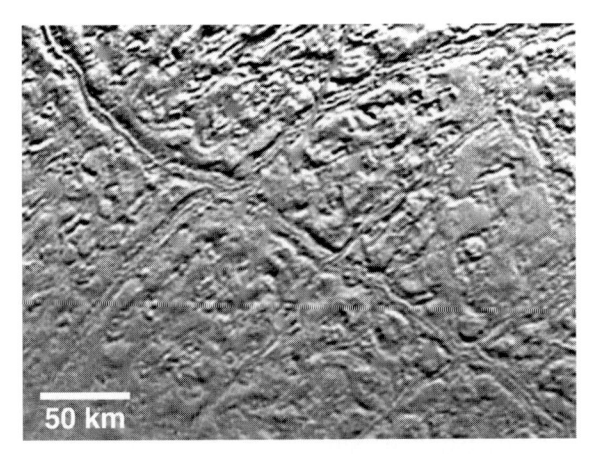

Figure 7.13. Neptune's satellite Triton exhibits several bands of sinuous ridges and double-ridge features reminiscent of those on Europa. This image shows the intersection of Slidr Sulci (running northwest to southeast) with a few smaller orthogonal ridge sets. The terrain surrounding the ridge sets is termed "cantaloupe terrain," composed of shallow quasi-circular depressions bounded by single ridges. North is to the top, and the illumination is from the bottom.

member of the binary was expelled and its place taken by the planet (Agnor and Hamilton, 2006). Of these, aerodynamic drag is expected to have occurred early in Neptune's history within a specific time period. Conditions for collisional and exchange capture are less time-sensitive, as Neptune migrated outwards due to encounters with material from the protoplanetary disk. Following capture, the orbit is thought to have circularized on a timescale of several hundred Myr to ~1 Gyr (Ross and Schubert, 1990; McKinnon *et al.*, 1995).

5.3.1. Ridges

Triton has a significant number of sinuous ridges (Figure 7.13), spanning a range of ages and degradation states. Typical ridges measure ~15–20 km in width, and some are continuous for ~800 km. A deep continuous axial depression ~5–10 km wide flanked by higher ridges results in a double-ridge morphology, and some ridges are bounded by shallow troughs ~20 km wide. Triton is the only other place in the solar system where Europa-like double ridges have been identified (subsection 5.1.1.3), and indeed these appear to be the dominant tectonic structure on Triton. The ridges on Triton are remarkably similar to Europan ridges, including single isolated troughs and double ridges, as well as much rarer triple and multi-crested ridges (Prockter *et al.*, 2005). Triton's ridges are typically many times wider than typical Europan ridges, but are morphologically subdued – the very limited available topography (Croft *et al.*, 1995) suggests that they are only a couple of hundred meters high, similar to Europan ridges.

Voyager-based models for ridge formation on Triton suggested cryovolcanic extrusion into graben (Smith *et al.*, 1989; Croft *et al.*, 1995). Investigations into ridge formation on Europa suggest that the graben model does not fit the observed morphologies, and given the strong resemblance of ridges on the two moons, it is reasonable that similar formation mechanisms are responsible on both. The shear heating model of ridge formation (Gaidos and Nimmo, 2000; Nimmo and Gaidos, 2002) was proposed for Triton's ridges by Prockter *et al.* (2005). They found that the magnitudes of stresses and heat fluxes required to generate ridges of the correct scale are comparable to predicted values generated during Triton's orbital evolution from a highly eccentric state. The much greater widths of Triton's ridges compared to ridges on Europa are likely due to the lower surface temperature, and thus a greater brittle–ductile transition depth.

The large-scale pattern of ridges and troughs on Triton is still not well understood. Equivocal results have come from attempts to compare the lineament patterns to plausible global stress models including despinning, orbital precession, and non-synchronous rotation (Collins and Schenk, 1994; Croft *et al.*, 1995; Prockter *et al.*, 2005).

If Triton's ridges do form by shear heating, the timescale becomes puzzling. Heat generated during capture could have easily melted Triton's interior (McKinnon *et al.*, 1995), probably enabling the surface to rapidly and completely overturn. Could the current surface be a relic of Triton's waning, tidally driven, geological activity? The ridges may not themselves be young, but the lack of impact craters and otherwise young surface suggests they formed relatively recently, which implies that capture was also a relatively recent event. Because the timescale for Triton's orbital evolution depends on the poorly known (and time-variable) internal mechanical layering of Triton and Neptune, the time of Triton's capture is not well constrained.

5.3.2. Tectonic interactions with cryovolcanic deposits

The surface of Triton exhibits many landforms that resemble terrestrial volcanic features (Croft *et al.*, 1995). These include cones 7–15 km in diameter, occurring individually or in clusters; chains of pits along ridges are similar to terrestrial tectonically controlled cinder cones and explosion pits; and circular to elongate depressions or pit paterae, occurring singly or in chains and $\sim$10–20 km in diameter, located in patches of smooth material 100–200 km in extent (Figure 7.14). These tend to follow regional tectonic trends and may be analogous to chains of explosion and collapse pits in terrestrial volcano–tectonic zones. Some of the most enigmatic features are the larger ring paterae, 50–100 km in scale, with an outer rim defined by a ring of coalescing pits. The pit and ring paterae along with their associated

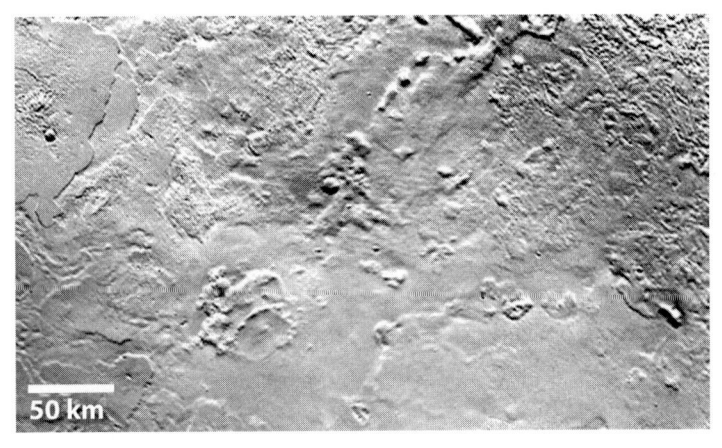

Figure 7.14. Image of plains on Triton, presumably emplaced by cryovolcanic flows. Some of the pit paterae occur in chains, and may be tectonically controlled. North is to the top left, and the illumination is from the bottom.

smooth deposits are interpreted to have formed as explosive cryovolcanic craters and deposits.

Triton's enigmatic "cantaloupe terrain" contains quasi-circular shallow depressions termed cavi, typically 25–35 km in diameter with slightly raised rims, giving the appearance of a cantaloupe rind (Figure 7.13; Croft *et al.*, 1995). Cavi have been suggested to represent cryovolcanic explosion craters, such as terrestrial maars, on the basis of their similar morphologies. However, the organized cellular pattern of the cantaloupe terrain has been proposed to closely resemble the expression of terrestrial salt diapirs, and the terrain may have formed due to diapirism resulting from gravity-driven overturn within an ice crust about 20 km thick (Schenk and Jackson, 1993). One possibility is that cryovolcanism on Triton layered more dense ices on top of less dense ice layers, leading to compositionally driven overturn of the ice layers. Cavi may be analogous to similar structures on Europa, which have also been proposed to have a diapiric origin.

5.3.3. Current activity

Voyager images of Triton's bright south polar cap revealed dark streaks, possibly methane converted to organic material by energetic particles and ultraviolet photons. This observation was remarkable given that Triton is thought to undergo a cycle of volatile deposition and sublimation from pole to pole on the cycle of 1 Triton year, or ~165 Earth years (Buratti *et al.*, 1994). This yearly cycle is expected to result in a meter or more of nitrogen, methane, and carbon dioxide frosts being sublimated from one pole and deposited on the other, thus the presence of the dark streaks implied they were very young. Searches of stereoscopic images of the south polar region clearly show plumes, up to 8 km in height with radii of up to 1 km, probably composed of dust and gas (Kirk *et al.*, 1995). The columns feed clouds of dark material that drifted with Triton's tenuous winds for more than one hundred kilometers.

Two models have been proposed to drive the plumes. One suggests explosive venting of nitrogen gas pressurized by solar heating (Kirk *et al.*, 1995). Triton's 38 K surface might be blanketed with transparent solid nitrogen. In this model, dark material lying immediately below the transparent layer is warmed by sunlight, undergoing a significant increase in temperature with respect to the surface and a corresponding increase in vapor pressure of the surrounding nitrogen. The highly pressurized nitrogen is trapped in pore spaces, then released to the surface through a vent, entraining dark material and lofting it into the atmosphere. The model suggests that a temperature increase of only 2° would be sufficient to propel the plumes to the observed altitudes. An alternative driving mechanism suggests that the heat source for the geysers comes from within the satellite, perhaps due to thermal convection in the underlying ice (Duxbury and Brown, 1997). This model seems consistent with Triton's young surface age and the extreme tidal heating predicted during capture of the satellite by Neptune.

6 Formerly active icy satellites

6.1 Ganymede

Ganymede is the largest satellite in the solar system, larger than the planet Mercury. It has the lowest normalized axial moment of inertia of any solid body known in the solar system, indicating that its mass is highly concentrated in the interior. Probably its interior consists of an iron core several hundred kilometers in radius, followed by a rocky mantle, with about 800 km of water ice on top (Anderson *et al.*, 1996). A molten iron core likely explains the internally generated magnetic dipole of Ganymede (Schubert *et al.*, 1996), the only satellite with its own magnetosphere. Variations in Ganymede's magnetic field also point to the existence of a subsurface ocean approximately 150 km below the surface, sandwiched between low pressure ice I at the surface and higher pressure phases of ice below (Kivelson *et al.*, 2002). The majority of its surface is dominated by tectonic features, but even the youngest of these features is overlain by younger impact craters, and is nominally 2 billion years old (Zahnle *et al.*, 2003), though there is considerable uncertainty in that age estimate. The surface of Ganymede is commonly divided into two broad categories, termed dark terrain and bright terrain. We will discuss the tectonics on both of these terrains, and then summarize the implications for tectonic driving mechanisms on Ganymede (see Pappalardo *et al.* (2004) for an up-to-date summary of all geology on Ganymede).

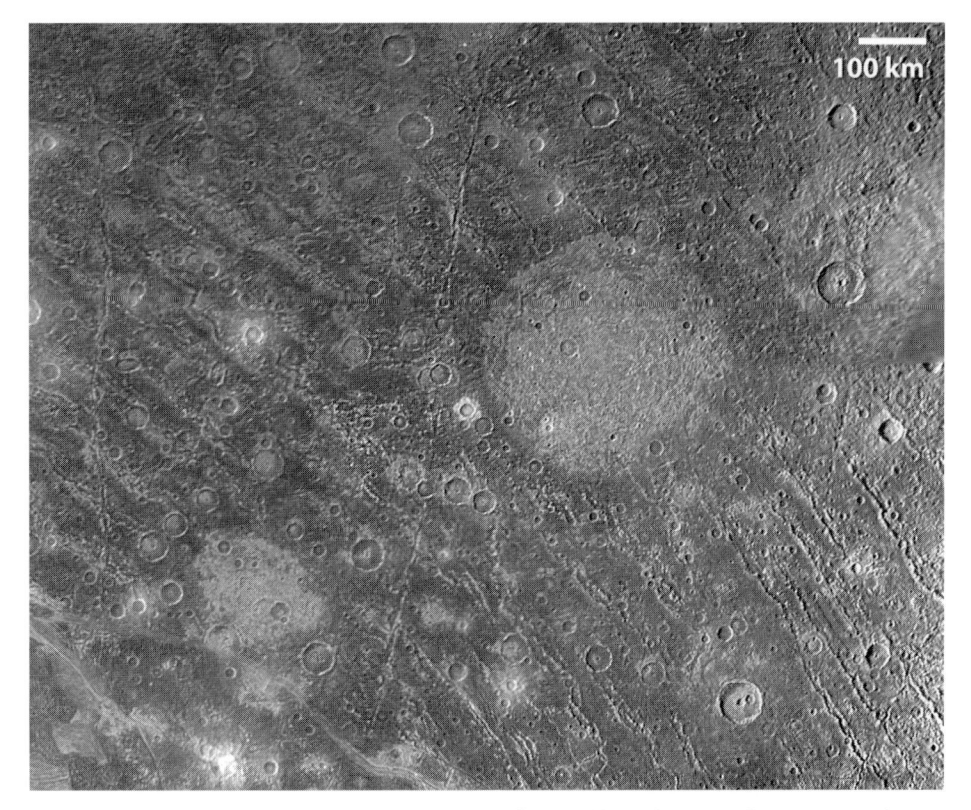

Figure 7.15. Ancient structures known as "furrows" arc from northwest to southeast in this area of Galileo Regio on Ganymede. North is to the top, and the illumination is from the left.

6.1.1. Dark terrain

Dark terrain covers one third of the surface of Ganymede and appears to be an ancient, perhaps primordial, surface. The surface is saturated with impact craters, and a surface layer of loose dark dust mantles much of the topography (Prockter et al., 1998). Viewed from a distance, many areas of dark terrain are dominated by sets of concentric ring arcs, termed furrows (Figure 7.15). Furrow sets can be thousands of kilometers across and are interpreted to be concentric fractures around ancient impact basins. As such, they are not endogenic tectonic features, but their characteristics do tell us something about the nature of the lithosphere in which they formed. Furrows must have formed in a lithosphere that was relatively thin (McKinnon and Melosh, 1980) compared to the present day, since more recent large impacts such as the basin Gilgamesh did not form these closely spaced features as they collapsed. Furrows are the oldest recognized feature on Ganymede, being cut by all other craters (Passey and Shoemaker, 1982), and thus giving us insight into an early period of higher heat flow and thinner lithosphere. The furrows themselves consist of two parallel ridges with a trough in between. They are interpreted to be graben-like features that have undergone topographic relaxation (McKinnon and Melosh, 1980). Nimmo and Pappalardo (2004) fit flexural models to the rift flank uplift on the furrow features and estimate that an ancient heat flux of 60–80 mW m^{-2} is necessary to fit the topography, corresponding to an elastic lithosphere 2–3 km thick. It is unknown, though, whether this represents the heat flux during the formation of the furrows or during the formation of bright terrain (see subsection 6.1.2), and how much the topography has viscously relaxed since the furrows formed.

From a distance, dark terrain also exhibits higher albedo streaks or irregular patches. Close-up Galileo images have resolved these to be concentrated areas of intense faulting. Motion along faults exposes bright ice below the dark surface, brightening the terrain. This may be one mode by which dark terrain changes into bright terrain (Prockter *et al.*, 2000b). In some areas of dark terrain, it appears that recent deformation has concentrated in former furrows, perhaps utilizing them as preexisting weaknesses in the lithosphere (Murchie *et al.*, 1986).

6.1.2 Bright terrain

The distinctive appearance of Ganymede results from the bright terrain that slices the dark terrain into discrete polygons. Within the bright terrain is a mosaic of crosscutting swaths and polygons of subparallel ridges and troughs (termed grooved terrain) or smooth bright plains (Figure 7.16). Grooved terrain exhibits several different morphologies of ridges and troughs, which appear to be related to the amount of strain accommodated in each polygon of the bright terrain. So far, no terrain with evidence for contractional strain has been identified. There is, however, abundant evidence for extensional strain all over the surface of Ganymede.

At the low-strain end of the strain spectrum are bright plains cut by parallel dark troughs and graben-like structures (Figure 7.17). Measurement of craters cut by grooves of this morphology show less than 5% extensional strain (Pappalardo and Collins, 2005). The origin of the smooth bright plains is unclear, though it has been suggested that they represent areas flooded by cryovolcanic flows (e.g., Shoemaker *et al.*, 1982), which is supported by the even, low topography of some of these areas (Schenk *et al.*, 2001b). However, no bright plains have been found without some form of ridges or troughs on their surfaces, so tectonism appears to be an essential part of bright terrain formation.

At the other end of the spectrum are areas of grooved terrain that appear to exhibit tilt-block normal faulting (Pappalardo *et al.*, 1998b). High-resolution observations of these areas show a landscape of parallel ridges and troughs with a triangular sawtooth cross section (Figure 7.18). In the tilt-block normal faulting model, triangular ridges are formed as the surface is cut apart by parallel normal faults, all

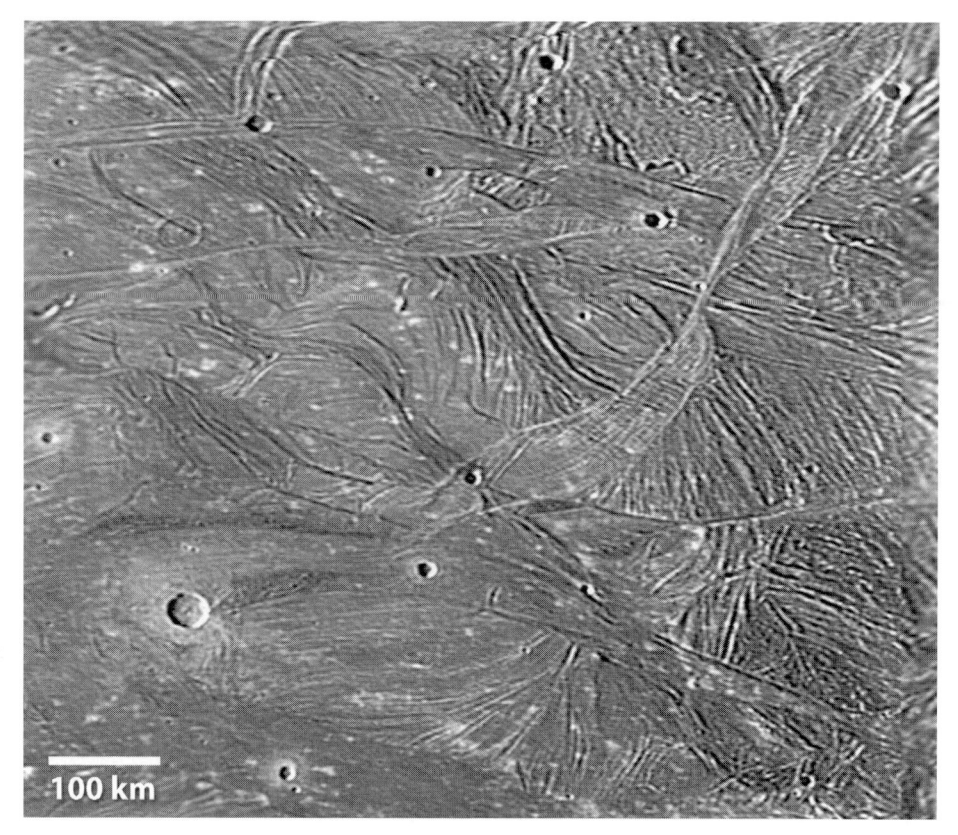

Figure 7.16. Broad-scale view of Ganymede grooved terrain, in Nun Sulci on the sub-Jovian hemisphere. North is to the top, and the illumination is from the left.

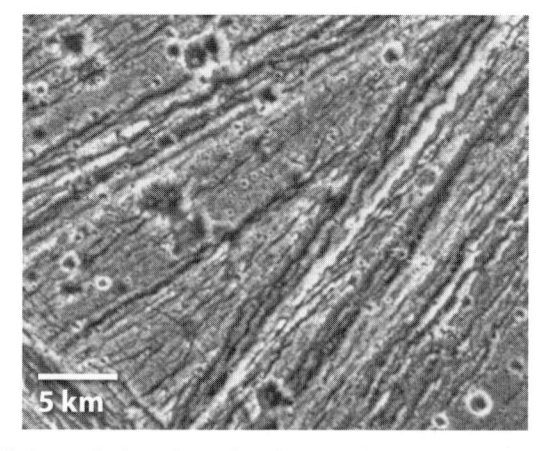

Figure 7.17. High-resolution view of graben sets in grooved terrain, Uruk Sulcus, Ganymede. North is to the top.

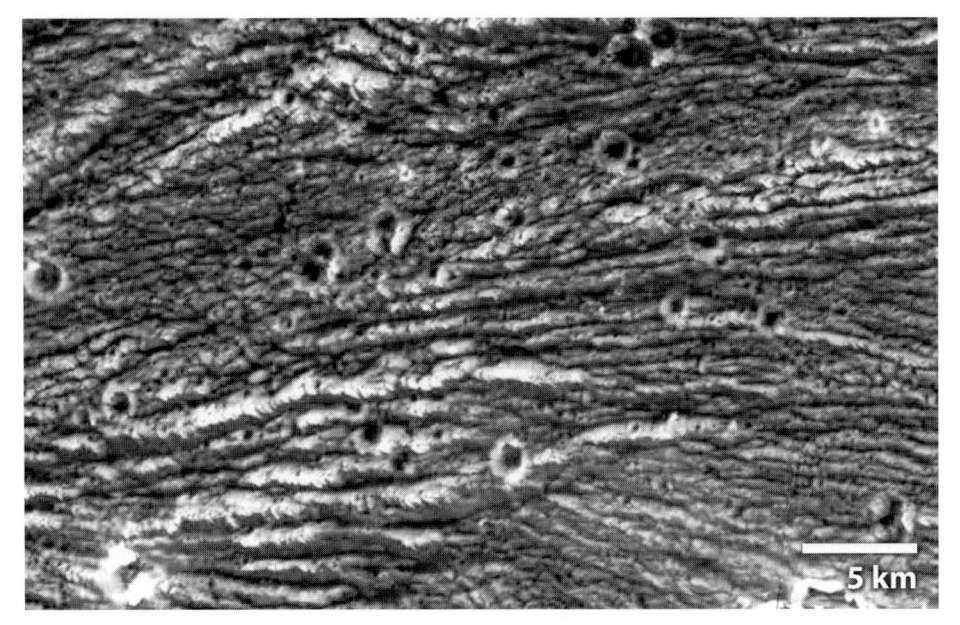

Figure 7.18. High-resolution view of ridges interpreted to be tilt blocks in grooved terrain, Uruk Sulcus, Ganymede. North is to the bottom right, and the illumination is almost overhead.

dipping in the same direction. As the terrain is pulled apart, motion along the fault exposes the fault scarp on one side of the ridge, while the original surface tilts back to form the other side of the ridge. These tilt-block ridges on Ganymede are typically ~1 km wide and 100–200 m high.

The geometry of this tilt-block faulting model was used by Collins *et al.* (1998a) to estimate that one region of grooved terrain had been pulled apart by about 50%, based on the throw on the fault scarps. In three other areas, Pappalardo and Collins (2005) found impact craters that had been cut by tilt-block faulting zones about 10–20 km in width. Using the craters as strain markers, they estimated that these sets of faults had accommodated from 50% up to 180% extensional strain. Some of these sets of faults had also accommodated a few kilometers of strike-slip motion in addition to the extension normal to the faults. The strain is high enough that no features from the preexisting surface can be recognized within the fault zone, a process termed tectonic resurfacing, which can wipe out craters and reset the surface age through tectonism alone.

In addition to extensional deformation, strike-slip motion is also observed as a component of Ganymede tectonics. Strike-slip motion was suspected from Voyager data based on the sigmoidal shapes of many small regions of grooved terrain, and offsets of background terrain features on either side of grooved terrain swaths

(Lucchitta, 1980; Murchie and Head, 1988). Higher resolution observations show several fault zones in which normal faults are organized into en echelon segments, indicating transtension (Pappalardo *et al.*, 1998b; Collins *et al.*, 1998b; DeRemer and Pappalardo, 2003). Two of the five fault zones measured by Pappalardo and Collins (2005) using craters as strain markers show significant levels of strike-slip motion along the faults.

In the tilt-block regions, the small-scale ridges and troughs described above are superimposed on broader undulating topography, with a ridge spacing (or wavelength) of ∼5 to 10 km (Patel *et al.*, 1999b) and an overall relief of about 500 m (Giese *et al.*, 1998; Squyres, 1981). This longer length scale of periodic deformation may be evidence of extensional instability of the brittle lithosphere over a ductile substrate (Fink and Fletcher, 1981; Herrick and Stevenson, 1990; Collins *et al.*, 1998a; Dombard and McKinnon, 2001). In such an instability, the lithosphere develops periodic pinches where extensional strain is preferentially concentrated; indeed, in these broad-scale valleys, the tilt blocks become smaller and may exhibit secondary faulting and greater strain. The existence of these lithospheric necks helps to constrain the properties of the lithosphere, since they are sensitive to the thickness of the lithosphere and the ductile properties of the substrate. Using this relationship, Dombard and McKinnon (2001) estimated that the grooved terrain formed at a strain rate of 10^{-16} to $10^{-14}\,\mathrm{s}^{-1}$ and a heat flux of 60–120 mW m^{-2}. However, more recent numerical simulations of extensional instabilities, using a different lithospheric strength envelope, have found that it may be more difficult to produce the observed ridges than the analytical models predict (Bland and Showman, 2007). More robust estimates should result from improved numerical models (e.g. Bland *et al.*, 2008a).

Using rift flank uplift at the edges of grooved terrain, Nimmo *et al.* (2002) estimated a heat flux of 100 mW m^{-2}, in good agreement with the extensional instability model. The heat fluxes from both these models indicate that grooved terrain formed with an elastic lithosphere about 1–2 km thick. In order for the observed faults to undergo shear failure through this lithosphere, stresses on the order of 1 MPa are required; the possible sources of this stress are the subject of the next section.

On Europa, smooth bands on the surface appear to be the result of complete spreading of the lithosphere (see subsection 5.1.1.3). Ganymede exhibits a few morphologically similar features, notably a 25-km wide swath of bright smooth material called Arbela Sulcus, cutting across the dark terrain of Nicholson Regio. Like the bands on Europa, this band can also be reconstructed by rotating one edge around a pole of rotation, matching preexisting features on the two sides (Head *et al.*, 2002). Though this feature is intriguing, it is still unclear how widespread lithospheric spreading is on Ganymede.

6.1.3 Implications for Ganymede evolution

Grooved terrain is the record of an active episode somewhere in the middle of Ganymede's history, but what could have triggered it? Any hypothesis for the driving mechanism behind groove formation must explain: (a) why it happened in the middle of Ganymede's history and not near the beginning, (b) how a sufficiently high amount of stress was generated to initiate fault failure, (c) the global, interconnected nature of grooved terrain faults, (d) the high levels of extensional strain observed, with no evidence found yet for contraction, and (e) the high heat flow inferred during the period of groove formation. It would be beneficial if the hypothesis also explained or related to other aspects of Ganymede, such as its molten iron core (e.g., Bland *et al.*, 2008b), and the distribution of craters on its surface.

Most of these points can be addressed by positing a heat pulse in Ganymede's interior. This could be the result of converting potential energy into heat as the satellite differentiates, or it could be due to enhanced tidal heating at some point in Ganymede's past. The three innermost Galilean satellites are currently locked in a 4:2:1 orbital resonance, called the Laplace resonance. As the satellites evolved towards this resonance, they may have passed through a Laplace-like resonance, which would have pumped up Ganymede's orbital eccentricity and caused enhanced tidal heating (Showman *et al.*, 1997). This tidal heating episode is sufficient to explain the inferred heat fluxes, and could also have warmed and melted some of the ice within Ganymede, generating a small amount of volume expansion (Showman *et al.*, 1997), which, in turn, caused surface extension. The tidal resonance hypothesis provides a natural explanation for how to delay the formation of grooved terrain.

If the heat pulse is due to differentiation, there is some question as to how it would remain undifferentiated for the initial part of its history (Friedson and Stevenson, 1983), although the mostly undifferentiated state of Callisto today (Anderson *et al.*, 2001) suggests this is a possibility. Differentiation, or completion of differentiation, could itself have been triggered by the tidal heating episode discussed above. One attractive aspect of differentiation is that since Ganymede is so large, there would be a large volume of high-pressure ice phases in the interior displaced to the outside, which would cause significant volume change (Squyres, 1980; Mueller and McKinnon, 1988). Volume change would cause isotropic tensile stress over the entire surface (see subsection 3.2.6), which may explain the large amount of extensional strain observed. Initial estimates of the amount of global expansion represented by all grooved terrain on Ganymede is significantly higher than would be expected from heating and melting alone, and is closer to the amount expected from interior differentiation (Collins, 2006).

Nonsynchronous rotation (subsection 3.2.2) is another possible source of stress for Ganymede grooved terrain, though the dominance of extension of Ganymede's surface argues that it may not be the primary source. The primary evidence for nonsynchronous rotation having occurred on Ganymede comes from its crater population. Synchronously rotating satellites accumulate more craters on the hemisphere that leads their orbital motion, but Ganymede's younger bright terrain shows a much weaker than expected asymmetry in crater density (Zahnle et al., 2001). The simplest explanation for this discrepancy is that Ganymede rotated nonsynchronously for a significant period of time, resulting in a uniform crater population. Later impacts contributed an anisotropic population, resulting in the diluted pattern we see today. Also, features called catenae on Ganymede are believed to be formed by comets split into fragments by a close pass by Jupiter (much like comet Shoemaker-Levy 9). All catenae should form on the Jupiter-facing hemisphere as the recently split comets leave the Jupiter system (Schenk et al., 1996), but on Ganymede a few of the catenae are found on the opposite hemisphere, indicating that it may have faced toward Jupiter for a finite time (Zahnle et al., 2001). True polar wander has also been proposed on Ganymede (Murchie and Head, 1986; Mohit et al., 2004), which could also explain the origin of the stresses and the crater asymmetry, but this finding has not been corroborated.

Ultimately, to distinguish among the hypotheses for grooved terrain formation, we need to compare the theoretical predictions to the details of the record of deformation in grooved terrain itself, which in turn will require careful mapping of the surface. The sparse nature of high-resolution data and the gaps that still remain in surface coverage mean that a definitive answer may have to wait for a new mission to the Jupiter system.

6.2 Miranda

Miranda is the smallest of the five major satellites of Uranus, with an average radius of only 236 km. The surface of Miranda (Figure 7.19) consists of cratered terrain crosscut by three "coronae," which are ovoidal to trapezoidal regions of low crater density, containing sets of ridges and troughs (Smith et al., 1986). The cratered terrain (Plescia, 1988; Stooke, 1991; Croft and Soderblom, 1991) is characterized by rolling topography punctuated by large muted craters and smaller sharp ones. Sharp and muted scarps and inward-facing scarp pairs, interpreted as normal fault scarps and graben, occur within the cratered terrain. The underpopulation of small craters and the muting of larger craters and scarps might be due to a large-scale mantling event, and dark material exposed in the walls of some fresh scarps and craters also argues for mantling (Croft and Soderblom, 1991). The presence of both muted (pre-mantling) and sharp (post-mantling) fault scarps indicates ongoing or multiple episodes of extensional tectonism during Miranda's period of endogenic activity. Crater counts suggest that Miranda's coronae may have been geologically active less than a billion years ago (Zahnle et al., 2003).

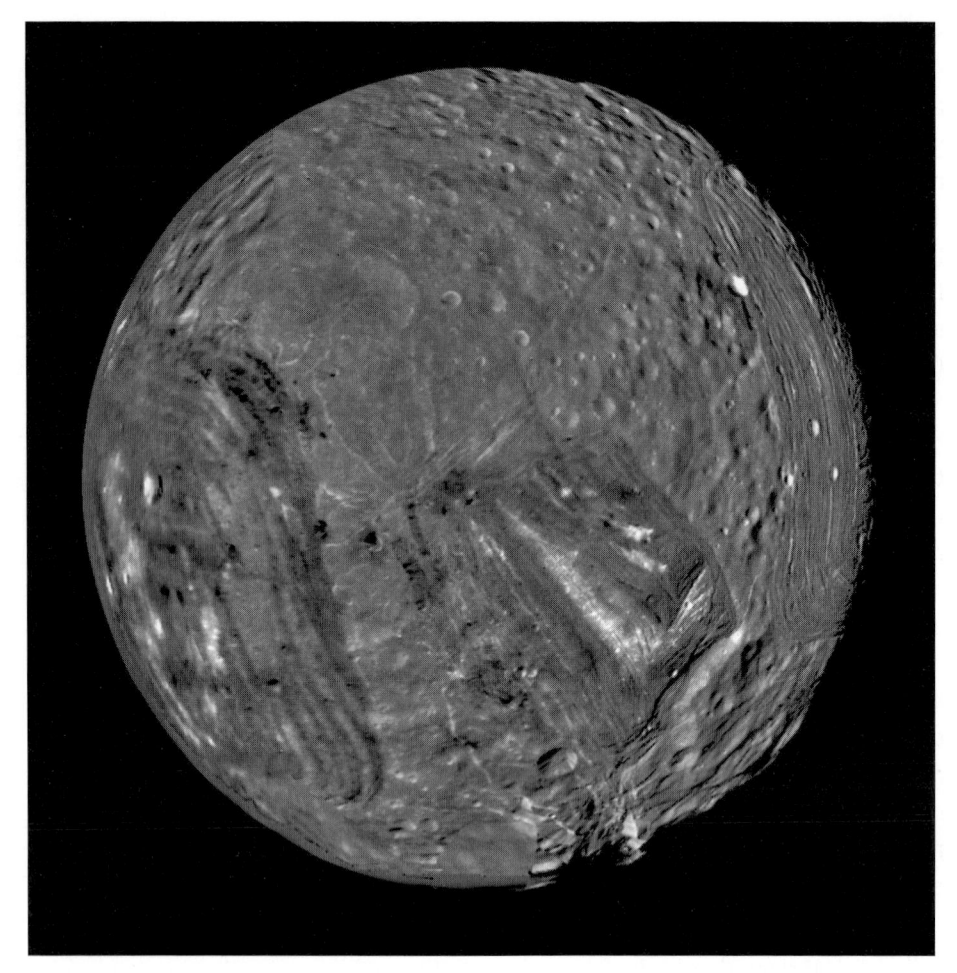

Figure 7.19. Broad-scale view of Miranda's surface, showing cratered terrain cut by the angular sets of faults known as coronae. The view is centered near the south pole.

The coronae are each comprised of an inner core and an outer belt, as defined by albedo and topographic variations (Smith et al., 1986; Greenberg et al., 1991; Pappalardo, 1994). The inner regions contain smooth material and/or intersecting ridges and troughs; the outer belts are predominantly comprised of distinct bands of subparallel ridges and troughs. Peculiarly, the coronae are "squared off," with relatively straight boundaries and rounded corners, reminiscent of a race track. The crater population and sharpness of topography within Miranda's coronae argues that corona formation postdated the event(s) that muted Miranda's large-scale topography (Plescia, 1988; Croft and Soderblom, 1991; Greenberg et al., 1991).

Upon the encounter of the Voyager spacecraft with Miranda the coronae were interpreted as manifestations of breakup by catastrophic impact and reaccretion of a partially differentiated proto-Miranda (Smith *et al.*, 1986). Janes and Melosh (1988) developed this idea into the "sinker" model, in which silicate-rich chunks of a shattered proto-Miranda sank toward the center of the reaccreting satellite. This would induce a downwelling wake, compressing the lithosphere above. Modeling by Janes and Melosh shows that resulting stresses would create a region of folds and/or thrusts with no preferred orientation, surrounded by an annulus of concentric folds. This tectonic pattern of a broad load on a small planetary body is distinct from that formed by a small load on a large planetary body, in that satellite curvature and membrane stress is an important factor in the tectonics of Miranda's coronae (Janes and Melosh, 1990).

Subsequently, a diapiric upwelling or "riser" model was proposed (Croft and Soderblom, 1991; Greenberg *et al.*, 1991; cf. McKinnon, 1988), in which coronae and their constituent ridges and troughs are surface manifestations of large-scale upwelling, perhaps associated with partial differentiation of Miranda. In this scenario, diapirs might pierce and replace the original surface to create coronae or might modify the original surface through extensional-tectonic deformation and extrusion (Greenberg *et al.*, 1991). In predicting the structures formed above a region of upwelling, this model simply requires a sign change from that of Janes and Melosh (1988), predicting a central region of disorganized extensional structures surrounded by a zone of concentric extensional faults (McKinnon, 1988; Janes and Melosh, 1990).

The origin of ridges and troughs within coronae is the principal constraint on models for the formation of coronae. If coronae were formed by downwelling currents, the ridge and trough terrain of their outer belts should be compressional in origin, expressed as folds or reverse faults (Janes and Melosh, 1988). If coronae were formed by upwelling, Miranda's ridge and trough terrain is predicted to be of extensional–tectonic origin (McKinnon, 1988; Janes and Melosh, 1990; Greenberg *et al.*, 1991), expressed as horst-and-graben structures or tilt blocks, potentially in combination with constructional fissure volcanism and/or intrusion.

A volcano-tectonic model for the evolution of coronae and their constituent ridges and troughs (Croft and Soderblom, 1991; Greenberg *et al.*, 1991) suggests that melt delivered to Miranda's subsurface erupted through preexisting fractures to create coronae. Consistent with the riser model, this suggests that many of the ridges and troughs within coronae formed by extrusion of viscous material along fissures. Schenk (1991) similarly concludes that many ridges within Elsinore and Inverness Coronae originated by linear extrusion of viscous volcanic material, while Jankowski and Squyres (1988) suggest that some ridges formed by solid-state emplacement of diapiric material.

The morphologies of scarps in Arden and Inverness Coronae, including limb profiles that show asymmetric steps (Figure 7.20), indicates that they were likely

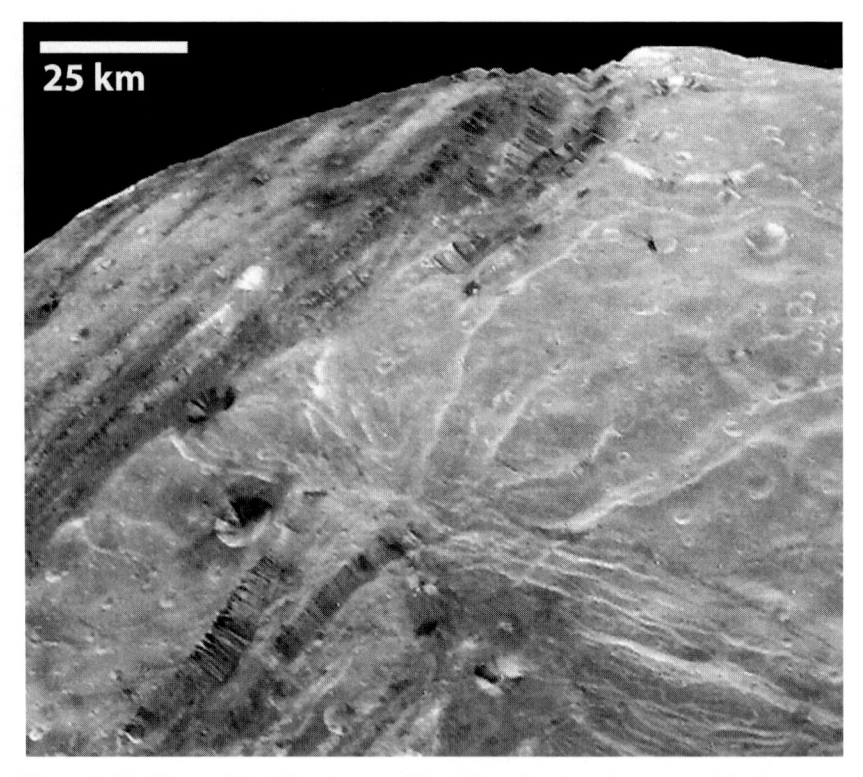

Figure 7.20. View of the Uranian satellite Miranda, showing normal fault scarps. The faults are formed in parallel sets of tilt blocks, and the sawtooth profile of one set of tilt blocks can be seen on the upper limb.

formed by normal faulting (Plescia, 1988; Thomas, 1988; Greenberg et al., 1991; Pappalardo, 1994; Pappalardo et al., 1997). Reconstruction of apparent tilt-block style normal faults in the outer belt of Arden Corona suggests that tens of percent extension has occurred, along faults with initial dips of ~50° (Pappalardo et al., 1997).

The weight of morphological evidence suggests that Miranda's coronae formed by extension and associated cryovolcanism, consistent with a riser model of corona formation. An upwelling origin of Miranda's coronae eliminates the need to invoke catastrophic breakup and reaccretion of the satellite as an explanation for its surface geology. Instead, coronae may represent the surface expression of broad diapirs rising within a relatively small satellite. The reason for the relatively straight sides and rounded corners of the coronae remains a mystery, but may be the result of structural control by more ancient structures. Miranda's relatively high current-day inclination is convincingly explained by passage through a tidal resonance with Umbriel, and passage through temporary resonances with Ariel and/or Umbriel would have tidally heated the moon to some degree (e.g., Dermott et al., 1988;

Figure 7.21. Global view of Ariel, showing ridged terrain crossing its surface.

Tittemore and Wisdom, 1990; Moons and Henrard, 1994; Peale, 1999). Perhaps it is this episode of tidal heating that is responsible for the renewal of tectonic activity we see on this tiny satellite.

6.3 Ariel

The surface of Ariel can be divided into three geologic units: cratered terrain, presumably the oldest material; smooth plains, displaying a relatively low crater density; and ridged terrain, characterized by bands of subparallel ridges and troughs (Plescia, 1987). Ridged terrain consists of 25- to 70-km wide swaths of east–west or northeast–southwest trending groups of parallel ridges and troughs. The ridges and troughs typically have spacing distances of 10–35 km and can be more than 100 km in length. Smaller scale ridges and troughs, with a regular spacing of about 5 km, are observed as well (Figure 7.21) (Nyffenegger and Consolmagno, 1988). Smooth material commonly occupies valley floors, in which case it exhibits a convex profile and can display medial ridges and/or troughs (Smith *et al.*, 1986; Jankowski and Squyres, 1988).

A variety of models have been invoked to account for the ridges and troughs on Ariel. Both large- and small-scale ridges and troughs have been hypothesized to be the product of normal faulting (Smith *et al.*, 1986; Nyffenegger and Consolmagno, 1988; Pappalardo, 1994). Some ridges within the ridged terrain might form by means of linear extrusion of viscous material (Ruzicka, 1988; Jankowski and Squyres, 1988). Smooth material occupying valley floors may have been emplaced as solid-state flows from linear vents (Jankowski and Squyres, 1988), perhaps mobilized by interstitial volatiles (Stevenson and Lunine, 1986). Medial troughs on the smooth material may be due to faulting, perhaps related to the opening of fissures (Smith *et al.*, 1986; Schenk, 1991), or they may be due to a "lava tube" style emplacement of smooth material (Croft and Soderblom, 1991). Isolated ridges in the smooth material may have formed as late stage extrusions (Smith *et al.*, 1986; Schenk, 1991). Extensional tectonics and extrusion both imply a tensile stress state in the satellite's lithosphere, perhaps induced by freezing of an initially molten interior (Smith *et al.*, 1986; Plescia, 1987).

6.4 Dione, Tethys, Rhea, and Titania

We have grouped together Dione, Tethys, and Rhea (satellites of Saturn) and Titania (satellite of Uranus) because they all have relatively minor amounts of tectonism on their surfaces compared to the preceding bodies, and the origin of their tectonic features is still somewhat mysterious. All of them have surfaces dominated by impact craters, with isolated tectonic features cutting across the cratered terrain. On all of these satellites, the dominant tectonic features are organized sets of scarps (probably normal fault scarps) and graben.

From Voyager data, linear zones of bright lineations, termed "wispy terrain," were mapped on Rhea's and Dione's surface (Smith *et al.*, 1981; Plescia, 1983). In neither case were these features seen clearly by Voyager, but Cassini has revealed those on Dione to be sets of faults (Wagner *et al.*, 2006; Moore and Schenk, 2007). The faults exhibit some bright scarps, and in some places are densely packed together in parallel groups (Figure 7.22). The main sets of faults on Dione appear to roughly follow a great circle, tilted with respect to the equator (Miller *et al.*, 2007), although there are sets that clearly deviate from this pattern. Cassini images of Dione also reveal north–south trending ridges along the western boundary of a resurfaced plain that might be compressional (Moore and Schenk, 2007). Graben and normal fault sets on Rhea, along with several broad ridges identified in both Voyager and Cassini data (Moore *et al.*, 1985; Moore and Schenk, 2007; Wagner *et al.*, 2007) trend dominantly north–south.

On Tethys, the main tectonic feature is a large, wide graben complex called Ithaca Chasma (Figure 7.23). Cassini mapping shows that the graben system is roughly 2–3 km deep with a raised rim up to 6 km high (Giese *et al.*, 2007; Moore and Schenk, 2007). Flexural modeling of the raised rim gives estimates of 16–20 km for the thickness of the elastic lithosphere during the formation of Ithaca Chasma (Giese *et al.*, 2007). Ithaca Chasma is of interest because it is offset only 15–20° from a great circle centered on the large Odysseus impact basin (whose

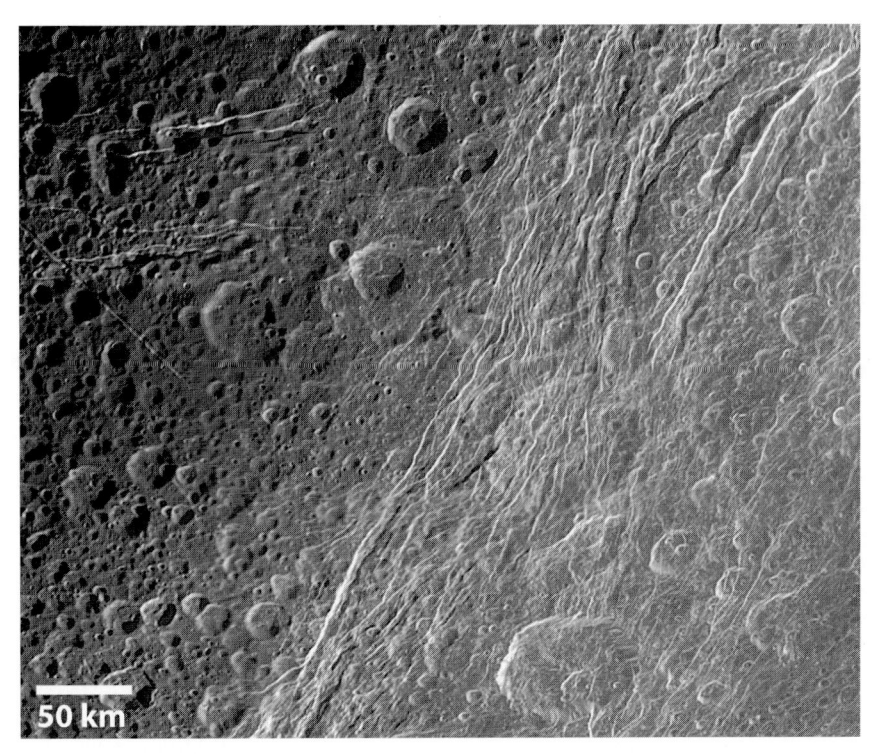

Figure 7.22. Closely packed sets of normal faults on Saturn's satellite Dione. Some faults are found in facing pairs, forming graben. North is to the top, and the illumination is from the right.

Figure 7.23. Ithaca Chasma on Saturn's satellite Tethys is interpreted to be a large graben complex. North is to the top, and illumination is from the bottom right.

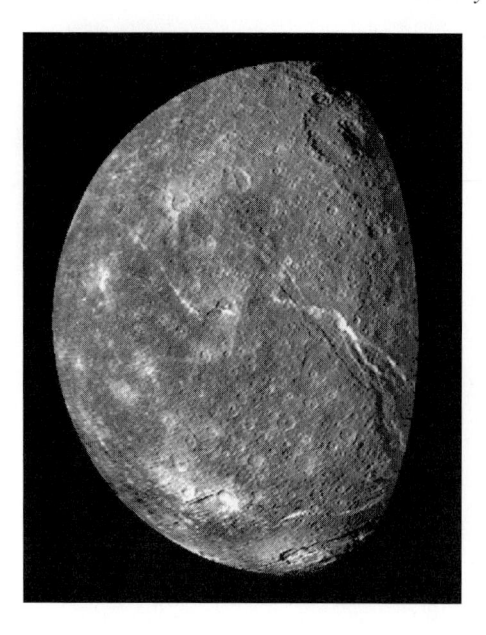

Figure 7.24. Global view of the Uranian satellite Titania, showing several graben-like structures crossing its surface.

diameter is roughly 0.4 the satellite radius). This has led to suggestions that the chasma is tectonically related to the basin (e.g., Moore and Ahern, 1983; Moore *et al.*, 2004a), though crater counts on the floor of Odysseus and the bottom of Ithaca Chasma indicate that the impact basin is younger (Giese *et al.*, 2007).

Finally, Titania exhibits a branching network of faults and graben, 20–50 km wide and 2–5 km deep (Figure 7.24), which cut across most of the craters on the surface (Smith *et al.*, 1986). The origin of these faults is unknown, due largely to the lack of a global image map of Titania.

These tectonic features, limited though they are, clearly imply something of the nature of the stress and thermal histories of these bodies. Freezing of water in the interior, the expansion of warming ice from radionuclide sources, been proposed for the expansion of these worlds' interiors (Pollack and Consolmagno, 1984; Hillier and Squyres, 1991; Castillo *et al.*, 2006). Reorientation of these bodies due to the mass asymmetries caused by large impact basins can produce stresses of $\sim$100 kPa, which could potentially leave a record of surface fracturing in response, depending on the state of the interior at the time (Nimmo and Matsuyama, 2007). Ongoing mapping and analysis of these satellites will elucidate these issues and give us a better insight into the origin of tectonic features on middle-sized icy worlds.

7 Satellites without widespread tectonic activity

7.1 Titan

Titan is by far the largest satellite of Saturn, close to Ganymede in size, and is distinguished by its massive, extended nitrogen–methane atmosphere. Titan is an active world, erasing craters from its surface at a geologically rapid pace. Only a handful of craters have been observed on the surface by the Cassini mission (Porco et al., 2005a; Stofan et al., 2006), though new ones are currently being discovered as our imaging coverage of the surface increases. Convincing evidence has been found for fluvial erosion features (Tomasko et al., 2005; Porco et al., 2005a; Stofan et al., 2006) and aeolian dunes (Stofan et al., 2006). Preliminary evidence has been found for cryovolcanic features (Stofan et al., 2006; Wall et al., 2009), but the evidence so far for tectonic activity on Titan is on shakier ground. Straight-sided features have been observed in near-infrared images (Porco et al., 2005a) but some of these have turned out to be fields of linear dunes (Lorenz et al., 2006). Eroded mountain chains have been observed in some of the radar data, but it is unclear what type of tectonic process formed them, or what the role of differential erosion is in determining the heights and shapes of these mountains (Radebaugh et al., 2007). Either Titan does not have active internally driven tectonic processes, or the erosional and depositional processes that modify Titan's surface (e.g., aeolian, fluvial) are so effective at masking and erasing tectonic features that we cannot clearly determine the nature of Titan's tectonics.

7.2 Callisto

Callisto, the outermost Galilean satellite, is similar in bulk properties to Ganymede, but its interior has not been differentiated as strongly into separate rock and ice layers. The dark, dusty surface is ancient and saturated with impact craters (see Moore et al., 2004b, for a full summary of Callisto geology). The most obvious tectonic features on Callisto are the concentric arcuate graben-like features that make up the multiring impact basins, such as Asgard and Valhalla (see subsection 3.3.4), somewhat similar to the furrows on Ganymede (subsection 6.1.1). Near the north pole, there is a group of narrow troughs, each several hundred kilometers long. They are oriented radial to a point on the surface, which may suggest an impact origin (Schenk, 1995), but the center of the system has not yet been imaged. If they are not impact related, they may be due to a late, slow, global expansion of Callisto, caused by a shutdown of internal convection (McKinnon, 2006). In this scenario, the colder poles would be favored locations for the tectonic expression of such late-stage expansion.

7.3 Mimas and Iapetus

Mimas and Iapetus are the innermost and outermost, respectively, of Saturn's major moons. Both have heavily cratered surfaces, and both show hints of a minor episode of ancient tectonic activity. On Mimas, Voyager data revealed a global pattern of linear troughs across the surface that could either be related to tidal stresses or to

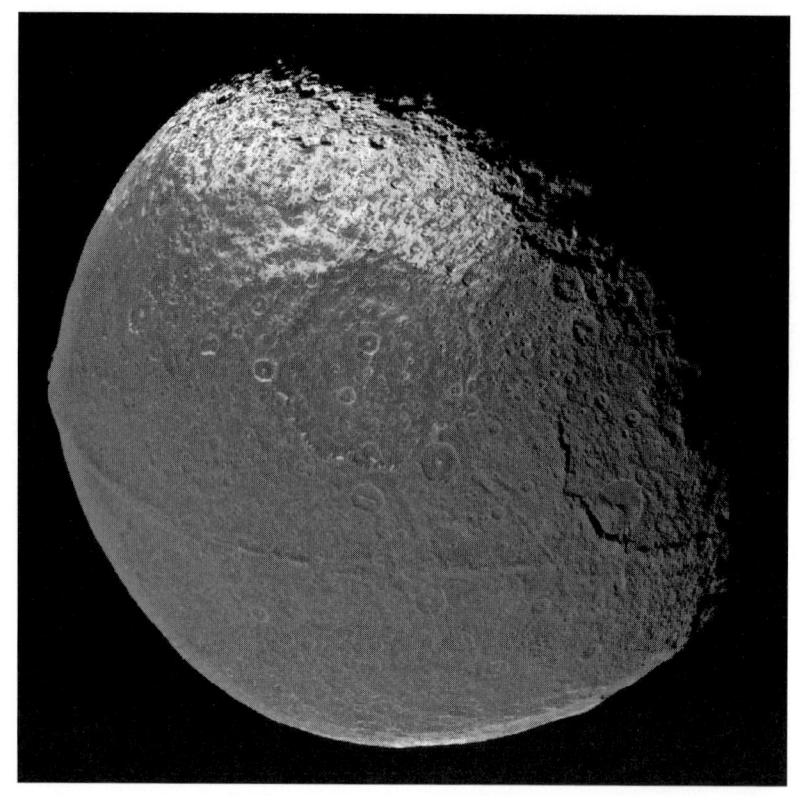

Figure 7.25. Global view of Saturn's satellite Iapetus, showing the enigmatic equatorial ridge curving around the center of the globe. The ridge is about 20 km wide and well over 1000 km long. Credit: NASA/JPL/Space Science Institute.

the large impact crater Herschel that dominates one side of Mimas (Moore et al., 2004a). Smith et al. (1981) suggested that large impacts could disrupt some of the inner satellites of Saturn. Large basins 400 to 600 km across are common on Iapetus and Rhea. Presumably, these impact events could induce incipient breakup fracturing on Iapetus, Rhea, and the smaller inner satellites, and Herschel on Mimas may be an example of this. Global mapping of recent Cassini images is ongoing to evaluate whether there is a link between the troughs and Herschel.

Other than scattered minor troughs and linear segments of crater walls, Iapetus displays only one major, possibly tectonic feature, but it is an impressive feature (Figure 7.25). A ridge up to 20 kilometers high runs exactly along the equator for more than one third of the satellite's circumference (Porco et al., 2005b). Its mode of origin and the origin of the stresses that may have formed it are unknown and present an intriguing mystery (Ip, 2006). It may well be linked to the highly oblate shape of this moon (Castillo-Rogez et al., 2007) in a manner that is presently unclear.

7.4 Other satellites

There are plenty of other satellites in the outer solar system that have not been mentioned in this chapter, but most of them are small and irregularly shaped. The larger satellites we have neglected here, such as Umbriel and Oberon (satellites of Uranus), are very poorly covered by current imaging data, and so nothing definitive can be said at this point. Pluto and its major satellite Charon will not be visited by spacecraft until the New Horizons flyby in 2015. The orbital evolution of this "binary planet" may have induced significant tidal stresses on their surfaces (Collins and Pappalardo, 2000), and it will be interesting to see if there are tectonic features on the outer frontier of our solar system.

8 Conclusions

As we close this chapter, let us consider how tectonics on the outer planet satellites differs from tectonics elsewhere. Three overarching conclusions are (in no particular order):

- *Tides are supremely important*: There are many factors influencing the tectonics of the outer planet satellites that are governed by the giant planets around which they orbit. The stress fields that lead to the formation and global pattern of tectonic features on these satellites are often controlled or strongly influenced by changes in the tidal figure of the satellite. The evolution of a satellite's orbit with time, which depends on dissipation within the giant planet and satellite and the orbital positions of a satellite's siblings, is also very important, leading to changes in tidal heating and diurnal stresses. As these orbital and tidal parameters change over time, we cannot assume that any satellite has remained in a steady state with respect to tidal stresses and energy input. The immense amount of recent and ongoing geological activity in the outer planet satellites, even the tiny ones, is a testament to the power of giant planet tides to drive geological activity.
- *Tectonic features in ice are interpretable from terrestrial experience (mostly)*: Despite major differences in material properties, many of the tectonic features, such as isolated normal faults, folds, graben, and tilt-block complexes, are easily recognizable in the thick icy crusts of the outer satellites. However, there are some active satellites with thin elastic lithospheres (Europa, the south pole of Enceladus, possibly Triton) that exhibit bizarre tectonic features that defy easy comparison with terrestrial analogues.
- *Extension is ubiquitous, contraction is hard to find*: There is evidence on almost every outer planet satellite for some type of extensional tectonic feature, but only in a few places is there good evidence for strike-slip motion or surface contraction. The prevalence of extensional features may be due to the relative

ease of lithospheric failure in tension as opposed to compression, or perhaps there is a deeper message. For example, it is becoming increasingly apparent that subsurface oceans may be common on icy satellites. Freezing of such an ocean and thickening of the floating ice shell can cause tensile stresses near the surface (see subsection 3.2.6).

Much work remains to be done to unravel the tectonic history and current behavior of the outer planet satellites, and we have only just begun to explore this region of our solar system. For most satellites, we still lack complete global imaging coverage at a resolution sufficient to distinguish tectonic features. Various tools used for exploring terrestrial tectonics, such as close-up fieldwork, subsurface electromagnetic sounding, global gravity fields, and global altimetry data, have started to be used on extraterrestrial bodies, notably Mars, but none of the outer planet satellites have had such attention lavished on them yet. Current proposals for a new flagship mission focused on one of the outer planet satellites would go a long way toward filling the data gap. Obtaining a global imaging, radar sounding, gravity, and altimetry dataset for one of these outer planet satellites would give us a more solid foundation for understanding the behavior of icy lithospheres in general and the power of tides to shape them.

Acknowledgments

We wish to thank Andrew Dombard and Simon Kattenhorn for their extraordinarily thorough and helpful reviews of this manuscript. The authors also wish to acknowledge support from the NASA Planetary Geology and Geophysics, Outer Planets Research, and Cassini Data Analysis programs. A sabbatical leave from Wheaton College for GCC contributed greatly to assembling and finishing this work. Portions of the work performed by RTP were carried out at the Jet Propulsion Laboratory, California Institute of Technology, under a contract with the National Aeronautics and Space Administration.

References

Agnor, C. B. and Hamilton, D. P. (2006). Neptune's capture of its moon Triton in a binary–planet gravitational encounter. *Nature*, **441**, 192–194.
Anderson, J. D., Lau, E. L., Sjogren, W. L., Schubert, G., and Moore, W. B. (1996). Gravitational constraints on the internal structure of Ganymede. *Nature*, **384**, 541–543.
Anderson, J. D., Jacobson, R. A., McElrath, T. P., Moore, W. B., Schubert, G., and Thomas, P. C. (2001). Shape, mean radius, gravity field, and interior structure of Callisto. *Icarus*, **153**, 157–161.
Aydin, A. (2006). Failure modes of the lineaments on Jupiter's moon, Europa: Implications for the evolution of its icy crust. *J. Struct. Geol.*, **28**, 2222–2236.

Bader, C. and Kattenhorn, S. A. (2007). Formation of ridge-type strike-slip faults on Europa (abs.). *Eos Trans. AGU*, **88(52)** (Fall Meet. Suppl.), P53B-1243.

Bagenal, F., Dowling, T. E., and McKinnon, W. B., eds. (2004). *Jupiter: The Planet, Satellites, and Magnetosphere*. New York: Cambridge University Press.

Barr, A. C. (2008). Mobile lid convection beneath Enceladus' south polar terrain. *J. Geophys. Res.*, **113**, E07009, doi: 10.1029/2008JE003114.

Barr, A. C. and McKinnon, W. B. (2007). Convection in ice I shells and mantles with self-consistent grain size. *J. Geophys. Res.*, **112**, E02012.

Barr, A. C. and Pappalardo, R. T. (2005). Onset of convection in the icy Galilean satellites: Influence of rheology. *J. Geophys. Res.*, **110**, E12005.

Bart, G. D., Turtle, E. P., Jaeger, W. L., Keszthelyi, L. P., and Greenberg, R. (2004). Ridges and tidal stress on Io. *Icarus*, **169**, 111–126.

Beeman, M., Durham, W. B., and Kirby, S. H. (1988). Friction of ice. *J. Geophys. Res.*, **93**, 7625–7633.

Benner, L. A. M. and McKinnon, W. B. (1995). Orbital behavior of captured satellites: The effect of solar gravity on Triton's postcapture orbit. *Icarus*, **114**, 1–20.

Best, M. G. (2003). *Igneous and Metamorphic Petrology*. Malden, MA: Blackwell.

Bergstralh, J. T., Miner, E. D., and Matthews, M. S., eds. (1991). *Uranus*. Tucson, AZ: University of Arizona Press.

Billings, S. E. and Kattenhorn, S. A. (2005). The great thickness debate: Ice shell thickness models for Europa and comparisons with estimates based on flexure at ridges. *Icarus*, **177**, 397–412.

Bills, B. G. (2005). Free and forced obliquities of the Galilean satellites of Jupiter. *Icarus*, **175**, 233–247.

Bland, M. T. and Showman, A. P. (2007). The formation of Ganymede's grooved terrain: Numerical modeling of extensional necking instabilities. *Icarus*, **189**, 439–456.

Bland, M. T., Beyer, R. A., and Showman, A. P. (2007). Unstable extension of Enceladus' lithosphere. *Icarus*, **192**, 92–105.

Bland, M. T., McKinnon, W. B., and Showman, A. P. (2008a). The formation of Ganymede's grooved terrain: Importance of strain weakening (abs.). Eos Trans. AGU (Fall Meet. Suppl.) P23A-1358.

Bland, M. T., Showman, A. P., and Tobie, G. (2008b). The production of Ganymede's magnetic field. *Icarus*, **198**, 384–399.

Bruesch, L. S. and Asphaug, E. (2004). Modeling global impact effects on middle-sized icy bodies: Applications to Saturn's moons. *Icarus*, **168**, 457–466.

Buck, W. R. (1991). Modes of continental lithospheric extension. *J. Geophys. Res.*, **96**, 20 161–20 178.

Buratti, B. J., Goguen, J. D., Gibson, J., and Mosher, J. (1994). Historical photometric evidence for volatile migration on Triton. *Icarus*, **110**, 303–314.

Burns, J. A. and Matthews, M. S., eds. (1986). *Satellites*. Tucson, AZ: University of Arizona Press.

Cassen, P., Reynolds, R. T., and Peale, S. J. (1979). Is there liquid water on Europa? *Geophys. Res. Lett.*, **6**, 731–734.

Castillo, J. C., Matson, D. L., Sotin, C., Johnson, T. V., Lunine, J. I., and Thomas, P. C. (2006). A new understanding of the internal evolution of saturnian icy satellites from Cassini observations (abs.). *Lunar Planet. Sci. Conf. XXXVII*, 2200. Houston, TX: Lunar and Planetary Institute (CD-ROM).

Castillo-Rogez, J., Matson, D., Sotin, C., Johnson, T., Lunine, J., and Thomas, P. (2007). Iapetus' geophysics: Rotation rate, shape, and equatorial ridge. *Icarus*, **190**, 179–202.

Chapman, C. R. and McKinnon, W. B. (1986). Cratering of planetary satellites. In *Satellites*, ed. J. A. Burns and M. S. Matthews. Tucson, AZ: University of Arizona Press, pp. 492–580.

Clark, B. E., Helfenstein, P., Veverka, J., Ockert-Bell, M., Sullivan, R. J., Geissler, P. E., Phillips, C. B., McEwen, A. S., Greeley, R., Neukum, G., Denk, T., and Klaasen, K. (1998). Multispectral terrain analysis of Europa from Galileo images. *Icarus*, **135**, 95–106.

Collins, G. C. (2006). Global expansion of Ganymede derived from strain measurements in grooved terrain (abs.). *Lunar Planet. Sci. Conf. XXXVII*, 2077. Houston, TX: Lunar and Planetary Institute (CD-ROM).

Collins, G. C. and Goodman, J. C. (2007). Enceladus' south polar sea. *Icarus*, **189**, 72–82.

Collins, G. C. and Pappalardo, R. T. (2000). Predicted stress patterns on Pluto and Charon due to their mutual orbital evolution (abs.). *Lunar Planet. Sci. Conf. XXXI*, 1035. Houston, TX: Lunar and Planetary Institute (CD-ROM).

Collins, G. C. and Schenk, P. (1994). Triton's lineaments: Complex morphology and stress patterns (abs.). *Lunar Planet. Sci. Conf. XXV*, 277–278. Houston, TX: Lunar and Planetary Institute (CD-ROM).

Collins, G. C., Head, J. W., and Pappalardo, R. T. (1998a). The role of extensional instability in creating Ganymede grooved terrain: Insights from Galileo high-resolution stereo imaging. *Geophys. Res. Lett.*, **25**, 233–236.

Collins, G. C., Head, J. W., and Pappalardo, R. T. (1998b). Geology of the Galileo G7 Nun Sulci target area, Ganymede (abs.). *Lunar Planet. Sci. Conf. XXIX*, 1755. Houston, TX: Lunar and Planetary Institute (CD-ROM).

Collins, G. C., Pappalardo, R. T., and Head, J. W. (1999). Surface stresses resulting from internal differentiation: Application to Ganymede tectonics (abs.). *Lunar Planet. Sci. Conf. XXX*, 1695. Houston, TX: Lunar and Planetary Institute (CD-ROM).

Cooper, J. F., Johnson, R. E., Mauk, B. H., Garrett, H. B., and Gehrels, N. (2001). Energetic ion and electron irradiation of the icy Galilean satellites. *Icarus*, **149**, 133–159.

Copernicus, N. (1543). *De Revolutionibus Orbium, Coelestium*. Nuremberg: Johannes Petrius.

Crawford, G. D. and Stevenson, D. J. (1988). Gas-driven water volcanism in the resurfacing of Europa. *Icarus*, **73**, 66–79.

Croft, S. K. and Soderblom, L. A. (1991). Geology of the uranian satellites. In *Uranus*, ed. J. T. Bergstralh, E. D. Miner and M. S. Matthews. Tucson, AZ: University of Arizona Press, pp. 561–628.

Croft, S. K., Kargel, J. S., Kirk, R. L., Moore, J. M., Schenk, P. M., and Strom, R. G. (1995). The geology of Triton. In *Neptune and Triton*, ed. D. P. Cruikshank. Tucson, AZ: University of Arizona Press, pp. 879–947.

Crown, D. A., Greeley, R., Craddock, R. A., and Schaber, G. G. (1992). Geologic map of Io. U.S. Geol. Surv. Misc. Invest. Ser., Map I-2209.

Cruikshank, D. P., ed. (1995). *Neptune and Triton*. Tucson, AZ: University of Arizona Press.

Cruikshank, D. P., Schmitt, B., Roush, T. L., Owen, T. C., Quirico, E., Geballe, T. R., de Bergh, C., Bartholomew, M. J., Dalle Ore, C. M., Douté, S., and Meier, R. (2000). Water ice on Triton. *Icarus*, **147**, 309–316.

De La Chapelle, S., Milsch, H., Castelnau, O., and Duval, P. (1999). Compressive creep of ice containing a liquid intergranular phase: Rate-controlling processes in the dislocation creep regime. *Geophys. Res. Lett.*, **26**, 251–254.

DeRemer, L. C., and Pappalardo, R. T. (2003). Manifestations of strike slip faulting on Ganymede (abs). *Lunar Planet. Sci. Conf. XXXIV*, 2033. Houston, TX: Lunar and Planetary Institute (CD-ROM).

Dermott, S. F., Malhotra, R., and Murray, C. D. (1988). Dynamics of the uranian and saturnian satellite systems: A chaotic route to melting Miranda? *Icarus*, **76**, 295–334.

Dobson, D. P., Meredith, P. G., and Boon, S. A. (2002). Simulation of subduction zone seismicity by dehydration of serpentine. *Science*, **298**, 1407–1410.

Dombard, A. J. and McKinnon, W. B. (2001). Formation of grooved terrain on Ganymede: Extensional instability mediated by cold, superplastic creep. *Icarus*, **154**, 321–336.

Dombard, A. J. and McKinnon, W. B. (2006a). Elastoviscoplastic relaxation of impact crater topography with application to Ganymede and Callisto. *J. Geophys. Res.*, **111**, E01001.

Dombard, A. J. and McKinnon, W. B. (2006b). Folding of Europa's icy lithosphere: An analysis of viscous-plastic buckling and subsequent topographic relaxation. *J. Struct. Geol.*, **28**, 2259–2269.

Durham, W. B. and Stern, L. A. (2001). Rheological properties of water ice: Applications to satellites of the outer planets. *Annu. Rev. Earth Planet. Sci.*, **29**, 295–330.

Durham, W. B., Kirby, S. H., and Stern, L. A. (1992). Effects of dispersed particulates on the rheology of water ice at planetary conditions. *J. Geophys. Res.*, **97**, 20 883–20 897.

Durham, W. B., Kirby, S. H., and Stern, L. A. (1993). Flow of ices in the ammonia-water system. *J. Geophys. Res.*, **98**, 17 667–17 682.

Durham, W. B., Kirby, S. H., and Stern, L. A. (1998). Rheology of planetary ices. In *Solar System Ices*, ed. B. Schmitt, C. de Bergh and M. Festou. Amsterdam: Kluwer.

Durham, W. B., Stern, L. A., and Kirby, S. H. (2001). Rheology of ice I at low stress and elevated confining pressure. *J. Geophys. Res.*, **106**, 11 031–11 042.

Durham, W. B., Stern, L. A., Kubo, T., and Kirby, S. H. (2005). Flow strength of highly hydrated Mg- and Na-sulfate hydrate salts, pure and in mixtures with water ice, with application to Europa. *J. Geophys. Res.*, **110**, E12010.

Duxbury, N. S. and Brown, R. H. (1997). The role of an internal heat source for the eruptive plumes on Triton. *Icarus*, **125**, 83–93.

Ellsworth, K. and Schubert, G. (1983). Saturn's icy satellites: Thermal and structural models. *Icarus*, **54**, 490–510.

Fagents, S. A., Greeley, R., Sullivan, R. J., Pappalardo, R. T., and Prockter, L. M. (2000). Cryomagmatic mechanisms for the formation of Rhadamanthys Linea, triple band margins, and other low-albedo features on Europa. *Icarus*, **144**, 54–88.

Figueredo, P. H. and Greeley, R. (2000). Geologic mapping of the northern leading hemisphere of Europa from Galileo solid-state imaging data. *J. Geophys. Res.*, **105**, 22 629–22 646.

Fink, J. H. and Fletcher, R. C. (1981). A mechanical analysis of extensional instability on Ganymede (abs.). *NASA TM-84211*, 51–53.

Finnerty, A. A., Ransford, G. A., Pieri, D. C., and Collerson, K. D. (1981). Is Europa surface cracking due to thermal evolution? *Nature*, **289**, 24–27.

Friedson, A. J. and Stevenson, D. J. (1983). Viscosity of rock-ice mixtures and applications to the evolution of icy satellites. *Icarus*, **56**, 1–14.

Frost, H. J. and Ashby, M. F. (1982). *Deformation-Mechanism Maps: The Plasticity and Creep of Metals and Ceramics*. New York: Pergamon Press.

Gaidos, E. J. and Nimmo, F. (2000). Tectonics and water on Europa. *Nature*, **405**, 637.

Galilei, G. (1610). *Siderius Nuncius*. Venice: Baglioni.

Gammon, P. H., Kiefte, H., Clouter, M. J., and Denner, W. W. (1983). Elastic constants of artificial and natural ice samples by Brillouin spectroscopy. *J. Glaciol.*, **29**, 433–460.

Gehrels, T. and Matthews, M. S., eds. (1984). *Saturn*. Tucson, AZ: University of Arizona Press.

Geissler, P. E., Greenberg, R., Hoppa, G., McEwen, A., Tufts, R., Phillips, C., Clark, B., Ockert-Bell, M., Helfenstein, P., Burns, J., Veverka, J., Sullivan, R., Greeley, R., Pappalardo, R. T., Head, J. W., Belton, M. J. S., and Denk, T. (1998a). Evolution of lineaments on Europa: Clues from Galileo multispectral imaging observations. *Icarus*, **135**, 107–126.

Geissler, P. E., Greenberg, R., Hoppa, G., Helfenstein, P., McEwen, A., Pappalardo, R., Tufts, R., Ockert-Bell, M., Sullivan, R., Greeley, R., Belton, M. J. S., Denk, T., Clark, B. E., Burns, J., and Veverka, J. (1998b). Evidence for non-synchronous rotation of Europa. *Nature*, **391**, 368.

Geissler, P., Greenberg, R., Hoppa, G. V., Tufts, B. R., and Milazzo, M. (1999). Rotation of lineaments in Europa's southern hemisphere (abs.). *Lunar Planet. Sci. Conf. XXX*, 1743. Houston, TX: Lunar and Planetary Institute (CD-ROM).

Giese, B., Oberst, J., Roatsch, T., Neukum, G., Head, J. W., and Pappalardo, R. T. (1998). The local topography of Uruk Sulcus and Galileo Regio obtained from stereo images. *Icarus*, **135**, 303–316.

Giese, B., Wagner, R., Neukum, G., and Sullivan, R. (1999). Doublet ridge formation on Europa: Evidence from topographic data. *Bull. AAS*, **31**, 62.08.

Giese, B., Wagner, R., Neukum, G., Helfenstein, P., and Thomas, P. C. (2007). Tethys: Lithospheric thickness and heat flux from flexurally supported topography at Ithaca Chasma. *Geophys. Res. Lett.*, **34**, L21203.

Goldreich, P. (1966). Final spin states of planets and satellites. *Astron. J.*, **71**, 1–7.

Goldreich, P., Murray, N., Longaretti, P. Y., and Banfield, D. (1989). Neptune's story. *Science*, **245**, 500–504.

Goldsby, D. L. and Kohlstedt, D. L. (2001). Superplastic deformation of ice: Experimental observations. *J. Geophys. Res.*, **106**, 11 017–11 030.

Golombek, M. P. and Banerdt, W. B. (1986). Early thermal profiles and lithospheric strength of Ganymede from extensional tectonic features. *Icarus*, **68**, 252–265.

Golombek, M. P. and Banerdt, W. B. (1990). Constraints on the subsurface structure of Europa. *Icarus*, **83**, 441–452.

Greeley, R., Sullivan, R., Coon, M. D., Geissler, P. E., Tufts, B. R., Head, J. W., Pappalardo, R. T., and Moore, J. M. (1998). Terrestrial sea ice morphology: Considerations for Europa. *Icarus*, **135**, 25–40.

Greeley, R., Collins, G. C., Spaun, N. A., Sullivan, R. J., Moore, J. M., Senske, D. A., Tufts, B. R., Johnson, T. V., Belton, M. J. S., and Tanaka, K. L. (2000). Geologic mapping of Europa. *J. Geophys. Res.*, **105**, 22 559–22 578.

Greeley, R., Chyba, C. F., Head, J. W., McCord, T. B., McKinnon, W. B., Pappalardo, R. T., and Figueredo, P. H. (2004). Geology of Europa. In *Jupiter: The Planet, Satellites, and Magnetosphere*, ed. F. Bagenal, T. E. Dowling and W. B. McKinnon. New York: Cambridge University Press, pp. 363–396.

Greenberg, R. (2004). The evil twin of Agenol: Tectonic convergence on Europa. *Icarus*, **167**, 313–319.

Greenberg, R. and Weidenschilling, S. J. (1984). How fast do Galilean satellites spin? *Icarus*, **58**, 186–196.

Greenberg, R., Croft, S. K., Janes, D. M., Kargel, J. S., Lebofsky, L. A., Lunine, J. I., Marcialis, R. L., Melosh, H. J., Ojakangas, G. W., and Strom, R. G. (1991). Miranda. In *Uranus*, ed. J. T. Bergstralh, E. D. Miner and M. S. Matthews. Tucson, AZ: University of Arizona Press, pp. 561–628.

Greenberg, R., Geissler, P., Hoppa, G., Tufts, B. R., Durda, D. D., Pappalardo, R., Head, J. W., Greeley, R., Sullivan, R., and Carr, M. H. (1998). Tectonic processes on Europa: Tidal stresses, mechanical response, and visible features. *Icarus*, **135**, 64–78.

Greenberg, R., Geissler, P., Hoppa, G., and Tufts, B. R. (2002). Tidal-tectonic processes and their implications for the character of Europa's icy crust. *Rev. Geophys.*, **40**, 1004.

Groenleer, J. M. and Kattenhorn, S. A. (2008). Cycloid crack sequences on Europa: Relationship to stress history and constraints on growth mechanics based on cusp angles. *Icarus*, **193**, 158–181.

Han, L. J. and Showman, A. P. (2005). Thermo-compositional convection in Europa's icy shell with salinity. *Geophys. Res. Lett.*, **32**, L20201.

Han, L. J. and Showman, A. P. (2008). Implications of shear heating and fracture zones for ridge formation on Europa. *Geophys. Res. Lett.*, **35**, L03202.

Head, J. W. (2000). RRR triple junctions on Europa: Clues to the nature of Europan crustal spreading processes (abs.). *Lunar Planet. Sci. Conf. XXXI*, 1286. Houston, TX: Lunar and Planetary Institute (CD-ROM).

Head, J. W., Pappalardo, R. T., and Sullivan, R. (1999). Europa: Morphological characteristics of ridges and triple bands from Galileo data (E4 and E6) and assessment of a linear diapirism model. *J. Geophys. Res.*, **104**, 24 223–24 236.

Head, J. W., Pappalardo, R., Collins, G., Belton, M. J. S., Giese, B., Wagner, R., Breneman, H., Spaun, N., Nixon, B., Neukum, G., and Moore, J. (2002). Evidence for Europa-like tectonic resurfacing styles on Ganymede. *Geophys. Res. Lett.*, **29**, doi:10.1029/2002GL015961.

Helfenstein, P., and Parmentier, E. M. (1980). Fractures on Europa: Possible response of an ice crust to tidal deformation. *Proc. Lunar Planet. Sci. Conf. 11*, 1987–1998.

Helfenstein, P. and Parmentier, E. M. (1983). Patterns of fracture and tidal stresses on Europa. *Icarus*, **53**, 415–430.

Helfenstein, P. and Parmentier, E. M. (1985). Patterns of fracture and tidal stresses due to nonsynchronous rotation: Implications for fracturing on Europa. *Icarus*, **61**, 175–184.

Helfenstein, P., Currier, N., Clark, B. E., Veverka, J., Bell, M., Sullivan, R., Klemaszewski, J., Greeley, R., Pappalardo, R. T., Head, J. W., Jones, T., Klaasen, K., Magee, K., Geissler, P., Greenberg, R., McEwen, A., Phillips, C., Colvin, T., Davies, M., Denk, T., Neukum, G., and Belton, M. J. S. (1998). Galileo observations of Europa's opposition effect. *Icarus*, **135**, 41–63.

Helfenstein, P., Thomas, P. C., Veverka, J., Rathbun, J., Perry, J., Turtle, E., Denk, T., Neukum, G., Roatsch, T., Wagner, R., Giese, B., Squyres, S., Burns, J., McEwen, A., Porco, C., and Johnson, T. V. (2006a). Patterns of fracture and tectonic convergence near the south pole of Enceladus (abs.). *Lunar Planet. Sci. Conf. XXXVII*, 2182. Houston, TX: Lunar and Planetary Institute (CD-ROM).

Helfenstein, P., Thomas, P. C., Veverka, J., Porco, C., Giese, B., Wagner, R., Roatsch, T., Denk, T., Neukum, G., and Turtle, E. (2006b). Surface geology and tectonism on Enceladus (abs.). *Eos Trans. AGU*, P22B-02.

Herrick, D. L. and Stevenson, D. J. (1990). Extensional and compressional instabilities in icy satellite lithospheres. *Icarus*, **85**, 191–204.

Hillier, J. and Squyres, S. W. (1991). Thermal stress tectonics on the satellites of Saturn and Uranus. *J. Geophys. Res.*, **96**, 15 665–15 674.

Hoppa, G. V., Tufts, B. R., Greenberg, R., and Geissler, P. (1999a). Strike-slip faults on Europa: Global shear patterns driven by tidal stress. *Icarus*, **141**, 287–298.

Hoppa, G. V., Tufts, B. R., Greenberg, R., and Geissler, P. E. (1999b). Formation of cycloidal features on Europa. *Science*, **285**, 1899–1902.

Hoppa, G., Greenberg, R., Geissler, P., Tufts, B. R., Plassmann, J., and Durda, D. D. (1999c). Rotation of Europa: Constraints from terminator and limb positions. *Icarus*, **137**, 341–347.

Hoppa, G. V., Tufts, B. R., Greenberg, R., and Geissler, P. E. (2000). Europa's sub-Jovian hemisphere from Galileo I25: Tectonic and chaotic surface features (abs.). *Lunar Planet. Sci. Conf. XXXI*, 1380. Houston, TX: Lunar and Planetary Institute (CD-ROM).

Hoppa, G. V., Tufts, B. R., Greenberg, R., Hurford, T. A., O'Brien, D. P., and Geissler, P. E. (2001). Europa's rate of rotation derived from the tectonic sequence in the Astypalaea region. *Icarus*, **153**, 208–213.

Hurford, T. A., Beyer, R. A., Schmidt, B., Preblich, B., Sarid, A. R., and Greenberg, R. (2005). Flexure of Europa's lithosphere due to ridge-loading. *Icarus*, **177**, 380–396.

Hurford, T. A., Helfenstein, P., Hoppa, G. V., Greenberg, R., and Bills, B. G. (2007). Eruptions arising from tidally controlled periodic openings of rifts on Enceladus. *Nature*, **447**, 292–294.

Hussmann, H., Sohl, F., and Spohn, T. (2006). Subsurface oceans and deep interiors of medium-sized outer planet satellites and large trans-neptunian objects. *Icarus*, **185**, 258–273.

Ip, W. H. (2006). On a ring of the equatorial ridge of Iapetus. *Geophys. Res. Lett.*, **33**, 16 203.

Ip, W. H., Kopp, A., Williams, D. J., McEntire, R. W., and Mauk, B. H. (2000). Magnetospheric ion sputtering: The case of Europa and its surface age. *Adv. Space Res.*, **26**, 1649–1652.

Jackson, J. (1989). Normal faulting in the upper continental crust: Observations from regions of active extension. *J. Struct. Geol.*, **11**, 15–36.

Jaeger, W., Turtle, E., Keszthelyi, L., Radebaugh, J., McEwen, A., and Pappalardo, R. (2003). Orogenic tectonism on Io. *J. Geophys. Res.*, **108**, doi:10.1029/2002JE001946.

Janes, D. M. and Melosh, H. J. (1988). Sinker tectonics: An approach to the surface of Miranda. *J. Geophys. Res.*, **93**, 3127–3143.

Janes, D. M. and Melosh, H. J. (1990). Tectonics of planetary loading: A general model and results. *J. Geophys. Res.*, **95**, 21 345–21 355.

Jankowski, D. G. and Squyres, S. W. (1988). Solid-state ice volcanism on the satellites of Uranus. *Science*, **241**, 1322–1325.

Jenyon, M. (1986). *Salt Tectonics*. New York: Elsevier.

Johnson, T. V. and Soderblom, L. A. (1982). Volcanic eruptions on Io: Implications for surface evolution and mass loss. In *Satellites of Jupiter*, ed. D. Morrison. Tucson, AZ: University of Arizona Press, pp. 634–646.

Kadel, S. D., Fagents, S. A., and Greeley, R. (1998). Trough-bounding ridge pairs on Europa: Considerations for an endogenic model of formation (abs.). *Lunar Planet. Sci. Conf. XXIX*, 1078. Houston, TX: Lunar and Planetary Institute (CD-ROM).

Kargel, J. S. and Pozio, S. (1996). The volcanic and tectonic history of Enceladus. *Icarus*, **119**, 385–404.

Kattenhorn, S. A. (2002). Nonsynchronous rotation evidence and fracture history in the Bright Plains region, Europa. *Icarus*, **157**, 490–506.

Kattenhorn, S. A. (2004). Strike-slip fault evolution on Europa: Evidence from tailcrack geometries. *Icarus*, **172**, 582–602.

Kattenhorn, S. A. and Marshall, S. T. (2006). Fault-induced perturbed stress fields and associated tensile and compressive deformation at fault tips in the ice shell of Europa: Implications for fault mechanics. *J. Struct. Geol.*, **28**, 2204–2221.

Kirchoff, M. R. (2006). *Mountain building on Io: An unsteady relationship between volcanism and tectonism.* Unpublished Ph.D. thesis, Washington University, St. Louis, MO.

Kirchoff, M. R. and McKinnon, W. B. (2009). Formation of mountains on Io: Variable volcanism and thermal stresses. *Icarus*, **201**, 598–614.

Kirk, R. L. and Stevenson, D. J. (1987). Thermal evolution of a differentiated Ganymede and implications for surface features. *Icarus*, **69**, 91–134.

Kirk, R. L., Soderblom, L. A., Brown, R. H., Keiffer, S. W., and Kargel, J. S. (1995). Triton's plumes: Discovery, characteristics, and models. In *Neptune and Triton*, ed. D. P. Cruikshank. Tucson, AZ: University of Arizona Press, pp. 949–989.

Kivelson, M. G., Khurana, K. K., and Volwerk, M. (2002). The permanent and inductive magnetic moments of Ganymede. *Icarus*, **157**, 507–522.

Lee, S., Pappalardo, R. T., and Makris, N. C. (2005). Mechanics of tidally driven fractures in Europa's ice shell. *Icarus*, **177**, 367–379.

Leith, A. C. and McKinnon, W. B. (1996). Is there evidence for polar wander on Europa? *Icarus*, **120**, 387–398.

Levison, H. F., Duncan, M. J., Zahnle, K., Holman, M., and Dones, L. (2000). Planetary impact rates from ecliptic comets. *Icarus*, **143**, 415–420.

Lorenz, R. D., and 39 others (2006). The sand seas of Titan: Cassini radar observations of longitudinal dunes. *Science*, **312**, 724–727.

Lucchitta, B. K. (1980). Grooved terrain on Ganymede. *Icarus*, **44**, 481–501.

Lucchitta, B. K. and Soderblom, L. A. (1982). Geology of Europa. In *Satellites of Jupiter*, ed. D. Morrison. Tucson, AZ: University of Arizona Press, pp. 521–555.

Macdonald, K. C. (1982). Mid-ocean ridges: Fine scale tectonic, volcanic, and hydrothermal processes within the plate boundary zone. *Annu. Rev. Earth Planet. Sci.*, **10**, 155.

Mackenzie, R. A., Iess, L., Tortora, P., and Rappaport, N. J. (2008). A non-hydrostatic Rhea. *Geophys. Res. Lett.*, **35**, L05204.

Maeno, N. and Arakawa, M. (2004). Adhesion shear theory of ice friction at low sliding velocities, combined with ice sintering. *J. Appl. Phys.*, **95**, 134–139.

Malin, M. C. and Pieri, D. C. (1986). Europa. In *Satellites*, ed. J. A. Burns and M. S. Matthews. Tucson, AZ: University of Arizona Press, pp. 689–717.

Manga, M. and Sinton, A. (2004). Formation of bands and ridges on Europa by cyclic deformation: Insights from analogue wax experiments. *J. Geophys. Res.*, **109**, E09001.

Manga, M. and Wang, C. Y. (2007). Pressurized oceans and the eruption of liquid water on Europa and Enceladus. *Geophys. Res. Lett.*, **34**, L07202.

Marcialis, R. and Greenberg, R. (1987). Warming of Miranda during chaotic rotation. *Nature*, **328**, 227–229.

Matsuyama, I. and Nimmo, F. (2007). Rotational stability of tidally deformed planetary bodies. *J. Geophys. Res.*, **112**, E11003.

Matsuyama, I. and Nimmo, F. (2008). Tectonic patterns on reoriented and despun planetary bodies. *Icarus*, **195**, 459–473.

McCarthy, C., Goldsby, D. L., and Cooper, R. F. (2007). Transient and steady-state creep responses of ice-I/magnesium sulfate hydrate eutectic aggregates (abs.). *Lunar Planet. Sci. Conf. XXXVIII*, 2429. Houston, TX: Lunar and Planetary Institute (CD-ROM).

McEwen, A. S. (1986). Tidal reorientation and the fracturing of Jupiter's moon Europa. *Nature*, **321**, 49–51.

McEwen, A. S., Keszthelyi, L. P., Lopes, R., Schenk, P. M., and Spencer, J. R. (2004). The lithosphere and surface of Io. In *Jupiter: The Planet, Satellites, and Magnetosphere*, ed. F. Bagenal, T. E. Dowling and W. B. McKinnon. New York: Cambridge University Press, pp. 307–328.

McKenzie, D., McKenzie, J. M., and Saunders, R. S. (1992). Dike emplacement on Venus and on Earth. *J. Geophys. Res.*, **97**, 15 977–15 990.

McKinnon, W. B. (1982). Tectonic deformation of Galileo Regio and limits to the planetary expansion of Ganymede. *Proc. Lunar Planet. Sci. 12*, 1585–1597.

McKinnon, W. B. (1988). Odd tectonics of a rebuilt moon. *Nature*, **333**, 701.

McKinnon, W. B. (1997). Mystery of Callisto: Is it undifferentiated? *Icarus*, **130**, 540–543.

McKinnon, W. B. (1998). Geodynamics of icy satellites. In *Solar System Ices*, ed. B. Schmitt, C. de Bergh and M. Festou. Amsterdam: Kluwer.

McKinnon, W. B. (1999). Convective instability in Europa's floating ice shell. *Geophys. Res. Lett.*, **26**, 951–954.

McKinnon, W. B. (2006). On convection in ice I shells of outer solar system bodies, with detailed application to Callisto. *Icarus*, **183**, 435–450.

McKinnon, W. B. and Leith, A. C. (1995). Gas drag and the orbital evolution of a captured Triton. *Icarus*, **118**, 392–413.

McKinnon, W. B. and Melosh, H. J. (1980). Evolution of planetary lithospheres: Evidence from multiringed structures on Ganymede and Callisto. *Icarus*, **44**, 454–471.

McKinnon, W. B., Lunine, J. I., and Banfield, D. (1995). Origin and evolution of Triton. In *Neptune and Triton*, ed. D. P. Cruikshank. Tucson, AZ: University of Arizona Press, pp. 807–877.

McKinnon, W. B., Schenk, P. M., and Dombard, A. J. (2001). Chaos on Io: A model for formation of mountain blocks by crustal heating, melting, and tilting. *Geology*, **29**, 103.

McNutt, M. K. (1984). Lithospheric flexure and thermal anomalies. *J. Geophys. Res.*, **89**, 11 180–11 194.

Melosh, H. J. (1975). Large impact craters and the Moon's orientation. *Earth Planet. Sci. Lett.*, **26**, 353–360.

Melosh, H. J. (1977). Global tectonics of a despun planet. *Icarus*, **31**, 221–243.

Melosh, H. J. (1980a). Tectonic patterns on a reoriented planet: Mars. *Icarus*, **44**, 745–751.

Melosh, H. J. (1980b). Tectonic patterns on a tidally distorted planet. *Icarus*, **43**, 334–337.

Melosh, H. J. and McKinnon, W. B. (1988). The tectonics of Mercury. In *Mercury*, ed. F. Vilas, C. R. Chapman and M. S. Matthews. Tucson, AZ: University of Arizona Press.

Melosh, H. J. and Turtle, E. P. (2004). Ridges on Europa: Origin by incremental ice-wedging (abs.). *Lunar Planet. Sci. Conf. XXXV*, 2029. Houston, TX: Lunar and Planetary Institute (CD-ROM).

Michaud, R. L., Pappalardo, R. T., and Collins, G. C. (2008). Pit chains on Enceladus: A discussion of their origin (abs.). *Lunar Planet. Sci. Conf. XXXIX*, 1678. Houston, TX: Lunar and Planetary Institute (CD-ROM).

Milazzo, M. P., Geissler, P. E., Greenberg, R., Keszthelyi, L. P., McEwen, A. S., Radebaugh, J., and Turtle, E. P. (2001). Non-synchronous rotation of Io? *Workshop on Jupiter: Planet, Satellites, and Magnetosphere*, 75–76.

Miller, D. J., Barnash, A. N., Bray, V. J., Turtle, E. P., Helfenstein, P., Squyres, S. W., and Rathbun, J. A. (2007). Interactions between impact craters and tectonic features on Enceladus and Dione. *Workshop on Ices, Oceans, and Fire*, 6007.

Mohit, P. S., Greenhagen, B. T., and McKinnon, W. B. (2004). Polar wander on Ganymede. *Bull. Am. Astron. Soc.*, **36**, 1084–1085.

Moons, M. and Henrard, J. (1994). Surfaces of section in the Miranda-Umbriel 3:1 inclination problem. *Celest. Mech. Dyn. Astron.*, **59**, 129–148.

Moore, J. M. and Ahern, J. L. (1983). The geology of Tethys. *J. Geophys. Res.*, **88**, A577-A584.

Moore, J. M. and Schenk, P. M. (2007). Topography of endogenic features on Saturnian mid-sized satellites (abs.). *Lunar Planet. Sci. Conf. XXXVIII*, 2136. Houston, TX: Lunar and Planetary Institute (CD-ROM).

Moore, J. M., Horner, V. M., and Greeley, R. (1985). The geomorphology of Rhea: Implications for geologic history and surface processes. *J. Geophys. Res.*, **90**, C785-C796.

Moore, J., McEwen, A., Albin, E., and Greeley, R. (1986). Topographic evidence for shield volcanism on Io. *Icarus*, **67**, 181–183.

Moore, J. M., Schenk, P. M., Bruesch, L. S., Asphaug, E. and McKinnon, W. B. (2004a). Large impact features on middle-sized icy satellites. *Icarus*, **171**, 421–443.

Moore, J. M., Chapman, C. R., Bierhaus, E. B., Greeley, R., Chuang, F. C., Klemaszewski, J., Clark, R. N., Dalton, J. B., Hibbitts, C. A., Schenk, P. M., Spencer, J. R., and Wagner, R. (2004b). Callisto. In *Jupiter: The Planet, Satellites, and Magnetosphere*, ed. F. Bagenal, T. E. Dowling and W. B. McKinnon. New York: Cambridge University Press, pp. 397–426.

Moore, W. B. (2006). Thermal equilibrium in Europa's ice shell. *Icarus*, **180**, 141–146.

Moore, W. B. and Schubert, G. (2003). The tidal response of Ganymede and Callisto with and without liquid water oceans. *Icarus*, **166**, 223–226.

Morrison, D., ed. (1982). *Satellites of Jupiter*. Tucson, AZ: University of Arizona Press.

Mueller, S. and McKinnon, W. B. (1988). Three-layered models of Ganymede and Callisto: Compositions, structures, and aspects of evolution. *Icarus*, **76**, 437–464.

Murchie, S. L. and Head, J. W. (1986). Global reorientation and its effect on tectonic patterns on Ganymede. *Geophys. Res. Lett.*, **13**, 345–348.

Murchie, S. L. and Head, J. W. (1988). Possible breakup of dark terrain on Ganymede by large-scale shear faulting. *J. Geophys. Res.*, **93**, 8795–8824.

Murchie, S. L., Head, J. W., Helfenstein, P., and Plescia, J. B. (1986). Terrain types and local-scale stratigraphy of grooved terrain on Ganymede. *J. Geophys. Res.*, **91**, E222-E238.

Murray, C. D. and Dermott, S. F. (1999). *Solar System Dynamics*. New York: Cambridge University Press.

Nimmo, F. (2004a). What is the Young's modulus of ice? (abs.). *Workshop on Europa's Icy Shell*, 7005.

Nimmo, F. (2004b). Stresses generated in cooling viscoelastic ice shells: Application to Europa. *J. Geophys. Res.*, **109**, E12001.

Nimmo, F. (2004c). Non-Newtonian topographic relaxation on Europa. *Icarus*, **168**, 205–208.

Nimmo, F. (2004d). Dynamics of rifting and modes of extension on icy satellites. *J. Geophys. Res.*, **109**, E01003.

Nimmo, F. and Gaidos, E. (2002). Strike-slip motion and double ridge formation on Europa. *J. Geophys. Res.*, **107**, doi:10.1029/2000JE001476.

Nimmo, F. and Manga, M. (2002). Causes, characteristics, and consequences of convective diapirism on Europa. *Geophys. Res. Lett.*, **29**, doi:10.1029/2002GL015754.

Nimmo, F. and Matsuyama, I. (2007). Reorientation of icy satellites by impact basins. *Geophys. Res. Lett.*, **34**, L19203.

Nimmo, F. and Pappalardo, R. T. (2004). Furrow flexure and ancient heat flux on Ganymede. *Geophys. Res. Lett.*, **31**, L19701.

Nimmo, F. and Pappalardo, R. T. (2006). Diapir-induced reorientation of Saturn's moon Enceladus. *Nature*, **441**, 614–616.

Nimmo, F. and Schenk, P. (2006). Normal faulting on Europa: Implications for ice shell properties. *J. Struct. Geol.*, **28**, 2194–2203.

Nimmo, F., Pappalardo, R. T., and Giese, B. (2002). Effective elastic thickness and heat flux estimates on Ganymede. *Geophys. Res. Lett.*, **29**, doi:10.1029/2001GL013976.

Nimmo, F., Pappalardo, R. T., and Giese, B. (2003a). On the origins of band topography, Europa. *Icarus*, **166**, 21–32.

Nimmo, F., Giese, B., and Pappalardo, R. T. (2003b). Estimates of Europa's ice shell thickness from elastically-supported topography. *Geophys. Res. Lett.*, **30**, doi:10.1029/2002GL016660.

Nimmo, F., Thomas, P. C., Pappalardo, R. T., and Moore, W. B. (2007a). The global shape of Europa: Constraints on lateral shell thickness variations. *Icarus*, **191**, 183–192.

Nimmo, F., Spencer, J. R., Pappalardo, R. T., and Mullen, M. E. (2007b). Shear heating as the origin of the plumes and heat flux on Enceladus. *Nature*, **447**, 289–291.

Nur, A. (1982). The origin of tensile fracture lineaments. *J. Struct. Geol.*, **4**, 31–40.

Nyffenegger, P. A. and Consolmagno, G. J. (1988). Tectonic episodes on Ariel: Evidence for an ancient thin crust (abs.). *Lunar Planet. Sci. Conf. XIX*, 873. Houston, TX: Lunar and Planetary Institute (CD-ROM).

Ojakangas, G. W. and Stevenson, D. J. (1989a). Thermal state of an ice shell on Europa. *Icarus*, **81**, 220–241.

Ojakangas, G. W. and Stevenson, D. J. (1989b). Polar wander of an ice shell on Europa. *Icarus*, **81**, 242–270.

O'Reilly, T. and Davies, G. (1981). Magma transport of heat on Io: A mechanism allowing a thick lithosphere. *Geophys. Res. Lett.*, **8**, 313–316.

Palguta, J., Anderson, J. D., Schubert, G., and Moore, W. B. (2006). Mass anomalies on Ganymede. *Icarus*, **180**, 428–441.

Pappalardo, R. T. (1994). *The origin and evolution of ridge and trough terrain and the geological history of Miranda*. Unpublished Ph.D. thesis, Arizona State University, Tempe, AZ.

Pappalardo, R. T. and Barr, A. C. (2004). The origin of domes on Europa: The role of thermally induced compositional diapirism. *Geophys. Res. Lett.*, **31**, L01701.

Pappalardo, R. T. and Collins, G. C. (2005). Strained craters on Ganymede. *J. Struct. Geol.*, **27**, 827–838.

Pappalardo, R. T. and Coon, M. D. (1996). A sea ice analogue for the surface of Europa (abs.). *Lunar Planet. Sci. Conf. XXVII*, 997–998. Houston, TX: Lunar and Planetary Institute (CD-ROM).

Pappalardo, R. T. and Sullivan, R. J. (1996). Evidence for separation across a gray band on Europa. *Icarus*, **123**, 557–567.

Pappalardo, R. T., Reynolds, S. J., and Greeley, R. (1997). Extensional tilt blocks on Miranda: Evidence for an upwelling origin of Arden Corona. *J. Geophys. Res.*, **102**, 13 369–13 379.

Pappalardo, R. T., Head, J. W., Greeley, R., Sullivan, R. J., Pilcher, C., Schubert, G., Moore, W. B., Carr, M. H., Moore, J. M., Belton, M. J. S., and Goldsby, D. L. (1998a). Geological evidence for solid-state convection in Europa's ice shell. *Nature*, **391**, 365–368.

Pappalardo, R. T., Head, J. W., Collins, G. C., Kirk, R. L., Neukum, G., Oberst, J., Giese, B., Greeley, R., Chapman, C. R., Helfenstein, P., Moore, J. M., McEwen, A., Tufts, B. R., Senske, D. A., Breneman, H. H., and Klaasen, K. (1998b). Grooved terrain on Ganymede: First results from Galileo high-resolution imaging. *Icarus*, **135**, 276–302.

Pappalardo, R. T., and 31 others (1999). Does Europa have a subsurface ocean? Evaluation of the geological evidence. *J. Geophys. Res.*, **104**, 24 015–24 056.

Pappalardo, R. T., Collins, G. C., Head, J. W., Helfenstein, P., McCord, T. B., Moore, J. M., Prockter, L. M., Schenk, P. M., and Spencer, J. R. (2004). Geology of Ganymede. In *Jupiter: The Planet, Satellites, and Magnetosphere*, ed. F. Bagenal, T. E. Dowling and W. B. McKinnon. New York: Cambridge University Press, pp. 363–396.

Passey, Q. R. (1983). Viscosity of the lithosphere of Enceladus. *Icarus*, **53**, 105–120.

Passey, Q. R. and Shoemaker, E. M. (1982). Craters and basins on Ganymede and Callisto: Morphological indicators of crustal evolution. In *Satellites of Jupiter*, ed. D. Morrison. Tucson, AZ: University of Arizona Press, pp. 379–434.

Patel, J. G., Pappalardo, R. T., Prockter, L. M., Collins, G. C., and Head, J. W. (1999a). Morphology of ridge and trough terrain on Europa: Fourier analysis and comparison to Ganymede (abs.). *Eos Trans. AGU*, P42A-12.

Patel, J. G., Pappalardo, R. T., Head, J. W., Collins, G. C., Hiesinger, H., and Sun, J. (1999b). Topographic wavelengths of Ganymede groove lanes from Fourier analysis of Galileo images. *J. Geophys. Res.*, **104**, 24 057–24 074.

Patterson, G. W. and Head, J. W. (2003). Crustal spreading on Europa: Inferring tectonic history from triple junction analysis (abs.). *Lunar Planet. Sci. Conf. XXXIV*, 1262. Houston, TX: Lunar and Planetary Institute (CD-ROM).

Patterson, G. W., Head, J. W., and Pappalardo, R. T. (2006). Plate motion on Europa and nonrigid behavior of the icy lithosphere: The Castalia Macula Region. *J. Struct. Geol.*, **28**, 2237–2258.

Peale, S. J. (1977). Rotation histories of the natural satellites. In *Planetary Satellites*, ed. J. A. Burns and M. S. Matthews. Tucson, AZ: University of Arizona Press, pp. 87–111.

Peale, S. J. (1999). Origin and evolution of the natural satellites. *Annu. Rev. Astron. Astrophys.*, **37**, 533–602.

Peale, S. J. (2003). Tidally induced volcanism. *Celest. Mech. Dyn. Astron.*, **87**, 129–155.

Peale, S., Cassen, P., and Reynolds, R. (1979). Melting of Io by tidal dissipation. *Science*, **203**, 892–894.

Person, M. J., Elliot, J. L., Gulbis, A. A. S., Pasachoff, J. M., Babcock, B. A., Souza, S. P., and Gangestad, J. (2006). Charon's radius and density from the combined data set of the 2005 July 11 occultation. *Astron. J.*, **132**, 1575–1580.

Petrenko, V. F. and Whitworth, R. W. (1999). *Physics of Ice*. Oxford: Oxford University Press.

Phillips, C. B., McEwen, A. S., Hoppa, G. V., Fagents, S. A., Greeley, R., Klemaszewski, J. E., Pappalardo, R. T., Klaasen, K. P., and Breneman, H. H. (2000). The search for current geologic activity on Europa. *J. Geophys. Res.*, **105**, 22 579–22 598.

Pieri, D. C. (1981). Lineament and polygon patterns on Europa. *Nature*, **289**, 17–21.

Plescia, J. B. (1983). The geology of Dione. *Icarus*, **56**, 255–277.

Plescia, J. B. (1987). Geological terrains and crater frequencies on Ariel. *Nature*, **327**, 201–204.

Plescia, J. B. (1988). Cratering history of Miranda: Implications for geologic processes. *Icarus*, **73**, 442–461.

Pollack, J. B. and Consolmagno, G. (1984). Origin and evolution of the Saturn system. In *Saturn*, ed. T. Gehrels and M. S. Matthews. Tucson, AZ: University of Arizona Press, pp. 811–866.

Porco, C. C., and 35 others (2005a). Imaging of Titan from the Cassini spacecraft. *Nature*, **434**, 159–168.

Porco, C. C., and 34 others (2005b). Cassini imaging science: Initial results on Phoebe and Iapetus. *Science*, **307**, 1237–1242.

Porco, C. C., and 24 others (2006). Cassini observes the active south pole of Enceladus. *Science*, **311**, 1393–1401.

Prockter, L. M. and Pappalardo, R. T. (2000). Folds on Europa: Implications for crustal cycling and accommodation of extension. *Science*, **289**, 941–944.

Prockter, L. M., Head, J. W., Pappalardo, R. T., Senske, D. A., Neukum, G., Wagner, R., Wolf, U., Oberst, J. O., Giese, B., Moore, J. M., Chapman, C. R., Helfenstein, P., Greeley, R., Breneman, H. H., and Belton, M. J. S. (1998). Dark terrain on Ganymede: Geological mapping and interpretation of Galileo Regio at high resolution. *Icarus*, **135**, 317–344.

Prockter, L. M., Pappalardo, R. T., and Head, J. W. (2000a). Strike-slip duplexing on Jupiter's icy moon Europa. *J. Geophys. Res.*, **105**, 9483–9488.

Prockter, L. M., Figueredo, P. H., Pappalardo, R. T., Head, J. W., and Collins, G. C. (2000b). Geology and mapping of dark terrain on Ganymede and implications for grooved terrain formation. *J. Geophys. Res.*, **105**, 22 519–22 540.

Prockter, L. M., Head, J. W., Pappalardo, R. T., Sullivan, R. L., Clifton, A. E., Giese, B., Wagner, R., and Neukum, G. (2002). Morphology of Europan bands at high resolution: A mid-ocean ridge-type rift mechanism. *J. Geophys. Res.*, **107**, doi:10.1029/2000JE001458.

Prockter, L. M., Nimmo, F., and Pappalardo, R. T. (2005). A shear heating origin for ridges on Triton. *Geophys. Res. Lett.*, **32**, L14202.

Quirico, E., Douté, S., Schmitt, B., de Bergh, C., Cruikshank, D. P., Owen, T. C., Geballe, T. R., and Roush, T. L. (1999). Composition, physical state, and distribution of ices at the surface of Triton. *Icarus*, **139**, 159–178.

Radebaugh, J., Keszthelyi, L., McEwen, A., Turtle, E., Jaeger, W., and Milazzo, M. (2001). Paterae on Io: A new type of volcanic caldera? *J. Geophys. Res.*, **106**, 33 005–33 020.

Radebaugh, J., Lorenz, R. D., Kirk, R. L., Lunine, J. I., Stofan, E. R., Lopes, R. M. C., and Wall, S. D. (2007). Mountains on Titan observed by Cassini radar. *Icarus*, **192**, 77–91.

Rist, M. A. (1997). High-stress ice fracture and friction. *J. Phys. Chem. B*, **101**, 6263–6266.

Ross, M. and Schubert, G. (1986). Tidal dissipation in a viscoelastic planet. *J. Geophys. Res.*, **91**, D447–D452.

Ross, M. and Schubert, G. (1989). Viscoelastic models of tidal heating in Enceladus. *Icarus*, **78**, 90–101.

Ross, M. N. and Schubert, G. (1990). The coupled orbital and thermal evolution of Triton. *Geophys. Res. Lett.*, **17**, 1749–1752.

Rubincam, D. P. (2003). Polar wander on Triton and Pluto due to volatile migration. *Icarus*, **163**, 469–478.

Ruzicka, A. (1988). The geology of Ariel (abs.). *Lunar Planet. Sci. Conf. XIX*, 1009. Houston, TX: Lunar and Planetary Institute (CD-ROM).

Sarid, A. R., Greenberg, R., Hoppa, G. V., Hurford, T. A., Tufts, B. R., and Geissler, P. (2002). Polar wander and surface convergence of Europa's ice shell: Evidence from a survey of strike-slip displacement. *Icarus*, **158**, 24–41.

Sarid, A. R., Greenberg, R., Hoppa, G. V., Geissler, P., and Preblich, B. (2004). Crack azimuths on Europa: Time sequence in the southern leading face. *Icarus*, **168**, 144–157.

Sarid, A. R., Greenberg, R., Hoppa, G. V., Brown, D. M., and Geissler, P. (2005). Crack azimuths on Europa: The G1 lineament sequence revisited. *Icarus*, **173**, 469–479.

Schenk, P. M. (1991). Fluid volcanism on Miranda and Ariel: Flow morphology and composition. *J. Geophys. Res.*, **96**, 1887–1906.

Schenk, P. M. (1995). The geology of Callisto. *J. Geophys. Res.*, **100**, 19 023–19 040.

Schenk, P. and Bulmer, M. (1998). Origin of mountains on Io by thrust faulting and large-scale mass movements. *Science*, **279**, 1514–1518.

Schenk, P. and Jackson, M. P. A. (1993). Diapirism on Triton: A record of crustal layering and instability. *Geology*, **21**, 299–302.

Schenk, P. M. and McKinnon, W. B. (1989). Fault offsets and lateral crustal movement on Europa: Evidence for a mobile ice shell. *Icarus*, **79**, 75–100.

Schenk, P. H. and Moore, J. M. (1995). Volcanic constructs on Ganymede and Enceladus: Topographic evidence from stereo images and photoclinometry. *J. Geophys. Res.*, **100**, 19 009–19 022.

Schenk, P. M. and Zahnle, K. (2007). On the negligible age of Triton's surface. *Icarus*, **192**, 135–149.

Schenk, P. M., Asphaug, E., McKinnon, W. B., Melosh, H. L., and Weissman, P. (1996). Cometary nuclei and tidal disruption: The geologic record of crater chains on Callisto and Ganymede. *Icarus*, **121**, 249–274.

Schenk, P., Wilson, R., Hargitai, H., McEwen, A., and Thomas, P. (2001a). The mountains of Io: Global and geologic perspectives from Voyager and Galileo. *J. Geophys. Res.*, **106**, 33 201–33 222.

Schenk, P. H., McKinnon, W. B., Gwynn, D., and Moore, J. M. (2001b). Flooding of Ganymede's bright terrains by low-viscosity water-ice lavas. *Nature*, **410**, 57–60.

Schenk, P., Wilson, R., and Davies, A. (2004a). Shield volcano topography and the rheology of lava flows on Io. *Icarus*, **169**, 98–110.

Schenk, P. M., Chapman, C. R., Zahnle, K., and Moore, J. M. (2004b). Ages and interiors: The cratering record of the Galilean satellites. In *Jupiter: The Planet, Satellites, and Magnetosphere*, ed. F. Bagenal, T. E. Dowling and W. B. McKinnon. New York: Cambridge University Press, pp. 427–456.

Schenk, P. M., Matsuyama, I., and Nimmo, F. (2008). True polar wander on Europa from global-scale small-circle depressions. *Nature*, **453**, 368–371.

Schmeltz, M., Rignot, E., and MacAyeal, D. (2002). Tidal flexure along ice-sheet margins: Comparisons of INSAR with an elastic-plate model. *Annals Glaciol.*, **34**, 202–208.

Schubert, G., Zhang, K., Kivelson, M. G., and Anderson, J. D. (1996). The magnetic field and internal structure of Ganymede. *Nature*, **384**, 544–545.

Schubert, G., Anderson, J. D., Spohn, T., and McKinnon, W. B. (2004). Interior composition, structure, and dynamics of the Galilean satellites. In *Jupiter: The Planet, Satellites, and Magnetosphere*, ed. F. Bagenal, T. E. Dowling and W. B. McKinnon. New York: Cambridge University Press, pp. 281–306.

Schubert, G., Anderson, J. D., Travis, B. J., and Palguta, J. (2007). Enceladus: Present internal structure and differentiation by early and long-term radiogenic heating. *Icarus*, **188**, 345–355.

Schulson, E. M. (2002). On the origin of a wedge crack within the icy crust of Europa. *J. Geophys. Res.*, **107**, doi:10.1029/2001JE001586.

Schulson, E. M. (2006). The fracture of water ice Ih: A short overview. *Meteorit. Planet. Sci.*, **41**, 1497–1508.

Segatz, M., Spohn, T., Ross, M. N., and Schubert, G. (1988). Tidal dissipation, surface heat flow, and figure of viscoelastic models of Io. *Icarus*, **75**, 187–206.

Shoemaker, E. M., Lucchitta, B. K., Wilhelms, D. E., Plescia, J. B., and Squyres, S. W. (1982). The geology of Ganymede. In *Satellites of Jupiter*, ed. D. Morrison. Tucson, AZ: University of Arizona Press, pp. 435–520.

Showman, A. P. and Han, L. J. (2005). Effects of plasticity on convection in an ice shell: Implications for Europa. *Icarus*, **177**, 425–437.

Showman, A. P., Stevenson, D. J., and Malhotra, R. (1997). Coupled orbital and thermal evolution of Ganymede. *Icarus*, **129**, 367–383.

Sklar, L. S., Polito, P., Zygielbaum, B., and Collins, G. C. (2008). Abrasion susceptibility of ultra-cold water ice: Preliminary measurements of abrasion rate, tensile strength, and elastic modulus (abs.). Science of Solar System Ices Workshop, 9076.

Smith, B. A., and 21 others (1979). The Jupiter system through the eyes of Voyager 1. *Science*, **204**, 951–972.

Smith, B. A., and 26 others (1981). Encounter with Saturn: Voyager 1 imaging science results. *Science*, **212**, 163–191.

Smith, B. A., and 28 others (1982). A new look at the Saturn system: The Voyager 2 images. *Science*, **215**, 505–537.

Smith, B. A., Soderblom, L. A., Beebe, R., Bliss, D., Brown, R. H., Collins, S. A., Boyce, J. M., Briggs, G. A., Brahic, A., Cuzzi, J. N., and Morrison, D. (1986). Voyager 2 in the Uranian system: Imaging science results. *Science*, **233**, 43–64.

Smith, B. A., Soderblom, L. A., Banfield, D., Barnet, C., Beebe, R. F., Basilevski, A. T., Bollinger, K., Boyce, J. M., Briggs, G. A., and Brahic, A. (1989). Voyager 2 at Neptune: Imaging science results. *Science*, **246**, 1422–1449.

Smith-Konter, B., and Pappalardo, R. T. (2008). Tidally driven stress accumulation and shear failure of Enceladus's tiger stripes. *Icarus*, **198**, 435–451.

Solomatov, V. S. (1995). Scaling of temperature- and stress-dependent viscosity convection. *Phys. Fluids*, **7**, 266–274.

Solomatov, V. S. and Moresi, L.-N. (2000). Scaling of time-dependent stagnant lid convection: Application to small-scale convection on Earth and other terrestrial planets. *J. Geophys. Res.*, **105**, 21 795–21 818.

Sotin, C., Head, J. W., and Tobie, G. (2002). Europa: Tidal heating of upwelling thermal plumes and the origin of lenticulae and chaos melting. *Geophys. Res. Lett.*, **29**, doi:10.1029/2001GL013844.

Spaun, N. A., Pappalardo, R. T., and Head, J. W. (2003). Evidence for shear failure in forming near-equatorial lineae on Europa. *J. Geophys. Res.*, **108**, doi:10.1029/2001JE001499.

Spencer, J. R., Pearl, J. C., Segura, M., Flasar, F. M., Mamoutkine, A., Romani, P., Buratti, B. J., Hendrix, A. R., Spilker, L. J., and Lopes, R. M. C. (2006). Cassini encounters Enceladus: Background and the discovery of a south polar hot spot. *Science*, **311**, 1401–1405.

Squyres, S. W. (1980). Volume changes in Ganymede and Callisto and the origin of grooved terrain. *Geophys. Res. Lett.*, **7**, 593–596.

Squyres, S. W. (1981). The topography of Ganymede's grooved terrain. *Icarus*, **46**, 156–168.

Squyres, S. W. and Croft, S. K. (1986). The tectonics of icy satellites. In *Satellites*, ed. J. A. Burns and M. S. Matthews. Tucson, AZ: University Arizona Press, pp. 293–341.

Squyres, S. W., Reynolds, R. T., Cassen, P. M., and Peale, S. J. (1983). The evolution of Enceladus. *Icarus*, **53**, 319–331.

Stempel, M. M. and Pappalardo, R. T. (2002). Lineament orientations through time near Europa's leading point: Implications for stress mechanisms and rotation of the icy shell (abs.). *Lunar Planet. Sci. Conf. XXXIII*, 1661. Houston, TX: Lunar and Planetary Institute (CD-ROM).

Stempel, M. M., Barr, A. C., and Pappalardo, R. T. (2005). Model constraints on the opening rates of bands on Europa. *Icarus*, **177**, 297–304.

Stern, S. A. and McKinnon, W. B. (2000). Triton's age and impactor population revisited in the light of Kuiper Belt fluxes: Evidence for small Kuiper Belt objects and recent geological activity. *Astron. J.*, **119**, 945–952.

Stevenson, D. (1996). Heterogeneous tidal deformation and geysers on Europa (abs.). *Europa Ocean Conf.*, 69–70.

Stevenson, D. J. and Lunine, J. I. (1986). Mobilization of cryogenic ice in outer solar system satellites. *Nature*, **323**, 46–48.

Stofan, E. R., and 35 others (2006). Mapping of Titan: Results from the first Titan radar passes. *Icarus*, **185**, 443–456.

Stooke, P. J. (1991). Geology of the inter corona regions of Miranda (abs.). *Lunar Planet. Sci. Conf. XXII*, 1341. Houston, TX: Lunar and Planetary Institute (CD-ROM).

Sullivan, R., Greeley, R., Homan, K., Klemaszewski, J., Belton, M. J. S., Carr, M. H., Chapman, C. R., Tufts, B. R., Head, J. W., Pappalardo, R., Moore, J., and Thomas, P. (1998). Episodic plate separation and fracture infill on the surface of Europa. *Nature*, **391**, 371–373.

Sullivan, R., Moore, J., and Pappalardo, R. (1999). Mass-wasting and slope evolution on Europa (abs.). *Lunar Planet. Sci. Conf. XXX*, 1747. Houston, TX: Lunar and Planetary Institute (CD-ROM).

Tackley, P. J. (2001). Convection in Io's asthenosphere: Redistribution of nonuniform tidal heating by mean flows. *J. Geophys. Res.*, **106**, 32 971–32 981.

Tackley, P., Schubert, G., Glatzmaier, G., Schenk, P., Ratcliff, J., Matas, J.-P. (2001). Three-dimensional simulations of mantle convection in Io. *Icarus*, **149**, 73–93.

Thomas, P. C. (1988). Radii, shapes, and topography of the satellites of Uranus from limb coordinates. *Icarus*, **73**, 427–441.

Thomas, P. C., Burns, J. A., Helfenstein, P., Squyres, S., Veverka, J., Porco, C., Turtle, E. P., McEwen, A., Denk, T., Giese, B., Roatsch, T., Johnson, T. V., and Jacobson, R. A. (2007). Shapes of saturnian icy satellites and their significance. *Icarus*, **190**, 573–584.

Tittemore, W. C. and Wisdom, J. (1990). Tidal evolution of the Uranian satellites: III. Evolution through the Miranda-Umbriel 3:1, Miranda-Ariel 5:3, and Ariel-Umbriel 2:1 mean-motion commensurabilities. *Icarus*, **85**, 394–443.

Tobie, G., Mocquet, A., and Sotin, C. (2005). Tidal dissipation within large icy satellites: Applications to Europa and Titan. *Icarus*, **177**, 534–549.

Tobie, G., Cadek, O., and Sotin, C. (2008). Solid tidal friction above a liquid water reservoir as the origin of the south pole hotspot on Enceladus. *Icarus*, **196**, 642–652.

Tomasko, M. G., and 39 others (2005). Rain, winds and haze during the Huygens probe's descent to Titan's surface. *Nature*, **438**, 765–778.

Tufts, B. R., Greenberg, R., Hoppa, G., and Geissler, P. (1999). Astypalaea Linea: A large-scale strike-slip fault on Europa. *Icarus*, **141**, 53–64.

Tufts, B. R., Greenberg, R., Hoppa, G., and Geissler, P. (2000). Lithospheric dilation on Europa. *Icarus*, **146**, 75–97.

Turcotte, D. L. (1983). Thermal stresses in planetary elastic lithospheres. *J. Geophys. Res.*, **88**, A585-A587.

Turcotte, D. L. and Schubert, G. (2002). *Geodynamics*. New York: Cambridge University Press.

Turcotte, D. L., Willemann, R. J., Haxby, W. F., and Norberry, J. (1981). Role of membrane stresses in the support of planetary topography. *J. Geophys. Res.*, **86**, 3951–3959.

Turtle, E., Jaeger, W., Keszthelyi, L., McEwen, A., Milazzo, M., Moore, J., Phillips, C., Radebaugh, J., Simonelli, D., Chuang, F., and Peter, S. (2001). Mountains on Io:

High-resolution Galileo observations, initial interpretations, and formation models. *J. Geophys. Res.*, **106**, 33 175–33 200.

Vaughan, D. G. (1995). Tidal flexure at ice shelf margins. *J. Geophys. Res.*, **100**, 6213–6224.

Wagner, R. J., Neukum, G., Giese, B., Roatsch, T., Wolf, U., and Denk, T. (2006). Geology, ages and topography of Saturn's satellite Dione observed by the Cassini ISS camera (abs.). *Lunar Planet. Sci. Conf. XXXVII*, 1805. Houston, TX: Lunar and Planetary Institute (CD-ROM).

Wagner, R. J., Neukum, G., Giese, B., Roatsch, T., and Wolf, U. (2007). The global geology of Rhea: Preliminary implications from the Cassini ISS data (abs.). *Lunar Planet. Sci. Conf. XXXVIII*, 1958. Houston, TX: Lunar and Planetary Institute (CD-ROM).

Wahr, J., Selvans, Z. A., Mullen, M. E., Barr, A. C., Collins, G. C., Selvans, M. M., and Pappalardo, R. T. (2009). Modeling stresses on satellites due to nonsynchronous rotation and orbital eccentricity using gravitational potential theory. *Icarus*, **200**, 188–206.

Walls, S. D., Lopes, R. M., Stofan, E. R., Wood, C. A., Radebaugh, J. L., Hörst, S. M., Stiles, B. W., Nelson, R. M., Kamp, L. W., Janssen, M. A., Lorenz, R. D., Lunine, J. I., Farr, T. G., Mitri, G., Paillou, P., Paganelli, F., and Mitchell, K. L. (2009). Cassini RADAR images at Hotei Arcus and western Xanadu, Titan: Evidence for geologically recent cryovolcanic activity. *Geophys. Res. Lett.*, **36**, L04203.

Watts, A. (2001). *Isostasy and Flexure of the Lithosphere*. New York: Cambridge University Press.

Watters, T. R., Schultz, R. A., and Robinson, M. S. (2000). Displacement–length relations of thrust faults associated with lobate scarps on Mercury and Mars: Comparison with terrestrial faults. *Geophys. Res. Lett.*, **27**, 3659–3662.

Weeks, W. F. and Cox, G. F. N. (1984). The mechanical properties of sea ice: A status report. *Ocean Sci. Eng.*, **9**, 135–198.

Willemann, R. J. (1984). Reorientation of planets with elastic lithospheres. *Icarus*, **60**, 701–709.

Wisdom, J. (2004). Spin-orbit secondary resonance dynamics of Enceladus. *Astron. J.*, **128**, 484–491.

Wyrick, D., Ferrill, D. A., Morris, A. P., Colton, S. L., and Sims, D. W. (2004). Distribution, morphology, and origins of Martian pit crater chains. *J. Geophys. Res.*, **109**, E06005.

Yoder, C. F. (1995). Astrometric and geodetic properties of Earth and the solar system. In *Global Earth Physics*, ed. T. J. Aherns. Washington, DC: AGU Press, pp. 1–32.

Zahnle, K., Dones, L., and Levison, H. F. (1998). Cratering rates on the Galilean satellites. *Icarus*, **136**, 202–222.

Zahnle, K., Schenk, P., Sobieszczyk, S., Dones, L., and Levison, H. F. (2001). Differential cratering of synchronously rotating satellites by ecliptic comets. *Icarus*, **153**, 111–129.

Zahnle, K., Schenk, P., Levison, H., and Dones, L. (2003). Cratering rates in the outer solar system. *Icarus*, **163**, 263–289.

Zuber, M. T. and Parmentier, E. M. (1984). Lithospheric stresses due to radiogenic heating of an ice-silicate planetary body: Implications for Ganymede's tectonic evolution. *J. Geophys. Res.*, **89**, B429-B437.

8

Planetary structural mapping

Kenneth L. Tanaka
U.S. Geological Survey, Flagstaff

Robert Anderson
Jet Propulsion Laboratory, California Institute of Technology, Pasadena

James M. Dohm
Department of Hydrology and Water Resources, University of Arizona, Tucson

Vicki L. Hansen
Department of Geological Sciences, University of Minnesota Duluth

George E. McGill
University of Massachusetts, Amherst

Robert T. Pappalardo
Jet Propulsion Laboratory, California Institute of Technology, Pasadena

Richard A. Schultz
Geomechanics – Rock Fracture Group, Department of Geological Sciences and Engineering, University of Nevada, Reno

and

Thomas R. Watters
Center for Earth and Planetary Studies, National Air and Space Museum, Smithsonian Institution, Washington, DC

Summary

As on Earth, other solid-surfaced planetary bodies in the solar system display landforms produced by tectonic activity, such as faults, folds, and fractures. These features are resolved in spacecraft observations directly or with techniques that extract topographic information from a diverse suite of data types, including radar

Planetary Tectonics, edited by Thomas R. Watters and Richard A. Schultz. Published by Cambridge University Press. Copyright © Cambridge University Press 2010.

backscatter and altimetry, visible and near-infrared images, and laser altimetry. Each dataset and technique has its strengths and limitations that govern how to optimally utilize and properly interpret the data and what sizes and aspects of features can be recognized. The ability to identify, discriminate, and map tectonic features also depends on the uniqueness of their form, on the morphologic complexity of the terrain in which the structures occur, and on obscuration of the features by erosion and burial processes. Geologic mapping of tectonic structures is valuable for interpretation of the surface strains and of the geologic histories associated with their formation, leading to possible clues about: (1) the types or sources of stress related to their formation, (2) the mechanical properties of the materials in which they formed, and (3) the evolution of the body's surface and interior where timing relationships can be determined. Formal mapping of tectonic structures has been performed and/or is in progress for Earth's Moon, the planets Mars, Mercury, and Venus, and the satellites of Jupiter (Callisto, Ganymede, Europa, and Io). Structures have also been recognized on some of the Saturnian (Titan, Dione, Rhea, Tethys, Iapetus, and Enceladus) and Uranian satellites (Miranda and Ariel) and Neptune's large moon, Triton. Of these, only Earth's Moon has provided rock samples that have been dated using radiometric techniques, thus constraining, in the best scenarios, the age of formation of specific structures. However, most of these bodies have a resurfacing history useful for relative structural history, which might be constrained by geologic mapping and, in some cases, by crater-density data. Because of the range in rheologic character represented by planetary crustal materials and in some cases exotic stress mechanisms acting upon them, planetary structures can include forms, relationships, and developmental patterns that are rare or non-existent on Earth.

1. Introduction

Structural mapping constitutes the fundamental approach to documenting the tectonic deformation of planetary surfaces in space and time. As on Earth, geologists characterize, map, and interpret rock materials and structures on planetary surfaces to interpret geologic histories from local to global scales. However, terrestrial mapping approaches need to be adapted to meet the needs and special challenges inherent in producing planetary geologic maps that rely on spacecraft data and lack ground truth. The philosophical basis and basic techniques used for all current planetary geologic mapping were developed by geologists studying the Moon prior to, during, and following lunar exploration using space vehicles. In particular, mappable units were defined by objective descriptive criteria that do not depend on genetic interpretations. Crustal history was determined by means of superposition and crosscutting relations, and by density of superposed craters. Wilhelms

(1972, 1990) provides a thorough exposition of this approach. Structures, especially craters, provide critical elements to this stratigraphic approach. Adequate data for mapping structures on a variety of planetary bodies have increased dramatically over the past few decades and now are available for Earth's Moon, Mercury, Mars, Venus, and the satellites of Jupiter, Saturn, and Neptune, as well as the asteroid Eros.

Tectonic structures result from deformation of crustal rock materials and may include individual or systems of joints, faults, folds, and combinations thereof in various size ranges (see Schultz *et al.*, Chapter 10). Such structures are expressed at the surface by a diversity of landforms that combine the morphology of the pre-deformed surface with that of the tectonic deformation. In addition, subsequent modification due to erosion, burial, and reactivation of deformation modifies the appearance of structural landforms. Morphology can be observed using suitable bases constructed from visible, infrared, and radar imaging data, as well as digital elevation models derived from laser altimetry, stereo photogrammetry and 2-D photoclinometry. Except in limited cases where good exposures and topographic information permit accurate measurement of stratigraphic offsets, the nature of deformation is inferred from a comparison between the planetary landform and the geologist's knowledge of Earth-based examples. Brittle extensional structures appear in spacecraft imagery as linear topographic elements with negative and/or positive relief, including normal faults, grabens, and rift systems that cut and commonly produce offsets in the strata (e.g., McGill, 1971; Wilhelms, 1987; Banerdt *et al.*, 1992; Schultz, 1991, 1999). Contractional structures, on the other hand, might be expressed as positive relief landforms, such as wrinkle ridges that overlie blind thrust faults (that do not break the planetary surface; e.g., Schultz, 2000a), surface-breaking thrust faults expressed by lobate scarps (see Watters and Nimmo, Chapter 2; Golombek and Phillips, Chapter 5), high-angle reverse faults expressed by high-relief ridges (see Watters and Nimmo, Chapter 2) and large mountain ranges that often mark the locations of buckled and/or over-thrusted crustal materials (e.g., Banerdt *et al.*, 1992; Schultz and Tanaka, 1994; Dohm and Tanaka, 1999; Schultz, 2000a). Strike-slip faults are found on several planets and satellites and display evidence of either brittle or ductile deformation, or both (e.g., Koenig and Aydin, 1998; Schultz, 1989, 1999; Watters, 1992; Schenk and McKinnon, 1989; Pappalardo *et al.*, 1999; Tuckwell and Ghail, 2003; Kumar, 2005; Okubo and Schultz, 2006).

Structural history is reconstructed among structures and materials such as lava flows, impact ejecta, and sedimentary deposits through crosscutting relations determined using all pertinent spacecraft information. This Earth-proven technique requires meticulous detective work. The spatial and temporal associations of tectonic structures among rock materials archive past geologic events, which can be thoughtfully deciphered through detailed geologic mapping. However, actual geologic histories are typically much more detailed and complex than can be fully realized and represented on a geologic map, and age correlations are commonly poorly constrained.

As with any archive of past events, evidence can be destroyed or buried. Therefore, mappers effectively reconstruct structural histories based on (1) collecting and synthesizing all available data and compiled geologic information, (2) clearly discussing applied methodologies and acknowledging critical assumptions and uncertainties, and (3) avoiding overinterpretation and bias. Geologic mapping, whether based on field investigations or remotely sensed data, is ultimately built on consistency arguments. Thus mappers' observations, assumptions, and interpretations should always be open to further scrutiny, including how interpretations imply relations about other units, structures, and history (e.g., Gilbert, 1886).

Absolute-age determinations, which depend on sizable returned samples, are only directly available for six locales on the Moon (e.g., Wilhelms, 1987). However, crater-density statistics are useful relative-age indicators for the more heavily cratered planetary bodies, because impact cratering is assumed to be a continuing, areally random process for most surfaces. The number of craters superposed on a surface provides an average relative age of the surface. For example, the density of superposed craters on the lunar highlands is much greater than the density of superposed craters on the lunar maria. If the flux of impacting objects was constant through time, an unrealistically young age for the maria and for post-mare craters is implied. Thus, it was realized that the flux was greater by orders of magnitude early in lunar (and solar system) history than in more recent time. Returned samples verified this inference. A controversy still exists, however, with regard to the nature and rate of the bolide flux decline. The radiometric ages of lunar basin ejecta seem to imply that most basin impacts occurred during a relatively brief interval of time, with essentially no surface ages older than about 4.1 Ga and none younger than about 3.8 Ga, with a major spike of impact events at about 3.9 Ga (Tera *et al.*, 1974; Kring and Cohen, 2002; Gomes *et al.*, 2005; Strom *et al.*, 2005). This "cataclysmic" model implies a very different history of the Moon during its first few hundred million years of existence than does the alternate model, which postulates a continuous, exponential decline in impact flux with time. The absence of direct evidence of basin events older than about 4.1 Ga may be due to complete destruction of evidence for older basins, including radiometric dates demonstrating associated basin-related activity, or to incomplete sampling dictated by the logistical constraints on the Apollo missions. Absolute ages for non-lunar planetary surfaces can be estimated, or modeled, based on extrapolations of impactor populations, cratering theory, the cratering-rate history determined for the Moon, and

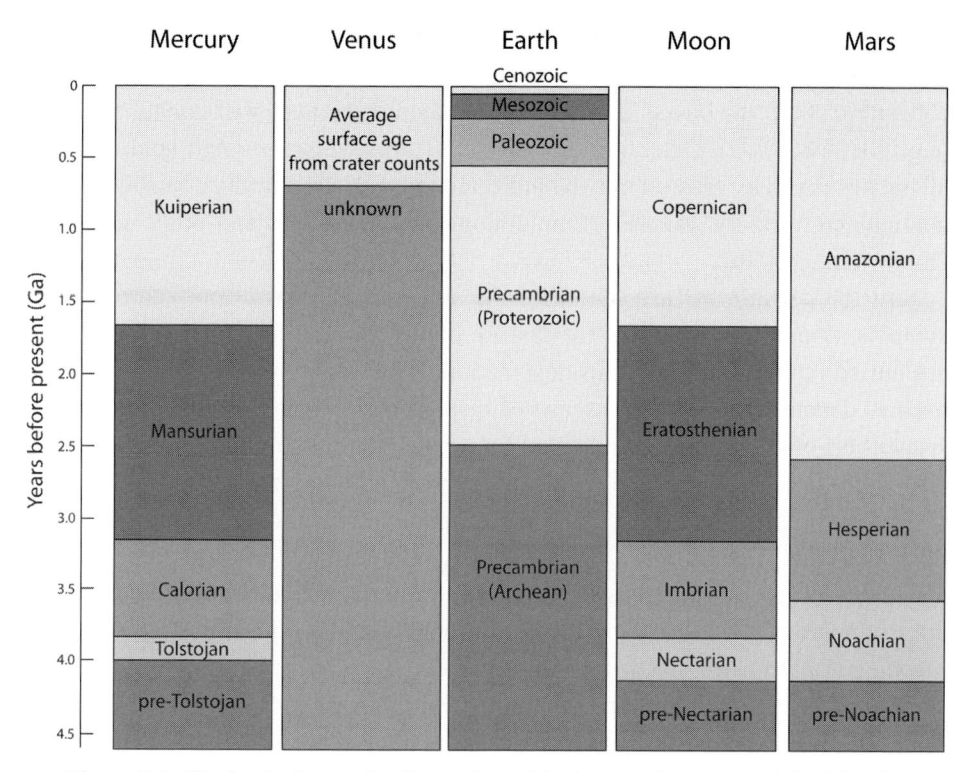

Figure 8.1. Geologic timescales for bodies of the inner solar system (after Tanaka and Hartmann, 2008).

other effects related to orbital distance from the Sun (e.g., Hartmann and Neukum, 2001). Based on this approach, absolute ages (having various degrees of uncertainty) for geologic epochs on the inner planets of the solar system are shown in Figure 8.1.

Maps that delineate structures of various ages can be used to characterize potential stress sources, strain magnitudes and history, and preexisting structural controls that may relate to episodes of local to global volcanism, tectonism, and impact cratering. Evidence for tectonic strain includes deformed features such as impact craters, which may hint at the type and amount of tectonism in the region or at the rheological nature of the impacted material, through assessing their geometric shape (Golombek *et al.*, 1996; Pappalardo and Collins, 2005). Tectonic structures and deformational processes may also control subsequent volcanic, intrusive, and hydrogeologic activity, including the diversion of surface and subsurface lavas and volatiles, as well as the discharge of subsurface fluids. Such activity can result in structurally controlled drainage pathways and canyon systems, aligned collapse features (e.g., pit crater chains), and volcanic constructs. Other indirect evidence for tectonism includes aligned mesas and valleys, such as those located in and surrounding large impact basins, where valleys and mesas radial to and concentric about the basins imply impact-related deformation of crustal materials (e.g., Melosh, 1989). Detection of major structures whose original landforms are all but destroyed by erosion may be indicated through structurally controlled linear anomalies in gravity, magnetic, mineralogic, geochemical, and other data (e.g., Dohm et al., 2007b).

Structural landforms are schematically shown on geologic maps through line symbols. Sometimes the lines are dashed, dotted, or queried to indicate that they are buried or have an uncertain designation. Additional symbols and colors are used to denote structure type, sense of displacement, and age. At times, closely spaced sets of structures are indicated by stipple patterns. In more recent mapping efforts, tectonic structures are digitized, enabling comparative analysis of tectonic structures in time and space using Geographic Information Systems (GIS) and other software (Tanaka et al., 1998; Dohm and Tanaka, 1999). Structure densities and orientations and spatial associations of structures with other geologic features are examples of what can be produced through GIS-based analysis of structural mapping (Dohm et al., 2001a).

Overall, planetary structural mapping provides unique insights into the deformational histories of planetary bodies. Structural mapping is integral to determining geologic histories at global to local scales, to evaluating geophysical, geochemical, and even in some cases climatic models relating to the time–space relations of the planet's interior, surface, and atmosphere (if present), to constraining surface rheology at various scales and through time, and to planning science objectives and designing experimental approaches for future planetary missions (e.g., Schulze-Makuch et al., 2007; Furfaro et al., 2007).

2 Mapping and dating structures with spacecraft data

2.1 Mapping with visible and infrared images

Images of planetary surfaces in visible and infrared wavelengths of adequate spatial resolution permit mapping and interpretation of surface landforms resulting from structural deformation (e.g., Avery and Berlin, 1992; Golombek, 1992; Lillesand and Kiefer, 1994; Bell et al., 1999). Data characteristics and quality are governed by planetary characteristics and the characteristics of the spacecraft camera systems. Following Lunar Orbiter photographs recorded on film in 1966–1967, Mariner, Viking, and Voyager spacecraft missions to Mars, Mercury, and the outer planetary satellites employed vidicon cameras that obtained digital images sensitive to light in the 350 to 650 nm wavelength range, with filter wheels providing broadband multi-spectral color imaging. Later, charge-coupled device (CCD) cameras, which expanded the sensitivity range to include near-ultraviolet to near-infrared wavelengths (200 to 1100 nm), have been used for missions to Mars, the Moon, the satellites of Jupiter, and asteroids (Mars Global Surveyor, Mars Odyssey, Mars Express, Clementine, Galileo, Giotto, and Near-Earth Asteroid Rendezvous).

Digital images are compiled as two-dimensional grids (i.e., raster arrays) composed of individual, square picture elements ("pixels"). Surface brightness is governed by the albedo of the surface (the ratio of reflected to incident surface energy within a defined wavelength range) and by illumination and surface geometries. Clouds can have a substantial albedo, thereby obscuring the underlying surface of planets with atmospheres. Images may be acquired in various geometries and illumination conditions. Some images, such as obtained by the Mars Orbiter Camera (MOC) and Thermal Emission Imaging System (THEMIS), primarily involved nadir (vertical) pointing of the camera, whereas other camera systems used extensive off-nadir pointing. Imaging campaigns that collect data at different seasons and times of day across the curving planetary surface capture variable directions of solar illumination and local surface changes. For Mars, seasonal haze and transient dust and CO_2 clouds may obscure surface viewing.

Higher resolution images provide correspondingly smaller areal coverage. In order to coherently map broad surfaces at highest resolutions, multiple images are combined into mosaics. Because lighting and atmospheric conditions and camera location, pointing, and color filter vary, photometric and geometric corrections are required. Spacecraft jitter and location uncertainties and camera noise effects also call for cosmetic improvements. However, some of these effects cannot be completely corrected through image processing techniques; for example, morning and afternoon lighting conditions will result in the reversal of hill-slope shading, which can result in confusing image mosaics. The more serious effects need to be recognized and documented to avoid mistaking them for surface features. In addition, geodetic control of the positioning of landforms varies according to available data and modeling. Thus, the mapped locations might include significant errors. This problem is especially evident in areas where the controls do not permit proper alignment of features between adjacent images in a mosaic.

Given the aforementioned limitations to image data, morphologic analysis can be performed where slope information can be gleaned. Both visual and thermal infrared images include intensity variations related to slope orientation vs. incident solar illumination, where sunlit slopes tend to be relatively bright (or warm in thermal infrared) and shaded (or cool) slopes tend to be dark. In addition, albedo variations of surface materials may highlight tectonic structures (e.g., structurally controlled deposits).

2.2 Mapping with radar images

Radar, radio detection and ranging – a form of active (i.e., it provides its own energy source) remote sensing – is able to operate generally independent of solar illumination or weather conditions. With the correct wavelength, radar "sees" through clouds or other media that might obstruct energy in visible light and other wavelengths. Radar works within the microwave band of the electromagnetic spectrum in the wavelength range from millimeters to 100 cm. Radar illuminates a terrain, detects return energy (radar return), and records an image. Three factors affect radar return: terrain slope (topography), terrain roughness (at the scale of the radar wavelength), and terrain electrical properties. The third factor, a function of composition and/or bulk density, is the hardest to quantify; we consider the first two in our brief discussion.

The military developed radar-imaging techniques in the 1950s. Geological applications began with image system declassification in the mid-to-late 1960s, and advanced in the 1970s with commercial radar image availability. Radar image analysis is largely similar to visible and infrared image analysis and primarily comprises "photogeologic" investigation of primary and secondary structures marked by topography or surface roughness. As such, geologic history interpretation must follow similar cautions (e.g., Hansen, 2000). Radar data are particularly useful for recognizing promontories or geomorphic features and linear ridges or troughs that represent fractures, faults, and folds as confirmed by field mapping. Radar data, acquired by orbiting spacecraft, can contribute to geological analysis of any solid planet surface. For example, NASA Magellan radar returned global coverage of breathtakingly detailed images of Venus' cloud-covered surface that form the basis of our understanding of Venus' tectonic and magmatic evolution (e.g., Solomon *et al.*, 1992; Ford *et al.*, 1993).

Although radar images resemble black-and-white photographs, radar's side-looking system requires a few simple considerations in interpretation. Oblique radar illumination produces strong signal returns and peaks that highlight surface roughness, which includes topographic highs (positive features), such as ridges, scarps, and promontories, and lows (negative features), which may include pits and impact craters, depending on shape, size (at radar resolution), and orientation with respect to radar illumination. Structures recognizable by their resulting landforms in radar images include: folds, grabens, faults, fractures, volcanic edifices and summit craters, impact craters, collapse depressions, and valley features such as Venusian valley networks (Komatsu *et al.*, 2001). Ground range images, a function of illumination direction and incidence angle and slope, form 2-D representations of 3-D surfaces. Radar artifacts, including radar shadow, foreshortening, and layover, can deter topographic interpretation or add geometric clues absent

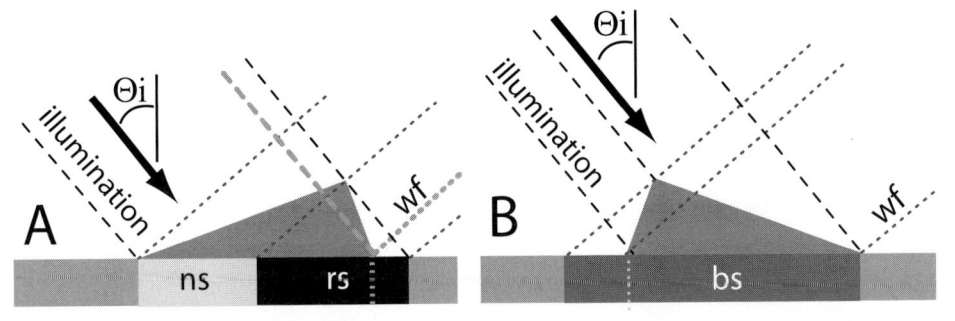

Figure 8.2. Radar ground-range images (basal strip) resulting from different topographic form, incidence angle (Θ_i), and illumination direction. Points on topographic form project parallel to wavefront (wf, perpendicular to illumination) to points on the ground-range image. Projected location and size of near slope (ns), back slope (bs) and radar shadow (rs) shown with shades of gray indicative of relative radar return and hence brightness. Gray lines show where surface locations would not be imaged. (A) Left illumination of asymmetric topographic form; foreshortening and radar shadow result in apparent symmetric shape. The point at the base of the near slope "projects" to its correct location; the high point projects forward of its true location; the base of the far slope "projects" to its correct location, but its presence is lost in radar shadow. The shallow near slope is imaged (ns), but it is "foreshortened" in the ground-range image; the entire far (steep) slope is lost in radar shadow. (B) Left illumination and an asymmetric topographic form with steep slope facing radar results in extreme foreshortening (called "layover"). Only the far slope is imaged because the high point projects forward, and the sharp break in slope is lost in extreme foreshortening, or layover. Although the near slope is lost to layover, none of the far slope is lost to radar shadow in this case.

in plan view imagery (Figure 8.2). Sharp breaks in slopes result in sharp changes in radar brightness in the ground range image, whereas gradual changes in slope yield gradual changes in radar brightness. Radar shadow, similar to visible light side-illumination shadowing, can hide breaks in slope. Foreshortening shortens the near slope or the slope facing radar illumination (Figure 8.2a). Layover, an extreme form of foreshortening, shortens the radar-facing slope so much that it disappears on the ground-range image, and only the back slope appears in the image (Figure 8.2b). Attention to radar illumination and wave-front geometry should yield consistent topographic interpretation among analysts, although geological interpretations of topography might be debated. Interpretation of material boundaries in radar imagery can be more controversial because radar backscatter could result from surface roughness (such as a pahoehoe flow versus aa flow surface) or from terrain electrical properties, or both. In addition, a single volcanic flow might change from pahoehoe to aa character resulting in a variable surface roughness that could yield a wavelength-dependent radar signature; yet the flow could have

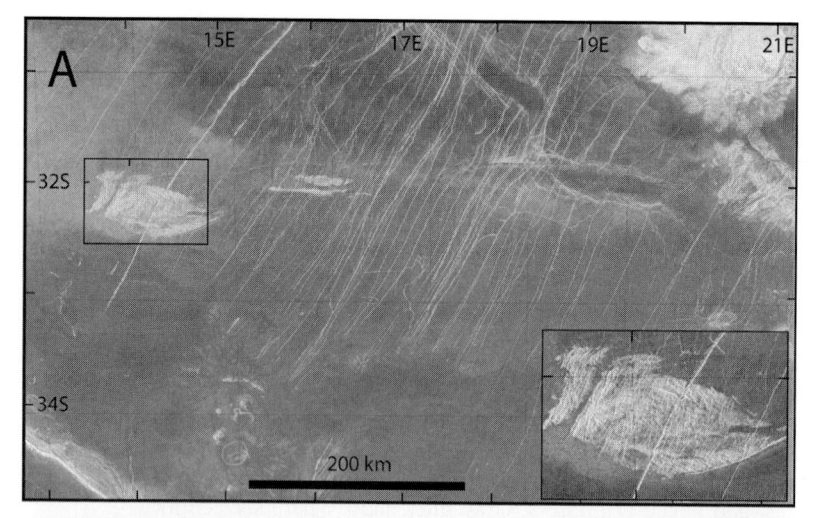

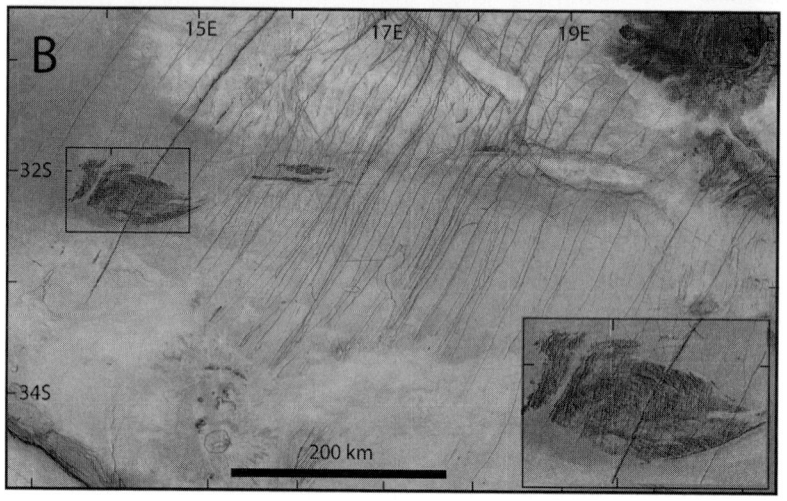

Figure 8.3. Comparison of left illumination SAR (A) and inverted SAR (B). Note impact crater in northeast corner of images and volcanic construct and northwest-trending fold belt in southwest corner. North–northeast trending fractures and pit chains are truncated, and thus post-dated, by flows in the southern sector. Insets show ribbon tessera terrain with west–northwest trending folds and north–northeast trending ribbon structures.

been emplaced as a single temporally equivalent geological unit. Thus, it is critical to consider radar interpretation separate from geological interpretation. Ford *et al.* (1993) provide an excellent review of radar interpretation with particular emphasis on Venus Magellan data.

Radar is exceptionally well suited for the identification of primary (emplacement related) or secondary (tectonic) structures that comprise 3-D topographic expression, while providing local to regional synoptic views of fracture, fault, and lineament patterns. Given that many secondary structures form lineaments, inverted or negative radar imagery commonly illustrates features more clearly because the human eye more easily differentiates dark lines on bright backgrounds rather than bright lines on dark backgrounds (Figure 8.3). Paired radar images with different incidence angles can be combined to construct stereo anaglyphs (e.g., Plaut, 1993), or radar data and complementary altimeter data can be combined to make synthetic stereo anaglyphs (Kirk *et al.*, 1992). Such data have been used with great success in Venus mapping (e.g., Hansen *et al.*, 1999, 2000).

The interpretation of geologic structures, material units, temporal relations among the structures and material units, and ultimately the geologic history using radar requires the same careful consideration that should be employed with any remote sensing dataset (e.g., Wilhelms, 1972, 1990; Tanaka *et al.*, 1994; Hansen, 2000). Radar data permit effective mapping of surface morphologies and structures (e.g., Ford *et al.*, 1989, 1993; Sabins, 1997). Geologic unit boundaries are, in many cases, clearly defined by radar backscatter intensity, as well as by crosscutting relations among materials and positive and negative features. Gradational contacts, or contacts not marked by radar distinctive features, can be difficult to trace. Termination of tectonic structures may represent the spatial limit of a deformation front, or termination might indicate the limit of younger on-lapping material. Structural facies boundaries represent a strain discontinuity related to a tectonic process, perhaps marked by a decrease in structural element density; younger burial is more likely if structural elements are abruptly truncated at a high angle to structural trend.

2.3 Mapping with topography

Topographic data also provide critical clues for tectonic landform analysis. Topographic data are critical to the proper characterization of the morphology and structural relief of tectonic landforms, which is required to identify the structure, to provide estimates of strain, and to formulate and test kinematic and mechanical models of the candidate structures (e.g., Schultz and Zuber, 1994; Schultz, 2000a; Okubo and Schultz, 2004, 2006). In addition, topographic data have revealed the presence and spatial extent of structures that are not readily detected in images.

Topographic data for tectonic landforms have been derived from images using a number of techniques. The most basic of these is shadow measurements. High spatial resolution topographic data can be obtained from single images using quantitative modeling of slope variations, or photoclinometry (see Hapke *et al.*, 1975; Davis and Soderblom, 1984; Kirk *et al.*, 2003). On Mars, photoclinometry has been applied to Viking, MOC, and other images for landforms in areas or conditions where albedo variations are minimal; relative errors in this technique can be below 5% (Davis and Soderblom, 1984; Tanaka and Davis, 1988). Another important means of obtaining high-resolution topographic data is from stereo imaging. Using digital stereo methods that employ automated and semi-automated matching algorithms, it is possible to obtain DEMs with spatial resolution and height accuracy limited only by the resolution and signal-to-noise ratio of the images (see Watters *et al.*, 1998; Cook *et al.*, 2000; Schenk and Bussey, 2004). These methods have also been used to derive topography from stereo radar images (see Herrick and Sharpton, 2000).

Topographic data for the terrestrial planets have also been obtained from Earth-based radar altimetry (see Downs *et al.*, 1982; Harmon *et al.*, 1986) and radar interferometry (Margot *et al.*, 1999). Radar interferometry compares two sets of radar altimetry data that yield differences in signal phase caused by elevation differences. Profiles derived from Earth-based radar altimetry have sufficient along-track spatial and vertical resolution to characterize individual tectonic features (see Watters and Nimmo, Chapter 2). These datasets provide regional context for more detailed topographic investigations of smaller-scale (local) structures, but have been largely superseded by measurement of planetary topography at regional and global scales by orbital spacecraft (such as for the Moon and Mars).

Global topographic data have been returned from radar and laser altimeters on orbiting spacecraft for Venus, the Moon, Mars, and asteroid 433 Eros (see Pettengill *et al.*, 1980; Ford and Pettengill, 1992; Zuber *et al.*, 1994; Smith *et al.*, 1999; Zuber *et al.*, 2000). For Venus and the Moon, current global topography datasets have high vertical but low spatial resolution. Although these data are important in determining the long-wavelength (i.e., regional) topography of deformed planetary surfaces, they generally lack the spatial resolution to resolve individual structural features. However, radar data can provide critical topographic constraints (e.g., Hansen, 2006). The Mars Orbiter Laser Altimeter (MOLA), in contrast, has returned some of the highest vertical and spatial (460 m to 115 m/pixel near the poles) resolution interpolated digital elevation models for any of the terrestrial planets thus far (Smith *et al.*, 1999, 2001; Zuber *et al.*, 2000; Okubo *et al.*, 2004). The high-resolution MOLA data were used to study Martian tectonic structures in unprecedented detail (e.g., Golombek *et al.*, 2001; Schultz and Lin, 2001; Schultz and Watters, 2001; Wilkins and Schultz, 2003; Okubo and Schultz, 2003, 2004, 2006; Schultz *et al.*, 2006; Golombek and Phillips, Chapter 4). These data have also revealed a previously undetected population of subdued wrinkle ridges in the northern lowlands of Mars, partly buried by sedimentary material (Withers and Neumann, 2001; Head *et al.*, 2002) or formed in weak materials (Tanaka *et al.*, 2003).

One of the challenges to mapping tectonic features using images and image mosaics is illumination bias introduced by the lighting geometry. Low-relief features in particular can be easily missed if the illumination direction is not favorable. Shaded-relief maps generated from DEMs with sufficient vertical and spatial resolution can be used to overcome illumination bias, because any illumination direction can be specified (Plate 23).

3 Structure mapping of planetary bodies

3.1 Moon

Important structures on Earth's Moon fall into three categories: impact craters, wrinkle ridges, and grabens (formerly called straight rilles; e.g., McGill, 1971; Masursky *et al.*, 1978; Golombek, 1979). Most of the extensional and contractional structures formed on the Moon were produced in association with impact basins (multiringed structures >300 km diameter on the Moon; e.g., Melosh, 1976, 1978, 1989; Solomon and Head, 1979, 1980; Golombek and McGill, 1983; Watters and Johnson, Chapter 4). Because of this relationship, wrinkle ridges and grabens received much attention in the earliest planetary structural mapping efforts (e.g., Schultz, 1976; Masursky *et al.*, 1978).

Until recent decades most geologists did not consider impact cratering an important geologic process. Terrestrial geologic experience indicated that sub-circular craters surrounded by material derived from the crater were volcanoes, and thus almost all geologists initially inferred that lunar craters were also volcanic. A notable early exception was Gilbert (1893), who recognized the importance of impact cratering in the evolution of the lunar crust. He also argued that the history of the Moon could be deciphered using techniques that are philosophically similar to those used by geologists studying the history of Earth. A major breakthrough in understanding was provided by the work of Baldwin (1963), who established a quantitative morphological continuum from bomb craters to lunar craters. By the mid 1960s, planetary scientists, with a small number of exceptions, considered this issue settled. The increasing realization that impact craters are not rare on Earth, and the analysis of returned lunar samples, finally ended this controversy.

Widespread ejecta deposits from large multiringed basins provide us with a relative timescale through use of classical superposition relations – younger ejecta deposits rest on older ejecta deposits. The lunar time–rock classification system is primarily based on the stratigraphy of basin deposits (Shoemaker and Hackman, 1962; Wilhelms, 1987). The interiors of many large multiringed basins are filled with relatively smooth, low-albedo material (the lunar maria), leading in the past to the concept that lunar maria is impact melt related directly to basin formation (e.g., Urey, 1952). But stratigraphic and crater-density relations indicate that the lunar maria are significantly younger than the impact basins that contain them (Figure 8.4). Indeed, returned samples demonstrate that the time gap between formation of lunar basins and their flooding by maria material spans hundreds of millions of years (e.g., Wilhelms, 1987). Within this stratigraphic framework, the formational ages of grabens and wrinkle ridges can be constrained by crosscutting and burial relations to outcrops of highland and maria materials.

Figure 8.4. Southeastern margin of Mare Imbrium. Montes Apenninus (MA) is a portion of the Imbrian basin rim; all features interior to this rim must be younger than the basin impact event. AB, Apennine Bench Formation; m, mare basalts; A, crater Archimedes (83 km in diameter); RB, Rima Bradley, an arcuate rille (graben). The Apennine Bench Formation is interpreted to be the oldest post-impact material; Archimedes ejecta and secondary craters are superposed on it, and Rima Bradley cuts it. Archimedes crater and Rima Bradley are, in turn, superposed by dark mare basalts. This sequence demonstrates that the mare basalts cannot be directly due to the Imbrian impact event (Wilhelms, 1987). Lunar Orbiter image IV-109-H3; center at 13.8°N, 3.6°W; north toward top.

Lunar wrinkle ridges generally consist of a broad, low arch from a few to a few tens of kilometers across, surmounted by a narrower, sinuous ridge that may be either symmetrically or asymmetrically placed on the arch (Figure 8.5). Hypotheses for the origin of these landforms fall into two categories: magmatic or tectonic (e.g., Maxwell *et al.*, 1975). Abundant evidence from the Moon and other planets favors a tectonic origin, with wrinkle ridges representing anticlines formed as the lunar crust warped upward due to displacement on a subsurface thrust fault (e.g., Howard and Muehlberger, 1973; Maxwell *et al.*, 1975; Schultz, 1976; Lucchitta, 1976, 1977; Plescia and Golombek, 1986; Sharpton and Head, 1988; Watters, 1988; Golombek *et al.*, 1991; Schultz, 2000a). Most lunar wrinkle ridges occur in lunar maria, a likely testament to mechanical layering of deformed materials. Some wrinkle ridges appear to be due to differential compaction over buried crater or basin rims; others most likely formed as a result of basin floor subsidence following extrusion of lunar maria, and thus reflect folding and faulting of the maria material. The common presence of wrinkle ridges within basins indicates that basin subsidence and deformation followed maria emplacement.

Rilles form narrow troughs that range up to hundreds of kilometers in length. Two types of rilles occur: sinuous (Figure 8.5) and straight to arcuate (Figure 8.4). Sinuous rilles generally begin at pits or other putative volcanic features, and thus

Figure 8.5. Interior of Oceanus Procellarum. Northwest of crater Nielsen (N) are typical wrinkle ridges (wr). The diameter of crater Nielsen is 10 km. Rimae Aristarchus (RA) and Vallis Schröteri (VS) are typical sinuous rilles, and are either collapsed lava tubes or lava channels. The material cut by the rilles, part of the Aristarchus plateau, is embayed by the younger mare basalts of Oceanus Procellarum. Lunar Orbiter image IV-158-H1; center at 42.0°N, 47.7°W; north toward top.

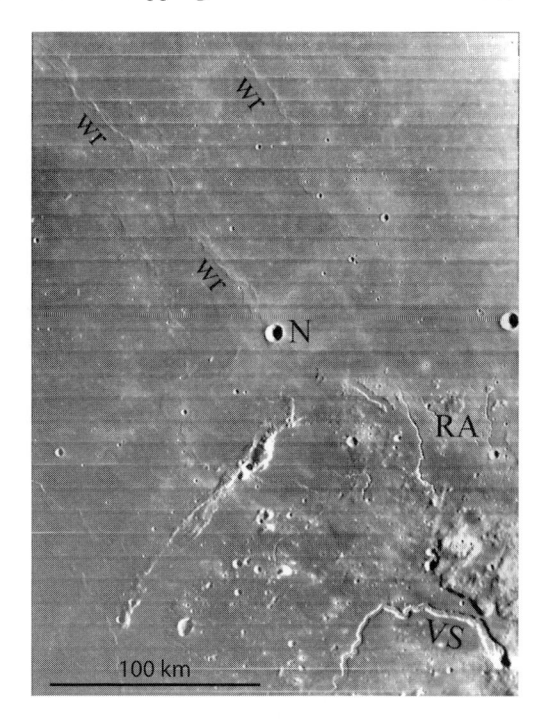

are believed to be analogous to terrestrial lava channels or collapsed lava tubes. The inferred volcanic source feature for a sinuous rille would be the same age as, or younger than, the materials crossed by the rille. Straight and arcuate rilles are grabens, as indicated by changes in rille width as it transects older topographic features, such as crater rims (McGill, 1971). These grabens commonly occur outside of and concentric to basins, and can be related to the same basin subsidence implied by intrabasin wrinkle ridges (e.g., Golombek and McGill, 1983). The three-dimensional geometry of straight and arcuate rilles has been used to estimate the thickness of the layer of impact ejecta (the megaregolith) that overlies local basement (Golombek, 1979; see Schultz et al., 2007, for a recent reanalysis and discussion).

A less common lunar tectonic landform are thrust fault scarps (Lucchitta, 1976; Howard and Muehlberger, 1973; Binder, 1982; Binder and Gunga, 1985). These scarps are often lobate and segmented, analogous in morphology to planetary lobate scarps (Watters and Johnson, Chapter 4). In contrast to the lobate scarps on Mercury and Mars (see Watters and Nimmo, Chapter 2; Golombek and Phillips, Chapter 5), lunar lobate scarps are much smaller scale structures (Binder and Gunga, 1985; Watters and Johnson, Chapter 4). Although lobate scarps on Mercury and Mars can have over a kilometer of relief, lunar scarps have a maximum relief of much less than 100 meters (Lucchitta, 1976; Howard and Muehlberger, 1973; Binder, 1982; Binder and Gunga, 1985; Watters and Johnson, Chapter 4), indicating lower local contractional strain accumulations. The lengths of the scarps are also small, with the longest segments reaching a maximum of only ~20 km (Binder and Gunga, 1985). In spite of their scale, lunar scarps are interpreted to be the result of thrust faulting (Lucchitta, 1976; Howard and Muehlberger, 1973; Binder, 1982; Binder and Gunga, 1985; Watters and Johnson, Chapter 4). Although the evidence of offset is not as dramatic as in the case of large-scale lobate scarps on Mercury and Mars, given the smaller surface gravity for the Moon that, in part, regulates the magnitude of fault offset (Schultz *et al.*, 2006), the morphology and the linkage between individual segments of the lunar scarps supports the interpretation that they are the surface expression of shallow thrust faults (Watters and Johnson, Chapter 4). The known lunar lobate scarps, as well as the mare ridges and grabens, have been mapped by Watters and Johnson (Chapter 4).

3.2 Mars

Although it has been more than 30 years since the Viking Orbiters imaged the surface of Mars, Mars' tectonic history continues to be a field of lively debate. Mars is dominated by two volcanic regions: the Tharsis magmatic complex of the western hemisphere (Dohm *et al.*, 2007a) and the Elysium rise of the eastern hemisphere (see Golombek and Phillips, Chapter 5). Large impact basins such as Utopia (McGill, 1989) and Hellas and the northern lowlands have also contributed to surface deformation. Tharsis is an enormous, high-standing region (roughly 25% of the surface area of the planet) capped by the solar system's largest shield volcanoes (e.g., Tanaka *et al.*, 1991; Banerdt *et al.*, 1992). This magmatic complex is the largest single tectonic and volcanic province on the terrestrial planets, with a rich history of geologic and tectonic activity that lasted throughout most of Martian geologic time (e.g., Solomon and Head, 1982; Dohm *et al.*, 2001b, 2007a; Phillips *et al.*, 2001).

As in the previous lunar work, stratigraphic mapping of regionally extensive terrains on Mars provides the context for defining the relative ages of major deformational events (e.g., Scott and Tanaka, 1986; Greeley and Guest, 1987; Tanaka, 1986, 1990; Tanaka *et al.*, 1991, 1992; Scott and Dohm, 1990, 1997; Dohm and Tanaka, 1999; Dohm *et al.*, 2001a,b; Anderson *et al.*, 2001; Knapmeyer *et al.*, 2006). In contrast to the Moon, however, internal processes have largely dominated the production of surface-breaking structures on Mars, with impact-basin tectonics playing a subsidiary role (e.g., Phillips *et al.*, 2001). Formation of the Martian crustal dichotomy (Watters *et al.*, 2007) and Tharsis occurred very early in the planet's history (e.g., Anderson *et al.*, 2001; Dohm *et al.*, 2001b, 2007a; Phillips *et al.*, 2001; Frey *et al.*, 2002; Nimmo and Tanaka, 2005; Frey, 2006), yet the structures from those Noachian times, as much as 4 Ga ago (Frey, 2006), are well preserved on Mars, in part due to the thin atmosphere that promotes slow rates of erosion and deposition relative to the Earth (e.g., Schultz, 1999) and in part due to the general lack of subsequent deformational events. On the other hand, other works indicate that the dichotomy could have evolved for a more extended period, possibly related to plate tectonics (Fairén and Dohm, 2004; Baker *et al.*, 2007). As a result, Martian structures have been used extensively to construct and test models for the internal evolution of the planet (e.g., Wise *et al.*, 1979; Banerdt *et al.*, 1982, 1992; Tanaka *et al.*, 1991; Márquez *et al.*, 2004; Dimitrova *et al.*, 2006; Baker *et al.*, 2007; Golombek and Phillips, Chapter 5).

The most common types of structures on Mars – grabens, thrust faults, and wrinkle ridges – demonstrate the predominance of extensional and contractional deformation, respectively, of the planet's crust or lithosphere (Plate 24; see Tanaka *et al.*, 1991; Scott and Dohm, 1997; Dohm and Tanaka, 1999; Schultz, 2000a; Dohm *et al.*, 2001a,b; Knapmeyer *et al.*, 2006; Schultz *et al.*, 2007; Golombek and Phillips, Chapter 5). However, Mars also displays clear evidence of strike-slip faulting (Forsythe and Zimbelman, 1988; Schultz, 1989; Okubo and Schultz, 2006). Recent work using high-resolution MOLA topography has identified the surface expression of igneous dikes in Tharsis (Schultz *et al.*, 2004; Goudy and Schultz, 2005).

The formation of Tharsis produced a vast system of grabens and wrinkle ridges that span the entire western hemisphere (Anderson *et al.*, 2001; Montesi and Zuber, 2003a,b). In contrast, grabens primarily accompanied the formation of Elysium (Carr, 1974; Wise *et al.*, 1979; Plescia and Saunders, 1982; Tanaka and Davis, 1988; Tanaka *et al.*, 1991; Banerdt *et al.*, 1992). Many of the grabens and wrinkle ridges associated with these regions, along with normal faults, dikes, and rifts, display a geometric relationship between the center of the magmatic-tectonic regions and the structures (Schultz, 1985; Watters and Maxwell, 1986; Tanaka *et al.*, 1991; Watters, 1993; Anderson *et al.*, 2001; Phillips *et al.*, 2001; Schultz *et al.*, 2006). Most of the structures identified within the Tharsis and Elysium regions are grabens (e.g., Tanaka and Davis, 1988; Tanaka *et al.*, 1991; Davis *et al.*, 1995; Anderson *et al.*, 2001). In places, grabens are associated with pit crater chains (e.g., Schultz, 1991; Okubo and Schultz, 2003; Wyrick *et al.*, 2004; Ferrill *et al.*, 2004; Goudy and Schultz, 2005). Rift-sized grabens, primarily located in the Tharsis region, have widths that generally range between 10 and 100 km and depths up to a few kilometers (e.g., Plate 25). Such large grabens are characterized by multiply faulted borders and floors (e.g., Plescia and Saunders, 1982; Dohm and Tanaka, 1999; Wilkins *et al.*, 2002; Wilkins and Schultz, 2003) and resemble complex terrestrial rift systems (Schultz, 1991, 1995; Banerdt *et al.*, 1992; Hauber and Kronberg, 2005). Other extensional structures include the Valles Marineris troughs (e.g., Blasius *et al.*, 1977; Lucchitta *et al.*, 1992; Mège and Masson, 1996; Schultz, 1991, 1995, 1998, 2000b; Wilkins and Schultz, 2003), structurally

Figure 8.6. Mariner 10 mosaic of Hero Rupes. Hero Rupes (58°S, 173°W) is one of the many large-scale lobate scarp thrust faults in the portion of the southern hemisphere imaged by Mariner 10.

controlled sapping channels (e.g., Davis *et al.*, 1995), and troughs of polygonal patterned ground (e.g., Pechmann, 1980; McGill, 1986; McGill and Hills, 1992; Buczkowski and McGill, 2002). Wrinkle ridges occur across the Martian surface (e.g., Chicarro *et al.*, 1985; Watters and Maxwell, 1986; Watters, 1988; Schultz, 2000a; Goudy *et al.*, 2005) and formed in association with impact basins, as on the Moon, and volcanotectonic provinces such as Tharsis, Thaumasia, and Elysium.

3.3 Mercury

In 1974 and 1975, the Mariner 10 spacecraft made three flybys of Mercury, returning over 2700 images of the eastern hemisphere, covering about 45% of the planet's surface (Strom, 1984). The resolution of images varied greatly from 100 to 4000 m/pixel. The three encounters of Mariner 10 occurred when the same hemisphere was illuminated. With the subsolar point located at about 0°N, 100°W, much of the imaged hemisphere had poor illumination geometry (near-nadir solar incidence) for the identification of tectonic landforms and morphologic features. In early 2008, MESSENGER became only the second spacecraft to visit Mercury. In the first of three flybys that will lead to Mercury orbit in 2011, MESSENGER imaged about 21% of the hemisphere unseen by Mariner 10 (Solomon *et al.*, 2008).

Landforms indicative of crustal shortening and extension are clearly evident on Mercury (Strom *et al.*, 1975; Melosh and McKinnon, 1988; Watters *et al.*, 2004;

Watters and Nimmo, Chapter 2). Lobate scarps, interpreted to be thrust faults, are the most widely distributed tectonic feature (Figure 8.6; Strom *et al.*, 1975; Melosh and McKinnon, 1988; Watters *et al.*, 1998, 2004; Solomon *et al.*, 2008). These faults occur throughout the imaged regions and deform the oldest intercrater plains and the youngest smooth plains (Watters and Nimmo, Chapter 2). A rare, related tectonic feature on Mercury are high-relief ridges. The maximum relief of high-relief ridges can exceed 1 km, and they appear to be spatially and temporally associated with lobate scarps. Deforming the same units as lobate scarps, in some cases, high-relief ridges transition into lobate scarps (Watters *et al.*, 2004; Watters and Nimmo, Chapter 2).

Wrinkle ridges are another common tectonic landform on Mercury, although they are not as broadly distributed as lobate scarps. Wrinkle ridges occur predominantly in smooth plains in the interior of the Caloris basin and in the smooth plains exterior to the basin (Strom *et al.*, 1975; Melosh and McKinnon, 1988; Watters *et al.*, 2005; Watters and Nimmo, Chapter 2).

Surprisingly, Mercury displays few extensional landforms in the imaged regions. Evidence of widespread extension is only found in the interior plains materials of the Caloris basin (Strom *et al.*, 1975; Melosh and McKinnon, 1988; Watters *et al.*, 2005; Murchie *et al.*, 2008; Watters and Nimmo, Chapter 2). Basin-radial and basin-concentric grabens form a remarkably complex pattern of extension (Murchie *et al.*, 2008; Watters and Nimmo, Chapter 2). This network of grabens crosscuts the wrinkle ridges in the Caloris basin (Strom *et al.*, 1975; Melosh and McKinnon, 1988; Watters *et al.*, 2005; Murchie *et al.*, 2008; Watters and Nimmo, Chapter 2).

Post Mariner 10 models for the origin of the tectonic stresses on Mercury involve global contraction due to secular cooling of the interior, tidal despinning, a combination of global contraction and tidal despinning, or a combination of global thermal contraction and the formation of the Caloris basin (Strom *et al.*, 1975; Cordell and Strom, 1977; Melosh and Dzurisin, 1978a,b; Pechmann and Melosh, 1979; Melosh and McKinnon, 1988; Thomas *et al.*, 1988; Thomas, 1997). These models predict distinctive patterns of tectonic features. Global contraction from slow thermal cooling results in global, horizontally isotropic compression (Solomon, 1976, 1977, 1978, 1979; Schubert *et al.*, 1988; Phillips and Solomon, 1997; Hauck *et al.*, 2004), predicting uniformly distributed, randomly oriented thrust faults. Tidal despinning predicts N–S oriented thrust faults in the equatorial zone and E–W normal faulting in the polar regions (Melosh and Dzurisin, 1978a,b; Melosh and McKinnon, 1988). Stresses related to the formation of the Caloris basin might result in Caloris-radial thrust faults (Thomas *et al.*, 1988; Thomas, 1997). Recent modeling suggests that mantle convection may be another important source of stress on Mercury (King, 2008). The spatial and temporal distribution of tectonic features is thus critical to constraining existing models for the origin of the stresses and future models that may emerge when the MESSENGER mission completes a global survey of Mercury (Solomon *et al.*, 2001, 2007).

Early efforts to map Mercury's tectonic features were based largely on the analysis of individual Mariner 10 image frames and hand-lain mosaics (see Strom *et al.*, 1975; Strom, 1984). Tectonic features were also mapped as part of the 1:5 000 000 geologic map series of Mercury (e.g., Schaber and McCauley, 1980) based on a series of shaded relief maps at the same scale (Davies and Batson, 1975; Davies *et al.*, 1978). Recent efforts to map tectonic features imaged by Mariner 10 involve digitization directly from image mosaics with improved radiometry and geometric rectification (Robinson *et al.*, 1999; Watters and Nimmo, Chapter 2). The less than ideal lighting geometry over much of the hemisphere imaged by Mariner 10 (incidence angles <45°) introduces an observational bias that must be considered (Watters *et al.*, 2004). In some areas imaged by Mariner 10, the limitations of the poor illumination geometry can be overcome by the availability of stereo coverage. Topography generated from stereo pairs helps to reveal tectonic landforms not easily detected in Mariner 10 images (see Watters *et al.*, 2001, 2002). The importance of illumination geometry in identifying tectonic landforms has been demonstrated by MESSENGER where previously undetected lobate scarps have been found near the Mariner 10 subsolar point (Solomon *et al.*, 2008; Watters and Nimmo, Chapter 2).

A tectonic map of lobate scarps imaged by Mariner 10 shows that they are unevenly distributed with preferred orientations, east-trending compressional tectonic features in the polar regions, and no dominant Caloris-radial pattern of lobate scarps (see Watters and Nimmo, Chapter 2). This suggests that some models for the global tectonic stresses do not fully account for the spatial distribution of the known tectonic landforms (see also Watters and Nimmo, Chapter 2, for a review of Mercury tectonics).

3.4 Venus

Venusian tectonic structures range from broad lowland basins to narrow linear features (e.g., Solomon *et al.*, 1992). The major global datasets for mapping structures include NASA Magellan mission high-resolution gravity, altimetry, and synthetic aperture radar (SAR) data. Given the long wavelength of gravity (resolves features >400 km) and altimetry (~20 km footprint) data, SAR imagery (~100 m/pixel) proves most useful for structural element identification. The morphology of Venusian structures is more readily visualized using inverted SAR images (Figures 8.7–8.10).

Regional geomorphic groups include lowland basins, linear deformation belts (~20–150 km wide, 1000's of km long) in either parallel or polygonal distribution,

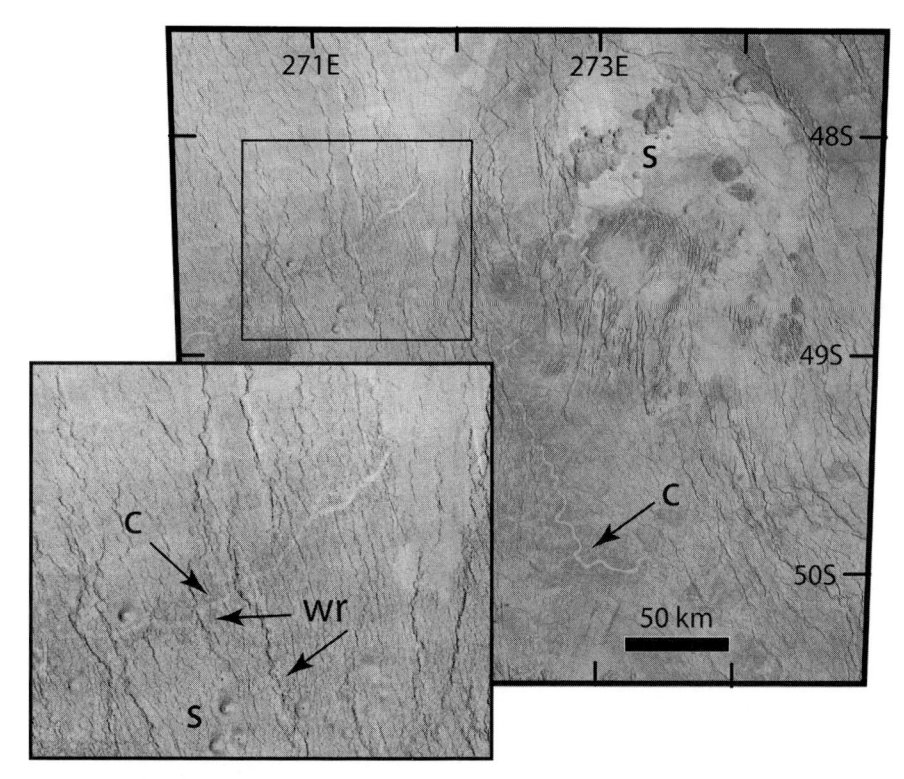

Figure 8.7. Shields and associated lava flows (e.g., Guest *et al.*, 1992; Addington, 2001; Hansen, 2005) in part of Helen Planitia variably bury kipukas of relatively high-standing deformed terrain in the NE corner. Variably spaced NNW-trending wrinkle ridges transect the area. Wrinkle ridges are less developed where shields and kipukas outcrop, either because shield lavas partly buried the wrinkle ridges or because mechanical differences between the host rocks favored wrinkle-ridge formation in the adjacent, lowland plains over the shield and kipuka terrains. Changes in wrinkle-ridge spacing could also reflect local differences in rheology or surface history; for example: (1) the area could have been deformed by closely spaced wrinkle ridges, then locally covered by a thin surface layer (NE corner of inset), followed by formation of widely spaced wrinkle ridges across the region; (2) closely and widely spaced wrinkle ridges could have formed synchronously, followed by local emplacement of a surface layer that covered closely spaced wrinkle ridges in the NE corner of the inset; and (3) surface flows of various layer thickness and coverage could have extended variably across the inset region; later formed wrinkle ridges (short and longer wavelength) could have formed generally synchronously, with wavelength reflecting the variable thickness (and/or strength) of individual surface units. Each of these histories can be accommodated within the context of the image data, and each should be considered as equally viable possibilities. Two channel segments (c) could represent a single channel (joining beneath the shield terrain) or two separate channels. Careful examination (see inset) indicates that a range of temporal relations among channel, shield (s) and wrinkle ridge (wr) formation are preserved. (Inverted Magellan SAR image mosaic; 75 m/pixel.)

372 *Planetary Tectonics*

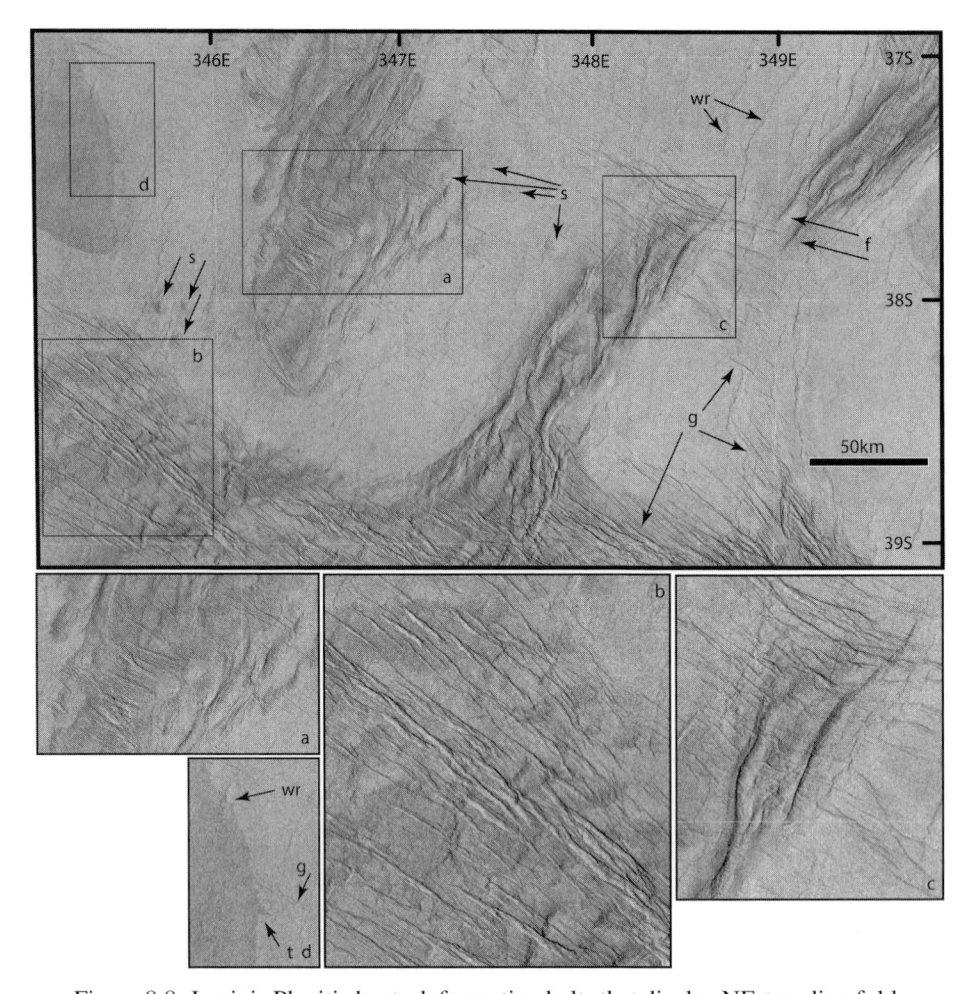

Figure 8.8. Lavinia Planitia hosts deformation belts that display NE-trending folds and NW-trending grabens. In the regions between deformation belts, wrinkle ridges parallel fold trends and grabens (or extension fractures) parallel graben trends in deformation belts. Note the gradational change in radar backscatter across fold crests and wrinkle ridges. Local radar backscatter boundaries show spatial relations with folds (a) and grabens (d) that indicate low-viscosity material embayed preexisting structural topography; elsewhere (b) radar backscatter boundaries show no spatial correlation with structural elements, indicating that backscatter boundaries pre-date fold and graben formation. In some cases, the change in radar backscatter is likely a function of structure topography and not related to different surface layers (e.g., c). Thus across this region, evidence for different surface histories can be gleaned. Although relatively clear embayment relations in the western part of inset a indicate that a low-viscosity surface layer was emplaced following both fold and graben formation, there is no evidence for material embayment following tectonism at location c. Here the distribution of folds and grabens appears to be a function of original strain partitioning, with strain intensity decreasing away from the deformation belt. At location d the embayed grabens presumably formed prior to the western surface unit, which predated local wrinkle-ridge formation. (Inverted Magellan SAR image mosaic; 75 m/pixel.)

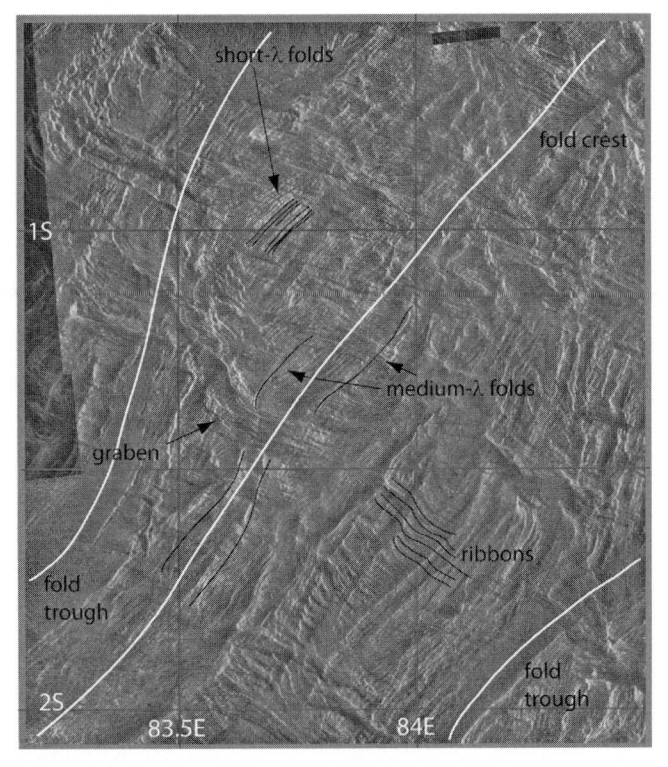

Figure 8.9. Part of Ovda Regio showing typical tessera terrain structures, including grabens, ribbons, and variable wavelength folds. White lines indicate troughs and crest of long-wavelength folds. See text for discussion. (Inverted Magellan SAR image mosaic; 75 m/pixel.)

large (~1000–2500 km diameter) quasi-circular domes (volcanic rises) and plateaus (crustal plateaus), linear troughs (up to ~400 km wide, 1000's of km long) called chasmata, and ~500 quasi-circular features (60–600 km diameter; Stofan et al., 1992, 2001) called coronae. Each of the regional geomorphic groups preserves a combination of individual structural elements. Smaller scale geomorphic features include ~970 impact craters (~1–280 km diameter; Phillips et al., 1992; Schaber et al., 1992; Herrick et al., 1997), medium to small volcanoes, relatively rare pancake domes, and tens to hundreds of thousands of 1–5-km diameter shields (Guest et al., 1992; Crumpler et al., 1997).

SAR's sensitivity to topography makes it an excellent tool for identifying primary landforms (i.e., intrinsic features of geologic map units) such as shield volcanoes, channels, levees, lava flow fronts, pit chains, and impact craters, as well as tectonic structures such as fault scarps, fold ridges, wrinkle ridges, and grabens (Figures 8.7–8.10; for additional discussion, see Ford et al., 1993). The relief of most of the smaller deformational structures is poorly expressed in SAR images, requiring

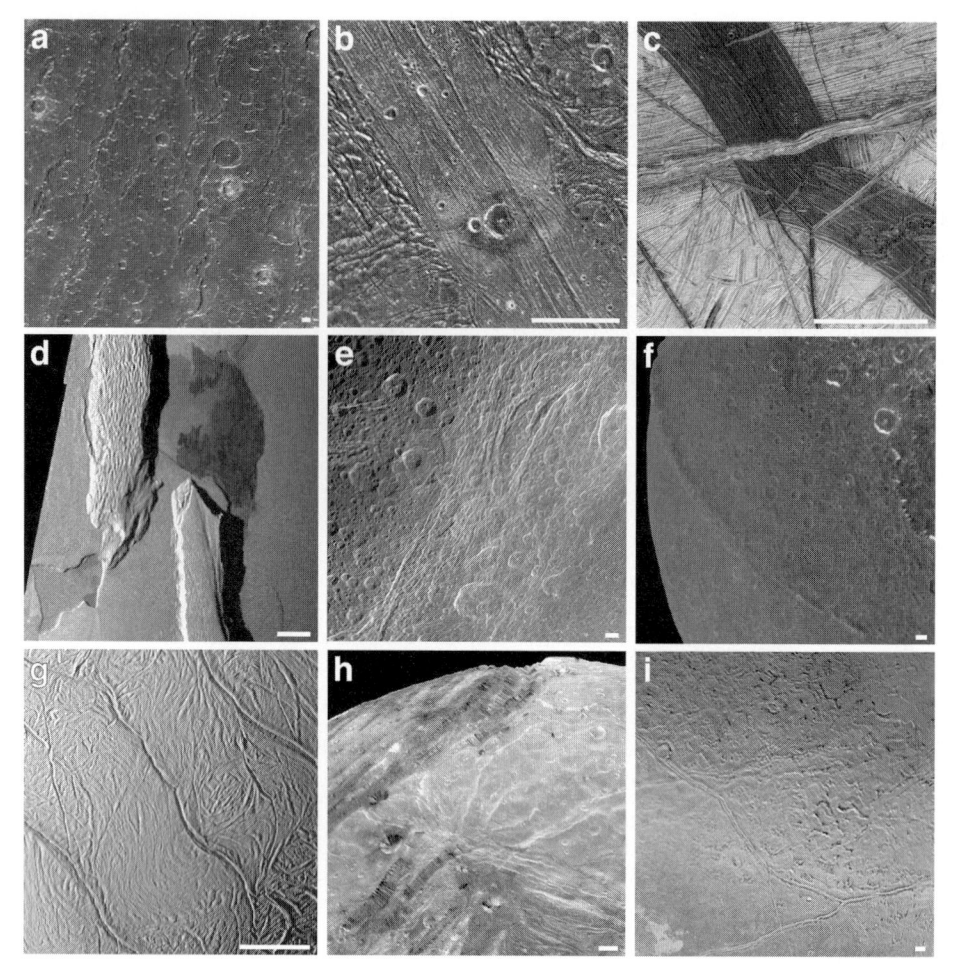

Figure 8.10. A montage of structures on several of the large and mid-sized outer planet satellites. Scale bars are each 20 km long. (a) Callisto: furrows surround the multiringed structure Valhalla. (b) Ganymede: grooved terrain of Byblus Sulcus is straddled by parallel and sub-orthogonal furrows in dark terrain. (c) Europa: the dark band Yelland Linea cuts older ridged plains and was subsequently overprinted by a prominent double ridge. (d) Io: mountains ~3.5 km tall adjacent to the dark-floored caldera Hi'iaka Patera. (e) Dione: fault scarps reveal brighter subsurface ice to create the satellite's wispy terrain. (f) Iapetus: a 20-km high ridge runs along the satellite's equator. (g) Enceladus: prominent fractures that cut across the satellite's south polar region are the sites of active venting. (h) Miranda: Inverness Corona (foreground) and Arden Corona (limb) are the sites of normal faulting, likely triggered by interior upwelling. (i) Triton: graben-like troughs in a region transitional from smooth terrain, which shows evidence for cryovolcanism, to the pitted "cantaloupe terrain," which may have formed by diapirism.

that their morphologic analysis relies heavily on their plan form. However, there is a rich record of their forms, orientations, distributions, and interactions with various features that assist with determination of their relative ages. In some cases, features display evidence of multiple episodes of activity. Figures 8.7 and 8.8 demonstrate this varied record and how multiple interpretations are commonly possible regarding relative ages of structures, and what mechanisms and factors control structural characteristics such as form, orientation, and spacing. Broad generalizations about their origin and age therefore cannot be made with certainty.

Tessera or tessera terrain (Figure 8.9) was originally called parquet terrain (in reference to its reticulate structural patterns). Tessera terrain, preserved in highland crustal plateaus and lowland regions and widely interpreted as collapsed plateaus (e.g., Ivanov and Head, 1996; Phillips and Hansen, 1994), consists of folds of multiple wavelengths, grabens, and variably flooded regions (Hansen and Willis 1996, 1998; Ghent and Hansen, 1999; Hansen, 2006). Discernible fold sets of common orientation have characteristic wavelengths that range from 0.3 km, the limit of effective SAR resolution, to ~100 km. Extensional structures, represented by long-aspect-ratio ribbon (steep-sided, graben-like) structures (Hansen and Willis, 1998) and shorter aspect ratio grabens, generally trend normal to fold crests. Cross-cutting relations, as well as mechanical arguments, illustrate that contractional and extensional structures formed broadly synchronously with the evolution of progressively longer wavelength structures with time (Hansen and Willis, 1998; Ghent and Hansen, 1999; Hansen *et al.*, 2000; Banks and Hansen, 2000; Hansen, 2006). Low-viscosity material, presumably lava, fills local structural lows; lava fill occurs both within troughs of long-wavelength folds and within structural lows along the crests and limbs of long-wavelength folds, indicating that flooding occurred throughout the development of the structural fabric (Hansen, 2006). Fold wavelength records progressively smaller amounts of shortening with a 0.3-km fold wavelength recording minimum shortening of ~30–40%, whereas 50- to 100-km wide folds record <1% layer shortening, indicating progressive fold formation of short- to long-wavelength folds with increasing layer thickness above a low-viscosity material, presumably lava (Hansen, 2006). For an alternative view of tessera terrain evolution, see Gilmore *et al.* (1998).

Ribbons (Figure 8.9) comprise alternating parallel ridges and troughs. Ribbons occur as suites of (listric?) normal faults with ramped trough terminations, and tensile fracture ribbons that show V-shaped terminations and matching opposing trough walls (Hansen and Willis, 1998).

Considerable progress has been made in mapping structures on Venus. Deformed terrains characterized by tessera, fold belts, fracture zones, and coronae have been delineated globally (e.g., Stofan *et al.*, 1992; Price and Suppe, 1995; Ivanov and Head, 1996). Although globally averaged crater statistics indicate that the mean crater densities of these terrains may vary, they provide little constraint for dating structures. Such inferences rely on crater statistics of the surfaces that host particular structures rather than the structures themselves; untestable assumptions include (1) the host surface was formed relatively quickly, and (2) the structures formed all at once, and immediately following the host surface. Therefore, the uncertainties in the crater statistics are large, permitting a wide range of plausible histories (Tanaka *et al.*, 1997; Campbell, 1999). The general paucity of reliable, spatially broad relative-age indicators along with complex geologic activity across the Venusian surface prevent construction of detailed structural histories at regional scales (e.g., Young and Hansen, 2003), as has been done for other planets and satellites such as the Moon and Mars. Thus resolving whether or not Venus underwent global catastrophic stratigraphic and structural events (e.g., Basilevsky and Head, 1998) or sporadic, regional activity at various times (Guest and Stofan, 1999), or a general evolution of spatially and temporally definable processes (e.g., Phillips and Hansen, 1998) will require detailed global mapping.

3.5 Outer planet satellites

The six large satellites (radius ≥ 1300 km) and 12 mid-sized satellites (radius ≥ 200 km) of the outer planets display a wide range of structures in diverse settings across four different planetary systems (Figure 8.10). Reconnaissance of the satellites was achieved by the Voyager 1 and 2 spacecraft, which surveyed all four planetary systems over the decade of 1979 to 1989 (with imaging resolutions lower than several hundred meters per pixel), followed by detailed and significantly higher resolution imaging (areas as high as several meters per pixel) by the Galileo spacecraft in orbit about Jupiter from 1995 to 2003, and the Cassini spacecraft, which began orbiting Saturn in 2004. A thorough review of the tectonics of outer planet satellites is provided by Collins *et al.* (Chapter 7).

Identification of structures on these satellites is complicated by materials, processes, and landforms that may be unfamiliar to the terrestrial geologist. With the notable exceptions of rocky Io and volatile-rich Triton, their surfaces are composed primarily of H_2O ice and various non-ice contaminants, whereas some contain other volatile components, such as CO_2 and SO_2. Unlike the terrestrial planets, global-scale tectonic processes are commonly linked to changes in satellite shape induced by tidal interactions with their parent planet, including despinning, orbital recession/procession, and reorientation relative to the tidal axes (nonsynchronous rotation or true polar wander). Volume change resulting from internal thermal evolution is also quite plausible, including thermal expansion or contraction, ice-phase transitions, or compositional differentiation. The variety of tectonic landforms is discussed below, organized by planetary system.

The only formal maps of the outer planet satellites generated to date are Voyager-based maps of the Galilean satellites of Jupiter; revisions to several of these maps are in progress, based on Galileo data. Structural mapping follows the definition and procedures outlined for the terrestrial planets, although some heavily tectonized material units (e.g., Ganymede's grooved terrain) necessarily are defined by their structural fabric (Wilhelms, 1990). The complexity of structural relationships combined with limitations of the available data make determination of relative ages difficult. Moreover, absolute age is poorly constrained, because the outer solar system impactor population has likely been different from that of the inner solar system, and it is difficult to extrapolate the current-day fluxes of observationally detected impactors back through time (Zahnle *et al.*, 2003; Schenk *et al.*, 2004).

Callisto. Of Jupiter's four Galilean satellites, Callisto's structural geology, dominated by exogenic processes, is by far the simplest (Moore *et al.*, 2004a). Callisto's impact craters show a variety of forms, from small bowl-shaped basins, through larger complex craters with central peaks, analogous to those on the Moon or Mercury. Unlike craters on the terrestrial planets, craters >35 km show central pits and those >60 km display central domes on both Callisto and Ganymede (Schenk *et al.*, 2004), perhaps due to the effect of subsurface ductile ice during the impact process (Schenk, 1993). Several large multiringed structures on Callisto (typified by Valhalla and Asgard) consist of sets of concentric ridges and/or scarps that are hundreds of kilometers in extent (Figure 8.10a) (Schenk and McKinnon, 1987). These probably surround the sites of large impact events that penetrated to a low-viscosity zone, plausibly a subsurface ocean >100 km beneath the surface, inducing surface extension that created concentric normal faults as the transient crater collapsed (McKinnon and Melosh, 1980; Schenk *et al.*, 2004). An induced magnetic field signature implies a briny ocean within Callisto today (Kivelson *et al.*, 1999), but the lack of endogenic activity means that this ocean is manifest at the surface probably only in the satellite's ancient impact-induced tectonics.

Ganymede. The surface of Ganymede (Pappalardo *et al.*, 2004) consists of about 1/3 dark terrain, which is heavily cratered and ancient, and 2/3 bright grooved terrain, which is less cratered and more recent (Figure 8.10b). Dark terrain contains arcuate systems of tectonic furrows, likely remnants of multiringed structures like those on Callisto (Schenk and McKinnon, 1987; Murchie and Head, 1988), and perhaps similarly an indication of a subsurface ocean (Schenk *et al.*, 2004); of less certain origin are semi-radial furrows also spatially associated with multiring structures. Ganymede's bright grooved terrain crosscuts the dark terrain in elongated swaths (sulci) that are pervasively tectonized in rift-like fashion (Shoemaker *et al.*, 1982; Pappalardo *et al.*, 2004). Two topographic "wavelengths" of structures occur within grooved terrain: (1) the broader (~8 km) scale is inferred to be related to extensional necking of the icy lithosphere above a ductile ice substrate

(Dombard and McKinnon, 2001), and (2) the finer (≤1 km) scale is inferred to have formed by tilt-block style normal faulting of stretched ice (Pappalardo *et al.*, 2004). Geometrical arguments and deformed impact craters indicate that extensional strain reaching tens of percent may be common in grooved terrain (Collins *et al.*, 1998; Pappalardo and Collins, 2005). A component of this strain includes strike-slip displacement (Pappalardo *et al.*, 2004).

Grooved terrain may be brightened and smoothed by a combination of faulting, which exposes bright ice from beneath a dark surface veneer of impactor debris, and icy volcanism (cryovolcanism), which resurfaces older dark terrain (Shoemaker *et al.*, 1982; Schenk *et al.*, 2001a; Pappalardo *et al.*, 2004). Scalloped depressions (paterae) are associated with some smooth lanes, and these may be caldera-like volcanotectonic features associated with cryovolcanism (Schenk *et al.*, 2001a). The extensional strain recorded within Ganymede's grooved terrain has been attributed to early differentiation and associated ice-phase changes and global expansion (Squyres, 1980), and attempts have been made to constrain the global degree of expansion that is consistent with geological observations (McKinnon, 1982; Golombek, 1982; Collins, 2006). It remains to be demonstrated how the inferred high degree of local extensional strain can be explained, given that no definitive examples of contractional structures have been identified. Although the age of Ganymede's grooved terrain is poorly constrained, the existence of intrinsic and induced magnetic fields at Ganymede indicate an active dynamo and a contemporary internal ocean (Kivelson *et al.*, 2002). It remains to be demonstrated whether grooved terrain might be an indirect manifestation of an internal ocean, such as related to convection in tidally heated ice above a global subsurface ocean, or of nonsynchronous rotation of a floating ice shell.

Europa. Europa's few large impact structures indicate that the satellite's average surface age is ∼60 Myr (Zahnle *et al.*, 2003; Schenk *et al.*, 2004), implying recent tectonic activity and associated resurfacing events. Analogous to those on Callisto and Ganymede, Europa's two known multiringed structures (Tyre and Pwyll) suggest transient crater penetration to a low-viscosity layer, plausibly a water ocean, ∼20 km below the icy surface (Schenk *et al.*, 2004). This ice shell thickness is consistent with models of basal tidal heating of a floating ice shell (Ojakangas and Stevenson, 1989; Moore, 2006). Europa's geology is complex and includes severely tectonized ridged plains that resemble a ball of string, with crosscutting troughs, double ridges, and bands (Figure 8.10c) (Greeley *et al.*, 2004). Linear to curvilinear or cycloidal troughs, many displaying raised flanks, are interpreted as near-vertical tensile cracks or shear fractures in Europa's ice shell (Greenberg *et al.*, 1998; Greeley *et al.*, 2004). Ubiquitous ridges are most commonly expressed as double ridges (a ridge pair with a medial trough), whereas some display a complex set of constituent subparallel ridges; modest strike-slip offset along these structures is common. Models of ridge formation include tidal squeezing, compression, dike intrusion, and localized shear heating (see Collins *et al.*, Chapter 7, for a review). Many ridges show diffuse dark flanks, called triple bands. Bands are linear to curvilinear features up to ~25 km wide, consisting of smooth, lineated, and/or hummocky interiors (Prockter *et al.*, 2002). In many cases, the surrounding terrain can be reconstructed, indicating that bands are sites where the brittle lithosphere was pulled apart and replaced by more mobile material from below. This resurfacing included cracking and/or faulting to form subparallel structures. Likely stress mechanisms for creating Europa's tectonic structures include: (1) diurnal stressing induced by radial and librational tides as Europa rapidly orbits Jupiter, (2) slow nonsynchronous rotation of Europa's ice shell relative to the satellite's tidal axes, and (3) true polar wander inducing change in the shape of the ice shell (Greeley *et al.*, 2004; Collins *et al.*, Chapter 7). The cycloidal shapes of many structures is convincing evidence of the action of diurnal stresses, where the effective speed of a propagating fracture is a close match to the diurnally rotating direction of tensile stress, with the fracture tracing an arc each Europa day (Hoppa *et al.*, 1999a). Moreover, rotating diurnal stresses can produce a preferred sense of strike-slip offset along faults in each hemisphere of Europa (Hoppa *et al.*, 1999b). Diurnal stresses are expected to be significant only if Europa's ice shell is decoupled from the rocky mantle by an ocean (Moore and Schubert, 2000). Therefore, cycloidal features are strong evidence for an ocean when the structures formed, and given the young age of the surface, this must be recently.

Europa's surface is marked by domes (indicative of tectonic uplift of the icy lithosphere) and pits (indicative of withdrawal of surface material and associated lithospheric downwarp). Scattered across the surface, reddish material has extruded to form spots and chaotic terrain, commonly associated with ridged plains deformation (Greeley *et al.*, 2004). In the largest and most dramatic examples (e.g., Conamara Chaos), older ridged plains are tectonically disrupted and blocks have rotated and translated in a hummocky matrix, and much of the original plains has been destroyed. Spots and chaos are attributed to whole-scale melting of Europa's ice shell (Greenberg *et al.*, 1999), or partial melting triggered by diapiric upwellings (Pappalardo *et al.*, 1999); however, the geophysical details of these models have yet to be satisfactorily understood (Nimmo and Giese, 2005). Crosscutting relationships among Europa's tectonic features and chaotic terrains indicate that structural deformation has become more narrowly focused through Europa's preserved history, and that chaotic terrain is overall more recent in the stratigraphy (Figueredo and Greeley, 2004), unless older chaos occurred but has been modified over time (Riley *et al.*, 2000). Recent chaos is consistent with recent ice shell thickening (Pappalardo *et al.*, 1999), perhaps related to secular changes in heat flux due to cyclical variations in orbital eccentricity (Hussmann and Spohn, 2004).

Io. Io is the most volcanically active body in the solar system, with a measured heat flux of $\sim 2\,\text{Wm}^2$, resulting from tidal heating of this rocky satellite (McEwen *et al.*, 2004). Its landscape is dominated by volcanic calderas (paterae), edifices, and flows, but ~3% of Io's surface consists of mountains that are likely tectonic in origin (Figure 8.10d) (McEwen *et al.*, 2004). Mountains are typically tens of kilometers across, average ~6 km tall, and reach heights of ~17 km. Many mountains are surrounded by debris aprons, evidence of mass wasting of loosely consolidated surface materials. Mountains potentially formed by large-scale up-thrusting of lithospheric blocks in response to global-scale compression, due to the high rate of volcanism and associated vertical loading and cycling (Schenk and Bulmer, 1998). Locally concentrated stress may have triggered mountain formation, such as through mantle diapirism (Turtle *et al.*, 2001; Jaeger *et al.*, 2003). Mountains are most abundant in two antipodal zones on the satellite, and are globally anticorrelated with volcanic centers (Schenk *et al.*, 2001b). Nonetheless, individual mountains and paterae are commonly associated, and some mountains show evidence of rifting, perhaps related to uplift (Jaeger *et al.*, 2003). Kilometer-scale ridges and troughs on the flanks of some mountains might be the expression of folds in a deformable surficial layer, perhaps driven by gravity sliding (Moore *et al.*, 2001). Similar small-scale structures on Io's plains may be contractional structures formed by diurnal tidal stress (Bart *et al.*, 2004).

Satellites of Saturn. As of this writing, the Cassini mission is changing our understanding of the tectonic histories of the Saturnian satellites, but few publications of the results yet exist. Smog-shrouded Titan is being revealed by radar imaging, and by imaging through near-infrared atmospheric windows using both Cassini's camera and infrared instrument (Elachi *et al.*, 2005; Porco *et al.*, 2005a; Sotin *et al.*, 2005). Several large impact craters occur on Titan, and lineaments and linear boundaries could be tectonic. Tectonic features occur on all of Saturn's middle-sized icy satellites. Voyager images revealed bright wispy terrains on both Dione and Rhea, which Cassini imaging has resolved as extensional fault scarps that expose relatively clean subsurface icy material (Figure 8.10e) (Johnson, 2005). Dione and Tethys exhibit ridges that may be contractional in origin (Moore and Ahern, 1983; Moore, 1984). Cassini imaging resolves the detailed internal fault structure of Ithaca Chasma on Tethys, a global-scale rift zone which roughly follows a great circle normal to the center of the impact basin Odysseus and thus may be related to the basin's formation or relaxation (Moore *et al.*, 2004b). Iapetus displays a 20-km high ridge that delineates the satellite's equator (Figure 8.10f), possibly indicating tidal despinning and associated equatorial radius decrease (Porco *et al.*, 2005b). Existing models of the thermal and corresponding tectonic evolution of the Saturnian satellites (Schubert *et al.*, 1986; Hillier and Squyres, 1991) will certainly be modified and improved upon, aiding interpretation of Cassini images.

Enceladus exhibits spectacular active vapor plumes escaping the satellite from prominent, warm tectonic structures in its extremely youthful and tectonically deformed south polar region (Figure 8.10g) (Porco et al., 2006; Spencer et al., 2006). Overall, it seems that intense tectonism (rather than cryovolcanism) may have resurfaced the youthful parts of Enceladus (Johnson, 2005). Tectonic interpretation of the south polar terrain suggests shortening of the satellite's spin axis (Porco et al., 2006), consistent with true polar wander of Enceladus to move a low-density internal mass anomaly to the spin axis (Nimmo and Pappalardo, 2006). Prominent tectonic structures in the south polar terrain, nicknamed "tiger stripes," may have formed as tension fractures and are believed to undergo shearing and open–close motions through the Enceladus tidal cycle, probably playing an important role in the production and release of the satellite's vapor plumes (Hurford et al., 2007; Nimmo et al., 2007; Spitale and Porco, 2007; Smith-Konter and Pappalardo, 2007). The mechanisms for tectonic deformation of Enceladus, and the possible role of an internal ocean or sea (Collins and Goodman, 2007; Nimmo et al., 2007), will continue to be important topics of investigation.

Satellites of Uranus. The Voyager 2 spacecraft passed closest to the innermost of the five major Uranian satellites, so imaging resolution is best at Miranda, while Titania and Oberon were not imaged sufficiently to resolve tectonic structures. Umbriel appears generally featureless and dark, but careful image processing reveals an ancient tectonic system (Helfenstein et al., 1989). Ariel displays a global-scale structural pattern of tectonic rifts, ridges, and troughs forming an obliquely intersecting pattern across much of the satellite's visible surface (Plescia, 1987). Some rift depressions appear to have been resurfaced by a viscous cryolava (Jankowski and Squyres, 1988). Abundant extensional tectonic structures on many of the Uranian satellites, as on the Saturnian satellites, are attributed to volume changes linked to thermal evolution (Schubert et al., 1986; Hillier and Squyres, 1991).

Despite its small size, Miranda exhibits striking structural geology (Figure 8.10h). Miranda displays three ovoidal regions termed "coronae" that are deformed by subparallel ridges and grooves. An early interpretation of the coronae suggested that they might be folds, which would be consistent with a "sinker" model, in which the satellite was disrupted early in its history by a large impact, with remnants of the proto-satellite's rocky core sinking downward through the reforming satellite and stirring downwelling currents to create the coronae above (Janes and Melosh, 1988). However, subsequent analyses showed that the coronae are likely extensional tectonic structures, including tilted blocks and cryovolcanic extrusions, more consistent with a model in which the coronae formed by tectonic deformation above large-scale upwelling plumes (Greenberg et al., 1991; Schenk, 1991; Pappalardo et al., 1997). Additional tectonic stress may have resulted from reorientation of Miranda in response to corona formation (Plescia, 1988).

Triton. The single large satellite of Neptune, Triton, has a surface rich in exotic ices (CO, CO_2, CH_4, and N_2), in addition to H_2O (Cruikshank et al., 2000). A variety of structures occur across its surface (Figure 8.10i), including troughs, double ridges, and active geysers, as well as morphological evidence for past cryovolcanism (Croft *et al.*, 1995). Because Triton has a retrograde orbit, it is generally believed that Triton is a captured body that originated as a Kuiper Belt object (Agnor and Hamilton, 2006). Triton's tectonic patterns may be influenced by its orbital procession as it spirals slowly inward toward Neptune, or by other global tectonic stresses (Croft *et al.*, 1995; Collins *et al.*, Chapter 7). Morphological comparison of Triton's ridges to smaller double ridges on Europa has prompted the suggestion that diurnal stressing and shear heating may have contributed to formation of Triton's ridges during the circularization of Triton's orbit following capture (Prockter *et al.*, 2005). Triton displays a unique region known informally as "cantaloupe terrain," where compositionally induced diapirism may have led to crustal overturn to form the pits (cavi) characteristic of this unusual region (Schenk and Jackson, 1993).

Overall, the diverse and sometimes bizarre tectonics of the outer planet satellites is inherently linked to satellite tides and the mechanical–rheological properties of ice (Collins *et al.*, Chapter 7). Our terrestrial experience is critical to their interpretation, yet is challenged by their unfamiliarity.

Acknowledgments

We would like to acknowledge Ron Greeley and Matt Golombek for their review comments that led to improvement of our chapter. Portions of this work performed by RA and RP were carried out at the Jet Propulsion Laboratory, California Institute of Technology, under a contract with the National Aeronautics and Space Administration. This work was also supported by the National Aeronautics and Space Administration under Grants issued through the Office of the Planetary Geology and Geophysics Program.

References

Addington, E. A. (2001). A stratigraphic study of small volcano clusters on Venus. *Icarus*, **149**, 16–36.

Agnor, C. B. and Hamilton, D. P. (2006). Neptune's capture of its moon Triton in a binary–planet gravitational encounter. *Nature*, **441**, 192–194.

Anderson, R. C., Dohm, J. M., Golombek, M. P., Haldemann, A., Franklin, B. J., Tanaka, K., Lias, J. and Peer, B. (2001). Significant centers of tectonic activity through time for the western hemisphere of Mars. *J. Geophys. Res.*, **106**, 20 563–20 585.

Avery, T. E. and Berlin, G. L. (1992). *Fundamentals of Remote Sensing and Airphoto Interpretation*, 5th edn. New York: Macmillan.

Baker, V. R., Maruyama, S., and Dohm, J. M. (2007). Tharsis superplume and the geological evolution of early Mars. In *Superplumes: Beyond Plate Tectonics*, eds. D. A. Yuen, S. Maruyama, S.-I. Karato and B. F. Windley. London: Springer, pp. 507–523.

Baldwin, R. B. (1963). *The Measure of the Moon.* Chicago: University of Chicago Press.

Banerdt, W. B., Phillips, R. J., Sleep, N. H., and Saunders, R. S. (1982). Thick-shell tectonics on one-plate planets: Applications to Mars. *J. Geophys. Res.*, **87**, 9723–9733.

Banerdt, W. B., Golombek, M. P. and Tanaka, K. L. (1992). Stress and tectonics on Mars. In *Mars*, eds. H. H. Kieffer, B. M. Jakosky, C. W. Snyder and M. S. Matthews. Tucson, AZ: University of Arizona Press, pp. 249–297.

Banks, B. K. and Hansen, V. L. (2000). Relative timing of crustal plateau magmatism and tectonism at Tellus Regio, Venus. *J. Geophys. Res.*, **105**, 17 655–17 668.

Bart, G. D., Turtle, E. P., Jaeger, W. L., Keszthelyi, L. P., and Greenberg, R. (2004). Ridges and tidal stress on Io. *Icarus*, **169**, 111–126.

Basilevsky, A. T. and Head, J. W. (1998). The geologic history of Venus: A stratigraphic view. *J. Geophys. Res.*, **103**, 8531–8544.

Bell, J. F., III, Campbell, B. A., and Robinson, M. S. (1999). Planetary geology. In *Remote Sensing for the Earth Sciences: Manual of Remote Sensing*, ed. A. N. Rencz. 3rd edn, Vol. 3. New York: John Wiley & Sons, pp. 509–563.

Binder, A. B. (1982). Post-Imbrian global lunar tectonism: Evidence for an initially totally molten moon. *Earth Moon Planets*, **26**, 117–133.

Binder, A. B. and Gunga, H. C. (1985). Young thrust-fault scarps in the highlands: Evidence for an initially totally molten Moon. *Icarus*, **63**, 421–441.

Blasius, K. R., Cutts, J. A., Guest, J. E., and Masursky, H. (1977). Geology of the Valles Marineris: First analysis of imaging from the Viking 1 orbiter primary mission. *J. Geophys. Res.*, **82**, 4067–4091.

Buczkowski, D. L. and McGill, G. E. (2002). Topography within circular grabens: Implications for polygon origin, Utopia Planitia, Mars. *Geophys. Res. Lett.*, **29**, doi:10.1029/2001GL014100.

Campbell, B. A. (1999). Surface formation rates and impact crater densities on Venus. *J. Geophys. Res.*, **104**, 21 951–21 955.

Carr, M. H. (1974). Tectonism and volcanism of the Tharsis region of Mars. *J. Geophys. Res.*, **79**, 3943–3949.

Chicarro, A. F., Schultz, P. H., and Masson, P. (1985). Global and regional ridge patterns on Mars. *Icarus*, **63**, 153–174.

Collins, G. C. (2006). Global expansion of Ganymede derived from strain measurements in grooved terrain (abs.). *Lunar Planet. Sci. Conf. XXXVII*, 2077. Houston, TX: Lunar and Planetary Institute (CD-Rom).

Collins, G. C. and Goodman, J. C. (2007). Enceladus' south polar sea. *Icarus*, **189**, 72–82.

Collins, G. C., Head III, J. W., and Pappalardo, R. T. (1998). Role of extensional instability in creating Ganymede grooved terrain: Insights from Galileo high-resolution stereo imaging. *Geophys. Res. Lett.*, **25**, 233–236.

Cook, A. C., Watters, T. R., Robinson, M. S., Spudis, P. D., and Bussey, D. B. J. (2000). Lunar polar topography derived from Clementine stereoimages. *J. Geophys. Res.*, **105**, 12 023–12 034.

Cordell, B. M. and Strom, R. G. (1977). Global tectonics of Mercury and the Moon. *Phys. Earth Planet. Inter.*, **15**, 146–155.

Croft, S. K., Kargel, J. S., Kirk, R. L., Moore, J. M., Schenk, P. M., and Strom, R. G. (1995). Geology of Triton. In *Neptune*, eds. J. T. Bergstralh *et al.* Tucson, AZ: University of Arizona Press, pp. 879–948.

Cruikshank, D. P., Schmitt, B., Roush, T. L., Owen, T. C., Quirico, E., Geballe, T. R., de Bergh, C., Bartholomew, M. J., Dalle Ore, C. M., Doute, S., and Meier, R. (2000). Water ice on Triton. *Icarus*, **147**, 309–316.

Crumpler, L. S., Aubele, J. C., Senske, D. A., Keddie, S. T., Magee, K. P., and Head, J. W. (1997). Volcanoes and centers of volcanism on Venus. In *Venus II: Geology, Geophysics, Atmosphere, and Solar Wind Environment*, eds. S. W. Bougher, D. M. Hunten and R. J. Phillips. Tucson, AZ: University of Arizona Press, pp. 697–756.

Davies, M. E. and Batson, R. M. (1975). Surface coordinates and cartography of Mercury. *J. Geophys. Res.*, **80**, 2417–2430.

Davies, M. E., Dwornik, S. E., Gault, D. E., and Strom, R. G. (1978). *Atlas of Mercury*, NASA Spec. Publ. SP-423.

Davis, P. A. and Soderblom, L. A. (1984). Modeling crater topography and albedo from monoscopic Viking Orbiter images: I. Methodology. *J. Geophys. Res.*, **89**, 9449–9457.

Davis, P. A., Tanaka, K. L., and Golombek, M. P. (1995). Topography of closed depressions, scarps, and grabens in the north Tharsis region of Mars: Implications for shallow crustal discontinuities and graben formation. *Icarus*, **114**, 403–422.

Dimitrova, L. L., Holt, W. E., Haines, A. J., and Schultz, R. A. (2006). Towards understanding the history and mechanisms of Martian faulting: The contribution of gravitational potential energy. *Geophys. Res. Lett.*, **33**, doi:10.1029/2005GL025307.

Dohm, J. M. and Tanaka, K. L. (1999). Geology of the Thaumasia region, Mars: Plateau development, valley origins, and magmatic evolution. *Planet. Space Sci.*, **47**, 411–431.

Dohm, J. M., Tanaka, K. L., and Hare, T. M. (2001a). Geologic map of the Thaumasia region of Mars. U.S. Geol. Surv. Misc. Invest. Ser. Map I-2650, scale 1:5 000 000.

Dohm, J. M., Ferris, J. C., Baker, V. R., Anderson, R. C., Hare, T. M., Strom, R. G., Barlow, N. G., Tanaka, K. L., Klemaszewski, J. E., and Scott, D. H. (2001b). Ancient drainage basin of the Tharsis region, Mars: Potential source for outflow channel systems and putative oceans or paleolakes. *J. Geophys. Res.*, **106**, 32 943–32 958.

Dohm, J. M., Baker, V. R., Maruyama, S., and Anderson, R. C. (2007a). Traits and evolution of the Tharsis superplume, Mars. In *Superplumes: Beyond Plate Tectonics*, eds. D. A. Yuen, S. Maruyama, S.-I. Karato and B. F. Windley. London: Springer, pp. 523–537.

Dohm, J. M., Barlow, N. G., Anderson, R. C., Williams, J.-P., Miyamoto, H., Ferris, J. C., Strom, R. G., Taylor, G. J., Fairén, A. G., Baker, V. R., Boynton, W. V., Keller, J. M., Kerry, K., Janes, D., Rodríguez, A., and Hare, T. M. (2007b). Possible ancient giant basin and related water enrichment in the Arabia Terra province, Mars. *Icarus*, doi:10.1016/j.Icarus.2007.03.006.

Dombard, A. J. and McKinnon, W. B. (2001). Formation of grooved terrain on Ganymede: Extensional instability mediated by cold, superplastic creep. *Icarus*, **154**, 321–336.

Downs, G. S., Mouginis-Mark, P. J., Zisk, S. H., and Thompson, T. W. (1982). New radar derived topography for the northern hemisphere of Mars. *J. Geophys. Res.*, **87**, 9747–9754.

Elachi, C., Wall, S., Allison, M., Anderson, Y., Boehmer, R., Callahan, P., Encrenaz, P., Flamini, E., Franceschetti, G., Gim, Y., Hamilton, G., Hensley, S., Janssen, M., Johnson, W., Kelleher, K. *et al.* (2005). Cassini radar views the surface of Titan. *Science*, **308**, 970–974.

Fairén, A. G. and Dohm, J. M. (2004). Age and origin of the lowlands of Mars. *Icarus*, **168**, 277–284.

Ferrill, D. A., Wyrick, D. Y., Morris, A. P., Sims, D. W., and Franklin, N. M. (2004). Dilational fault slip and pit chain formation on Mars. *GSA Today*, **14**, 4–12.

Figueredo, P. H. and Greeley, R. (2004). Resurfacing history of Europa from pole-to-pole geological mapping. *Icarus*, **167**, 287–312.

Ford, P. G. and Pettengill, G. H. (1992). Venus topography and kilometer-scale slopes. *J. Geophys. Res.*, **97**, 13 103–13 114.

Ford, J. P., Blom, R. G., Crisp, J. A., Elachi, C., Farr, T. G., Saunders, R. S., Theilig, E. E., Wall, S. D., and Yewell, S. B. (1989). *Spaceborne Radar Observations: A Guide for Magellan Radar-Image Analysis*. JPL Publ. 89–41. Pasadena, CA: Jet Propulsion Laboratory, 126pp.

Ford, J. P., Plaut, J. J., Weitz, C. M., Farr, T. G., Senske, D. A., Stofan, E. R., Michaels, G., and Parker, T. J. (1993). *Guide to Magellan Image Interpretation*. JPL Publ. 93–24. Pasadena, CA: Jet Propulsion Laboratory, 148pp.

Forsythe, R. D. and Zimbelman, J. R. (1988). Is the Gordii Dorsum escarpment on Mars an exhumed transcurrent fault? *Nature*, **336**, 143–146.

Frey, H. V. (2006). Impact constraints on, and a chronology for, major events in early Mars history. *J. Geophys. Res.*, **111**, doi:10.1029/2005JE002449.

Frey, H., Roark, J. H., Shockey, K. M., Frey, E. L., and Sakimoto, S. H. E. (2002). Ancient lowlands on Mars. *Geophys. Res. Lett.*, **29**, 1384, doi:10.1029/2001GL013832.

Furfaro, R., Dohm, J. M., Fink, W., Kargel, J. S., Schulze-Makuch, D., Fairén, A. G., Ferre, P. T., Palmero-Rodriguez, A., Baker, V. R., Hare, T. M., Tarbell, M., Miyamoto, H. H., and Komatsu, G. (2007). The search for life beyond Earth through fuzzy expert systems. *Planet. Space Sci.*, **56**, 448–472.

Ghent, R. R. and Hansen, V. L. (1999). Structural and kinematic analysis of eastern Ovda Regio, Venus: Implications for crustal plateau formation. *Icarus*, **139**, 116–136.

Gilbert, G. K. (1886). The inculcation of scientific method by example, with an illustration drawn from the Quaternary of Utah. *Am. J. Sci.*, **31** (3rd Series), 284–299.

Gilbert, G. K. (1893). The Moon's face, a study of the origin of its features. *Philos. Soc. Wash. Bull.*, **12**, 241–292.

Gilmore, M. S., Collins, G. C., Ivanov, M. A., Marinangeli, L., and Head, J. W. (1998). Style and sequence of extensional structures in tessera terrain, Venus. *J. Geophys. Res.*, **103**, 16 813–16 840.

Golombek, M. P. (1979). Structural analysis of lunar grabens and the shallow crustal structure of the Moon. *J. Geophys. Res.*, **84**, 4657–4666.

Golombek, M. P. (1982). Constraints on the expansion of Ganymede and the thickness of the lithosphere (abs.). *Proc. Lunar Planet. Sci. Conf. 13. J. Geophys. Res.*, **87**, A77–A83.

Golombek, M. P. (1992). Planetary tectonic processes, terrestrial planets. In *The Astronomy and Astrophysics Encyclopedia*, ed. S. P. Maran. New York: Van Nostrand Reinhold, pp. 544–546.

Golombek, M. P. and McGill, G. E. (1983). Grabens, basin tectonics, and the maximum total expansion of the Moon. *J. Geophys. Res.*, **88**, 3563–3578.

Golombek, M. P., Plescia, J. B., and Franklin, B. J. (1991). Faulting and folding in the formation of planetary wrinkle ridges. *Proc. Lunar Planet. Sci. Conf. 21*, 679–693.

Golombek, M. P., Tanaka, K. L., and Franklin, B. J. (1996). Extension across Tempe Terra, Mars, from measurements of fault scarp widths and deformed craters. *J. Geophys. Res.*, **101**, 26 119–26 130.

Golombek, M. P., Anderson, F. S., and Zuber, M. T. (2001). Martian wrinkle ridge topography: Evidence for subsurface faults from MOLA. *J. Geophys. Res.*, **106**, 23 811–23 821.

Gomes, R., Levison, H. F., Tsiganis, K., and Morbidelli, A. (2005). Origin of the cataclysmic late heavy bombardment period of the terrestrial planets. *Nature*, **435**, 466–469.

Goudy, C. L. and Schultz, R. A. (2005). Dike intrusions beneath grabens south of Arsia Mons, Mars. *Geophys. Res. Lett.*, **32**, doi:10.1029/2004GL021977.

Goudy, C. L., Schultz, R. A., and Gregg, T. K. P. (2005). Coulomb stress changes in Hesperia Planum, Mars, reveal regional thrust fault reactivation. *J. Geophys. Res.*, **110**, doi:10.1029/2004JE002293.

Greeley, R. and Guest, J. E. (1987). Geologic map of the eastern equatorial region of Mars. U.S. Geol. Surv. Misc. Invest. Ser., Map I-1802-B, scale 1:15 000 000.

Greeley, R., Chyba, C., Head, J. W., McCord, T., McKinnon, W. B., and Pappalardo, R. T. (2004). Geology of Europa. In *Jupiter: The Planet, Satellites and Magnetosphere*, eds. F. Bagenal, T. E. Dowling and W. B. McKinnon. Tucson, AZ: University of Arizona Press, pp. 329–362.

Greenberg, R., Croft, S. K., Janes, D. M., Kargel, J. S., Lebofsky, L. A., Lunine, J. I., Marcialis, R. L., Melosh, H. J., Ojakangas, G. W., and Strom, R. G. (1991). Miranda. In *Uranus*, eds. J. T. Bergstralh *et al*. Tucson, AZ: University of Arizona Press, pp. 693–735.

Greenberg, R., Geissler, P. E., Hoppa, G., Tufts, B. R., Durda, D. D., Pappalardo, R., Head, J. W., Greeley, R., Sullivan, R., and Carr, M. H. (1998). Tectonic processes on Europa: Tidal stresses, mechanical response, and visible features. *Icarus*, **135**, 64–78.

Greenberg, R., Hoppa, G. V., Tufts, B. R., Geissler, P. E., and Reilly, J. (1999). Chaos on Europa. *Icarus*, **141**, 263–286.

Guest, J. E. and Stofan, E. R. (1999). A new view of the stratigraphic history of Venus. *Icarus*, **139**, 55–66.

Guest, J. E., Bulmer, M. H., Aubele, J. C., Beratan, K., Greeley, R., Head, J. W., Michaels, G., Weitz, C., and Wiles, C. (1992). Small volcanic edifices and volcanism in the plains on Venus. *J. Geophys. Res.*, **97**, 15 949–15 966.

Hansen, V. L. (2000). Geologic mapping of tectonic planets. *Earth Planet. Sci. Lett.*, **176**, 527–542.

Hansen, V. L. (2005). Venus's shield terrain. *Geol. Soc. Am. Bull.*, **117**, 808–822.

Hansen, V. L. (2006). Geologic constraints on crustal plateau surface histories, Venus: The lava pond and bolide impact hypotheses. *J. Geophys. Res.*, **111**, doi:10.1029/2006JE002714.

Hansen, V. L. and Willis, J. J. (1996). Structural analysis of a sampling of tesserae: Implications for Venus geodynamics. *Icarus*, **123**, 296–312.

Hansen, V. L. and Willis, J. J. (1998). Ribbon terrain formation, southwestern Fortuna Tessera, Venus: Implications for lithosphere evolution. *Icarus*, **132**, 321–343.

Hansen, V. L., Banks, B. K., and Ghent, R. R. (1999). Tessera terrain and crustal plateaus, Venus. *Geology*, **27**, 1071–1074.

Hansen, V. L., Phillips, R. J., Willis, J. J., and Ghent, R. R. (2000). Structures in tessera terrain, Venus: Issues and answers. *J. Geophys. Res.*, **105**, 4135–4152.

Hapke, B., Danielson, E., Klaasen, K., and Wilson, L. (1975). Photometric observations of Mercury from Mariner 10. *J. Geophys. Res.*, **80**, 2431–2443.

Harmon, J. K., Campbell, D. B., Bindschadler, K. L., Head, J. W., and Shapiro, I. I. (1986). Radar altimetry of Mercury: A preliminary analysis. *J. Geophys. Res.*, **91**, 385–401.

Hartmann, W. K. and Neukum, G. (2001). Cratering chronology and evolution of Mars. *Space Sci. Rev.*, **96**, 165–194.

Hauber, E. and Kronberg, P. (2005). The large Thaumasia graben on Mars: Is it a rift? *J. Geophys. Res.*, **110**, doi:10.1029/2005JE002407.

Hauck, S. A., Dombard, A. J., Phillips, R. J., and Solomon, S. C. (2004). Internal and tectonic evolution of Mercury. *Earth Planet. Sci. Lett.*, **222**, 713–728.

Head, J. W., Kreslavsky, M. A., and Pratt, S. (2002). Northern lowlands of Mars: Evidence for widespread volcanic flooding and tectonic deformation in the Hesperian period. *J. Geophys. Res.*, **107**, doi:10.1029/2000JE001445.

Helfenstein, P., Thomas, P. C., and Veverka, J. (1989). Evidence from Voyager II photometry for early resurfacing of Umbriel. *Nature*, **338**, 324–326.

Herrick, R. R. and Sharpton, V. L. (2000). Implications from stereo-derived topography of Venusian impact craters. *J. Geophys. Res.*, **105**, 20 245–20 262.

Herrick, R. R., Sharpton, V. L., Malin, M. C., Lyons, S. N., and Feely, K. (1997). Morphology and morphometry of impact craters. In *Venus II: Geology, Geophysics, Atmosphere, and Solar Wind Environment*, eds. S. W. Bougher, D. M. Hunten and R. J. Phillips. Tucson, AZ: University of Arizona Press, pp. 1015–1046.

Hillier, J. and Squyres, S. W. (1991). Thermal stress tectonics on the satellites of Saturn and Uranus. *J. Geophys. Res.*, **96**, 15 665–15 674.

Hoppa, G. V., Tufts, B. R., Greenberg, R., and Geissler, P. E. (1999a). Formation of cycloidal features on Europa. *Science*, **285**, 1899–1902.

Hoppa, G., Tufts, B. R., Greenberg, R., and Geissler, P. (1999b). Strike-slip faults on Europa: Global shear patterns driven by tidal stress. *Icarus*, **141**, 287–298.

Howard, K. A. and Muehlberger, W. R. (1973). Lunar thrust faults in the Taurus-Littrow region. In *Apollo 17 Prel. Sci. Rep.*, NASA SP-330, 31–12 to 31–21.

Hurford, T. A., Helfenstein, P., Hoppa, G. V., Greenberg, R., and Bills, B. G. (2007). Eruptions arising from tidally controlled periodic openings of rifts on Enceladus. *Nature*, **447**, 292–294.

Hussmann, H. and Spohn, T. (2004). Coupled thermal and orbital evolution of Europa and Io. *Icarus*, **171**, 391–410.

Ivanov, M. A. and Head, J. W. (1996). Tessera terrain on Venus: A survey of the global distribution, characteristics, and relation to surrounding units from Magellan data. *J. Geophys. Res.*, **101**, 14 861–14 908.

Jaeger, W. L., Turtle, E. P., Keszthelyi, L. P., Radebaugh, J., McEwen, A. S., and Pappalardo, R. T. (2003). Orogenic tectonism on Io. *J. Geophys. Res.*, **108**, doi:10.1029/2002JE001946.

Janes, D. M. and Melosh, H. J. (1988). Sinker tectonics: An approach to the surface of Miranda. *J. Geophys. Res.*, **93**, 3127–3143.

Jankowski, D. J. and Squyres, S. W. (1988). Solid-state ice volcanism on the satellites of Uranus. *Science*, **241**, 1322–1325.

Johnson, T. V. (2005). Geology of the icy satellites. *Space Sci. Rev.*, **116**, 401–420.

King, S. D. (2008). Pattern of lobate scarps on Mercury's surface reproduced by a model of mantle convection. *Nature Geosciences*, **1**, 229–232.

Kirk, R. L., Soderblom, L. A., and Lee, E. L. (1992). Enhanced visualization for interpretation of Magellan radar data: Supplement to the Magellan Special Issue. *J. Geophys. Res.*, **97**, 16 371–16 381.

Kirk, R. L., Howington-Kraus, E., Redding, B., Galuszka, D., Hare, T. M., Archinal, B. A., Soderblom, L. A., and Barrett, J. M. (2003). High-resolution topomapping of candidate MER landing sites with Mars Orbiter Camera narrow-angle images. *J. Geophys. Res.*, **108**, doi:10.1029/2003JE002131.

Kivelson, M. G., Khurana, K. K., Stevenson, D. J., Bennett, L., Joy, S., Russell, C. T., Walker, R. J., Zimmer, C., and Polanskey, C. (1999). Europa and Callisto: Induced or intrinsic fields in a periodically varying plasma environment. *J. Geophys. Res.*, **104**, 4609–4625.

Kivelson, M. G., Khurana, K. K., and Volwerk, M. (2002). The permanent and inductive magnetic moments of Ganymede. *Icarus*, **157**, 507–522.

Knapmeyer, M., Oberst, J., Hauber, E., Wählisch, M., Deuchler, C., and Wagner, R. (2006). Working models for spatial distribution and level of Mars' seismicity. *J. Geophys. Res.*, **111**, E11006, doi:10.1029/2006JE002708.

Koenig, E. and Aydin, A. (1998). Evidence for large-scale strike-slip faulting on Venus. *Geology*, **26**, 551–554.

Komatsu, G., Gulick, V. C., and Baker, V. R. (2001). Valley networks on Venus. *Geomorphology*, **37**, 225–240.

Kring, D. A. and Cohen, B. A. (2002). Cataclysmic bombardment throughout the inner solar system 3.9–4.0 Ga. *J. Geophys. Res.*, **107**, doi:10.1029/2001JE001529.

Kumar, P. S. (2005). An alternative kinematic interpretation of Thetis Boundary Shear Zone, Venus: Evidence for strike-slip ductile duplexes. *J. Geophys. Res.*, **110**, doi:10.1029/2004JE002387.

Lillesand, T. M. and Kiefer, R. W. (1994). *Remote Sensing and Image Interpretation*, 3rd edn. New York: John Wiley & Sons.

Lucchitta, B. K. (1976). Mare ridges and highland scarps: Results of vertical tectonism? *Proc. Lunar Sci. Conf. 7*, 2761–2782.

Lucchitta, B. K. (1977). Topography, structure, and mare ridges in southern Mare Imbrium and northern Oceanus Procellarum. *Proc. Lunar Sci. Conf. 8*, 2691–2703.

Lucchitta, B. K., McEwen, A. S., Clow, C. D., Geissler, R. B., Singer, R. B., Schultz, R. A., and Squyres, S. W. (1992). The canyon system on Mars. In *Mars*, eds. H. H. Kieffer, B. M. Jakosky, C. W. Snyder and M. S. Matthews. Tucson, AZ: University of Arizona Press, pp. 453–492.

Margot, J. L., Campbell, D. B., Jurgens, R. F., and Slade, M. A. (1999). Topography of the lunar poles from radar interferometry: A survey of cold trap locations. *Science*, **284**, 1658–1660.

Márquez, A., Fernández, C., Anguita, F., Farelo, A., Anguita, J., and de la Casa, M.-A. (2004). New evidence for a volcanically, tectonically, and climatically active Mars. *Icarus*, **172**, 573–581.

Masursky, H., Colton, G. W., and El-Baz, F. (eds.) (1978). *Apollo Over the Moon: A View From Orbit*, NASA SP-362. Washington, DC: NASA Scientific and Technical Information Office.

Maxwell, T. A., El-Baz, F., and Ward, S. H. (1975). Distribution, morphology, and origin of ridges and arches in Mare Serenitatis. *Geol. Soc. Am. Bull.*, **86**, 1273–1278.

McEwen, A. S., Kezthelyi, L., Lopes, R., Schenk, P., and Spencer, J. (2004). The lithosphere and surface of Io. In *Jupiter: The Planet, Satellites and Magnetosphere*, eds. F. Bagenal, T. E. Dowling and W. B. McKinnon. Tucson, AZ: University of Arizona Press, pp. 307–328.

McGill, G. E. (1971). Attitude of fractures bounding straight and arcuate lunar rilles. *Icarus*, **14**, 53–58.

McGill, G. E. (1986). The giant polygons of Utopia, northern Martian plains. *Geophys. Res. Lett.*, **13**, 705–708.

McGill, G. E. (1989). Buried topography of Utopia, Mars: Persistence of a giant impact depression. *J. Geophys. Res.*, **94**, 2753–2759.

McGill, G. E. and Hills, L. S. (1992). Origin of giant Martian polygons. *J. Geophys. Res.*, **97**, 2633–2647.

McKinnon, W. B. (1982). Tectonic deformation of Galileo Regio and limits to the planetary expansion of Ganymede. *Proc. Lunar Planet. Sci. Conf. 12*, 1585–1597.

McKinnon, W. B. and Melosh, H. J. (1980). Evolution of planetary lithospheres: Evidence from multiring basins on Ganymede and Callisto. *Icarus*, **44**, 454–471.

Mége, D. and Masson, P. (1996). Amounts of crustal stretching in Valles Marineris, Mars. *Planet. Space Sci.*, **44**, 749–781.

Melosh, H. J. (1976). On the origin of fractures radial to lunar basins. *Proc. Lunar Sci. Conf.* 7, 2967–2982.

Melosh, H. J. (1978). The tectonics of mascon loading. *Proc. Lunar Planet. Sci. Conf. 9*, 3513–3525.

Melosh, H. J. (1989). *Impact Cratering: A Geologic Process.* New York: Oxford University Press.

Melosh, H. J. and Dzurisin, D. (1978a). Tectonic implications for gravity structure of Caloris basin, Mercury. *Icarus*, **33**, 141–144.

Melosh, H. J. and Dzurisin, D. (1978b). Mercurian global tectonics: A consequence of tidal despinning? *Icarus*, **35**, 227–236.

Melosh, H. J. and McKinnon, W. B. (1988). The tectonics of Mercury. In *Mercury*, eds. F. Vilas, C. R. Chapman and M. S. Matthews. Tucson, AZ: University of Arizona Press, pp. 374–400.

Moore, J. M. (1984). The tectonic and volcanic history of Dione. *Icarus*, **59**, 205–220.

Moore, J. M. and Ahern, J. L. (1983). The geology of Tethys. *J. Geophys. Res.*, **88**, A577–A584.

Moore, J. M., Sullivan, R. J., Chuang, F. C., Head, J. W., McEwen, A. S., Milazzo, M. P., Nixon, B. E., Pappalardo, R. T., Schenk, P. M., and Turtle, E. P. (2001). Landform degradation and slope processes on Io: The Galileo view. *J. Geophys. Res.*, **106**, 33 223–33 240.

Moore, J. M., Chapman, C. R., Chapman., C., Bierhaus, E., Greeley, R., Chuang, F., Klemaszewski, J., Clark, R., Dalton, J., Hibbitts, C., Schenk, P., Spencer, J., and Wagner, R. (2004a). Callisto. In *Jupiter: The Planet, Satellites and Magnetosphere*, eds. F. Bagenal, T. E. Dowling and W. B. McKinnon. Tucson, AZ: University of Arizona Press, pp. 397–426.

Moore, J. M., Schenk, P. M., Bruesch, L. S., Asphaug, E., and McKinnon, W. B. (2004b). Large impact features on middle-sized icy satellites. *Icarus*, **171**, 421–443.

Moore, W. B. (2006). Thermal equilibrium in Europa's ice shell. *Icarus*, **180**, 141–146.

Moore, W. B. and Schubert, G. (2000). The tidal response of Europa. *Icarus*, **147**, 317–319.

Montesi, L. G. J. and Zuber, M. T. (2003a). Spacing of faults at the scale of the lithosphere and localization instability: 1. Theory. *J. Geophys. Res.*, **108**, doi:10.1029/2002JB001923.

Montesi, L. G. J. and Zuber, M. T. (2003b). Clues to the lithospheric structure of Mars from wrinkle ridge sets and localization instability. *J. Geophys. Res.*, **108**, doi:10.1029/2002JE001974.

Murchie, S. L. and Head, J. W. (1988). Possible breakup of dark terrain on Ganymede by large-scale shear faulting. *J. Geophys. Res.*, **93**, 8795–8824.

Murchie, S. L., Watters, T. R., Robinson, M. S., Head, J. W., Strom, R. G., Chapman, C. R., Solomon, S. C., McClintock, W. E., Prockter, L. M., Domingue, D. L., and Blewett, D. T. (2008). Geology of the Caloris Basin, Mercury: A new view from MESSENGER. *Science*, **321**, 73–76.

Nimmo, F. and Giese, B. (2005). Thermal and topographic tests of Europa chaos formation models from Galileo E15 observations. *Icarus*, **177**, 327–340.

Nimmo, F. and Pappalardo, R. T. (2006). Diapir-induced reorientation of Saturn's moon Enceladus. *Nature*, **441**, 614–616.

Nimmo, F. and Tanaka, K. (2005). Early crustal evolution of Mars. *Annu. Rev. Earth Planet. Sci.*, **33**, 133–161.

Nimmo, F., Spencer, J. R., Pappalardo, R. T., and Mullen, M. E. (2007). Shear heating as the origin of the plumes and heat flux on Enceladus. *Nature*, **447**, 289–291.

Ojakangas, G. W. and Stevenson, D. J. (1989). Thermal state of an ice shell on Europa. *Icarus*, **81**, 220–241.

Okubo, C. H. and Martel, S. J. (1998). Pit crater formation on Kilauea volcano, Hawaii. *J. Volcan. Geotherm. Res.*, **86**, 1–18.

Okubo, C. H. and Schultz, R. A. (2003). Thrust fault vergence directions on Mars: A foundation for investigating global-scale Tharsis-driven tectonics. *Geophys. Res. Lett.*, **30**, doi:10.1029/2003GL018664.

Okubo, C. H. and Schultz, R. A. (2004). Mechanical stratigraphy in the western equatorial region of Mars based on thrust fault-related fold topography and implications for near-surface volatile reservoirs. *Geol. Soc. Am. Bull*, **116**, 594–605.

Okubo, C. H. and Schultz, R. A. (2006). Variability in Early Amazonian Tharsis stress state based on wrinkle ridges and strike-slip faulting. *J. Struct. Geol.*, **28**, 2169–2181.

Okubo, C. H., Schultz, R. A., and Stefanelli, G. S. (2004). Gridding Mars Orbiter Laser Altimeter data with GMT: Effects of pixel size and interpolation methods on DEM integrity. *Computers Geosci.*, **30**, 59–72.

Pappalardo, R. T. and Collins, G. C. (2005). Strained craters on Ganymede. *J. Struct. Geol.*, **27**, 827–838.

Pappalardo, R. T., Reynolds, S. J., and Greeley, R. (1997). Extensional tilt blocks on Miranda: Evidence for an upwelling origin of Arden Corona. *J. Geophys. Res.*, **102**, 13 369–13 379.

Pappalardo, R. T. *et al.* (1999). Does Europa have a subsurface ocean? Evaluation of the geological evidence. *J. Geophys. Res.*, **104**, 24 015–24 055.

Pappalardo, R. T., Collins, G. C., Head, J. W., Helfenstein, P., McCord, T., Moore, J. M., Prockter, L. M., Schenk, P. M., and Spencer, J. R. (2004). Geology of Ganymede. In *Jupiter: The Planet, Satellites and Magnetosphere*, eds. F. Bagenal, T. E. Dowling and W. B. McKinnon. Tucson, AZ: University of Arizona Press, pp. 363–396.

Pechmann, J. C. (1980). The origin of polygonal troughs on the northern plains of Mars. *Icarus*, **42**, 185–210.

Pechmann, J. B. and Melosh, H. J. (1979). Global fracture patterns of a despun planet application to Mercury. *Icarus*, **38**, 243–250.

Pettengill, G. H., Eliason, E., Ford, P. G., Loriot, G. B., Masursky, H., and McGill, G. E. (1980). Pioneer Venus radar results: Altimetry and surface properties. *J. Geophys. Res.*, **85**, 8261–8270.

Phillips, R. J. and Hansen, V. L. (1994). Tectonic and magmatic evolution of Venus. *Annu. Rev. Earth Planet. Sci.*, **22**, 597–654.

Phillips, R. J. and Hansen, V. L. (1998). Geological evolution of Venus: Rises, plains, plumes, and plateaus. *Science*, **279**, 1492–1497.

Phillips, R. J. and Solomon, S. C. (1997). Compressional strain history of Mercury (abs.). *Lunar Planet. Sci. Conf. XXVIII*, 1107–1108.

Phillips, R. J., Raubertas, R. F., Arvidson, R. E., Sarkar, I. C., Herrick, R. R., Izenberg, N., and Grimm, R. E. (1992). Impact craters and Venus resurfacing history. *J. Geophys. Res.*, **97**, 15 923–15 948.

Phillips, R. J., Zuber, M. T., Solomon, S. C., Golombek, M. P., Jakosky, B. M., Banerdt, W. B., Smith, D. E., Williams, R. M. E., Hynek, B. M., Aharonson, O., and Hauck II, S. A. (2001). Ancient geodynamics and global-scale hydrology on Mars. *Science*, **291**, 2587–2591.

Plaut, J. J. (1993). Stereo imaging. In *Guide to Magellan Image Interpretation*, eds. J. P. Ford, J. J. Plaut, C. M. Weitz, T. G. Farr, D. A. Senske, E. R. Stofan, G. Michaels and

T. J. Parker. JPL Publ. 93–24. Pasadena, CA: NASA and Jet Propulsion Laboratory, pp. 33–37.

Plescia, J. B. (1987). Geologic terrains and crater frequencies on Ariel. *Nature*, **327**, 201–204.

Plescia, J. B. (1988). Cratering history of Miranda: Implications for geologic processes. *Icarus*, **73**, 442–461.

Plescia, J. B. and Golombek, M. P. (1986). Origin of planetary wrinkle ridges based on the study of terrestrial analogs. *Geol. Soc. Am. Bull.*, **97**, 1289–1299.

Plescia, J. B. and Saunders, R. S. (1982). Tectonic history of the Tharsis Region, Mars. *J. Geophys. Res.*, **87**, 9775–9791.

Polit, A. T. (2005). Influence of mechanical stratigraphy and strain on the displacement–length scaling of normal faults on Mars, 2005. M.S. thesis, University of Nevada, Reno.

Porco, C. C., Baker, E., Barbara, J., Beurle, K., Brahic, A., Burns, J. A., Charnoz, S., Cooper, N., Dawson, D. D., Del Genio, A. D., Tilmann, D., Dones, L., Dyudina, U., Evans, M. W., Fussner, S. *et al.* (2005a). Imaging of Titan from the Cassini spacecraft. *Nature*, **434**, 159.

Porco, C. C., Baker, E., Barbara, J., Beurle, K., Brahic, A., Burns, J. A., Charnoz, S., Cooper, N., Dawson, D. D., Del Genio, A. D., Denk, T., Dones, L., Dyudina, U., Evans, M. W., Giese, B. *et al.* (2005b). Cassini imaging science: Initial results on Phoebe and Iapetus. *Science*, **307**, 1237–1242.

Porco C. C., Helfenstein, P., Thomas, P. C., Ingersoll, A. P., Wisdom, J., West, R., Neukum, G., Denk, T., Wagner, R., Roatsch, T., Kieffer, S., Turtle, E., McEwen, A., Johnson, T. B., Rathbun, J. *et al.* (2006). Cassini observes the active south pole of Enceladus. *Science*, **311**, 1393–1401.

Price, M. and Suppe, J. (1995). Constraints on the resurfacing history of Venus from the hypsometry and distribution of volcanism, tectonism, and impact craters. *Earth, Moon and Planets*, **71**, 99–145.

Prockter, L. M., Head, J. W., Pappalardo, R. T., Sullivan, R. L., Clifton, A. E., Giese, B., Wagner, R., and Neukum, G. (2002). Morphology of Europan bands at high resolution: A mid-ocean ridge-type rift mechanism. *J. Geophys. Res.*, **107**, doi:10.1029/2000JE001458.

Prockter, L. M., Pappalardo, R. T., and Nimmo, F. (2005). A shear heating origin for ridges on Triton. *Geophys. Res. Lett.*, **32**, doi:10.1029/2005GL022832.

Riley, J., Hoppa, G. V., Greenberg, R., Tufts, B. R., and Geissler, P. (2000). Distribution of chaotic terrain on Europa. *J. Geophys. Res.*, **105**, 22 599–22 615.

Robinson, M. S., Davies, M. E., Colvin, T. R., and Edwards, K. E. (1999). A revised control network for Mercury. *J. Geophys. Res.*, **104**, 30 847–30 852.

Sabins, F. F. (1997). *Remote Sensing: Principles and Interpretation.* New York: W.H. Freeman and Company.

Schaber, G. G. and McCauley, J. F. (1980). Geologic map of the Tolstoj quadrangle of Mercury. U.S. Geol. Surv. Misc. Invest. Ser., Map I-1199, scale 1:5 000 000.

Schaber, G. G., Strom, R. G., Moore, H. J., Soderblom, L. A., Kirk, R. L., Chadwick, D. J., Dawson, D. D., Gaddis, L. A., Boyce, J. M., and Russell, J. (1992). Geology and distribution of impact craters on Venus: What are they telling us? *J. Geophys. Res.*, **97**, 13 257–13 302.

Schenk, P. M. (1991). Fluid volcanism on Miranda and Ariel: Flow morphology and composition. *J. Geophys. Res.*, **96**, 1887–1906.

Schenk, P. M. (1993). Central pit and dome craters: Exposing the interiors of Ganymede and Callisto. *J. Geophys. Res.*, **98**, 7475–7498.

Schenk, P. M. and Bulmer, M. H. (1998). Origin of mountains on Io by thrust faulting and large-scale mass movements. *Science*, **279**, 1514–1517.

Schenk, P. M. and Bussey, D. B. J. (2004). Galileo stereo topography of the lunar north polar region. *Geophys. Res. Lett.*, **31**, doi:10.1029/2004GL021197.

Schenk, P. M. and Jackson, M. P. A. (1993). Diapirism on Triton: A record of crustal layering and instability. *Geology*, **21**, 299–302.

Schenk, P. M. and McKinnon, W. B. (1987). Ring geometry on Ganymede and Callisto. *Icarus*, **72**, 209–234.

Schenk, P. M. and McKinnon, W. B. (1989). Fault offsets and lateral crustal movement on Europa: Evidence for a mobile ice shell. *Icarus*, **79**, 75–100.

Schenk, P. M., McKinnon, W. B., Gwynn, D., and Moore, J. M. (2001a). Flooding of Ganymede's bright terrains by low-viscosity water-ice lavas. *Nature*, **410**, 57–60.

Schenk, P., Hargitai, H., Wilson, R., McEwen, A., and Thomas, P. (2001b). The mountains of Io: Global and geological perspectives from Voyager and Galileo. *J. Geophys. Res.*, **106**, 33 201–33 222.

Schenk, P. M., Chapman, C. R., Zahnle, K., and Moore, J. M. (2004). Ages and interiors: The cratering record of the Galilean satellites. In *Jupiter: The Planet, Satellites and Magnetosphere*, eds. F. Bagenal, T. E. Dowling and W. B. McKinnon. Tucson, AZ: University of Arizona Press, pp. 427–456.

Schubert, G., Spohn, T., and Reynolds, R. T. (1986). Thermal histories and internal structures of the moons of the solar system. In *Satellites*, eds. J. A. Burns and M. S. Matthews. Tucson, AZ: University of Arizona Press, pp. 224–292.

Schultz, P. H. (1976). *Moon Morphology*. Austin: University of Texas Press.

Schultz, R. A. (1985). Assessment of global and regional tectonic models for faulting in the ancient terrains of Mars. *J. Geophys. Res.*, **90**, 7849–7860. (Correction to Schultz, R. A. Assessment of global and regional tectonic models for faulting in the ancient terrains of Mars. *J. Geophys. Res.*, **91**, 12 861–12 863, 1986.)

Schultz, R. A. (1989). Strike-slip faulting of ridged plains near Valles Marineris, Mars. *Nature*, **341**, 424–426.

Schultz, R. A. (1991). Structural development of Coprates Chasma and western Ophir Planum. *J. Geophys. Res.*, **96**, 22 777–22 792.

Schultz, R. A. (1995). Gradients in extension and strain at Valles Marineris. *Planet. Space Sci.*, **43**, 1561–1566.

Schultz, R. A. (1998). Multiple-process origin of Valles Marineris basins and troughs. *Planet. Space Sci.*, **46**, 827–834.

Schultz, R. A. (1999). Understanding the process of faulting: Selected challenges and opportunities at the edge of the 21st century. *J. Struct. Geol.*, **21**, 985–993.

Schultz, R. A. (2000a). Localization of bedding-plane slip and backthrust faults above blind faults: Keys to wrinkle ridge structure. *J. Geophys. Res.*, **105**, 12 035–12 052.

Schultz, R. A. (2000b). Fault-population statistics at the Valles Marineris Extensional Province, Mars: Implications for segment linkage, crustal strains, and its geodynamical development. *Tectonophysics*, **316**, 169–193.

Schultz, R. A. and Lin, J. (2001). Three-dimensional normal faulting models of Valles Marineris, Mars, and geodynamic implications. *J. Geophys. Res.*, **106**, 16 549–16 566.

Schultz, R. A. and Tanaka, K. L. (1994). Lithospheric-scale buckling and thrust structures on Mars: The Coprates rise and south Tharsis ridge belt. *J. Geophys. Res.*, **99**, 8371–8385.

Schultz, R. A. and Watters, T. R. (2001). Forward mechanical modeling of the Amenthes Rupes thrust fault on Mars. *Geophys. Res. Lett.*, **28**, 4659–4662.

Schultz, R. A. and Zuber, M. T. (1994). Observations, models, and mechanisms of failure of surface rocks surrounding planetary surface loads. *J. Geophys. Res.*, **99**, 14 691–14 702.

Schultz, R. A., Okubo, C. H., Goudy, C. L., and Wilkins, S. J. (2004). Igneous dikes on Mars revealed by MOLA topography. *Geology*, **32**, 889–892.

Schultz, R. A., Okubo, C. H., and Wilkins, S. J. (2006). Displacement–length scaling relations for faults on the terrestrial planets. *J. Struct. Geol.*, **28**, 2182–2193.

Schultz, R. A., Moore, J. M., Grosfils, E. B., Tanaka, K. L., and Mège, D. (2007). The Canyonlands model for planetary grabens: Revised physical basis and implications. In *The Geology of Mars: Evidence from Earth-Based Analogues*, eds. M. G. Chapman and I. P. Skilling. Cambridge: Cambridge University Press, pp. 371–399.

Schulze-Makuch, D., Dohm, J. M., Fan, C., Fairén, A. G., Rodriguez, J. A. P., Baker, V. R., and Fink, W. (2007). Exploration of hydrothermal targets on Mars. *Icarus*, **189**, 308–324.

Scott, D. H. and Dohm, J. M. (1990). Chronology and global distribution of fault and ridge systems on Mars. *Proc. Lunar Planet. Sci. Conf. 20*, 487–501.

Scott, D. H. and Dohm, J. M. (1997). Mars structural geology and tectonics. In *Encyclopedia of Planetary Sciences*. New York: Van Nostrand Reinhold, pp. 461–463.

Scott, D. H. and Tanaka, K. L. (1986). Geologic map of the western equatorial region of Mars. U.S. Geol. Surv. Misc. Invest. Ser. Map I-1802-A, scale 1:15 000 000.

Sharpton, V. L. and Head, J. W. (1988). Lunar mare ridges: Analysis of ridge-crater intersections and implications for the tectonic origin of mare ridges. *Proc. Lunar Planet. Sci. Conf. 18*, 307–317.

Shoemaker, E. M. and Hackman, R. J. (1962). Stratigraphic basis for a lunar timescale. In *The Moon*, eds. Z. Kopal and Z. K. Mikhailov. London: Academic Press, pp. 289–300.

Shoemaker, E. M., Lucchitta, B. K., Plescia, J. B., Squyres, S. W., and Wilhelms, D. E. (1982). The geology of Ganymede. In *Satellites of Jupiter*, ed. D. Morrison. Tucson, AZ: University of Arizona Press, pp. 435–520.

Simons, M., Solomon, S. C., and Hager, B. H. (1997). Localization of gravity and topography: Constraints on the tectonics and mantle dynamics of Venus. *Geophys. J. Int.*, **131**, 24–44.

Smith, D. E., Zuber, M. T., Solomon, S. C., Phillips, R. J., Head, J. W., Garvin, J. B., Banerdt, W. B., Muhleman, D. O., Pettingill, G. H., Neumann, G. A., Lemoine, F. G., Abshire, J. B., Aharonson, O., Brown, C. D., Hauck, S. A., II et al. (1999). The global topography of Mars and implications for surface evolution. *Science*, **284**, 1495–1503.

Smith, D. E., Zuber, M. T., Frey, H. V., Garvin, J. B., Head, J. W., Muhleman, D. O., Pettingill, G. H., Phillips, R. J., Solomon, S. C., Zwally, H. J., Banerdt, W. B., Duxbury, T. C., Golombek, M. P., Lemoine, F. G., Neumann, G. A., et al. (2001). Mars Orbiter Laser Altimeter (MOLA): Experiment summary after the first year of global mapping of Mars. *J. Geophys. Res.*, **106**, 23 689–23 722.

Smith-Konter, B. and Pappalardo, R. T. (2008). Tidally driven stress accumulation and shear failure of Enceladus's tiger stripes. *Icarus*, **198**, 435–451.

Solomon, S. C. (1976). Some aspects of core formation in Mercury. *Icarus*, **28**, 509–521.

Solomon, S. C. (1977). The relationship between crustal tectonics and internal evolution in the Moon and Mercury. *Phys. Earth Planet. Inter.*, **15**, 135–145.

Solomon, S. C. (1978). On volcanism and thermal tectonics on one-plate planets. *Geophys. Res. Lett.*, **5**, 461–464.

Solomon, S. C. (1979). Formation, history and energetics of cores in the terrestrial planets. *Phys. Earth Planet. Inter.*, **19**, 168–182.

Solomon, S. C. and Head, J. W. (1979). Vertical movements in mare basins: Relation to mare emplacement, basin tectonics, and lunar thermal history. *J. Geophys. Res.*, **84**, 1667–1682.

Solomon, S. C. and Head, J. W. (1980). Lunar mascon basins: Lava filling, tectonics, and evolution of the lithosphere. *Rev. Geophys.*, **18**, 107–141.

Solomon, S. C. and Head, J. W. (1982). Evolution of the Tharsis province of Mars: The importance of heterogeneous lithospheric thickness and volcanic construction. *J. Geophys. Res.*, **87**, 9755–9774.

Solomon, S. C., Smrekar, S. E., Bindschadler, D. L., Grimm, R. E., Kaula, W. M., McGill, G. E., Phillips, R. J., Saunders, R. S., Schubert, G., Squyres, S. W., and Stofan, E. R. (1992). Venus tectonics: An overview of Magellan observations. *J. Geophys. Res.*, **97**, 13 199–13 255.

Solomon, S. C., McNutt, R. L., Gold, R. E., Acuña, M. H., Baker, D. N., Boynton, W. V., Chapman, C. R., Cheng, A. F., Gloeckler, G., Head, J. W., Krimigis, S. M., McClintock, W. E., Murchie, S. L., Peale, S. J., Philips, R. J., Robinson, M. S., Slavin, J. A., Smith, D. E., Strom, R. G., Trombka, J. I., and Zuber, M. T. (2001). The MESSENGER Mission to Mercury: Scientific objectives and implementation. *Planet. Space Sci.*, **49**, 1445–1465.

Solomon, S. C., McNutt, R. L., Jr., Watters, T. R., Lawrence, D. J., Feldman, W. C., Head, J. W., Krimigis, S. M., Murchie, S. L., Phillips, R. J., Slavin, J. A., and Zuber, M. T. (2008). Return to Mercury: A global perspective on MESSENGER's first Mercury flyby. *Science*, **321**, 59–62.

Sotin, C., Jaumann, R., Buratti, B. J., Brown, R. H., Clark, R. N., Soderblom, L. A., Baines, K. H., Bellucci, G., Bibring, J.-P., Capaccioni, F., Cerroni, P., Combes, M., Coradini, A., Cruikshank, D. P., Drossart, P. *et al.* (2005). Release of volatiles from a possible cryovolcano from near-infrared imaging of Titan. *Nature*, **435**, 786–789.

Spencer, J. R., Pearl, J. C., Segura, M., Flasar, F. M., Mamoutkine, A., Romani, P., Buratti, B. J., Hendrix, A. R., Spilker, L. J., and Lopes, R. M. C. (2006). Cassini encounters Enceladus: Background and the discovery of a south polar hot spot. *Science*, **311**, 1401–1405.

Spitale, J. and Porco, C. (2007). Association of the jets of Enceladus with the warmest regions on its south-polar fractures. *Science*, **449**, 695–697.

Squyres, S. W. (1980). Volume changes in Ganymede and Callisto and the origin of grooved terrain. *Geophys. Res. Lett.*, **7**, 593–596.

Stofan, E. R., Sharpton, V. L., Schubert, G., Baer, G., Bindschadler, D. L., Janes, D. M., and Squyres, S. W. (1992). Global distribution and characteristics of coronae and related features on Venus: Implications for origin and relation to mantle processes. *J. Geophys. Res.*, **97**, 13 347–13 378.

Stofan, E. R., Smrekar, S. E., Tapper, S. W., Guest, J. E., and Grindrod, P. M. (2001). Preliminary analysis of an expanded corona database for Venus. *Geophys. Res. Lett.*, **28**, 4267–4270.

Strom, R. G. (1984). Mercury. In *The Geology of the Terrestrial Planets*, ed. M. H. Carr. NASA SP-469. Washington, DC: U.S. Government Printing Office, pp. 13–55.

Strom, R. G., Trask, N. J., and Guest, J. E. (1975). Tectonism and volcanism on Mercury. *J. Geophys. Res.*, **80**, 2478–2507.

Strom, R. G., Malhotra, R., Ito, T., Yoshida, F., and Kring, D. A. (2005). The origin of planetary impactors in the inner solar system. *Science*, **309**, 1847–1850.

Tanaka, K. L. (1986). The stratigraphy of Mars. In Proceedings of the 17th Lunar and Planetary Science Conference. *J. Geophys. Res.*, Supplement., pt. 1, **91**, E139-E158.

Tanaka, K. L. (1990). Tectonic history of the Alba Patera-Ceraunius Fossae region of Mars. *Proc. Lunar Planet. Sci. Conf.* **20**, 515–523.

Tanaka, K. L. and Davis, P. A. (1988). Tectonic history of the Syria Planum province of Mars. *J. Geophys. Res.*, **93**, 14 893–14 917.

Tanaka, K. L. and Hartmann, W. K. (2008). The planetary timescale. In *The Concise Geologic Time Scale*, eds. J. G. Ogg, G. M. Ogg and F. M. Gradstein. New York: Cambridge University Press, pp. 13–22.

Tanaka, K. L., Golombek, M. P., and Banerdt, W. B. (1991). Reconciling stress and structural histories of the Tharsis region of Mars. *J. Geophys. Res.*, **96**, 15 617–15 633.

Tanaka, K. L., Scott, D. H., and Greeley, R. (1992). Global stratigraphy. In *Mars*, eds. H. H. Kieffer, B. M. Jakosky, C. W. Snyder and M. S. Matthews. Tucson, AZ: University of Arizona Press, pp. 345–382.

Tanaka, K. T., Moore, H. J., Schaber, G. G., Chapman, M. G., Stofan, E. R., Campbell, D. B., Davis, P. A., Guest, J. E., McGill, G. E., Rogers, P. G., Saunders, R. S., and Zimbelman, J. R. (1994). The Venus geologic mappers' handbook. U.S. Geol. Surv. Open-File Rep. 94–438, 66 pp.

Tanaka, K. L., Senske, D. A., Price, M., and Kirk, R. L. (1997). Physiography, geomorphic/geologic mapping, and stratigraphy of Venus. In *Venus II: Geology, Geophysics, Atmosphere, and Solar Wind Environment*, eds. S. W. Bougher, D. M. Hunten and R. J. Phillips. Tucson, AZ: University of Arizona Press, pp. 667–694.

Tanaka, K. L., Dohm, J. M., Lias, J. H., and Hare, T. M. (1998). Erosional valleys in the Thaumasia region of Mars: Hydrothermal and seismic origins. *J. Geophys. Res.*, **103**, 31 407–31 419.

Tanaka, K. L., Skinner Jr., J. A., Hare, T. M., Joyal, T., and Wenker, A. (2003). Resurfacing history of the northern plains of Mars based on geologic mapping of Mars Global Surveyor data. *J. Geophys. Res.*, **108**, doi:10.1029/2002JE001908.

Tera, F., Papanastassiou, D. A., and Wasserburg, G. J. (1974). Isotopic evidence for a terminal lunar cataclysm. *Earth Planet. Sci. Lett.*, **22**, 1–21.

Thomas, P. G. (1997). Are there other tectonics than tidal despinning, global contraction and Caloris-related events on Mercury? A review of questions and problems. *Planet. Space Sci.*, **45**, 3–13.

Thomas, P. G., Masson, P., and Fleitout, L. (1988). Tectonic history of Mercury. In *Mercury*, ed. F. Vilas, C. R. Chapman and M. S. Matthews. Tucson, AZ: University of Arizona Press.

Tuckwell, G. W. and Ghail, R. C. (2003). A 400-km-scale strike-slip zone near the boundary of Thetis Regio, Venus. *Earth Planet. Sci. Lett.*, **211**, 45–45.

Turtle, E. P., Jaeger, W. L., Keszthelyi, L. P., McEwen, A. S., Milazzo, M., Moore, J., Phillips, C. B., Radebaugh, J., Simonelli, D., Chuang, F., and Schuster, P. (2001). Galileo SSI Team, Mountains on Io: High-resolution Galileo observations, initial interpretations, and formation models. *J. Geophys. Res.*, **106**, 33 175–33 200.

Urey, H. C. (1952). *The Planets*. New Haven, CT: Yale University Press.

Watters, T. R. (1988). Wrinkle ridge assemblages on the terrestrial planets. *J. Geophys. Res.*, **93**, 10 236–10 254.

Watters, T. R. (1992). A system of tectonic features common to Earth, Mars, and Venus. *Geology*, **20**, 609–612.

Watters, T. R. (1993). Compressional tectonism on Mars. *J. Geophys. Res.*, **98**, 17 049–17 060.

Watters, T. R. (2004). Elastic dislocation modeling of wrinkle ridges on Mars. *Icarus*, **171**, 284–294.

Watters, T. R. and Maxwell, T. A. (1986). Orientation, relative age, and extent of the Tharsis plateau ridge system. *J. Geophys. Res.*, **91**, 8113–8125.

Watters, T. R., Robinson, M. S., and Cook, A. C. (1998). Topography of lobate scarps on Mercury: New constraints on the planet's contraction. *Geology*, **26**, 991–994.

Watters, T. R., Robinson, M. S., and Cook, A. C. (2001). Large-scale lobate scarps in the southern hemisphere of Mercury. *Planet. Space Sci.*, **49**, 1523–1530.

Watters, T. R., Schultz, R. A., Robinson, M. S., and Cook, A. C. (2002). The mechanical and thermal structure of Mercury's early lithosphere. *Geophys. Res. Lett.*, **29**, 1542.

Watters, T. R., Robinson, M. S., Bina, C. R., and Spudis, P. D. (2004). Thrust faults and the global contraction of Mercury. *Geophys. Res. Lett.*, **31**, doi:10.1029/2003GL019171.

Watters, T. R., Nimmo, F., and Robinson, M. S. (2005). Extensional troughs in the Caloris Basin of Mercury: Evidence of lateral crustal flow. *Geology*, **33**, doi:10.1130/G21678.

Watters, T. R., McGovern, P. J., and Irwin III, R. P. (2007). Hemispheres apart: The crustal dichotomy on Mars. *Annu. Rev. Earth Planet. Sci.*, **35**, 621–652.

Wilhelms, D. E. (1972). Geologic mapping of the second planet. U.S. Geol. Surv. Interagency Report, Astrogeology **55**, 36 pp.

Wilhelms, D. E. (1987). The geologic history of the Moon. U.S. Geol. Surv., Prof. Paper 1348.

Wilhelms, D. E. (1990). Geologic mapping. In *Planetary Mapping*, eds. R. Greeley and R. M. Batson. New York: Cambridge University Press, pp. 208–260.

Wilkins, S. J., Schultz, R. A., Anderson, R. C., Dohm, J. M., and Dawers, N. C. (2002). Deformation rates from faulting at the Tempe Terra extensional province, Mars. *Geophys. Res. Lett.*, **29**, doi:10.1029/2002GL015391.

Wilkins, S. J. and Schultz, R. A. (2003). Cross faults in extensional settings: Stress triggering, displacements localization, and implications for the origin of blunt troughs at Valles Marineris. *J. Geophys. Res.*, **108**, doi:10.1029/2002JE001968.

Wise, D. U., Golombek, M. P., and McGill, G. E. (1979). Tharsis province of Mars: Geologic sequence, geometry, and a deformation mechanism. *Icarus*, **38**, 456–472.

Withers, P. and Neumann, G. A. (2001). Enigmatic northern plains of Mars. *Nature*, **410**, 651.

Wyrick, D., Ferrill, D. A., Morris, A. P., Colton, S. L., and Sims, D. W. (2004). Distribution, morphology, and origins of Martian pit crater chains. *J. Geophys. Res.*, **109**, doi:06010.01029/02004JE002240.

Young, D. A. and Hansen, V. L. (2003). Geologic map of the Rusalka quadrangle (V-25), Venus. U.S. Geol. Surv. Invest. Ser., Map I-2783, scale 1:5 000 000.

Zahnle, K., Schenk, P., Levison, H., and Dones, L. (2003). Cratering rates in the outer solar system. *Icarus*, **163**, 263–289.

Zuber, M. T., Smith, D. E., Lemoine, F. G., and Neumann, G. A. (1994). The shape and internal structure of the Moon from the Clementine mission. *Science*, **266**, 1839–1843.

Zuber, M. T., Solomon, S. C., Phillips, R. J., Smith, D. E., Tyler, G. L., Aharonson, O., Balmino, G., Banerdt, W. B., Head, J. W., Johnson, C. L., Lemoine, F. G., McGovern, P. J., Neumann, G. A., Rowlands, D. D., and Zhong, S. (2000). Internal structure and early thermal evolution of Mars from Mars Global Surveyor topography and gravity. *Science*, **287**, 1788–1793.

9

Strength and deformation of planetary lithospheres

David L. Kohlstedt
Department of Geology and Geophysics, University of Minnesota, Minneapolis

and

Stephen J. Mackwell
Lunar and Planetary Institute, Houston

Summary

Robotic missions to destinations throughout our solar system have illuminated in increasing detail evidence of past and present tectonics combined with manifestations of internal dynamics. Interpretation of observations, such as sustenance of high mountains on Venus for potentially hundreds of millions of years, formation of the grooved terrain on the surface of Ganymede, and tidally driven tectonics and volcanism on Io, requires the application of realistic constitutive equations describing the rheological properties for the materials that constitute the crusts and interiors of these planetary bodies. Appropriate flow laws can only be derived from careful experimental studies under conditions that may be reliably extrapolated to those believed to exist on and in the planetary body under consideration. In addition, knowledge of the appropriate rheological behavior may, coupled with measurements made from orbiting satellites, enable the determination of geophysical properties, such as heat flow, that are otherwise not quantifiable without an expensive surface mission. In this chapter, we review the current state of knowledge of the rheological properties of materials appropriate to understanding tectonic behavior and interior dynamics for the terrestrial planets as well as the major Jovian satellites. We then discuss the utility of experimentally constrained constitutive equations in understanding large-scale processes on Venus, Mars, Europa, Ganymede and Io.

1 Introduction

Historically, much of our understanding of the deformation behavior of planetary materials derives from experimental investigations undertaken to explore the mechanical properties of minerals and rocks as related to tectonic processes on

Planetary Tectonics, edited by Thomas R. Watters and Richard A. Schultz. Published by Cambridge University Press. Copyright © Cambridge University Press 2010.

our own planet, Earth. Since the early 1980s, however, a few laboratory studies of rock strength have been undertaken specifically to address questions raised in the planetary community. In this chapter on strength and deformation of planetary lithospheres, we draw on examples from recent research related to the rheological properties of two planets and three planetary satellites. From these case studies and the larger body of literature for Earth, we develop the understanding of rock deformation necessary to analyze quantitatively the dynamic behavior of planetary bodies.

This chapter is laid out in four sections. In the first, we introduce the basic mechanisms and governing equations for flow of crystalline materials by examining the case of ice deformation within the context of Jupiter's icy satellites, Europa and Ganymede. In the second, we explore laboratory constraints on lithospheric strength, and we describe the concept of a strength envelope, which defines an upper limit to rock strength as a function of depth as applied to Earth's twin, Venus. In the third, we extend the constitutive flow equations to emphasize the role of water in the mechanical properties of nominally anhydrous silicate minerals within the framework of possible importance to the early history of Mars. In the fourth, we amplify on the effect of partial melting on the strength of rocks, a topic of central importance to deformation of Jupiter's moon, Io. Taken together, these four sections provide an integrated overview of the rheological behavior of rocks within the framework of the flow and strength of the lithospheres of planetary bodies.

1.1 Flow of rocks: Europa and Ganymede

The need for experimentally determined flow laws for water ice in understanding and interpreting planetary features and processes has become acute in the past several decades. Observations of the icy Galilean satellites from Voyager 1 and 2 in addition to Galileo reveal surfaces that have evolved significantly over time, with evidence of recent tectonic and potentially volcanic activity, as well as impact cratering. Europa and Ganymede are also believed to have icy shells overlaying liquid water layers, where conditions may be appropriate for the development of life. Experimentally determined rheological properties can be used to constrain the processes that formed these surface features, to model the ability of the surface to retain tectonic or impact structures, and even to allow determination of the thickness of the icy shells. Rheological (flow) laws are also useful in modeling the water-ice polar caps on Mars, and potentially provide useful constraints on flow of an icy regolith needed to understand the formation of features such as the large slumps evident on the flanks of Olympus Mons.

On Earth, the rheological properties of ice dictate the flow of glaciers and ice sheets. Hence, several laboratory and numerous field studies have explored in some detail the flow behavior of ice I, the low-pressure phase of ice relevant to planetary bodies. This work has been extended with deformation experiments on ice in the broader context of planetary tectonics, first undertaken by Durham and colleagues (for a review, see Durham and Stern, 2001). In their ground-breaking work, these researchers developed a cryogenic high-pressure apparatus that allowed them to deform relatively coarse grained ice at reasonably high stresses entirely within the plastic flow regime, without the complications introduced by fracturing along grain boundaries that limited the applicability of results from earlier studies.

1.2 Mechanisms of deformation

As with other crystalline materials, ice deforms by several different mechanisms involving zero-dimensional (point), one-dimensional (line), and two-dimensional (planar) defects. That is, plastic deformation occurs by diffusion of ions, motion of dislocations, and sliding along grain boundaries. Broadly speaking, deformation mechanisms can be divided into two categories, those that are controlled by diffusion of ions (diffusion creep) and those that are governed by propagation (glide, climb, and cross slip) of dislocations (dislocation creep). In both cases, the relative motion of neighboring grains by ductile movement along their common interface (i.e., grain boundary sliding), a process regulated by ionic diffusion or dislocation motion, can be an essential process. For a given material such as ice, each deformation mechanism is characterized by a unique dependence of strain rate (viscosity) on parameters such as stress, grain size, temperature, and pressure. Hence, each mechanism will dominate over a limited range of these conditions. For example, as discussed below, diffusion creep tends to be most important in fine-grained rocks at low differential stresses, while dislocation creep often dictates the rate of deformation in coarse-grained rocks at larger differential stresses.

The flow behavior of a crystalline material is generally expressed in terms of its viscosity, η. In general, viscosity is a function of the imposed differential stress, σ, temperature, T, confining pressure, P, elements of the microstructure, S, and activities, a, or fugacities, f, of the chemical components expressed as (e.g., Evans and Kohlstedt, 1995)

$$\eta = \eta(\sigma, T, P, S_1, S_2, \ldots, a_1, a_2, \ldots, f_1, f_2, \ldots). \tag{9.1}$$

Microstructural elements include parameters such as grain size, subgrain size, texture and fabric, dislocation density, and, if more than one phase is present, their proportions, morphology, and distributions. For silicate minerals or rocks, chemical activities and fugacities of components include the activity of silica, a_{SiO_2}, and iron oxide, a_{FeO}, and fugacities include oxygen fugacity, f_{O_2}, and water fugacity, f_{H_2O}. In experimental studies of the flow behavior of ice and silicate rocks, deformation results are frequently expressed in terms of strain rate, $\dot{\varepsilon}$, rather than viscosity, with the two quantities related by

$$\eta \equiv \frac{\sigma}{\dot{\varepsilon}}. \qquad (9.2)$$

1.2.1 Deformation by ionic diffusion coupled with grain boundary sliding

Diffusion creep can be divided into two regimes based on whether the diffusive flux is dominated by transport along grain boundaries or through grain matrixes (grain interiors). Both processes yield Newtonian viscosities, that is, viscosities that are independent of differential stress. Restated in terms of strain rate, the strain rate is linearly proportional to differential stress. Additionally, in both diffusion creep regimes, strain rate increases as the characteristic diffusion distance decreases, that is, with decreasing grain size. The strain rate – stress – grain size relationship thus takes the general form

$$\dot{\varepsilon}_{\text{diff}} \propto \frac{\sigma}{d^p}. \qquad (9.3)$$

Models predict a grain size exponent in Equation (9.3) of $p = 2$ if grain matrix (interior) diffusion dominates the diffusion flux (Nabarro, 1948; Herring, 1950) and $p = 3$ if grain boundary diffusion governs (Coble, 1963). These original formulations of the flow laws for diffusion creep considered deformation of a single spherical grain. Subsequent analyses have emphasized the importance of grain boundary sliding in polycrystalline materials as an essential part of diffusion creep (Liftshitz, 1963; Raj and Ashby, 1971). If grain boundaries are too weak to support shear stresses, the creep process can be considered to be one in which grain boundary sliding is accommodated by diffusion both through grain matrixes (interiors) and along grain boundaries. As is common, we refer to this deformation mechanism of grain boundary sliding with diffusion accommodation simply as diffusion creep.

Since grain boundary (gb) and grain matrix (gm) diffusion are independent/parallel processes, Equation (9.3) can be written to include the dependence of strain rate on the grain matrix and grain boundary diffusivities, D_{gm} and D_{gb} (Raj and Ashby, 1971):

$$\dot{\varepsilon}_{\text{diff}} = 14 \left(\frac{\sigma V_m}{RT} \right) \left(D_{\text{gm}} + \pi \frac{\delta}{d} D_{\text{gb}} \right) \left(\frac{1}{d^2} \right), \qquad (9.4)$$

where V_m is the molar volume, R is the gas constant, and δ is the grain boundary width; δ is on the order of 1 nm for most materials important in planetary applications (e.g., Carter and Sass, 1981; Ricoult and Kohlstedt, 1983). In diffusion creep, grain size and grain boundary width correspond to microstructural elements introduced in Equation (9.1).

The primary effects of temperature and pressure on creep rate (viscosity) enter through the grain matrix and grain boundary diffusivities both of which have the form

$$D_{gm/gb} = D^o_{gm/gb} \exp\left(-\frac{E_{gm/gb} + PV_{gm/gb}}{RT}\right), \qquad (9.5)$$

where $E_{gm/gb}$ and $V_{gm/gb}$ are the activation energy and activation volume, respectively, for grain matrix or grain boundary diffusion. With decreasing grain size, the contribution of grain boundary diffusion to deformation becomes more significant than the contribution due to grain matrix diffusion. This point is captured in Equation (9.4) since $\dot{\varepsilon}_{gb} \propto 1/d^3$, while $\dot{\varepsilon}_{gm} \propto 1/d^2$. Furthermore, based on Equation (9.5), grain boundary diffusion contributes more to the creep rate than does grain matrix diffusion at low temperatures since normally $E_{gb} < E_{gm}$.

1.2.2 Deformation by dislocation processes

Several distinct mechanisms for deformation by dislocation motion – which takes place by glide, climb and cross slip – have been considered in the materials science literature; compilations and descriptions of dislocation creep mechanisms are recorded in Poirier (1985) and Evans and Kohlstedt (1995). In this chapter, we will primarily discuss glide, in which dislocations move in their slip or glide plane, and climb, in which dislocations require diffusion to move normal to their slip plane. Broadly, dislocation mechanisms of deformation can be divided into low-temperature processes in which dislocation glide dominates and high-temperature mechanisms in which dislocation climb controls the rate of deformation (Weertman, 1999). Laboratory and field studies of the rheological behavior of ice I have generally emphasized the latter in analyses of deformation results.

At high temperatures (roughly $T \geq T_m/2$), dislocation climb as well as glide and cross slip contribute to deformation. The bulk of the strain results from dislocation glide, while the strain rate is controlled by climb or cross slip. Flow laws in this dislocation creep regime are generally written in the form of a power law (Weertman, 1999):

$$\dot{\varepsilon}_{disl} = A_{disl}\, \sigma^n \exp\left(-\frac{E_{disl} + PV_{disl}}{RT}\right). \qquad (9.6)$$

In the dislocation creep regime, the stress exponent typically lies in the range $3 \leq n \leq 5$. Strain rate does not depend on grain size, provided that there are a sufficient number of slip systems to fulfill the von Mises (1928) criterion (discussed below) or, at least, the relaxed von Mises criterion (Paterson, 1969). In this case, viscosity is non-Newtonian, that is, a function of differential stress. For many materials, the activation energy for dislocation creep is equal to the activation energy for grain matrix diffusion; this observation holds for ice for which the activation energy for dislocation creep is equal to that for self-diffusion of either hydrogen or oxygen (Weertman, 1983). Therefore, the strain rate is written as a function of the grain matrix diffusion coefficient

$$\dot{\varepsilon} \propto D_{gm} \propto \exp\left(-\frac{E_{gm} + PV_{gm}}{RT}\right), \tag{9.7}$$

suggestive of an important role of dislocation climb in controlling the rate of plastic deformation (Weertman, 1968, 1999).

For ice and rock samples with irregular, three-dimensional grain shapes to deform homogeneously by dislocation creep without dilatation resulting from the formation of cracks or voids, several slip systems must simultaneously operate in all of the grains in the aggregate. Based on purely geometric arguments, von Mises (1928) demonstrated that five independent slip systems must operate within each mineral grain for homogeneous deformation. The bulk of the strain can be accommodated by glide on two or three of the slip systems, while the other slip systems take up only minor amounts of strain in order to minimize strain heterogeneity and avoid development of cracks and voids. In this case, the aggregate strength is expected to lie intermediate between the strengths of the weakest and the strongest of the required slip systems and to have constitutive parameters dominated by the strongest required slip system. If deformation occurs by climb as well as glide, three slip systems suffice for homogeneous deformation (Groves and Kelly, 1969). Similarly, processes such as grain boundary and grain matrix diffusion as well as grain boundary sliding often serve as a further accommodation process, hence necessitating fewer active slip systems. If grain boundary sliding contributes to the deformation, the aggregate flow law may show a dependence on grain size, consistent with the enhanced role of this process at finer grain sizes.

Many metals and some ceramics are composed of crystalline grains with high symmetry, such that the von Mises criterion is easily met. However, in most rocks, the crystalline grains have relatively low symmetry and, hence, the number of active slip systems is limited. Thus, despite the relative weakness of the easiest slip system in a mineral, the aggregate strength may be very high due to the necessity for dislocation glide on an energetically unfavorable slip system. For example, in ice, slip is significantly easier on the basal plane than on non-basal slip systems. Activation of non-basal slip in single crystals requires stresses a factor of 10 to 100 greater than required for the basal slip system. The strength of polycrystalline ice with a coarse grain size ($\sim$1–10 mm) lies between that of single crystals oriented for basal slip, with dislocations gliding on the basal plane of ice, and those oriented for non-basal slip, as illustrated in the log–log plot of strain rate as a function of differential stress in Figure 9.1.

Figure 9.1. Log–log plot of strain rate as a function of differential stress, illustrating the strength of coarse-grained polycrystalline ice relative to that of single crystals of ice oriented for slip on the basal plane and single crystals of ice oriented for slip on non-basal planes. Note that the differential stress required to deform polycrystalline samples at a given strain rate lies between the values for slip on the hard and easy slip systems in ice. Comparison of the slopes, that is, the stress exponent n, in Equation (9.6), of these three lines suggests that slip on non-basal systems controls the rate of deformation, while the relative positions of the lines indicates that basal slip contributes significantly to flow. Modified from Goldsby and Kohlstedt (2001).

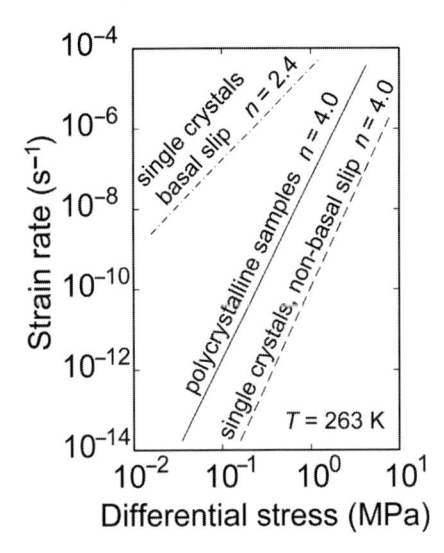

The significant contrast in strength of the various slip systems in minerals such as ice means that strain-producing, non-dislocation processes may become active at stresses below those required to activate the stronger slip systems. In particular, grain boundary sliding can be very important in ice and rock samples deforming by dislocation creep. As the rate of grain boundary sliding depends on the size of the sliding interface, this phenomenon displays a dependence on grain size, with finer grain sizes favoring the activation of grain boundary sliding. Grain boundary sliding is also often characterized by a smaller value of the stress exponent than observed in dislocation creep. The resultant flow law will otherwise closely resemble a standard flow law for dislocation (power-law) creep.

1.2.3 Deformation by dislocation processes coupled with grain boundary sliding

As in the diffusion creep regime, grain boundary sliding can contribute significantly to deformation in the dislocation creep regime, such that it is appropriate to discuss a creep regime in which dislocation motion is accompanied by grain boundary sliding. In detail, grain boundary sliding in the dislocation creep regime differs from that in the diffusion creep regime due to movement of dislocations along grain boundaries; hence, in this regime, grain boundary sliding is sometimes referred to as enhanced or "stimulated" grain boundary sliding (Kaibyshev, 1992). In this regime, termed here the dislocation – grain boundary sliding (disl–gbs) regime, strain rate is again a function of grain size, as well as a non-linear function of differential stress:

$$\dot{\varepsilon}_{\text{disl-gbs}} = A_{\text{disl-gbs}} \frac{\sigma^n}{d^p} \exp\left(-\frac{E_{\text{disl-gbs}} + PV_{\text{disl-gbs}}}{RT}\right). \quad (9.8)$$

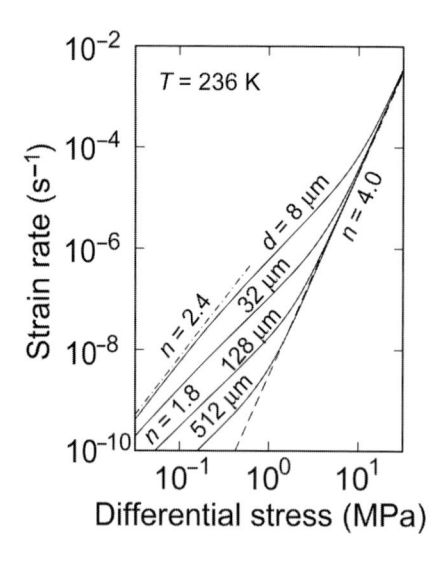

Figure 9.2. Log–log plot of viscosity as a function of differential stress for fine-grained polycrystalline ice with grain sizes of 8, 32, 128, and 512 μm. The flow law for coarse-grained ice deforming in the dislocation creep regime is indicated by the dashed line of slope $n = 4.0$, while the flow law for single crystals of ice oriented for slip on the basal plane is included as a dot-dashed line of slope $n = 2.4$. Between these two bounds, polycrystalline ice flows by grain boundary sliding accommodated by dislocation movement in the dislocation – grain boundary sliding regime characterized by a slope of $n = 1.8$.

Models for this creep regime yield values of $n = 2$, $p = 2$ and $n = 3$, $p = 1$ (Gifkins, 1972; Langdon, 1994), where the former values of n and p were derived for the case in which grains are free of subgrains (sometimes stated as "grain size smaller than subgrain size"), while the latter pair of values applies when the grain size is larger than the subgrain size. Laboratory experiments demonstrate that flow of fine-grained ice is, in fact, characterized by $n = 1.8$ and $p = 1.4$ (Goldsby and Kohlstedt, 2001; Durham et al., 2001). As illustrated in Figure 9.2, the dislocation – grain boundary sliding regime dominates flow at low differential stresses and finer grain sizes, while dislocation creep dominates at higher differential stresses and coarser grain sizes. As grain size increases, the transition from dislocation creep to dislocation – grain boundary sliding creep moves to lower differential stresses. At the low differential stresses (≤ 0.1 MPa) appropriate for glaciers, the flow behavior of coarse-grained ice observed in *in situ* field studies is dominated by the dislocation – grain boundary sliding flow law determined from laboratory experiments (Goldsby, 2006).

1.2.4 Creep of ice I

A constitutive equation describing the creep of ice thus consists of at least three flow regimes. Diffusion creep, dislocation creep, and dislocation – grain boundary sliding creep operate largely independently. Thus, the constitutive equation has the form

$$\dot{\varepsilon} = \dot{\varepsilon}_{\text{diff}} + \dot{\varepsilon}_{\text{disl-gbs}} + \dot{\varepsilon}_{\text{disl}}. \tag{9.9a}$$

As diffusion creep appears to operate at differential stresses and grain sizes well below those explored in the laboratory and those appropriate for flow in glacial and planetary ice bodies, we eliminate this term from Equation (9.9a). In addition, Goldsby and Kohlstedt (2001) have argued that the dislocation – grain boundary sliding regime can be viewed as the operation of two interdependent/serial processes: basal glide and grain boundary sliding. Therefore, in Figure 9.2, at a given grain size and strain rate, the mechanism of deformation changes from dislocation creep ($n = 4.0$) to grain boundary sliding controlled creep ($n = 1.8$). Goldsby (2006) has suggested that for $d > 1$ mm, $\sigma > 10^{-4}$ MPa, and $T > 220$ K, the flow of ice can be well described by simplifying Equation (9.9a) to

$$\dot{\varepsilon} \approx \dot{\varepsilon}_{\text{disl-gbs}} + \dot{\varepsilon}_{\text{disl}}, \tag{9.9b}$$

where $n = 1.8$, $p = 1.4$ in the dislocation – grain boundary sliding regime and $n = 4.0$, $p = 0$ in the dislocation creep regime.

Deformation experiments undertaken to study the flow of ice with applications to satellites of the outer planets have built on an extensive literature in the glaciology field (see references in Durham and Stern, 2001; Goldsby, 2006; Iverson, 2006). In turn, recent experiments on ice designed to investigate the tectonic behavior of icy satellites, such as Europa and Ganymede, have impacted investigations of flow of ice sheets and glaciers (e.g., Cuffey et al., 2000a,b). Prior to the late 1990s, the results of much of the work on the rheological properties of ice were interpreted in terms of creep in the dislocation regime as given by an equation of the form of Equation (9.6). In this regime, flow of ice was earlier described with a stress exponent of $n = 3$ with $E = 139$ kJ/mol, following the work of Glen (1952, 1955). An extensive series of experiments by Durham et al. (1983, 1997) at high pressures, however, demonstrated that values of $n = 4.0$ and $E = 61$ kJ/mol are more appropriate in the dislocation creep regime.

Within the past decade, a concerted effort has been made to explore the diffusion creep regime of ice. Since diffusion creep is expected to dominate over dislocation creep in fine-grained samples, cylinders of ice I were cycled from ice I to ice II and then back to ice I by increasing and then rapidly decreasing pressure (Stern et al., 1997). Nucleation of new grains upon crossing the phase boundary resulted in ice grains a few microns across. To explore the effect of grain size on creep rate, samples were systematically heated for prescribed times to promote grain growth. While these experiments failed to identify a diffusion creep regime with $n = 1$ and $p = 3$, they did encounter an extended creep regime within which strain rate was markedly dependent on grain size (Goldsby and Kohlstedt, 2001; Durham et al., 2001). As described above, this grain size sensitive creep regime is characterized by robust values for the stress exponent of $n = 1.8$ and the grain size exponent of $p = 1.4$ with an activation energy of 49 kJ/mol. The presence of numerous four-grain junctions in deformed samples (as opposed to three-grain junctions found in textural equilibrium) suggests that grain boundary sliding and grain switching

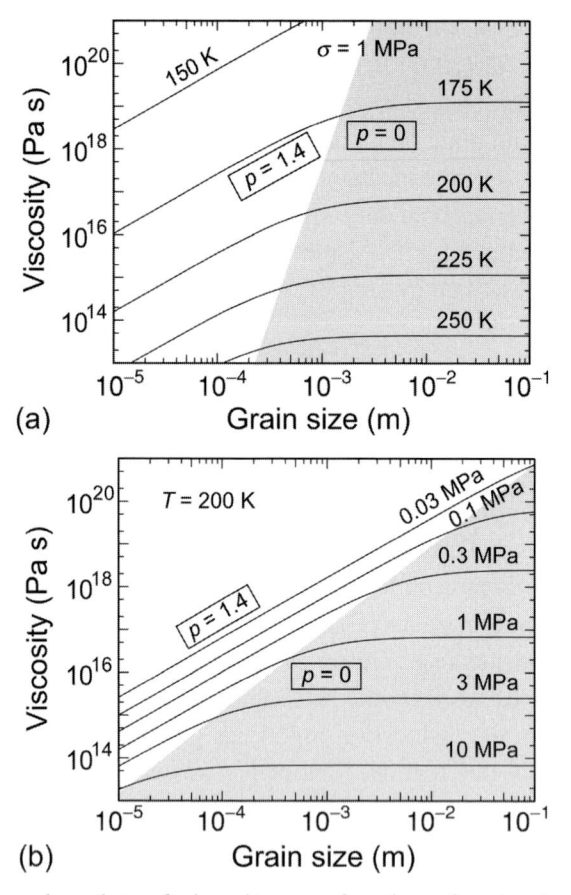

Figure 9.3. Log–log plots of viscosity as a function of grain size at (a) fixed differential stress of 1 MPa and (b) fixed temperature of 200 K. Dislocation creep dominates the viscosity at large grain sizes (shaded regions on the right of (a) and (b)), while dislocation – grain boundary sliding creep dominates at smaller grain sizes. Strain rate is independent of grain size ($p = 0$) in the former regime, while strain rate increases with decreasing grain size in the latter regime ($p = 1.4$). In (a), the transition between the two regimes occurs at smaller grain size with increasing temperature. In (b), the transition between the two regimes occurs at smaller grain size with increasing differential stress.

events played an important role. Hence, this regime was identified as one in which dislocation creep is accompanied by grain boundary sliding, as described by Equation (9.8).

The effects of temperature, differential stress, and grain size on the flow of ice due to the combined effects of dislocation creep and dislocation – grain boundary sliding as described by Equation (9.9b) are illustrated in the log–log plots of viscosity as a function of grain size in Figures 9.3a and 9.3b. These figures explore the effect of activation energy (temperature dependence), stress exponent

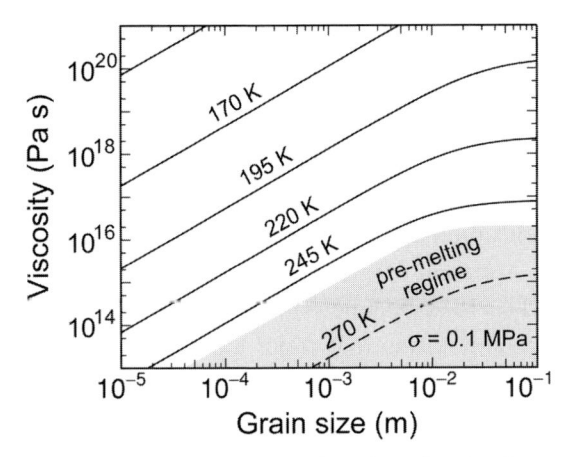

Figure 9.4. Log–log plot of strain rate as a function of grain size at a constant differential stress of 0.1 MPa for temperatures both below and above the pre-melting temperature of ∼255 K. The temperature interval between constant-temperature curves is 25 K. With increasing temperature between 170 and 245 K, the contours become progressively closer. However, the spacing between the curve for 245 K and the curve for 270 K is significantly wider than that between the curves for 220 K and 245 K, reflecting the marked increase in activation energy from ∼50 to ∼180 kJ/mol in crossing the pre-melting point.

(stress dependence), and grain size exponent (grain size dependence) on creep rate. In both Figures 9.3a and 9.3b, the contribution to the plastic flow from the grain-size sensitive, dislocation – grain boundary sliding creep regime increases with decreasing grain size. Based on Figure 9.3a, the dislocation creep regime extends to lower grain size as temperature increases, reflecting the higher activation energy in this regime. Based on Figure 9.3b, the dislocation creep regime extends to smaller grain sizes at higher stress because the stress exponent is larger in the dislocation creep regime ($n = 4.0$) than in the dislocation – grain boundary sliding regime ($n = 1.8$).

Above ∼255 K, the flow behavior of ice is dramatically affected by the apparent onset of pre-melting at grain boundaries (Duval *et al.*, 1983; Dash *et al.*, 1995). Deformation of ice occurs more rapidly for a given grain size and temperature than predicted from experimental data obtained at lower temperatures. The disproportionately large decrease in viscosity at temperatures above 255 K is clear in the log–log plot of viscosity as a function of grain size in Figure 9.4. The apparent activation energy at $T > 255$ K increases to ∼180 kJ/mol in the diffusion creep regime and to ∼190 kJ/mol in the dislocation – grain boundary sliding creep regime (Goldsby and Kohlstedt, 2001). A detailed summary of flow law parameters is presented in Table 1 of Durham and Stern (2001) and in Goldsby (2006).

1.3 Application to Europa and Ganymede

The constitutive equation describing the flow of ice I has been used in analyses of the tectonic behavior of Europa and Ganymede. Two problems related to the topography on Ganymede have been analyzed based on the combined flow laws for dislocation and dislocation – grain boundary sliding creep in ice I. First, the upper limit on the relaxation time for impact craters was calculated based on flow by solid-state creep. Apparent retention ages for craters on Ganymede are ≤1 Gyr (Passey and Shoemaker, 1982; Zahnle *et al.*, 1998). Finite element models of crater relaxation lead to the conclusion that crater topography can be sustained against viscous relaxation for billions of years, consistent with observation (Dombard and McKinnon, 2000). Second, formation of grooved terrain on Ganymede as a result of extensional plastic instability has been examined. Early analysis of this problem based on dislocation flow of ice led to the conclusion that a necking instability could not explain the ridge–trough topographic relief (Herrick and Stevenson, 1990). However, models including the grain-size sensitive, dislocation – grain boundary sliding creep regime yielded instabilities with topographic wavelengths consistent with those observed for the grooved terrain on Ganymede (Dombard and McKinnon, 2001).

Laboratory-determined flow properties of ice also provide constraints on the tectonic behavior of the ice shell on Europa (McKinnon, 1999; Ruiz and Tejero, 2000, 2003; Barr *et al.*, 2004). Models have examined the possibility of convection in a thin ice shell as a mechanism for transporting heat efficiently to the surface, as opposed to heat flow entirely by conduction. The results of these models depend critically on the flow laws used to characterize ice deformation. For example, Ruiz and Tejero (2003) conclude that convection could occur in the ice shell on Europa, accounting for the high heat flow associated with tidal heating, only if ice deforms primarily by the dislocation – grain boundary sliding mechanism. These researchers further conclude that the thickness of the ice shell is in the range ~15 to 50 km, consistent with morphology and size of impact craters (Schenk, 2002).

2 Strength envelopes: Venus

In the dynamics of Earth and other planetary bodies, many of the most fundamental processes that we seek to understand involve solid-state deformation of minerals and rocks. To model this behavior adequately, we must understand the processes involved in deformation as appropriate to the planetary setting and, when possible, utilize experimentally derived data to constrain the models. Application of experimental results is not always feasible due to limitations of the models, computational constraints, or uncertainty about the nature and conditions in the regions under consideration. However, utilization of experimentally derived mechanical data in problems of planetary dynamics and tectonics will usually provide a greater understanding than simplified models, whether or not a perfect fit to observation is obtained. Models incorporating experimental results that clearly do not fit observational data also provide important constraints on planetary processes.

Application of experimentally derived and field-correlated mechanical property data for minerals and rocks to processes in planetary interiors requires an understanding of both the nature of the planetary body and the limitations of the experiments themselves. Injudicious application of experimental data in planetary models generally leads to inappropriate conclusions and, ultimately, to an underestimation of the true merit of experimental work. At the simplest level, experimental measurements can be applied to issues of rheological behavior through illustrations of strength distributions in the lithosphere, such as the strength envelope concept (Sibson, 1977; Goetze and Evans, 1979; Brace and Kohlstedt, 1980; Kohlstedt et al., 1995). In such diagrams, the strength of rock is plotted against depth in the planetary body, under the assumptions of uniform lateral distribution of deformation, constant rate of deformation with depth, and a specific dependence of temperature on depth. Such models, though simple, allow an appreciation of issues such as depth of the seismogenic zone, integrated strength of the lithosphere, identification of zones of maximum strength in the lithosphere, distribution of vertically stratified zones of weakness, and potential detachment of crust and mantle. However, these simplified models include assumptions that will not hold in all environments. For example, rheological properties describing plastic flow, which define the behavior at depths greater than a few kilometers, are strongly temperature dependent, making the selection of an appropriate temperature profile critical. Also, the use of a depth-independent strain rate on the assumption of uniform lateral distribution of deformation ignores the abundant evidence of plastic shear zones in the continental crust of Earth. Water also has a major effect on mechanical properties for all deformation processes and with all lithologies. However, water may be present only episodically, and fluid pressures may vary significantly both laterally and vertically.

In addition, most rocks within the interior of the terrestrial planets are polyphase aggregates, composed of several minerals in differing volumetric ratios, each with characteristic morphologies and mechanical properties. Aggregate behavior will be a complex integration of the properties of individual phases with contributions due to interactions between phases (for example, interphase boundaries have very different properties than grain boundaries). For deformation in the convecting upper mantle of the terrestrial planets, dislocation creep can be reasonably approximated by the behavior of polycrystalline olivine (Zimmerman and Kohlstedt, 2004), as olivine is the predominant and weakest major mineral. However, such simplifications are not always possible, and rocks of gabbroic composition have rheologies that require a volumetric averaging of the behavior of the pyroxene and plagioclase feldspar components (Mackwell *et al.*, 1998). Consequently, strength envelopes based on single mineral flow behavior may not adequately model flow at depth. In summary, application of strength envelopes must be made judiciously with as complete an understanding of the thermal, tectonic, hydrologic, and lithologic setting as is possible, and using the appropriate flow laws.

2.1 The strength envelope model

2.1.1 Brittle deformation

In strength envelope models, deformation in the most shallow portion of the lithosphere is governed by sliding on existing fault surfaces, as described by Byerlee's law (Byerlee, 1978). For a fault with a normal at 60° from the maximum principal stress, σ_1, Byerlee's law can be written as (Sibson, 1974; Brace and Kohlstedt, 1980; McGarr *et al.*, 1982; McGarr, 1984; Kohlstedt *et al.*, 1995)

$$\sigma_1 - P_P = 4.9(\sigma_3 - P_P) \qquad \sigma_3 - P_P < 100 \text{ MPa} \qquad (9.10a)$$

$$\sigma_1 - P_P = 3.1(\sigma_3 - P_P) + 210 \qquad \sigma_3 - P_P > 100 \text{ MPa}, \qquad (9.10b)$$

where σ_3, the minimum principal stress, equals the lithostatic pressure, and P_P is the pore pressure. The stress supported by a rock mass deforming by frictional sliding depends predominantly on the lithostatic pressure, is largely independent of temperature and strain rate, and, to first order, does not vary substantially between rock types. Of course, the maximum principal stress that can be supported by frictional sliding is limited. If the fracture strength, σ_f, is reached, by an increase either in the local stress or in the pore fluid pressure, the rock may deform by cataclasis, rather than localized sliding on preexisting faults (see e.g., Kohlstedt *et al.*, 1995).

2.1.2 Semi-brittle deformation

Semi-brittle deformation becomes important at stresses lower than the fracture stress when dislocation glide becomes activated at points of stress concentration and/or when fluid-assisted grain boundary diffusion (pressure solution creep) allows some relaxation of the stress. While some attempts have been made to quantify the boundary between fully brittle (frictional sliding or fracture) and semi-brittle (brittle plus plastic) behavior (e.g., Chester, 1988), we will follow the observation-based assumption from Kohlstedt *et al.* (1995) that semi-brittle behavior is initiated when the plastic flow strength of the rock is about five times the frictional strength. This transition stress is often referred to as the brittle–ductile transition (BDT).

As depth increases, increasing temperature enables a transition from semi-brittle processes to fully plastic flow, which is only possible if the lithostatic pressure, σ_3, is sufficient to suppress brittle processes. This stress-induced transition (referred to as the Goetze criterion, C. Goetze, private communication, 1975; Kohlstedt et al., 1995) occurs approximately when

$$\sigma_1 - \sigma_3 = (\sigma_3 - P_P). \tag{9.11}$$

This criterion defines the maximum differential stress $(\sigma_1 - \sigma_3)$ that can be supported by a rock through purely plastic processes and defines the brittle–plastic transition (BPT). Between the BDT and the BPT, the strength of the rocks is rather ill-defined, as a range of brittle and plastic processes all contribute to deformation (Evans and Kohlstedt, 1995; Kohlstedt et al., 1995). In this region, the strength envelope defined by the end-member behaviors alone will provide only an upper limit on lithospheric strength, particularly at the highest strength regions. Deformation is likely to be localized, as often observed in the field, for example, in the form of mylonites (e.g., Sibson, 1977).

2.1.3 Plastic flow

At depths at which the temperature is high enough for purely plastic processes to dominate, deformation of the rocks is controlled by micro-mechanical mechanisms such as dislocation glide, dislocation climb, and diffusion creep (see, e.g., Poirier, 1985; Kohlstedt et al., 1995). In this regime at lower temperatures and higher stresses, dislocation glide dominates; dislocations within individual mineral grains are constrained to move along crystallographically controlled glide planes. At higher temperatures, diffusion becomes fast enough to enable diffusion creep and to promote dislocation creep processes that involve both climb and glide of dislocations, with diffusion-controlled climb rate-limiting deformation. For rocks with relatively large grain sizes, dislocation creep is likely to control aggregate behavior, whereas for rocks with finer grain sizes, diffusion creep is favored. As dislocations are crystallographically controlled and move on glide planes, homogeneous deformation of an aggregate by dislocation processes requires motion of dislocations on multiple slip systems, as delineated by von Mises (1928) and discussed in the sections on Europa and Mars. As the minerals that comprise the dominant rock types in the interior of terrestrial planets generally lack a sufficient number of easy slip systems for deformation to occur homogeneously, some additional component of strain accommodation, such as grain boundary sliding or diffusion, is usually required. Both diffusion and dislocation creep processes are thermally activated and can be described by a flow law (see section on Europa and, e.g., Kohlstedt et al., 1995) with an Arrhenius dependence of strain rate on temperature and pressure, as described by Equations (9.5–9.8). Strain rate may also depend on other aspects of the thermodynamic environment, such as oxygen fugacity (for iron-bearing silicates), activity of component oxides, and water fugacity. For example, in the dislocation creep regime, mantle olivine deforms a factor of ~8 faster when oxygen fugacity is buffered near Fe/FeO than when buffered at Ni/NiO (Keefner *et al.*, 2005).

2.1.4 Water weakening

Water is known to produce a weakening of rock throughout Earth's interior in the brittle, brittle–plastic, and plastic deformation regimes. At shallow depths, weakening results from a decrease in the loading across fault surfaces due to the fluid pressure; at greater depths it may result from weakening of fault surfaces due to the presence of platy hydrous minerals with easy cleavage planes parallel to the fault plane, or due to decreases in the intracrystalline or intercrystalline strength of the minerals/rocks themselves. In particular, the presence of water has been observed to reduce the high-temperature strength of most silicates, including quartzite (Gleason and Tullis, 1995; Post *et al.*, 1996), feldspar (Rybacki and Dresen, 2000), pyroxene (Ross and Nielsen, 1978; Bystricky and Mackwell, 2001; Chen *et al.*, 2006), and olivine (Chopra and Paterson, 1984; Mackwell *et al.*, 1985; Karato *et al.*, 1986; Mei and Kohlstedt, 2000a,b; Karato and Jung, 2003; see also section on Mars). This weakening effect has been correlated to increased rates of dislocation climb within mineral grains, as well as enhanced rates of intracrystalline (grain matrix) diffusion and grain boundary diffusion. In addition, increased water activity enhances grain growth and recovery processes such as recrystallization, which also affect the mechanical behavior of a rock mass. Thus, any model for deformation of the interior of a planetary body must allow for the possible role of water.

2.1.5 Strength envelope for oceanic lithosphere

Oceanic crust on Earth is composed of rocks derived from basalt; as such, mechanical properties appropriate to basaltic or gabbroic rocks should provide a reasonable approximation to the behavior of this region. However, the ocean crust is relatively thin (~6 km), and quite cold, so that deformation is primarily brittle and occurs by frictional sliding on systems of faults. Fluid circulation occurs to significant depths (Ranero *et al.*, 2003), decreasing the effective pressure within the fault zones. Also, the presence of aqueous fluids results in the formation of hydrous minerals, such as serpentine, which lower the coefficient of friction within the fault zone (Escartin *et al.*, 1997).

Even well into the oceanic upper mantle, deformation by fully plastic processes is not anticipated on the basis of experimental studies of deformation in mantle rocks. It is worth noting that seismic studies have recognized the presence of

Figure 9.5. Strength envelope plotting rock strength as a function of depth in the Earth for typical oceanic lithosphere deforming at a strain rate of 10^{-14} s^{-1}. An oceanic geotherm from Turcotte and Schubert (1982) for 60 m.y. lithosphere is assumed. The basaltic composition crust of 6 km thickness deforms by frictional sliding, modeled using Byerlee's law. The crust overlies a dry mantle lithosphere, which extends to 80 km depth and is modeled using rheological properties for dry olivine. The zone between approximately 10 and 38 km (the dotted line) is characterized by semi-brittle behavior. Below 80 km, a wet olivine rheology is used to model the asthenosphere, following Hirth and Kohlstedt (1996). The inset shows the contrast in strength between stiff dry lithosphere and convecting wet asthenosphere.

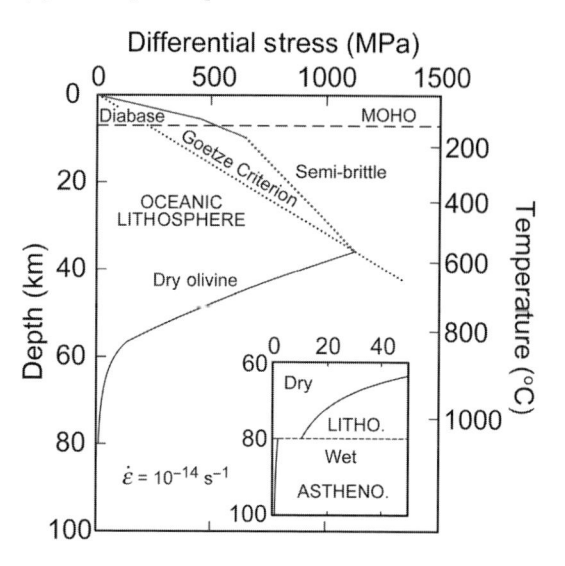

what appear to be shear zones in the uppermost oceanic mantle (e.g., Reston, 1990; Flack and Klemperer, 1990; Ranero et al., 2003). According to the model of Hirth and Kohlstedt (1996) for formation of new oceanic crust at mid-ocean ridges, partial melting of convecting mantle below the ridge results in depletion of the water content of the nominally anhydrous mineral grains due to the strong partitioning of water into the melt phase (see also Karato, 1986; Morgan, 1997). This partitioning results in an olivine-dominated, dry residuum. Thus, the oceanic lithosphere is composed of a mostly dry basaltic crust overlying a dry peridotitic upper mantle, which can be modeled approximately by the rheological properties of dry olivine since olivine is the most abundant and the weakest mineral (Zimmerman and Kohlstedt, 2004). The underlying asthenosphere is part of the convective upper mantle and can be described by the rheological properties of wet olivine. Timescales for diffusion of water from the asthenosphere into the dry lithosphere are slow relative to the rate of plate motion (Hirth and Kohlstedt, 1996).

We constructed a strength envelope diagram in Figure 9.5 for deformation of the oceanic lithosphere under the assumptions of constant rate of deformation across the lithospheric column and no localization in the plastic deformation regime. We used the oceanic geotherm for 60-m.y. old lithosphere from Turcotte and Schubert (1982, pp. 163–167). The brittle upper lithosphere, incorporating mostly the crust, was modeled using a frictional sliding law (Byerlee, 1978) with hydrostatic pore pressure. A substantial semi-brittle regime is bounded above by the criterion that some plastic accommodation process will become activated when the stress required for frictional sliding reaches $\sim 1/5$ of the plastic flow stress for the rock. The deeper bound on the semi-brittle regime is defined by the Goetze criterion (Equation 9.11), which states that brittle processes will contribute to bulk

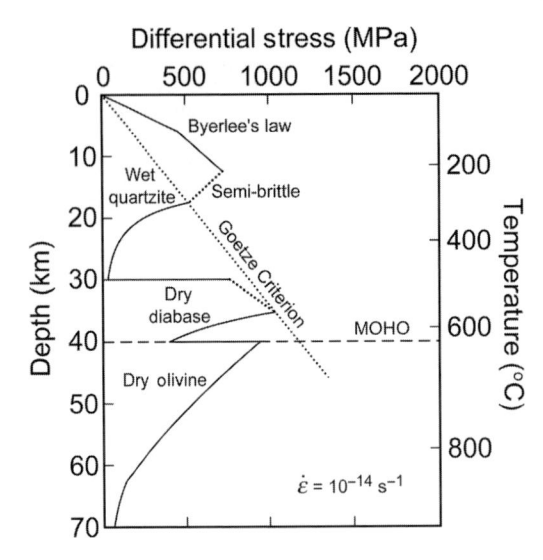

Figure 9.6. Strength envelope plot showing rock strength as a function of depth in the Earth for a model continental lithosphere deforming at a strain rate of 10^{-14} s^{-1}. A continental geotherm from Chapman (1986) for a surface heat flow of 60 mW/m was assumed. An upper crust of wet quartzite deforms at shallow depths by frictional sliding, and greater depths by dislocation creep. A dry lower crust (at amphibolite–granulite metamorphic conditions) composed of gabbroic composition rocks is modeled using rheological properties for dry diabase. Deformation in the olivine-rich mantle lithosphere is modeled by dislocation glide of dry olivine to about 60 km depth and by dislocation creep of dry olivine at greater depths. Semi-brittle regions exist in both the upper crust and lower crust; these regions are bounded below by the Goetze criterion.

deformation as long as the differential stress is greater than the overburden pressure. The actual rock strength between these bounds is not well constrained by laboratory experiments or field observations.

The plastic upper lithospheric mantle was modeled using a low-temperature deformation (dislocation glide) model for olivine (Goetze, 1978; Evans and Goetze, 1979) at temperatures ≤800 °C and a high-temperature dry dunite (dislocation, power law) flow law (Chopra and Paterson, 1984) at higher temperatures. An activation volume of 15×10^{-6} m^3/mol was used for deformation in both regimes. In the inset in Figure 9.5, a marked decrease in strength occurs in the transition from the dry, more rigid lithosphere to the wet, convecting asthenosphere, as noted by Hirth and Kohlstedt (1996).

2.1.6 Strength envelope for continental lithosphere

In a simplified lithology for Earth's continental lithosphere, an upper crust composed of wet quartzite overlies a dry lower crust (assuming granulite facies metamorphism) of gabbroic rocks, which overlies a wet or dry uppermost mantle composed mostly of olivine-rich rock (dunite). In cratonic settings, the lithospheric upper mantle is likely to be dry, consistent with the long residence time for cratonic keels over the history of Earth. Near active margins, a wet olivine rheology may be more appropriate.

Following this simplified lithology and assuming constant strain rate throughout the lithospheric column, as well as no localization in the plastic regime, we used a continental geotherm for a surface heat flow of 60 mW/m (Chapman, 1986) to plot a strength envelope for the continental lithosphere in Figure 9.6. In this figure, the uppermost brittle crust is described by a frictional sliding law (Byerlee, 1978) with pore fluid pressure equal to the hydrostatic pressure. The fully plastic portion of the upper crust is modeled by deformation of wet quartzite (Gleason and Tullis, 1995). The plastic lower crust is modeled by creep of dry diabase (Mackwell et al., 1995, 1998); the lower continental crust is generally believed to be comparatively dry due to amphibolite–granulite facies metamorphic conditions. For the uppermost mantle lithosphere, we use the low-temperature, dislocation glide model for deformation of olivine (Goetze, 1978; Evans and Goetze, 1979) and the high-temperature, dry dislocation flow law for dunite (Chopra and Paterson, 1984). Again, we assumed a value for the activation volume of 15×10^{-6} m^3/mol for plastic deformation of olivine. Semi-brittle behavior is observed at mid- and lower-crustal conditions, as indicated by the dotted lines in Figure 9.6.

2.1.7 Localization

Even within the fully plastic regime, deformation may become localized with the development of shear zones, as is commonly observed even at depths greater than 10–15 km (Ramsey, 1980). Montési and Zuber (2002) and Montési and Hirth (2003) have analyzed such localization in terms of a feedback loop involving grain size reduction that results in strain weakening. Their basic argument is that dynamic recrystallization reduces the grain size, causing a rock to weaken by enhancing the contributions of grain size sensitive creep mechanisms (i.e., diffusion creep and/or dislocation – grain boundary sliding creep) as well as by reducing the dislocation density within the grains. A key element of their analyses is the evolution of grain size, which is controlled by a competition between grain size reduction due to dynamic recrystallization and grain size increase due to grain growth. This trade-off between grain growth and dynamic recrystallization provides a natural regulator for stress buildup in the lithosphere; as stress increases, grain size in the deforming (shear) zone decreases, allowing easier deformation through dislocation – grain boundary sliding and/or diffusion creep processes and, thus, causing the stress to drop. Such a feedback system tends to stabilize preexisting plastic shear zones.

Support for this mechanism for localization can be drawn from experimental deformation results on clinopyroxene aggregates and olivine aggregates, both of which exhibit a decrease in strength (at a given strain rate and temperature) with decreasing grain size, well within the dislocation creep regime (Bystricky et al., 2000; Bystricky and Mackwell, 2001; Hirth and Kohlstedt, 2003). These observations are consistent with deformation by a dislocation – grain boundary sliding process. For olivine under laboratory conditions, sample strength has been observed to decrease by a factor of about three in going from the dislocation creep regime to the dislocation – grain boundary sliding regime. Although localization will result in higher strain rates in the zone of deformation relative to pervasive flow, the reduced strength of the material in the shear zone should more than offset these higher rates of deformation, stabilizing or enhancing localization. Shear displacements along plastic shear zones may also promote increased localization due to shear heating in regions of greatest strain rate. In addition, the presence of a fluid phase can lead to strain localization as discussed in our section on Io (Hier-Majumder and Kohlstedt, 2006; Kohlstedt and Holtzman, 2009).

2.1.8 Lithospheric deformation: the global picture

Taking all of the various aspects of deformation in Earth's lithosphere into consideration, we develop cartoons to represent a variety of plate tectonic scenarios likely to exist on Earth, as illustrated in Plate 26. In the first illustration, oceanic lithosphere (albeit shown with exaggerated crustal thickness) with a relatively dry upper crust overlies a dry, strong lower crust; uppermost mantle deforming by dislocation glide overlies a region that flows by power-law dislocation creep. The model in the second illustration is similar but with a wet lower crust, as might be found in the back-arc setting, where hydrous fluids can permeate into the lower crust from dewatering of the underlying slab. The model for a dry continental craton in the third illustration has a dry, strong crust overlying a dry, strong upper mantle, which may also be distinct mineralogically from the underlying wet convecting mantle (asthenosphere). While only illustrations, the sketches lay out the distribution and nature of strain that is possible in various geologic settings on Earth. Localization of deformation occurs not only in the brittle field but also in the semi-brittle regime and even in the plastic deformation regime of the lower crust and uppermost mantle.

2.2 Application to Venus

2.2.1 The role of water

On Earth, the rheological properties of the lithosphere are strongly influenced by even trace amounts of water dissolved in nominally anhydrous silicate minerals. This effect, often termed "water weakening" or "hydrolytic weakening," is developed in more detail in the section on Mars. In contrast to Earth, the lithosphere on Venus is expected to be thoroughly dehydrated due to high surface temperatures and extensive volcanic degassing, which are likely to have also gradually dehydrated the Venusian interior (Kaula, 1990). While the process of crustal formation probably generated a largely dry crust, subsequent plutonic activity may have resulted in episodic flushing of fluids along fault zones and associated production of hydrous minerals. Although sustained pore fluid pressure along faults seems unlikely, the presence of hydrous minerals may promote some localized weakness within the fault zone.

The comparative dryness of the Venusian lithosphere will affect the rheological behavior throughout; in the shallow crust, the pore pressure will be zero, so that

Equations (9.10a) and (9.10b) can be simplified to

$$\sigma_1 - \sigma_3 = 3.9\sigma_3 \quad \sigma_3 < 100 \,\text{MPa} \tag{9.12a}$$
$$\sigma_1 - \sigma_3 = 2.1\sigma_3 + 210 \quad \sigma_3 > 100 \,\text{MPa}, \tag{9.12b}$$

increasing the stress required for sliding on faults relative to the water-present situation. At greater depths, the absence of fluid and consequent lack of pressure-solution deformation processes will also strengthen the semi-brittle regime. The effective absence of water associated with faults may not only strengthen them but also inhibit their formation, as noted by Regenauer-Lieb *et al.* (2001) and Regenauer-Lieb and Kohl (2003).

At temperatures sufficiently high for fully plastic processes to occur, deformation of the crustal and mantle components of the lithosphere is likely to occur by dislocation creep or dislocation – grain boundary sliding processes. Constraints on crustal rheology come from the experimental work of Mackwell *et al.* (1995, 1998) on rocks of basaltic composition. These experiments were performed under dry conditions on samples of natural diabase extracted from basaltic dikes in Maryland and South Carolina, USA. The samples contain mostly plagioclase feldspar and pyroxene (augite and hypersthene), with minor oxide and hydrous minerals; prior to deformation, the samples were heat treated at high temperature to remove hydrous components. The measured strengths for the diabase are intermediate between those for plagioclase (Shelton and Tullis, 1981; Rybacki and Dresen, 2000; Dimanov *et al.*, 2003) and pyroxene (Bystricky and Mackwell, 2001), with most deformation accommodated by dislocation creep within the plagioclase feldspar grains. The more plagioclase-rich Columbia diabase is distinctly weaker than the more pyroxene-rich Maryland diabase under otherwise identical deformation conditions. In addition, dry diabase is not much lower in strength than olivine-rich rocks deformed under dry conditions (e.g., Chopra and Paterson, 1984; Karato *et al.*, 1986; Mei and Kohlstedt, 2000b).

As noted above, the formation of basaltic crust on Venus over time, with the resultant dehydration of the olivine-rich residuum, combined with the lack of an effective path for returning water into the Venusian interior, has probably resulted in a depletion of water in the interior of this planet. Such a loss of volatiles from the interior would likely be even more severe if a cyclic pattern of planetary resurfacing occurred, as envisioned by Turcotte (1993, 1995). A dehydrated mantle will be characterized by a significantly higher viscosity than on Earth due to the absence of water weakening of the major minerals under conditions that are otherwise similar. For example, at uppermost mantle conditions on Earth, the viscosity decrease due to water pressure equal to lithostatic pressure would be equivalent to a reduction in temperature of $\sim$100 K (about a factor of ten in viscosity). Thus, in order for a mostly dehydrated Venus to dissipate internal heat, especially given the high surface temperature, it is likely that the interior of Venus is hotter than the interior

418 *Planetary Tectonics*

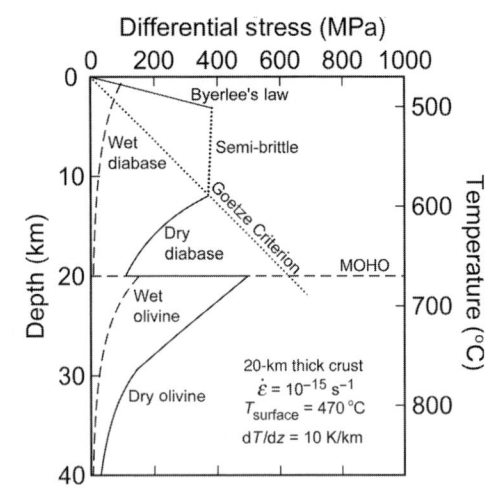

Figure 9.7. Strength envelope plotting rock strength as a function of depth in Venus for a lithosphere deforming at a strain rate of 10^{-15} s^{-1}. We assumed a surface temperature of 470 °C, a thermal gradient of 10 K/km, and a crustal thickness of 20 km, conditions believed to be appropriate for Venus. A dry crust of basaltic composition was modeled at shallow depths by a frictional sliding law (Byerlee's law) and at greater depths by rheological properties for dry diabase deforming by dislocation creep. A significant region of semi-brittle behavior controls deformation over much of the crust. Rheology properties for wet diabase, included for comparison, predict an unrealistically weak crust for Venus. The mantle lithosphere, which is presumed to be strongly depleted in water, is modeled using rheological properties for dislocation glide (to about 30 km depth) and dislocation creep (at greater depths) for olivine under dry conditions. Due to long-term water loss from the mantle, no wet asthenosphere is expected on Venus.

of Earth. Solomatov and Moresi (2000) have used convection theory to argue that the mantle of Venus should be ~200 K hotter than the mantle of Earth, a difference in temperature that is more than sufficient to compensate for the lack of water weakening in the interior. Nonetheless, the model for a dry, stiff lithosphere on Earth overlying a wet asthenosphere (Hirth and Kohlstedt, 1996) does not translate to Venus, where the lack of water means that an anomalously weak asthenospheric layer is absent. As a result, while the lithospheric plates on the surface of Earth move largely independently of the convecting mantle, lithospheric deformation and mantle convection on Venus will be strongly mechanically coupled. The prediction of strong coupling is consistent with that previously deduced by Herrick and Phillips (1992) and Bindschadler *et al.* (1992) from Magellan observations of tectonic structures on Venus.

2.2.2 Strength envelopes for Venus

The strength envelope in Figure 9.7 is plotted for conditions believed to be appropriate for the lithosphere of Venus, with a crustal thickness of 20 km, a strain rate of 10^{-15} s^{-1}, a surface temperature of 470 °C, and a geothermal gradient of 10 K/km. It is noteworthy that previous creep tests on diabase samples from the same or similar localities (Caristan, 1982; Shelton and Tullis, 1981) were significantly affected by small amounts of water present in the samples in the form of hydrous minerals that likely formed during near-surface weathering. As is evident in Figure 9.7, dried samples of diabase (Mackwell *et al.*, 1995, 1998) yield viscosities that are much larger than those predicted from the previous work and are more consistent with the observations of long-term passive maintenance of topography on Venus.

Several general inferences can be made on the basis of strength envelopes for Venus, such as the one in Figure 9.7. Much of the strength of the lithosphere resides in the crust, due to the high strength of basaltic-composition rocks under dry conditions. Passive maintenance of topography may thus be possible for at least tens of millions of years. The strength contrast between crust and mantle is quite modest, so that no lower crustal weak zone (the jelly sandwich model) is predicted based on experimental work on dry systems. Thus, surface tectonics will likely be coupled to strain throughout the lithosphere. Unlike the case for Earth, a wet, weak asthenosphere is absent so that convection and lithospheric deformation will be strongly coupled. In addition, regional and planetary-scale tectonics will likely directly reflect underlying mantle processes such as convection.

It must, however, be noted that the choice of parameters in Figure 9.7 is only poorly constrained by observation, with significant uncertainty in crustal thickness, strain rate, and geothermal gradient. To assess in general terms the effects of variations in crustal thickness, strain rate, and geothermal gradient, we can contrast the strength envelopes in Figure 9.8 with that in Figure 9.7. Such strength envelopes, which are based on experimentally determined rheological properties, lead to several general observations. (1) By comparing Figure 9.8a with Figure 9.7, the main effect of increasing crustal thickness (from 20 km to 30 km) is that more of the lithospheric strength resides in the crustal region; otherwise, the general conclusions based on Figure 9.7 still hold. (2) As demonstrated in Figure 9.8b, an increase in strain rate results in a much stronger lithosphere but otherwise has no effect on the conclusions drawn from Figure 9.7, as the relationship between strain rate and stress in the plastic regime is very similar for olivine and diabase. (3) As manifested in Figure 9.8c, a decrease in thermal gradient in the lithosphere markedly increases the strength of the lithosphere but otherwise does not affect the overall conclusions, as the activation energies for deformation of olivine and diabase are not so dissimilar. It should also be noted that, while a single thermal gradient throughout the lithosphere is clearly an oversimplification, alternative models lack any stronger basis in observation. In fact, one of the strongest constraints on the thermal gradient comes from modeling the thermal profile on the

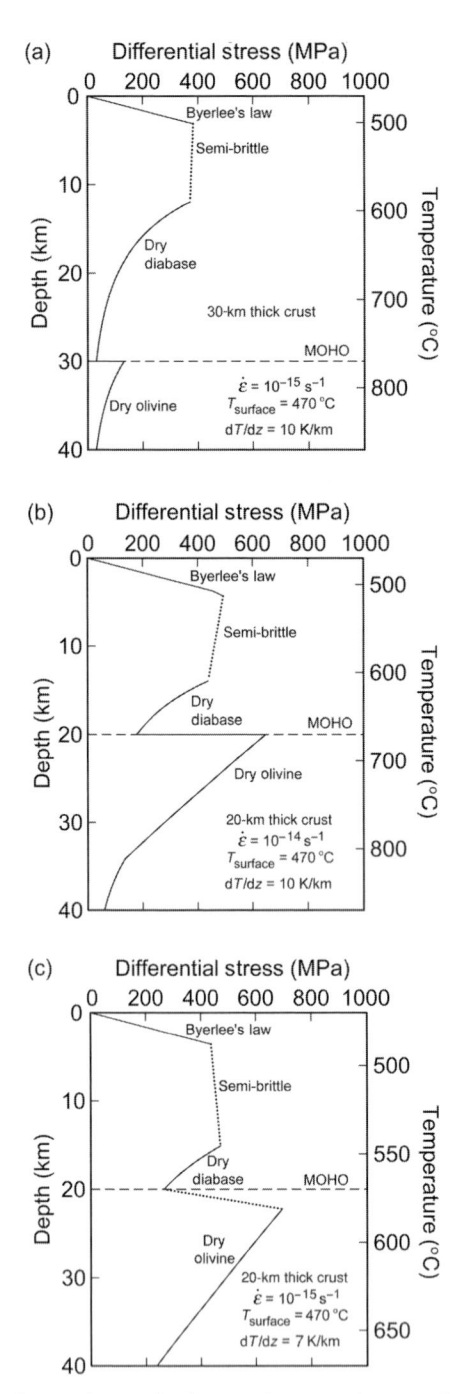

Figure 9.8. Strength envelope plotting rock strength as a function of depth in Venus to illustrate the effects of variation in (a) crustal thickness, (b) strain rate, and (c) thermal gradient by comparison to the boundary conditions used in Figure 9.7. Increasing crustal thickness resulted in a somewhat weaker lithosphere overall, while both increased strain rate and decreased thermal gradient strengthened the lithosphere. However, despite these changes in overall strength, all models still predict a strong lithosphere and strong mechanical coupling between crustal and mantle regions.

basis of observation combined with experimental measurements of deformation behavior (Phillips *et al.*, 1997).

Overall, the strength envelopes for Venus based on experimental deformation of dry diabase and olivine are consistent with our current knowledge of the history of this planet, in contrast to models based on rheological properties for undried rocks. These strength envelopes provide the basis for arguments in favor of stagnant lid dynamics for Venus and Mars, and they give substance to the conclusion that recycling of water into terrestrial planetary interiors is necessary for the maintenance of plate tectonics. Without such a return path for water, planetary interiors gradually dehydrate, making convective processes more difficult and potentially necessitating an increase in internal temperature to maintain effective dissipation of internal heat. Removal of volatiles from the interior of a terrestrial planet may also result in the loss of an anomalously weak asthenospheric layer and the promotion of a stronger coupling between the convecting mantle and the relatively rigid lithosphere.

3 Water weakening: Mars

Early in the planet's history, the lithosphere of Mars may have been significantly weaker than the lithosphere of Earth. The primary difference in rheological behavior is related to the difference in Fe content of mantle rocks, with the Martian mantle being about twice as rich in Fe as Earth's mantle (Rubie *et al.*, 2004). Iron influences the strength of rocks in two ways. Based on experience with olivine, strength decreases with increasing Fe content (e.g., Ricoult and Kohlstedt, 1985; Bai *et al.*, 1991). In addition, water solubility and hence water weakening are more pronounced in Fe-rich samples than in Fe-poor rocks (Zhao *et al.*, 2004).

To develop the theoretical framework necessary to understand one of the mechanisms by which water can weaken silicate minerals and rocks, we return to the description of diffusion-controlled dislocation creep and examine the diffusion coefficient in more detail. One of the often-cited models is that of Weertman (1968), in which dislocation sources emit dislocations that glide on parallel planes. The edge portions of the dislocations glide until two on nearby planes trap one another. To allow other dislocations emitted by the same sources to continue to glide, the edge dislocations climb toward each other and annihilate. Hence, a great deal of the strain is accomplished by glide, but the strain rate is controlled by climb. The resulting strain rate is proportional to the rate of diffusion, as described by Equations (9.6) and (9.7).

To discuss the effect of water on the rheological properties of olivine, we start with a description of its role in relaxing the von Mises (1928) criterion of homogeneous deformation requiring five independent slip systems. Next we examine the dependence of strain rate (viscosity) on water concentration and then introduce a mechanism by which water might weaken nominally anhydrous minerals. Finally, we explore the possible implication of water on the viscosity of the Martian mantle.

3.1 Deformation under hydrous conditions

3.1.1 The von Mises criterion and the role of water

For the orthorhombic mineral olivine at high temperature and low stress, the easiest slip systems are (010)[100] and (001)[100], both involving dislocations with the shortest Burgers vector, [100]. The notation (010)[100] indicates that the slip plane is (010) and the slip direction [100]. Activation of the (010)[001] system requires stresses roughly a factor of three greater than for the systems with [100] Burgers vectors (Bai *et al.*, 1991). While the (100)[001] system has been activated in experiments investigating the (001)[100] slip system, it has not been possible to quantify its behavior, other than to note that slip systems involving dislocations with [001] Burgers vectors are more difficult to activate than the slip systems involving dislocations with [100] Burgers vectors (Durham *et al.*, 1985). However, by analogy to slip with the [100] Burgers vector, it is likely that (100)[001] may be comparable to or slightly stronger than (010)[001]. Under anhydrous deformation conditions, the strength of coarse-grained dunite lies near that of the strongest slip system, (010)[001], as illustrated in Figure 9.9, indicating that at least some activation of this slip system is required to deform the aggregate. In this case, three independent slip systems are activated of the five required by the von Mises criterion for homogeneous deformation of an aggregate; the von Mises criterion is relaxed because deformation is not rigorously homogeneous and some strain is undoubtedly taken up by diffusion. In contrast, the strength of fine-grained dunite lies near that for the weakest slip system, (010)[100]. In addition, the strain rate is grain size dependent, suggestive of a significant role of grain boundary sliding while still well within the dislocation creep regime, as described by Equation (9.8) (Wang, 2002; Hirth and Kohlstedt, 2003).

A clear role for water in de-emphasizing the limitations imposed by the von Mises criterion can be shown for olivine-rich rocks. At experimental conditions with $f_{H_2O} \approx 300$ MPa, water decreases the strength of olivine single crystals oriented to favor slip on each of the slip systems by a factor of ~2, as demonstrated by comparing results plotted in Figure 9.9 for anhydrous conditions with those plotted in Figure 9.10 for hydrous conditions (Mackwell *et al.*, 1985). This weakening effect is correlated with an increase in dislocation climb, likely reflecting enhanced diffusion rates for silicon in olivine under wet conditions (Mackwell *et al.*, 1985; Costa and Chakraborty, 2008). At the same time, the strength of coarse-grained

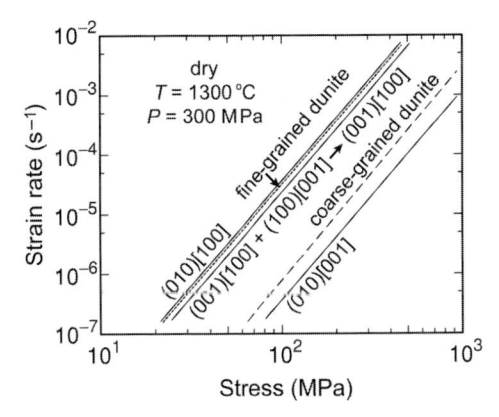

Strength and deformation of planetary lithospheres 423

Figure 9.9. Log–log plot of strain rate as a function of differential stress for olivine single crystals (solid lines), as well as fine-grained (short-dash line) and coarse-grained (long-dash line) dunite deformed under anhydrous conditions. The olivine single crystals are oriented to maximize the resolved shear stress and thus slip on the (010)[100], (001)[100] plus (100)[001], and (010)[001] slip systems. Deformation of crystals oriented to activate the (001)[100] slip system plus the (100)[001] slip system is dominated by the former. Note the proximity of the fine-grained dunite flow behavior to that of the weak slip system (010)[100] and the proximity of the coarse-grained dunite flow behavior to that of the hardest slip system, (010)[001].

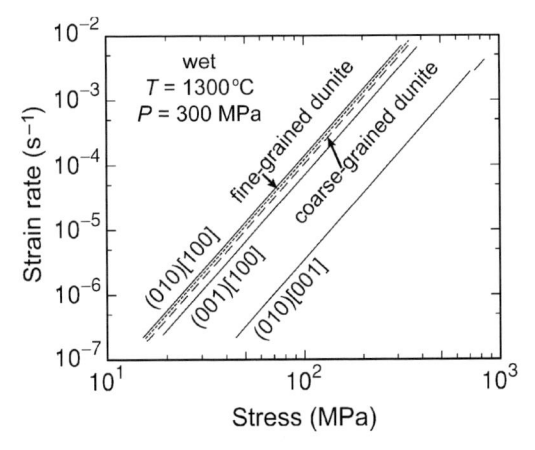

Figure 9.10. Log–log plot of strain rate as a function of differential stress for olivine single crystals (solid lines), as well as fine-grained (short-dash line) and coarse-grained (long-dash line) dunite deformed under hydrous conditions. The olivine single crystals are oriented to maximize the resolved shear stress and thus slip on the (010)[100], (001)[100] plus (100)[001], and (010)[001] slip systems. Deformation of crystals oriented to activate the (001)[100] slip system plus the (100)[001] slip system is dominated by the former. Note the proximity of both the fine-grained and the coarse-grained dunite flow behavior to that of the weakest slip system, (010)[100].

dunite exhibits a factor of >4 weakening effect (Chopra and Paterson, 1984), falling near the strength of the easiest slip systems, (010)[100] and (001)[100]; under hydrous conditions, the strengths of coarse-grained and fine-grained dunite are similar, as illustrated in Figure 9.10. Thus, in the presence of water, the need for slip on the (010)[001] slip system is decreased and only two independent slip systems appear to be required for plastic deformation of the aggregate. The reduction in the number of independent slip systems required by the von Mises criterion from five to two probably reflects the increased rates of grain matrix diffusion and, hence, of dislocation climb (Mackwell et al., 1985) as well as relaxation of the constraint of homogeneous deformation throughout the sample.

3.1.2 Comparison of deformation under anhydrous and hydrous conditions

Deformation experiments on polycrystalline samples of olivine demonstrate that strain rate increases approximately linearly with increasing water fugacity (Mei and Kohlstedt, 2000a,b). The flow law under hydrous conditions can then be expressed as a modified form of Equation (9.6) as

$$\dot{\varepsilon} = A f_{H_2O}^r \frac{\sigma^n}{d^p} \exp\left(-\frac{E + PV}{RT}\right), \qquad (9.13a)$$

or in terms of viscosity as

$$\eta = \frac{1}{A} \frac{d^p}{f_{H_2O}^r \sigma^{n-1}} \exp\left(\frac{E + PV}{RT}\right). \qquad (9.13b)$$

In the diffusion creep regime, $r \approx 1$ (Mei and Kohlstedt, 2000a), while in the dislocation creep regime, $r \approx 1$ to 1.2 (Mei and Kohlstedt, 2000b; Karato and Jung, 2003). The stress exponent $n = 1$ and grain size exponent $p = 3$ in the diffusion creep regime, while $n = 3.5$ and $p = 0$ in the dislocation creep regime (Hirth and Kohlstedt, 2003). The viscosity of dunite as a function of pressure under hydrous conditions is compared to that for dunite under anhydrous conditions in Figure 9.11, using the flow parameters reported by Karato and Jung (2003). The viscosity of samples deformed under anhydrous conditions increases with increasing pressure due to the explicit pressure dependence in Equation (9.6). In contrast, the viscosity of samples deformed under hydrous conditions initially decreases rapidly and then more gradually with increasing pressure due to competition between the implicit effect of water fugacity, which increases with increasing pressure under water-saturated conditions (Pitzer and Sterner, 1994), and the explicit effect of pressure in Equation (9.13b). If this plot were extended beyond a pressure of $\sim$3 GPa for the samples deformed under water-saturated conditions, the strain rate would begin to increase with increasing pressure; this region has yet to be investigated experimentally.

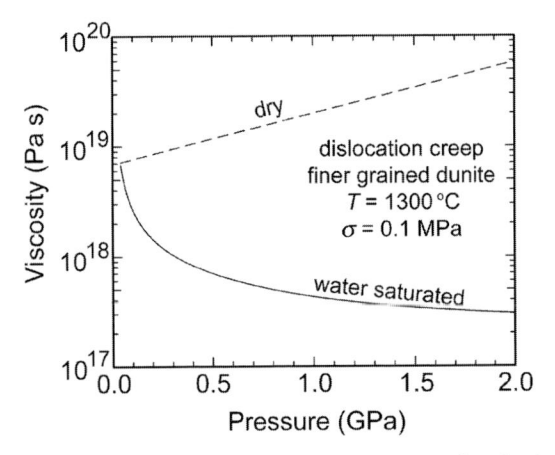

Figure 9.11. Semi-log plot of viscosity versus pressure for dunite in the dislocation creep regime. Under anhydrous (dry) conditions, viscosity increases with increasing pressure based on an exponential dependence of strain rate on pressure, Equation (9.6). Under hydrous (wet) conditions, viscosity decreases with increasing pressure due to the approximately linear dependence of strain rate on water fugacity, Equation (9.13b), since water fugacity increases with increasing pressure under water-saturated conditions (e.g., Pitzer and Sterner, 1994). This implicit effect of water fugacity on viscosity with increasing pressure is offset to some degree by the explicit exponential increase in viscosity with increasing pressure in Equation (9.13b).

3.1.3 The importance of point defects in ionic diffusion and diffusion-controlled creep

The key issue in discussing the effect of water on diffusion in nominally anhydrous minerals is the mechanism by which protons, introduced into grains in the presence of water, affect the diffusion coefficients for the constituent ions. Ionic diffusivities are determined by the concentration and diffusivity (i.e., mobility) of the point defects (vacancies and interstitials) involved in the diffusion process. As the most completely studied mineral in this regard is olivine, $(Mg_{1-x}Fe_x)_2SiO_4$, we will focus on it. If diffusion of a specific ion, for example Si, occurs dominantly by a vacancy mechanism, then the ionic self-diffusivity is given by

$$D_{Si} = X_{V_{Si}} D_{V_{Si}}, \qquad (9.14)$$

where D_{Si} is the diffusion coefficient for Si, $X_{V_{Si}}$ is the concentration of silicon vacancies, and $D_{V_{Si}}$ is the diffusion coefficient of silicon vacancies. Since the concentration of silicon vacancies is very small ($\sim 10^{-12}$ to $\sim 10^{-6}$), it is clear that $D_{Si} \ll D_{V_{Si}}$. For minerals composed of several ions, self-diffusivities for each of the ions will have the form given by Equation (9.14).

In minerals composed of more than one ion, the strain rate described by Equation (9.4) for diffusion creep or Equation (9.7) for dislocation creep is determined by the slowest diffusing species traveling along its fastest path, either through grain matrices (interiors) or along grain boundaries. This situation arises because the fluxes of all of the species that make up the crystalline grains are coupled in order to maintain stoichiometry (Ruoff, 1965; Poirier, 1985; Dimos *et al.*, 1988; Jaoul, 1990; Schmalzried, 1995). In the case of olivine, $D_{Me} \gg D_O > D_{Si}$ (Me = octahedrally coordinated metal cations: Buening and Buseck, 1973; Misener, 1974; Hermeling and Schmalzried, 1984; Nakamura and Schmalzried, 1984; Jaoul *et al.*, 1995; Chakraborty, 1997; O: Gerard and Jaoul, 1989; Ryerson *et al.*, 1989; Si: Houlier *et al.*, 1990; Dohmen *et al.*, 2002), such that the effective diffusion coefficient, D_{eff}, that limits the rate of deformation is (Dimos *et al.*, 1988)

$$D_{eff} = \frac{D_O D_{Si}}{D_O + 4D_{Si}} \approx D_{Si}. \quad (9.15)$$

In Equation (9.1), the dependencies of viscosity (strain rate) on the activities and fugacities of the chemical components enter through the concentrations of point defects. In the specific case described by Equation (9.15), these point defects are vacancies and interstitials on the silicon sublattice. The defect diffusivities (mobilities) are not significantly affected by the activities or fugacities of chemical components (Dieckmann and Schmalzried, 1977a,b; Dieckmann *et al.*, 1978; Schmalzried, 1981, pp. 174–175; Shewmon, 1983, pp. 155–160; Nakamura and Schmalzried, 1984). Therefore, the question to be answered is "how does water affect the concentration of vacancies and interstitials on the silicon sublattice in olivine?"

3.1.4 Point defects in olivine under anhydrous conditions

As the basis for addressing this question, it is instructive to first consider diffusion in olivine under anhydrous conditions. Two steps are needed in order to determine the dependencies of the concentrations of point defects on component activities and fugacities. The fact that Fe is present not only as Fe^{2+} but also in small concentrations ($\sim 10^{-3}$–10^{-5}) as Fe^{3+} is critical. First, it is necessary to specify the charge neutrality condition, which for olivine is (e.g., Sockel, 1974; Greskovich and Schmalzried, 1970; Schmalzried, 1978)

$$[Fe^{\bullet}_{Me}] = 2[V''_{Me}]; \quad (9.16)$$

the point defects involved in the charge neutrality condition are referred to as the majority defects. The Kröger–Vink notation (Kröger and Vink, 1956) is introduced in discussing point defect chemistry to specify the species, site (subscript), and charge (superscript). The square brackets denote concentration relative to the formula unit or lattice molecule and will be used interchangeably with the symbol X. As an example, $X_{V''_{Me}} \equiv [V''_{Me}]$. Charge is referenced to the perfect lattice, with a dot indicating a positive charge and a slash denoting a negative charge, both relative to the neutral charge of a normally occupied site, as indicated by an ×. In general, the charge neutrality condition is not *a priori* known for any particular mineral and must be determined experimentally by studying properties such as thermogravimetry, ionic conductivity, electrical conductivity, and point defect relaxation kinetics.

Second, defect reactions must be written in terms of structural elements of the ideal crystal, point defects in the ideal crystal treated as quasi-particles, and chemical components of the system (Schmalzried, 1981 pp. 37–57, 1995 pp. 27–37). It is useful to explore a point defect reaction involving the majority point defects, that is, Fe^{3+} on a metal cation Me site, $Fe^{\bullet}_{Me}$, and a vacant metal cation site, V''_{Me}. Numerous point defect reactions can be written that involve these two point defects, but we focus on one that involves just these two defects:

$$\frac{1}{2}O_2(srg) + MeSiO_3(srg) + Me^{\times}_{Me} + 2Fe^{\times}_{Me} = 2Fe^{\bullet}_{Me} + V''_{Me} + Me_2SiO_4(srg), \quad (9.17)$$

where srg designates a site of repeatable growth, such as a dislocation or grain boundary. The law of mass action applied to Equation (9.17) yields

$$\frac{[Fe^{\bullet}_{Me}]^2 [V''_{Me}] a_{ol}}{a_{en} [Fe^{\times}_{Me}]^2 [Me^{\times}_{Me}] (f_{O_2}/f^o_{O_2})^{1/2}} = K_{17}(T, P, X_{Fa}), \quad (9.18)$$

where Fa indicates fayalite, Fe_2SiO_4, ol indicates olivine, Me_2SiO_4, en denotes enstatite, $MeSiO_3$, and $f^o_{O_2}$ is the reference state oxygen fugacity. In writing Equation (9.18), it is assumed that the concentrations of the point defects are dilute so that their activities can be replaced by their concentrations. If Equations (9.16) and (9.18) are combined with $a_{ol} = 1$, since olivine is present, and $[Me^{\times}_{Me}] \approx 1$, as it differs from unity only by the concentration of vacancies on the metal cation sublattice, $\sim 10^{-4}$, then

$$[Fe^{\bullet}_{Me}] = 2[V''_{Me}] = 2^{1/3}[Fe^{\times}_{Me}]^{2/3}(f_{O_2}/f^o_{O_2})^{1/6} a_{en}^{1/3} K_{17}^{1/3}(T, P, X_{Fa}). \quad (9.19)$$

The dependence of the concentration of Si vacancies is now obtained by considering a reaction involving Si vacancies and one or more of the point defects already considered. For example,

$$O_2(srg) + Me_2SiO_4(srg) + Si^{\times}_{Si} + 4Fe^{\times}_{Me} = 4Fe^{\bullet}_{Me} + V''''_{Si} + 2MeSiO_3(srg). \quad (9.20)$$

The law of mass action yields

$$[V''''_{Si}] = [Fe^{\bullet}_{Me}]^{-4}[Fe^{\times}_{Me}]^4 (f_{O_2}/f^o_{O_2}) a_{en}^{-2} K_{20}(T, P, X_{Fa})$$
$$= [Fe^{\times}_{Me}]^{4/3}(f_{O_2}/f^o_{O_2})^{1/3} a_{en}^{-10/3} \frac{K_{20}(T, P, X_{Fa})}{2^{4/3} K_{17}^{4/3}(T, P, X_{Fa})}, \quad (9.21)$$

where again the relations $a_{ol} = 1$ and $[Si_{Si}^\times] \approx 1$ are used; the second equality in Equation (9.21) is obtained by using the relation for $[Fe_{Me}^\bullet]$ from Equation (9.19). Therefore, if Si diffuses by a vacancy mechanism and as vacancy diffusivity is not significantly affected by the activities or fugacities of chemical components,

$$D_{Si} \propto [V_{Si}''''] \propto [Fe_{Me}^\times]^{4/3} f_{O_2}^{1/3} a_{en}^{-10/3}, \qquad (9.22)$$

where the oxygen fugacity and enstatite activity dependencies of diffusivity and thus of diffusion-controlled creep rate are shown explicitly (see Equation 9.1).

Tracer diffusion experiments of Si, the slowest diffusing species through grain interiors in olivine, suggest that the rate of silicon diffusion decreases with increasing oxygen fugacity; specifically, $D_{Si} \propto f_{O_2}^{-0.19}$ (Houlier et al., 1990). This inverse dependence of Si diffusivity on oxygen fugacity has been interpreted as indicating that Si is diffusing by an interstitial mechanism rather than by a vacancy mechanism, since $[Si_i^{\bullet\bullet\bullet\bullet}] \propto [V_{Si}'''']^{-1}$ (also see Equation 9.22). In contrast, creep experiments on olivine single crystals and olivine-rich rocks reveal that strain rate increases with increasing oxygen fugacity (Kohlstedt and Hornack, 1981; Kohlstedt and Ricoult, 1984; Jaoul, 1990; Bai et al., 1991; Jin et al., 1994; Keefner et al., 2005), suggesting that diffusion of Si may not be limiting the rate of creep. Recent diffusion experiments carried out under anhydrous conditions, however, demonstrate two important points: first, $D_O \gg D_{Si}$, and second, the activation energy for self-diffusion of Si of ~530 kJ/mol (Dohmen et al., 2002) is essentially identical to the activation energy for high-temperature creep of olivine (Chopra and Paterson, 1984; Hirth and Kohlstedt, 2003). This value of the activation energy for Si diffusion is almost a factor of two larger than the value reported previously (Houlier et al., 1990) and approximately a factor of two larger than the activation energies for self-diffusion of O (Dohmen et al., 2002) and Me (Chakraborty, 1997). Dohmen et al. (2002) note that the analysis of the Si diffusion data by Houlier et al. (1990) was seriously affected by very short diffusion distances, which resulted in an underestimation of the activation energy by almost a factor of two (290 versus 530 kJ/mol) and an associated large uncertainty in the dependence of Si diffusivity on oxygen fugacity. For the present example, therefore, we assert that the high-temperature rate of creep of olivine-rich rocks is controlled by Si diffusing by a vacancy mechanism, but the reader must be aware that not all alternative models can be safely rejected (for further discussion, see Kohlstedt, 2006, 2007).

3.1.5 Point defects in olivine under hydrous conditions

With the introduction of water into the surrounding environment, protons diffuse rapidly into nominally anhydrous minerals (Mackwell and Kohlstedt, 1990; Kohlstedt and Mackwell, 1998, 1999; Wang and Zhang, 1996; Hercule and Ingrin, 1999; Carpenter Woods et al., 2000; Woods, 2000; Stalder and Skogby, 2002;

Blanchard and Ingrin, 2004) and influence the point defect chemistry (i.e., point defect concentrations) and thus diffusivities of all of the ionic species. Although our understanding of the point defect chemistry of olivine and other nominally anhydrous silicate minerals under hydrous conditions is far from complete, certain aspects are well constrained by experiment and theory.

Theoretical calculations by Brodholt and Refson (2000) and Braithwaite *et al.* (2003) for olivine under hydrous conditions reveal the importance of point defect associates formed between one or more hydroxyl ions, $(OH)_O^{\bullet}$, and a silicon and/or a metal cation vacancy. Both conclude that the most energetically favorable mechanism for introducing protons into olivine is by the defect associate $\{V_{Si}'''' - 4(OH)_O^{\bullet}\}^{\times}$. Here, the curly brackets, { }, denote a defect associate, which is often written as $(4H)_{Si}^{\times}$, indicating four protons in close proximity to a vacant Si site, the so-called hydrogarnet defect. While this shorthand is convenient, the physical meaning is best described as a defect associate, which avoids the implicit suggestion that protons lie physically within the vacant site. Braithwaite *et al.* (2003), however, argue that the defect associate $\{V_{Si}'''' - 4(OH)_O^{\bullet}\}^{\times}$, should yield an O–H stretching band that is, in fact, not observed in infrared spectra. Therefore, they conclude that the most common water-derived point defect involving a silicon vacancy is the defect associate $\{V_{Si}'''' - 3(OH)_O^{\bullet}\}'$, stabilized by a charge compensating defect such as $Fe_{Me}^{\bullet}$.

In this contribution, we explore the possible role of the defect associate $\{V_{Si}'''' - 3(OH)_O^{\bullet}\}'$ in the charge neutrality conditions. However, we emphasize the importance of $(OH)_O^{\bullet}$ rather than $Fe_{Me}^{\bullet}$ as the stabilizing, charge-compensating defect because the concentration of $(OH)_O^{\bullet}$ increases systematically with increasing water fugacity and is not limited by the total amount of Fe present in the system. As for the anhydrous case, we start by writing the charge neutrality condition:

$$[(OH)_O^{\bullet}] = [\{V_{Si}'''' - 3(OH)_O^{\bullet}\}'] \equiv [(3H)_{Si}']. \quad (9.23)$$

The dependence of the concentrations of these two defects on water fugacity can be obtained by examining a reaction involving both of them:

$$2H_2O(srg) + Me_2SiO_4(srg) + Si_{Si}^{\times} + O_O^{\times} = (OH)_O^{\bullet} + (3H)_{Si}' + 2MeSiO_3(srg). \quad (9.24)$$

The law of mass action applied to Equation (9.24) combined with Equation (9.23) yields

$$[\{V_{Si}'''' - 3(OH)_O^{\bullet}\}'] = [(OH)_O^{\bullet}] = (f_{H_2O}/f_{H_2O}^o)^1 a_{en}^{-1} K_{24}^{1/2}(T, K, X_{Fa}), \quad (9.25)$$

where $f_{H_2O}^o$ is the reference state water fugacity. One can write similar equations for defect associates in which 1, 2 or 4 protons are associated with a silicon vacancy. However, based on the work of Braithwaite *et al.* (2003), the $(3H)_{Si}'$ defects are the

Si defects present in greatest concentration. The total concentration of Si vacancies is then expressed as

$$\begin{aligned}X_{V_{Si}}^{total} &= [V_{Si}''''] + [\{V_{Si}'''' - (OH)_O^\bullet\}'''] + [\{V_{Si}'''' - 2(OH)_O^\bullet\}''] \\ &\quad + [\{V_{Si}'''' - 3(OH)_O^\bullet\}'] + [\{V_{Si}'''' - 4(OH)_O^\bullet\}^\times] \\ &\approx [\{V_{Si}'''' - 3(OH)_O^\bullet\}'] \\ &\propto f_{H_2O}^1.\end{aligned} \quad (9.26)$$

Since silicon vacancies and protons diffuse much more rapidly than silicon ions, defect associates such as $\{V_{Si}'''' - 3(OH)_O^\bullet\}'$ should be able to diffuse quickly and contribute directly to Si diffusion (Brodholt and Refson, 2000; Wang et al., 2004; Hier-Majumder et al., 2005); a silicon atom can hop into such a defect, displacing the protons, which move into the site vacated by the silicon. In this regard, it is instructive to recall that hydrogen ions in nominally anhydrous minerals can be thought of in three different but related ways. First, strictly speaking, hydrogen ions occupy interstitial sites denoted by $H_i^\bullet$. Second, hydrogen ions reside in close proximity to oxygen ions as indicated by the notation $(OH)_O^\bullet$ and detected as OH-stretching bands in infrared spectra. Third, hydrogen ions are protons, $p^\bullet$, which diffuse much faster than Me, O, and Si. In reality, hydrogen ions in a silicate lattice are all of these, as interstitial hydrogens are generally bonded into the oxygen lattice and are relatively easily mobilized as protons. From the diffusion perspective, therefore, it is useful to replace the defect associate $\{V_{Si}'''' - 3(OH)_O^\bullet\}'$ with $\{V_{Si}'''' - 3p^\bullet\}'$. Although the water fugacity dependence of Si diffusion has yet to be determined experimentally, diffusion experiments demonstrate that Si diffuses a factor of >100 times faster under hydrous conditions (1300 °C, 2 GPa) than under anhydrous conditions (1300 °C, 0.1 MPa) (Costa and Chakraborty, 2008) due effectively to an increase in the number of silicon vacancies that are stabilized due to the presence of protons. It should be noted that, even under hydrous conditions, Me diffusion is still orders of magnitude faster than Si or O diffusion (Hier-Majumder et al., 2005), so that $D_{Me} > D_O \geq D_{Si}$ under both anhydrous and hydrous conditions.

In the previous studies, it was proposed that deformation of olivine under hydrous conditions is rate-limited by diffusion of Si by an interstitial mechanism (Mei and Kohlstedt, 2000a,b; Karato and Jung, 2003). However, neither study took into account defect associates such as the one described by Equations (9.25) and (9.26). Here, we emphasize the importance of defect associates formed between protons and Si vacancies, based on theoretical calculations that predict that the concentration of these defect associates increases rapidly with increasing water concentration (Brodholt and Refson, 2000; Braithwaite et al., 2003). Therefore, under hydrous conditions, the approximate linear dependence of strain rate on water fugacity in Equation (9.13) is consistent with climb-controlled creep in

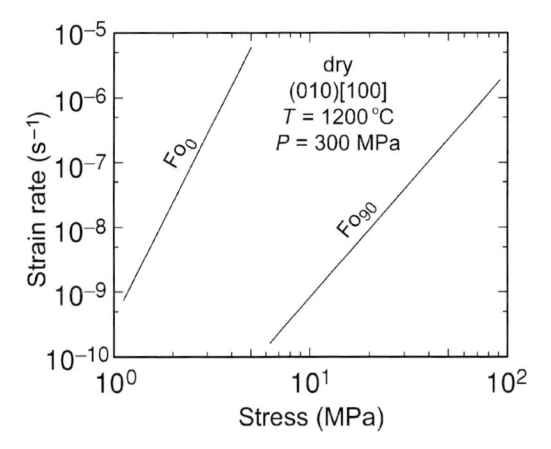

Figure 9.12. Comparison of strain rate as a function of differential stress for single crystals of Fo$_0$ (Ricoult and Kohlstedt, 1985) and Fo$_{90}$ (Bai *et al.*, 1991). At a given strain rate, fayalite crystals are over an order of magnitude weaker than Fe-poor samples.

which the climb rate is governed by the slowest diffusing species, Si, diffusing by a vacancy mechanism, Equations (9.25) and (9.26).

3.2 Application to Mars

Analyses of mantle minerals contained within Martian meteorites indicate that the Martian interior is characterized by olivines with a higher iron content (about Fo$_{80}$) than on Earth (about Fo$_{90}$), and by mantle silicates that have water contents not so different from terrestrial examples, although there is some debate about the true mantle water budget for the interior of Mars. The viscosity of the mantle of Mars compared to that of Earth will be influenced by two factors. First, deformation experiments on single crystals of olivine demonstrate that, at a given temperature and strain rate, strength decreases with increasing Fe content (e.g., compare Ricoult and Kohlstedt, 1985 with Bai *et al.*, 1991). This point is illustrated in Figure 9.12, which compares the flow law for Fo$_{90}$ with that for Fo$_0$. This general type of behavior is anticipated from the difference in solidus temperature, T_s, of these crystals with $T_s(\text{Fo}_{90}) \approx 2130\,\text{K}$ and $T_s(\text{Fo}_0) \approx 1478\,\text{K}$ combined with the difference in iron content and thus in point defect chemistry. A common scaling law, used to compare the rate of deformation of materials with the same structure but different melting points, is based on the solidus temperature through the activation energy. Specifically, the activation energy in Equation (9.6) is replaced such that $E/R = gT_s$, where the value of g depends upon the system of interest (Weertman, 1970; Weertman and Weertman, 1975). A value of $E = 540$ kJ/mol provides a good description of the temperature dependence of creep rate for single crystals of Fo$_{90}$

(Bai et al., 1991), that is, g ≈ 30. For this value of g, the predicted difference in strain rate for Fo$_{90}$ and Fo$_0$ at 1200 °C is ~6 × 10^5, a value in reasonable agreement with that obtained by comparison of the creep data for these compositions in Figure 9.12. If we now take this approach to predict the relative strain rates at fixed stress for a temperature of 1200 °C, we obtain a factor of 2 to 3 faster strain rate for Fo$_{80}$ than for Fo$_{90}$ at a given differential stress. (Note that recently published experimental results demonstrate that this difference in strain rate is roughly correct; however, a detailed examination of the data reveals that scaling based on solidus temperature alone is not sufficient to predict fully the experimental results (Zhao et al., 2009).)

Second, in addition to the direct effect of Fe on viscosity, water weakening is anticipated to increase with increasing Fe content because the solubility of protons/hydrogen increases with increasing Fe concentration in olivine. Based on laboratory experiments (Zhao et al., 2004), the solubility of hydrogen, C_H, can be expressed as

$$C_H \propto f_{H_2O}^1 \exp\left(\frac{\beta X_{Fa}}{RT}\right), \qquad (9.27)$$

with $\beta \approx 97$ kJ/mol. This dependence of proton/hydrogen solubility in olivine enters through the Gibbs free energy for incorporation of protons into olivine, a quantity that is sensitive to the Fe:Mg ratio. Given the linear dependence of hydrogen concentration on water fugacity in Equation (9.27), the strain rate in Equation (9.13a) can be written as

$$\dot{\varepsilon} \propto C_H^r \exp\left(\frac{r\beta X_{Fa}}{RT}\right). \qquad (9.28)$$

At 1200 °C, the effect of increased water solubility in Fo$_{80}$ crystals relative to that in Fo$_{90}$ crystals results in a factor of 2.2 ($r = 1$) to 2.6 ($r = 1.2$) increase in strain rate. Therefore, in addition to the direct effect of more Fe-rich olivine being weaker than less Fe-rich olivine, the indirect effect of increased water solubility in Martian olivine introduces an additional weakening effect. Combined, we anticipate a factor of ≥5 difference between the viscosity of the Martian mantle and Earth's mantle (assuming water saturation in both bodies and similar temperature distributions), as illustrated in Figure 9.13. Of course, neither the temperature nor the water content of the Martian interior is well constrained, so that this estimate is a maximum potential contrast in viscosity for Earth versus Mars. If present-day Mars is significantly cooler than Earth, as one might argue based on its smaller size, and if the interior water content is similar to or less than in Earth, then the Martian interior might be much more viscous.

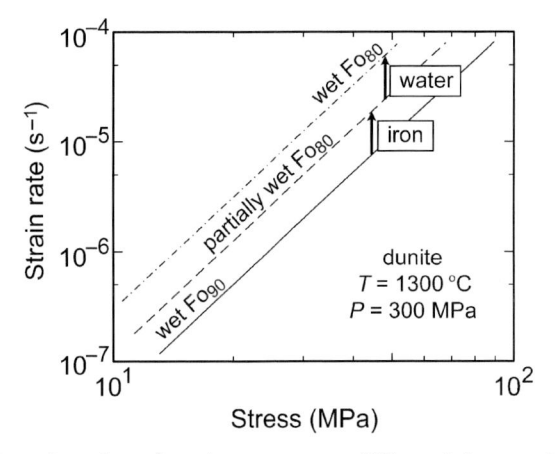

Strength and deformation of planetary lithospheres 433

Figure 9.13. Log–log plot of strain rate versus differential stress illustrating the effect on creep rate behavior of increasing iron content in olivine. Based on Figure 9.12, at a given stress, the strain rate increases as iron content increases. In addition, water solubility increases with increasing iron content, thus further increasing the strain rate.

It is interesting to speculate about the potential effect on the tectonic history of Mars if the Martian mantle is weaker than Earth's mantle. Present-day Mars lacks plate tectonics, and the only evidence consistent with an episode of plate tectonics earlier in its history is the poorly understood magnetic reversal structures recorded in the southern highlands (Connerney, 1999). Thus, Martian mantle dynamics have been dominated by stagnant-lid convection over much of its history (Reese *et al.*, 1998), rather than Earth-like plate tectonics for which heat dissipation occurs predominantly through subduction of cold oceanic lithosphere. In stagnant-lid mode, most heat dissipation occurs through volcanic activity and, potentially, through crustal delamination. On Mars, abundant evidence of plumes exists in the formation of Tharsis Rise and in the very long effusive history of the massive volcanoes, including Olympus Mons and the Tharsis volcanoes.

While few true constraints exist on the evolution of a stagnant-lid planet as a function of mantle water content, it seems reasonable to assume that lower viscosities resulting from iron enrichment and higher water content than Earth would favor more active convection and effective heat transport from the deep interior to near the surface. A feedback system involving increased internal temperatures driving convective vigor, and consequent thermal erosion of the lithosphere, would tend to maintain a lithospheric thickness that permitted sufficient heat loss to stabilize the interior dynamics of the planet. The observation of a thick lithosphere on Mars would appear to argue that the mantle viscosity is higher than on Earth. Hence, even though the iron content of the Martian interior is relatively high, mantle viscosity is actually higher than on Earth because the temperature is lower and/or the water content is depleted in the Martian interior, perhaps due to early water loss. Such early water loss and gradually decreasing temperature might allow development of a stagnant lid following an early period of plate tectonics.

4 Partial melting: Io

On Earth, the rheological behavior of partially molten rocks has been studied largely in the context of deformation and melt transport at mid-ocean ridges and in mantle plumes, where viscosities on the order of 10^{18} to 10^{20} Pa s are expected (Hirth and Kohlstedt, 1996; Ito *et al.*, 2003). Deformation experiments on partially molten lherzolites (Cooper and Kohlstedt, 1984, 1986; Bussod and Christie, 1991; Zimmerman and Kohlstedt, 2004), harzburgites (Bai *et al.*, 1997), and mixtures of olivine plus mid-ocean ridge basalt (MORB) (Hirth and Kohlstedt, 1995a,b) demonstrate only a modest reduction in viscosity with the addition of a small amount of melt; for a melt fraction $\varphi = 0.05$, viscosity decreases by a factor of ~4 relative to melt-free samples. On Io, the most volcanically active body in our solar system, magma transports tidally generated heat out of the mantle to the surface. Models of heat dissipation of tidal energy needed to balance the surface heat flux on Io require relatively low viscosities, as low as 10^8 to 10^{12} Pa s if dissipation occurs dominantly in a low-viscosity asthenosphere or on the order of 10^{16} to 10^{17} Pa s if dissipation takes place mantle wide (Segatz *et al.*, 1988; Tackley *et al.*, 2001). These lower values for viscosity require very large melt fractions, while the higher values necessitate very high temperatures and/or melt contents significantly greater than present in partially molten rocks beneath mid-ocean ridges on Earth.

4.1 Deformation of partially molten rocks

4.1.1 Effect of melt on deformation

The influence of melt on the deformation behavior of partially molten rocks depends critically on the grain-scale distribution of the melt phase (e.g., Kohlstedt, 1992, 2002; Xu *et al.*, 2004). The melt distribution, in turn, is usually described in terms of two parameters, the dihedral angle, θ, and the melt fraction, φ (von Bargen and Waff, 1986). The dihedral angle, the angle between two crystalline grains in contact with a liquid phase, is illustrated in Figures 9.14 and 9.15. Physically, the dihedral angle is a measure of the solid–solid interfacial energy, γ_{ss}, (e.g., grain boundary energy) relative to the solid–liquid interfacial energy, γ_{sl}:

$$\cos\left(\frac{\theta}{2}\right) = \frac{\gamma_{ss}}{2\gamma_{sl}}. \tag{9.29}$$

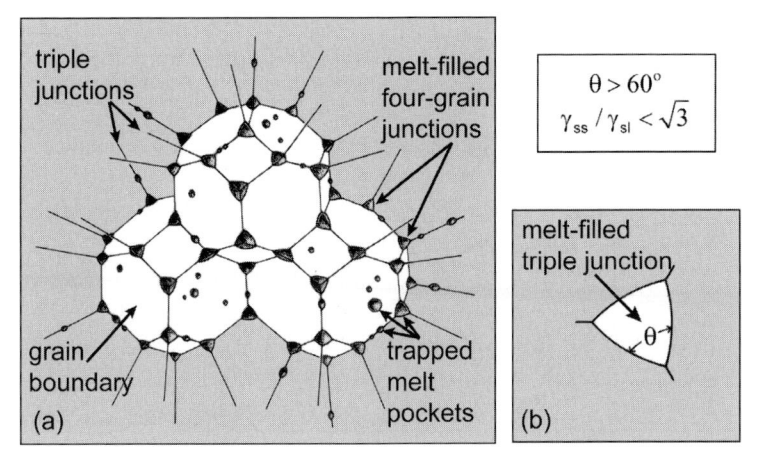

Figure 9.14. Melt distribution for solid–liquid system in which $\theta > 60°$. Melt is confined to isolated pockets, at least below a critical melt fraction. (a) Three-dimensional drawing of melt distribution illustrating trapped melt at four-grain junctions, in three-grain junctions, and on grain boundaries. (b) Two-dimensional cross-sectional view through melt pocket trapped along triple junction in (a). Modified from Lee et al. (1991).

In terms of dihedral angle or, equivalently, interfacial energies, three distinct regimes can be defined. (1) For solid–liquid systems in which the solid–liquid interfacial energy is relatively large (i.e., $\gamma_{ss} < \sqrt{3}\,\gamma_{sl}$), grain boundaries are favored over solid–liquid interfaces. Such systems are characterized by a large dihedral angle, $\theta > 60°$, with the fluid confined to isolated pockets, at least below a critical melt fraction (von Bargen and Waff, 1986), as illustrated in Figure 9.14. Examples of such systems involving silicate rocks include hydrous fluids, at least at modest pressures and high temperatures (Watson and Brenan, 1987; Brenan and Watson, 1988; Lee et al., 1991; Holness and Graham, 1991; Holness, 1993; Mibe et al., 1998; Hier-Majumder and Kohlstedt, 2006), and metal-sulfide melts, at least at low to modest pressures (Minarik et al., 1996; Ballhaus and Ellis, 1996; Gaetani and Grove, 1999; Hustoft and Kohlstedt, 2006). (2) For solid–liquid systems in which the solid–liquid energy is relatively small (i.e., $\gamma_{sl} \leq \gamma_{ss}/2$), the fluid forms a thin film along all of the grain boundaries, and the solid–fluid system is characterized by $\theta = 0°$. Dihedral angles approaching zero have been observed in some analogue systems such as camphor-benzoic acid eutectic (Takei, 2000; Takei and Shimizu, 2003) and in some high-temperature structural ceramic materials (Lange et al., 1980; Clarke, 1987). (3) For intermediate conditions with $0 < \theta \leq 60°$ (i.e., $\sqrt{3}\,\gamma_{sl} \leq \gamma_{ss} < 2\gamma_{sl}$), liquids form interconnected networks along triple junctions, as illustrated in Figure 9.15. Partially molten silicates frequently fall into this category (Waff and Bulau, 1979; Toramaru and Fujii, 1986), as do important analogue

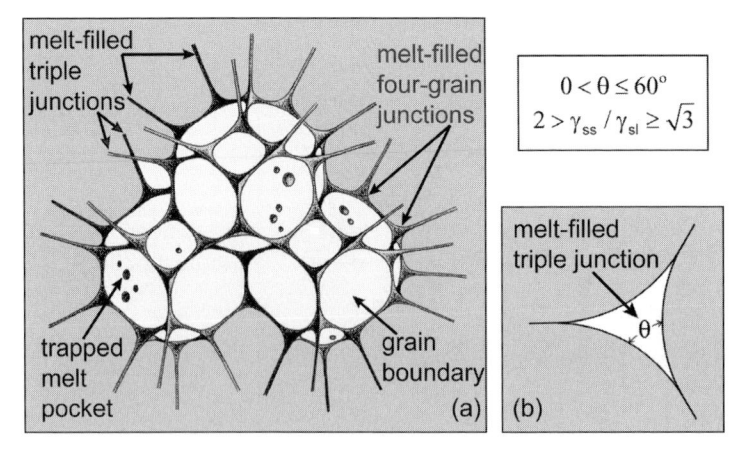

Figure 9.15. Melt distribution for solid–liquid system in which $0 < \theta \leq 60°$. Melt is distributed along all triple junctions and in all four-grain junctions, even at very small melt fractions. (a) Three-dimensional drawing of melt distribution illustrating melt tubules along triple junctions passing through four-grain junctions. Melt can be trapped on grain boundaries. (b) Two-dimensional cross-sectional view through melt-filled triple junction viewed in three dimensions in (a). Modified from Lee *et al.* (1991).

materials used to study the effect of a liquid phase on acoustic properties (Takei, 2000).

In a recent series of papers, Takei (1998, 2000, 2001, 2002) has emphasized characterization of melt distribution in terms of the wetness parameter, ψ,

$$\psi = \frac{A_{sl}}{A_{sl} + 2A_{ss}}, \qquad (9.30)$$

where A_{sl} and A_{ss} are the solid–liquid and solid–solid interfacial areas; in brief, ψ is the ratio of the wetted area to the total interfacial area for each grain. In her papers, Takei discusses the wetness as the link between microscopic melt distribution and macroscopic properties. Hence, wetness can be taken as a state parameter, such as introduced in Equation (9.1). In an isotropic solid–liquid system at thermodynamic equilibrium, wetness incorporates both the dihedral angle and melt fraction (e.g., von Bargen and Waff, 1986); as the dihedral angle decreases and/or melt fraction increases, ψ increases. Wetness is an important parameter for characterizing melt distribution because, unlike the combination of dihedral angle and melt fraction, wetness can be determined for solid–liquid systems that have anisotropic interfacial energies (i.e., cannot be described by a single dihedral angle). In addition, wetness can be measured, either directly from micrographs or indirectly from properties such as seismic velocities, under non-hydrostatic (i.e., deformation) conditions (Takei, 2001).

For each of the three types of melt morphology noted above, the influence of the fluid phase of the viscosity of the fluid–rock system is different. (i) In the case of a non-wetting liquid at small melt fractions in which liquid is confined to isolated pockets, the liquid only modestly influences viscosity. The primary effect of the liquid phase is to reduce the amount of grain-to-grain contact and thus increase the local stress. Since the fluid is not interconnected, it cannot significantly enhance diffusion mass transport (Chen and Argon, 1979; Tharp, 1983). (ii) If the liquid phase totally wets the grain boundaries, the rates of ionic diffusion are generally significantly enhanced relative to rates of diffusion along melt-free grain boundaries. As a result, the contribution of diffusion creep in which grain boundary transport dominates is greatly enhanced. This effect could be expressed by replacing D_{gb} in Equation (9.4) with D_{film}, where typically $D_{film} \gg D_{gb}$ (e.g., Elliot, 1973; Rutter, 1976; Raj, 1982). It should be noted that it is the flux of ions along grain boundaries that is important in determining the rate of diffusion creep, not simply the grain boundary diffusivity. Hence, if the solubilities of ions that compose the crystalline grains are negligible in the film, then the deformation rate will not be significantly enhanced (Gust et al., 1993), since flux is proportional to the product of diffusivity times concentration. (iii) If the fluid phase is restricted to triple junctions, it influences the deformation behavior of partially molten rocks in at least two ways (Cooper et al., 1989). First, the presence of melt reduces the load-bearing area of a rock, thus increasing the local differential stress. Second, because diffusion through melt is significantly faster than diffusion through melt-free grain boundaries, the presence of melt in triple junctions effectively reduces the distance that ions have to diffuse in traveling from regions of compression to regions of tension (or lower compressive stress), thus increasing the contribution of diffusion creep.

Flow laws developed to describe creep of melt-bearing rocks have the general form

$$\dot{\varepsilon}_\varphi \equiv \dot{\varepsilon}(\theta,\varphi) = \dot{\varepsilon}_{\varphi=0}\mathcal{F}(\theta,\varphi), \tag{9.31}$$

where $\dot{\varepsilon}_{\varphi=0}$, the strain rate of a melt-free sample, takes on the appropriate form from the equations written above for diffusion and dislocation creep. Based on an analysis of diffusion creep rate-limited by grain boundary diffusion with a melt distribution for an isotropic melt–solid system, $\mathcal{F}(\theta,\varphi)$ is given by (Cooper et al., 1989)

$$\mathcal{F}(\theta,\varphi) = \left(\frac{1}{1-(\Delta d'/d')}\right)^2_{sc} \left(\frac{1}{1-(\Delta d'/d')}\right)^2_{se} = \left(\frac{1}{1-\varphi^{1/2}g(\theta)}\right)^4. \tag{9.32}$$

In this analysis, a fraction of the grain boundary contact, $\Delta d'/d'$, is replaced by melt. In Equation (9.32), the subscripts sc and se indicate the effects of strain rate enhancement due to rapid diffusion (short-circuit, sc, diffusion) through the melt phase, and of local stress enhancement (se) due to the reduction of grain-to-grain contact on creep rate, respectively. The term $g(\theta)$ is given by

$$g(\theta) = \frac{1.06 \sin(30° - (\theta/2))}{\left(\frac{1+\cos\theta}{\sqrt{3}} - \sin\theta - \frac{\pi}{90}(30° - (\theta/2))\right)^{1/2}}. \tag{9.33}$$

A recent extension of this model for diffusion creep of partially molten rocks from two dimensions to three dimensions, including the observed anisotropic wetting of grains, has been initiated by Takei and Holtzman (2009a,b,c). In this 3-D model, the viscosity of a partially molten rock is expressed in terms of grain boundary wetness. This model for deformation in the diffusion creep regime predicts a factor of five decrease in viscosity with a very small increase in melt fraction from $\varphi = 0$ to $\varphi = 0.001$, a result not anticipated by the Cooper-Kohlstedt model. This factor of five decrease in viscosity associated with an increase in φ from 0 to 0.001 has yet to be verified experimentally. One possible explanation for this observation is that, in general, samples synthesized for physical property measurements contain a trace amount of melt. This melt forms due to impurities introduced from the crushing and grinding processes or due to inclusions within the olivine crystals used to prepare the powders from which samples are fabricated (Faul and Jackson, 2006). Therefore, truly melt-free samples ($\varphi = 0$) have not for the most part formed the baseline for examining the effect of melt on creep of mantle rocks.

In reality, melt has a larger effect on strain rate than that predicted by Equations (9.31) to (9.33), in part at least because melt wets not only all of the triple junctions but also some fraction of the grain boundaries due to the anisotropic wetting properties of most crystalline materials (e.g., Cooper and Kohlstedt, 1982; Waff and Faul, 1992; Hirth and Kohlstedt, 1995a), as illustrated with the transmission electron micrograph in Figure 9.16. In addition, the fraction of grain boundaries wetted by melt increases with increasing melt fraction (Hirth and Kohlstedt, 1995a) and, as discussed below, melt distribution is markedly affected by deformation (Kohlstedt and Zimmerman, 1996; Zimmerman et al., 1999; Holtzman et al., 2003a,b). Also, the presence of melt at triple junctions helps to mitigate the buildup of stresses at these three grain junctions, thus relaxing constraints on grain boundary sliding (Hirth and Kohlstedt, 1995b). Although these factors have yet to be incorporated into a comprehensive model for deformation, the dependence of creep rate on the melt fraction of partially molten ultramafic rocks is reasonably well described by the empirical relation (Kelemen et al., 1997)

$$\mathcal{F}(\theta,\varphi) = \exp(\alpha(\theta)\varphi), \tag{9.34}$$

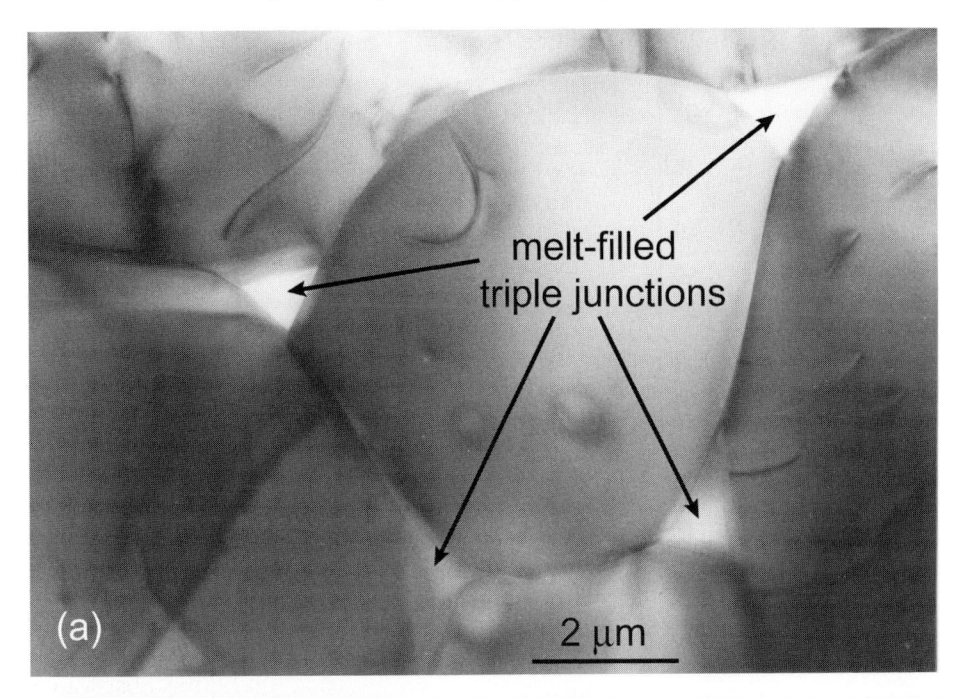

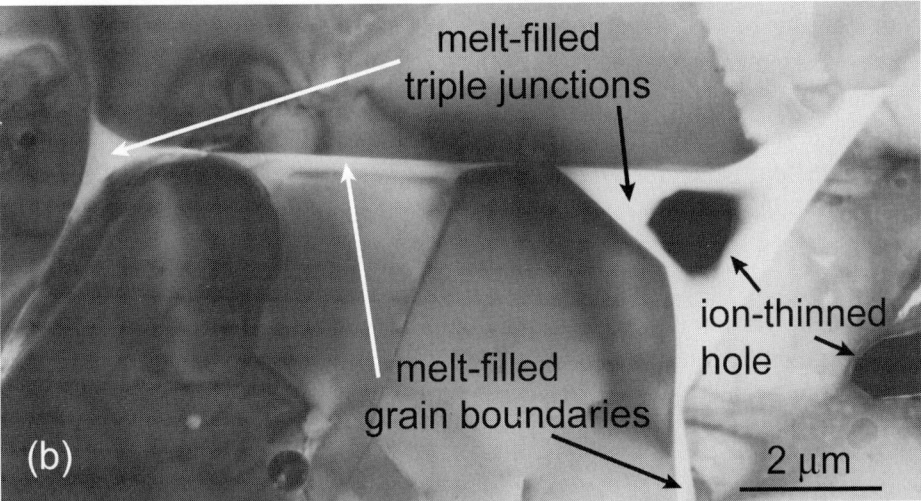

Figure 9.16. Transmission electron micrographs of melt distribution in partially molten aggregate of olivine plus mid-ocean ridge basalt. Melt is present in all triple junctions, as well as along some grain boundaries. From Kohlstedt (2002) with permission from MSA.

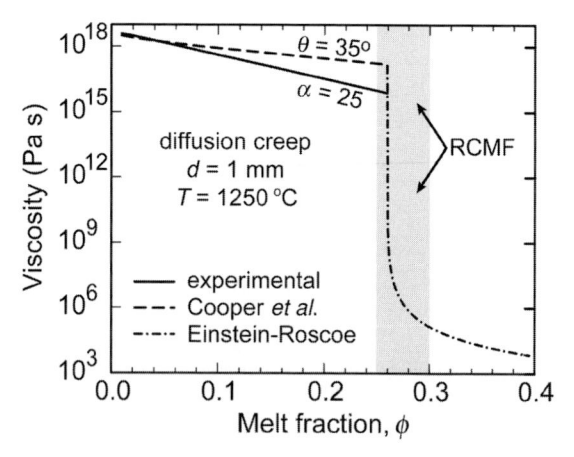

Figure 9.17. Semi-log plot of viscosity as a function of melt fraction for the diffusion creep regime. The dashed line is based on Equations (9.31) and (9.32) for $\theta = 35°$. The solid line is an empirical fit of experimental data to Equation (9.34) with $\alpha = 25$. The dot-dashed line is an extrapolation to higher melt fractions using the Einstein-Roscoe relationship in Equation (9.35). The shaded region identifies the rheologically critical melt fraction for the olivine plus basalt system based on experiments by Scott and Kohlstedt (2006).

with the value of $\alpha(\theta)$ being somewhat larger in the dislocation creep regime than in the diffusion creep regime (Mei et al., 2002; Zimmerman and Kohlstedt, 2004). It should be noted that the recent 3-D model of Takei and Holtzman (2009a) also provides a good fit to the experimental data for deformation of partially molten rocks in the diffusion creep regime, at least for $\varphi \geq 0.01$.

At very high melt fractions, a transition occurs from deformation that is largely controlled by a framework of grains to one governed by a suspension of grains in a melt. In the olivine plus basalt system, this transition occurs for a rheologically critical melt fraction (RCMF) in the range $0.25 \leq \varphi \leq 0.30$ (Scott and Kohlstedt, 2006). At melt fractions greater than the RCMF, deformation is often described by the Einstein-Roscoe equation (Einstein, 1906, 1911; Roscoe, 1952)

$$\eta_{\text{rock}} = \frac{\eta_{\text{melt}}}{(1.35\varphi - 0.35)^{2.5}}, \qquad (9.35)$$

which leads to a very rapid decrease in viscosity with increasing melt fraction. The decrease in viscosity with increasing melt fraction is illustrated in Figure 9.17 for the diffusion creep regime. At melt fractions up to $\varphi \approx 0.05$, the model for diffusion creep given by Equations (9.31–9.33) and the empirical relation described by Equation (9.34) yield similar results. At higher melt fractions, however, the viscosity of partially molten rocks decreases more rapidly than predicted by this model. Above $\varphi \approx 0.26$, viscosity decreases very quickly with increasing melt fraction as a transition occurs from framework-like to suspension-like behavior.

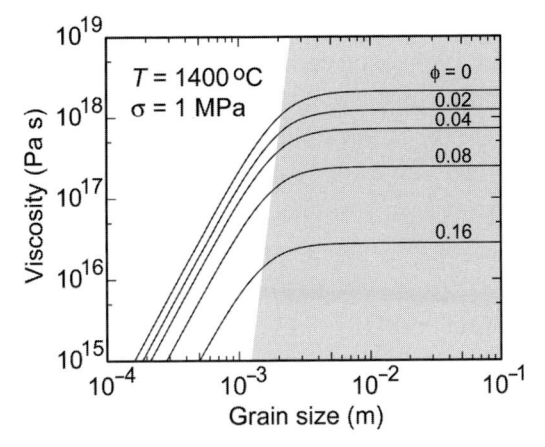

Figure 9.18. Log–log plot of viscosity as a function of grain size at 1400 °C and a stress of 1 MPa for several melt fractions based on Equation (9.34), with values of $\alpha_{\text{diff}} = 21$ in the diffusion creep regime and $\alpha_{\text{disl}} = 27$ in the dislocation creep regime (shaded region), based on the laboratory experiments of Zimmerman and Kohlstedt (2004) on partially molten lherzolite samples. Since $\alpha_{\text{diff}} < \alpha_{\text{disl}}$, the boundary between the two creep regimes moves toward smaller grain size with increasing melt fraction. In the diffusion creep regime, viscosity increases as the cube of the grain size.

The dependence of viscosity on melt fraction based on laboratory experiments and described by Equation (9.34) is further explored in Figure 9.18. Flow law parameters were taken from an experimental study of creep of partially molten lherzolite (Zimmerman and Kohlstedt, 2004). In the diffusion creep regime, $\alpha_{\text{diff}} = 21$, $Q = 370$ kJ/mol, $n = 1$ and $p = 3$, while in the dislocation creep regime $\alpha_{\text{disl}} = 27$, $Q = 550$ kJ/mol, $n = 4.3$ and $p = 0$. Results are shown for a temperature of 1400 °C and a differential stress of 1 MPa. For a melt-free sample, the viscosity is $\sim 10^{18}$ Pa s for a grain size ≥ 1 mm and decreases with decreasing grain size below ~ 1 mm as diffusion creep becomes increasingly important. In the dislocation creep regime in which the viscosity is stress dependent, the viscosity will increase to $\sim 10^{20}$ Pa s if the differential stress is decreased to 0.3 MPa. This range of viscosity is in good agreement with values calculated from observation of postglacial rebound (e.g., Peltier, 1998; Kaufmann and Lambeck, 2002) and analyses of the geoid on Earth (Craig and McKenzie, 1986; Hager, 1991; Panasyuk and Hager, 2000).

4.1.2 Effect of deformation/stress on melt distribution

Not only does a small amount of melt weaken a partially molten rock, but also the distribution of melt changes in response to an applied stress. The scale at which redistribution of melt in a deforming rock takes place is governed by the compaction length, δ_c, that is, the distance over which melt flow and matrix deformation are coupled, such that pressure gradients can develop in the fluid, leading to melt segregation (McKenzie, 1984; Scott and Stevenson, 1986). Here we use a simplified form of the expression for compaction length in terms of the shear viscosity of the partially molten rock, η, the permeability of the partially molten rock, k, and the viscosity of the melt, μ. The bulk viscosity is assumed to be smaller than or comparable to the shear viscosity, so that

$$\delta_c \approx \sqrt{k \frac{4/3\eta}{\mu}}. \qquad (9.36)$$

In experiments, melt segregation occurs if the compaction length is less than or on the order of the sample thickness, that is, $\delta_c \leq \delta_t$, and does not occur in samples for which $\delta_c \gg \delta_t$. However, even if $\delta_c \gg \delta_t$, melt reorients in response to the applied stress, even though large-scale segregation does not occur. This point is illustrated directly in samples of olivine plus a few percent basalt sheared to large strain, as shown in Figure 9.19 (Kohlstedt and Zimmerman, 1996; Zimmerman *et al.*, 1999). In these samples, the compaction length is ~10 mm, while the sample thickness is <1 mm (Holtzman *et al.*, 2003a). Stress-induced anisotropy of a partially molten analogue material of unknown compaction length has also been identified *in situ* using ultrasonic shear waves (Takei, 2001).

If $\delta_c \leq \delta_t$, melt segregates into a network of melt-rich bands that are separated by melt-depleted lenses, as illustrated in Figure 9.20, an optical micrograph of a sample deformed to high strain in simple shear. Before deformation, melt was homogeneously distributed, similar to the melt distribution in Figure 9.19a, with a starting melt fraction of $\varphi \approx 0.04$. After deformation, the anastomosing network of bands contains a relatively large melt fraction, $\varphi \approx 0.2$, while the melt-depleted lenses contain a much smaller melt fraction, $\varphi \approx 0.01$. One of the striking features of the melt-rich bands is that they are oriented, on average, ~20° to the shear plane but antithetic to the shear direction. This average orientation of the bands does not increase with increasing shear strain but remains, on average, approximately constant. Since the bands must rotate with the flow field of the matrix, they reach a maximum angle of ~30°, beyond which the melt pressure in the bands increases, forcing flow in the resulting pressure gradient. Melt can then flow from the bands at higher angles to those at lower angles through the network of high-permeability, melt-rich pathways.

Segregation of melt into melt-rich bands during high-strain experiments has a pronounced influence on both viscosity and permeability of the rock. As melt fraction rises in the melt-rich bands, extracting melt from neighboring regions, the viscosity of the bands decreases thus localizing deformation into narrow melt-rich shear zones. Likewise, as the melt fraction in the bands increases, permeability

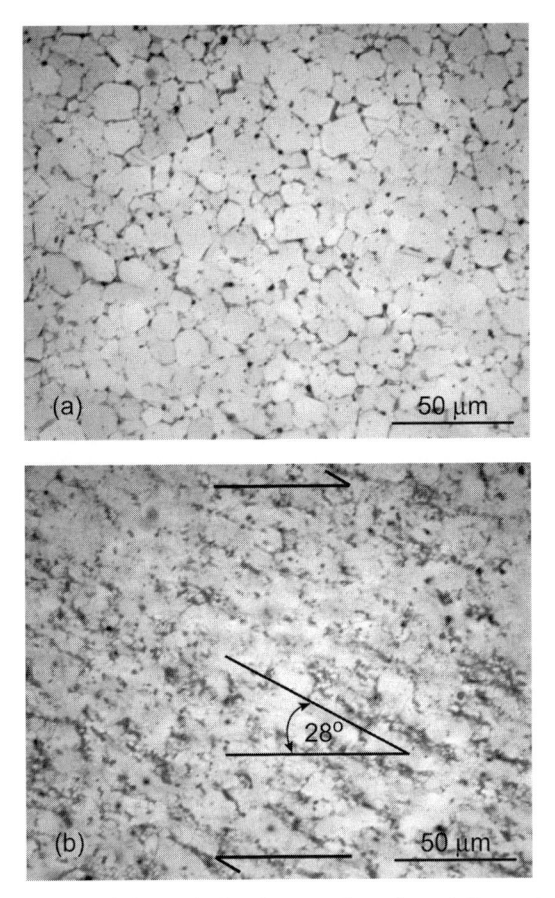

Figure 9.19. Reflected light optical micrographs of partially molten samples of olivine plus basalt. (a) In a hydrostatically annealed sample, melt is distributed in triple junctions, as well as along some grain boundaries. (b) In a sheared sample, melt is aligned ∼28° to the shear plane and antithetic to the shear direction.

increases since permeability scales approximately as melt fraction squared or cubed (Turcotte and Schubert, 1982, pp. 383–384). Hence, the fluid-rich bands provide paths for rapid transport of melt and aqueous fluids from deep within the mantle to the surface (Daines and Kohlstedt, 1997; Holtzman *et al.*, 2003a,b; Hier-Majumder and Kohlstedt, 2006).

4.2 Application to Io

Io is characterized by extensive volcanic activity and an associated high surface heat flux, which is approximately 35 times that of Earth (Keszthelyi and McEwen, 1997; Lopes *et al.*, 2001). On Io, internal heat is generated by tidal dissipation associated with its Laplace resonance with Europa and Ganymede as the

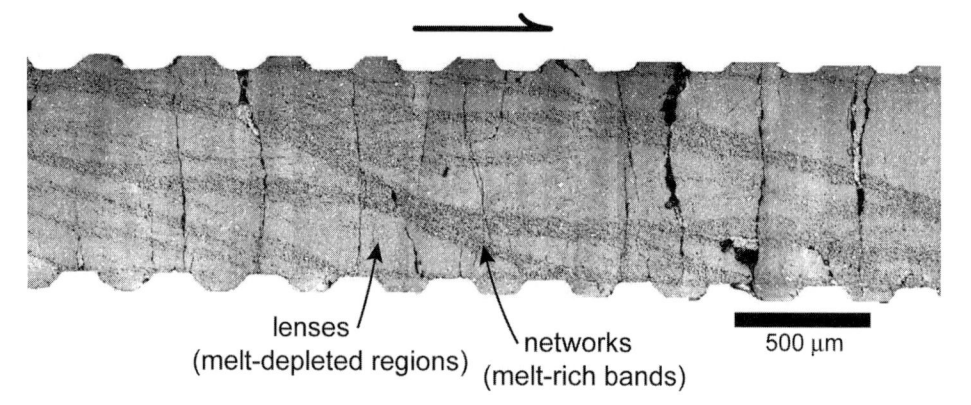

Figure 9.20. Reflected light optical micrograph of sample of olivine plus basalt plus chromite deformed in simple shear to a strain of ∼3. The melt-rich bands are dark, while the melt-depleted lenses are light. The vertical cracks developed during cooling and depressurization due to the mismatch in elastic constants and thermal expansion between the samples and the tungsten pistons. The grooves along the top and bottom of the sample were formed by the pistons, which were serrated to prevent slip between the sample and the pistons.

three satellites orbit Jupiter. Two models have been proposed to account for this large amount of heating within Io. In the first, tidal heat is dissipated throughout the entire mantle with a viscosity of $\sim 10^{16}$–10^{17} Pa s; in this model, most of the tidal heating occurs near the core-mantle boundary (Ross and Schubert, 1985, 1986; Segatz et al., 1988). Such a model appears to require that the entire mantle remains just above the solidus temperature throughout its depth in order to maintain these low viscosities. In the second model, heat is primarily dissipated in a thin, ∼100-km thick asthenosphere with a viscosity of $\sim 10^{8}$–10^{12} Pa s (Segatz et al., 1988).

A viscosity of $< 10^{12}$ Pa s cannot be obtained at 1400 °C without a melt fraction exceeding the rheologically critical melt fraction, as illustrated in Figures 9.17 and 9.18. Even at a temperature of 1600 °C, Figure 9.21, the temperature of the hottest lavas erupting on Io, viscosity is $> 10^{13}$ Pa s unless the melt fraction exceeds 0.15 or the grain size is less than ∼10 μm. It is unlikely that grain size will be that small at such a high temperature for which grain growth should be rapid. The differential stress in such a system is also expected to be small, such that the dynamically recrystallized grain size should be relatively large (∼10 mm for $\sigma =$ 1 MPa; van der Wal et al., 1993). As pointed out by Tackley et al. (2001), if the melt fraction is large, gravity-driven compaction and associated melt extraction will almost certainly occur, thus forcing an increase in viscosity by reducing the melt fraction. Such extraction will be further enhanced in a convective regime for which melt segregation due to deformation will significantly increase permeability

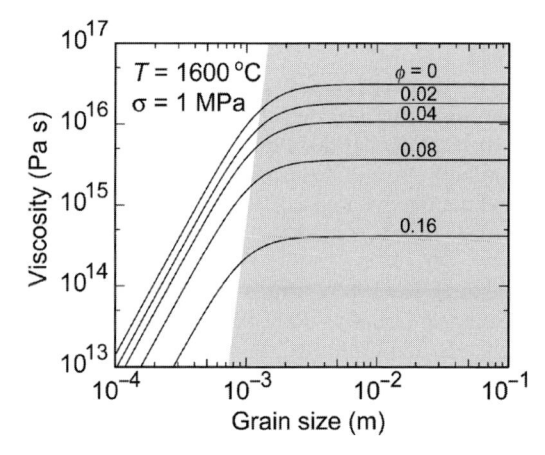

Figure 9.21. Log–log plot of viscosity as a function of grain size at 1600 °C and a stress of 1 MPa for several melt fractions based on Equation (9.34), with the same values for creep parameters as used in Figure 9.17.

(Holtzman *et al.*, 2003a,b). Hence, mantle viscosity cannot be less than $\sim 10^{16}$ Pa s, and an extremely low viscosity asthenosphere appears to be unsustainable. Thus, experimental constraints favor models of whole mantle convection on Io at high temperatures (~ 1600 °C) and relatively low melt fractions (<0.01).

Acknowledgments

The authors thank Ben Holtzman for his intellectual input and artistic skill in preparing Plate 26. The authors greatly appreciate the detailed critique provided by Laurent Montési. Support from NASA grants NNG04G173G and NNX07AP68G through the Planetary Geology and Geophysics Program to DLK and under CAN-NCC5–679 to SJM is gratefully acknowledged. Support from NSF grants OCE-0648020 (deformation of partially molten rocks) and EAR-0439747 (deformation of nominally anhydrous minerals) to DLK is also gratefully acknowledged. This chapter is LPI contribution #1240.

References

Bai, Q., Mackwell, S. J., and Kohlstedt, D. L. (1991). High-temperature creep of olivine single crystals: 1. Mechanical results for buffered samples. *J. Geophys. Res.*, **96**, 2441–2463.

Bai, Q., Jin, Z., and Green, H. W. (1997). Experimental investigation of partially molten peridotite at upper mantle pressure and temperature. In *Deformation Enhanced Fluid Transport in the Earth's Crust and Mantle*, ed. M. Holness. London: Chapman & Hall.

Ballhaus, C. and Ellis, D. J. (1996). Mobility of core melts during Earth's accretion. *Earth Planet. Sci. Lett.*, **143**, 137–145.

Barr, A. C., Pappalardo, R. T., and Zhong, S. (2004). Convective instability in ice I with non-Newtonian rheology: Application to the icy Galilean satellites. *J. Geophys. Res.*, **109**, E12008, doi:10.1029/2004JE002296.

Bindschadler, D. L., Schubert, G., and Kaula, W. M. (1992). Coldspots and hotspots: Global tectonics and mantle dynamics of Venus. *J. Geophys. Res.*, **97**, 13 495–13 532.

Blanchard, M. and Ingrin, J. (2004). Kinetics of deuteration in pyrope. *Eur. J. Mineral.*, **16**, 567–576.

Brace, W. F. and Kohlstedt, D. L. (1980). Limits on lithospheric stress imposed by laboratory experiments. *J. Geophys. Res.*, **85**, 6248–6252.

Braithwaite, J. S., Wright, K., and Catlow, C. R. A. (2003). A theoretical study of the energetics and IR frequencies of hydroxyl defects in forsterite. *J. Geophys. Res.*, **108**, 2284, doi:10.1029/2002JB002126.

Brenan, J. M. and Watson, E. B. (1988). Fluids in the lithosphere: 2. Experimental constraints on CO_2 transport in dunite and quartzite at elevated P-T conditions with implications for mantle and crustal decarbonation processes. *Earth Planet. Sci. Lett.*, **91**, 141–158.

Brodholt, J. P. and Refson, K. (2000). An *ab initio* study of hydrogen in forsterite and a possible mechanism for hydrolytic weakening. *J. Geophys. Res.*, **105**, 18 977–18 992.

Buening, D. K. and Buseck, P. R. (1973). Fe–Mg lattice diffusion in olivine. *J. Geophys. Res.*, **78**, 6852–6862.

Bussod, G. Y. and Christie, J. M. (1991). Textural development and melt topology in spinel lherzolite experimentally deformed at hypersolidus conditions. *J. Petrol., Spec. Vol.*, 17–39.

Byerlee, J. D. (1978). Friction of rocks. *Pure Appl. Geophys.*, **116**, 615–626.

Bystricky, M., Kunze, K., Burlini, L., and Burg, J.-P. (2000). High shear strain of olivine aggregates: Rheological and seismic consequences. *Science*, **290**, 1564–1567.

Bystricky, M. and Mackwell, S. (2001). Creep of dry clinopyroxene aggregates. *J. Geophys. Res.*, **106**, 13 443–13 454.

Caristan, Y. (1982). The transition from high temperature creep to fracture in Maryland diabase. *J. Geophys. Res.*, **87**, 6781–6790.

Carpenter Woods, S., Mackwell, S., and Dyar, D. (2000). Hydrogen in diopside: Diffusion profiles. *Amer. Min.*, **85**, 480–487.

Carter, C. B. and Sass, S. L. (1981). Electron diffraction and microscopy techniques for studying grain-boundary structure. *J. Am. Ceram. Soc.*, **64**, 335–345.

Chakraborty, S. (1997). Rates and mechanisms of Fe-Mg interdiffusion in olivine at 980° to 1300°C. *J. Geophys. Res.*, **102**, 12 317–12 331.

Chapman, D. S. (1986). Thermal gradients in the continental crust. In *The Nature of the Continental Crust*, ed. J. B. Dawson, D. A. Carswell, J. Hall and K. H. Wedepohl, *Spec. Publ. Geol. Soc. London*, **24**, 63–70.

Chen, I. W. and Argon, A. S. (1979). Steady state power-law creep in heterogeneous alloys with microstructures. *Acta Metall.*, **27**, 785–791.

Chen, S., Hiraga, T., and Kohlstedt, D. L. (2006). Water weakening of clinopyroxene in the dislocation creep regime. *J. Geophys. Res.*, **111**, B08203, doi:10.1029/2005JB003885.

Chester, F. M. (1988). The brittle ductile transition in a deformation-mechanism map for halite. *Tectonophys.*, **154**, 125–136.

Chopra, P. N. and Paterson, M. S. (1984). The role of water in the deformation of dunite. *J. Geophys. Res.*, **89**, 7861–7876.

Clarke, D. R. (1987). On the equilibrium thickness of intergranular glass phases in ceramic materials. *J. Am. Ceram. Soc.*, **70**, 15–22.
Coble, R. (1963). A model for boundary diffusion controlled creep in polycrystalline materials. *J. Appl. Phys.*, **34**, 1679–1682.
Connerney, J. E. P., Acuña, M. H., Wasilewski, P. J., Ness, N. F., Rème, H., Mazelle, C., Vignes, D., Lin, R. P., Mitchell, D. L., and Cloutier, P. A. (1999). Magnetic lineations in the ancient crust of Mars. *Science*, **284**, 794–798.
Cooper, R. F. and Kohlstedt, D. L. (1982). Interfacial energies in the olivine–basalt system. In *High-Pressure Research in Geophysics, Advances in Earth and Planetary Sciences*, Vol. 12, ed. S. Akimota and M. H. Manghnani, Center for Academic Publications Japan, Tokyo, pp. 217–228.
Cooper, R. F. and Kohlstedt, D. L. (1984). Solution-precipitation enhanced creep of partially molten olivine-basalt aggregates during hot-pressing. *Tectonophys.*, **107**, 207–233.
Cooper, R. F. and Kohlstedt, D. L. (1986). Rheology and structure of olivine-basalt partial melts. *J. Geophys. Res.*, **91**, 9315–9323.
Cooper, R. F., Kohlstedt, D. L., and Chyung, C. K. (1989). Solution-precipitation enhanced creep in solid-liquid aggregates which display a non-zero dihedral angle. *Acta Metall.*, **37**, 1759–1771.
Costa, R. and Chakraborty, S. (2008). The effect of water on Si and O diffusion rates in olivine and implications for transport properties and processes in the upper mantle. *Phys. Earth Planet. Inter.* **166**, 11–29, doi:10.1016/j.pepi.2007.10.006.
Craig, C. H. and McKenzie, D. (1986). The existence of a thin low-viscosity layer beneath the lithosphere. *Earth Planet. Sci. Lett.*, **78**, 420–426.
Cuffey, K. M., Thorsteinsson, T., and Waddington, E. D. (2000a). A renewed argument for crystal size control of ice sheet strain rates. *J. Geophys. Res.*, **105**, 27 889–27 894.
Cuffey, K. M., Conway, H., Gades, A., Hallet, B., Raymond, C. F., and Whitlow, S. (2000b). Deformation properties of subfreezing glacier ice: Role of crystal size, chemical impurities, and rock particles inferred from in situ measurements. *J. Geophys. Res.*, **105**, 27 895–27 915.
Daines, M. J. and Kohlstedt, D. L. (1997). Influence of deformation on melt topology in peridotites. *J. Geophys. Res.*, **102**, 10 257–10 271.
Dash, J. G., Fu, H. Y., and Wettlaufer, J. S. (1995). The premelting of ice and its environmental consequences. *Rep. Prog. Phys.*, **58**, 115–167.
Dieckmann, R. and Schmalzried, H. (1977a). Defects and cation diffusion in magnetite (I). *Ber. Bunsenges. Phys. Chem.*, **81**, 344–347.
Dieckmann, R. and Schmalzried, H. (1977b). Defects and cation diffusion in magnetite (II). *Ber. Bunsenges. Phys. Chem.*, **81**, 414–419.
Dieckmann, R., Mason, T. O., Hodge, J. D., and Schmalzried, H. (1978). Defects and cation diffusion in magnetite (III). Tracer diffusion of foreign cations as a function of temperature and oxygen potential. *Ber. Bunsenges. Phys. Chem.*, **82**, 778–783.
Dimanov, A., Lavie, M. P., Dresen, G., Ingrin, J., and Jaoul, O. (2003). Creep of polycrystalline anorthite and diopside. *J. Geophys. Res.*, **108**, 2061, doi:10.1029/2002JB001815.
Dimos, D., Wolfenstine, J., and Kohlstedt, D. L. (1988). Kinetic demixing and decomposition of multicomponent oxides due to a nonhydrostatic stress. *Acta Met.*, **36**, 1543–1552.
Dohmen, R., Chakraborty, S., and Becker, H.-W. (2002). Si and O diffusion in olivine and implications for characterizing plastic flow in the mantle. *Geophys. Res. Lett.*, **29**, 2030, doi:10.1029/2002GL015480.

Dombard, A. J. and McKinnon, W. B. (2000). Long-term retention of impact crater topography on Ganymede. *Geophys. Res. Lett.*, **27**, 3663–3666.

Dombard, A. J. and McKinnon, W. B. (2001). Formation of grooved terrain on Ganymede: Extensional instability mediated by cold, superplastic creep. *Icarus*, **154**, 321–336.

Durham, W. B. and Stern, L. A. (2001). Rheological properties of water ice: Applications to satellites of the outer planets. *Annu. Rev. Earth Planet. Sci.*, **29**, 295–330.

Durham, W. B., Heard, H. C., and Kirby, S. H. (1983). Experimental deformation of polycrystalline H_2O ice at high pressure and low temperature: Preliminary results. *J. Geophys. Res.*, **88**, 377–392.

Durham, W. B., Ricoult, D. L., and Kohlstedt, D. L. (1985). Interaction of slip systems in olivine. In *Point Defects in Minerals*, ed. R. N. Schock, Washington, DC: American Geophysical Union, pp. 185–193.

Durham, W. B., Kirby, S. H., and Stern, L. A. (1997). Creep of water ices at planetary conditions: A compilation. *J. Geophys. Res.*, **102**, 16 293–16 302.

Durham, W. B., Kirby, S. H., and Stern, L. A. (2001). Rheology of ice I at low stress and elevated confining pressure. *J. Geophys. Res.*, **106**, 11 031–11 042.

Duval, P., Ashby, M. F., and Anderman, I. (1983). Rate-controlling processes in the creep of polycrystalline ice. *J. Phys. Chem.*, **87**, 4066–4074.

Einstein, A. (1906). Eine neue Bestimmung der Molekuldimensionen. *Annu. Phys.*, **19**, 289–306.

Einstein, A. (1911). Berichtigung zu meiner Arbeit: eine neue Bestimmung der Molekuldimensionen. *Annu. Phys.*, **34**, 591–592.

Elliot, D. (1973). Diffusion flow laws in metamorphic rocks. *Geol. Soc. Am. Bull.*, **84**, 2645–2664.

Escartin, J., Hirth, G., and Evans, B. (1997). Effects of serpentinization on the lithospheric strength and style of normal faulting at slow-spreading ridges. *Earth Planet. Sci. Lett.*, **151**, 181–189.

Evans, B. and Goetze, C. (1979). The temperature variation of hardness of olivine and its implications for polycrystalline yield stress. *J. Geophys. Res.*, **84**, 5505–5524.

Evans, B. and Kohlstedt, D. L. (1995). Rheology of rocks. In *Rock Physics and Phase Relations: A Handbook of Physical Constants*, ed. T. J. Ahrens, Washington, DC: American Geophysical Union, pp. 148–165.

Faul, U. and Jackson, I. (2006). The effect of melt on the creep strength of polycrystalline olivine (abs.). *Eos Trans. AGU*, **87**, Fall Meet. Suppl., MR11B-0129.

Flack, C. A. and Klemperer, S. L. (1990). Reflections from mantle fault zones around the British Isles. *Geology*, **18**, 528–532.

Gaetani, G. A. and Grove, T. L. (1999). Wetting of olivine by sulfide melt: Implications for Re/Os ratios in mantle peridotite and late-stage core formation. *Earth Planet. Sci. Lett.*, **169**, 147–163.

Gerard, O. and Jaoul, O. (1989). Oxygen diffusion in San Carlos olivine. *J. Geophys. Res.*, **94**, 4119–4128.

Gifkins, R. C. (1972). Grain boundary sliding and its accommodation during creep and superplasticity. *Metall. Trans.*, **7A**, 1225–1232.

Glen, J. W. (1952). Experiments on the deformation of ice. *J. Glaciol.*, **2**, 111–114.

Glen, J. W. (1955). The creep of polycrystalline ice. *Proc. R. Soc. Lond. Ser. A*, **228**, 519–538.

Gleason, G. C. and Tullis, J. (1995). A flow law for dislocation creep of quartz aggregates determined with the molten salt cell. *Tectonophysics*, **247**, 1–23.

Goetze, C. (1978). The mechanisms of creep in olivine. *Philos. Trans. R. Soc. Lond. A*, **288**, 99–119.

Goetze, C. and Evans, B. (1979). Stress and temperature in the bending lithosphere as constrained by experimental rock mechanics. *Geophys. J. R. Astron. Soc.*, **59**, 463–478.

Goldsby, D. L. (2006). Superplastic flow of ice relevant to glacier and ice sheet mechanics. In *Glacier Science and Environmental Change*, ed. P. Knight, Oxford, Blackwell Publishing, pp. 308–314.

Goldsby, D. L. and Kohlstedt, D. L. (2001). Superplastic flow of ice: Experimental observations. *J. Geophys. Res.*, **106**, 11 017–11 030.

Greskovich, C. and Schmalzried, H. (1970). Non-stoichiometry and electronic defects in Co_2SiO_4 and in $CoAl_2O_4$-$MgAl_2O_4$ crystalline solutions. *J. Phys. Chem. Solids*, **31**, 639–646.

Groves, G. W. and Kelly, A. (1969). Change of shape due to dislocation climb. *Philos. Mag.*, **19**, 977–986.

Gust, M., Goo, G., Wolfenstine, J., and Mecartney, M. (1993). Influence of amorphous grain boundary phases on the superplastic behavior of 3-mol%-yttria- stabilized tetragonal zirconia polycrystals (3Y-TZP). *J. Am. Ceram. Soc.*, **76**, 1681–1690.

Hager, B. H. (1991). Mantle viscosity: A comparison of models from postglacial rebound and from the geoid, plate driving forces, and advected heat flux. In *Glacial Isostasy, Sea-Level and Mantle Rheology*, ed. R. Sabadini *et al.*, Dordrecht: Kluwer Academic Publishers, pp. 493–513.

Hercule, S. and Ingrin, J., (1999). Hydrogen in diopside: Diffusion, kinetics of extraction-incorporation, and solubility. *Am. Min.*, **84**, 1577–1587.

Hermeling, J. and Schmalzried, H. (1984). Tracer diffusion of the Fe cations in olivine $(Fe_xMg_{1-x})_2SiO_4$ (III). *Phys. Chem. Miner.*, **11**, 161–166.

Herrick, R. R. and Phillips, R. J. (1992). Geological correlations with the interior density structure of Venus. *J. Geophys. Res.*, **97**, 16 017–16 034.

Herrick, D. L. and Stevenson, D. J. (1990). Extensional and compressional instabilities in icy satellite lithospheres. *Icarus*, **85**, 191–204.

Herring, C. (1950). Diffusional viscosity of a polycrystalline solid. *J. Appl. Phys.*, **21**, 437–445.

Hier-Majumder, S. and Kohlstedt, D. L. (2006). Role of dynamic grain boundary wetting in fluid circulation beneath volcanic arcs. *Geophys. Res. Lett.*, **33**, L08305, doi:10.1029/2006GL025716.

Hier-Majumder, S., Anderson, I. M., and Kohlstedt, D. L. (2005). Influence of protons on Fe-Mg interdiffusion in olivine. *J. Geophys. Res.*, **110**, B02202, doi:10.1029/2004JB003292.

Hirth, G. and Kohlstedt, D. L. (1995a). Experimental constraints on the dynamics of the partially molten upper mantle: Deformation in the diffusion creep regime. *J. Geophys. Res.*, **100**, 1981–2001.

Hirth, G. and Kohlstedt, D. L. (1995b). Experimental constraints on the dynamics of the partially molten upper mantle: Deformation in the dislocation creep regime. *J. Geophys. Res.*, **100**, 15 441–15 449.

Hirth, G. and Kohlstedt, D. L. (1996). Water in the oceanic upper mantle: Implications for rheology, melt extraction and the evolution of the lithosphere. *Earth Planet. Sci. Lett.*, **144**, 93–108.

Hirth, G. and Kohlstedt, D. L. (2003). Rheology of the upper mantle and the mantle wedge: A view from the experimentalists. In *Inside the Subduction Factory*, Geophysical Monograph 138, ed. J. Eiler, Washington, D.C., American Geophysical Union, pp. 83–105.

Holness, M. B. (1993). Temperature and pressure dependence of quartz-aqueous fluid dihedral angles: The control of adsorbed H_2O on the permeability of quartzites. *Earth Planet. Sci. Lett.*, **117**, 363–377.

Holness, M. B. and Graham, C. M. (1991). Equilibrium dihedral angles in the system H_2O-CO_2-NaCl-calcite, and implications for fluid flow during metamorphism. *Contrib. Mineral. Petrol.*, **108**, 368–383.

Holtzman, B. K., Groebner, N. J., Zimmerman, M. E., Ginsberg, S. B., and Kohlstedt, D. L. (2003a). Deformation-driven melt segregation in partially molten rocks. *Geochem., Geophys., Geosyst.*, **4**, 8607, doi:10.1029/2001GC000258.

Holtzman, B. K., Kohlstedt, D. L., Zimmerman, M. E., Heidelbach, F., Hiraga, T., and Hustoft, J. (2003b). Melt segregation and strain partitioning: Implications for seismic anisotropy and mantle flow. *Science*, **301**, 1227–1230.

Houlier, B., Cheraghmakani, M., and Jaoul, O. (1990). Silicon diffusion in San Carlos olivine. *Phys. Earth Planet. Inter.*, **62**, 329–340.

Hustoft, J. W. and Kohlstedt, D. L. (2006). Metal-silicate segregation in deforming dunitic rocks. *Geochem., Geophys., Geosyst.*, **7**, Q02001, doi:10.1029/2005GC001048.

Ito, G., Lin, J., and Graham, D. (2003). Observational and theoretical studies of the dynamics of mantle plume-mid-ocean ridge interaction. *Rev. Geophys.*, **41**, 1017, doi:10.1029/2002RG000117.

Iverson, N. R. (2006). Laboratory experiments in glaciology. In *Glacier Science and Environmental Change*, ed. P. Knight, Oxford, Blackwell Publishing, pp. 449–458.

Jaoul, O. (1990). Multicomponent diffusion and creep in olivine. *J. Geophys. Res.*, **95**, 17 631–17 642.

Jaoul, O., Bertran-Alvarez, Y., Liebermann, R. C., and Price, G. D. (1995). Fe-Mg interdiffusion in olivine up to 9 GPa at $T = 600$–$900\,°C$: Experimental data and comparison with defect calculations. *Phys. Earth Planet. Inter.*, **89**, 199–218.

Jin, Z. M., Bai, Q., and Kohlstedt, D. L. (1994). Creep of olivine crystals from four localities. *Phys. Earth Planet. Inter.*, **82**, 55–64.

Kaibyshev, O. (1992). *Superplasticity of Alloys, Intermetallides, and Ceramics*. New York, Springer-Verlag.

Karato, S.-I. (1986). Does partial melting reduce the creep strength of the upper mantle? *Nature*, **319**, 309–310.

Karato, S.-I. and Jung, H. (2003). Effects of pressure on high-temperature dislocation creep in olivine. *Philos. Mag.*, **83**, 401–414.

Karato, S.-I., Paterson, M. S., and Fitz Gerald, J. D. (1986). Rheology of synthetic olivine aggregates: Influence of grain size and water. *J. Geophys. Res.*, **91**, 8151–8176.

Kaufmann, G. and Lambeck, K. (2002). Glacial isostatic adjustment and the radial viscosity profile from inverse modeling. *J. Geophys. Res.*, **107**, 2280, doi:10.1029/2001JB000941.

Kaula, W. M. (1990). Venus: A contrast in evolution to Earth. *Science*, **247**, 1191–1196.

Keefner, J. W., Mackwell, S. J., and Kohlstedt, D. L. (2005). Dunite viscosity dependence on oxygen fugacity (abs.). *Lunar Planet. Sci. Conf. XXXVI*, 1915.

Kelemen, P. B., Hirth, G., Shimizu, N., Spiegelman, M., and Dick, H. J. B. (1997). A review of melt migration processes in the adiabatically upwelling mantle beneath spreading ridges. *Philos. Trans. R. Soc. Lond. A*, **355**, 283–318.

Keszthelyi, L. and McEwen, A. (1997). Magmatic differentiation of Io. *Icarus*, **130**, 437–448.

Kohlstedt, D. L. (1992). Structure, rheology and permeability of partially molten rocks at low melt fractions. In *Mantle Flow and Melt Generation at Mid-Ocean Ridges*,

Monograph 71, ed. J. Phipps-Morgan, D. K. Blackman and J. M. Sinton. Washington, DC: American Geophysical Union. pp. 103–121.

Kohlstedt, D. L. (2002). Partial melting and deformation. In *Plastic Deformation in Minerals and Rocks*, ed. S. I. Karato and H. R. Wenk. Reviews in Mineralogy and Geochemistry, Vol. 51, Mineralogical Society of America, pp. 105–125.

Kohlstedt, D. L. (2006). Water and rock deformation: The case for and against a climb-controlled creep rate. In *Water in Nominally Anhydrous Minerals*, ed. H. Keppler and J. R. Smyth. Reviews in Mineralogy and Geochemistry, Vol. 62, ser. ed. J. J. Rosso, Mineralogical Society of America, pp. 377–396.

Kohlstedt, D. L. (2007). Properties of rocks and minerals: constitutive equations, rheological behavior, and viscosity of rocks. In *Treatise on Geophysics*, ed. G. Schubert. Vol. 2.14. Oxford: Elsevier, pp. 389–417.

Kohlstedt, D. L. and Holtzman, B. K. (2009). Shearing melt out of the Earth: An experimentalist's perspective on the influence of deformation on melt extraction. *Annu. Rev. Earth Planet. Sci.*, **37**, 561–593, doi:10.1146/annurev.earth.031208.100104.

Kohlstedt, D. L. and Hornack, P. (1981). The effect of oxygen partial pressure on creep in olivine. In *Anelasticity in the Earth, Geodynamic Series*, 4, ed. F. D. Stacey, M. S. Paterson and A. Nicolas. Washington, American Geophysical Union, pp. 101–107.

Kohlstedt, D. L. and Mackwell, S. J. (1998). Diffusion of hydrogen and intrinsic point defects in olivine. *Z. Phys. Chem.*, **207**, 147–162.

Kohlstedt, D. L. and Mackwell, S. J. (1999). Solubility and diffusion of 'water' in silicate minerals. In *Microscopic Processes in Minerals*, ed. K. Wright and C. R. A. Catlow, NATO-ASI Series. Dordrecht, Kluwer Academic Publisher, pp. 539–559.

Kohlstedt, D. L. and Ricoult, D. L. (1984). High-temperature creep of olivines. In *Deformation of Ceramics II*, ed. R. E. Tressler and R. C. Bradt. New York, Plenum Publishing, pp. 251–280.

Kohlstedt, D. L. and Zimmerman, M. E. (1996). Rheology of partially molten mantle rocks. *Annu. Rev. Earth Planet. Sci.*, **24**, 41–62.

Kohlstedt, D. L., Evans, B., and Mackwell, S. J. (1995). Strength of the lithosphere: Constraints imposed by laboratory experiments. *J. Geophys. Res.*, **100**, 17 587–17 602.

Kröger, F. A. and Vink, H. J. (1956). Relation between the concentration of imperfections in crystalline solids. In *Solid State Physics 3*, ed. F. Seitz and D. Turnball. New York, Academic Press, pp. 367–435.

Langdon, T. G. (1994). A unified approach to grain boundary sliding in creep and superplasticity. *Acta Met.*, **42**, 2437–2443.

Lange, F. F., Davis, B. I., and Clarke, D. R. (1980). Compressive creep of Si_3N_4/MgO alloys. Part 1: Effect of composition. *J. Mater. Sci.*, **15**, 601–610.

Lee, V., Mackwell, S. J., and Brantley, S. L. (1991). The effect of fluid chemistry on wetting textures in novaculite. *J. Geophys. Res.*, **96**, 10 023–10 037.

Liftshitz, I. M. (1963). On the theory of diffusion-viscous flow of polycrystalline bodies. *Soviet Phys. JETP*, **17**, 909–920.

Lopes, R. M. C., Kamp, L. W., Douté, S., Smythe, W. D., Carlson, R. W., McEwen, A. S., Geissler, P. E., Kieffer, S. W., Leader, F. E., Davies, A. G., Barbinis, E., Mehlman, R., Segura, M., Shirley, J., and Soderblom, L. A. (2001). Io in the near-infrared: NIMS results from the Galileo flybys in 1999 and 2000. *J. Geophys. Res.*, **106**, 33 053–33 078.

Mackwell, S. J. and Kohlstedt, D. L. (1990). Diffusion of hydrogen in olivine: Implications for water in the mantle. *J. Geophys. Res.*, **95**, 5079–5088.

Mackwell, S. J., Kohlstedt, D. L., and Paterson, M. S. (1985). The role of water in the deformation of olivine single crystals. *J. Geophys. Res.*, **90**, 11 319–11 333.

Mackwell, S. J., Zimmerman, M., Kohlstedt, D. L., and Scherber, D. (1995). Experimental deformation of dry Columbia diabase: Implications for tectonics on Venus. In *Proceedings of the 35th U.S. Symposium on Rock Mechanics*, ed. J. J. K. Daemen and R. A. Schultz, pp. 207–214.

Mackwell, S. J., Zimmerman, M. E., and Kohlstedt, D. L. (1998). High-temperature deformation of dry diabase with application to tectonics on Venus. *J. Geophys. Res.*, **103**, 975–984.

McGarr, A. (1984). Scaling of ground motion parameters, state of stress, and focal depth. *J. Geophys. Res.*, **89**, 6969–6979.

McGarr, A., Zoback, M. D., and Hanks, T. C. (1982). Implications of an elastic analysis of in situ stress measurements near the San Andreas fault. *J. Geophys. Res.*, **87**, 7797–7806.

McKenzie, D. (1984). The generation and compaction of partially molten rock. *J. Petrol.*, **25**, 713–765.

McKinnon, W. B. (1999). Convective instability in Europa's floating ice shell. *Geophys. Res. Lett.*, **26**, 951–954.

Mei, S. and Kohlstedt, D. L. (2000a). Influence of water on plastic deformation of olivine: 1. Diffusion creep regime. *J. Geophys. Res.*, **105**, 21 457–21 469.

Mei, S. and Kohlstedt, D. L. (2000b). Influence of water on plastic deformation of olivine: 2. Dislocation creep regime. *J. Geophys. Res.*, **105**, 21 471–21 481.

Mei, S., Bai, W., Hiraga, T., and Kohlstedt, D. L. (2002). Influence of water on plastic deformation of olivine-basalt aggregates. *Earth Planet. Sci. Lett.*, **201**, 491–507.

Mibe, K., Fujii, T., and Yasuda, A. (1998). Connectivity of aqueous fluid in the Earth's upper mantle. *Geophys. Res. Lett.*, **25**, 1233–1236.

Minarik, W. G., Ryerson, F. J., and Watson, E. B. (1996). Textural entrapment of core-forming melts. *Science*, **272**, 530–533.

Misener, D. J. (1974). Cationic diffusion in olivine to 1400 °C and 35 kbar. In *Geochemical Transport and Kinetics*, ed. A. W. Hofmann, B. J. Giletti, H. S. Yoder Jr. and R. A. Yund. Washington, DC: Carnegie Institution of Washington, pp. 117–129.

Montési, L. G. J. and Hirth, G. (2003). Grain size evolution and the rheology of ductile shear zones: From laboratory experiments to postseismic creep. *Earth Planet. Sci. Lett.*, **211**, 97–110.

Montési, L. G. J. and Zuber, M. T. (2002). A unified description of localization for application to large-scale tectonics. *J. Geophys. Res.*, **107**, doi:10.1029/2001JB000465.

Nabarro, F. (1948). Deformation of crystals by the motion of single ions. In *Report on a Conference on the Strength of Solids*. London, Physical Society, pp. 75–90.

Nakamura, A. and Schmalzried, H. (1984). On the Fe^{2+}-Mg^{2+} interdiffusion in olivine (II). *Ber. Bunsenges. Phys. Chem.*, **88**, 140–145.

Panasyuk, S. V. and Hager, B. H. (2000). Inversion for mantle viscosity profiles constrained by dynamic topography and the geoid, and their estimated errors. *Geophys. J. Int.*, **143**, 821–836.

Passey, Q. R. and Schoemaker, E. M. (1982). Craters and basins on Ganymede and Callisto: Morphological indicators of crustal evolution. In *Satellites of Jupiter*, ed. D. Morrison and M. S. Matthews. Tucson, University of Arizona Press, pp. 379–434.

Paterson, M. S. (1969). The ductility of rocks. In *Physics of Strength and Plasticity*, ed. A. S. Argon. Cambridge, MA, MIT Press, pp. 377–392.

Peltier, W. R. (1998). Global glacial isostasy and relative sea level: Implications for solid earth geophysics and climate system dynamics. In *Dynamics of the Ice Age Earth*, ed. P. Wu. Switzerland: Trans Tech Publications, pp. 17–54.

Phillips, R. J., Johnson, C. L., Mackwell, S. J., Morgan, P., Sandwell, D. T., and Zuber, M. T. (1997). Lithospheric mechanics and dynamics of Venus. In *Venus II*, ed. S. W. Bougher, D. M. Hunten and R. J. Phillips. Tucson, AZ: University of Arizona Press, pp. 1163–1204.

Morgan, J. P. (1997). The generation of a compositional lithosphere by mid-ocean ridge melting and its effect on subsequent off-axis hotspot upwelling and melting. *Earth Planet. Sci. Lett.*, **146**, 213–232.

Pitzer, K. S. and Sterner, S. M. (1994). Equations of state valid continuously from zero to extreme pressures for H_2O and CO_2. *J. Chem. Phys.*, **101**, 3111–3116.

Poirier, J.-P. (1985). *Creep of Crystals: High-temperature Deformation Processes in Metals, Ceramics and Minerals*. Cambridge, Cambridge University Press.

Post, A. D., Tullis, J., and Yund, R. A. (1996). Effects of chemical environment on dislocation creep of quartzite. *J. Geophys. Res.*, **101**, 22 143–22 155.

Raj, R. (1982). Creep in polycrystalline aggregates by matter transport through a liquid phase. *J. Geophys. Res.*, **87**, 4731–4739.

Raj, R. and Ashby, M. F. (1971). On grain boundary sliding and diffusional creep. *Metall. Trans.*, **2**, 1113–1127.

Ramsey, J. G. (1980). Shear zone geometry: A review. *J. Structural Geol.*, **2**, 83–99.

Ranero, C. R., Phipps Morgan, J., McIntosh, K., and Reichert, C. (2003). Bending-related faulting and mantle serpentinization at the Middle America Trench. *Nature*, **425**, 367–373.

Reese, C. C., Solomatov, V. S., and Moresi, L.-N. (1998). Heat transport efficiency for stagnant lid convection with dislocation viscosity: Application to Mars and Venus. *J. Geophys. Res.*, **103**, 13 643–13 658.

Regenauer-Lieb, K. and Kohl, T. (2003). Water solubility and diffusivity in olivine: Its role for planetary tectonics. *Mineral. Mag.*, **67**, 697–717.

Regenauer-Lieb, K., Yuen, D. A., and Branlund, J. (2001). The initiation of subduction: Criticality by addition of water? *Science*, **294**, 578–580.

Reston, T. J. (1990). Mantle shear zones and the evolution of the northern North Sea basin. *Geology*, **18**, 272–275.

Ricoult, D. L. and Kohlstedt, D. L. (1983). Structural width of low-angle grain boundaries in olivine. *Phys. Chem. Minerals*, **9**, 133–138.

Ricoult, D. L. and Kohlstedt, D. L. (1985). Creep of Co_2SiO_4 and Fe_2SiO_4 crystals in a controlled thermodynamic environment. *Philos. Mag. A*, **51**, 79–93.

Roscoe, R. (1952). The viscosity of suspensions of rigid spheres. *Brit. J. Appl. Phys.*, **3**, 267–269.

Ross, J. V. and Nielsen, K. C. (1978). High-temperature flow of wet polycrystalline enstatite. *Tectonophys.*, **44**, 233–261.

Ross, M. and Schubert, G. (1985). Tidally forced viscous heating in a partially molten Io. *Icarus*, **64**, 391–400.

Ross, M. and Schubert, G. (1986). Tidal dissipation in a viscoelastic planet. *J. Geophys. Res.*, **91**, 447–452.

Rubie, D. C., Gessmann, C. K., and Frost, D. J. (2004). Partitioning of oxygen during core formation on the Earth and Mars. *Nature*, **429**, 58–61.

Ruiz, J. and Tejero, R. (2000). Heat flows through the ice lithosphere of Europa. *Geophys. Res. Lett.*, **105**, 29 283–29 289.

Ruiz, J. and Tejero, R. (2003). Heat flow, lenticulae spacing, and possibility of convection in the ice shell of Europa. *Icarus*, **162**, 362–373.

Ruoff, A. L. (1965). Mass transfer problems in ionic crystals with charge neutrality. *J. Appl. Phys.*, **36**, 2903–2907.

Rutter, E. H. (1976). The kinetics of rock deformation by pressure solution. *Philos. Trans. R. Soc. Lond. A*283, 203–219.

Rybacki, E. and Dresen, G. (2000). Dislocation and diffusion creep of synthetic anorthite aggregates. *J. Geophys. Res.*, **105**, 26 017–26 036.

Ryerson, F. J., Durham, W. B., Cherniak, D. J., and Lanford, W. A. (1989). Oxygen diffusion in olivine: Effect of oxygen fugacity and implications for creep. *J. Geophys. Res.*, **94**, 4105–4118.

Schenk, P. M. (2002). Thickness constraints on the icy shells of Galilean satellites from a comparison of crater shapes. *Nature*, **417**, 419–421.

Schmalzried, H. (1978). Reactivity and point defects of double oxides with emphasis on simple silicates. *Phys. Chem. Minerals*, **2**, 279–294.

Schmalzried, H. (1981). *Solid State Reactions*. Weinheim, Verlag Chemie, pp. 37–57; 174–175.

Schamzlried, H. (1995). *Chemical Kinetics of Solids*. New York, VCH Publishers, pp. 27–37.

Scott, D. R. and Stevenson, D. J. (1986). Magma ascent by porous flow. *J. Geophys. Res.*, **91**, 9283–9296.

Scott, T. and Kohlstedt, D. L. (2006). The effect of large melt fraction on the deformation behavior of peridotite. *Earth Planet. Sci. Lett.*, **246**, 177–187.

Segatz, M., Spohn, T., Ross, M. N., and Schubert, G. (1988). Tidal dissipation, surface heat flow, and figure of viscoelastic models of Io. *Icarus*, **75**, 187–206.

Shelton, G. and Tullis, J. (1981). Experimental flow laws for crustal rocks (abs.). *Eos Trans. AGU*, **62**, 396.

Shewmon, P. G. (1983). *Diffusion in Solids*. Jenks, OK, J. Williams Book Company, pp. 155–160.

Sibson, R. H. (1974). Frictional constraints on thrust, wrench and normal faults. *Nature*, **249**, 542–544.

Sibson, R. H. (1977). Fault rocks and fault mechanisms. *J. Geol. Soc. London*, **133**, 191–213.

Sockel, H. G. (1974). Defect structure and electrical conductivity of crystalline ferrous silicate. In *Defects and Transport in Oxides*, ed. M. S. Seltzer and R. I. Jaffe. New York, Plenum Press, pp. 341–354.

Solomatov, V. S. and Moresi, L.-N. (2000). Scaling of time-dependent stagnant lid convection: Application to small-scale convection on Earth and other terrestrial planets. *J. Geophys. Res.*, **105**, 21 795–21 818, doi:10.1029/2000JB900197.

Stalder, R. and Skogby, H. (2002). Hydrogen incorporation in enstatite. *Eur. J. Mineral.*, **14**, 1139–1144.

Stern, L. A., Durham, W. B., and Kirby, S. H. (1997). Grain-sized-induced weakening of H_2O ices I and II and associated anisotropic recrystallization. *J. Geophys. Res.*, **102**, 5313–5325.

Tackley, P., Schubert, G., Glatzmaier, G. A., Schenk, P., Ratcliff, J. T., and Matas, J.-P. (2001). Three-dimensional simulations of mantle convection in Io. *Icarus*, **149**, 79–93.

Takei, Y. (1998). Constitutive mechanical relations of solid-liquid composites in terms of grain-boundary contiguity. *J. Geophys. Res.*, **103**, 18 183–18 203.

Takei, Y. (2000). Acoustic properties of partially molten media studied on a simple binary system with a controllable dihedral angle. *J. Geophys. Res.*, **105**, 16 665–16 682.

Takei, Y. (2001). Stress-induced anisotropy of partially molten media inferred from experimental deformation of a simple binary system under acoustic monitoring. *J. Geophys. Res.*, **106**, 567–588.

Takei, Y. (2002). Effect of pore geometry on V_P/V_S: From equilibrium geometry to crack. *J. Geophys. Res.*, **107**(B21), 2043, 10.1029/2001JB000522.

Takei, Y. and Holtzman, B. K. (2009a). Viscous constitutive relations of solid–liquid composites in terms of grain-boundary contiguity: I. Grain boundary diffusion-control model. *J. Geophys. Res.*, **114**, B06205, doi:10.1029/2008JB005850.

Takei, Y. and Holtzman, B. K. (2009b). Viscous constitutive relations of solid–liquid composites in terms of grain-boundary contiguity: II. Compositional model for small melt fractions. *J. Geophys. Res.*, **114**, B06206, doi:10.1029/2008JB005851.

Takei, Y. and Holtzman, B. K. (2009c). Viscous constitutive relations of solid–liquid composites in terms of grain-boundary contiguity: III. Causes and consequences of viscous anisotropy. *J. Geophys. Res.*, **114**, B06207, doi:10.1029/2008JB005852.

Takei, Y. and Shimizu, I. (2003). The effects of liquid composition, temperature, and pressure on the equilibrium dihedral angles of binary solid–liquid systems inferred from a lattice-like model. *Phys. Earth Planet. Inter.*, **139**, 225–242.

Tharp, T. M. (1983). Analogies between the high-temperature deformation of polyphase rocks and the mechanical behavior of porous powder metal. *Tectonophys.*, **96**, T1-T11.

Toramaru, A. and Fujii, N. (1986). Connectivity of melt phase in a partially molten peridotite. *J. Geophys. Res.*, **91**, 9239–9252.

Turcotte, D. L. (1993). An episodic hypothesis for Venusian tectonics. *J. Geophys. Res.*, **98**, 17 061–17 068.

Turcotte, D. L. (1995). How does Venus lose heat? *J. Geophys. Res.*, **100**, 16 931–16 940.

Turcotte, D. L. and Schubert, G. (1982). *Geodynamics: Applications of Continuum Physics to Geological Problems*. New York, John Wiley, pp. 163–167; 383–384.

van der Wal, D., Chopra, P. N., Drury, M., and Fitz Gerald, J. D. (1993). Relationships between dynamically recrystallized grain size and deformation conditions in experimentally deformed olivine rocks. *Geophys. Res. Lett.*, **20**, 1479–1482.

von Bargen, N. and Waff, H. S. (1986). Permeabilities, interfacial areas and curvatures of partially molten systems: Results of numerical computations of equilibrium microstructures. *J. Geophys. Res.*, **91**, 9261–9276.

von Mises, R. (1928). Mechanik der plastischen Formänderung von Kristallen. *Z. Angew. Math. Mech.*, **8**, 161–185.

Waff, H. S. and Bulau, J. R. (1979). Equilibrium fluid distribution in an ultramafic partial melt under hydrostatic stress conditions. *J. Geophys. Res.*, **84**, 6109–6114.

Waff, H. S. and Faul, U. H. (1992). Effects of crystalline anisotropy on fluid distribution in ultramafic partial melts. *J. Geophys. Res.*, **97**, 9003–9014.

Wang, Z. (2002). Effect of pressure and water on the kinetics properties of olivine, Ph.D. thesis, University of Minnesota.

Wang, Z., Hiraga, T., and Kohlstedt, D. L. (2004). Effect of H^+ on Fe-Mg interdiffusion in olivine, $(Mg,Fe)_2SiO_4$. *Appl. Phys. Lett.*, **85**, 209–211.

Wang, L. and Zhang, Y. (1996). Diffusion of the hydrous component in garnet. *Am. Min.*, **81**, 706–718.

Watson, E. B. and Brenan, J. M. (1987). Fluids in the lithosphere: 1. Experimentally determined wetting characteristics of CO_2-H_2O fluids and their implications for fluid transport, host-rock physical properties, and fluid inclusion formation. *Earth Planet. Sci. Lett.*, **85**, 497–515.

Weertman, J. (1968). Dislocation climb theory of steady-state creep. *Trans. Am. Soc. Metals*, **61**, 681–694.

Weertman, J. (1970). The creep strength of the Earth's mantle. *Rev. Geophys. Space Phys.*, **8**, 145–168.

Weertman, J. (1983). Creep deformation of ice. *Annu. Rev. Earth Planet. Sci.*, **11**, 215–240.

Weertman, J. (1999). Microstructural mechanisms in creep. In *Mechanics and Materials: Fundamentals and Linkages*, ed. M. A. Meyers, R. W. Armstrong and H. Kirchner. New York: John Wiley and Sons, pp. 451–488.

Weertman, J. and Weertman, J. R. (1975). High temperature creep of rock and mantle viscosity. *Annu. Rev. Earth Planet. Sci.*, **3**, 293–315.

Woods, S. (2000). The kinetics of hydrogen diffusion in single crystal enstatite. Ph.D. thesis, Pennsylvania State University.

Xu, Y., Zimmerman, M. E., and Kohlstedt, D. L. (2004). Deformation behavior of partially molten mantle rocks. In *Rheology and Deformation of the Lithosphere at Continental Margins*. MARGINS Theoretical and Experimental Earth Science Series, Vol. I. ed. G. D. Karner, N. W. Driscoll, B. Taylor and D. L. Kohlstedt. Columbia University Press, pp. 284–310.

Zahnle, K., Dones, L., and Levison, H. F. (1998). Cratering rates on Galilean satellites. *Icarus*, **136**, 202–222.

Zhao, Y. H., Ginsberg, S. G., and Kohlstedt, D. L. (2004). Solubility of hydrogen in olivine: Effects of temperature and Fe content. *Contrib. Mineral. Petrol.*, **147**, 155–161, doi:10.1007/s00410-003-0524-4.

Zhao, Y.-H., Zimmerman, M. E., and Kohlstedt, D. L. (2009). Effect of iron content on the creep behavior of olivine: 1. Anhydrous conditions, *Earth Planet. Sci. Lett.* **287**, 229–240, doi:10.1016/j.epsl.2009.08.006.

Zimmerman, M. E. and Kohlstedt, D. L. (2004). Rheological properties of partially molten lherzolite. *J. Petrol.*, **45**, 275–298.

Zimmerman, M. E., Zhang, S., Kohlstedt, D. L., and Karato, S. (1999). Melt distribution in mantle rocks deformed in shear. *Geophys. Res. Lett.*, **26**, 1505–1508.

10

Fault populations

Richard A. Schultz
Geomechanics – Rock Fracture Group, Department of Geological Sciences and Engineering, University of Nevada, Reno

Roger Soliva
Université Montpellier II, Département des Sciences de la Terre et de l'Environnement, France

Chris H. Okubo
U.S. Geological Survey, Flagstaff

and

Daniel Mège
*Laboratoire de Planetologie et Geodynamique, UFR des Sciences et Techniques
Université de Nantes, France*

Summary

Faults have been identified beyond the Earth on many other planets, satellites, and asteroids in the solar system, with normal and thrust faults being most common. Faults on these bodies exhibit the same attributes of fault geometry, displacement–length scaling, interaction and linkage, topography, and strain accommodation as terrestrial faults, indicating common processes despite differences in environmental conditions, such as planetary gravity, surface temperature, and tectonic driving mechanism. Widespread extensional strain on planetary bodies is manifested as arrays and populations of normal faults and grabens having soft-linked and hard-linked segments and relay structures that are virtually indistinguishable from their Earth-based counterparts. Strike-slip faults on Mars and Europa exhibit classic and diagnostic elements such as rhombohedral push-up ranges in their echelon stepovers and contractional and extensional structures located in their near-tip quadrants. Planetary thrust faults associated with regional contractional strains occur as surface-breaking structures, known as lobate scarps, or as blind faults beneath an anticlinal fold at the surface, known as a wrinkle ridge. Analysis of faults and fault

---

Planetary Tectonics, edited by Thomas R. Watters and Richard A. Schultz. Published by Cambridge University Press. Copyright © Cambridge University Press 2010.

populations can reveal insight into the evolution of planetary surfaces that cannot be gained from other techniques. For example, measurements of fault-plane dip angles provide information on the frictional strength of the faulted lithosphere. The depth of faulting, and potentially, paleogeothermal gradients and seismic moments, can be obtained by analysis of the topographic changes associated with faulting. Because the sense of fault displacement (normal, strike-slip, or thrust) is related to the local and regional stress states, fault dip angle and displacement characteristics can provide values for crustal strength and magnitudes of stress and strain in map view and at depth while the fault population was active. Statistical characterization of fault-population attributes, such as spacing, length, and displacement, provides an exciting and productive avenue for exploring the mechanical stratigraphy, fault restriction, partitioning of strain between small and large faults, and the processes of fault growth over a wide range of scales that are useful for defining or testing geodynamic models of lithospheric and planetary evolution.

1 Introduction

Faults on the Earth or other planetary bodies rarely occur as solitary entities. Instead, they occur as members of a set, array, network, or population. In a population, faults display wide variation in their primary characteristics, such as length, displacement, and spacing. However, these characteristics do not occur at random. All of the faults' characteristics depend on one another, so that knowledge of one or two key characteristics can provide insight into the values and relationships among the others.

In this chapter we first define the common fault geometries and then review the stress states in a planetary lithosphere that are associated with faults, using the conditions in the Earth's crust as a reference. We then briefly explore some of the main characteristics of fault populations, again using examples from Earth since these have been investigated in the most detail. Because topographic data are becoming more widely available for planetary fault populations, we show how measurements of the structural topography generated by faulting can reveal information about properties of the faults and of the faulted lithosphere. Last, we show how strains can be calculated for planetary fault populations, and end with a summary of challenges for future work on these exciting issues.

2 Faulted planetary lithospheres

Faults have been documented on nearly every geologic surface in the solar system. Normal faults and grabens are probably the most common and are found on Mercury (Watters *et al.*, Chapter 2), Venus (McGill *et al.*, Chapter 3), the Moon (Watters and

Johnson, Chapter 4), Mars (Golombek and Phillips, Chapter 5), Europa, Ganymede, and several smaller icy satellites of the outer planets including Tethys, Dione, and Miranda (Collins *et al.*, Chapter 7). Thrust faults have been identified on Mercury, Venus, the Moon, and Mars (e.g., Suppe and Connors, 1992; Williams *et al.*, 1994; Solomon *et al.*, 2008; and chapters in this volume). Strike-slip faults have been identified on Mars (e.g., Schultz, 1989; Okubo and Schultz, 2006b; Andrews-Hanna *et al.*, 2008) and on the icy satellite Europa that shows large lateral displacements, such as those found at terrestrial transform plate boundaries (Schenk and McKinnon, 1989; Kattenhorn and Marshall, 2006). Individual dilatant cracks (joints) and deformation bands (Aydin *et al.*, 2006; Fossen *et al.*, 2007) have both been identified on Mars (Okubo and McEwen, 2007; Okubo *et al.*, 2008a) and perhaps Europa (Aydin, 2006), and the presence of subsurface igneous dikes has been inferred on Mars from surface topographic data (Schultz *et al.*, 2004). In this chapter we focus on faults on the planets and satellites.

2.1 Definition and geometries of faults

The terminology of geologic structures such as joints, faults, and deformation bands has recently been reassessed and streamlined by Schultz and Fossen (2008). Following this terminology, a **fault** is a sharp structural discontinuity defined by its slip planes (surfaces of discontinuous displacement) and related structures including fault core and damage zones (e.g., cracks, deformation bands, slip surfaces, and other structural discontinuities) that formed at any stage in the evolution of the structure. Commonly associated structures such as drag or faulted fault-propagation folds are associated elements not included in the term fault, although clay smearing or other early forms of strain localization may be included.

A **fault set** is a collection of faults that have some element in common, such as age, length, spacing, type, or orientation. A **fault array** is a fault set in which all faults are genetically related to each other (i.e., same deformational event or rock type). A **fault zone** is a narrow array of relatively closely spaced faults having similar strikes. A **fault system** is a spatially extensive array in which the faults interact mechanically. A **fault population** is a system comprised of all faults having the full range of lengths, spacings, displacement distributions, and other characteristics that record the progressive evolution of the faulted domain. Populations of faults, as well as joints (Segall, 1984a) and deformation bands (Fossen *et al.*, 2007), are said to be **self-organizing** (e.g., Sornette *et al.*, 1990) in the sense that their physical, geometric, and statistical characteristics evolve with increasing deformation of the region (e.g., Cowie *et al.*, 1995; Cladouhos and Marrett, 1996; Ackermann *et al.*, 2001).

Terminology of normal faults

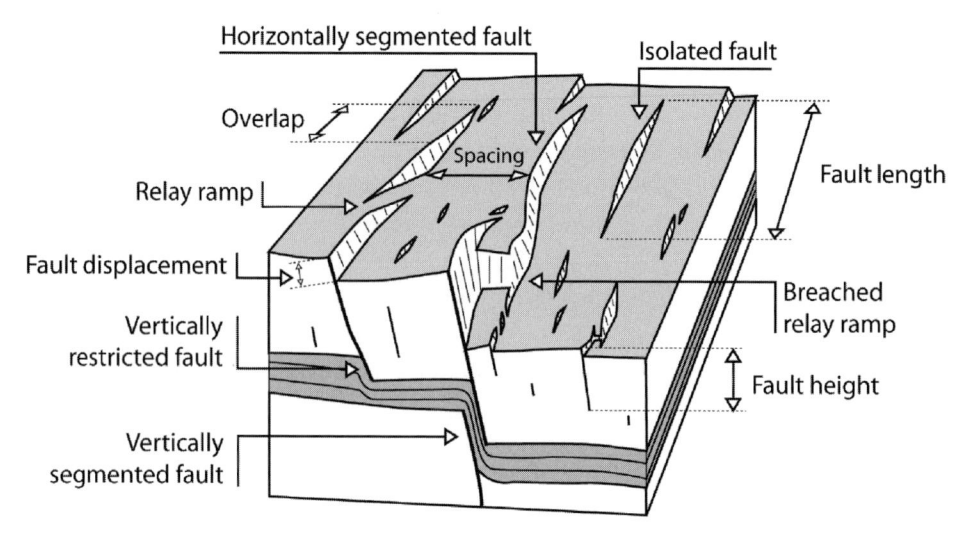

Figure 10.1. Block diagram showing the main geometric characteristics of a surface-breaking fault population. Although normal faults are shown, the descriptions are also applicable to surface-breaking strike-slip and thrust faults on the Earth and other planets and satellites.

Fault systems are composed of "isolated faults" and "segmented faults" (Figure 10.1). **Isolated faults** are defined as faults showing no evidence of significant mechanical interaction with other nearby or surrounding faults (Willemse, 1997; Gupta and Scholz, 2000a; Soliva and Benedicto, 2004), i.e., without relay zones or breaching (e.g., Davison, 1994) allowing transfer of displacement to another fault. A **segmented fault** is composed of two or more non-colinear overlapping fault segments that are arranged in echelon patterns (see Davison, 1994). Fault segments are separated by relay zones, or **stepovers** (Aydin, 1988), which are defined as the rock volume between overlapping (echelon) fault tips in which the fault segments interact through their stress fields. This interaction results in a transfer of displacement between the fault segments, an increase of fault-end displacement gradient that is accommodated by continuous deformation, and distortion of the rock volume located between the two fault segments. A fault can be segmented in three dimensions (3-D; vertically and horizontally, Figure 10.1), i.e., containing vertical, horizontal and obliquely oriented relay zones leading to very simplistic (elliptical or rectangular shapes) to more complex fault geometries (Kattenhorn and Pollard, 1999, 2001; Walsh et al., 2003; Benedicto et al., 2003). A segmented fault can therefore be composed of fault segments that are breached (connected by cross-faults, or "hard-linked") or not (echelon or "soft-linked"). A linked

Table 10.1. *Effective lithostatic stress gradients and rock-mass depths for terrestrial planets*

Planet/Satellite	Gravity, g (m s^{-2})	Dry lithostat σ_v (MPa km^{-1})	Wet lithostat σ_v (MPa km^{-1})	Depth of rock-mass zone, z_0 (km)
Mercury	3.78	10.6	–	2.6–5.2
Venus	8.8	24.6	–	1.1–2.2
Earth	9.8	–	17.6	1–2
Moon	1.62	4.5	–	6–12
Mars	3.7	10.4	6.7	2.6–5.3

Assumes $\sigma_v = \rho(1-\lambda)gz$ with $\rho = 2800$ kg m^{-3} (dry crustal rock). Values calculated and shown where dry or wet conditions can be reasonably inferred. Approximate values for z_0 for Mercury, Venus, Moon, and Mars for the depth range of 1–2 km calculated for depths on those bodies corresponding to σ_v on Earth for dry basalt taken to be at 1–2 km depths.

(formerly segmented) fault is called "**kinematically coherent**" (Willemse *et al.*, 1996) because it acts as a single mechanical break.

2.2 Stress states and faulting

The reference stress state for a planetary lithosphere can be inferred from measurements of *in situ* stress within the Earth's crust. Subsurface stresses are, in general, compressive (e.g., McGarr and Gay, 1978; Brown and Hoek, 1978; Engelder, 1993, pp. 10–15; Plumb, 1994; Zoback *et al.*, 2003), except perhaps for rare exceptions due to subsurface inhomogeneities (e.g., lava tubes, faults) or for locations close to the surface, where one of the principal stresses may be tensile. The vertical stress magnitude, or "lithostat," is given by $\sigma_v = \rho(1 - \lambda)gz$, in which ρ is the average density of rock, λ is the Hubbert-Rubey pore-fluid pressure ratio with $\lambda = P_{\text{water}}/\rho_{\text{rock}}$ (Hubbert and Rubey, 1959; Suppe, 1985, p. 300; Price and Cosgrove, 1990, p. 68; Weijermars, 1997, pp. 42, 98–99), g is gravitational acceleration at the planetary surface, and z is the depth below the surface (McGarr and Gay, 1978; Zoback *et al.*, 2003), with compressive stresses taken in this chapter to be positive numbers. Using values of $\rho = 2800$ kg m^{-3} and either dry or hydrostatic pore-water conditions ($\lambda = 1/\rho_{\text{rock}} \sim 0.4$), as would be the case for the effective principal stresses in the Earth (e.g., Suppe, 1985; Engelder, 1993) and, perhaps at times, for Mars, the calculated lithostats are listed in Table 10.1. These gradients in effective vertical stress σ_v are well documented for the Earth (e.g., Brown and Hoek, 1978; McGarr and Gay, 1978).

Classical rock mechanics treatments suggest values for the minimum horizontal stress of approximately one-third of the lithostatic value based on the Poisson response of an ideal intact linearly-elastic unconfined rock in the horizontal direction (e.g., Jaeger and Cook, 1979; Jaeger et al., 2007; see Suppe, 1985, for the "Earth pressure coefficient"). Measurements of *in situ* stress in the Earth's crust demonstrate instead, however, that the magnitudes of the horizontal principal stresses are controlled by the frictional resistance of the fractured planetary lithosphere (e.g., Zoback et al., 2003). As originally developed by Goetz and Evans (1979) and Brace and Kohlstedt (1980) in the context of lithospheric strength envelopes for the Earth, the horizontal principal stresses are limited to about one-third to one-fifth of the dry ($\lambda = 0$) or effective ($\lambda > 0$) lithostat, with greater principal-stress differences (or principal-stress ratios) leading to faulting (see Kohlstedt et al., 1995, and Kohlstedt and Mackwell, Chapter 9). As a result, the dry or effective principal stresses that drive faulting are all compressive (e.g., Jaeger et al., 2007, p. 74), as was shown more than a half-century ago in E. M. Anderson's fault classification scheme (Anderson, 1951; Figure 10.1), so that all three types of faults – normal, strike-slip, and thrust – can be regarded as compressive structures that also shear (see also Sibson, 1974; Marone, 1998; Scholz, 1998).

The critical (minimum) value of the remote (dry or effective) principal stresses for faulting of a planetary lithosphere to be achieved is then given most simply by the Coulomb criterion for frictional slip (Jaeger and Cook, 1979, p. 97; Price and Cosgrove, 1990, p. 26)

$$\sigma_1 = \sigma_c + q\sigma_3, \qquad (10.1)$$

in which σ_c is the unconfined compressive strength of the rock mass (Bieniawski, 1989; Schultz, 1995, 1996) and $q = ([\mu^2 + 1]^{0.5} + \mu)^2$ with μ being the average static (or maximum; see Marone, 1998) friction coefficient of lithospheric rocks. Typical values of static and dynamic friction coefficients for crustal rocks on the Earth are $\mu = 0.2$–0.8 (Paterson and Wong, 2005, pp. 166–170; Jaeger et al., 2007, p. 70), with strength given by values of static friction at the high end of the range. Setting $\mu = 0.6$ (corresponding to a representative angle of friction for the rock of $\varphi = \tan^{-1}(\mu) = 31°$; see Sibson, 1994), $q = 3.12$ and $\sigma_c = 3.5$ MPa (assuming a typically small value of cohesion for the near-surface rock mass of $C_0 = 1.0$ MPa; see Hoek, 1983; Schultz, 1993, 1996; Hoek and Brown, 1997). Typical ranges of friction coefficient μ of 0.4–0.85 lead to values of $\varphi = 22$–40 and $q = 2.2$–4.68, respectively. For a given value of vertical stress or depth, the maximum (dry or effective) compressive principal stress must therefore be at least 2–5 times larger than the value of the minimum compressive principal stress for normal, strike-slip, or thrust faulting to initiate in a planetary lithosphere. This critical value defines the brittle (Byerlee) frictional strength of planetary rocks having icy or silicate compositions (e.g., Sibson, 1974; Brace and Kohlstedt, 1980; Kohlstedt et al., 1995; Scholz, 2002, pp. 146–155; Kohlstedt and Mackwell, Chapter 9).

The fault dip angle is related to the friction coefficient (or angle) of the faulted planetary lithosphere. Noting that $q = \tan^2(\theta_{opt})$, the optimum dip angle θ_{opt} is given by (e.g., Jaeger and Cook, 1979)

$$\theta_{opt} = \left(45° + \frac{\phi}{2}\right) = \left[90° - \frac{\tan^{-1}\left(\frac{1}{\mu}\right)}{2}\right], \quad (10.2)$$

where θ_{opt} is the angle between σ_1 and the normal to the optimum slip plane. This relationship assumes that one of the (dry or effective) principal stresses is vertical, which is a common occurrence in the Earth (e.g., McGarr and Gay, 1978) and likely in other planets and satellites as well. For a friction coefficient of $\mu = 0.6$ (corresponding to a friction angle $\varphi = 30.5°$), the optimum fault dip angle for a normal fault would be 60.5°; a thrust fault would be oriented according to σ_1 being horizontal, resulting in an optimum dip angle of 29.5°. These values are in accord with the measured dip angles of many large steeply dipping terrestrial faults (Sibson, 1994) that may be modified (either steepened or shallowed) during the progressive deformation of a faulted domain.

At the planetary surface and shallow subsurface, however, faults can dip at initial angles that are steeper than the optimum angle (e.g., McGill and Stromquist, 1979; Gudmundsson, 1992; Moore and Schultz, 1999; McGill et al., 2000; Ferrill and Morris, 2003) because of the pressure and depth dependence of frictional strength in the near surface (e.g., Hoek, 1983; Schultz, 1995) and differences in the initial failure mechanism of near-surface strata (e.g., Gudmundsson, 1992; Schultz, 1996; Peacock, 2002; Crider and Peacock, 2004). Sometimes called the "rock-mass zone" (Schultz, 1993), this region of locally greater effective friction coefficient extends from the planetary surface down to depths of $\sim$1–2 km on the Earth, corresponding approximately to depths on the planets and satellites where the vertical principal compressive stress $\sigma_1 < 10$–35 MPa (with specific values depending on the dry or wet rock density; see Table 10.1). Within this near-surface zone, rock-mass strength is well approximated by the Hoek-Brown criterion (Hoek and Brown, 1980; Hoek, 1983, 1990; Brady and Brown, 1993, pp. 132–135; Franklin, 1993) which is given by

$$\sigma_1 = \sigma_3 + \sqrt{m\sigma_c\sigma_3 + s\sigma_c^2}, \quad (10.3)$$

in which m and s are non-dimensional parameters that describe the friction and degree of fracturing of the rock mass and σ_c is the unconfined compressive strength of the intact planetary lithospheric rock material (i.e., its lithology such as basalt or tuff). Values of the parameters are given by the sources cited above, as well as Schultz (1993, 1995, 1996); the criterion has been applied to terrestrial and planetary faulting by Angelier (1994), Schultz (1993, 1995, 1996, 2002), Schultz and Zuber (1994), Schultz and Watters (1995), Ferrill and Morris (2003), Schultz *et al.* (2004, 2006), Okubo and Schultz (2004), Neuffer and Schultz (2006), and Andrews-Hanna *et al.* (2008). Stress models for prediction of the types and locations of planetary faults that do not incorporate a criterion for rock-mass strength such as Equations (10.1) or (10.3) (e.g., Banerdt *et al.*, 1992; Freed *et al.*, 2001; Golombek and Phillips, Chapter 5) potentially can correctly predict the observed faults (especially strike-slip) when the lithospheric strength is explicitly included (Schultz and Zuber, 1994; Andrews-Hanna *et al.*, 2008).

In structural geology, the change in length ΔL between two points in a rock normalized by the original length L_0 between them is referred to variously as the extension, elongation, linear strain, or normal strain. The sign of this quantity, computed by using $\varepsilon_n = \Delta L/L_0$, is taken to be positive for an increase in length (extension) or negative for a length decrease (contraction or shortening). In this chapter we refer to ε_n as the *normal strain* (a component of the local strain tensor), following the convention from rock mechanics (e.g., Means, 1976, p. 152; Jaeger *et al.*, 2007, p. 43), noting that it applies to penetrative deformation at the particular scale of interest (e.g. Pappalardo and Collins, 2005). For geometrically sparser fault populations, the normal strain ε_n in a given direction (i.e., the horizontal planetary surface normal to fault strike) can be calculated by summing the geometric fault moments as described in Section 5 (see Equation 10.12) below.

Anderson's (1951) classification scheme for faults succinctly associates the three main fault types (normal, thrust and strike-slip) with the 3-D regional stress states needed to drive the required sense of slip along optimally oriented surfaces. Anderson's fault classification scheme is shown in Figure 10.2. With one principal stress vertical (σ_v), the other two are necessarily horizontal (σ_H and σ_h; e.g., McGarr and Gay, 1978; Angelier, 1994). In order of decreasing compressive stress magnitude, the dry or effective principal stresses in a planetary crust are $\sigma_1 > \sigma_2 > \sigma_3$ and $\sigma_H > \sigma_h$. The fault's strike is defined to be parallel to σ_2 (the intermediate principal stress; Sibson, 1974), using the assumption that only the extreme (maximum and minimum) principal stresses are important for driving frictional sliding in a planetary lithosphere (σ_1 and σ_3; e.g., Paterson and Wong, 2005, pp. 35–38). This correspondence between fault strike and σ_2 is commonly observed in nature when the magnitude of normal strain parallel to the fault, ε_2, is negligibly small (i.e., a two-dimensional strain field; see Reches, 1978, 1983; Aydin and Reches, 1982; Krantz, 1988, 1989; Figure 10.2).

For normal faulting, the maximum dry or effective principal stress σ_1 is oriented vertically, denoted the vertical stress σ_v; with the minimum remote dry or effective principal stress σ_h being horizontal (σ_3), the remote stress state for normal faulting and grabens in a planetary crust is given for typical values of friction coefficient ($\mu = 0.6$–0.85) by $\sigma_v = 3$–$5\,\sigma_h$. For thrust faulting, on the other hand,

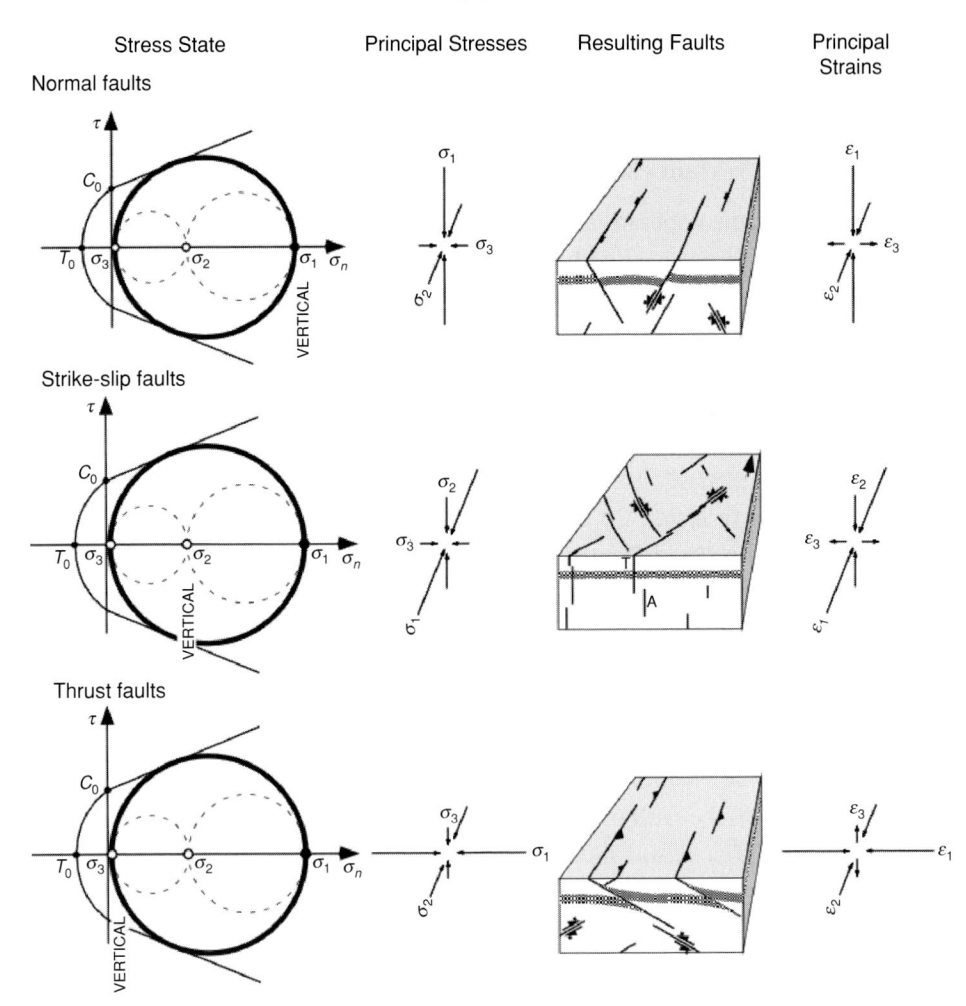

Figure 10.2. The Anderson (1951) classification scheme for faults based on the orientations of the remote (regional) principal stresses relative to the planetary surface. The principal normal strains are also shown (right-hand column); note the change in sign of normal strain ε_1 for extension (normal faults) and ε_3 for strike-slip and contraction (thrust faults). This normal strain, with the opposite sense of the other two, is the "odd axis" of Krantz (1988). Its extensional sense is required when a rock mass deforms with constant volume, as is approximately the case for planetary lithospheres.

σ_1 is horizontal and σ_3 is vertical, so that $\sigma_H = 3\text{--}5\,\sigma_v$. For strike-slip faulting, σ_2 is vertical, so that $\sigma_1 = \sigma_H = 3$ to $5\,\sigma_h$. Fault sets on a planetary surface are prima facie evidence that the state of stress in a planetary crust was given approximately by one of these three expressions. The magnitudes of the resulting strains, however, are related to the magnitude of displacement that has accumulated along the faults in the population, as well as the sizes and spatial relationships between the faults (e.g., Segall, 1984a; Gupta and Scholz, 2000b; Schultz, 2003a; see Section 5).

3 Main characteristics of fault populations

The analyses of fault populations began with Earth examples, so the first salient works and main references cited here are for terrestrial fault systems. The characteristics and processes of fault system development (e.g., McGill and Stromquist, 1979; Davison, 1994) described in this section are observed as well in planetary fault systems (e.g., Muehlberger, 1974; Lucchitta, 1976; Sharpton and Head, 1988; Banerdt *et al.*, 1992; McGill, 1993; Schultz and Fori, 1996; Mège and Masson, 1996; Schultz, 1991, 1997, 1999, 2000a,b; Koenig and Aydin, 1998; Mangold *et al.*, 1998; Watters *et al.*, 1998; Wilkins and Schultz, 2003; Okubo and Schultz, 2003, 2006b; Goudy *et al.*, 2005; Hauber and Kronberg, 2005; Kattenhorn and Marshall, 2006; Kiefer and Swafford, 2006; Knapmeyer *et al.*, 2006), although the rheologies and characteristics of the lithospheric strength envelopes for those bodies differ in detail from those for the Earth (see Kohlstedt and Mackwell, Chapter 9).

3.1 Fault system morphology

A fault population can be quantitatively described by using a series of geometrical attributes inherent to the fault pattern (see Figure 10.1). Fault *displacement*, i.e., the net slip along the fault (also called the fault "offset"), is an important geometrical attribute since it provides information on fault kinematics and the amount of strain accommodated by the fault. In the absence of three-dimensional data from the fault surface (e.g., Nicol *et al.*, 1996; Willemse, 1997; Kattenhorn and Pollard, 2001; Wilkins and Schultz, 2005), the continuous measure of fault displacement along fault trace (the fault's "displacement distribution" or "displacement profile") can be obtained by measuring the displacement of preexisting markers, such as bedding or impact craters, either in a horizontal plane (such as the planetary surface) or in a vertical plane (such as a cross-sectional exposure of the fault; Wilkins and Gross, 2002). On the Earth and other planets and satellites, with many surface-breaking faults but rarer cross-sectional exposures, displacement distributions along the faults' horizontal traces (called the fault "length") are more commonly measured and reported. In addition, however, displacement distributions are generally easier to obtain along normal faults, especially along their horizontal lengths (e.g., Dawers *et al.*, 1993), than along strike-slip faults (e.g., Peacock and Sanderson, 1995), for which horizontal markers would be needed, or thrust faults (e.g., Davis *et al.*, 2005), where folding and related deformation can complicate the displacement distribution. This is the reason why fault population analyses have been emphasized for normal faulting environments and also why the following text in this chapter will be based on normal fault populations.

Three other main geometrical attributes used in fault population studies are *length*, *spacing* and *overlap* (Figure 10.1). The **length** of a fault is defined by the distance along the fault trace between the fault tips (where fault offsets decrease to zero) measured along a horizontal surface. Fault **spacing** is the horizontal distance normal to fault strike between two faults. Fault **overlap** is the horizontal distance parallel to fault strike along which two faults overstep (i.e., in the relay ramp between two normal faults (e.g., Davison, 1994; Moore and Schultz, 1999; Schultz *et al.*, 2007) or thrust faults (Aydin, 1988; Davis *et al.*, 2005), or the length of a pull-apart or push-up range (Aydin and Nur, 1982; Schultz, 1989; Aydin and Schultz, 1990; Aydin *et al.*, 1990) along a pair of echelon strike-slip faults). These geometrical attributes are important for quantitatively describing the geometry of both the relay zones and the overall fault population itself.

Much attention has been devoted to potential measurement biases on these geometrical attributes (e.g., Marrett *et al.*, 1999; Ackermann *et al.*, 2001; Soliva and Schultz, 2008; and references therein). Two classes of bias can be defined as "natural bias" and "detection bias." Natural bias results from natural geologic processes, such as fault scarp erosion and basin in-filling, that lead to underestimates of fault length, displacement, overlap, and spacing. Detection biases are inherent to the particular data acquisition method (e.g., field photographs, aerial or satellite images, digital elevation models (DEMs); see Priest (1993) for a comprehensive and quantitative treatment of detection biases). Faults with lengths that exceed the dimensions of the measurement area are underestimated, introducing an upper bias referred to as "censoring," whereas image resolution, for example, may lead to undercounting of small faults, introducing a lower bias known as "truncation." Similarly, measurements of fault lengths and displacements are limited by the spatial and vertical resolution of a DEM (e.g., Hooper *et al.*, 2003).

The formation of the largest faults and the distribution of strain appear as widely variable in normal fault systems. Two end-member cases can be identified: (1) localized fault systems, with a few large faults accumulating around 50% of the total strain accommodated by the population and a large number of small faults (with a complementary strain) (Figure 10.3a and 10.3c), and (2) distributed fault systems, with strain regularly distributed along evenly spaced faults having a characteristic length scale (Figure 10.3b and 10.3d). These fault system geometries, which are a function of several factors, including deformation rate, stress transmission mode, rheology of the lithospheric strength envelopes including stratification, strain magnitude, and properties inherent to the faults and their physical characteristics (see Section 3.3), can be identified and then described precisely by using the fault population statistics.

468   *Planetary Tectonics*

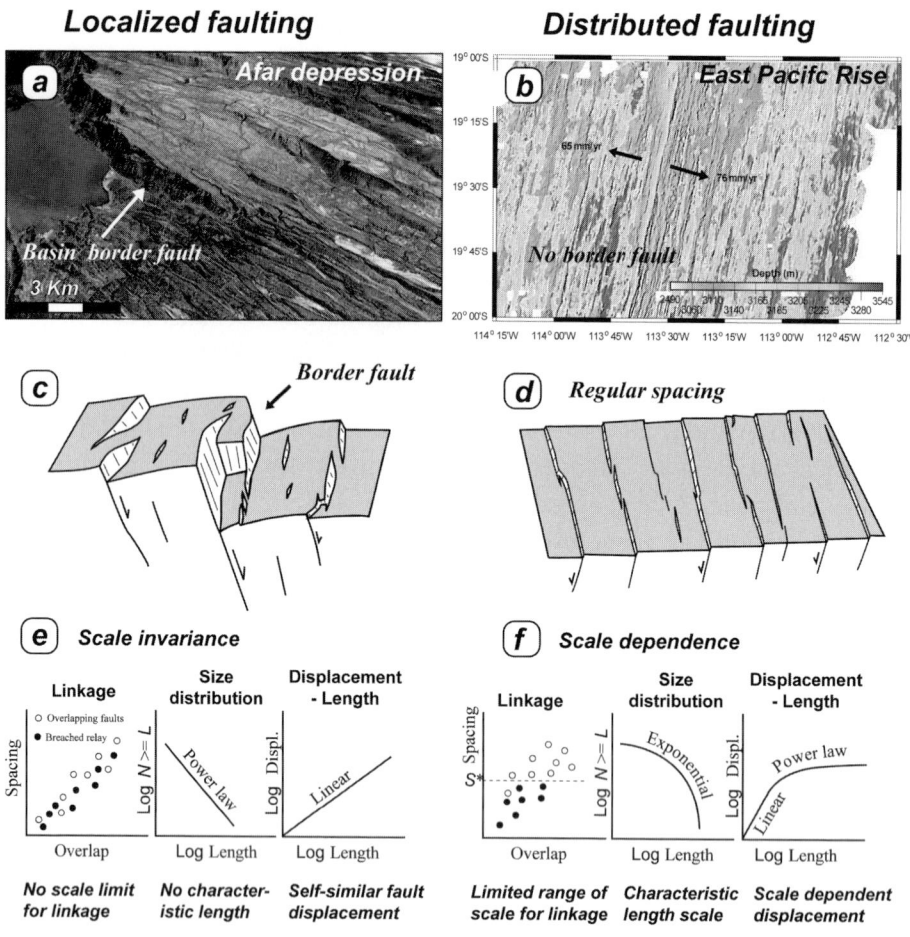

Figure 10.3. Two end-member cases of fault population geometries, after Soliva and Schultz (2008). (a) Normal fault population with localized faulting along relatively few large faults in the Afar depression. (b) Normal fault population from the East Pacific Rise, with distributed faulting along many regularly spaced faults of small and subequal displacement. Figure parts (c) and (d) are schematic views of the fault population geometry of the cases presented in (a) and (b). Figure parts (e) and (f) are the statistical properties specific to each of these cases.

3.2 Statistical properties

Statistical analysis applied to fault patterns was developed mainly in the 1990s in order to: (1) decipher quantitatively fault and fault-population growth, and (2) predict the fault morphology. For these two reasons, research within the Earth Science community was undertaken to quantify the geometry of faults in as wide a scale range as possible to provide measured dimensions and displacements of faults as tests of various fault growth scaling laws (see summary by Cowie *et al.*, 1996). As mentioned previously, normal fault systems were thoroughly analyzed because of their generally clear expression of the displacement distribution (i.e., topography) along their surface traces.

3.2.1 D–L scaling

The first scaling law studied on multiple fault populations is the *maximum displacement–length* relation (D_{max}–L, or more compactly, D–L). Since displacements accumulate along faults during their lateral and down-dip growth, or "propagation," this relation is intended to describe quantitatively, from a simplified mechanical basis, how the faults grow. By analyzing different fault populations separately, the data show that this relation can be explained in log–log space by the following equation (e.g., Scholz and Cowie, 1990; Cowie and Scholz, 1992a,b; Clark and Cox, 1996):

$$D_{max} = \gamma L^n. \quad (10.4)$$

The parameter γ is called the "scaling factor" (Cowie and Scholz, 1992b) or a "characteristic shear strain" (Watterson, 1986), and the power-law exponent n describes the rate of displacement accumulation relative to the fault length L.

The slope of individual fault populations across the full range of lengths and fault types available was shown to be approximately $n = 1.0$ (Scholz and Cowie, 1990; Gudmundsson and Bäckström, 1991; Cowie and Scholz, 1992a; Dawers *et al.*, 1993; Schlische *et al.*, 1996; Clark and Cox, 1996). Work has also shown, however, that a single relation of the form of Equation (10.4) – with a single unique value of γ – cannot represent all the data from every fault population when all are plotted together (Figure 10.4) (Clark and Cox, 1996; Wibberley *et al.*, 1999; Schultz and Fossen, 2002; Soliva *et al.*, 2005; Schultz *et al.*, 2006, 2008). Instead, each fault population has its own particular scaling law, principally with its own intercept γ that is associated with several factors, including lithology, fault geometry, frictional properties, and stress states. In detail, the distinctiveness of individual fault populations is revealed by variability of the values of γ and n. For example, the variability of these parameters between various fault systems shown from the Earth in Figure 10.3 ($0.538 < n < 2$, and for $n = 1$, $0.0001 < \gamma < 0.6$) suggests that some of the processes acting on fault growth on a given planet or satellite that can modify γ and n are scale dependent, with others related to particular fault geometries within the population (see also Schultz, 1999):

- Host-rock rigidity, as for example soft sediments in the subsurface (Muraoka and Kamata, 1983; Wibberley *et al.*, 1999; Gudmundsson, 2004),
- Friction of the fault zone, as for example the transition from deformation band (cm to m scale) to faults (m to km scale) in sandstones (Wibberley *et al.*, 2000; Fossen *et al.*, 2007),

470   *Planetary Tectonics*

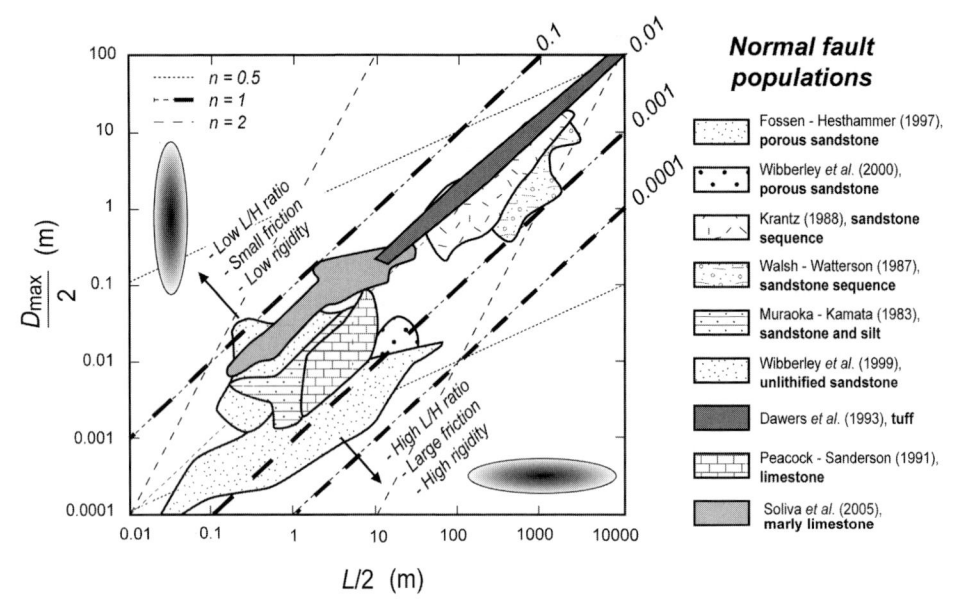

Figure 10.4. Log–log diagram of maximum displacement – length data for terrestrial normal faults, drawn using the convention of $D_{\max}/2$ and fault half-lengths ($L/2$) sometimes used in fault-population studies, following Wibberley *et al.* (1999) and others. Note the wide variation between data groups, especially for small scales. Lines with different exponents n from Equation (3) are reported and labeled. Principal factors that influence the D_{max}/L ratio deduced from field studies and rock fracture mechanics are noted on the diagram including fault aspect ratio (L/H), shown as shaded tipline ellipses.

- Propagation in layered sequences, as for example faults confined to particular layers and vertically restricted by subjacent and superjacent shale layers (Schultz and Fossen, 2002; Wilkins and Gross, 2002; Soliva *et al.*, 2005),
- Fault initiation, for example the transition from fracture opening to faulting (Gudmundsson, 1992; Peacock, 2002; Crider and Peacock, 2004), and potentially,
- The rheology of the lithospheric strength envelopes (Cowie, 1998; Bellahsen *et al.*, 2003; Soliva and Schultz, 2008).

As a result, the displacement–length scaling relations for a particular fault population can only be understood once the details of fault geometry, interaction and linkage, rock type, mechanical stratigraphy, and geodynamic context are documented and utilized.

3.2.2 Length distribution

Lengths of seismic (earthquake) ruptures were studied in the 1980s and subsequently associated with the faults in order to quantify the long-term fault population strain (e.g., Scholz and Cowie, 1990). One of the main purposes of these early

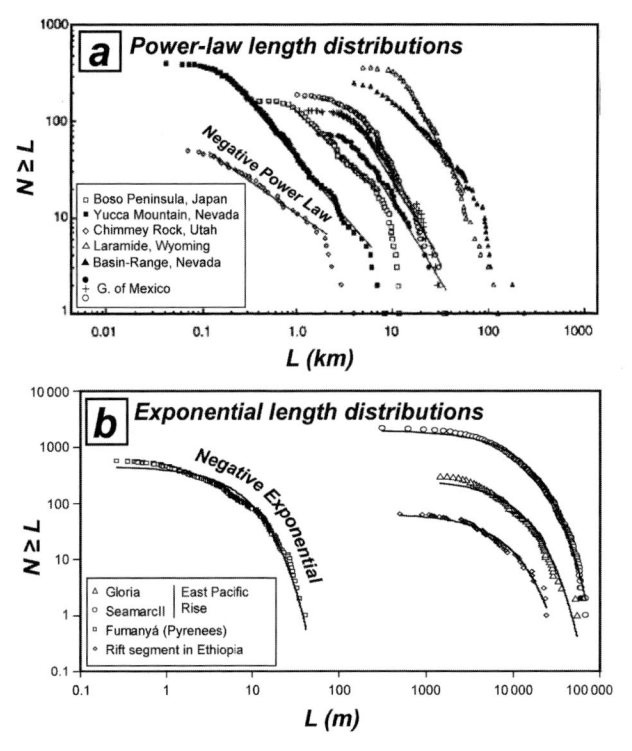

Figure 10.5. Example of characteristic length distributions observed on terrestrial fault populations. (a) Negative power-law length distributions (also called scale-invariant populations). (b) Negative exponential length distributions (scale-dependent populations).

studies was to discuss the relative contribution of larger and smaller faults in the same population, which has implications for strain calculations using remote sensing data from the Earth, as well as from the planets and satellites. A series of measurements of fault populations in the Earth's crust exhibited a negative power-law length distribution on cumulative frequency diagrams (Marrett and Allmendinger, 1991; Walsh et al., 1991; Scholz et al., 1993), with a negative power-law exponent, c, varying from ~0.5 to ~2 (Figure 10.5a). Similar results were found for Martian fault populations (Schultz and Fori, 1996; Schultz, 2000a). This power-law (or approximately "fractal") distribution reflects strain localized mainly along a few large faults, which themselves contribute up to ~50% of the population moment and strain accommodation for the case of a typical (and fractal) power-law exponent $c = 2$ (Kakimi, 1980; Villemin and Sunwoo, 1987; Scholz and Cowie, 1990), with the remainder of the moment and strain distributed on the smaller faults in a complementary proportion (Walsh et al., 1991).

This behavior has been interpreted to be the result of the long-term stability and self-similarity of the stress-shadowing process (or elimination process for joints; Aydin and DeGraff, 1988) that controls fault propagation, clustering, and therefore linkage in the whole fault population (see also Cladouhos and Marrett, 1996). However, the assertion that a fault population is self-similar requires a single value of c that remains constant throughout its development, which is not borne out in nature. Kakimi (1980) suggested that the "fractal dimension" of a given fault population varies with strain magnitude, i.e., have steeper slopes (larger c) when total strains are smaller, and have shallower slopes (smaller c) when total strains are greater, a result verified in numerical experiments by Cladouhos and Marrett (1996), for earthquakes by Wesnousky (1999), and for faults on Mars by Wilkins et al. (2002). The variation in the magnitude of fault scaling parameters means that the term "self-similar" may not strictly apply, except perhaps to a particular snapshot of a fault population's development (e.g., see Tchalenko, 1970, for an example).

Alternatively, both field examples and analogue modeling have shown that fault populations involving strain distributed along regularly spaced faults are generally characterized by negative exponential relations and show a characteristic length scale (e.g., Cowie et al., 1994; Ackermann et al., 2001) (Figure 10.5b). The common aspect of these fault populations is that they grow within a single mechanical layer or unit in which the faults are vertically confined. The confinement of the faults within the layer (also called fault "restriction," e.g., Nicol et al., 1996; Schultz and Fossen, 2002) limits the horizontal extent of fault interaction through their stress fields to a nearly constant value (Soliva et al., 2006), similar to stratabound joints whose regular spacings scale with the layer thickness (Bai and Pollard, 2000). It appears that the fault population reaches a stage with a characteristic length (Ackerman et al., 2001) that can evolve to a maximum length if the layer is "saturated" (Soliva et al., 2005), i.e., when the fault spacing stops evolving and the spacing then stabilizes at a constant value.

3.2.3 Spacing

Fault spacing is a sensitive response to the stress field within the fault population (e.g., Cowie and Roberts, 2001; Roberts et al., 2004; Soliva et al., 2006). Fault spacing, which is dependent on fault displacement magnitude and distribution (Crider and Pollard, 1998; Cowie and Roberts, 2001; Soliva and Benedicto, 2004), is linearly related to fault overlap when the fault-length distributions are described by power laws and when D_{max}–L scaling is linear (i.e., a scale-independent, non-restricted fault population, Figures 10.3a and 10.5a; see also Segall and Pollard, 1983, for analogous spacing relationships in nonrestricted joint populations,

Figure 10.6. (a) Log–log diagram of relay displacement vs. fault spacing, including different published datasets over a large scale range. Gray straight line is the maximum value of relay displacement to separation ratio (D/S) for the data composed only of open relays, with equation labeled. Black straight line is the minimum value of D/S for the data composed of fully breached relay, with equation labeled. (b) Log–log diagram of fault overlap vs. spacing, including different published datasets (gray surfaces) over a large scale range. See Soliva and Benedicto (2004) for the source of data.

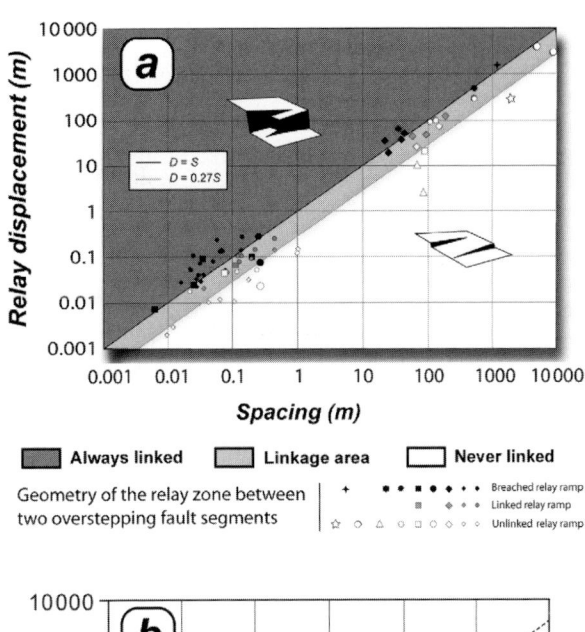

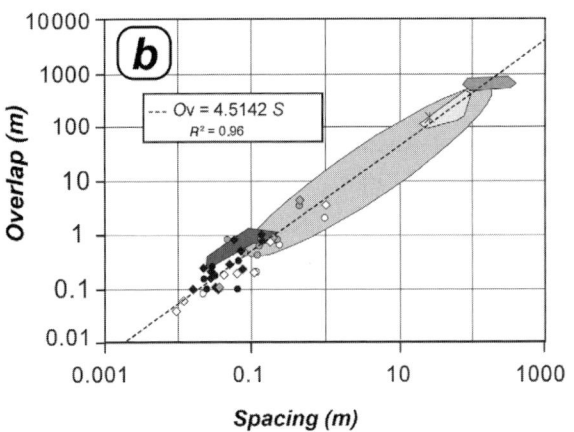

and Olson, 1993, for spacing in restricted joint populations). This fault-length-dependent spacing relationship implies that rocks can support long-term and wide-ranging fault interactions over a broad range of scales (observed from 1 mm to 100 km) (Aydin and Nur, 1982; Peacock, 2003) (Figure 10.6).

On the other hand, fault systems that are characterized by exponential length distributions (Figures 10.3b and 10.5b) generally show strain distributed along regularly spaced faults (i.e., a lognormal distribution on the length–frequency diagram; e.g., Ackermann et al., 2001; Soliva and Schultz, 2008; Figure 10.7). As discussed in the previous section, regular fault spacing is due to the limited horizontal extent of the shear stress reduction (or "shadow") zone around the vertically restricted faults that is, in turn, a function of the short and constant fault height in the population (Soliva et al., 2006). This effect also limits the maximum

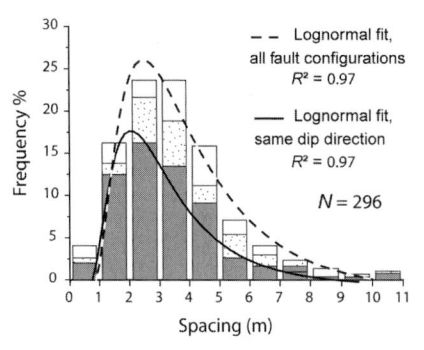

Figure 10.7. Histogram showing the frequency of fault spacing along scan lines crossing a fault population. N is the number of detected intersections between the faults and the scan lines. Spacing between faults having the same dip direction, in horst, and in graben configurations are distinguished. Broken and solid lines are logarithmic-normal fits for all configurations and for faults of the same dip direction, respectively. Least-squares coefficients (R^2) are labeled.

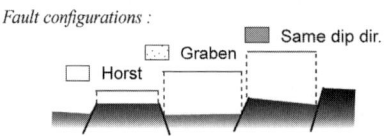

distance for strong fault interaction, thereby controlling the dimensions of relay ramps and eventual fault linkage (Soliva and Benedicto, 2004). This behavior is not consistent with self-similar fault segmentation, but instead is related in a scale-dependent manner to the thickness of the mechanical unit in which the faults are confined (see Ackermann *et al.* (2001) and Soliva *et al.* (2006) for normal faults, Schultz and Fossen (2002) for deformation bands, and Hu and Evans (1989) and Bai and Pollard (2000) for joint sets).

3.3 Mechanisms of fault growth

Fault geometries are frequently analyzed using Linear Elastic Fracture Mechanics (LEFM) (e.g., Pollard and Segall, 1987; Walsh and Watterson, 1988; Pollard and Fletcher, 2005) although some are better matched by using post-yield fracture mechanics (PYFM) (e.g., Cowie and Scholz, 1992b; Schultz and Fossen, 2002) or "symmetric linear stress distribution" (Bürgmann *et al.*, 1994; Schultz *et al.*, 2006) models. Figure 10.8 summarizes these three quasi-static models. In each of these models, the host rock (taken to be either two-dimensional or three-dimensional in extent) having an approximately homogeneous linear elastic behavior contains a shear displacement–discontinuity (the fault) subject to the far-field, remote, "regional" tectonic stresses and the constitutive relations of the fault (i.e., a constant or variable value of friction along the fault).

In the LEFM model, a constant stress drop (or "driving stress") across the fault produces an elliptical distribution displacement along a straight planar fault, and unrealistically large (infinite or "singular") local stress concentration at the fault tips (Figures 10.8b and c). In the PYFM model, cohesive-frictional end zones are defined that represent the inelastic processes (such as microcracking and fault-tip

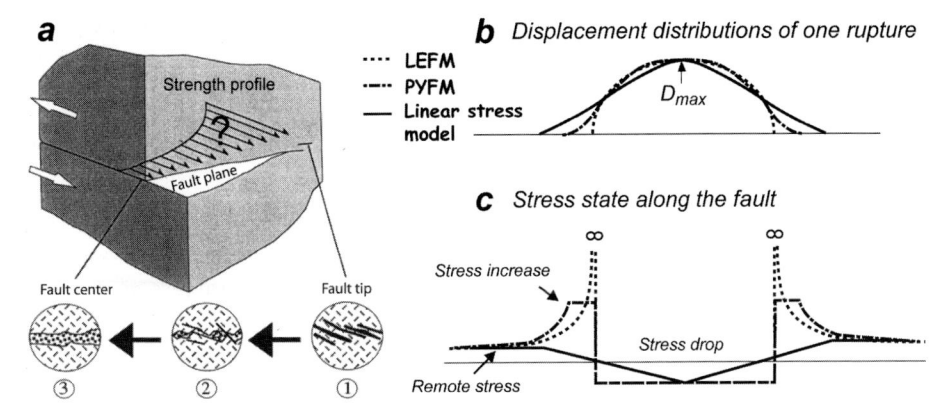

Figure 10.8. Mechanical models of shear rupture along a fault surface. (a) Schematic representation of the fault zone from the tip to the fault center showing the evolution of the fault rock damage and suggesting the evolution of the frictional properties (after Cowie and Scholz, 1992b). (b) Displacement profiles predicted by three mechanical models. (c) Resulting stress distributions along the fault plane. See text for discussion.

growth) along and around the fault terminations (Figure 10.8a, see the fault tip). This model therefore integrates the concept of rock yield strength within a larger volume than possible for the LEFM approach, limiting the amount of local stress increase at fault tips to this strength (with values several to several tens of MPa) and producing a bell-shaped displacement distribution along the fault (Figures 10.8b and c; Cowie and Scholz, 1992b; Cooke, 1997; Martel, 1997; Martel and Boger, 1998). In the "symmetric linear stress distribution" model, a linear variation of frictional strength is prescribed along the fault, from a lower (weaker) value at fault center to a larger (stronger) value at the fault tips. This approach, which implies a non-constant stress drop along the fault, aims to simulate a variation in constitutive relations, or "maturity," along the fault in which the fault-zone material or gouge is more mature and less resistant to slip near the fault center. This model produces a linear displacement distribution of displacement along the fault, as commonly observed (e.g., Manighetti *et al.*, 2001, 2005), and corresponding patterns of stress changes off the fault as inferred from stress-triggering studies (e.g., Cowie and Roberts, 2001; Roberts *et al.*, 2004) (Figures 10.8b and c).

Work based on these three fault models reveals the importance of four principal sets of parameters:

- Remote stress state
- Host-rock mechanical properties
- Fault geometry
- Friction and the constitutive relations along the fault

The remote stress state in 3-D governs the initial sense of fault displacement (normal, strike-slip, or thrust) and also the displacement magnitude via the differential or driving stress (e.g., Cowie and Scholz, 1992b; Bürgmann et al., 1994). It therefore exerts a primary influence on the average value of D_{max}/L for a given fault population (Schultz and Fossen, 2002; Schultz et al., 2006). However, the remote stresses are frequently difficult to estimate for inactive fault populations or from planetary observations, and they can be estimated only in a few terrestrial cases where outcrop conditions allow measurements of parameters such as 3-D fault geometry, friction, and material properties (e.g., see Scholz, 2002).

Material properties of the rock surrounding a fault, such as its stiffness or rigidity (as expressed principally by its Young's or shear moduli), near-tip yield strength (Scholz and Lawler, 2004), and viscosity (Bellahsen et al., 2003), are also key factors that modulate fault displacement (Walsh and Watterson, 1988; Cowie and Scholz, 1992b; Bürgmann et al., 1994; Wibberley et al., 1999; Gudmundsson, 2004). This is particularly due to the wide variety of rock mechanical properties that promote a large range of possible values for rocks (e.g., shear modulus, $0.5\,\text{GPa} < G < 50\,\text{GPa}$) from laboratory testing (Hatheway and Kiersch, 1989).

Fault tipline (the line defined by fault surface termination, i.e., where displacement equals zero; Davison, 1994) geometry is an important characteristic that also controls displacement distribution and magnitudes (Cowie et al., 1992b; Willemse, 1997; Schultz and Fossen, 2002). Moreover, the morphology of the fault surface is also important. For example, corrugations of the fault surface resulting from rock heterogeneity or fault linkage during its evolution (e.g., Schultz and Balasko, 2003; Okubo and Schultz, 2006a) can facilitate, or inhibit, displacement with respect to the slip direction (conservative and non-conservative barriers, respectively, in the sense of King and Yielding, 1984).

Fault friction can be thought of as a function of the normal stress and friction coefficient for the fault surface and has been integrated into all three fault growth models discussed above (Figure 10.8c). These models are largely consistent with field observations that show variations in meter-scale fault segmentation geometry, cataclastic fault-rock textures, and fault-rock type from the tips to the center of a fault (e.g., Caine et al., 1996; Wibberley et al., 2000) (Figure 10.8a). Frictional resistance (friction coefficient times the normal stress, plus cohesion if any) along faults modifies the displacement magnitude and can affect the displacement distribution along a fault (e.g., Aydin and Schultz, 1990; Schultz and Aydin, 1990; Aydin et al., 1990; Schultz, 1992; Kattenhorn and Pollard, 1999; Figure 10.8b). Based on these models and fault rock observations, it can be concluded that in porous siliciclastic rocks friction can influence the slope of the D_{max}–L scaling relation for some fault sets (Bürgmann et al., 1994; Wibberley et al., 1999). Because the magnitude of normal stress resolved on fault planes is related to planetary gravity

g, the scaling relations for fault populations on other planets and satellites (having smaller values of g than Earth) differ systematically in the value of their scaling intercepts γ throughout the solar system (Schultz et al., 2006), as discussed below. Other potential factors such as far-field extension rate (in terrestrial oceanic fast- vs. slow-spreading centers or continental rifts) or modes of slip event accumulation (Gutenberg-Richter vs. "characteristic," e.g., Wesnousky, 1994, 1999; Scholz, 2002) are probably of importance for fault population development but are not yet clearly demonstrated with planetary examples and theory.

3.3.1 Fault slip and 3-D propagation

Propagation of a fault requires a critical value of near-tip displacement gradient leading to an amount of fault-tip stress equal to the rock's local yield strength (Cowie and Scholz, 1992b; Bürgmann et al., 1994; Gupta and Scholz, 2000a; Scholz and Lawler, 2004). If the near-tip stress reaches the shear yield strength, the rock fails there by macroscopic shearing, and displacement accumulates along the various parts of the lengthening fault. LEFM models predict an infinitely large value of near-tip stress at the tip of a fault (Figure 10.8) and unambiguously predict a scaling exponent of $n = 0.5$ that is inconsistent with the data compiled in Figure 10.4 (Scholz, 2002, p. 116; Olson, 2003; Schultz et al., 2008).

The two other classes of fault-growth models discussed above (PYFM and the symmetric linear displacement model; Figure 10.8) are consistent with a linear D_{max}–L scaling ($n = 1$) because they avoid producing a near-tip singularity. In these two models, γ is a function of (1) elastic properties; (2) driving stress; (3) shear yield strength; and (4) fault aspect ratio (L/H; see Figure 10.1). These approaches, implicitly or explicitly, consider "radial" or "proportional" fault growth (fault propagation having approximately the same rates down-dip and horizontally) and predict a range of fault displacement profiles from bell-shaped to linear (Figure 10.8b). The growth of such an isolated fault can produce nearly circular or elliptical tipline shapes (e.g., Nicol et al., 1996; Martel and Boger, 1998) if the rock strength is comparable around the fault tipline. In layered rocks, H can remain constant during fault growth if the tipline is restricted by a lithologic or rheological barrier (i.e., "vertical restriction" in Figure 10.1) (Scholz, 1997; Schultz and Fossen, 2002). In this case, the slope of the fault-population exponent changes from $n = 1$ in the earlier, non-restricted, proportional growth phase, to $n < 1$ as the faults grow laterally while being restricted vertically (Schultz and Fossen, 2002; Soliva et al., 2005; Fossen and Gabrielsen, 2005, p. 161; Figure 10.4).

3.3.2 Interaction and linkage

Fault interaction and linkage are a major process leading to fault growth (Peacock and Sanderson, 1991; Dawers and Anders, 1995; Mansfield and Cartwright, 1996;

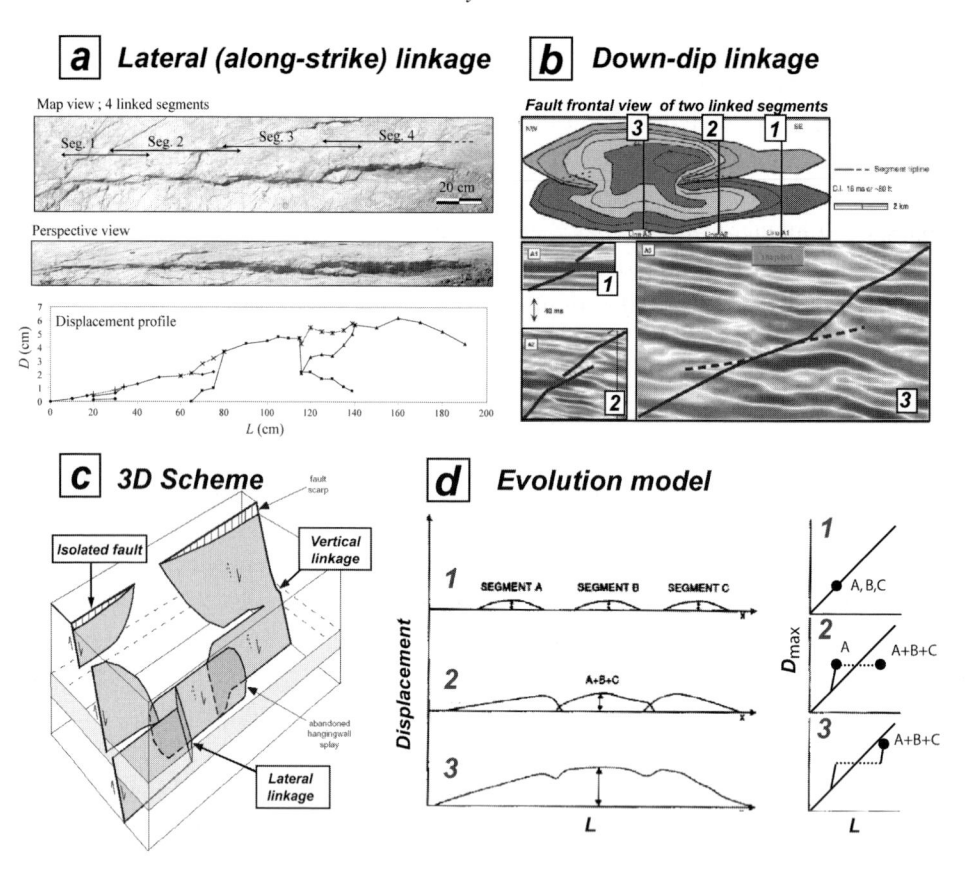

Figure 10.9. 3-D geometry and evolution of segmented normal faults. (a) Geometry of lateral linkage and associated displacement distribution. (b) Three-dimensional (3-D) geometry of vertical linkage and associated 3-D displacement distribution. (c) 3-D representation of segmented faults. (d) Displacement evolution model.

Crider and Pollard, 1998; Cowie and Roberts, 2001). Field data and theory have shown that two initially isolated fault segments can interact through their stress fields as they grow, eventually linking across their relay zones in 3-D (e.g., Segall and Pollard, 1980; Figures 10.9a, b and c). During the first step of fault interaction, the increase of shear stress around the stepover, or relay zone, leads to a transfer of displacement on one or each segment if both are actively slipping, leading to an increase of displacement gradient along the interacting fault ends. This interaction promotes an increase of the D_{max}/L ratio, which ultimately can lead to an abrupt increase in length by linkage of the temporarily over-displaced fault segments and a subsequent period of fault displacement recovery for the newly linked fault (Figure 10.9d). When fault linkage and displacement readjustment are achieved, the resulting linked segmented fault can behave as a new larger kinematically coherent fault having a D_{max}/L ratio consistent with non-linked isolated faults (e.g., Cowie and Roberts, 2001). These perturbations of fault displacement, due to the short-range mechanical interactions between the closely-spaced fault segments, can explain a large component of the scatter observed on D_{max}–L diagrams (Figure 10.4). Fault interaction and linkage also control other fault population characteristics, such as: (1) fault length distribution (Cladouhos and Marrett, 1996), (2) fault spacing (Soliva *et al.*, 2006), geometry of syntectonic basins and deposits (e.g., Gawthorpe and Hurst, 1993), and (3) slip and sedimentation rate (e.g., Ravnas and Bondevic, 1997; Cowie and Roberts, 2001).

3.3.3 Whole fault system development

The compiled datasets shown in Figure 10.4 give a synoptic view of the D_{max}–L scaling relationships of normal faults observed at the Earth's surface. These data show a scatter of the D_{max}/L ratio (from 10^{-3} to 4×10^{-2}) for faults of $L < 200$ m. This suggests that at a small scale (relative to the mechanical unit thicknesses typical of stratified igneous or sedimentary sequences), fault displacement is greatly influenced by both the lithological discontinuities (acting on fault shapes) and the rheology (stiffness or rigidity, friction) of each rock type. This wide D_{max}–L variability is possible because of the small dimension of the faults with respect to the mechanical unit thicknesses, allowing the faults to be sensitive to the specific rheology of each mechanical unit. In contrast, if the fault dimension is large enough with respect to the lithological stratification (for example 1 km long for mechanical units of meter-scale thickness), displacement must then be controlled by the average rheology of the entire bounding layered sequence. This seems to be a reasonable explanation for at least some of the scale dependence of the D_{max}–L data variability (e.g., Soliva *et al.*, 2005; see the large scatter for small faults compared to large faults in Figure 10.4). Therefore, regardless of the fault initiation process (see Crider and Peacock, 2004) or other factors such as propagation rates (e.g., Walsh and Watterson, 1987; Peacock and Sanderson, 1996), it is improbable that a fault will grow with a consistently linear D_{max}–L behavior (i.e., without change of slope) from the centimeter to the kilometer scale in layered sequences of contrasting lithologies.

To understand fault population growth and scaling, Gupta and Scholz (2000a) calculated the perturbation of the maximum Coulomb shear stress (King *et al.*, 1994; Harris, 1998) around a series of faults. They showed that interaction and subsequent linkage develop preferentially for similar spacing/overlap ratios independent of the scale of observation, where D_{max}–L scaling, fault aspect ratio, and the tipline geometry are scale invariant. Their work suggests that a self-similar segmentation geometry is mechanically possible in a fault population if the shear stress perturbation around the faults scales linearly with their horizontal lengths.

This scale-invariant process of linkage allows the formation of very large faults by linkage of smaller growing segments (e.g., Cowie *et al.*, 1995), therefore allowing large strain localization along just a few faults (corresponding to the first end-member case discussed above, Figure 10.3a). The increase of fault size (1) enlarges the rock volume of reduced stress that shadows the activity of smaller faults, and (2) allows the development of the largest faults, which promotes an approximately fractal geometry (or scale-invariant negative power-law length distributions) of the fault population (Sornette *et al.*, 1990; Cladouhos and Marrett, 1996).

Because rock-mass characteristics such as rheological contrasts in layered stratigraphies can control the geometry of one fault, they can therefore control fault interactions throughout the entire fault population (e.g., Soliva *et al.*, 2006). The population can thereby change from localized to distributed strain (e.g., see the two end-member cases shown in Figure 10.3). This case concerns populations of active faults that are confined within a layer of given thickness. Here, faults grow horizontally with their vertical extent being limited, or restricted, by adjacent layers that act as mechanical barriers (e.g., Scholz, 1997), leading to regularly spaced faults. In this regime, faults are no longer self-similar in displacement distribution since they are vertically restricted, and instead generally exhibit flat-topped displacement profiles (Ackermann *et al.*, 2001; Soliva *et al.*, 2006). The scaling of restricted faults is well explained by non-linear growth paths on the D_{max}–L diagram (i.e., 3-D PYFM conditions with constant fault height; Schultz and Fossen, 2002; Soliva *et al.*, 2005; Figure 10.10c). This growth sequence has been observed in fault populations over a wide range of scales and structural contexts (Cowie *et al.*, 1994; Carbotte and Macdonald, 1994; dePolo, 1998; Poulimenos, 2000; Manighetti *et al.*, 2001; Bohnenstiehl and Carbotte, 2001; Polit, 2005; Soliva and Schultz, 2008; Polit *et al.*, 2009). Cowie *et al.* (1994) describe crustal-scale fault populations in oceanic lithosphere at the East Pacific Rise and Soliva and Schultz (2008) along the Main Ethiopian Rift, where much of the strain is distributed on nearly evenly spaced faults (Figure 10.5b). At the East Pacific Rise, the fault population has been interpreted to indicate growth within (confined to) the oceanic brittle crust (Cowie, 1998; Bohnenstiehl and Kleinrock, 1999; Garel *et al.*, 2002), whereas at the Main Ethiopian Rift the faults seem confined to competent basalts. These faults also show non-linear D_{max}–L scaling with a significant decrease in the D_{max}/L ratio with increasing fault length, i.e., $n < 1$ (Cowie *et al.*, 1994; Manighetti *et al.*, 2001).

3.4 Scaling relations for planetary faults

Precision measurements of the maximum displacement ("offset," D_{max}) and map lengths L of surface-breaking faults on Mars and Mercury demonstrate that less

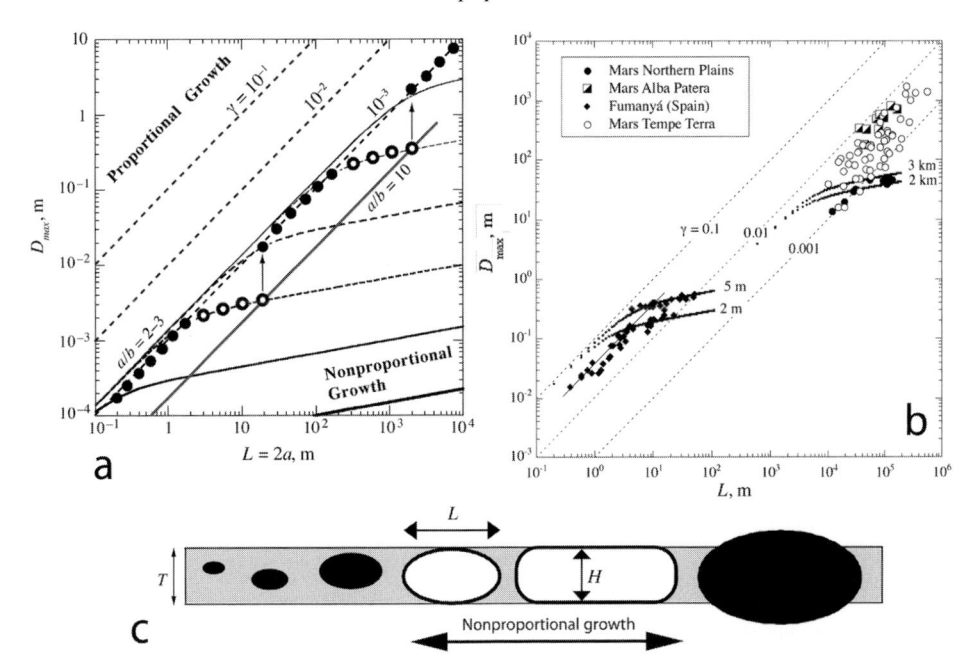

Figure 10.10. 3-D displacement–length scaling relations and the growth of stratigraphically restricted faults. (a) Fault growth paths on the D_{max}–L diagram (after Schultz and Fossen, 2002) showing stair-step trajectory of alternating proportional (linear, filled symbols) and non-proportional (restricted, open symbols) fault growth. (b) Examples of restricted fault populations on Earth (normal faults from Fumanyá in the southeast Pyrenees, after Soliva et al., 2005) and Mars (graben-bounding normal faults from the northern plains, after Polit, 2005, and Polit et al., 2009). (c) Cross-sectional fault geometries shown schematically for each part of the growth sequence. Filled and open symbols for fault-shape ellipses as in (a).

displacement per unit length is accumulated along faults on these planets than along terrestrial ones. For example, normal faults from Tempe Terra (Mars) and thrust faults from Arabia Terra (Mars) show D_{max}/L ratios of 6.7×10^{-3} (Wilkins et al., 2002; Watters, 2003) and 6×10^{-3} (Watters et al., 1998), respectively. Thrust faults from Mercury also show D_{max}/L ratios of 6.5×10^{-3} (Watters et al., 2000, 2002; Watters and Nimmo, Chapter 2). The fault populations discussed here currently lack evidence for significant restriction, although many of their characteristics such as displacement distributions that could suggest restriction remain to be investigated; in contrast, a set of restricted grabens from the Tharsis area of Mars (Polit, 2005; Polit et al., 2009) are discussed below (see Figure 10.10c). Typical values for terrestrial faults (normal, strike-slip, or thrust) are ~ 1–5×10^{-2} (see the recent compilations by Schultz et al., 2006, 2008). Currently, topographic data of sufficient accuracy and resolution to assess D_{max}–L scaling of faults are available only for Mars and Mercury.

The data for Martian normal faults, such as those on Tempe Terra (Wilkins et al., 2002), are systematically offset to smaller values of displacement by a factor of about five from the terrestrial data (Figures 10.10b and 10.11a). A similar offset is observed for thrust faults on both Mars and Mercury (Plate 25b). Detailed examination of the Martian and Mercurian faults indicates that the smaller D_{max}/L ratios result from smaller displacements (accurately measured from topographic data; e.g., Watters et al., 1998, 2000, 2002; Wilkins et al., 2002) rather than an overestimation of fault lengths by the same factor of five.

Schultz et al. (2006) found that the D_{max}/L ratio for non-restricted faults depends on three primary factors: stiffness of the rock surrounding the faults (Young's modulus or shear modulus (rigidity)), shear driving stress, and yield strength, with all three of these primary factors being influenced to various degrees by planetary gravity g. For the same conditions of rock type (e.g., basaltic rock mass), fault type (normal), and fluid-saturated crustal rocks (i.e., "wet" conditions with $\lambda = 0.36$–0.4), g reduces D_{max} for Martian faults, relative to terrestrial ones, by $g_{Mars}/g_{Earth} = 0.38$ (via the driving stress term). Yield strength in shear scales with gravity, with the strength of the Martian basaltic rock mass being approximately one-half of the corresponding terrestrial one. Modulus decreases with decreasing g as a result of reduced overburden stress, to a normalized value of ~ 0.84 for the (wet) Martian case. The combined effect of g on all three key factors is a reduction in D_{max}/L of about a factor of 5–6, consistent with the data for normal and thrust faults from the literature (e.g., Clark and Cox, 1996; Schultz et al., 2006).

Restricted faults have only recently been recognized in planetary datasets (Figure 10.10b) and the implications of this class of fault for the stratigraphy, seismology, and tectonics of planets and satellites is as important for those bodies as for the Earth itself (see discussion by Knapmeyer et al., 2006). Fault restriction can be identified in terrestrial and planetary datasets by using one or more diagnostic techniques, including quantitative examinations of the fault-related topography (Soliva et al., 2005; Polit et al., 2009), spacing (e.g., Soliva et al., 2006), D_{max}–L ratios (e.g., Soliva and Benedicto, 2005; Polit et al., 2009), relay-ramp dimensions (e.g. Soliva and Benedicto, 2004), and length–frequency data (e.g., Gupta and Scholz, 2000b; Soliva and Schultz, 2008). Stratigraphically restricted faults represent snapshots of the progressive growth of fault systems in layered sequences and their strain magnitudes can be computed by using the well-known equations for "large faults" (see Section 5).

Assessment of D_{max}–L scaling relations of faults on the Moon, Venus, and icy satellites of the outer solar system is currently hindered by large uncertainties in measurements of displacement (due to low-resolution, or unavailable, topographic data) and, to a lesser extent, length (due to imaging data having coarse spatial resolution). The results summarized here (Figure 10.11) suggest that faults on Venus

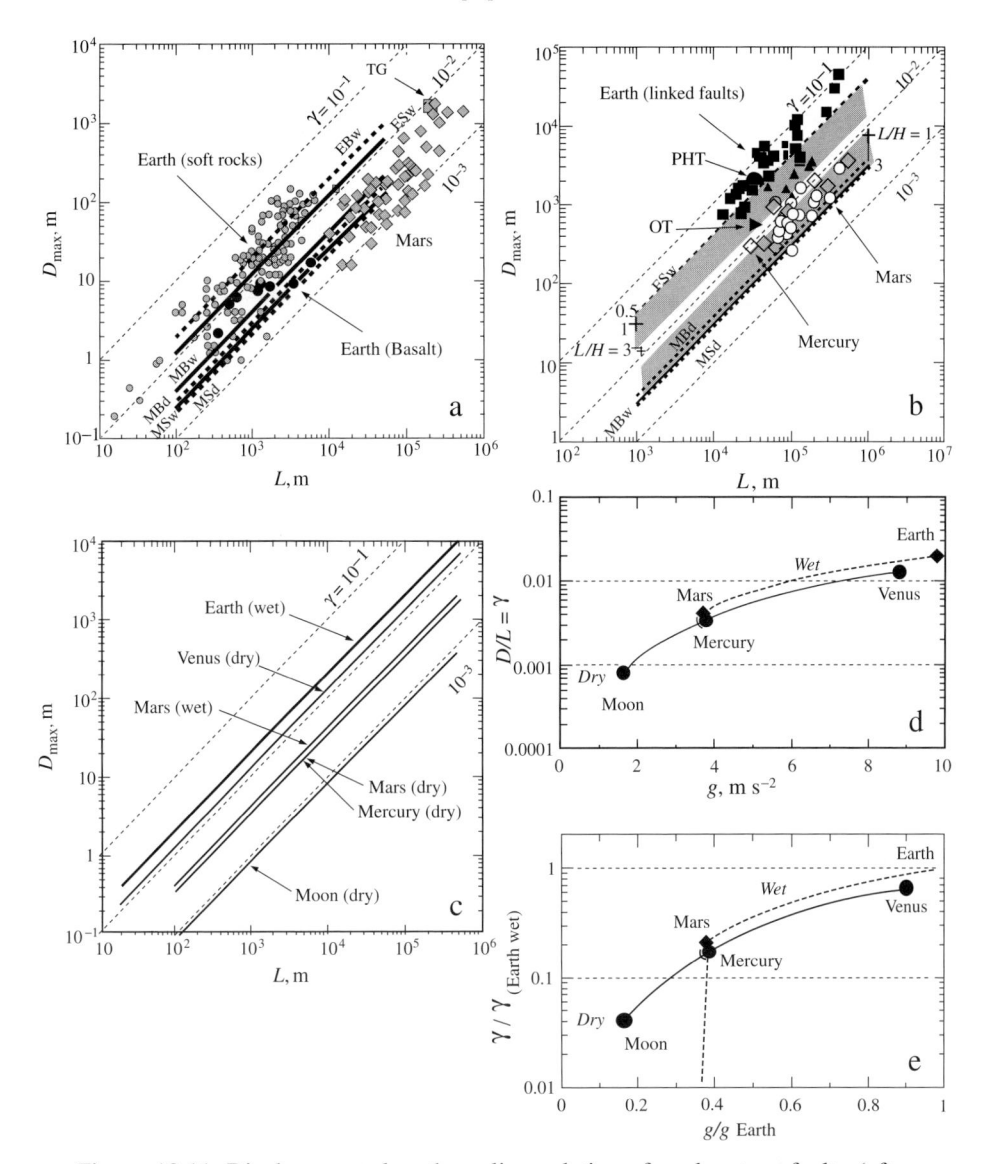

Figure 10.11. Displacement–length scaling relations for planetary faults (after Schultz et al., 2006). (a) D_{max}–L data for normal faults from Earth (gray circles, sandstone and non-welded tuff; black circles, basalt) and Mars (gray diamonds, Tempe Terra; gray square, Thaumasia graben, linked faults, 'TG'); data from Schlische et al. (1996), Wilkins et al. (2002), and Hauber and Kronberg (2005). Calculated scaling relations (see Schultz et al., 2006 for parameters): EBw, Earth basaltic rock mass with wet conditions; ESw, Earth sandstone rock mass with wet conditions; MBw, Mars basaltic rock mass with wet conditions; MBd, Mars basaltic rock mass with dry conditions; MSw, Mars sandstone rock mass with wet conditions; MSd, Mars sandstone rock mass with dry conditions. (b) D_{max}–L data for thrust faults from Earth (black squares and triangles), Mars (open circles) and

484 Planetary Tectonics

(see McGill *et al.*, Chapter 3) should accumulate somewhat smaller displacements than their terrestrial counterparts given an ∼10% reduction in gravity ($g = 8.8$ m s^{-2}) relative to the Earth. Faults on the icy satellites of Jupiter and Saturn (see Collins *et al.*, Chapter 7) probably also scale with gravity, with particular values of the D_{max}/L ratio depending on appropriate values of near-tip ice strength and ice stiffness (e.g., Nimmo and Schenk, 2006) along with reduced surface gravities of those satellites. Lunar faults will likely scale with its smaller surface gravity as well, with faults that cut highland regolith (which has significantly smaller values of modulus than does basalt) exhibiting larger displacements than those that cut mare basalts, all other factors being equal (see Watters and Johnson, Chapter 4).

Comparisons of displacement–length scaling between planets and satellites should also be made on a consistent basis for faults that do not cut through the mechanical or thermal lithosphere, so that flexure or tilting of faulted blocks does not contribute to increased values of offset (e.g., Cowie and Scholz, 1992b; Nimmo and Schenk, 2006). Additionally, faults should be isolated from other, nearby faults (i.e., not segments from a fault zone or rift) and not be stratigraphically restricted to ensure the clearest comparison with terrestrial and other data that are collected following these guidelines. Because fault-related strains depend on the D_{max}/L ratio along with the fault density (Gupta and Scholz, 2000b; Schultz, 2003a), the average strain accommodated by faulting at the surface of a planetary body, for the same style of tectonic domain, may generally decrease as a function of gravity.

4 Fault-related topography

The topographic signature of a fault at the planetary surface reveals its geometry and characteristics in the subsurface, as demonstrated from many terrestrial studies (e.g., Ma and Kusznir, 1993; Willemse, 1997; Niño *et al.*, 1998; Soliva and Benedicto, 2005). For example, the magnitude and distribution of uplift along normal

---

←

Mercury (gray diamonds); data from Elliott (1976), black squares; Mége and Riedel (2001), black triangles; Shaw *et al.* (2002), black circle (Puente Hills Blind-Thrust System, 'PHT'); Davis *et al.* (2005), right-pointing black triangle (Ostler Thrust, 'OT'); Watters *et al.* (2000, 2002); and Watters (2003). Calculated scaling relations labeled as in (a) but with $L/H = 0.5$ for terrestrial thrust faults with lower ticks at $L/H = 1.0$ and 3.0 (upper shaded region in the figure), and $L/H = 3.0$ for Martian and Mercurian thrust faults with upper tick at $L/H = 1.0$ (lower shaded region). (c) Predicted values of D_{max}/L for smaller planets and satellites. All curves calculated for normal faults assuming $L/H = 3$, $N = 3000$, and basaltic rock-mass parameters. (d) Summary of D_{max}–L scaling for terrestrial planets calculated for wet basaltic crusts (dashed curves) and dry basaltic crusts (solid curves). (e) D_{max}–L scaling values for smaller planets normalized by (wet) terrestrial ones.

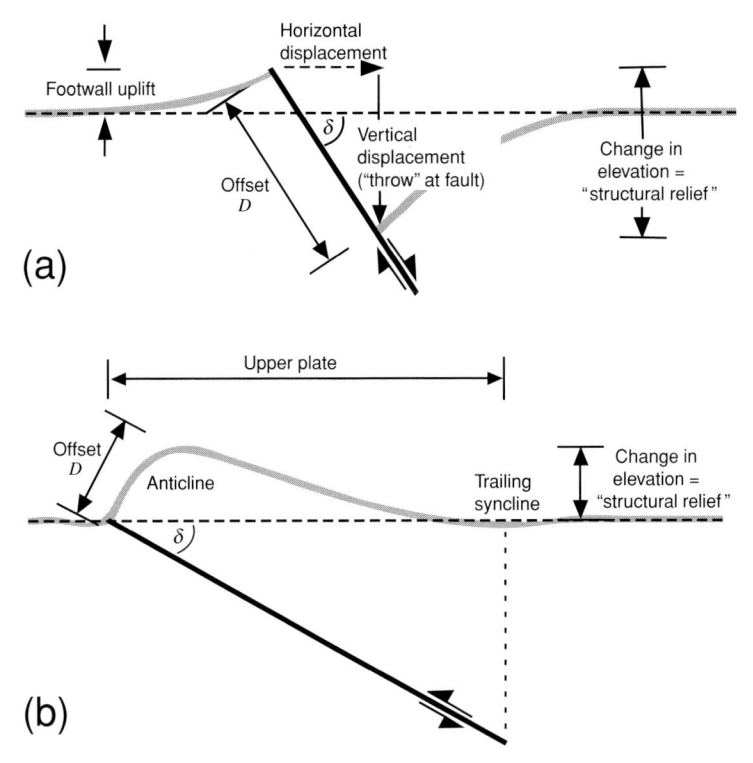

Figure 10.12. Relationships between cumulative fault offset D, fault dip angle δ, and structural topography, shown in cross section, due to deformation of the planetary surface by the faulting for (a) normal faults (after Schultz and Lin, 2001) and (b) thrust faults (after Schultz, 2000b).

faults (i.e., the small footwall uplift on normal fault or graben flanks; Weissel and Karner, 1989) and thrust faults (i.e., the major uplift on the upper plate called "lobate ridges" on the Moon, Mars, and Mercury, Figure 10.12; Niño et al., 1998; Cohen, 1999; Schultz, 2000b; Ma and Kusznir, 2003) is a function of the map length and down-dip height of an individual fault. Topographic uplift across faults is also a reliable indicator of the subsurface fault geometry on Mars (e.g., Schultz, 1999, 2000b; Schultz and Lin, 2001; Schultz and Watters, 2001; Watters et al., 2002; Wilkins et al., 2002; Wilkins and Schultz, 2003; Okubo and Schultz, 2003, 2004, 2006a; Polit et al., 2009), where outstanding high-resolution topographic data currently exist.

Because erosion and degradation of topography is relatively slow on Mars, fault-related topography is well expressed, especially for younger faults. However, even Noachian thrust faults (with ages ~4 Ga; see Figure 8.1 of Tanaka et al., Chapter 8) have topography that is sufficiently well preserved to reveal the subsurface details (e.g., Schultz, 2003b; Okubo and Schultz, 2003, 2004; Goudy et al., 2005; Grott

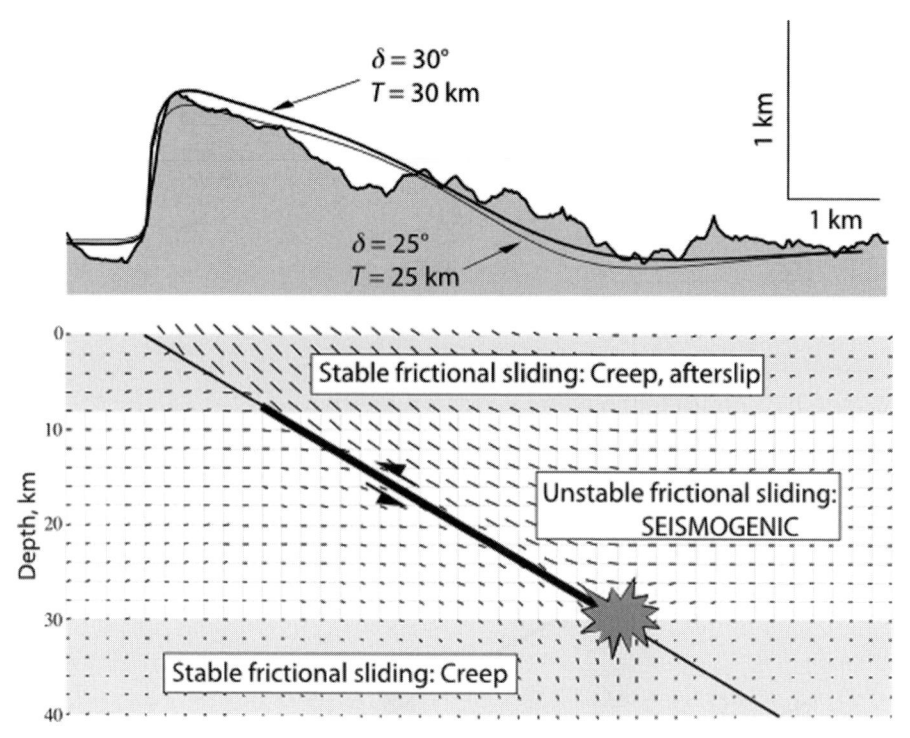

Figure 10.13. Topography measured across Amenthes Rupes, a Martian surface-breaking thrust fault, along with the topography predicted from 1.5 km of slip along the thrust fault (after Schultz and Watters, 2001). Lower panel shows the predicted displacement trajectories in the Martian lithosphere associated with the Amenthes Rupes thrust fault, with the orientation and length of the tick marks indicating the predicted local direction and magnitude of displacements; the largest values occur above the thrust fault (the "upper plate"). Regimes of Martian frictional stability shown as shaded regions and labeled; star indicates maximum depth of seismic rupture along the fault (after Schultz, 2003b).

et al., 2006). For example, forward mechanical models of the topography, both boundary element and finite element, produced by normal faults (Schultz and Lin, 2001; Hauber and Kronberg, 2005) and thrust faults (Schultz, 2000b; Schultz and Watters, 2001; Watters *et al.*, 2002; Okubo and Schultz, 2004; Grott *et al.*, 2006) demonstrate how topographic profiles across faults on Mars, and also Mercury, can be used to accurately determine the dip angle and depth of faulting (Figure 10.13). These models calculate the displacements on faults subject to a specified set of conditions, including remote tectonic stresses, fault geometry and constitutive relations such as frictional strength, and material properties of the surrounding rock, and then calculate the associated topographic changes of the planetary surface (see Schultz and Aydin, 1990; Schultz, 1992; and Okubo and Schultz, 2004, for details of the boundary element method and appropriate parameters used in the program FAULT to make these calculations). The topographic changes along Martian strike-slip faults (Schultz, 1989; Okubo and Schultz, 2006b) provide an additional avenue for exploring fault geometry and lithospheric stress states (e.g., Andrews-Hanna et al., 2008).

Investigation of MOLA profiles has also revealed evidence of igneous dikes below certain Martian grabens (Schultz et al., 2004) by detection of the subtle yet diagnostic topographic signature (e.g., Rubin and Pollard, 1988; Mastin and Pollard, 1988; Rubin, 1992) produced above a dike at the planetary surface (see also Goudy and Schultz, 2005). An example is shown in Figure 10.14. Perhaps counterintuitively at first thought, the rock directly above the dike tip is neither displaced nor extended to any large degree, but instead, the planetary surface on either side of a dike is displaced upward and outward, forming the characteristic pair of gentle topographic swells shown in Figure 10.14c (and noted, for example, by Rubin and Pollard, 1988). In contrast, the surface topography associated with slip along two inward-dipping normal faults is elevated yet concave-upward in the footwall (Rubin and Pollard, 1988; Weissel and Karner, 1989; Schultz and Lin, 2001; see Figure 10.12a) and decays more rapidly with distance away from the fault than does the topographic rise produced by dike inflation (Figure 10.14c). The several distinctive characteristics of the topographic signatures of normal faults and subsurface dikes, apparent in Figures 10.12a and 10.14c, permit the identification of the type of extensional structure beneath a planetary surface (i.e., dike or fault).

The flanking topographic uplifts above a dike also correspond to the locations of increased horizontal tensile stresses, noted previously, for example, by Williams (1957) and Delaney et al. (1986) and related to bending of the rock there. Given sufficient bending, ground cracks and two inward-dipping normal faults can nucleate at the topographic crests and propagate downward, forming a structural graben above the dike (e.g., Rubin and Pollard, 1988; Mastin and Pollard, 1988; Rubin, 1992; Schultz, 1996; Figure 10.14b) whose width scales with the depth to the dike top (Rubin and Pollard, 1988; Mastin and Pollard, 1988; Schultz, 1996; Okubo and Martel, 1998; Schultz et al., 2004). The predicted displacement trajectories in the Martian lithosphere associated with inflation of the dike (Figure 10.14d) indicate that most of the deformation of a planetary lithosphere occurs closest to the dike, with the magnitude of deformation decreasing away from it, consistent with previous work on terrestrial dike-related topographic changes. The displacement magnitudes in the lithosphere scale with the magma pressure and inversely with lithospheric stiffness (Young's modulus). Assessment of the subsurface structure in areas of planetary volcanotectonic activity is critical to evaluating the relationships, for example, between regional extension, faulting, and dike intrusion (e.g., Grosfils and Head, 1994; Koenig and Pollard, 1998; Ernst et al., 2001; Wilson and Head, 2002; Mège et al., 2003; Schultz et al., 2004) and between groundwater discharge

488 *Planetary Tectonics*

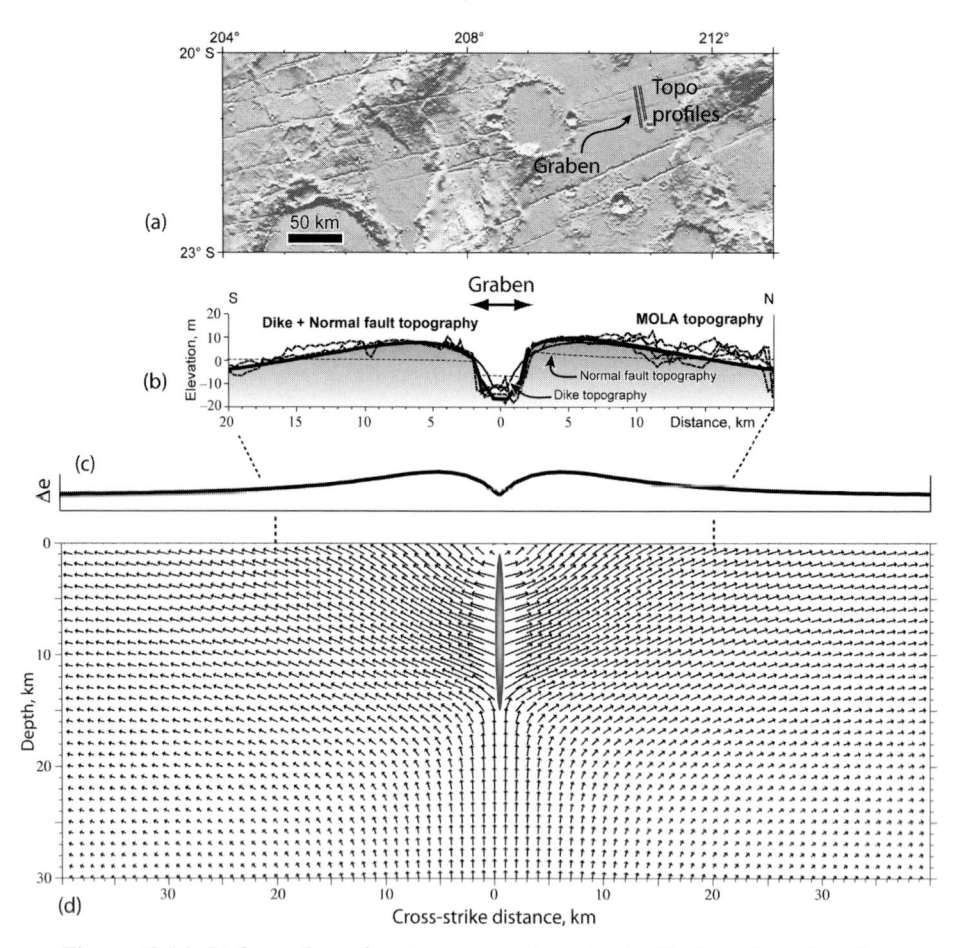

Figure 10.14. Deformation of a planetary surface due to dilation of a subsurface igneous dike, following Schultz *et al.* (2004). (a) Shaded relief image showing several northeast-trending grabens in the Tharsis region of Mars; the locations of four MOLA topographic profiles oriented normal to one of these grabens are indicated. (b) Topographic slice across the graben shown in (a) showing the four MOLA profiles (heavy dashed lines) and three predictions of structural topography: uplift due only to a subsurface dike (smooth curve; parameters given in Schultz *et al.*, 2004), uplift due only to the graben-bounding normal faults (fine dashed curve), and the sum of dike and fault topographies (bold curve). The location of the graben at the crest of regional dike-related topography is indicated. (c) Predicted surface topography above a dike. (d) The predicted displacement trajectories in the Martian lithosphere associated with inflation of a dike due to magma pressure, shown following Figure 10.13 but here with arrowheads.

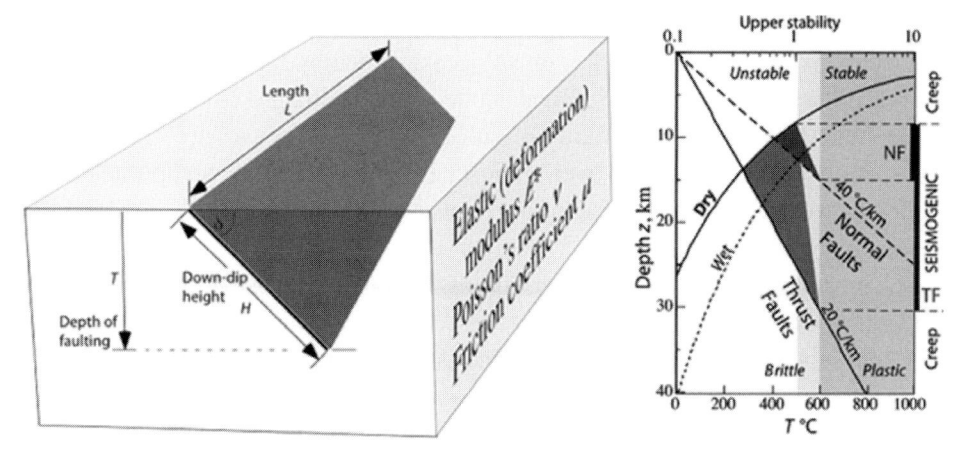

Figure 10.15. The depth of faulting T is related to the down-dip height H of the coseismic displacement distribution on a planetary fault (left panel, after Schultz and Lin, 2001) and, in turn, to the 600 °C isotherm for mafic rocks (right panel, after Schultz, 2003b), appropriate to most silicate planetary bodies.

in Martian outflow channels and the associated dike-related grabens (Hanna and Phillips, 2006).

The maximum depth of faulting T in a planetary lithosphere of mafic composition is defined approximately by the 600 °C isotherm (Abercrombie and Ekström, 2001; Grott et al., 2006; Knapmeyer et al., 2006), which is associated with the lower stability transition between unstable (seismogenic) frictional sliding above and stable sliding (creep) below (e.g., Tse and Rice, 1986; Scholz, 1998). Using the best-fit value of $T = 30$ km (Schultz and Watters, 2001; Grott et al., 2006) for the faulted domain at Amenthes Rupes in Arabia Terra (eastern Mars), the paleogeothermal gradient during Martian thrust faulting there was approximately 20 °C km^{-1} (assuming a surface temperature of $\sim$0 °C and an approximately linear gradient). Down-dip portions of the Martian thrust faults, deeper than 30 km, would tend to slip stably and would contribute only small components to the surface topography, given their greater depth below the surface (e.g., Cohen, 1999). On the other hand, Martian normal faults in Tempe Terra and Alba Patera attain depths of $\sim$15 km (Wilkins et al., 2002; Polit et al., 2009), implying a paleogeothermal gradient there of about 40 °C km^{-1} (Figure 10.15).

The upper (shallow) limit of seismogenic slip is related to the upper stability transition (Marone, 1998; Scholz, 1998), above which fault zone material (such as gouge) is velocity strengthening (Marone and Scholz, 1988). This upper stability transition is pressure dependent and independent of fault type (Scholz, 1998). By scaling the values used for terrestrial faults (3–4 km: Cowie et al., 1994; Scholz, 1998, and hydrostatic pore-fluid conditions) to Martian conditions ($g = 3.7$ m s^{-2}), frictional sliding along Martian faults should be conditionally stable (barring large perturbations, such as marsquakes on subjacent or nearby fault segments, or rapid healing processes; see Scholz, 1998) at depths shallower than ~8–10 km for a "wet" lithosphere (hydrostatic pore-fluid pressure) or ~5–7 km for a "dry" lithosphere. An active hydrologic system ("wet" lithosphere), along with slow slip rates along the faults, would promote healing of the fault zone, leading to decreasing depth for the upper stability transition.

Seismogenic (unstable) frictional sliding along the largest thrust fault in the Amenthes Rupes population (Schultz, 2003b) should have occurred primarily between depths of 8 and 30 km (with the depth of the lower stability transition corresponding to the likely marsquake nucleation depth; Figure 10.15). Using the depth range obtained above for unstable frictional sliding, ~82% of the total moment release and strain associated with the Martian thrust fault population was seismogenic (assuming $L/H = 3$; 80% for $L/H = 2$). The fraction of seismogenic strain for a given Martian fault population will decrease for smaller and less deeply penetrating surface-breaking faults given that the upper ~8 km globally should remain largely devoid of nucleating marsquakes along normal, strike-slip, or thrust faults.

5 Strain

The strain signature associated with the three fault types is well known (e.g., Reches, 1978, 1983; Krantz, 1988, 1989), as shown in Figure 10.2. In an extending tectonic domain with coaxial stress–strain relations, the vertical principal stress is the lithostat and the two horizontal principal stresses are smaller in magnitude but still compressive, as shown by *in situ* stress measurements on the Earth (McGarr and Gay, 1978; Brown and Hoek, 1978; Plumb, 1994; Zoback *et al.*, 2003). The domain extends in the direction of the least horizontal principal stress and thins vertically, producing an extensional normal strain horizontally and a contractional (thinning) normal strain vertically. For thrust faulting, the maximum principal stress is horizontal and the lithostat becomes the least principal stress (both are, of course, compressive, as is the intermediate (horizontal) principal stress), leading to lithospheric thickening (with an extensional vertical normal strain) and horizontal shortening normal to the maximum (horizontal) principal stress. For strike-slip faulting, the lithostat serves as the intermediate (compressive) principal stress, and the maximum and minimum (compressive) principal stresses are horizontal, leading to a contractional normal strain perpendicular to the maximum horizontal principal stress direction and an extensional normal strain perpendicular to the minimum horizontal principal stress direction. Shear strains can also be calculated for these fault types, as well as the more complicated, spatially varying, inhomogeneous

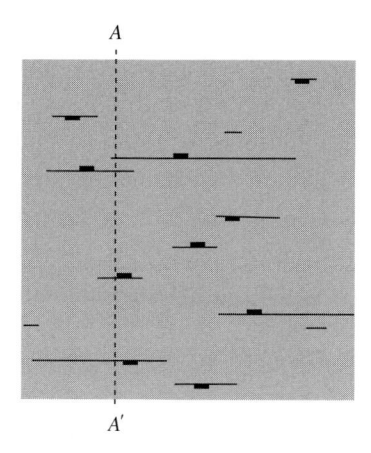

Figure 10.16. A one-dimensional (1-D) sampling traverse (A–A′) across a fault population. This sparse fault population suggests how the value of strain depends on where the traverse is taken.

displacement and strain fields that are particularly significant within a few fault lengths of a fault (e.g., Chinnery, 1961; Barnett *et al.*, 1987; Ma and Kusznir, 1993).

An extensive literature exists on how the amount of strain accommodated by a population of faults can be quantified (see Kostrov, 1974; Segall, 1984a,b; Wojtal, 1989; Scholz and Cowie, 1990; Marrett and Allmendinger, 1991; Westaway, 1994; Scholz, 1997; and Borgos *et al.*, 2000, for representative approaches). Using the approach of a simple horizontal one-dimensional (1-D, line) traverse across a deformed region (e.g., Golombek *et al.*, 1996; see Figure 10.16), the amount of displacement across each fault is first measured from topographic data and then corrected to take into account only the component of displacement parallel to the traverse (i.e., correct for strike and dip; Peacock and Sanderson, 1993; Scholz, 1997; see below). For closely spaced faults (called "penetrative deformation") it may be easier to trace the offset of a passive marker from one side to the other, instead of measuring all the fault displacements; see Pappalardo and Collins (2005) for a calculation of the strains along a dense, closely-spaced fault population on Ganymede. However, many planetary fault populations tend to be sparse – that is, faults that are widely spaced relative to their lengths (e.g., Segall, 1984a; Barnett *et al.*, 1987). Strains measured along a 1-D traverse can therefore miss many small faults. In addition, measurements of fault offset will likely not be made at the positions of maximum fault displacement for all faults transected by the traverse.

Instead of using a 1-D line traverse for calculating strain, an alternative approach to obtaining fault-related normal strain parallels that from seismology, i.e., relating incremental displacements accumulated along rupture patches during an earthquake and the total (or cumulative) displacements accumulated along faults (e.g., Segall,

1984a; Scholz and Cowie, 1990; Scholz, 1997). Any fault has three characteristic dimensions, including length L (defined again as its horizontal dimension), maximum D_{max} or average D displacement (usually located near the fault's midpoint), and height H (measured normal to length, along the fault plane, in the vertical or down-dip direction). The procedure for calculating the average horizontal normal strain for a deforming region, for example, is straightforward and summarized in this section. Specifically, the three variables (L, D, and H) for each fault in the population are assessed (see Figure 10.10c), summed, and then divided by the dimensions of the deforming region.

First the geometric moment M_g is calculated, which is given by

$$M_g = DLH \quad (10.5)$$

and which is defined by *average* displacement D, fault length L, and down-dip fault height H, with units of m³ (King, 1978; Scholz and Cowie, 1990; Ben-Zion, 2001). D is the average offset along the fault (not D_{max}), measured in the plane of the fault; it is not the component in a horizontal plane, as will be needed later for the horizontal normal strains for extension or contraction. The geometric moment represents the volume of deformed rock associated with a fault population. As a fault grows in size, its surface area increases; because a fault's displacement scales with L, the geometric moment M_g increases as a fault grows in size and in displacement.

The amount of deformation attributed to each fault is given by a related scalar quantity, the quasi-static fault moment M_f (see Pollard and Segall, 1987, p. 302):

$$M_f = GAD = GM_g = GDLH, \quad (10.6)$$

where G is the shear modulus of the surrounding rock mass (where $G = E/[2(1 - \nu^2)]$), A is the surface area of the fault as defined by its shape (length L times height H), and D is the average (relative) displacement across the fault. M_f has units of MJ (joules $\times 10^6$) for values of modulus in 10^6 Pa and L, H, and D in meters. The quasi-static fault moment represents the total energy consumed by the rock mass in producing the fault displacements within the region.

The work done by faulting, as recorded in the measured fault displacements, W_f, is the sum of the quasi-static fault moments for all faults in a region:

$$W_f = \sum_{i=1}^{N} (M_f)_i, \quad (10.7)$$

where M_{fi} is the quasi-static moment for each fault and N is the total number of faults in the region. The work also has units of energy (MJ) or, equivalently, 10^6 N m. W_f does not explicitly depend on the size of the deforming region that contains the faults, and it also neglects the generally much smaller contributions of processes such as fault formation. Although there are some implicit relationships between the quantities in Equations (10.6) and (10.7) and region size (e.g., *A* may be limited by stratal or crustal thickness (Scholz and Cowie, 1990; Westaway, 1994; see also Figures 10.10c and 10.13), and *D* and *G* may depend on scale and driving stress (Cowie and Scholz, 1992b; Schultz *et al.*, 2006)), the total work done by faulting (Equation (10.7) represents a convenient method for quantifying the role of faulting in lithospheric deformation.

Fault strain is a tensor quantity, with components such as normal and shear strain in various directions. For example, the total strain accommodated by a population of normal faults will have a component of extensional normal strain, perpendicular to the average strike of the faults (their "extension direction"), another component of extensional normal strain parallel to the fault strike (which will be small for most cases; e.g., Krantz, 1988), and a component of contractional normal strain in the vertical direction, corresponding to crustal thinning (e.g., Wilkins *et al.*, 2002). Similarly, a thrust fault population will have a component of contractional normal strain perpendicular to the average strike of the faults (their shortening or "vergence" direction), another component of contractional normal strain parallel to the fault strike, and a component of extensional normal strain in the vertical direction, corresponding to crustal thickening (e.g., Schultz, 2000b).

The desired components of the strain tensor can be obtained by using either of two methods. First, all the information needed for Equation (10.6) – the geometric fault moment – can be specified, along with fault dip, fault strike, and displacement direction (rake) for each fault. The component of interest can be obtained by solving Kostrov's (1974) equation

$$\varepsilon_{kl} = \frac{1}{2V} \sum_{i=1}^{N} \left(M_f\right)_i, \tag{10.8}$$

as outlined by Aki and Richards (1980, pp. 117–118) and which has been used extensively in seismotectonics and structural geology (e.g., Molnar, 1983; Scholz and Cowie, 1990; Westaway, 1992; Scholz, 1997; Scholz, 2002, pp. 306–309; see also Wilkins *et al.*, 2002; Schultz, 2003a,b; Knapmeyer *et al.*, 2006; Dimitrova *et al.*, 2006, for applications to planetary fault populations). Alternatively, measurements along a traverse can be taken, corrected explicitly for fault strike and rake, and then substituted into a simpler set of strain equations that already have the dip correction incorporated into them (e.g., Scholz, 1997). This second, simpler method for obtaining the horizontal normal strain perpendicular to the average strike of a set of faults, which is probably the most important and widely used quantity in planetary fault population studies, is outlined next, although both methods will produce the same results.

The strike correction, for normal or thrust faults, is the component of $D_{\max}$ in a particular horizontal direction (e.g., Priest, 1993, pp. 96–97; Peacock and Sanderson, 1993). This is obtained by calculating the component of fault displacement D_s along the direction of interest, such as a traverse line (such as one perpendicular to the average strike of a set of faults), by using

$$D_s = D_{\max} |\cos(\Delta\psi)|, \tag{10.9}$$

in which $\Delta\psi = $ (strike of fault ψ minus the strike of traverse ψ_T). The component of horizontal displacement along the traverse direction is given by

$$D_{s,d} = D_{\max} |\cos(\Delta\psi)| \cos\delta, \tag{10.10}$$

which includes the dip correction (given by the last term in Equation (10.10)).

The other correction that must be made to the displacement data is to reduce the value of $D_{\max}$ to an average value of displacement for the fault. The average displacement D is used in fault-set inversions for paleostresses (e.g., Marrett and Allmendinger, 1990; Angelier, 1994), as well as for fault-related strain (e.g., Scholz and Cowie, 1990; Scholz, 1997). The average displacement $D = \kappa D_{\max}$, where κ is a fraction of the maximum displacement $D_{\max}$, depending on the specific displacement distribution along the fault. Scholz and Cowie (1990) assumed a value of $\kappa = 0.5$. Dawers et al. (1993) obtained values of κ for small normal faults in Bishop Tuff of 0.61; Moore and Schultz (1999) found values of κ between 0.3 and 0.7 for normal faults from Canyonlands National Park. A fault having a linear displacement profile has $\kappa = 0.5$, whereas one with an ideal elliptical profile (assuming LEFM conditions) has $\kappa = 0.7854$.

Using these three corrections to $D_{\max}$, the horizontal normal strain due to a particular fault can be calculated. The horizontal normal strain ε_n is M_g normalized by the appropriate dimension of the faulted region having thickness T, horizontal area A, and volume $V = TA$. For "small" faults (e.g., Scholz and Cowie, 1990; Scholz, 1997) (Figure 10.17), $H_i < T/\sin\delta_i$; for "large" faults, $H_i = T/\sin\delta_i = H_0$, so the horizontal normal strain (assuming constant fault dip angles) is obtained from Kostrov's equation (10.8) explicitly as (e.g., Scholz, 1997)

$$\varepsilon_n = \frac{\sin 2\delta}{2V} \sum_{i=1}^{N} [D_i L_i H_i]$$
$$\varepsilon_n = \frac{\sin 2\delta}{2AT} \sum_{i=1}^{N} \left[D_i L_i \frac{T}{\sin\delta}\right]. \tag{10.11}$$

The first of Equation (10.11) is for small faults, the second is the approximate limiting value for large faults. "Large" faults in a population, as discussed in this section, are considered to be vertically restricted; "small" faults in a population

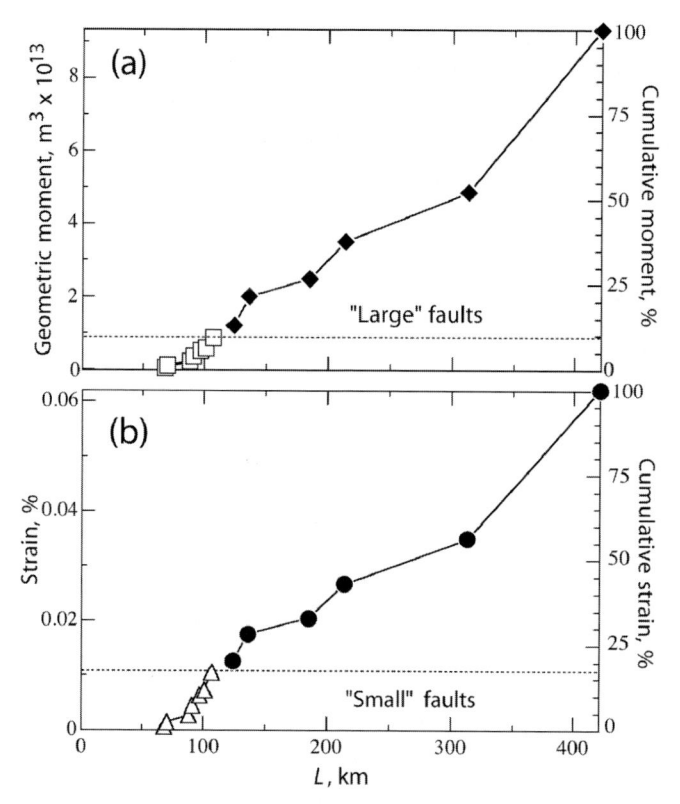

Figure 10.17. (a) Geometric moment and (b) contractional horizontal normal strain (shown as absolute values) calculated for thrust faults within the Amenthes Rupes population of eastern Mars (after Schultz, 2003b).

are unrestricted (Figure 10.10c). Using the trigonometric substitution $\sin 2\delta = 2\sin\delta\cos\delta$ and collecting terms, the horizontal normal strain is written as

$$\varepsilon_n = \frac{\sin\delta\cos\delta}{V}\sum_{i=1}^{N}[D_i L_i H_i]$$
$$\varepsilon_n = \frac{\cos\delta}{A}\sum_{i=1}^{N}[D_i L_i], \quad (10.12)$$

in which δ is fault dip angle and D is the average displacement on a particular fault (using Equation (10.8) and the correction for average displacement from $D_{\max}$); again, the first of Equation (10.12) is for small faults, the second is for large faults. The sign of D must be specified for these equations, using $D > 0$ for normal faults and $D < 0$ for thrust faults. Using this convention, extensional normal strain will be positive and contractional normal will be negative.

In these equations, M_f is calculated for the component of the complete moment tensor for the population in the horizontal plane (i.e., the planetary surface) and normal to fault strike (e.g., Aki and Richards, 1980, pp. 117–118; Scholz, 1997). These equations are thus defined for normal or thrust faults, with pure dip-slip rakes of 90°, and provide the horizontal normal strain (extension for normal faults, contraction for thrust faults) accommodated by the fault population perpendicular to its strike and in the horizontal plane. Analogous equations can be defined for the shear strain accommodated in the horizontal plane by a population of strike-slip faults, although this fault type is comparatively rare on planetary surfaces except for Earth and Europa.

The **vertical normal strain** associated with dip-slip faulting is given by (Aki and Richards, 1980, pp. 117–118)

$$\varepsilon_v = -\frac{\sin \delta \cos \delta}{V} \sum_{i=1}^{N} [D_i L_i H_i]$$
$$\varepsilon_v = -\frac{\cos \delta}{A} \sum_{i=1}^{N} [D_i L_i]$$
(10.13)

(the first expression of Equation (10.13) is again for small faults, the second is the approximate limiting value for large faults). Note the sign change relative to the horizontal fault-normal strain: this is the "odd axis" strain of Krantz (1988, 1989; see Figure 10.2). For normal faulting, this strain component quantifies the amount of lithospheric **thinning**, whereas for thrust faulting, the amount of **thickening** of the faulted section.

Direct calculation of the horizontal normal strain due to dip-slip faulting in any dataset by using Equation (10.12) provides a straightforward measure of the extensional or contractional strain associated by a tectonic event. Similarly, the vertical normal strain (thinning or thickening) of a faulted section can be obtained easily from the same set of measurements. As an example, the moments and contractional horizontal normal strain for the Amenthes Rupes thrust fault population in eastern Mars were calculated from the MOLA topography (Figure 10.15; Schultz, 2003b; Grott et al., 2006). The figure reveals the incremental increases in both quantities due to each fault, as well as the dominant effect of the largest faults in the population.

6 Challenges and future work

Future advances in the study of planetary fault populations can be made through a number of techniques, including the use of high-resolution topography. Digital elevation models (DEMs) derived from stereo observations, and to some extent photoclinometry, have the potential to reveal a wealth of information on the geometry and slip distributions of faults. DEMs with postings of one to tens of meters spacing can be constructed from Mars from imagery acquired by the Mars Orbiter Camera,

High Resolution Stereo Camera, Context imager and High Resolution Imaging Science Experiment (HiRISE) camera (Kirk *et al.*, 2003; Neukum *et al.*, 2004; Williams *et al.*, 2004; Jaumann *et al.*, 2005; Kronberg *et al.*, 2007). While the public availability of preprocessed DEMs is currently limited, the stereo image data are widely available through NASA's Planetary Data System (http://pds.jpl.nasa.gov) and processing of these image data into DEMs can be achieved with standard software (Albertz *et al.*, 2005; Kirk *et al.*, 2007).

The potential insight that can be gained from high-resolution DEMs of fault populations is well worth the effort of processing these data. Although much work has been accomplished from study of fault-related topography measured by photoclinometry, radar, and MOLA data, analysis of DEMs based on more recent datasets will help to extend the current state of knowledge. Significantly, high-resolution DEMs can help to quantify the geometries and slip distributions of planetary faults that are as short as a few kilometers in length (e.g., Okubo *et al.*, 2008b; Polit *et al.*, 2009), and help to extend current understanding of populations of faults at comparable length scales (e.g., Figure 10.10b). High-resolution DEMs can also help to resolve key spatial details (such as fault-tip displacements and cross-cutting relationships) of the longer faults that have been previously examined in lower resolution planetary datasets. Images having high spatial resolution from HiRISE, for example, are themselves useful for planetary tectonic studies as they are revealing, for the first time, joints (Okubo and McEwen, 2007) and deformation bands (Okubo *et al.*, 2008a) on Mars.

Quantification of the geometries and displacements of planetary fault populations can reveal significant insight into the evolution of planetary surfaces. Measurements of fault-plane dip angles reveal effective fault frictional strengths (Equation (10.2)), providing a means of inferring crustal properties. Further, the sense of fault displacement provides insight into the causative stress state (Figure 10.2). Together, fault dip angle and displacement characteristics can provide important constraints, such as crustal strength and magnitudes of stress and strain in 3-D, for the time span over which a particular fault population was active. In this way, analyses of planetary fault populations of different spatial and temporal distributions will be an important source of boundary conditions for geodynamic models of lithospheric evolution (Grott and Breuer, 2008), as well as interpretations of the geologic history of planetary surfaces.

Acknowledgments

Reviews by Steve Wojtal and Ken Tanaka improved the clarity and flow of the chapter. RAS was supported by grants from NASA's Planetary Geology and Geophysics Program and NASA's Mars Data Analysis Program. DM was supported by grants from CNRS/INSU's Programme National de Planétologie.

References

Abercrombie, R. W. and Ekström, G. (2001). Earthquake slip on oceanic transform faults. *Nature*, **410**, 74–76.

Ackermann, R. V., Schlische, R. W., and Withjack, M. O. (2001). The geometric and statistical evolution of normal fault systems: An experimental study of the effects of mechanical layer thickness on scaling laws. *J. Struct. Geol.*, **23**, 1803–1819.

Aki, K. and Richards, P. G. (1980). *Quantitative Seismology: Theory and Methods*. Vol. I. San Francisco: W. H. Freeman.

Albertz, J., Dorninger, P., Dorrer, E., Ebner, H., Gehrke, S., Giese, B., Gwinner, K., Heipke, C., Howington-Kraus, E., Kirk, R. L., Lehmann, H., Mayer, H., Muller, J.-P., Oberst, J., Ostrovskiy, A., Renter, J., Reznik, S., Schmidt, R., Scholten, F., Spiegel, M., Stilla, U., Wählisch, M., Neukum G., Attwenger, M., Barrett, J., and Casley, S. (2005). HRSC on Mars Express: Photogrammetric and cartographic research. *Photogramm. Eng. Remote Sens.*, **71**, 1153–1166.

Anderson, E. M. (1951). *The Dynamics of Faulting and Dyke Formation, with Applications to Britain*. Edinburgh, Oliver & Boyd.

Andrews-Hanna, J. C., Zuber, M. T., and Hauck II, S. A. (2008). Strike-slip faults on Mars: Observations and implications for global tectonics and geodynamics. *J. Geophys. Res.*, **113**, E08002, doi:10.1029/2007JE002980.

Angelier, J. (1994). Fault slip analysis and paleostress reconstruction. In *Continental Deformation*, ed. P. L. Hancock. New York, Pergamon, pp. 53–100.

Aydin, A. (1988). Discontinuities along thrust faults and the cleavage duplexes. In *Geometries and Mechanisms of Thrusting, with Special Reference to the Appalachians*, ed. G. Mitra and S. Wojtal. *Geol. Soc. Am. Spec. Pap.*, **222**, 223–232.

Aydin, A. (2006). Failure modes of the lineaments on Jupiter's moon, Europa: Implications for the origin of its icy crust. *J. Struct. Geol.*, **28**, 2222–2236.

Aydin, A. and DeGraff, J. M. (1988). Evolution of polygonal fracture patterns in lava flows. *Science*, **239**, 471–476.

Aydin, A. and Nur, A. (1982). Evolution of pull-apart basins and their scale independence. *Tectonics*, **1**, 11–21.

Aydin, A. and Reches, Z. (1982). Number and orientation of fault sets in the field and in experiments. *Geology*, **10**, 107–112.

Aydin, A. and Schultz, R. A. (1990). Effect of mechanical interaction on the development of strike-slip faults with echelon patterns. *J. Struct. Geol.*, **12**, 123–129.

Aydin, A., Schultz, R. A., and Campagna, D. (1990). Fault-normal dilation in pull-apart basins: Implications for the relationship between strike-slip faults and volcanic activity. *Ann. Tectoni.*, **4**, 45–52.

Aydin, A., Borja, R. I., and Eichhubl, P. (2006). Geological and mathematical framework for failure modes in granular rock. *J. Struct. Geol.*, **28**, 83–98.

Bai, T. and Pollard, D. D. (2000). Fracture spacing in layered rocks: A new explanation based on the stress transition. *J. Struct. Geol.*, **22**, 43–57.

Banerdt, W. B., Golombek, M. P., and Tanaka, K. L. (1992). Stress and tectonics on Mars. In *Mars*, ed. H. H. Kieffer, B. M. Jakosky, C. W. Snyder and M. S. Matthews. Tucson AZ: University of Arizona Press, pp. 249–297.

Barnett, J. A. M., Mortimer, J., Rippon, J. H., Walsh, J. J., and Watterson, J. (1987). Displacement geometry in the volume containing a single normal fault. *Am. Assoc. Petrol. Geol. Bull.*, **71**, 925–937.

Bellahsen, N., Daniel, J.-M., Bollinger, L., and Burov, E. (2003). Influence of viscous layers on growth of normal faults: Insights from experimental and numerical models. *J. Struct. Geol.*, **25**, 1471–1485.

Benedicto, A., Schultz, R., and Soliva, R. (2003). Layer thickness and the shape of faults. *Geophys. Res. Lett.*, **30**, 2076, 10.1029/2003GL018237.

Ben-Zion, Y. (2001). On quantification of the earthquake source. *Seismol. Res. Lett.*, **72**, 151–152.

Bieniawski, Z. T. (1989). *Engineering Rock Mass Classifications: A Complete Manual for Engineers and Geologists in Mining, Civil, and Petroleum Engineering*. New York, Wiley.

Bohnenstiehl, D. R. and Carbotte, S. M. (2001). Faulting patterns near 19°30'S on the East Pacific Rise: Fault formation and growth at a superfast spreading center. *Geochem., Geophys., Geosyst.*, **2**, 2001GC000156.

Bohnenstiehl, D. R. and Kleinrock, M. C. (1999). Faulting and fault scaling on the median valley floor of the Trans-Atlantic Geotraverse (TAG) segment, 26°N on the Mid-Atlantic Ridge. *J. Geophys. Res.*, **104**, 29 351–29 364.

Borgos, H. G., Cowie, P. A., and Dawers, N. H. (2000). Practicalities of extrapolating one-dimensional fault and fracture size-frequency distributions to higher-dimensional samples. *J. Geophys. Res.*, **105**, 28 377–28 391.

Brace, W. F. and Kohlstedt, D. L. (1980). Limits on lithospheric stress imposed by laboratory experiments. *J. Geophys. Res.*, **85**, 6248–6252.

Brady, B. H. G. and Brown, E. T. (1993). *Rock Mechanics for Underground Mining*. London, Chapman and Hall.

Brown, E. T. and Hoek, E. (1978). Trends in relationships between measured in-situ stresses and depth. *Int. J. Rock Mech. Min. Sci. Geomech. Abs.*, **15**, 211–215.

Bürgmann, R., Pollard, D. D., and Martel, S. J. (1994). Slip distributions on faults: Effects of stress gradients, inelastic deformation, heterogeneous host-rock stiffness, and fault interaction. *J. Struct. Geol.*, **16**, 1675–1690.

Caine, J. S., Evans, J. P., and Forster, C. B. (1996). Fault zone architecture and permeability structure. *Geology*, **24**, 1025–1028.

Carbotte, S. M. and Macdonald, K. C. (1994). Comparison of seafloor tectonic fabric at intermediate, fast, and super fast spreading ridges: Influence of spreading rate, plate motions, and ridge segmentation on fault patterns. *J. Geophys. Res.*, **99**, 13 609–13 631.

Chinnery, M. A. (1961). The deformation of the ground around surface faults. *Seismol. Soc. Am. Bull.*, **51**, 355–372.

Cladouhos, T. T. and Marrett, R. (1996). Are fault growth and linkage models consistent with power-law distributions of fault lengths? *J. Struct. Geol.*, **16**, 281–293.

Clark, R. M. and Cox, S. J. D. (1996). A modern regression approach to determining fault displacement–length relationships. *J. Struct. Geol.*, **18**, 147–152.

Cohen, S. C. (1999). Numerical models of crustal deformation in seismic fault zones. *Adv. Geophys.*, **41**, 133–231.

Cooke, M. L. (1997). Fracture localization along faults with spatially varying friction. *J. Geophys. Res.*, **102**, 22 425–22 434.

Cowie, P. A. (1998). Normal fault growth in three dimensions in continental and oceanic crust. In *Faulting and Magmatism at Mid-Ocean Ridges*, ed. R. Buck, P. Delaney, J. Karson and Y. Lagabrielle. AGU Monograph 106, pp. 325–348.

Cowie, P. A. and Roberts, G. P. (2001). Constraining slip rates and spacing for active normal faults. *J. Struct. Geol.*, **23**, 1901–1915.

Cowie, P. A. and Scholz, C. H. (1992a). Displacement–length scaling relationships for faults: Data synthesis and discussion. *J. Struct. Geol.*, **14**, 1149–1156.

Cowie, P. A. and Scholz, C. H. (1992b). Physical explanation for the displacement–length relationship of faults using a post-yield fracture mechanics model. *J. Struct. Geol.*, **14**, 1133–1148.

Cowie, P. A., Malinverno, A., Ryan, W. B. F., and Edwards, M. H. (1994). Quantitative fault studies on the East Pacific Rise: A comparison of sonar imaging techniques. *J. Geophys. Res.*, **99**, 15 205–15 218.

Cowie, P. A., Sornette, D., and Vanneste, C. (1995). Multifractal scaling properties of a growing fault population. *Geophys. J. Inter.*, **122**, 457–469.

Cowie, P. A., Knipe, R. J., and Main, I. G. (1996). Introduction to the Special Issue. *J. Struct. Geol.*, **28** (2/3), v–xi.

Crider, J. G. and Peacock, D. C. P. (2004). Initiation of brittle faults in the upper crust: A review of field observations. *J. Struct. Geol.*, **26**, 691–707.

Crider, J. G. and Pollard, D. D. (1998). Fault linkage: Three-dimensional mechanical interaction between echelon normal faults. *J. Geophys. Res.*, **103**, 24 373–24 391.

Davis, K., Burbank, D. W., Fisher, D., Wallace, S., and Nobes, D. (2005). Thrust-fault growth and segment linkage in the active Ostler fault zone, New Zealand. *J. Struct. Geol.*, **27**, 1528–1546.

Davison, I. (1994). Linked fault systems: Extensional, strike-slip and contractional. In *Continental Deformation*, ed. P. L. Hancock. New York, Pergamon, pp. 121–142.

Dawers, N. H. and Anders, M. H. (1995). Displacement–length scaling and fault linkage. *J. Struct. Geol.*, **17**, 607–614.

Dawers, N. H., Anders, M. H., and Scholz, C. H. (1993). Growth of normal faults: Displacement–length scaling. *Geology*, **21**, 1107–1110.

Delaney, P. T., Pollard, D. D., Ziony, J. I., and McKee, E. H. (1986). Field relations between dikes and joints: Emplacement processes and paleostress analysis. *J. Geophys. Res.*, **91**, 4920–4938.

dePolo, C. M. (1998). A reconnaissance technique for estimating the slip rates of normal-slip faults in the Great Basin, and application to faults in Nevada, USA. Ph.D. thesis, University of Nevada, Reno.

Dimitrova, L. L., Holt, W. E., Haines, A. J., and Schultz, R. A. (2006). Towards understanding the history and mechanisms of Martian faulting: The contribution of gravitational potential energy. *Geophys. Res. Lett.*, **33**, L08202, 10.1029/2005GL025307.

Elliott, D. (1976). The energy balance and deformation mechanism of thrust sheets. *Philos. Trans. R. Soc. Lond.*, **A283**, 289–312.

Engelder, T. (1993). *Stress Regimes in the Lithosphere*. Princeton, NJ, Princeton University Press.

Ernst, R. E., Grosfils, E. B., and Mège, D. (2001). Giant dike swarms: Earth, Venus, and Mars. *Annu. Rev. Earth Planet. Sci.*, **29**, 489–534.

Ferrill, D. A. and Morris, A. P. (2003). Dilational normal faults. *J. Struct. Geol.*, **25**, 183–196.

Fossen, H. and Gabrielsen, R. H. (2005). *Strukturgeologi*. Bergen, Norway, Fagbokforlaget.

Fossen, H., Schultz, R. A., Shipton, Z. K., and Mair, K. (2007). Deformation bands in sandstone: A review. *J. Geol. Soc. Lond.*, **164**, 755–769.

Franklin, J. A. (1993). Empirical design and rock mass characterization. In *Comprehensive Rock Engineering*, ed. J. A. Hudson. Vol. 2, ed. C. Fairhurst. New York. Pergamon Press, pp. 795–806.

Freed, A. M., Melosh, H. J., and Solomon, S. C. (2001). Tectonics of mascon loading: Resolution of the strike-slip faulting paradox. *J. Geophys. Res.*, **106**, 20 603–20 620.

Gawthorpe, R. L. and Hurst, J. M. (1993). Transfer zone in extensional basins: Their structural style and influence on drainage development and stratigraphy. *J. Geol. Soc. Lond.*, **150**, 1137–1152.

Garel, E., Dautiel, O., and Lagabrielle, Y. (2002). Deformation processes at fast to ultra-fast oceanic spreading axes: Mechanical approach. *Tectonophy.*, **346**, 223–246.

Goetze, C. and Evans, B. (1979). Stress and temperature in the bending lithosphere as constrained by experimental rock mechanics. *Geophys. J. R. Astron. Soc.*, **59**, 463–478.

Golombek, M. P., Tanaka, K. L., and Franklin, B. J. (1996). Extension across Tempe Terra, Mars, from measurements of fault scarp widths and deformed craters. *J. Geophys. Res.*, **99**, 23 163–23 171.

Goudy, C. L. and Schultz, R. A. (2005). Dike intrusions beneath grabens south of Arsia Mons, Mars. *Geophys. Res. Lett.*, **32**, 5, 10.1029/2004GL021977.

Goudy, C. L., Schultz, R. A., and Gregg, T. K. P. (2005). Coulomb stress changes in Hesperia Planum, Mars, reveal regional thrust fault reactivation. *J. Geophys. Res.*, **110**, E10005, 10.1029/2004JE002293.

Grosfils, E. and Head, J. W. (1994). Emplacement of a radiating dike swarm in western Vinmara Planitia, Venus: Interpretation of the regional stress field orientation and subsurface magmatic configuration. *Earth, Moon and Planets*, **66**, 153–171.

Grott, M. and Breuer, D. (2008). The evolution of the Martian elastic lithosphere and implications for crustal and mantle rheology. *Icarus*, **193**, 503–515.

Grott, M., Hauber, E., Werner, S. C., Kronberg, P., and Neukum, G. (2006). Mechanical modeling of thrust faults in the Thaumasia region, Mars, and implications for the Noachian heat flux. *Icarus*, **186**, 517–526.

Gudmundsson, A. (1992). Formation and growth of normal faults at the divergent plate boundary in Iceland. *Terra Nova*, **4**, 464–471.

Gudmundsson, A. (2004). Effects of Young's modulus on fault displacement. *Comptes Rendus Geosci.*, **336**, 85–92.

Gudmundsson, A. and Bäckström, K. (1991). Structure and development of the Sveeinagja graben, Northeast Iceland. *Tectonophys.*, **200**, 111–125.

Gupta, A. and Scholz, C. H. (2000a). A model of normal fault interaction based on observations and theory. *J. Struct. Geol.*, **22**, 865–879.

Gupta, A. and Scholz, C. H. (2000b). Brittle strain regime transition in the Afar depression: Implications for fault growth and seafloor spreading. *Geology*, **28**, 1078–1090.

Hanna, J. C. and Phillips, R. J. (2006). Tectonic pressurization of aquifers in the formation of Mangala and Athabasca Valles, Mars. *J. Geophys. Res.*, **111**, E03003, doi: 10.1029/2005JE002546.

Harris, R. A. (1998). Introduction to special section: Stress triggers, stress shadows, and implications. *J. Geophys. Res.*, **103**, 24 347–24 358.

Hatheway, A. W. and Kiersch, G. A. (1989). Engineering properties of rock. In *Practical Handbook of Physical Properties of Rocks and Minerals*, ed. R. S. Carmichael, Boca Raton, Fl: CRC Press, pp. 672–715.

Hauber, E. and Kronberg, P. (2005). The large Thaumasia graben on Mars: Is it a rift? *J. Geophys. Res.*, **110**, E07003, 10.1029/2005JE002407.

Hoek, E. (1983). Strength of jointed rock masses. *Géotechnique*, **33**, 187–223.

Hoek, E. (1990). Estimating Mohr-Coulomb friction and cohesion from the Hoek-Brown failure criterion (abs.). *Int. J. Rock Mech. Min. Sci. Geomech. Abs.*, **27**, 227–229.

Hoek, E. and Brown, E. T. (1980). Empirical strength criterion for rock masses. *J. Geotech. Eng. Div. Am. Soc. Civ. Eng.*, **106**, 1013–1035.

Hoek, E. and Brown, E. T. (1997). Practical estimates of rock mass strength. *Int. J. Rock Mech. Min. Sci.*, **34**, 1165–1186.

Hooper, D. M., Bursik, M. I., and Webb, F. H. (2003). Application of high-resolution, interferometric DEMs to geomorphic studies of fault scarps, Fish Lake Valley, Nevada, California, USA. *Remote Sens. Environ.*, **84**, 255–267.

Hu, M. S. and Evans, A. G. (1989). The cracking and decohesion of thin films on ductile substrate. *Acta Mater.*, **37**, 917–925.

Hubbert, M. K. and Rubey, W. W. (1959). Role of fluid pressure in mechanics of overthrust faulting: I. Mechanics of fluid-filled porous solids and its application to overthrust faulting. *Geol. Soc. Am. Bull.*, **70**, 115–166.

Jaeger, J. C. and Cook, N. G. W. (1979). *Fundamentals of Rock Mechanics*. 3rd edn. New York, Chapman and Hall.

Jaeger, J. C., Cook, N. G. W., and Zimmerman, R. W. (2007). *Fundamentals of Rock Mechanics*. 4th edn. Oxford, Blackwell.

Jaumann, R., Reiss, D., Frei, S., Neukum, G., Scholten, F., Gwinner, K., Roatsch, T., Matz, K.-D., Mertens, V., Hauber, E., Hoffmann, H., Köhler, U., Head, J. W., Hiesinger, H., and Carr, M. H. (2005). Interior channels in Martian valleys: Constraints on fluvial erosion by measurements of the Mars Express High Resolution Stereo Camera. *Geophys. Res. Lett.*, **32**, L16203, 10.1029/2005GL023415.

Kakimi, T. (1980). Magnitude-frequency relation for displacement of minor faults and its significance in crustal deformation. *Bull. Geol. Surv. Jap.*, **31**, 467–487.

Kattenhorn, S. A. and Marshall, S. T. (2006). Fault-induced perturbed stress fields and associated tensile and compressive deformation at fault tips in the ice shell of Europa: Implications for fault mechanics. *J. Struct. Geol.*, **28**, 2204–2221.

Kattenhorn, S. A. and Pollard, D. D. (1999). Is lithostatic loading important for the slip behavior and evolution of normal faults in the Earth's crust? *J. Geophys. Res.*, **104**, 28 879–28 898.

Kattenhorn, S. A. and Pollard, D. D. (2001). Integrating 3D seismic data, field analogs and mechanical models in the analysis of segmented normal faults in the Wytch Farm oil field, southern England. *Am. Assoc. Petrol. Geol. Bull.*, **85**, 1183–1210.

Kiefer, W. S. and Swafford, L. C. (2006). Topographic analysis of Devana Chasma, Venus: Implications for rift system segmentation and propagation. *J. Struct. Geol.*, **28**, 2144–2155.

King, G. C. P. (1978). Geological faulting: Fracture, creep and strain. *Philos. Trans. R. Soc. Lond.*, **A288**, 197–212.

King, G. and Yielding, G. (1984). The evolution of a thrust fault system: Processes of rupture initiation, propagation and termination in the 1980 El Asnam (Algeria) earthquake. *Geophys. J. R. Astron. Soc.*, **77**, 915–933.

King, G. C. P., Stein, R. S., and Lin, J. (1994). Static stress changes and the triggering of earthquakes. *Seismol. Soc. Am. Bull.*, **84**, 935–953.

Kirk, R. L., Howington-Kraus, E., Redding, B., Galuszka, D., Hare, T. M., Archinal, B. A., Soderblom, L. A., and Barrett, J. M. (2003). High-resolution topomapping of candidate MER landing sites with Mars Orbiter Camera narrow-angle images. *J. Geophy. Res.*, **108**, 8088, 10.1029/2003JE002131.

Kirk, R. L., Howington-Kraus, E., Rosiek, M. R., Cook, D., Anderson, J., Becker, K., Archinal, B. A., Keszthelyi, L., King, R., McEwen, A. S., and the HiRISE Team (2007). Ultrahigh resolution topographic mapping of Mars with HiRISE stereo

images: Methods and first results (abs.). *Seventh International Conference on Mars*, 3381.
Knapmeyer, M., Oberst, J., Hauber, E., Wählisch, M., Deuchler, C., and Wagner, R. (2006). Working models for spatial distribution and level of Mars' seismicity. *J. Geophys. Res.*, **111**, E11006, 10.1029/2006JE002708.
Koenig, E. and Aydin, A. (1998). Evidence for large-scale strike-slip faulting on Venus. *Geology*, **26**, 551–554.
Koenig, E. and Pollard, D. D. (1998). Mapping and modeling of radial fracture patterns on Venus. *J. Geophys. Res.*, **103**, 15 183–15 202.
Kohlstedt, D. L., Evans, B., and Mackwell, S. J. (1995). Strength of the lithosphere: Constraints imposed by laboratory experiments. *J. Geophys. Res.*, **100**, 17 587–17 602.
Kostrov, B. (1974). Seismic moment and energy of earthquakes, and seismic flow of rock. *Izvestiya, Phys. Solid Earth*, **13**, 13–21.
Krantz, R. W. (1988). Multiple fault sets and three-dimensional strain: Theory and application. *J. Struct. Geol.*, **10**, 225–237.
Krantz, R. W. (1989). Orthorhombic fault patterns: The odd axis model and slip vector orientations. *Tectonics*, **8**, 483–495.
Kronberg, P., Hauber, E., Grott, M., Werner, S. C., Schäfer, T., Gwinner, K., Giese, B., Masson, P., and Neukum, G. (2007). Acheron Fossae, Mars: Tectonic rifting, volcanism, and implications for lithospheric thickness. *J. Geophys. Res.*, **112**, E04005, 10.1029/2006JE002780.
Lucchitta, B. K. (1976). Mare ridges and related highland scarps: Results of vertical tectonism? *Proc. Lunar Sci. Conf.* **7**, 2761–2782.
Ma, X. Q. and Kusznir, N. J. (2003). Modelling of near-field subsurface displacements for generalized faults and fault arrays. *J. Struct. Geol.*, **15**, 1471–1484.
Mangold, N., Allemand, P., and Thomas, P. G. (1998). Wrinkle ridges of Mars: Structural analysis and evidence for shallow deformation controlled by ice-rich décollements. *Planet. Space Sci.*, **46**, 345–356.
Manighetti, I., King, G. C. P., Gaudemer, Y., Scholz, C. H., and Doubre, C. (2001). Slip accumulation and lateral propagation of active normal faults in Afar. *J. Geophys. Res.*, **106**, 13 667–13 696.
Manighetti, I., Campillo, M., Sammis, C., Mai, P. M., and King, G. (2005). Evidence for self-similar, triangular slip distributions on earthquakes: Implications for earthquake and fault mechanics. *J. Geophys. Res.*, **110**, B05302, doi:10.1029/2004JB003174.
Mansfield, C. S. and Cartwright, J. A. (1996). High resolution fault displacement mapping from three-dimensional seismic data: Evidence for dip linkage during fault growth. *J. Struct. Geol.*, **18**, 249–263.
Marone, C. (1998). Laboratory-derived friction laws and their application to seismic faulting. *Annu. Rev. Earth Planet. Sci.*, **26**, 643–696.
Marone, C. and Scholz, C. H. (1988). The depth of seismic faulting and the transition from stable to instable slip regimes. *Geophys. Res. Lett.*, **15**, 621–624.
Marrett, R. and Allmendinger, R. W. (1990). Kinematic analysis of fault slip data. *J. Struct. Geol.*, **12**, 973–986.
Marrett, R. and Allmendinger, R. W. (1991). Estimates of strain due to brittle faulting: Sampling of fault populations. *J. Struct. Geol.*, **13**, 735–738.
Marrett, R., Orteg, O. J., and Kelsey, J. M. (1999). Extent of power-law scaling for natural fractures in rocks. *Geology*, **27**, 799–802.
Martel, S. J. (1997). Effects of cohesive zones on small faults and implications for secondary fracturing and fault trace geometry. *J. Struct. Geol.*, **19**, 835–847.

Martel, S. J. and Boger, W. A. (1998). Geometry and mechanics of secondary fracturing around small three-dimensional faults in granitic rock. *J. Geophys. Res.*, **103**, 21 299–21 314.

Mastin, L. G. and Pollard, D. D. (1988). Surface deformation and shallow dike intrusion processes at Inyo Craters, Long Valley, California. *J. Geophys. Res.*, **93**, 13 221–13 235.

McGarr, A. and Gay, N. C. (1978). State of stress in the Earth's crust. *Ann. Rev. Earth Planet. Sci.*, **6**, 405–436.

McGill, G. E. (1993). Wrinkle ridges, stress domains, and kinematics of Venusian plains. *Geophys. Res. Lett.*, **20**, 2407–2410.

McGill, G. E. and Stromquist, A. W. (1979). The grabens of Canyonlands National Park, Utah: Geometry, mechanics, and kinematics. *J. Geophys. Res.*, **84**, 4547–4563.

McGill, G. E., Schultz, R. A., and Moore, J. M. (2000). Fault growth by segment linkage: An explanation for scatter in maximum displacement and trace length data from the Canyonlands Grabens of SE Utah: discussion. *J. Struct. Geol.*, **22**, 135–140.

Means, W. D. (1976). *Stress and Strain: Basic Concepts of Continuum Mechanics for Geologists*. New York, Springer-Verlag.

Mége, D. and Masson, P. (1996). Amounts of crustal stretching in Valles Marineris, Mars. *Planet. Space Sci.*, **44**, 749–782.

Mége, D. and Riedel, S. P. (2001). A method for estimating 2D wrinkle ridge strain from application of fault displacement scaling to the Yakima folds, Washington. *Geophys. Res. Lett.*, **28**, 3545–3548.

Mége, D., Cook, A. C., Garel, E., Lagabrielle, Y., and Cormier, M.-H. (2003). Volcanic rifting at Martian graben. *J. Geophys. Res.*, **108**, 5044, doi:10.1029/2002JE001852.

Molnar, P. (1983). Average regional strain due to slip on numerous faults of different orientations. *J. Geophys. Res.*, **88**, 6430–6432.

Moore, J. M. and Schultz, R. A. (1999). Processes of faulting in jointed rocks of Canyonlands National Park, Utah. *Geol. Soc. Am. Bull.*, **111**, 808–822.

Muehlberger, W. R. (1974). Structural history of southeastern Mare Serenitatis and adjacent highlands. *Proc. Lunar Sci. Conf. 5*, 101–110.

Neuffer, D. P. and Schultz, R. A. (2006). Mechanisms of slope failure in Valles Marineris, Mars. *Q. J. Eng. Geol. Hydrogeol.*, **39**, 227–240.

Neukum, G., Jaumann, R., Hoffmann, H., Hauber, E., Head, J. W., Basilevsky, A. T., Ivanov, B. A., Werner, S. C., van Gasselt, S., Murray, J. B., McCord, T., and the HRSC Co-Investigator Team (2004). Recent and episodic volcanic and glacial activity on Mars revealed by the High Resolution Stereo Camera. *Nature*, **432**, 971–979.

Nicol, A., Watterson, J., Walsh, J. J., and Childs, C. (1996). The shapes, major axis orientations and displacement patterns of fault surfaces. *J. Struct. Geol.*, **18**, 235–248.

Nimmo, F. and Schenk, P. (2006). Normal faulting on Europa: Implications for ice shell properties. *J. Struct. Geol.*, **28**, 2194–2203.

Niño, F., Philip, H., and Chéry, J. (1998). The role of bed-parallel slip in the formation of blind thrust faults. *J. Struct. Geol.*, **20**, 503–516.

Okubo, C. H. and Martel, S. J. (1998). Pit crater formation on Kilauea volcano, Hawaii. *J. Volcanol. Geotherm. Res.*, **86**, 1–18.

Okubo, C. H. and McEwen, A. S. (2007). Fracture controlled paleo-fluid flow in Candor Chasma, Mars. *Science*, **315**, 983–985.

Okubo, C. H. and Schultz, R. A. (2003). Thrust fault vergence directions on Mars: A foundation for investigating global-scale Tharsis-driven tectonics. *Geophys. Res. Lett.*, **30**, 2154, 10.1029/2003GL018664.

Okubo, C. H. and Schultz, R. A. (2004). Mechanical stratigraphy in the western equatorial region of Mars based on thrust fault-related fold topography and implications for near-surface volatile reservoirs. *Geol. Soc. Am. Bull.*, **116**, 594–605.

Okubo, C. H. and Schultz, R. A. (2006a). Near-tip stress rotation and the development of deformation band stepover geometries in mode II. *Geol. Soc. Am. Bull.*, **118**, 343–348.

Okubo, C. H. and Schultz, R. A. (2006b). Variability in Early Amazonian Tharsis stress state based on wrinkle ridges and strike-slip faulting. *J. Struct. Geol.*, **28**, 2169–2181.

Okubo, C. H., Schultz, R. A., Chan, M. A., Komatsu, G., and the HiRISE Team (2008a). Deformation band clusters on Mars and implications for subsurface fluid flow. *Geol. Soc. Am. Bull.*, **121** 474–482.

Okubo, C. H., Lewis, K. L., McEwen, A. S., Kirk, R. L., and the HiRISE Team (2008b). Relative age of interior layered deposits in southwest Candor Chasma based on high-resolution structural mapping. *J. Geophys. Res.*, **113**, E12002, doi:10.1029/2008JE003181.

Olson, J. E. (1993). Joint pattern development: Effects of subcritical crack growth and mechanical crack interaction. *J. Geophys. Res.*, **98**, 12 251–12 265.

Olson, J. E. (2003). Sublinear scaling of fracture aperture versus length: An exception or the rule? *J. Geophys. Res.*, **108**, 2413, doi:10.1029/2001JB000419.

Pappalardo, R. T. and Collins, G. C. (2005). Strained craters on Ganymede. *J. Struct. Geol.*, **27**, 827–838.

Paterson, M. S. and Wong, T.-F. (2005). *Experimental Rock Deformation: The Brittle Field*, 2nd edn. Berlin, Springer.

Peacock, D. C. P. (2002). Propagation, interaction and linkage in normal fault systems. *Earth-Sci. Rev.*, **58**, 121–142.

Peacock, D. C. P. (2003). Scaling of transfer zones in British Isles. *J. Struct. Geol.*, **25**, 1561–1567.

Peacock, D. C. P. and Sanderson, D. J. (1991). Displacement, segment linkage and relay ramps in normal fault zones. *J. Struct. Geol.*, **13**, 721–733.

Peacock, D. C. P. and Sanderson, D. J. (1993). Estimating strain from fault slip using a line sample. *J. Struct. Geol.*, **15**, 1513–1516.

Peacock, D. C. P. and Sanderson, D. J. (1995). Strike-slip relay ramps. *J. Struct. Geol.*, **17**, 1351–1360.

Peacock, D. C. P. and Sanderson, D. J. (1996). Effects of propagation rate on displacement variations along faults. *J. Struct. Geol.*, **18**, 311–320.

Plumb, R. A. (1994). Variations of the least horizontal stress magnitude in sedimentary basins. In *Rock Mechanics: Models and Measurements, Challenges from Industry*, ed. P. Nelson and S. E. Laubach. Rotterdam, Balkema, pp. 71–77.

Polit, A. T. (2005). Influence of mechanical stratigraphy and strain on the displacement–length scaling of normal faults on Mars. M. S. thesis, University of Nevada, Reno.

Polit, A. T., Schultz, R. A., and Soliva, R. (2009). Geometry, displacement–length scaling, and extensional strain of normal faults on Mars with inferences on mechanical stratigraphy of the Martian crust. *J. Struct. Geol.*, **31**, 662–673.

Pollard, D. D. and Fletcher, R. C. (2005). *Fundamentals of Structural Geology*. Cambridge: Cambridge University Press.

Pollard, D. D. and Segall, P. (1987). Theoretical displacements and stresses near fractures in rock: With applications to faults, joints, dikes, and solution surfaces. In *Fracture Mechanics of Rock*, ed. B. K. Atkinson. New York, Academic Press, pp. 277–349.

Poulimenos, G. (2000). Scaling properties of normal fault populations in the western Corinth Graben, Greece: Implications for fault growth in large strain settings. *J. Struct. Geol.*, **27**, 307–322.

Price, N. J. and Cosgrove, J. W. (1990). *Analysis of Geological Structures*. Cambridge, Cambridge University Press.

Priest, S. D. (1993). *Discontinuity Analysis for Rock Engineering*. New York, Chapman and Hall.

Ravnas, R. and Bondevik, K. (1997). Architecture and controls on the Bathonian Kimmeridgian shallow-marine syn-rift wedges of the Oseberg-Brage area, northern North Sea. *Basin Res.*, **9**, 197–226.

Reches, Z. (1978). Analysis of faulting in three-dimensional strain field. *Tectonophys.*, **47**, 109–129.

Reches, Z. (1983). Faulting of rocks in three-dimensional strain fields II. Theoretical analysis. *Tectonophys.*, **95**, 133–156.

Roberts, G. P., Cowie, P., Papanikolaou, I., and Michetti, A. M. (2004). Fault scaling relationships, deformation rates and seismic hazards: An example from the Lazio-Abruzzo Apennines, central Italy. *J. Struct. Geol.*, **26**, 377–398.

Rubin, A. M. (1992). Dike-induced faulting and graben subsidence in volcanic rift zones. *J. Geophys. Res.*, **97**, 1839–1858.

Rubin, A. M. and Pollard, D. D. (1988). Dike-induced faulting in rift zones of Iceland and Afar. *Geology*, **16**, 413–417.

Schenk, P. and McKinnon, W. B. (1989). Fault offsets and lateral crustal movement on Europa: Evidence for a mobile ice shell. *Icarus*, **79**, 75–100.

Schlische, R. W., Young, S. S., Ackerman, R. V., and Gupta, A. (1996). Geometry and scaling relations of a population of very small rift related normal faults. *Geology*, **24**, 683–686.

Scholz, C. H. (1997). Earthquake and fault populations and the calculation of brittle strain. *Geowiss.*, **15**, 124–130.

Scholz, C. H. (1998). Earthquakes and friction laws. *Nature*, **391**, 37–42.

Scholz, C. H. (2002). *The Mechanics of Earthquakes and Faulting*. 2nd edn. Cambridge: Cambridge University Press.

Scholz, C. H. and Cowie, P. A. (1990). Determination of total strain from faulting using slip measurements. *Nature*, **346**, 837–838.

Scholz, C. H. and Lawler, T. M. (2004). Slip tapers at the tips of faults and earthquake ruptures. *Geophys. Res. Lett.*, **31**, L21609, 10.1029/2004GL021030.

Scholz, C. H., Dawers, N. H., Yu, J.-Z., and Anders, M. H. (1993). Fault growth and fault scaling laws: Preliminary results. *J. Geophys. Res.*, **98**, 21 951–21 961.

Schultz, R. A. (1989). Strike-slip faulting of ridged plains near Valles Marineris, Mars. *Nature*, **341**, 424–426.

Schultz, R. A. (1991). Structural development of Coprates Chasma and western Ophir Planum, central Valles Marineris rift, Mars. *J. Geophys. Res.*, **96**, 22 777–22 792.

Schultz, R. A. (1992). Mechanics of curved slip surfaces in rock. *Eng. Anal. Bound. Elem.*, **10**, 147–154.

Schultz, R. A. (1993). Brittle strength of basaltic rock masses with application to Venus. *J. Geophys. Res.*, **98**, 10 883–10 895.

Schultz, R. A. (1995). Limits on strength and deformation properties of jointed basaltic rock masses. *Rock Mech. Rock Eng.*, **28**, 1–15.

Schultz, R. A. (1996). Relative scale and the strength and deformability of rock masses. *J. Struct. Geol.*, **18**, 1139–1149.

Schultz, R. A. (1997). Displacement–length scaling for terrestrial and Martian faults: Implications for Valles Marineris and shallow planetary grabens. *J. Geophys. Res.*, **102**, 12 009–12 015.

Schultz, R. A. (1999). Understanding the process of faulting: Selected challenges and opportunities at the edge of the 21st century. *J. Struct. Geol.*, **21**, 985–993.

Schultz, R. A. (2000a). Fault-population statistics at the Valles Marineris Extensional Province, Mars: Implications for segment linkage, crustal strains, and its geodynamical development. *Tectonophys.*, **316**, 169–193.

Schultz, R. A. (2000b). Localization of bedding plane slip and backthrust faults above blind thrust faults: Keys to wrinkle ridge structure. *J. Geophys. Res.*, **105**, 12 035–12 052.

Schultz, R. A. (2002). Stability of rock slopes in Valles Marineris, Mars. *Geophys. Res. Lett.*, **30**, 1932, 10.1029/2002GL015728.

Schultz, R. A. (2003a). A method to relate initial elastic stress to fault population strains. *Geophys. Res. Lett.*, **30**, 1593, 10.1029/2002GL016681.

Schultz, R. A. (2003b). Seismotectonics of the Amenthes Rupes thrust fault population, Mars. *Geophys. Res. Lett.*, **30**, 1303, 10.1029/2002GL016475.

Schultz, R. A. and Aydin, A. (1990). Formation of interior basins associated with curved faults in Alaska. *Tectonics*, **9**, 1387–1407.

Schultz, R. A. and Balasko, C. M. (2003). Growth of deformation bands into echelon and ladder geometries. *Geophys. Res. Lett.*, **30**, 2033, 10.1029/2003GL018449.

Schultz, R. A. and Fori, A. N. (1996). Fault-length statistics and implications of graben sets at Candor Mensa, Mars. *J. Struct. Geol.*, **18**, 373–383.

Schultz, R. A. and Fossen, H. (2002). Displacement–length scaling in three dimensions: The importance of aspect ratio and application to deformation bands. *J. Struct. Geol.*, **24**, 1389–1411.

Schultz, R. A. and Fossen, H. (2008). Terminology for structural discontinuities. *Am. Assoc. Petrol. Geol. Bull.*, **92**, 853–867.

Schultz, R. A. and Lin, J. (2001). Three-dimensional normal faulting models of Valles Marineris, Mars, and geodynamic implications. *J. Geophys. Res.*, **106**, 16 549–16 566.

Schultz, R. A. and Watters, T. R. (2001). Forward mechanical modeling of the Amenthes Rupes thrust fault on Mars. *Geophys. Res. Lett.*, **28**, 4659–4662.

Schultz, R. A. and Zuber, M. T. (1994). Observations, models, and mechanisms of failure of surface rocks surrounding planetry surface loads. *J. Geophys. Res.*, **99**, 14 691–14 702.

Schultz, R. A., Okubo, C. H., Goudy, C. L., and Wilkins, S. J. (2004). Igneous dikes on Mars revealed by MOLA topography. *Geology*, **32**, 889–892.

Schultz, R. A., Okubo, C. H., and Wilkins, S. J. (2006). Displacement–length scaling relations for faults on the terrestrial planets. *J. Struct. Geol.*, **28**, 2182–2193.

Schultz, R. A., Moore, J. M., Grosfils, E. B., Tanaka, K. L., and Mège, D. (2007). The Canyonlands model for planetary grabens: Revised physical basis and implications. In *The Geology of Mars: Evidence from Earth-Based Analogues*, ed. M. G. Chapman. Cambridge: Cambridge University Press, pp. 371–399.

Schultz, R. A., Soliva, R., Fossen, H., Okubo, C. H., and Reeves, D. M. (2008). Dependence of displacement–length scaling relations for fractures and deformation bands on the volumetric changes across them. *J. Struct. Geol.* **30**, 1405–1411.

Segall, P. (1984a). Formation and growth of extensional fracture sets. *Geol. Soc. Am. Bull.*, **95**, 454–462.

Segall, P. (1984b). Rate-dependent extensional deformation resulting from crack growth in rock. *J. Geophys. Res.*, **89**, 4185–4195.

Segall, P. and Pollard, D. D. (1980). Mechanics of discontinuous faults. *J. Geophys. Res.*, **85**, 4337–4350.

Segall, P. and Pollard, D. D. (1983). Joint formation in granitic rock of the Sierra Nevada. *Geol. Soc. Am. Bull.*, **94**, 563–575.

Sharpton, V. L. and Head, J. W. (1988). Lunar mare ridges: Analysis of ridge-crater intersections and implications for the tectonic origin of mare ridges. *Proc. Lunar Planet. Sci. Conf.* 18, 307–317.

Shaw, J. H., Plesch, A., Dolan, J. F., Pratt, T. L., and Fiore, P. (2002). Puente Hills blind-thrust system, Los Angeles, California. *Seismol. Soc. Am. Bull.*, **92**, 2946–2960.

Sibson, R. H. (1974). Frictional constraints on thrust, wrench and normal faults. *Nature*, **249**, 542–544.

Sibson, R. H. (1994). An assessment of field evidence for 'Byerlee' friction. *Pure Appl. Geophys.*, **142**, 645–662.

Soliva, R. and Benedicto, A. (2004). A linkage criterion for segmented normal faults. *J. Struct. Geol.*, **26**, 2251–2267.

Soliva, R. and Benedicto, A. (2005). Geometry, scaling relations and spacing of vertically restricted normal faults. *J. Struct. Geol.*, **27**, 317–325.

Soliva, R. and Schultz, R. A. (2008). Distributed and localized faulting in extensional settings: Insight from the North Ethiopian Rift – Afar transition area. *Tectonics*, **27**, TC2003, doi:10.1029/2007TC002148.

Soliva, R., Schultz, R. A., and Benedicto, A. (2005). Three-dimensional displacement–length scaling and maximum dimension of normal faults in layered rocks. *Geophys. Res. Lett.*, **32**, L16302, 10.1029/2005GL023007.

Soliva, R., Benedicto, A., and Maerten, L. (2006). Spacing and linkage of confined normal faults: Importance of mechanical thickness. *J. Geophys. Res.*, **110**, B01402, 10.1029/2004JB003507.

Solomon, S. C., McNutt, R. L., Jr., Watters, T. R., Lawrence, D. J., Feldman, W. C., Head, J. W., Krimigis, S. M., Murchie, S. L., Phillips, R. J., Slavin, J. A., and Zuber, M. T. (2008). Return to Mercury: A global perspective on MESSENGER's first Mercury flyby. *Science*, **321**, 59–62.

Sornette, A., Davy P., and Sornette, D. (1990). Growth of fractal fault patterns. *Phys. Rev. Lett.*, **65**, 2266–2269.

Suppe, J. (1985). *Principles of Structural Geology*. Englewood Cliffs, NJ, Prentice-Hall.

Suppe, J. and Connors, C. (1992). Critical-taper wedge mechanics of fold-and-thrust belts on Venus: Initial results from Magellan. *J. Geophys. Res.*, **97**, 13 545–13 561.

Tchalenko, J. S. (1970). Similarities between shear zones of different magnitudes. *Geol. Soc. Am. Bull.*, **81**, 1625–1640.

Tse, S. T. and Rice, J. R. (1986). Crustal earthquake instability in relation to the variation of frictional slip properties. *J. Geophys. Res.*, **91**, 9452–9472.

Villemin, T. and Sunwoo, C. (1987). Distribution logarithmique self similaire des rejets et longueurs de failles: Exemple du bassin Houiller Lorrain. *Compte Rendu Acad. Sci., Série II*, **305**, 1309–1312.

Walsh, J. J. and Watterson, J. (1987). Distribution of cumulative displacement and seismic slip on a single normal fault surface. *J. Struct. Geol.*, **9**, 1039–1046.

Walsh, J. J. and Watterson, J. (1988). Analysis of the relationship between displacements and dimensions of faults. *J. Struct. Geol.*, **10**, 239–247.

Walsh, J. J., Watterson, J., and Yielding, G. (1991). The importance of small-scale faulting in regional extension. *Nature*, **351**, 391–393.

Walsh, J. J., Bailey, W. R., Childs, C., Nicol, A., and Bonson, C. G. (2003). Formation of segmented normal faults: A 3-D perspective. *J. Struct. Geol.*, **25**, 1251–1262.

Watters, T. R. (2003). Thrust faults along the dichotomy boundary in the eastern hemisphere of Mars. *J. Geophys. Res.*, **108**, 5054, 10.1029/2002JE001934.

Watters, T. R., Robinson, M. S., and Cook, A. C. (1998). Topography of lobate scarps on Mercury: New constraints on the planet's contraction. *Geology*, **26**, 991–994.

Watters, T. R., Schultz, R. A., and Robinson, M. S. (2000). Displacement–length scaling relations of thrust faults associated with lobate scarps on Mercury and Mars: Comparison with terrestrial faults. *Geophys. Res. Lett.*, **27**, 3659–3662.

Watters, T. R., Schultz, R. A., Robinson, M. S., and Cook, A. C. (2002). The mechanical and thermal structure of Mercury's early lithosphere. *Geophys. Res. Lett.*, **29**, 10.1029/2001GL014308.

Watterson, J. (1986). Fault dimensions, displacements and growth. *Pure Appl. Geophys.*, **124**, 365–373.

Weijermars, R. (1997). *Principles of Rock Mechanics*. Amsterdam: Alboran Science Publishing.

Weissel, J. K. and Karner, G. D. (1989). Flexural uplift of rift flanks due to mechanical unloading of the lithosphere during extension. *J. Geophys. Res.*, **94**, 13 919–13 950.

Wesnousky, S. G. (1994). The Gutenberg-Richter or characteristic earthquake distribution, which is it? *Seismol. Soc. Am. Bull.*, **84**, 1940–1959.

Wesnousky, S. G. (1999). Crustal deformation processes and the stability of the Gutenberg-Richter relationship. *Seismol. Soc. Am. Bull.*, **89**, 1131–1137.

Westaway, R. (1992). Seismic moment summation for historical earthquakes in Italy: Tectonic implications. *J. Geophys. Res.*, **97**, 15 437–15 464.

Westaway, R. (1994). Quantitative analysis of populations of small faults. *J. Struct. Geol.*, **16**, 1259–1273.

Wibberley, C. A. J., Petit, J.-P., and Rives, T. (1999). Mechanics of high displacement gradient faulting prior to lithification. *J. Struct. Geol.*, **21**, 251–257.

Wibberley, C. A. J., Petit, J.-P., and Rives, T. (2000). Mechanics of cataclastic 'deformation band' faulting in high-porosity sandstone, Provence. *Comptes Rendus Acad. Sci., Paris*, **331**, 419–425.

Wilkins, S. J. and Gross, M. R. (2002). Normal fault growth in layered rocks at Split Mountain, Utah: Influence of mechanical stratigraphy on dip linkage, fault restriction and fault scaling. *J. Struct. Geol.*, **24**, 1413–1429 (erratum, *J. Struct. Geol.*, **24**, 2007).

Wilkins, S. J. and Schultz, R. A. (2003). Cross faults in extensional settings: Stress triggering, displacement localization, and implications for the origin of blunt troughs in Valles Marineris, Mars. *J. Geophys. Res.*, **108**, 5056, 10.1029/ 2002JE001968.

Wilkins, S. J. and Schultz, R. A. (2005). 3D cohesive end-zone model for source scaling of strike-slip interplate earthquakes. *Seismol. Soc. Am. Bull.*, **95**, 2232–2258.

Wilkins, S. J., Schultz, R. A., Anderson, R. C., Dohm, J. M., and Dawers, N. C. (2002). Deformation rates from faulting at the Tempe Terra extensional province, Mars. *Geophys. Res. Lett.*, **29**, 1884, 10.1029/2002GL015391.

Willemse, E. J. M. (1997). Segmented normal faults: Correspondence between three-dimensional mechanical models and field data. *J. Geophys. Res.*, **102**, 675–692.

Willemse, E. J. M., Pollard, D. D., and Aydin, A. (1996). Three-dimensional analyses of slip distributions on normal fault arrays with consequences for fault scaling. *J. Struct. Geol.*, **18**, 295–309.

Williams, C. A., Connors, C., Dahlen, F. A., Price, E. J., and Suppe, J. (1994). Effect of the brittle-ductile transition on the topography of compressive mountain belts on the Earth and Venus. *J. Geophys. Res.*, **99**, 19 947–19 974.

Williams, M. L. (1957). On the stress distribution at the base of a stationary crack. *J. Appl. Mech.*, **24**, 109–114.

Wilson, L. and Head, J. W. (2002). Tharsis-radial graben system as the surface manifestation of plume-related dike intrusion complexes: Models and implications. *J. Geophys. Res.*, **107**, 5057, doi:10.1029/2001JE001593.

Wojtal, S. (1989). Measuring displacement gradients and strains in faulted rocks. *J. Struct. Geol.*, **11**, 669–678.

Zoback, M. D., Barton, C. A., Brudy, M., Castillo, D. A., Finkbeiner, T., Grollimund, B. R., Moos, D. B., Peska, P., Ward, C. D., and Wiprut, D. J. (2003). Determination of stress orientation and magnitude in deep wells. *Int. J. Rock Mech. Min. Sci.*, **40**, 1049–1076.

Index

f indicates figures, *t* indicates tables.

Accretion, small bodies 240
Activation energy 413, 417–419
Adventure Rupes, Mercury 20–21, 23, 26
Alba Patera, Mars 192, 212, 489
Albedo 11
Alpha Regio, Venus 97
Altimetry 18–19
Amazonian (Martian time scale) 184–186, 188, 353f
Amenthes Rupes, Mercury 26, 28, 495f, 496
Analog, terrestrial 32, 49, 51
Andal-Coleridge basin, Mercury 29
Anderson's fault classification 462, 464
Angle, friction (*see* Friction angle, fault)
Angle, incidence 19, 23–24
Annulus, coronae, Venus 85
Anomaly, remnant magnetic 190–191
Anticline (*see also* Fault, thrust) 4, 7, 153, 302, 364
Anticrack (*see also* Deformation band) 459
Antoniadi Dorsa, Mercury 30–32
Aphrodite Terra, Venus 89, 97
Apollo spacecraft mission 127
Apollodorus crater, Mercury 41
Arabia Terra, Mars 189–190, 192, 481, 489
Arch (*see also* Ridge, wrinkle) 17, 33–35, 87, 91–93, 123, 128–130, 364
Archean, Earth 106
Arden Corona, Miranda 326
Arecibo, Earth-based radar system 18, 21–22, 35, 37f
Argyre basin, Mars 193
Ariel 326–328, 381
Array, fault 459
Asgard basin, Callisto 331, 374
Aspect ratio, fault 27, 149
Associates, point defect 425–429
Asteroids 233, 235–240, 244–248, 250, 258–259
  Binary 244–245
  Near-Earth 233, 245
  Trojan 245
Asthenosphere 12, 109, 413–414, 416, 418, 444–445
Astrolabe Rupes, Mercury 29

Astypalaea Linea, Europa 300, 302
Aureole deposit, Mars 201
Average displacement (*see* Displacement, average fault)

Back-arc setting, Earth 416
Bands 9, 324, 327, 378–379
  Deformation (*see* Deformation band)
  Melt-rich 442, 444
  Pull-apart 301, 304
  Shear 299–300
  Smooth 321
  Triple 297
Basalt 32, 124–128, 442
Basin
  Multi-ring impact 363
  Pull-apart (*see also* Fault, strike-slip) 467
Beagle Rupes, Mercury 19–20, 25, 28
Beethoven basin, Mercury 42, 43
Belt, mountain (*see* Deformation, contractional, mountain belt)
Bias, in statistical data
  Censoring 467
  Detection 467
  Measurement 467
  Natural 467
  Truncation 467
Blind thrust fault (*see* Fault, thrust, blind)
Bombardment, heavy 16, 25, 63, 66, 152, 171, 187
Bombardment, terminal (*see* Bombardment, heavy)
Boosaule Montes, Io 289
Brackett graben, Moon 138–139
Bramante-Schubert basin, Mercury 29
Bright terrain (*see* Terrain, bright, Ganymede)
Brittle-ductile transition (*see* Transition, brittle-ductile; Strength Envelope; Goetze criterion)
Buckling (*see* Deformation, buckling)
Bulge, tidal 4, 67, 273–278, 281
Burgers vector 422
Byblus Sulcus, Ganymede 374

511

Byerlee's law (*see* Byerlee's rule)
Byerlee's rule 268, 292

Calderas, Io 287, 287f
Calisto Fossae, Eros 8, 242, 250, 256
Callisto 276t, 374f, 377
Calorian (Mercurian time scale) 64, 66, 70, 353f
Caloris basin, Mercury 4, 15–17, 25, 36–44, 50–56, 59, 64, 67, 70–71, 134, 143, 150, 369
Camera, vidicon 11, 356
Canadian Rocky Mountains, Earth 49
Cantaloupe terrain, Triton 10, 313–315, 374, 382
Cap, polar 191, 315, 398
Cassini spacecraft mission 236, 376
Cataclasis 410
Cavi 315, 382
Cavity, transient 53
Ceramics 402
Chain
  Cycloidal 306
  Pit crater 195, 355, 366
Channel, outflow, Mars 186, 188–189, 195
Chaotic terrain, Europa 379
Charge-coupled device (CCD) imaging system 357
Chasma 5, 82, 86, 94, 102–105, 109, 203, 373
Chasmata (*see* Chasma)
Claritas Fossae, Mars 188, 203–204, 208, 212
Clathrate 270
Clementine spacecraft mission 127
Climb, dislocation 401–402
Coefficient of friction (*see* Friction, coefficient of)
Cohesion (*see also* Coulomb criterion, Byerlee's rule) 462
Collisions, small bodies 235, 237
Columbia Hills, Mars 185, 217
Columbia Plateau, Earth 32, 51, 132, 150, 153
Comet nuclei 236
Comets 235–239
Compaction length 441–442
Compensation 57, 95, 99
Compressional
  Stress (*see* Stress, compressive)
  Structure (*see* Structure, contractional)
Compressive structure (*see* Structure, contractional)
Conamara Chaos, Europa 379
Constitutive equation 404, 408
Contraction, global 4, 6, 17, 54, 56, 63–66, 69, 70–71, 121, 153, 170–171, 213, 217, 369
Convection 109–111, 283–285, 408, 418, 433
Convection, mantle 4, 45, 62, 68–70, 111, 367, 416, 444
Cooling, planetary (*see* Contraction, global)
Coprates Chasma, Mars 203
Coprates rise, Mars 203
Core
  Fault 459
  Planetary 70, 192
Corona 5, 82, 102, 323–325, 371–372, 374, 382
Coulomb criterion (*see also* Byerlee's rule) 278, 462
Crack (*see* Joint)
Crack, cycloidal 304

Crater, impact 188, 199f, 306, 309–312, 314, 316, 320, 328, 331–332, 354, 356, 363
  Floor-fractured 141
  Shape 253
  Statistics 354, 375
Craton, dry continental 416
Creep 268–269, 399, 403, 404, 405, 406, 408
  Coble 400
  Diffusion 269, 399, 400–401, 403–405, 409, 411, 415, 424–425, 437, 438, 440
  Dislocation 269, 399–402
  Extended regime 405
  Nabarro 400
  Rate 401, 405–406, 428, 431, 433, 437–438
  Stress exponent 401, 403–404, 406, 424
Crosscutting relations 25, 43, 352–353, 361
Crust 12–14, 15–18, 183–185, 187, 188–190, 458, 461
  Basaltic 413, 417–418, 484
  Terrestrial oceanic 301
Crustal shortening (*see* Deformation, contractional)
Cryovolcanism 311–312, 315, 326, 374, 378, 381–382

Damage zone, fault 459
Data, topographic 10, 11, 18, 19–20, 127, 361, 458–459
Defect, point 425–426
Defects 269, 399, 425–429
Deformation band 459, 469, 474, 497
Deformation
  Brittle 268, 270
  Buckling 50, 88, 135, 153
  Compressional (*see* Deformation, contractional)
  Contractional 52, 65, 170, 490, 492
    Fold-and-thrust belt 308
    Mountain belt 88, 100–102
  Distributed 86–87
  Ductile 269–270
  Elastic 268
  Extensional
    Fracture belt 5, 82, 86, 92–93, 103–104
    Ridge belt 91–92, 101, 107–108, 206
  Ice 269–270, 402–404
  Localized (*see* Localization, strain; Rift)
  Newtonian 284, 400, 402
  Penetrative 86, 89, 464, 491
  Plastic 298, 399, 402, 412, 413, 415–416, 424
  Self-organizing 459
  Semi-brittle 409–418
  Volume change 282–283
Deimos 233, 234f, 238–239
DEM (*see* Digital Elevation Model)
Demagnetization, thermal 193–194
Depression, quasi-circular (QCD) 188
Despinning 17, 61–63, 67, 71, 279f, 281
Devana Chasma, Venus 104
Diabase
  Columbia 417
  Dry 414–415, 417
  Maryland 417
Diapirism 298–299, 315

Dichotomy
    Crustal 366
    Global, Mars 186, 190, 214
Differentiation 245
Digital Elevation Model (DEM) 18, 353, 362, 467
Dike, igneous 367, 487–488
Dione 328–330, 352, 375, 380
Dip angle, fault 458, 463, 485–486, 494–495, 497
Discontinuity
    Mechanical 137, 195
    Structural, defined 471
Discovery Rupes, Mercury 19–21, 23, 26, 28–29, 31–32, 65
Dislocation
    Climb 401–402, 411, 421, 422, 424
    Cross slip 399
    Glide 401–402, 411–412, 413–414, 416
Displacement-length ratio (*see* Ratio, displacement-length)
Displacement-length scaling relations (*see* Scaling, displacement-length)
Displacement
    Distribution
        Fault 466, 475–476: Bell-shaped 475; Flat-topped 480; Linear 475; Gradient, near-tip 477
    On fault 472
Distribution, melt 434–436, 439
D_{max}/L ratio (*see* Ratio, displacement-length)
Dome 294, 373, 377, 379
Dorsum (*see* Deformation, extensional)
Dorsa Aldrovandi, Moon 133–134, 136, 148
Dorsa Lister, Moon 135
Dorsum Buckland, Moon 131–132
Dorsum Nicol, Moon 130–132, 139
Driving stress (*see* Stress, driving)
Ductile deformation (*see* Deformation, ductile)
Dune, eolian 331
Dunite 414–415, 422–425
Dynamo 69, 193, 214–215, 378

Earth, geologic time scale 353f
East Pacific Rise, Earth 468f, 480
Echelon (*see* Pattern, echelon)
Eistla Regio, Venus 89, 95–96
Elysium Mons, Mars 205
Elysium, Mars 188, 192t, 205, 207, 366–368
Embayment relation 25, 37–38
Enceladus 10, 265, 276t, 283, 307–312, 333, 352, 374, 380–381
End zone, fault 474
Endeavour Rupes, Mercury 21, 23t
Eros 8, 233–234, 238–245, 248–254, 256–257, 352–353, 362
Erosion
    Fluvial, Titan 216, 331
    On Venus 81–83, 103
Europa 9–10, 264–265, 276t, 280, 294–307, 374, 378–379, 398, 405, 457, 459, 496
Evaporite 185

Evolution
    Crustal 105–106
    Thermal 47
Extension (*see* Deformation, extensional)
Extension, crustal 9–10, 136–137, 152
Extensional strain (*see* Strain, extensional)
Exterior plains, Mercury 36, 63

Failure criteria (*see* Byerlee's rule; Coulomb criterion; Hoek–Brown criterion; Flow law)
Fault
    Definition of 459–461
    Geometry
        Conjugate 137
        Hard-linked (*see also* Linkage, fault) 460
        Isolated 460, 477
        Kinematically coherent 473
        Large 51, 477, 480–482, 494–496
        Listric 26
        Master 137, 195
        Restricted 472, 476, 479–481, 482, 494–495
        Segmented (*see also* Interaction, mechanical; Linkage, fault) 460–461
        Small 467, 479, 494, 495
        Soft-linked (*see also* Interaction, mechanical) 460
Fault type
    Normal 482
    Reverse (*see* Fault, thrust)
    Strike-slip 27, 464
    Thrust (*see also* Wrinkle ridge) 3–4, 6, 8, 10, 15–17, 19–20, 25–28, 32, 49, 51, 197–199, 202, 213, 279f, 281, 290, 292, 353, 364–369, 458–460, 462–467, 481–489, 493–495
        Blind 51, 183, 353
        Mars 183, 197
        Mercury 15–17, 19–20, 25–28
        Moon 121–122, 126–127, 134–135, 143–148
        Outer planet satellites 292, 302
        Small bodies 251–252
        Surface-breaking 3, 15, 17, 19, 353
        Venus 88
Faulting
    Depth of 26, 46, 51–52, 170, 486, 489
    Localized 468
    Tilt-block 318, 320
    Work done by 493
Flexure 43, 285–286
Flow law 397–398, 401, 404, 407–408, 437
Flow, lateral crustal, Mercury 54–55, 60, 71
Fold belt (*see* Deformation, contractional, fold-and-thrust belt)
Folding (*see* Folds)
Folds 1, 10, 51, 66, 92, 97, 101–102, 197–198, 294, 302, 325, 351, 353, 358, 372–373, 375, 380–381, 459
Folds, Yakima, Earth 51
Fossae Plinius, Moon 138–139, 166
Fracture (*see* Joint, Fault)
Fracture belt (*see* Deformation, extensional, fracture belt)

Fracture mechanics
   Linear elastic 295, 474
   Post-yield 474
Fragmentation 237, 239–240, 246, 251
Friction angle, fault 463
Friction, coefficient of 462–463
Frictional strength (see Strength, frictional; Byerlee's rule)
Fugacity 399, 412, 424–425, 427–430, 432
Furrow 317

Galilean satellites (see Satellites, Galilean)
Galileo spacecraft mission 266, 376
Ganymede 9, 267, 274, 276t, 352, 397–398, 405, 408
Gaspra 8, 234, 239t, 245, 249–250, 257–258
Geoid 186, 190, 208, 210, 441
Geotherm, terrestrial continental 414
Geotherm, terrestrial oceanic 413
Gertjon Corona, Venus 94
Geyser, icy satellite 264, 265
Gilgamesh, Ganymede 316, 317
Glacier 396, 404–405
Glide, dislocation 399, 401–402, 410–411, 414–416, 418f, 421
Goetze criterion 411, 413–414
Goldstone radar system, Earth 34–35
Gouge, fault 475, 489
Graben (see also Fault, normal)
   Circumferential 58–60, 208–209
   Mars 183–184, 188, 194–196
   Mercury 15–17, 19, 39, 41–45
   Moon 125, 136–141
   Outer planet satellites 288f, 314, 318–319
   Small bodies 250
   Venus 82, 86, 88f, 90, 92–93, 97–98
Gradient, displacement, fault 460
Gradient, thermal 47, 55, 418–420
Grain size 399–408, 415, 422, 424
Gravity 122, 160–168, 482, 484
Grid, tectonic 39
Groove 8, 238, 247, 249–250, 252–258, 318–323, 377–378, 442
Groundwater 185
Growth, fault 458, 474–477

Half-space, elastic 26
Healing, fault 490
Heat flow (see Heat flux)
Heat flux 15, 45, 47–49, 54–55, 62, 64, 69, 71, 185, 194, 314, 318, 321–322, 379, 434, 443
Heating, shear 299
Height, fault 473, 480, 489, 492
Helen Planitia, Venus 371
Hellas, Mars 187, 189–190, 192–193, 205–206
Hemisphere
   Leading 302, 308
   Trailing 308
Hero Rupes, Mercury 28–29
Herschel, Mimas 332
Hesperia Planum, Mars 36, 205

Hesperian (Martian time scale) 183–186, 188–190, 192t, 200–207, 211, 213, 216–217, 353f
Hi'iaka Patera, Io 374
High Resolution Imaging Science Experiment (HiRISE) 187
High Resolution Stereo Camera (HRSC) 187
Highlands, southern, Mars 183, 186–188
Hinks Dorsum, Eros 251
HiRISE (see High Resolution Imaging Science Experiment)
History, thermal 17–18
Hoek-Brown criterion 463
Hotspot, Venus 85, 94, 109–110
HRSC (see High Resolution Stereo Camera)
Hyabusa spacecraft mission 239t, 247
Hypsometry 84

Iapetus 265, 274, 276t, 331–332, 352, 374, 380
Ice, water 9–10, 12, 246, 264–267, 270
Ida 234, 238–240, 243, 245, 247–251, 253, 256–257
Image, digital 356–357
Impact, giant 190
Inertia, moment of 57, 160, 192, 248, 308
Initiation, fault 469
Inlier, Venus 82, 96, 101, 141
Interaction, mechanical (see also Linkage, fault) 459–460
Inverness Corona, Miranda 374f
Io 9, 12, 264–266, 272, 275–276t, 283, 286–294, 352, 372, 374f, 376, 379, 397–398, 434–445
Iron 421, 431, 433
Ishtar Terra, Venus 96, 99–102
Isidis, Mars 188, 190
Isotherm 45–46, 48f, 489f
Ithaca Chasma, Tethys 328–330, 380
Itokawa spacecraft mission 239t, 247

Joint 90, 253, 351, 472–474
Jupiter 245, 352–353

Kostrov's equation 494
Kuiper belt objects 8, 245

Lakshmi Planum, Venus 100–101
Landform, tectonic 361, 365
Landslide 201, 289, 294
Lavinia Planitia, Venus 88f, 91–93, 372f
Layer, stratigraphic or mechanical 240–241, 244, 271, 472, 480
Lee-Lincoln scarp, Moon 122–123, 145–146, 148
Length, of structures
   Cumulative 23–25, 62, 152
   Distribution
      Negative exponential 472
      Negative power-law 480
   Fault 51, 149, 467, 470–471, 479
Libration, obliquity 275
Life 398
Linkage, fault 52, 470, 472, 474, 476
Lithosphere 12–13, 45–46, 109, 110, 153, 171, 418f, 420f, 458–466, 486f, 487, 488f, 490

Seismogenic (*see also* Transition, lower stability) 26, 421, 501, 502
  Terrestrial oceanic 301, 477
Lithostat (*see* Stress, lithostatic)
Littrow graben, Moon 140
Littrow ridge, Moon 133, 134f, 136, 148
Lobate scarp (*see* Scarp, lobate)
Localization, strain 89, 415–416, 459
Lowlands, northern, Mars 190, 193, 206, 213–215, 217
Lunae Planum, Mars 202

Magellan spacecraft mission 358, 360, 370, 371f, 372f, 373f
Magnetic field 69, 83, 192–194, 307, 316, 377
Main Ethiopian Rift, Earth 480
Mantle 45–46, 48f, 409, 410–421
Mapping
  Geologic 11
  Planetary geologic 11, 352
  Structural 352
  Symbols 356
Mare Humorum, Moon 137f, 147, 166–167
Mare Imbrium, Moon 130, 140, 166–167, 364f
Mare ridge (*see* Ridge, mare)
Mare Serenitatis, Moon 124, 129–136, 138, 140f, 144, 148, 153, 164, 166
Mare, lunar 33, 39, 43
Mariner 10 spacecraft mission 16–20, 368f
Mariner 9 spacecraft mission 187
Mars 12, 19, 22, 26, 28, 30, 32–33, 35–36, 41–42, 50, 55, 62, 69, 87, 89, 124–125, 128, 130, 136, 147, 151, 171, 183–217, 364–366
Mars Express spacecraft mission 185–187, 357
Mars, geologic time scale 353f
Mars Global Surveyor (MGS) spacecraft mission 186, 210, 357
Mars Odyssey spacecraft mission 186, 191, 357
Mars Orbiter Camera (MOC) 187
Mars Orbiter Laser Altimeter (MOLA) 186, 362
Mars Reconnaissance Orbiter (MRO) spacecraft mission 185–186
Marsquake 490
Mascon (Mass Concentration) 6, 43–45, 54, 56, 59, 71, 122, 125, 127, 153, 162, 164–167, 171, 205, 209
Mass wasting 83, 380
Material
  Polycrystalline 400
  Silicate 12, 235, 267
Mathilde 234f, 239–241, 246, 256
Maxwell Montes, Venus 83, 100–102
Maxwell time 270, 272, 275, 280
Measurement, shadow length 18, 33
Mechanism, vacancy 425
Megaregolith 137, 171, 365
Melt 434
Melting, partial 434–445
Memnonia Fossae, Mars 196f
Mercury 3, 15, 124, 353, 366, 480–482
Mercury Dual Imaging System (MDIS) 16

Mercury, geologic time scale 353f
Mercury Laser Altimeter (MLA) 19
Meridiani Planum, Mars 185, 217
MESSENGER spacecraft mission 19–20, 24, 38, 368
Metals 402
Meteorite 233, 235
Mimas 276t, 331–332
Mineral, hydrous 412–413, 416
Miranda 265, 276t, 311, 323–326, 352, 374f, 381
Mirni Rupes, Mercury 29
Model, kinematic 27, 137
Modeling, dislocation 27–28
Modulus, shear (*see* Rigidity)
Moment
  Geometric 492
  Population 471
  Quasi-static fault 492
  Seismic 159
Mongibello Mons, Io 290f
Montes Riphaeus, Moon 146–147
Moon 1–3, 6–7, 9–10, 12, 16, 19, 32, 35, 37, 53–55, 59, 62, 87, 121–171
Moon, geologic time scale 353f
Moonquake 6, 121–122, 126–127, 154–159, 162, 168–169, 172
  Deep 121, 126
  Shallow 126, 158–159
Morozov scarp, Moon 143f
Morphology 353, 361
Motion
  Dislocation 399
  Plate 413
Mylonite, formed along faults 411

Near Earth Asteroid Rendezvous (NEAR) spacecraft mission 8, 233, 239t, 241–242, 251–252
Neptune 312–316
Noachian (Martian time scale) 183–185, 187–194, 197, 200–209, 211–212, 214–217, 353f
Noctis Labyrinthus, Mars 184, 188, 202–203, 211
Normal fault (*see* Fault, normal)
Northern plains (*see* Lowlands, northern, Mars)
Nun Sulci, Ganymede 319f

Ocean
  Magma 169, 171
  Subsurface 265, 284, 297–298, 307
Oceanus Procellarum, Moon 129, 146–147, 152, 167, 172, 365f
Odin Planitia, Mercury 33, 36–37
Odysseus, Tethys 328, 330, 380
Offset (*see* Displacement, on fault)
Olivine 412–413, 438–443, 444
  Dry 413
  Polycrystalline 409
  Wet 413
Olympus Mons, Mars 184, 188, 192t, 201, 398, 433
Oort cloud 237
Organics 235–236
Ovda Regio, Venus 97f, 99, 373f
Overlap, fault 467, 472

Pantheon Fossae, Mercury 41, 56
Parameter, wetness 436
Parga Chasma, Venus 94f
Paterae 287, 293, 314, 378–380
Pattern
   Echelon 43, 90, 125, 321, 460
   Fault 465f, 467
   Polygonal 41–42, 52, 86, 90, 141, 368
Permeability 442, 443, 444
Phases, high-pressure 267, 322
Phobos 233–234, 238–239, 241, 249, 250f, 253–256
Phoebe 236, 239t
Photoclinometry 18, 21–23, 42–43, 195, 200, 353, 361
Phyllosilicates 185, 217, 236
Pile, rubble 245–248
Pit crater chain (*see* Chain, pit crater)
Pits 201, 249, 253–254, 308–314, 358, 364, 377, 379, 382
Plains
   Intercrater, Mercury 24
   Ridged 188
   Venus 86
   Volcanic 32
Planets, terrestrial 32, 45
Plastic flow (*see* Deformation, plastic)
Plate boundaries, transform 458
Plate tectonics 2–3, 84–86, 109–111, 185
Plate, upper (*see also* Fault, thrust) 26, 485–486
Plateau, crustal 82, 86, 99–101
Plateaus 365f, 373, 375
Plume, mantle 5, 7, 86, 95–96, 99, 208, 293
Poisson's ratio 26, 49, 57–58, 66
Pola Regio, Ida 257
Polar wander, true 376, 379, 381
Population
   Fault 457
   Joint 473
Porosity 241, 246, 248, 268, 296, 299
Precambrian, Earth 98, 353f
Pressure solution 410
Pressure
   Pore fluid 410, 415, 416, 461
   Hydrostatic 489
   Lithostatic 13, 410
   Magma 487
   Overburden (*see* Pressure, lithostatic)
Profile, displacement (*see* Distribution, displacement)
Propagation, fault 459, 472, 477
Properties
   Mechanical 2, 12–13, 148–149
   Rheological 266–267, 397–398
Pwyll, Europa 378

Quartzite, wet 414–415

Rabelais crater, Mercury 29
Rabelais Dorsum, Mercury 31–32, 31f
Radar 86–89, 90–91, 96–97, 107–108, 358–361
Radar, synthetic aperture 11, 370
Radionuclide 236, 245
Raditladi basin, Mercury 41

Rahe Dorsum, Eros 8, 242–243, 242f, 251–252, 251f, 252f, 256–257
Rameau crater, Mercury 26
Ramp, relay 467, 474, 482
Range, push-up (*see also* Fault, strike-slip) 457, 467
Ratio, displacement-length 13–14, 197
Rebound, postglacial 441
Recrystallization 270, 412, 415
Regolith 8, 66, 135, 148, 155, 161, 170–171, 238–241, 241f, 249, 253–254, 258–259, 308, 398, 484
Resolution Rupes, Mercury 20–21, 23t, 28–29, 31–32
Resolution, image 368, 376, 381, 467
Resonance, Laplace 322, 443
Resurfacing, global, Venus 85, 109–111
Rhea 265, 267, 274, 276t, 328–329, 332, 352, 380
Rheology 47–48, 48f, 266–272
Ridge
   Belt (*see* Deformation, extensional, ridge belt)
   Complex 295–297, 295f, 299, 305f
   Cycloid 304–305, 305f
   Double 295f, 296–298, 305f, 313f
   High-relief 4, 17, 30–32, 44, 353, 369
   Mare, Moon 123–129, 129f, 131, 134–136
   Mid-ocean, terrestrial 52, 413, 434, 439f
   Narrow 33
Ridge, wrinkle (*see also* Faulting, thrust) 15–17, 32–39, 86–89, 123–125, 128–136, 146–149, 152–153, 164–168, 189, 197–199, 198f
   Mars 199, 205, 207, 213, 217
   Mercury 33, 39–40, 50–52, 59–60
   Moon 124–125, 128–136, 146–149, 152–153, 164
   Venus 82, 86–91
Ridged plains (*see* Plains, ridged)
Rift 94, 104, 183–184, 195, 202–204, 210, 367, 380
Rigidity 62–64, 273, 476
Rille (*see also* Graben) 6, 121–125, 129f, 136, 138t, 143, 152, 163, 363–365
   Concentric 164
   Linear 125, 138t
   Sinuous 125, 364–365, 365f
Rima Ariadaeus, Moon 136f
Rimae Hippalus, Moon 137, 137f
Rings, impact basin 29
Rise, volcanic 82, 85, 94–96, 94f, 373
Rock
   Basaltic 12
   Mass (*see also* Hoek–Brown criterion) 410, 461–465, 461t, 465f, 480
   Polycrystalline 409, 424
Rotation
   Nonsynchronous 10, 278–280, 293–294, 300, 303–304, 323, 376, 378–379
   Synchronous 255, 273, 281–282
Rupes Recta (Straight Wall), Moon 141–143

Santa Maria Rupes, Mercury 20–23
Satellites
   Galilean 294, 322, 331, 377, 398
   Icy 9–10, 270–272, 398, 405, 484
   Outer planet 9–10, 264, 376–382

Index

Scaling, displacement-length 469–470, 481f, 483f, 484
Scarp
  Highland 126
  Lobate 6, 15, 20–30, 23t, 52, 61, 61f, 65–71, 143f, 365–366
    Mars 199f
    Mercury 19–30
    Moon 122, 127, 143–144, 146–148, 150–153, 170–172
Schiaparelli Dorsum, Mercury 33
Sedna Planitia, Venus 87f
Seismicity 127, 154–159, 163–169, 172–173
Serpentine 412
Set, fault 459
Shamshu Patera, Io 287
Shear
  Failure (*see* Coulomb criterion)
  Modulus (*see* Rigidity)
  Strength (*see* Byerlee's rule) 252
  Zone 415–416
Sheet, ice 398, 405
Shell, floating ice 267, 275, 278, 280, 285
Shield field, Venus 91, 106–107
Shock wave 237
Shoemaker–Levy 9 238
Shulamite Corona, Venus 94f
Singularity, stress 477
Sliding
  Grain boundary 269, 399–407
  On pre-existing faults (*see* Byerlee's rule) 410
  Stimulated grain boundary 403
Slip system 402, 422, 423f, 424
Slip, net (*see* Displacement, on fault)
Slope
  Regional 93
  Talus 104
Small bodies 8–9, 233
Solis Planum, Mars 192t, 202
Spacing
  Fault 458–459, 467, 472–474, 479
  Fractures, Venus 90–92
  Of structures 96
Stepover, fault (*see also* Ramp, relay) 457, 460, 490
Stickney, Phobos 254–256
Strain 490–496
  Partitioning 458
  Rate 48f, 63, 237, 268–270, 277f, 280, 301, 399, 403f, 405, 407f, 419, 420f, 421, 422, 423f, 424–426, 428, 430, 431f, 432, 433f, 438
  Compressional (*see* Strain, contractional) 51, 54, 58, 205, 213–214
  Contractional 7, 8, 17, 65, 170–172, 213–214, 298, 318, 366, 457, 496
  Distributed (*see* Deformation, distributed) 16, 86–91, 467, 480
  Extensional 52, 55, 59, 211, 302, 308, 318, 320–322, 378, 457
  Fault 493
  Horizontal normal (*see also* Strain, extensional) 492–496, 495f

Localization (*see* Localization, strain) 50, 89, 416, 459, 480
Odd-axis model 465f, 496
Regional (*see also* Stress state, remote) 51–52, 202, 211
Vertical normal 490, 496
Stratigraphy 14, 363, 379, 458, 470, 482
Stratigraphy, mechanical 14, 458, 470
Strength
  Envelope (*see also* Byerlee's rule, Flow law) 191, 398, 408–410, 413, 413f, 418f, 418–419, 420f, 421, 462, 466–467, 470
  Frictional (*see also* Coulomb criterion 410, 458, 462–463, 475, 497
  Lithospheric (*see also* Byerlee's rule; Hoek-Brown criterion) 398, 411, 419, 462, 464, 466–467
  Shear (*see* Byerlee's rule; Coulomb criterion; Hoek–Brown criterion; Flow law)
  Tensile 238, 246, 268, 303, 306
  Yield 9, 85, 109, 110, 148, 475–477, 482
Stress
  Compressional (*see* Stress, compressive)
  Compressive 16, 62, 71, 88–90, 92, 166, 207, 210, 283, 291–292, 437, 461, 463–464
  Concentration 474
  Differential 67–68, 269, 281, 399–404, 404f, 406f, 411, 414, 423f, 431f, 432, 433f, 437, 441, 444
  Diurnal 277, 280, 294, 297, 304–306, 308, 311, 333, 379, 382
  Driving 296, 474, 476–477, 482, 493
  Drop (*see* Stress, driving)
  Effective lithostatic 461t
  Effective principal 462, 464
  In situ 13, 490
  Membrane 184, 207, 209, 274, 325
  Principal (*see* Stress, effective principal)
  Regional (*see* Stress state, remote)
  Remote 49, 476, 486
  Thermal 54, 67, 89, 91, 126, 169–170, 238–239, 292, 293
Stress states and faulting (*see also* Byerlee's rule) 458, 461–465
Strike-slip fault (*see* Fault, strike-slip)
Stripe, tiger 10, 307, 309–312, 310f, 381
Structure, ribbon 97, 360f
Subduction 5, 105, 109–110, 209, 286, 291, 433
Subsidence 10, 36–37, 45, 53–54, 56, 58, 60, 64, 71, 95, 124, 166, 206, 292–294, 311, 364
Sulci 311, 313f, 319f, 377
Sulfur 65, 286
Superposition 32, 44, 89, 121, 152–153, 213, 252, 363
Sveinsdóttir basin, Mercury 20f, 25
Syncline (*see also* Folds) 26, 302
Syria Planum, Mars 183, 188, 203–204
System, fault 458–459

Tailcrack (*see also* Wing crack; Fault, strike-slip) 306
Tectonics
  Planetary 1–2
  Stagnant lid 110, 284, 434
  Vertical 85–86

Tempe rift, Mars 202
Tempe Terra, Mars 183–184, 201–202, 212t, 481–482, 483f, 489
Tempel-1 233, 239t, 244f
Temperature 401, 406f, 407f, 410, 411, 415, 417, 418f, 419, 421, 432, 433, 456
Temperature, homologous 266, 272
Tension fracture (*see* Joint)
Terminator 23–24
Terrain
    Bright, Ganymede 318–321
    Dark 316, 317–318, 321, 374f, 377, 378
    Grooved 308, 318, 323, 319f, 374, 377, 408
    Heavily cratered 187–188
Tessera 82, 86, 88, 96–99, 99–112, 97f, 373f, 375
Tethys 265, 276t, 328, 329f, 352, 380, 459
Tharsis, Mars 101, 194, 481, 488, (these are ok)
Thaumasia rift, Mars 203–204
Thaumasia, Mars 183–184, 188
Themis Regio, Venus 94f, 96, 102
Thermal Emission Imaging System (THEMIS) 187
Thickness
    Crustal 62, 69–70
    Effective elastic 45, 48, 50, 63, 160, 271f, 271, 273, 297
    Elastic (*see* Thickness, effective elastic)
    Lithospheric 45–47, 48f, 63, 103, 164, 297, 433
Thrust fault (*see* Fault, thrust)
Tides, diurnal 275, 277–278, 277f
Time scales, geologic, terrestrial planet 355f
Time, relaxation 408
Tipline, fault 470f, 476, 477
Tir Planitia, Mercury 32–37, 34t, 35f, 37f
Titan 9, 11, 267, 276t, 283, 330–331, 352, 380
Titania 265, 276t, 328–329, 330f, 381
Tolstoj, Mercury 43
Tolstojan (Mercurian time scale) 25, 28–29, 44, 70, 353f
Topography
    Fault-related 13–14, 244, 457, 482, 484–490
    Stereo derived 19, 130
Townsend Dorsum, Ida 243, 251
Transition
    Brittle-ductile 15, 45, 50, 207, 314, 410
    Brittle-plastic 410, 411
    Lower stability 490
    Upper stability 489, 490
Transform fault (*see* Fault, strike-slip)
Traverse, line 494

Triton 10, 264–265, 267, 276t, 281, 312–316, 333, 352, 376, 381–382
Trough 188, 189, 191, 195, 202–203, 206, 242f, 249–251, 254, 295f, 296, 297, 299, 301, 311, 312, 318, 321, 324, 325, 327, 332, 358, 367, 375f, 380, 382
    Cycloidal 378
    Extensional 2, 4, 40f, 50, 52, 54, 60, 133f, 136
Tube, lava 125, 328, 365f, 461
Tyre, Europa 378

Umbriel 276t, 326, 333
Uplift, footwall 485f, 485
Uruk Sulcus, Ganymede 319f, 320f
Utopia, Mars 42, 188, 366

Valhalla, Callisto 331, 374f, 377
Valles Marineris, Mars 7, 184, 186, 188, 191, 195, 202–203, 210–212, 212t, 367
Venus 4–5, 370–376, 397, 398, 408, 416–419, 420f, 461t
Venus, bulk properties 82
Venus, geologic time scale 353f
Vergence, thrust fault 144, 493
Vienna Regio, Ida 243f, 256
Viking spacecraft mission 356
Volatiles 236, 238, 253, 259, 298, 328, 355, 417, 421
Volcanism 5, 9, 53, 59, 205, 208, 286, 290–294, 325, 355, 378, 397
von Mises criterion 401–402, 422

Wander, polar 16, 67, 70, 192–194, 279f, 280–281
Water 2, 9, 10, 12, 13, 82, 100, 103, 109, 110, 151, 184–186, 212, 214, 236, 246, 253, 270, 272, 274, 282–283, 286, 298, 301, 378, 407–408, 409, 410, 413, 416–418, 418f, 421, 422, 425f, 429
Water weakening (*see* Weakening, hydrolytic)
Weakening, hydrolytic 416
Width, fault (*see* Height, fault)
Wing crack (*see also* Faulting, strike-slip) 300, 363–364, 365f, 367, 368f, 369, 371f
Wrinkle ridge (*see* Ridge, wrinkle)

Yelland Linea, Europa 372
Yield strength (*see* Strength, yield)
Young's modulus 57, 58f, 267–268, 271f, 274, 276t, 295, 487

Zone, fault 320–321, 412, 416, 459, 475f, 484, 490